Advances in Mathematical Fluid Mechanics

For further volumes:
http://www.springer.com/series/5032

Birkhäuser

Advances in Mathematical Fluid Mechanics

Advances in Mathematical Fluid Mechanics is a forum for the publication of high quality monographs, or collections of works, on the mathematical theory of fluid mechanics, with special regards to the Navier-Stokes equations. Its mathematical aims and scope are similar to those of the *Journal of Mathematical Fluid Mechanics*. In particular, mathematical aspects of computational methods and of applications to science and engineering are welcome as an important part of the theory. So also are works in related areas of mathematics that have a direct bearing on fluid mechanics.

The monographs and collections of works published here may be written in a more expository style than is usual for research journals, with the intention of reaching a wide audience. Collections of review articles will also be sought from time to time.

Tomáš Bodnár • Giovanni P. Galdi •
Šárka Nečasová
Editors

Fluid-Structure Interaction and Biomedical Applications

Birkhäuser

Editors
Tomáš Bodnár
Faculty of Mechanical Engineering
Department of Technical Mathematics
Czech Technical University in Prague
Prague, Czech Republic

Giovanni P. Galdi
Swanson School of Engineering
University of Pittsburgh
Pittsburgh, PA, USA

Šárka Nečasová
Institute of Mathematics
Academy of Sciences of the Czech Republic
Prague, Czech Republic

ISBN 978-3-0348-0821-7 ISBN 978-3-0348-0822-4 (eBook)
DOI 10.1007/978-3-0348-0822-4
Springer Basel Heidelberg New York Dordrecht London

Library of Congress Control Number: 2014943951

Mathematics Subject Classification (2010): 74F10, 35Q35, 35Q30, 76N10, 60G35, 93E10, 92C55, 76Z05, 74L15, 92C45, 65C20, 65M60, 65M08, 65M12

Printed on acid-free paper

Springer is part of Springer Science+Business Media (www.springer.com)

Preface

Fluid–structure interaction (FSI) is among the most popular and intriguing problems in applied sciences. Their range of interest covers many significant areas of research including industrial applications, such as the manufacturing of short-fiber composites, biological applications, such as separation of macromolecules by electrophoresis, DNA sequencing, and blood flow, as well as animal locomotion and damage of structures.

The focus of this book is to present, in a methodical way, updated and comprehensive descriptions and analysis of some among the most relevant problems in the framework of FSI, with a special view to biomedical applications. More specifically, this work, organized in chapters, is dedicated to as many different fundamental aspects of FSI from different perspectives, that comprehend a mathematical analysis of basic questions, like well-posedness of the relevant initial and boundary value problems, as well as the modeling and the numerical simulation of a number of fundamental phenomena related to human biology. These latter include blood flow in arteries and veins, blood coagulation, and speech modeling.

The book is organized in seven chapters presenting in detail selected subjects:

1. *Chapter 1: Mathematical and Numerical Analysis of Some FSI Problems*—Some specific existence and numerical results applied to a 2D/1D fluid–structure coupled model, for an incompressible fluid and a thin elastic structure are presented. Special focus is on mathematical and numerical difficulties arising when studying this kind of problems, such as the geometrical nonlinearities or the added mass effect. In particular the link between the strategies of proof of weak or strong solutions and the possible algorithms to discretize these type of coupled problems is pointed out.
2. *Chapter 2: Fluid–Structure Interaction in Hemodynamics: Modeling, Analysis, and Numerical Simulation*—To study the interaction between blood flow and cardiovascular tissue a stable and efficient FSI solver is an indispensable tool. The multi-physics nature of this class of problems suggests the use of partitioned, modular algorithms based on an operator splitting approach that would separate the different physics in the problem. Such a scheme is presented, to be used not

only in computations but also to prove existence of weak solutions to this class of problems. Particular attention is paid to multi-physics FSI problems involving structures consisting of multiple layers.

3. *Chapter 3: Hyperbolic–Parabolic Coupling and the Occurrence of Resonance in Partially Dissipative Systems*—Elastic solids, when subjected to a time-periodic load of frequency ω, may respond with a drastic increase of the magnitude of basic kinematic and dynamic quantities, such as displacement, velocity, and energy.Objective of the presented analysis is to investigate whether the interaction of an elastic solid with a dissipative agent can affect and possibly prevent the occurrence of resonance. This problem is studied in a broad class of dynamical systems (called partially dissipative), whose dynamics is governed by strongly continuous semigroups of contractions. For such systems the sharp necessary and sufficient conditions for the occurrence of resonance are provided.
4. *Chapter 4: Topics in the Mathematical Theory of Interactions of Incompressible Viscous Fluid with Rigid Bodies*—The recent results devoted to the interactions between a set of rigid bodies and a surrounding viscous fluid, the whole system filling a container, are reviewed. It is assumed that the motion of (resp.) the rigid bodies is governed by the incompressible Navier–Stokes equations (resp. Newton laws), and that velocities and stress tensors are continuous at the fluid/body interfaces. The well-posedness of the associated Cauchy problem, with a specific eye towards the handling of contact between bodies or between one body and the container boundary, is addressed.
5. *Chapter 5: Numerical Simulation of Fluid–Structure Interaction Problems of Flow in Vocal Folds*—Here the main attention is paid to the mathematical description of a corresponding problem and to the description of the applied numerical methods. The mathematical description consists of the elasticity equations describing the motion of an elastic structure, and the air flow modeled by the Navier–Stokes equations. The solution of dynamic elasticity equations is realized with the aid of conforming finite elements or the elastic structure motion is modeled by a simplified model of vibrating rigid body. Both compressible and incompressible fluid models are considered. The approximation of flow in moving domains is treated with the aid of the arbitrary Lagrangian–Eulerian method.
6. *Chapter 6: Data Assimilation in Cardiovascular Fluid–Structure Interaction Problems*—This chapter provides an introduction to methods for data assimilation, mostly developed in fields like meteorology, applied to computational hemodynamics. The focus is mainly on two of them: methods based on stochastic arguments (Kalman filtering) and variational methods. Some examples are addressed that have been approached with different techniques, in particular the estimation of vascular compliance from displacement measures.
7. *Chapter 7: Mathematical Models for Blood Coagulation*—The historical exposure of the development of classical coagulation modeling theories is followed by a basic overview of blood coagulation biochemistry. The recent developments of cell-based models are explained in detail to demonstrate the current shift from the classical cascade/waterfall models. This phenomenological overview

is followed by a survey of available mathematical concepts used to describe the blood coagulation process at various spatial scales including some of the related biophysical phenomena.

We believe that the variety of the topics along with the different approaches used to tackle the corresponding problems will help the reader to have a global and updated view on the latest results on the subject and on the relevant open questions. For the same reason we expect the book to become a friendly companion to scientists having their main research focus in diverse disciplines, such as mathematics, physics, mathematical biology, bioengineering, and medicine.

Prague, Czech Republic — Tomáš Bodnár
Pittsburgh, PA, USA — Giovanni Paolo Galdi
Prague, Czech Republic — Šárka Nečasová
February 2014

Contents

1 Mathematical and Numerical Analysis of Some FSI Problems 1
Céline Grandmont, Mária Lukáčová-Medvid'ová, and Šárka Nečasová
1.1 Introduction 1
1.1.1 Variational Formulation and Energy Estimates 6
1.1.2 Difficulties 12
1.2 Mathematical Analysis 13
1.2.1 A Linear Simplified Problem 16
1.2.2 Existence of Weak Solutions 19
1.2.3 Existence of Strong Solutions 31
1.2.4 Non-Newtonian Shear-Dependent Fluid 36
1.3 Numerical Analysis 50
1.3.1 Explicit Scheme and Added Mass Effect 52
1.3.2 A Semi-Implicit Scheme 53
1.3.3 Kinematically Coupled Schemes 58
1.4 Conclusions 69
References 69

2 Fluid–Structure Interaction in Hemodynamics: Modeling, Analysis, and Numerical Simulation 79
Sunčica Čanić, Boris Muha, and Martina Bukač
2.1 Introduction 80
2.2 Mathematical Models of Arterial Walls 83
2.2.1 Elastodynamics of Thin Structures 84
2.2.2 Elastodynamics of Structures with Finite Thickness ("Thick Structures") 100
2.3 A Benchmark Problem 104
2.3.1 The Model Equations 104
2.3.2 The Coupling Conditions 106
2.3.3 The Boundary and Initial Conditions 107

2.4 FSI Literature Review 109
2.4.1 Literature on Analysis of FSI Problems 110
2.4.2 Literature on Numerical Simulation of FSI Problems 111
2.5 Solution Framework 113
2.5.1 The Energy of the Coupled Problem 114
2.5.2 ALE Formulation 118
2.5.3 The Splitting Scheme: General Framework 120
2.5.4 A Modified Splitting Scheme Achieving Higher Accuracy 124
2.6 Existence of a Weak Solution 125
2.6.1 Problem Definition 126
2.6.2 The Energy of the Coupled Problem 129
2.6.3 The ALE Formulation and Lie Splitting 130
2.6.4 Weak Solutions 134
2.6.5 Approximate Solutions 138
2.6.6 Convergence of Approximate Solutions 144
2.6.7 The Limiting Problem and Weak Solution 157
2.6.8 Construction of the Appropriate Test Functions 157
2.7 Numerical Simulation 167
2.7.1 Problem Definition 167
2.7.2 The Energy of the Coupled Problem 170
2.7.3 The ALE Formulation 171
2.7.4 Weak Formulation of FSI Problem (2.177)–(2.191) 173
2.7.5 Numerical Implementation of the Splitting Scheme 174
2.7.6 Discretized Scheme in Weak Form 177
2.7.7 Numerical Examples 180
2.8 Conclusions 187
References 188

3 Hyperbolic–Parabolic Coupling and the Occurrence of Resonance in Partially Dissipative Systems 197
Giovanni Paolo Galdi, Mahdi Mohebbi, Rana Zakerzadeh, and Paolo Zunino
3.1 Introduction 197
3.2 Resonance in a Linearized Elastic Solid 202
3.3 An Interesting Case Study: Linear Thermoelasticity 206
3.3.1 The One-Dimensional Case 208
3.3.2 The Higher Dimensional Case 213
3.4 An Abstract Approach 217
3.5 Some Applications 231
3.5.1 Three-Dimensional Linear Thermoelasticity (Revisited) 231
3.5.2 Linear Magnetoelasticity 233
3.5.3 A Liquid–Structure Interaction Problem Showing Generic Absence of Resonance 237

3.5.4 Interaction of a Viscous Liquid with a Thin Structure: The Flat Case 241
3.5.5 Interaction of a Viscous Liquid with a Thin Structure: The Curved Case 244
3.6 Numerical Experiments 249
References 254

4 Topics in the Mathematical Theory of Interactions of Incompressible Viscous Fluid with Rigid Bodies 257
Matthieu Hillairet
4.1 Introduction 257
4.2 Basic Equations 258
4.2.1 Notations 258
4.2.2 Equations of Motion 260
4.2.3 Outline of the Paper 262
4.3 Existence and Uniqueness for the Initial Boundary-Value Problems 262
4.3.1 Many Bodies in a Container: Weak Solutions 263
4.3.2 Motion of One Body in a Bounded Domain: Classical Solutions 281
4.4 The Contact Issue 292
4.4.1 Blow-Up Criterion for Classical Solutions 292
4.4.2 On Weak Solutions with Contact 299
4.4.3 Contact vs No-Contact 305
References 318

5 Numerical Simulation of Fluid–Structure Interaction Problems with Applications to Flow in Vocal Folds 321
Miloslav Feistauer, Petr Sváček, and Jaromír Horáček
5.1 Introduction 322
5.2 Incompressible Flow in Time-Dependent Domains 327
5.2.1 Incompressible Navier–Stokes Equations 328
5.2.2 Arbitrary Lagrangian–Eulerian Method 329
5.2.3 Numerical Approximation of the Incompressible Navier–Stokes Equations 330
5.2.4 Numerical Solution of the Nonlinear Discrete Problem ... 335
5.3 Structural Models 335
5.3.1 Aeroelastic Model of Vocal Folds Vibration with Two Degrees of Freedom 335
5.3.2 Dynamic Elasticity Problem 344
5.3.3 Finite Element Space Discretization of the Elasticity Problem 345
5.3.4 Time Discretization of the Structural Problem 347
5.4 Coupled FSI Problems 348
5.4.1 Coupled Problem of Incompressible Flow and Vocal Fold Rigid Body Model 348

5.4.2 Coupled Problem of Incompressible Flow Problem and Elastic Structure ... 351
5.5 Construction of the ALE Mapping ... 352
5.5.1 Analytical Definition of the ALE Mapping for Two Degrees of Freedom ... 352
5.5.2 Artificial Elasticity Problem ... 353
5.6 FSI Algorithms ... 355
5.6.1 Algorithms for Interaction of Fluid Flow and Vibrating Rigid Body ... 355
5.6.2 Algorithms for Interaction of Fluid and Elastic Structure ... 356
5.7 Numerical Examples ... 358
5.7.1 Interaction of Incompressible Flow with Rigid Body Models of Vocal Folds ... 358
5.7.2 Interaction of Incompressible Flow with Elastic Model of Vocal Folds ... 366
5.8 Interaction of Compressible Flow with Elastic Structure ... 370
5.8.1 Compressible Navier–Stokes Equations ... 371
5.8.2 ALE Form of Compressible Navier–Stokes Equations ... 372
5.9 Discretization of Viscous Compressible Flow ... 372
5.9.1 Discontinuous Galerkin Space Semidiscretization ... 373
5.9.2 Time Discretization by the BDF Method ... 375
5.9.3 Semi-implicit BDF Scheme ... 376
5.9.4 Realization of the Boundary Conditions ... 377
5.9.5 Stabilization by the Local Artificial Viscosity ... 379
5.9.6 Coupling Procedure ... 381
5.10 Numerical Results Obtained by the Discontinuous Galerkin Method ... 382
5.10.1 Flow in a Channel with Prescribed Motion of Walls ... 382
5.10.2 Interaction of Compressible Flow with a Model of Elastic Vocal Folds ... 384
5.11 Conclusion ... 387
5.11.1 Discussion of the Results ... 387
5.11.2 Open Problems and Topics for Future Work ... 388
References ... 388

6 Data Assimilation in Cardiovascular Fluid–Structure Interaction Problems: An Introduction ... 395
Luca Bertagna, Marta D'Elia, Mauro Perego, and Alessandro Veneziani
6.1 Preliminaries ... 396
6.1.1 An Introductory Example ... 400
6.2 Probabilistic Approach ... 402
6.2.1 Basic Notation and Concepts ... 404
6.2.2 Minimum Variance and Other In-Out Estimators ... 406

6.2.3 The Kalman Filter for Linear Problems 412
6.2.4 Extension of the Kalman Filter to Nonlinear Problems 423
6.2.5 The Unscented Kalman Filter 426
6.3 Deterministic Variational Assimilation Methods 428
6.3.1 Least Squares Estimators 430
6.3.2 Constrained Minimization Problems with PDEs: A Simple Working Example 432
6.3.3 Reducing the Costs Via Solution Reduction 445
6.4 Some Applications of Data Assimilation in Hemodynamics Problems 449
6.4.1 Assimilation of Velocity Measures in Blood Flow Simulations 449
6.4.2 Estimation of the Arterial Compliance from Measurements of Displacement: An Inverse Fluid–Structure Interaction Problem 460
6.5 Conclusions 476
References 477

7 Mathematical Models for Blood Coagulation 483
Tomáš Bodnár, Antonio Fasano, and Adélia Sequeira
7.1 Introduction 483
7.2 Historical Remarks 484
7.3 Cells and Proteins Intervening in the Formation and Dissolution of Clots 489
7.3.1 Blood Cells 489
7.3.2 Platelet Regulators 491
7.3.3 The Coagulation "Factors" 492
7.3.4 Fibrinolysis Factors 496
7.3.5 Factors Inhibitors 496
7.4 The Cell-Based Model for Secondary Hemostasis 498
7.4.1 Secondary Hemostasis 498
7.4.2 Fibrinolysis (Fig. 7.8) 503
7.5 Bleeding Disorders 504
7.5.1 Platelet-Related Bleeding Disorders 504
7.5.2 Disorders Due to Coagulation Factors Deficiency 506
7.5.3 Disorders Due to Proteins C, S, Z or to Vitamin K Deficiency 507
7.5.4 Deep Vein (or Venous) Thrombosis (DVT) 507
7.5.5 Heart Arrhythmia and Thromboembolism 508
7.5.6 Coagulation on Artificial Surfaces 508
7.5.7 Disseminated Intravascular Coagulation 509
7.6 The 3-Pathway Cascade Model 509

7.7 Mathematical Description of Coagulation Models 511
7.7.1 Scale-Based Classification of Coagulation Models 512
7.7.2 Feature-Based Classification of Coagulation Models 538
7.8 Conclusions and Remarks 556
References 557

Chapter 1
Mathematical and Numerical Analysis of Some FSI Problems

Céline Grandmont, Mária Lukáčová-Medvid'ová, and Šárka Nečasová

Abstract In this chapter we deal with some specific existence and numerical results applied to a 2D/1D fluid–structure coupled model, for an incompressible fluid and a thin elastic structure. We will try to underline some of the mathematical and numerical difficulties that one may face when studying this kind of problems such as the geometrical nonlinearities or the added mass effect. In particular we will point out the link between the strategies of proof of weak or strong solutions and the possible algorithms to discretize these type of coupled problems.

Keywords ALE method • Elastic structure • Existence of the weak solution • Fluid–structure interaction • Geometric conservation laws • Hemodynamics • Kinematic coupling algorithm • Navier–Stokes equations • Non-Newtonian shear-dependent fluids • Operator splitting

MSC2010: 35Q35, 74F10, 76Z05, 76M10, 76M12, 74S05, 74S10

1.1 Introduction

Many physical phenomena deal with a fluid interacting with a moving rigid or deformable structure. These kinds of problems have a lot of important applications, for instance, in aeroelasticity, biomechanics, hydroelasticity, sedimentation, etc. From the analytical point of view as well as from the numerical point of view they have been studied extensively over the past years. The purpose of the present chapter is to present an overview of some of the analytical and numerical difficulties that may be encountered when dealing with fluid–structure interaction problems and how one can deal with these difficulties. We will focus on the case where the fluid is a viscous incompressible fluid and interacts with a thin elastic structure located on one part of the fluid boundary domain. We will assume that the deformations

T. Bodnár et al. (eds.), *Fluid-Structure Interaction and Biomedical Applications*, Advances in Mathematical Fluid Mechanics, DOI 10.1007/978-3-0348-0822-4_1,

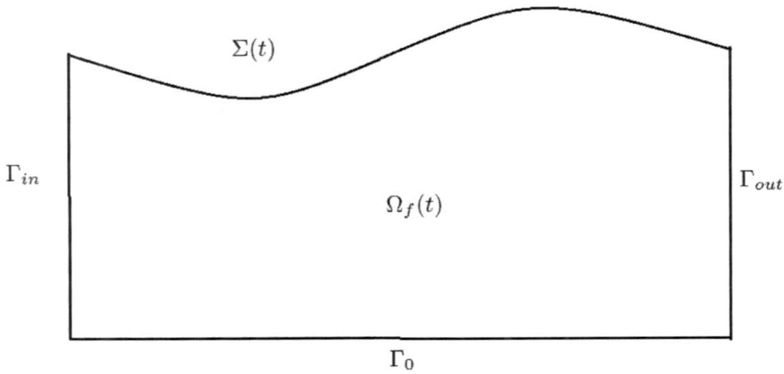

Fig. 1.1 Description of the fluid domain $\Omega_f(t)$

of the fluid domain cannot be neglected and which consequently depends on the structure displacement. The fluid evolves according to the structure displacement itself resulting from the fluid force. We restrict ourselves to the two-dimensional case. Note that the model we will consider in what follows can be viewed as a first model to describe the blood flow in large arteries [138].

Let us start by setting the full nonlinear coupled problem we will study in what follows. The fluid is described by the unsteady Navier–Stokes equations (or by incompressible non-Newtonian fluid system), whereas the structure will be a thin linear elastic structure. We consider $\Omega_f(t)$ the domain occupied by the fluid at time t, $\Omega_f(t) \subset \mathbb{R}^2$. Furthermore, we assume that the fluid boundary can be decomposed into four parts: $\partial\Omega_f(t) = \Sigma(t) \cup \overline{\Gamma_0} \cup \overline{\Gamma_{\text{in}}} \cup \overline{\Gamma}_{\text{out}}$, see Fig. 1.1.

The first part $\Sigma(t)$ denotes the elastic wall. The boundary $\Sigma(t)$ is consequently the deformed configuration of the structure which is an unknown of the problem depending on the structure displacement, whereas Γ_0, Γ_{in}, Γ_{out} are fixed given boundaries where different kind of boundary conditions could be applied. We denote by $\hat{\Omega}_f$ and $\hat{\Sigma}$ the reference configurations of the fluid and of the structure, respectively. Here we consider that $\hat{\Sigma}$ is flat and equal to $(0, L) \times \{R\}$ and that $\hat{\Omega}_f = (0, L)\times(0, R)$. The behavior of the structure is described by the displacement of each point of the reference configuration. Then each point $\hat{\boldsymbol{x}}$ of $\hat{\Sigma}$ occupies at time t the position $\boldsymbol{x}(t) = \hat{\boldsymbol{x}} + \boldsymbol{d}(\hat{\boldsymbol{x}}, t)$ where $\boldsymbol{d}$ denotes the displacement of the structure that satisfies the constitutive equations that describe the structure motion.

Concerning the fluid, most of the descriptions are Eulerian ones. The unknowns (velocity, pressure) are evaluated at each time and at each point of the physical domain. The resolution of the fluid part, if one considers a Newtonian flow, then consists in: *finding* $(\boldsymbol{u}, p)$ *defined over* $\Omega_f(t)$ *such that*

$$\rho_f \partial_t \boldsymbol{u} + \rho_f(\boldsymbol{u}.\nabla)\boldsymbol{u} - \mu\Delta\boldsymbol{u} + \nabla p = \boldsymbol{f}, \quad \text{in } \Omega_f(t), \tag{1.1}$$

$$\operatorname{div} \boldsymbol{u} = 0, \quad \text{in } \Omega_f(t), \tag{1.2}$$

where $\boldsymbol{u}$ denotes the fluid velocity, p its pressure, ν its viscosity, ρ_f its density (assumed to be constant). Moreover $\boldsymbol{f}$ denotes a given exterior load. Equations (1.1) stand for the momentum conservation whereas (1.2) traduces the incompressibility constraint. These equations are completed with initial data

$$\boldsymbol{u}(t=0,\cdot) = \boldsymbol{u}_0(\cdot), \quad \text{in } \Omega_f(0), \tag{1.3}$$

and boundary conditions.

On the boundaries which are outside the fluid–structure interface, i.e., on $\Gamma_f = \partial\Omega_f(t) \setminus \Sigma(t)$, one can consider different kind of boundary conditions. For instance, one could impose homogeneous Dirichlet condition over Γ_f:

$$\boldsymbol{u} = \boldsymbol{0}, \quad \text{on } \Gamma_f. \tag{1.4}$$

This type of boundary conditions means that the fluid is enclosed in the cavity $\Omega_f(t)$. It is clear that in the most of the applications, other kind of boundary conditions are to be considered. For instance, in the case of blood flow modeling or in the case of the air flow in the respiratory system where one considers only a portion of the arterial network, or a portion of the bronchial tree, some parts of Γ_f such that Γ_{in} and Γ_{out} may represent artificial boundaries. If measurement of the velocity is available (which is rather unlikely) over Γ_{in} and Γ_{out} one may impose given Dirichlet boundary conditions. Otherwise we may also have access to some pressure measurements and then impose the Neumann boundary conditions that write

$$\boldsymbol{\sigma}_f(\boldsymbol{u}, p) \cdot \boldsymbol{n} = -p_{\text{in}}\boldsymbol{n}, \quad \text{on } \Gamma_{\text{in}}, \tag{1.5}$$

$$\boldsymbol{\sigma}_f(\boldsymbol{u}, p) \cdot \boldsymbol{n} = -p_{\text{out}}\boldsymbol{n}, \quad \text{on } \Gamma_{\text{out}}, \tag{1.6}$$

completed with

$$\boldsymbol{u} = \boldsymbol{0}, \quad \text{on } \Gamma_0 = \Gamma_f \setminus (\Gamma_{\text{in}} \cup \Gamma_{\text{out}}), \tag{1.7}$$

where $\boldsymbol{\sigma}_f(\boldsymbol{u}, p) = -p\boldsymbol{I} + 2\nu\mathbb{D}(\boldsymbol{u})$ denotes the fluid stress tensor, $\mathbb{D}(\boldsymbol{u})$ stands for the fluid strain tensor, $\mathbb{D}(\boldsymbol{u}) = \frac{1}{2}(\nabla\boldsymbol{u} + (\nabla\boldsymbol{u})^T)$, and $\boldsymbol{n}$ denotes the exterior unit normal to the considered boundary.

Note that many other boundary conditions could be imposed, for instance

$$\boldsymbol{\sigma}_f(\boldsymbol{u}, p) \cdot \boldsymbol{n} \cdot \boldsymbol{n} = -p_{\text{in}}, \boldsymbol{u} \cdot \boldsymbol{\tau} = 0, \quad \text{on } \Gamma_{\text{in}}, \tag{1.8}$$

$$\boldsymbol{\sigma}_f(\boldsymbol{u}, p) \cdot \boldsymbol{n} \cdot \boldsymbol{n} = -p_{\text{out}}, \boldsymbol{u} \cdot \boldsymbol{\tau} = 0, \quad \text{on } \Gamma_{\text{out}}, \tag{1.9}$$

or boundary conditions on the pressure only (see [32]), or a prescribe velocity profile of unknown intensity (see [93]).

One could also consider simplified reduced models that represent the part of the fluid–structure domain that have been cut. These models will then be coupled to the Navier–Stokes system. We refer to [103] for a theoretical study of the steady and unsteady Navier–Stokes system with a given prescribed flux, or average pressure. We also refer to [69–71, 138, 140, 157] for the case of blood flows and $3D/0D$ coupling or $3D/1D$ coupling and to [9] for the case of air flow modeling.

Concerning the structural part, we consider here a thin elastic structure whose displacement $\boldsymbol{d}$ is decomposed into two components: ξ the longitudinal one and η the radial or transversal one. For the sake of simplicity we assume that ξ and η satisfy two linear decoupled equations, which correspond to a rod or beam model. One could consider also more complex model such as shell models (Koiter, Nagdhi, etc.) where ξ and η are coupled, see, e.g., Bukač et al. [21], Raghu et al. [142], Čanić et al. [24], and the references therein.

The considered equations satisfied by $\boldsymbol{d}(t,x,R) = (\xi(t,x), \eta(t,x))^T$ in the reference configuration $(0,L)$ are:

$$\rho_s e \partial_{tt}\xi - \alpha_1 \partial_{xx}\xi - \beta_1 \partial_{xx}\partial_t \xi = -(\boldsymbol{T}_f)_1 + g_1, \text{ on } (0,L), \tag{1.10}$$

$$\rho_s e \partial_{tt}\eta + \gamma \partial_x^4 \eta - \alpha_2 \partial_{xx}\eta - \beta_2 \partial_{xx}\partial_t \eta = -(\boldsymbol{T}_f)_2 + g_2, \text{ on } (0,L), \tag{1.11}$$

where ρ_s denotes the density of the structure (assumed to be constant), e the thickness of the elastic wall. The positive constants γ, $\alpha_i, i = 1,2$ denote mechanical constants depending on the elastic properties of the media and the $\beta_i, i = 1,2$ some additional viscous damping. Two types of load act on the structure: the load $\boldsymbol{T}_f$ coming from the fluid, that will be defined later on, and a given external surfacic load denoted $\boldsymbol{g}$. Note that one could add to (1.11) the term $-\frac{2}{3}\rho_s e^3 \partial_{xx}\partial_{tt}\eta$ that represents the inertia of rotation.

These equations have to be completed with the initial conditions:

$$\begin{aligned} &\xi(t=0,\cdot) = \xi_0(\cdot), \eta(t=0,\cdot) = \eta_0(\cdot) \\ &\partial_t \xi(t=0,\cdot) = \xi_1(\cdot), \partial_t \eta(t=0,\cdot) = \eta_1(\cdot) \text{ on } (0,L), \end{aligned} \tag{1.12}$$

and by the boundary conditions. We can assume here that the rod is clamped and thus prescribe homogeneous Dirichlet boundary conditions, even if these type of conditions are not reflecting reality in the case of blood flow modeling. Thus we have, for instance

$$\xi(0) = \xi(L) = 0 \text{ and } \eta(0) = \eta(L) = 0 \tag{1.13}$$

$$\partial_x \eta(0) = \partial_x \eta(L) = 0, \text{ whenever } \gamma > 0. \tag{1.14}$$

The next question that we aim to study is:

What are the coupling conditions between the fluid and the structure?

First, we will assume that the fluid sticks to the elastic boundary and consequently the fluid velocity and the structure velocity are equal at the interface. That writes, taking into account the fact that the fluid velocity is written in the Eulerian coordinates and the structure velocity in the Lagrangian ones:

$$\boldsymbol{u}(t, \hat{\boldsymbol{x}} + \boldsymbol{d}(t, \hat{\boldsymbol{x}})) = \partial_t \boldsymbol{d}(t, \hat{\boldsymbol{x}}), \text{ for } (t, \hat{\boldsymbol{x}}) \in (0, T) \times \hat{\Sigma}. \tag{1.15}$$

This can be rewritten in the following form:

$$\boldsymbol{u}(t, x + \xi(t, x), R + \eta(t, x)) = (\partial_t \xi(t, x), \partial_t \eta(t, x))^T \text{ for } (t, x) \in (0, T) \times (0, L).$$

The second coupling condition traduces the action–reaction principle. In the case where one considers the coupling of a fluid with a thick structure this principle leads to the equality of the normal component of the fluid and structure stress tensors. Here the structure is thin and consequently the action of the fluid on the structure appears in the right hand side of the structure equations as a surfacic load. This load $\boldsymbol{T}_f$ can be defined in a variational way by:

$$\int_{\Sigma(t)} (\boldsymbol{\sigma}_f(\boldsymbol{u}, p) \cdot \boldsymbol{n}) \cdot \boldsymbol{v} d\sigma_t = \int_{\hat{\Sigma}} \boldsymbol{T}_f \cdot \overline{\boldsymbol{v}} d\sigma, \quad \forall \boldsymbol{v} \tag{1.16}$$

with $\overline{\boldsymbol{v}}(t, \hat{\boldsymbol{x}}) = \boldsymbol{v}(t, \hat{\boldsymbol{x}} + \boldsymbol{d}(t, \hat{\boldsymbol{x}}))$ for each $\hat{\boldsymbol{x}} \in \hat{\Sigma}$ and $\boldsymbol{n}$ denoting the exterior unit normal to the deformed wall $\Sigma(t)$. The quantity $\boldsymbol{T}_f$ represents the surface force, written in the reference configuration, applied by to fluid on the structure.

Note that other types of coupling conditions have been considered in the literature. For instance in [100], the existence of weak solution is studied in the case, where $\gamma > 0$ and $\xi = 0$. In their study the kinematic condition simply ensures that the deformable boundary is impermeable.

In order to rewrite $\boldsymbol{T}_f$ in a strong way, let us introduce $\boldsymbol{\phi}_f$ a deformation of the fluid domain that maps $[0, T] \times \hat{\Omega}_f$ into $\cup_t \{t\} \times \Omega_f(t)$ and such that $\boldsymbol{\phi}_f(t, \hat{\boldsymbol{x}}) = \hat{\boldsymbol{x}} + \boldsymbol{d}(t, \hat{\boldsymbol{x}})$ for $(t, \hat{\boldsymbol{x}}) \in [0, T] \times \hat{\Sigma}$. Let $\boldsymbol{F}_f$ be the deformation gradient and J_f the Jacobian of $\boldsymbol{\phi}_f$. The equality (1.16) writes

$$\boldsymbol{T}_f = J_f \boldsymbol{\sigma}_f(\boldsymbol{u}, p) \cdot (\boldsymbol{F}_f)^{-T} \cdot \hat{\boldsymbol{n}}, \tag{1.17}$$

where $\hat{\boldsymbol{n}}$ denotes the exterior unit normal to the reference interface $\hat{\Sigma}$.

Remark 1.1. The previous definition of $\boldsymbol{T}_f$ depends only on the displacement $\boldsymbol{d}$ of the elastic wall (which is a part of the fluid boundary) and not on the particular chosen extension $\boldsymbol{\phi}_f$.

The third coupling condition comes from the fact that the fluid equations are written in the Eulerian coordinates and consequently the fluid domain $\Omega_f(t) = \boldsymbol{\phi}_f(t, \hat{\Omega}_f)$ is unknown and depends on the displacement of the structure.

Remark 1.2. Whenever we consider homogeneous boundary conditions for the fluid (1.4) or mixed boundary conditions (1.5), (1.6), (1.7), the pressure is always defined in a unique way and not up to an additional constant. This is clear in the case when Neumann boundary conditions are prescribed. In the case of Dirichlet boundary conditions the average pressure is determined by adding an additional constraint to the system that states that, since the fluid is incompressible, the total volume of the fluid cavity is preserved. In our case it writes:

$$\int_0^L \partial_t \eta(1 + \partial_x \xi) - \int_0^L \partial_t \xi \partial_x \eta = 0, \tag{1.18}$$

which is equivalent to

$$\int_0^L \eta(1 + \partial_x \xi) = C, \tag{1.19}$$

where C is a given constant.

From the mathematical point of view the average pressure is the Lagrange multiplier associated with (1.18) or (1.19).

We are now well prepared to write the variational formulation of the coupled problem and derive, at least formally, a priori energy estimates satisfied by any smooth enough solution.

1.1.1 Variational Formulation and Energy Estimates

The following calculations are formal and based on the assumption that the solution $(\boldsymbol{u}, p, \boldsymbol{d})$ of the system and the chosen test functions are regular enough.

Let $\boldsymbol{v}$ be a fluid test function satisfying $\boldsymbol{v} = \boldsymbol{0}$ on Γ_f in the case of homogeneous boundary conditions or $\boldsymbol{v} = \boldsymbol{0}$ on Γ_0 in the case of mixed Neumann–Dirichlet boundary conditions. Let $(b_1, b_2)^T$ be a structure test function satisfying the same homogeneous Dirichlet boundary conditions (1.13) as $(\xi, \eta)^T$.

We multiply the fluid conservation of momentum (1.1) by $\boldsymbol{v}$ and integrate over the fluid domain $\Omega_f(t)$. In the same way, we multiply the structure equations (1.10), (1.11), respectively, by b_1 and b_2 and integrate over the structure reference domain $(0, L)$. After integration by parts and by summing up the fluid and the structure contributions, we obtain in the case where $\boldsymbol{v} = \boldsymbol{0}$ over Γ_f

$$\begin{aligned}
&\int_{\Omega_f(t)} \rho_f \partial_t \boldsymbol{u} \cdot \boldsymbol{v} + \int_{\Omega_f(t)} \rho_f (\boldsymbol{u} \cdot \nabla)\boldsymbol{u} \cdot \boldsymbol{v} + \nu \int_{\Omega_f(t)} \mathbb{D}(\boldsymbol{u}) : \mathbb{D}(\boldsymbol{v}) \\
&\quad - \int_{\Omega_f(t)} p \operatorname{div} \boldsymbol{v} - \int_{\Sigma(t)} \boldsymbol{\sigma}_f(\boldsymbol{u}, p) \cdot \boldsymbol{n} \cdot \boldsymbol{v} \\
&+ \int_0^L \rho_s e \, \partial_{tt} \xi \, b_1 + \alpha_1 \int_0^L \partial_x \xi \partial_x b_1 + \beta_1 \int_0^L \partial_x \partial_t \xi \, \partial_x b_1
\end{aligned}$$

$$+\int_0^L \rho_s e\,\partial_{tt}\eta\, b_2 + \gamma\int_0^L \partial_{xx}\eta\,\partial_{xx}b_2 + \alpha_2\int_0^L \partial_x\eta\,\partial_x b_2 + \beta_2\int_0^L \partial_x\partial_t\eta\,\partial_x b_2$$
$$= -\int_{\hat{\Sigma}} \boldsymbol{T}_f\cdot\boldsymbol{b} + \int_{\Omega_f(t)} \boldsymbol{f}\cdot\boldsymbol{v} + \int_0^L g_1 b_1 + \int_0^L g_2 b_2,$$
$$\forall(\boldsymbol{v},\boldsymbol{b}), \boldsymbol{b}(\hat{\boldsymbol{x}}) = (b_1(x), b_2(x))^T\,\forall\hat{\boldsymbol{x}} = (x,R)\in\hat{\Sigma}, \boldsymbol{v} = \boldsymbol{0} \text{ on } \Gamma_f.$$

Note that the terms at the fluid–structure interface are

$$\int_{\Sigma(t)} \boldsymbol{\sigma}_f(\boldsymbol{u},p)\cdot\boldsymbol{n}\cdot\boldsymbol{v} - \int_{\hat{\Sigma}} \boldsymbol{T}_f\cdot\boldsymbol{b} = \int_{\hat{\Sigma}} \boldsymbol{T}_f\cdot(\overline{\boldsymbol{v}}-\boldsymbol{b}), \tag{1.20}$$

thanks to the definition of $\boldsymbol{T}_f$. By choosing the test functions satisfying the kinematic condition:
$\boldsymbol{b}(t,\hat{\boldsymbol{x}}) = \boldsymbol{v}(t,\hat{\boldsymbol{x}}+\boldsymbol{d}(t,\hat{\boldsymbol{x}}))$ for $\hat{\boldsymbol{x}}\in\hat{\Sigma}$ and by using the definition (1.16) of $\boldsymbol{T}_f$, we obtain the following variational formulation of the problem:

$$\int_{\Omega_f(t)} \rho_f\partial_t\boldsymbol{u}\cdot\boldsymbol{v} + \int_{\Omega_f(t)} \rho_f(\boldsymbol{u}\cdot\nabla)\boldsymbol{u}\cdot\boldsymbol{v} + \nu\int_{\Omega_f(t)} \mathbb{D}(\boldsymbol{u}):\mathbb{D}(\boldsymbol{v}) - \int_{\Omega_f(t)} p\,\mathrm{div}\boldsymbol{v}$$
$$+\int_0^L \rho_s e\,\partial_{tt}\xi\, b_1 + \alpha_1\int_0^L \partial_x\xi\partial_x b_1 + \beta_1\int_0^L \partial_x\partial_t\xi\,\partial_x b_1$$
$$+\int_0^L \rho_s e\,\partial_{tt}\eta\, b_2 + \gamma\int_0^L \partial_{xx}\eta\,\partial_{xx}b_2 + \alpha_2\int_0^L \partial_x\eta\,\partial_x b_2 + \beta_2\int_0^L \partial_x\partial_t\eta\,\partial_x b_2$$
$$= \int_{\Omega_f(t)} \boldsymbol{f}\cdot\boldsymbol{v} + \int_{\hat{\Sigma}} \boldsymbol{g}\cdot\boldsymbol{b},$$
$$\forall(\boldsymbol{v},\boldsymbol{b}) \text{ such that } \boldsymbol{v} = \boldsymbol{0} \text{ on } \Gamma_f,$$
$$\boldsymbol{b}(t,\hat{\boldsymbol{x}}) = (b_1(t,x), b_2(t,x))^T = \boldsymbol{v}(t,\hat{\boldsymbol{x}}+\boldsymbol{d}(t,\hat{\boldsymbol{x}})) \text{ for } \hat{\boldsymbol{x}} = (x,R)\in\hat{\Sigma}. \tag{1.21}$$

We have to underline the fact that the previous weak formulation is not standard since the test functions depend on time and on the unknown solution. This is due to the difference between the Lagrangian formulation for the structure motion and the Eulerian one for the fluid description.

In the case where we have mixed Neumann–Dirichlet boundary conditions, we obtain

$$\int_{\Omega_f(t)} \rho_f\partial_t\boldsymbol{u}\cdot\boldsymbol{v} + \int_{\Omega_f(t)} \rho_f(\boldsymbol{u}\cdot\nabla)\boldsymbol{u}\cdot\boldsymbol{v} + \nu\int_{\Omega_f(t)} \mathbb{D}(\boldsymbol{u}):\mathbb{D}(\boldsymbol{v})$$
$$-\int_{\Omega_f(t)} p\,\mathrm{div}\boldsymbol{v} - \int_{\Sigma(t)} \boldsymbol{\sigma}_f(\boldsymbol{u},p)\cdot\boldsymbol{n}\cdot\boldsymbol{v}$$
$$-\int_{\Gamma_{\mathrm{in}}} \boldsymbol{\sigma}_f(\boldsymbol{u},p)\cdot\boldsymbol{n}\cdot\boldsymbol{v} - \int_{\Gamma_{\mathrm{out}}} \boldsymbol{\sigma}_f(\boldsymbol{u},p)\cdot\boldsymbol{n}\cdot\boldsymbol{v}$$

$$
\begin{aligned}
&+\int_0^L \rho_s e\, \partial_{tt}\xi\, b_1 + \alpha_1 \int_0^L \partial_x \xi \partial_x b_1 + \beta_1 \int_0^L \partial_x \partial_t \xi\, \partial_x b_1 \\
&+\int_0^L \rho_s e\, \partial_{tt}\eta\, b_2 + \gamma \int_0^L \partial_{xx}\eta\, \partial_{xx} b_2 + \alpha_2 \int_0^L \partial_x \eta\, \partial_x b_2 + \beta_2 \int_0^L \partial_x \partial_t \eta\, \partial_x b_2 \\
&= -\int_{\hat{\Sigma}} \boldsymbol{T}_f \cdot \boldsymbol{b} + \int_{\Omega_f(t)} \boldsymbol{f} \cdot \boldsymbol{v} + \int_0^L g_1 b_1 + \int_0^L g_2 b_2, \\
&\forall (\boldsymbol{v}, \boldsymbol{b}), \boldsymbol{b}(\hat{\boldsymbol{x}}) = (b_1(x), b_2(x))^T \text{ for } \hat{\boldsymbol{x}} = (x, R) \in \hat{\Sigma}, \boldsymbol{v} = \boldsymbol{0} \text{ on } \Gamma_0.
\end{aligned}
$$

Taking into account the Neumann boundary conditions on Γ_{in} and Γ_{out} we get

$$
\begin{aligned}
&\int_{\Omega_f(t)} \rho_f \partial_t \boldsymbol{u} \cdot \boldsymbol{v} + \int_{\Omega_f(t)} \rho_f (\boldsymbol{u} \cdot \nabla)\boldsymbol{u} \cdot \boldsymbol{v} + \nu \int_{\Omega_f(t)} \mathbb{D}(\boldsymbol{u}) : \mathbb{D}(\boldsymbol{v}) - \int_{\Omega_f(t)} p \operatorname{div} \boldsymbol{v} \\
&+\int_0^L \rho_s e\, \partial_{tt}\xi\, b_1 + \alpha_1 \int_0^L \partial_x \xi \partial_x b_1 + \beta_1 \int_0^L \partial_x \partial_t \xi\, \partial_x b_1 \\
&+\int_0^L \rho_s e\, \partial_{tt}\eta\, b_2 + \gamma \int_0^L \partial_{xx}\eta\, \partial_{xx} b_2 + \alpha_2 \int_0^L \partial_x \eta\, \partial_x b_2 + \beta_2 \int_0^L \partial_x \partial_t \eta\, \partial_x b_2 \\
&= \int_{\Omega_f(t)} \boldsymbol{f} \cdot \boldsymbol{v} - \int_{\Gamma_{\text{out}}} p_{\text{out}} \boldsymbol{v} \cdot \boldsymbol{n} - \int_{\Gamma_{\text{in}}} p_{\text{in}} \boldsymbol{v} \cdot \boldsymbol{n} + \int_{\hat{\Sigma}} \boldsymbol{g} \cdot \boldsymbol{b}, \\
&\forall (\boldsymbol{v}, \boldsymbol{b}) \text{ such that } \boldsymbol{v} = \boldsymbol{0} \text{ on } \Gamma_0, \\
&\boldsymbol{b}(t, \hat{\boldsymbol{x}}) = (b_1(t, x), b_2(t, x))^T = \boldsymbol{v}(t, \hat{\boldsymbol{x}} + \boldsymbol{d}(t, \hat{\boldsymbol{x}})) \text{ for } \hat{\boldsymbol{x}} = (x, R) \in \hat{\Sigma}.
\end{aligned}
\tag{1.22}
$$

Energy Estimates

In order to obtain the energy estimates satisfied by $(\boldsymbol{u}, p)$ we choose the fluid velocity $\boldsymbol{u}$ and the structure velocity $\partial_t \boldsymbol{d}$ as test functions in the previous variational formulations (1.21) or (1.22). These test functions are admissible since they satisfy the kinematic condition (1.15). Thanks to the fluid incompressibility, in the case of homogeneous Dirichlet boundary conditions on Γ_f and since the interface $\Sigma(t)$ moves at the structure velocity which is equal to the fluid velocity, we note that due to the Reynolds transport theorem it holds

$$
\int_{\Omega_f(t)} \rho_f \partial_t \boldsymbol{u} \cdot \boldsymbol{u} + \int_{\Omega_f(t)} \rho_f (\boldsymbol{u} \cdot \nabla)\boldsymbol{u} \cdot \boldsymbol{u} = \frac{d}{dt} \int_{\Omega_f(t)} \frac{\rho_f}{2} |\boldsymbol{u}|^2. \tag{1.23}
$$

Consequently the kinetic energy of the fluid appears leading to the following energy balance:

$$\underbrace{\int_{\Omega_f(t)} \frac{\rho_f}{2}|\boldsymbol{u}|^2(t)}_{\text{Fluid kinetic energy}} + \underbrace{\nu \int_0^t \int_{\Omega_f(t)} |\mathbb{D}(\boldsymbol{u})|^2}_{\text{Fluid dissipation}} + \underbrace{\int_0^L \frac{\rho_s e}{2}\left(|\partial_t \xi|^2 + |\partial_t \eta|^2\right)}_{\text{Kinetic energy of the structure}}$$
$$+ \underbrace{\frac{\alpha_1}{2}\int_0^L |\partial_x \xi|^2 + \frac{\gamma}{2}\int_0^L |\partial_{xx}\eta|^2 + \frac{\alpha_2}{2}\int_0^L |\partial_x \eta|^2}_{\text{Mechanical energy of the structure}}$$
$$+ \underbrace{\frac{\beta_1}{2}\int_0^t\int_0^L |\partial_x \partial_t \xi|^2 + \frac{\beta_2}{2}\int_0^t\int_0^L |\partial_x \partial_t \eta|^2}_{\text{Structure dissipation}}$$
$$= \underbrace{\int_0^t \int_{\Omega_f(s)} \boldsymbol{f}\cdot\boldsymbol{u} + \int_0^t \int_{\hat{\Sigma}} \boldsymbol{g}\cdot\partial_t \boldsymbol{d}}_{\text{Power of the exterior forces}} + \underbrace{\int_{\Omega_f(0)} \frac{\rho_f}{2}|\boldsymbol{u}_0|^2}_{\text{Initial energy of the fluid } = E_f(0)}$$
$$+ \underbrace{\int_0^L \frac{\rho_s e}{2}(|\xi_1|^2 + |\eta_1|^2) + \frac{\alpha_1}{2}\int_0^L |\partial_x \xi_0|^2 + \frac{\gamma}{2}\int_0^L |\partial_{xx}\eta_0|^2 + \frac{\alpha_2}{2}\int_0^L |\partial_x \eta_0|^2}_{\text{Initial energy of the structure } = E_s(0)} \tag{1.24}$$

Remark 1.3. In the case of an elastic or hyperelastic thick structure, the energy takes the same form except for the mechanical energy of the structure which depends on the considered material.

The energy balance (1.24) implies, by using the Gronwall lemma, that for $t \leq T$

$$\int_{\Omega_f(t)} \frac{\rho_f}{2}|\boldsymbol{u}|^2(t) + \nu \int_0^t \int_{\Omega_f(s)} |\mathbb{D}(\boldsymbol{u})|^2 + \int_0^L \frac{\rho_s e}{2}\left(|\partial_t \xi|^2(t) + |\partial_t \eta|^2(t)\right)$$
$$+ \frac{\alpha_1}{2}\int_0^L |\partial_x \xi|^2(t) + \frac{\gamma}{2}\int_0^L |\partial_{xx}\eta|^2(t) + \frac{\alpha_2}{2}\int_0^L |\partial_x \eta|^2(t)$$
$$+ \frac{\beta_1}{2}\int_0^t\int_0^L |\partial_x \partial_t \xi|^2 + \frac{\beta_2}{2}\int_0^t\int_0^L |\partial_x \partial_t \eta|^2$$
$$\leq e^t \left(E_f(0) + E_s(0) + C\int_0^t e^{t-s}\left(\int_{\Omega_f(s)} |\boldsymbol{f}|^2 + \int_0^L |\boldsymbol{g}|^2 \right)\right). \tag{1.25}$$

Note that here we have not used the dissipation coming from the structure to obtain this energy estimate and this is the reason why we have a constant in time whose behavior is exponential. Consequently the previous estimates is also valid for $\beta_1 = \beta_2 = 0$. By taking advantage of both the dissipation of the fluid and the structure one obtains an estimate of the form

$$C\left(E_f(0) + E_s(0) + C\int_0^t \left(\int_{\Omega_f(s)} |\boldsymbol{f}|^2 + \int_0^L |\boldsymbol{g}|^2\right)\right),$$

where C does not depend on time. Consequently in any case, we obtain that, at least formally, if

$$\begin{cases} \boldsymbol{f} \in L^2((0,T)\times\mathbb{R}^2),\ \boldsymbol{g} \in L^2((0,T)\times(0,L)), \\ \xi_0 \in H^1(0,L), \\ \eta_0 \in H^2(0,L) \text{ if } \gamma > 0,\ \eta_0 \in H^1(0,L) \text{ otherwise}, \\ \xi_1 \in L^2(0,L),\ \eta_1 \in L^2(0,L), \end{cases}$$

then

$$\begin{cases} \boldsymbol{u} \in L^\infty(0,T;L^2(\Omega_f(t))), \\ \nabla\boldsymbol{u} \in L^2(0,T;L^2(\Omega_f(t))), \\ \xi \in W^{1,\infty}(0,T;L^2(0,L)) \cap L^2(0,T;H^1(0,L)), \\ \eta \in W^{1,\infty}(0,T;L^2(0,L)) \cap L^2(0,T;H^2(0,L)),\ \text{if } \gamma > 0, \\ \eta \in W^{1,\infty}(0,T;L^2(0,L)) \cap L^2(0,T;H^1(0,L)) \text{ otherwise}, \\ \text{if } \beta_i > 0, i = 1,2,\ \partial_t\xi \in L^2(0,T;H^1(0,L)),\ \partial_t\eta \in L^2(0,T;H^1(0,L)). \end{cases}$$

In the case when the Neumann boundary conditions are prescribed on Γ_{in} and Γ_{out}, then the Newton equality (1.23) has to be modified and the flux of kinetic energy appears at the artificial boundaries:

$$\begin{aligned} \int_{\Omega_f(t)} \rho_f \partial_t\boldsymbol{u}\cdot\boldsymbol{u} + \int_{\Omega_f(t)} \rho_f(\boldsymbol{u}\cdot\nabla)\boldsymbol{u}\cdot\boldsymbol{u} &= \frac{d}{dt}\int_{\Omega_f(t)} \frac{\rho_f}{2}|\boldsymbol{u}|^2 \\ &+ \int_{\Gamma_{\text{in}}} \rho_f\frac{|\boldsymbol{u}|^2}{2}\boldsymbol{u}\cdot\boldsymbol{n} + \int_{\Gamma_{\text{out}}} \rho_f\frac{|\boldsymbol{u}|^2}{2}\boldsymbol{u}\cdot\boldsymbol{n}. \end{aligned} \tag{1.26}$$

Consequently the energy balance becomes

$$\underbrace{\int_{\Omega_f(t)} \frac{\rho_f}{2}|\boldsymbol{u}|^2(t)}_{\text{Fluid kinetic energy}} + \underbrace{\nu\int_0^t\int_{\Omega_f(t)} |\mathbb{D}(\boldsymbol{u})|^2}_{\text{Fluid dissipation}} + \underbrace{\int_0^L \frac{\rho_s e}{2}\left(|\partial_t\xi|^2(t) + |\partial_t\eta|^2(t)\right)}_{\text{Kinetic energy of the structure}}$$

$$
\underbrace{+\frac{\alpha_1}{2}\int_0^L |\partial_x \xi|^2(t) + \frac{\gamma}{2}\int_0^L |\partial_{xx}\eta|^2(t) + \frac{\alpha_2}{2}\int_0^L |\partial_x \eta|^2(t)}_{\text{Mechanical energy of the structure}}
$$

$$
\underbrace{+\frac{\beta_1}{2}\int_0^t\int_0^L |\partial_x\partial_t \xi|^2 + \frac{\beta_2}{2}\int_0^t\int_0^L |\partial_x\partial_t \eta|^2}_{\text{Structure dissipation}}
$$

$$
= \underbrace{\int_0^t\int_{\Omega_f(s)} \boldsymbol{f}\cdot\boldsymbol{u} + \int_0^t\int_{\hat{\Sigma}} \boldsymbol{g}\cdot\partial_t \boldsymbol{d}}_{\text{Power of the exterior forces}} + \underbrace{\int_{\Omega_f(0)} \frac{\rho_f}{2}|\boldsymbol{u}_0|^2}_{\text{Initial energy of the fluid } =E_f(0)}
$$

$$
\underbrace{+\int_0^L \frac{\rho_s e}{2}(|\xi_1|^2 + |\eta_1|^2) + \frac{\alpha_1}{2}\int_0^L |\partial_x \xi_0|^2 + \frac{\gamma}{2}\int_0^L |\partial_{xx}\eta_0|^2 + \frac{\alpha_2}{2}\int_0^L |\partial_x \eta_0|^2}_{\text{Initial energy of the structure } =E_s(0)}
$$

$$
\underbrace{-\int_0^t\int_{\Gamma_{\text{in}}} \rho_f \frac{|\boldsymbol{u}|^2(t)}{2}\boldsymbol{u}(t)\cdot\boldsymbol{n} - \int_0^t\int_{\Gamma_{\text{out}}} \rho_f \frac{|\boldsymbol{u}|^2(t)}{2}\boldsymbol{u}(t)\cdot\boldsymbol{n}}_{\text{Flux of fluid kinetic energy at the interfaces}} \tag{1.27}
$$

These two last additional terms have an undetermined sign and could not be easily estimated by the fluid energy. In dimension 2 one could obtain an energy estimate locally in time and for small initial data but it is not feasible in dimension 3.

We refer to [9, 103, 138] for existence results in the case of Navier–Stokes in a given domain with Neumann type boundary conditions. Note that in these papers the solutions are strong and not weak. Since no energy estimate could be derived, a way to get rid of this difficulty is to modify the initial problem and impose:

$$
\boldsymbol{\sigma}_f\left(\boldsymbol{u}, p + \frac{\rho_f}{2}|\boldsymbol{u}|^2\right)\cdot\boldsymbol{n} = -p_{\text{in}}\boldsymbol{n} \quad \text{on } \Gamma_{\text{in}}, \tag{1.28}
$$

$$
\boldsymbol{\sigma}_f\left(\boldsymbol{u}, p + \frac{\rho_f}{2}|\boldsymbol{u}|^2\right)\cdot\boldsymbol{n} = -p_{\text{out}}\boldsymbol{n} \quad \text{on } \Gamma_{\text{out}}, \tag{1.29}
$$

where p is replaced by the total pressure $p + \frac{\rho_f}{2}|\boldsymbol{u}|^2$. By doing so one obtains an energy estimate and one can hope to prove existence of solutions in the energy space (see [93, 109–111, 121, 129], where such kind of boundary conditions are considered). These type of boundary conditions will also be used in Sect. 1.2.4. Note moreover that imposing the Neumann boundary conditions (1.5), (1.6) for the Navier–Stokes system leads also to numerical issues since the energy entering the computational domain is not controlled, which may induce numerical instabilities [94, 113, 126].

1.1.2 *Difficulties*

This type of coupled problem raises challenging questions from the analytical and numerical point of view, that we will try to review, at least some of them, in the present chapter. But first of all let us describe which type of theoretical and numerical difficulties one face when studying the existence of solutions or simulating these coupled system.

Mathematical Difficulties

Considering the Cauchy problem and consequently the well-posedness of fluid–structure interaction problems the main difficulties are mainly of three types:

- The full system is nonlinear. There are two types of nonlinearities: the nonlinear convective terms of the Navier–Stokes system and the geometrical nonlinearities due to the fact that the fluid equations are set in an unknown domain depending on the structure displacement. In the case of the non-Newtonian fluids we have in addition nonlinearities in the diffusive term.

  One of the key points here is to define a functional framework that enables to give a sense of traces of fluid velocity-fields on moving boundaries. Such function spaces have to be designed keeping in mind that classical methods for proving existence of weak solutions require a fortiori to be able to prove compactness of any bounded family of solutions. In our case, if one considers weak solutions, the a priori bounds imply that ξ is in $H^1(0, L)$ for a.e t and η is in $H^2(0, L)$ for a.e t if $\gamma > 0$, $H^1(0, L)$ otherwise. Note that these regularities are not sufficient for $\Omega_f(t)$ to be a Lipschitz domain. So we need to work with more regular solutions or to consider only transversal motion (i.e., $\xi = 0$).
- The full unsteady system involves two different kinds of PDE. The fluid has a parabolic behavior whereas, in the case of undamped elasticity (i.e., $\beta_i = 0$, $i = 1, 2$), the structure equations are hyperbolic. Consequently, one has a parabolic–hyperbolic coupling inducing a gap between the fluid regularities and the structure regularities.
- The incompressibility of the fluid may be difficult to handle. In particular the pressure load applied by the fluid on the structure may not be estimated easily in suitable spaces. This difficulty is linked to the added mass effect that may also imply numerical instabilities.

Numerical Difficulties

From a numerical point of view one of the difficulties arises from the geometric nonlinearity. The question is then how to follow the moving interface. We will not focus on this aspect here, but many methods have been developed to get rid of the difficulties linked to the moving domain. In particular, one can refer to [43] for the ALE method, [85] for the fictitious domain method, or [133] for the immersed boundary method.

We will focus here on the fluid–structure coupling. Indeed, even if we consider a linear coupled problem, one major issue is how to efficiently discretize in time the coupling conditions (1.15) and (1.17)? The *implicit* (or *strongly coupled schemes*) are stable since they preserve the energy balance at the interface. The *explicit* (or *staggered*) *scheme* are cheaper but do not preserve the energy balance at the interface and may lead to numerical instabilities in particular in the case of strong added mass effect of the fluid on the structure (i.e., ρ_f close to ρ_s), cf. also Proposition 1.1. In order to balance out the energy at the interface and to stabilize explicitly coupled schemes typically inner sub-iterations are needed, see, e.g., [25, 37, 63, 69, 72, 130, 131, 140], and the references therein. Semi-implicit schemes have been introduced based on the implicit treatment of the added mass effect and the explicit treatment of the viscous stress [63]. In the case of a thin structure there is however also another strategy to solve fluid–structure interaction problem without inner sub-iterations. The so-called kinematically coupled schemes treat implicitly only the hydrodynamic fluid–structure coupling (i.e., the added mass effect), whereas the contribution of the elastic structure is treated explicitly. We refer, e.g., to [23, 56, 57, 62, 98, 110, 121] and to our more detailed discussion in Sect. 1.3. See also [82].

The goal of this chapter is to enlighten some of these difficulties, some strategies to overcome them and to present few open questions. Consequently we will review some of the existence results that can be found, as well as some of the numerical schemes. We will only give the key ideas and steps of these results and the reader may refer to corresponding papers for the details.

1.2 Mathematical Analysis

Last years, existence of weak or strong solutions for fluid–structure coupled problems have been the object of numerous researches.

A vast majority of works concern a rigid solid moving in a viscous incompressible Newtonian fluid whose behavior is described by the equations of Navier–Stokes (historically, the weak formulation of the problem of the motion of rigid bodies in viscous fluids has been introduced and studied in [112], and further in [33, 39, 40, 75, 89, 107, 108, 146, 147, 149, 153, 154] for existence of weak or strong solutions). Note that, in these cases, the displacement of the structure remains regular enough and that we have a parabolic-ODE coupling. We refer to Chap. 2 for further details.

In these problems a challenging point is the existence of collisions. Let us first mention that in the case of compressible fluids this problem has been clarified in [52, Lemma 3.1, Corollary 3.1] where it is proven that solids are allowed to touch but not to penetrate one another unless they did so at the initial time. In the incompressible case the situation is different.

In [102, 104] a no-collision result is proven when there is only one body in a bounded two-dimensional cavity. Later on, the result was extended to the three-dimensional situation in [105]. The case of grazing collisions was studied in [106]. In [52] (see p. 287) the case of a rigid sphere surrounded by an incompressible viscous fluid inside a cavity was considered and a "paradoxical" solution to the subsequent problem in which the sphere sticks to the ceiling of the cavity without falling down was constructed. Moreover in [151], collisions, if any, are proved to occur with zero relative *translational* velocity as soon as the boundaries of the rigid objects are smooth and the gradient of the underlying velocity field is square integrable—a hypothesis satisfied by any Newtonian fluid flow of finite energy. The possibility or impossibility of collisions in a viscous fluid is related to the properties of the velocity gradient. A simple argument reveals that the velocity gradient must become singular (unbounded) at the contact point since otherwise the streamlines would be well defined, in particular, they could never meet each other.

In [45], inspired by Feireisl et al. [54], the motion of several rigid bodies in a non-Newtonian fluid of power-law type (see Chap. 1 of [123] for details) is considered. It is shown that not only that a weak solution exists but also that collisions cannot occur in such viscous fluids. The question of the influence of the smoothness of boundary on the existence of collisions was recently investigated in [78] and [77]. Moreover slip boundary conditions at the fluid–structure interface have been considered in [79] and [80] where, respectively, existence of weak solutions is proven up to collision and it is showed that collision may occur. Recently, the existence result was extended up to the contact [28] but for a slightly different problem.

Concerning the motion of rigid bodies in a viscous compressible fluid let us mention [40], then [52], where the existence of global-in-time weak solutions was proved. This case was extended to case with self-gravitation force in [46]. See also regularity results [17].

Comparing with results in the incompressible case, there is no restriction on the existence time, regardless of possible collisions of two or more bodies or contact of a body with the boundary.

In [53] the problem of the *long time behavior* of global-in-time solutions is addressed. The authors restrict themselves to the simplest situation of a rigid *ball* in a viscous fluid occupying a two-dimensional bounded domain. Assuming there is a body force (gravity) acting in the vertical direction, they show that the rigid body approaches, as time tends to infinity, a static state when the body touches the boundary. Thus the contact albeit possibly absent in any finite instant must occur in the asymptotic regime in the long run.

Concerning investigation of motion of bodies in inviscid case one can refer to [83, 84].

Concerning an elastic structure evolving in incompressible flow, we can refer to [41] and [19] where the structure is described by a finite number of eigenmodes or to [10, 16, 119, 120] for an artificially damped elastic structure. In [15] the case of a compressible fluid was also considered. Note that, in these cases, the displacement of the structure remains regular enough and that we have either a parabolic–parabolic or a parabolic-ODE coupling that enables to consider strong solutions as well as weak solutions.

Concerning three-dimensional elastic structure very few results are available. One can refer to [76, 87, 152] in the steady state case and [34, 35, 114] for the full unsteady case. In the later works the authors consider the existence of strong solutions for small enough data locally in time. Note that unrealistic compatibility conditions are required for these existence results and that some drawbacks are known to exist in the proof of [34] (they surely may be overcome but to the price of tedious adaptations).

To be complete, let us mention that strong existence of the motion of elastic bodies in a viscous compressible case was recently investigated in [18].

Concerning the fluid-beam or fluid-plate coupled system one considers in this chapter, the 2D/1D steady state case has been considered in [86] for homogeneous Dirichlet boundary conditions on the boundary Γ_f that is not the fluid–structure interface. Existence of a unique regular enough solution is obtained for small enough applied forces. To our knowledge this is the only work where both the transverse and the longitudinal displacement are considered. In the unsteady framework the only studied case, so far, is the case $\xi = 0$. One may refer to [26] where a 3D/2D coupling is studied and where the structure is a damped plate in flexion (i.e., $\xi = 0$, $\gamma > 0$) and to [88] in the case of a plate in flexion (i.e., $\xi = 0$, $\gamma > 0$ and $\beta_2 = 0$). The previous results deal with the existence of weak solutions, i.e., in the energy spaces. Note that in these problems the displacement of the structure is only in H^2 in space which is not sufficient to imply a Lipschitz regularity of the fluid domain $\Omega_f(t)$. Nevertheless $\Omega_f(t)$ is at least C^0 and taking the advantage of the only transverse motion of the elastic interface one can prove that at least one weak solution exists. These results also apply in the case of a 2D/1D coupled problem with $\xi = 0$ and $\gamma = \beta_2 = 0$ as we will see in Sect. 1.2.2. More recently [129] have considered the 2D/1D coupling with $\xi = 0$, $\gamma > 0$ and $\beta_2 \geq 0$ (note that in this case the fluid domain is Lipschitz) and involving also Neumann type boundary conditions (1.28), (1.29). The authors have given an alternative proof of existence of a weak solutions based on ideas coming from numerical schemes introduced in [98] and further developed in [56, 57, 62, 110, 121]. The proof is then based on a numerical scheme where a splitting strategy is used for the structure part, emphasizing the link between stable numerical schemes and strategies to prove existence of solution. We would like to point out yet another theoretical results concerning the existence of weak solution of the fluid–structure interaction problem based on numerical techniques. In [111] the interaction of a non-Newtonian fluid with the viscoelastic membrane was studied, for inflow/outflow boundary conditions (1.28), (1.29) are applied. The proof is based analogously as the numerical scheme described in [109] on the artificial compressibility approach as well as on the global iterations with respect to the moving domain, cf. also [110]. Such a link to numerics is also present in [119, 120] where the existence of strong solutions for 3D/2D, or 2D/1D coupled problem involving a damped elastic structure is studied (see also [10] for 2D/1D strong solutions). The proofs of [119, 120] are based on a splitting strategy for the Stokes system and on an implicit treatment of the added mass effect. Also we would like to mention [99] where the continuity of the solution with respect to the data was investigated a 2D/1D context and for $\gamma > 0$. The study of the existence of weak solution for unsteady fluid–structure interaction problem for shear-thickening flow was investigated in [111]. This yields the existence of at least one weak solution of the fully coupled unsteady fluid–structure interaction between the non-Newtonian shear-dependent fluid and the elastic string. Note that in this case an additional viscous term is added $\partial_x^4 \partial_t \eta$. For more details see Sect. 1.2.4. Moreover the coupling of a Newtonian (or resp. a generalized Newtonian flow) and a linearly elastic Koiter shell has recently been studied in [118], (resp. [117]). In these studies the mid-surface of the structure is not flat anymore and other types of arguments to prove compactness than the ones we will develop here are used. To complete the references see [49, 68, 90, 91, 139, 141].

The aim of this section is to present some known results on existence of a solution of fluid-beam (or rod) coupled problems and to show how these problems could be approximated or decoupled, how compactness results could be derived. The section is organized as follows: we will consider the unsteady problem and review some of the existing results that can be found on the problem. Note that we will only consider the case where $\xi = 0$ which is the case that has been treated in the literature so far. In a first part we will explain on a simplified linear problem the so-called added mass effect and why it may lead to some mathematical (and numerical) difficulties. Then we will review some results of existence of weak and strong solutions. In particular, we will see how to prove existence of weak solutions and how one can obtain compactness of a sequence of approximated solutions in the case of a damped structure first and then in the undamped case. Next the general ideas of the proof of existence of strong solution will be developed. We will explain the decoupling strategy used. This strategy enables to exploit the properties of each sub-problem but impose to treat carefully the so-called added mass effect. This theoretical section is concluded by presenting the existence result for the interaction of a non-Newtonian fluid with a viscoelastic structure. The proof inherits some techniques from numerical simulations, such as the artificial compressibility and the global iterations with respect to the moving domain (i.e., the Schauder fixed point theorem). We will underline which kind of additional difficulties arise due to the non-Newtonian behavior of the fluid.

1.2.1 A Linear Simplified Problem

Let us consider the following toy problem that has been introduced in [25] to illustrate the role played by the added mass effect on the numerical stability of explicit schemes for fluid–structure interaction problems. For the fluid we consider a perfect, inviscid flow described by the following equations:

$$\rho_f \partial_t \boldsymbol{u} + \nabla p = \boldsymbol{0}, \quad \text{in } \hat{\Omega}_f, \tag{1.30}$$

$$\operatorname{div} \boldsymbol{u} = 0, \quad \text{in } \hat{\Omega}_f, \tag{1.31}$$

$$p = p_{\text{in}}, \quad \text{on } \Gamma_{\text{in}}, \tag{1.32}$$

$$p = p_{\text{out}}, \quad \text{on } \Gamma_{\text{out}}, \tag{1.33}$$

$$\boldsymbol{u} \cdot \boldsymbol{n} = \boldsymbol{0}, \quad \text{on } \Gamma_0. \tag{1.34}$$

Note that we neglect the fluid domain variations and the equations are set in the reference configuration $\hat{\Omega}_f$. Concerning the structure part we consider a thin clamped elastic structure but whose displacement is only vertical and whose transverse component η satisfied, for instance

$$\rho_s e \partial_{tt} \eta - \alpha_2 \partial_{xx} \eta - \beta_2 \partial_{xx} \partial_t \eta = p, \text{ on } (0, L). \tag{1.35}$$

Moreover since the fluid is inviscid the kinematic condition at the interface writes

$$\boldsymbol{u} \cdot \boldsymbol{n} = \partial_t \eta, \text{ on } \hat{\Sigma}, \tag{1.36}$$

Note that here all geometrical and convective nonlinearities have been omitted. We will assume that all the quantities are regular enough to justify the following derivations. Due to the incompressibility constraint (1.31), by taking the divergence of the fluid equation (1.30) we obtain that the pressure satisfies

$$-\Delta p = 0, \quad \text{in } \hat{\Omega}_f, \tag{1.37}$$

$$\frac{\partial p}{\partial \boldsymbol{n}} = -\rho_f \partial_t \boldsymbol{u} \cdot \boldsymbol{n}, \quad \text{on } \hat{\Sigma} \tag{1.38}$$

$$\frac{\partial p}{\partial \boldsymbol{n}} = 0, \quad \text{on } \Gamma_0, \tag{1.39}$$

together with Dirichlet boundary conditions (1.32), (1.33) on Γ_{in} and Γ_{out}.

Due to (1.36), the boundary condition (1.38) writes

$$\frac{\partial p}{\partial \boldsymbol{n}} = -\rho_f \partial_{tt} \eta \quad \text{on } \hat{\Sigma}. \tag{1.40}$$

Consequently we can rewrite the pressure load applied by the fluid on the structure as $p = q - \rho_f \mathcal{M}(\partial_{tt} \eta)$ where q satisfies:

$$-\Delta q = 0, \quad \text{in } \hat{\Omega}_f, \tag{1.41}$$

$$\frac{\partial q}{\partial \boldsymbol{n}} = 0, \quad \text{on } \Gamma_0 \cup \hat{\Sigma}, \tag{1.42}$$

$$q = p_{\text{in}}, \quad \text{on } \Gamma_{\text{in}}, \tag{1.43}$$

$$q = p_{\text{out}}, \quad \text{on } \Gamma_{\text{out}}, \tag{1.44}$$

and $\mathcal{M}$ is the Neumann to Dirichlet operator that associates with any v on Γ_f, the trace over Γ_f of p solution of

$$-\Delta p = 0, \quad \text{in } \hat{\Omega}_f, \tag{1.45}$$

$$\frac{\partial p}{\partial \boldsymbol{n}} = v, \quad \text{on } \hat{\Sigma}, \tag{1.46}$$

$$\frac{\partial p}{\partial \boldsymbol{n}} = 0, \quad \text{on } \Gamma_f, \tag{1.47}$$

$$p = 0, \quad \text{on } \Gamma_{\text{in}} \cup \Gamma_{\text{out}}. \tag{1.48}$$

The mapping $\mathcal{M}$ is continuous from $H^{-1/2}(\hat{\Sigma})$ into $H^{1/2}(\hat{\Sigma})$ and is a compact, self-adjoint, positive operator on $L^2(\hat{\Sigma})$. Consequently the coupled system can be written only in terms of η which satisfies:

$$(\rho_s e\mathcal{I} + \rho_f \mathcal{M})\partial_{tt}\eta - \alpha_2 \partial_{xx}\eta - \beta_2 \partial_{xx}\partial_t \eta = q, \text{ on } (0, L),$$

where $\mathcal{I}$ is the identity mapping. We note that the action of the fluid on the structure can be seen as an added mass through the operator $\mathcal{M}$. Now, one can prove easily that, if $\boldsymbol{f} \in L^2(0, T; L^2(\Omega_f))$, with initial conditions in well-chosen spaces, there exists a weak solution $\eta \in C^1(0, T; H_0^1((0, L))$ of this problem. Nevertheless if one tries to prove existence of a solution by naively decoupling the fluid from the structure, one will be able to do it only in the case where the structure density is large enough compared to the fluid one. Note first that since the structure is mono-dimensional, the natural way to decouple the fluid from the structure is to solve the fluid equations with a given velocity on the fluid–structure boundary. That gives a stress which is applied to the elastic media. Finally one obtains a new structure velocity. The fixed point strategy consists then in associating with a given displacement η_1 and displacement η_2 solution of

$$\rho_s e\partial_{tt}\eta_2 - \alpha_2 \partial_{xx}\eta_2 - \beta_2 \partial_{xx}\partial_t \eta_2 = q - \rho_f \mathcal{M}(\partial_{tt}\eta_1), \text{ on } (0, L).$$

Here one can clearly see that η_2 will have the same regularity in time as η_1. Consequently the fixed point mapping will not be a contraction even locally in time unless the structure density is large enough with respect to the fluid density. This restriction can be found in [89] but is only due to the choice of strategy of proof. That underlines the fact if one wants to decouple the fluid from the structure the added mass effect has to be treated "implicitly" (see [119, 120] in the context of existence results and [1–3, 25, 55, 64] in the context of numerical schemes) or other strategies have to be thought about (see [129] in the context of existence result and [56, 57, 62, 98] in the context of numerical schemes).

In the next subsection, we will consider the existence of weak and strong solutions for the unsteady problem in the case where $\xi = 0$. In the first part, adapted from [26, 88], existence of weak solutions is tackled and the global fluid–structure formulation is preserved to avoid the added mass effect (only the geometrical part being decoupled). In a second part, we explain the strategy used in [119, 120], the proof being based on a suitable decoupling of the fluid–structure problem where the added mass effect is kept implicit.

1.2.2 Existence of Weak Solutions

In this subsection we will consider the case of the full nonlinear coupled problem but with $\xi = 0$. Consequently the fluid domain is a subgraph defined by:

$$\Omega_f(t) = \Omega_\eta(t) = \{(t, x, y) \in \mathbb{R}^2, x \in (0, L),\ 0 < y < R + \eta(t, x)\}.$$

The displacement η will satisfied a wave equation (thus $\gamma = 0$ in (1.11)). In a first part we will consider an additional viscous term ($\beta_2 > 0$) then we study the vanishing viscosity limit ($\beta_2 \to 0$) and we prove that there exists at least one weak solution of the fluid–membrane coupled problem. Moreover to simplify we will consider fluid boundary conditions that allow to obtain energy estimates and consequently impose homogeneous Dirichlet boundary conditions (1.4). Note one could also impose (1.28), (1.29).

The Damped Wave Equation: $\beta_2 > 0$

An energy estimate of the same type as (1.25) shows that one can look for an elastic displacement η in $W^{1,\infty}(0, T; L^2(0, L)) \cap H^1(0, T; H_0^1(0, L))$. Nevertheless this regularity implies that the fluid–structure interface is continuous for all time but not Lipschitz. Consequently the set $\Omega_\eta(t)$ is an open set (till $R + \eta(t, x) > 0$) which not Lipschitz. One of the first questions is then to properly define the functional spaces to which the fluid velocity belongs to and to rigorously define the fluid trace velocity on the moving interface.

Remark 1.4. In the case where one considers a beam equation (i.e., $\gamma > 0$) the energy estimates give that the elastic displacement belongs to $L^\infty(0, T; H_0^2(0, L))$. Thus the mapping $\boldsymbol{\phi}_f$ is a C^1 diffeomorphism as long as $R + \eta > 0$. And consequently $\Omega_\eta(t)$ is Lipschitz.

In our case, we introduce the open set $\Omega_\eta^T = \cup_{t\in(0,T)}\Omega_\eta(t) \times \{t\}$ and we define

$$L^2(0, T; H^1(\Omega_\eta(t))) = \left\{v \in L^2(\Omega_\eta^T),\ \nabla v \in L^2(\Omega_\eta^T)\right\},$$

$$L^2(0, T; H_0^1(\Omega_\eta(t))) = \overline{\mathcal{D}(\Omega_\eta^T)}^{L^2(0,T;H^1(\Omega_\eta(t)))},$$

$$\mathcal{V}_f = \left\{\boldsymbol{v} \in C^1\left(\overline{\Omega_\eta^T}\right),\ \mathrm{div}\boldsymbol{v} = 0,\ \boldsymbol{v} = \mathbf{0} \text{ on } (0, T) \times \Gamma_f\right\},$$

and

$$V_f = \overline{\mathcal{V}_\delta}^{L^2(0,T;H^1(\Omega_\eta(t)))}.$$

Concerning the sense to give to the trace of the fluid velocity on the moving interface, we can easily give a sense in $L^2(0, L)$ to $\boldsymbol{u}(t, x, R + \eta(t, x))$. Indeed the elastic displacement being only transversal and the fluid velocity being H^1 in space, we have:

$$\boldsymbol{u}(t, x, R + \eta(t, x)) = \int_0^{R+\eta(t,x)} \partial_z \boldsymbol{u}(t, x, s)ds.$$

This specificity of our coupled problem is strongly used in [26]. In particular it enables to build divergence free lifting of structure test functions by noting that $(0, b(x))^T$ with $b \in H^1(0, L)$ belongs to $H^1(\Omega_\eta(t))$ and is divergence free. Moreover the only transverse motions imply that $\Omega_\eta(t)$ is a subgraph. Density properties, Korn inequality follow (even if the fluid domain is not Lipschitz) as well as the possibility to contract the fluid domain with respect to the vertical direction.

The weak solutions that are considered are defined by (by choosing all the constants except the viscosity to be equal to 1)

- $\boldsymbol{u} \in V_f \cap L^\infty(0, T; L^2(\Omega_\eta(t)),$
- $\eta \in W^{1,\infty}(0, T; L^2(0, L)) \cap H^1(0, T; H_0^1(0, L)),$
- $\boldsymbol{u}(t, x, R + \eta(t, x)) = (0, \partial_t \eta(t, x))^T$, for a.e. t, on $(0, L)$
- For all $(\boldsymbol{\phi}, b) \in \mathcal{V}_f \times C^1([0, T]; H_0^1(0, L))$ such that $\boldsymbol{\phi}(t, x, R + \eta(t, x)) = (0, b(t, x))^T$, $(t, x) \in [0, T] \times (0, L)$, we have for a.e. t

$$\begin{aligned}
&\int_{\Omega_\eta(t)} \boldsymbol{u}(t)\cdot\boldsymbol{\phi}(t) - \int_0^t \int_{\Omega_\eta(s)} \boldsymbol{u}\cdot\partial_t\boldsymbol{\phi} + 2\nu \int_0^t \int_{\Omega_\eta(s)} \nabla\boldsymbol{u} : \nabla\boldsymbol{\phi}\\
&\quad + \int_0^t \int_{\Omega_\eta(s)} (\boldsymbol{u}\cdot\nabla)\boldsymbol{u}\cdot\boldsymbol{\phi} - \int_0^t \int_0^L (\partial_t\eta)^2 b + \int_0^L \partial_t\eta(t)\, b(t)\\
&\quad - \int_0^t \int_0^L \partial_t\eta\, \partial_t b + \beta_2 \int_0^t \int_0^L \partial_x\partial_t\eta\, \partial_x b + \int_0^t \int_0^L \partial_x\eta\, \partial_x b\\
&= \int_0^t \int_{\Omega_\eta(s)} \boldsymbol{f}\cdot\boldsymbol{\phi} + \int_0^t \int_0^L g\, b + \int_{\Omega_\eta(0)} \boldsymbol{u}_0\boldsymbol{\phi}(0) + \int_0^L \eta_1 b(0). \qquad (1.49)
\end{aligned}$$

Here we have integrated by parts both in space and in time making a new convection term appearing: $\int_0^t \int_0^L (\partial_t\eta)^2 b$. Moreover the fluid dissipation involves only the gradient of $\boldsymbol{u}$ and not its symmetric part. This simplification comes, once again, from the only traverse motion of the elastic structure. Indeed using the equality of the velocity at the interface and the fluid incompressibility leads to

$$(2\mathbb{D}(\boldsymbol{u}) \cdot \boldsymbol{n})_2 = (\nabla \boldsymbol{u} \cdot \boldsymbol{n})_2, \qquad \text{on } \Sigma(t).$$

This identity enables to simplify the weak formulation.

As noted in Remark 1.2, the volume of the fluid domain has to be preserved during the deformations. Unlike the case where $\xi \neq 0$, the condition satisfied by the structure displacement is linear and η should satisfy $\int_0^L \partial_t \eta = 0$. This simplification comes, once again, from the only transverse motion.

Let us assume that the data satisfy the following compatibility conditions:

$$\begin{cases} \min\limits_{[0,L]}(R + \eta_0) > 0, \\ \operatorname{div} \boldsymbol{u}_0 = 0 \text{ in } \Omega_\eta(0), \\ \boldsymbol{u}_0 \cdot \boldsymbol{n} = 0 \text{ on } \Gamma^f, \\ \boldsymbol{u}_0(t, x, R + \eta_0(x)) \cdot \hat{\boldsymbol{n}} = (0, \eta_1)^T \cdot \hat{\boldsymbol{n}} \text{ on } (0, L), \\ \displaystyle\int_0^L \eta_1 = 0. \end{cases} \tag{1.50}$$

The existence theorem can then be enunciated as follows:

Theorem 1.1. *By assuming that the data satisfy*

$$\begin{aligned} &\boldsymbol{f} \in L^2((0,T) \times \mathbb{R}^2), \; g \in L^2((0,T) \times (0,L)), \\ &\boldsymbol{u}_0 \in L^2(\Omega_\eta(0)), \; \eta_1 \in L^2(0,L), \; \eta_0 \in H_0^1(0,L) \end{aligned} \tag{1.51}$$

and verify the compatibility conditions (1.50), there exists at least one weak solution $(\boldsymbol{u}, \eta)$ *to the coupled problem as long as the elastic structure does not touch the bottom of the fluid cavity. Moreover this solution satisfies the energy estimate.*

The proof of this result follows a standard scheme:

- Construction of approximated solutions that satisfy energy bounds;
- Derivation of additional bounds, since the energy bounds are not sufficient to obtain compactness to pass to the limit in the nonlinear terms;
- Passage to the limit.

The first step is to build suitable approximated solutions. As we have seen for the toy problem, to naively decouple the fluid system from the structure equation may lead to difficulties due to the added mass effect. Consequently we will not decouple the fluid from the structure. To get rid of the difficulties of the unsmooth fluid domain and convective nonlinearities, the fluid domain and the convective velocity will be regularized. By doing so, and if we want to conserve a solution satisfying the energy bound, one has either to consider a regularization process imposing that the regularized fluid convective velocity matches the velocity of the domain or to modify the fluid–structure variational formulation (1.49). The first alternative has been used in [39, 41] and relies on a representation of the fluid and structure velocities. Nevertheless it requires more regularity of the structure motion than the one we have here. We then modify the variational formulation by noting

$$\int_{\Omega_\eta(t)} (\boldsymbol{u}\cdot\nabla\boldsymbol{u})\boldsymbol{\phi} = \frac{1}{2}\int_{\Omega_\eta(t)} (\boldsymbol{u}\cdot\nabla\boldsymbol{u})\boldsymbol{\phi} - \frac{1}{2}\int_{\Omega_\eta(t)} (\boldsymbol{u}\cdot\nabla\boldsymbol{\phi})\boldsymbol{u} - \frac{1}{2}\int_0^L (\partial_t\eta)^2 b,$$

with $\boldsymbol{\phi}(x, R+\eta(x)) = (0, b(x))^T$ on $(0, L)$. The convective term is then transformed as follows:

$$\frac{1}{2}\int_{\Omega_\eta^*(t)} (\boldsymbol{u}^*\cdot\nabla\boldsymbol{u})\boldsymbol{\phi} - \frac{1}{2}\int_{\Omega_\eta^*(t)} (\boldsymbol{u}^*\cdot\nabla\boldsymbol{\phi})\boldsymbol{u} - \frac{1}{2}\int_0^L \partial_t\eta^*\partial_t\eta b,$$

where the asterisk denotes regularized quantities.

The modified variational formulation then writes, with $\Omega\eta^*(t)$ denoting the regularized domain:

$$\begin{aligned}
&\int_{\Omega\eta^*(t)} \boldsymbol{u}(t)\cdot\boldsymbol{\phi}(t) - \int_0^t\int_{\Omega\eta^*(s)} \boldsymbol{u}\cdot\partial_t\boldsymbol{\phi} + 2\nu\int_0^t\int_{\Omega_\eta^*(s)} \nabla\boldsymbol{u}:\nabla\boldsymbol{\phi}\\
&\quad \frac{1}{2}\int_0^t\int_{\Omega_\eta^*(s)} (\boldsymbol{u}^*\cdot\nabla\boldsymbol{u})\boldsymbol{\phi} - \frac{1}{2}\int_0^t\int_{\Omega_\eta^*(s)} (\boldsymbol{u}^*\cdot\nabla\boldsymbol{\phi})\boldsymbol{u} + \frac{1}{2}\int_0^t\int_0^L \partial_t\eta^*\partial_t\eta b\\
&\quad -\int_0^t\int_0^L \partial_t\eta\,\partial_t b + \beta_2\int_0^t\int_0^L \partial_x\partial_t\eta\,\partial_x b + \int_0^t\int_0^L \partial_x\eta\,\partial_x b\\
&\quad = \int_0^t\int_{\Omega_\eta^*(s)} \boldsymbol{f}\cdot\boldsymbol{\phi} + \int_0^t\int_0^L g\,b + \int_{\Omega_\eta^*(0)} \boldsymbol{u}_0\boldsymbol{\phi}(0) + \int_0^L \eta_1 b(0),
\end{aligned} \tag{1.52}$$

for test functions satisfying div $\boldsymbol{\phi} = 0$ and $\boldsymbol{\phi}(t, x, R+\eta^*(t, x)) = (0, b(t, x))^T$, for $x \in (0, L)$.

Then by taking ($\boldsymbol{\phi} = \boldsymbol{u}, b = \partial_t\eta$) as test functions we recover the fluid kinetic energy since

$$\begin{aligned}
&\int_{\Omega_\eta^*(t)} |\boldsymbol{u}(t)|^2 - \int_0^t\int_{\Omega_\eta^*(s)} \boldsymbol{u}\cdot\partial_t\boldsymbol{u} + \frac{1}{2}\int_0^t\int_{\Omega_\eta^*(s)} (\boldsymbol{u}^*\cdot\nabla\boldsymbol{u})\boldsymbol{u} - \frac{1}{2}\int_0^t\int_{\Omega_\eta^*(s)} (\boldsymbol{u}^*\cdot\nabla\boldsymbol{u})\boldsymbol{u}\\
&\quad + \frac{1}{2}\int_0^t\int_0^L \partial_t\eta^*(\partial_t\eta)^2 = \frac{1}{2}\int_{\Omega_\eta^*(t)} |\boldsymbol{u}(t)|^2 - \frac{1}{2}\int_{\Omega_\eta^*(0)} |\boldsymbol{u}(0)|^2.
\end{aligned}$$

The existence of at least one weak solution of the approximated problem (1.52) relies on a fixed point argument: we consider a given motion of the fluid domain δ and a given convective fluid velocity $\boldsymbol{v}$, that we regularized and we denote by δ^* and

$\boldsymbol{v}^*$ these regularizations. Denoting by

$$\Omega_\delta(t) = \{(x, y) \in \mathbb{R}^2, x \in (0, L),\ 0 < y < R + \delta(t, x)\}$$

we solve the following linearized-approximated problem:

$$\begin{aligned}
&\int_{\Omega_{\delta^*}(t)} \boldsymbol{u}(t) \cdot \boldsymbol{\phi}(t) - \int_0^t \int_{\Omega_{\delta^*}(s)} \boldsymbol{u} \cdot \partial_t \boldsymbol{\phi} + 2\nu \int_0^t \int_{\Omega_{\delta^*}(s)} \nabla \boldsymbol{u} : \nabla \boldsymbol{\phi} \\
&\frac{1}{2}\int_0^t \int_{\Omega_{\delta^*}(s)} (\boldsymbol{v}^* \cdot \nabla \boldsymbol{u})\boldsymbol{\phi} - \frac{1}{2}\int_0^t \int_{\Omega_\delta^*(s)} (\boldsymbol{v}^* \cdot \nabla \boldsymbol{\phi})\boldsymbol{u} + \frac{1}{2}\int_0^t \int_0^L \partial_t \delta^* \partial_t \eta b \\
&\quad - \int_0^t \int_0^L \partial_t \eta\, \partial_t b + \beta_2 \int_0^t \int_0^L \partial_x \partial_t \eta\, \partial_x b + \int_0^t \int_0^L \partial_x \eta\, \partial_x b \\
&\quad = \int_0^t \int_{\Omega_{\delta^*}(s)} \boldsymbol{f} \cdot \boldsymbol{\phi} + \int_0^t \int_0^L g\, b + \int_{\Omega_{\delta^*}(0)} \boldsymbol{u}_0 \boldsymbol{\phi}(0) + \int_0^L \eta_1 b(0), \qquad (1.53)
\end{aligned}$$

for test functions satisfying $\operatorname{div} \boldsymbol{\phi} = 0$ and $\boldsymbol{\phi}(t, x, R + \delta^*(t, x)) = (0, b(t, x))^T$, for $x \in (0, L)$. Note that now the test functions do not depend on the solution anymore, the problem is linear and the fluid domain motion is smooth. Yet the fluid and the structure equations are still coupled. One can then easily prove that there exists a unique solution to this problem thanks to the Galerkin method for instance. Consequently we are able to associate with $(\boldsymbol{v}, \delta)$ a new velocity $\boldsymbol{u}$ and a new elastic displacement η. The idea to prove the compactness for the Schauder fixed point procedure is to take $(\partial_t \boldsymbol{u}, \partial_{tt}\eta)$ as test functions that will lead to L^2_{tx} bounds of these quantities and then to use the elliptic regularity (note that we are in the case where $\beta_2 > 0$). We remark nevertheless that $(\partial_t \boldsymbol{u}, \partial_{tt}\eta)$ are not admissible test functions. Yet we can easily modify the fluid acceleration to obtain a fluid test function that is divergence free and is equal to the structure acceleration at the interface. Let us emphasize that this construction uses the regularized domain motion. Moreover the L^2_{tx} bounds on $(\partial_t \boldsymbol{u}, \partial_{tt}\eta)$ also require the regularization step. Consequently the additional estimates depend on this regularization parameter.

Once the existence of a solution of the approximated problem is obtained, we need to pass to the limit as the regularization parameter tends to zero. The energy estimates lead to uniform bounds of the approximated solution in $V_f^* \cap L^\infty(0, T; L^2(\Omega_\eta^*(t)))$ for the fluid velocity and in $W^{1,\infty}(0, T; L^2(0, L)) \cap H^1(0, T; H_0^1(0, L))$ for the structure displacement. In particular, we obtain the uniform convergence in C^0 of the sequence of interface displacement that enables to pass to the limit in the sequence of fluid domains. Nevertheless, these bounds are not sufficient to obtain the desired strong convergences. Some additional bounds are needed to obtain the L^2_{tx} compactness of the velocities. This compactness in L^2_{tx} is indeed needed to pass to the limit in the nonlinear convective terms such as $\frac{1}{2}\int_0^t \int_{\Omega_\eta^*(s)} (\boldsymbol{u}^* \cdot \nabla \boldsymbol{u})\boldsymbol{\phi} - \frac{1}{2}\int_0^t \int_{\Omega_\eta^*(s)} (\boldsymbol{u}^* \cdot \nabla \boldsymbol{\phi})\boldsymbol{u}$ or $\frac{1}{2}\int_0^t \int_0^L \partial_t \eta^* \partial_t \eta b$. The estimates obtained at the previous step cannot be used since they depend on the regularization parameter. It is here a key step in the proofs of existence of weak solutions. One may try to apply the Aubin's lemma to prove the desired compactness. Nevertheless it is not straightforward to apply this lemma in the case of divergence free functions defined on moving time-dependent domains. We can refer to [74, 145] for incompressible Navier–Stokes in moving domains or to [39, 40, 146] for existence weak solutions of fluid solid coupled problem.

One option to obtain compactness is to study quantities like: $\|\boldsymbol{u}(t+h)-\boldsymbol{u}(t)\|_{L^2_{tx}}$ and $\|\partial_t\eta(t+h)-\partial_t\eta(t)\|_{L^2_{tx}}$. Indeed, we are going to use the following lemma that characterizes the compact sets of $L^p(0,T;X)$ where X is a Banach space (see [150]).

Lemma 1.1. *Let X be a Banach space and $F \hookrightarrow L^q(0,T;X)$ with $1 \le q < \infty$. Then F is a relatively compact set of $L^q(0,T;X)$ if and only if*

i) $\left\{\int_{t_1}^{t_2} f(t)dt,\, f \in F\right\}$ *is relatively compact in X, $\forall 0 < t_1 < t_2 < T$*
ii) $\|f(t+h)-f(t)\|_{L^q(0,T;X)} \longrightarrow 0$ *as h goes to zero, uniformly with respect to f in F.*

We will now apply Lemma 1.1 to the sequence $F = (\overline{\boldsymbol{u}}, \partial_t\eta)$, indexed by the regularization parameter, $q = 2$ and $X = L^2(B) \times L^2(0,L)$. Note that here we have introduced a set B that contains all the fluid domains $\Omega^*_\eta(t)$ for any $t \in (0,T)$ and we have extended the fluid velocity by defining

$$\overline{\boldsymbol{u}} = \begin{cases} \boldsymbol{u} \text{ in } \Omega^*_\eta(t) \\ (0,\partial_t\eta)^T \text{ in } B \setminus \Omega^*_\eta(t). \end{cases} \tag{1.54}$$

Remark 1.5. This extension relies strongly on the fact that there is only the transverse motion of the elastic structure. Note moreover that, for this reason, $\overline{\boldsymbol{u}}$ is divergence free and that if $\partial_t\eta$ is in H^1 in space (which is the case since $\beta_2 > 0$) this extension is also in $H^1(B)$ in space.

The first point i) of Lemma 1.1 is clearly satisfied thanks to energy estimates and we have to verify the second point. Given any $h > 0$, we denote $g^-(t,\cdot) = g(t-h,\cdot)$ and $g^+(t,\cdot) = g(t+h,\cdot)$. The assertion $ii)$ is a consequence of the following lemma:

Lemma 1.2. *Let $T > 0$ such that* $\min_{[0,T]\times[0,L]}(1+\eta) \ge \alpha > 0$. *Then for all $h > 0$ small enough, we have*

$$\int_0^T\int_B \rho|\overline{\boldsymbol{u}} - (\overline{\boldsymbol{u}})^-|^2 + \int_0^T\int_0^L (\partial_t\eta - \partial_t\eta^-)^2 \le C\sqrt{h}, \tag{1.55}$$

and

$$\int_0^T\int_B |\rho\overline{\boldsymbol{u}} - \rho^-(\overline{\boldsymbol{u}})^-|^2 \le C\sqrt{h}, \tag{1.56}$$

with η extended by η_0 for $t < 0$ (hence $\partial_t \eta$ extended by 0) and $\overline{\boldsymbol{u}}$ extended by 0 for $t < 0$, and where $\rho(t)$ denotes the characteristic function of $\Omega^*_\eta(t)$. These estimates are uniform with respect to the regularization parameter.

The idea to prove this lemma is to choose test functions in the variational formulation (1.52) that looks like $\int_{t-h}^t \boldsymbol{u}$ and $\int_{t-h}^t \partial_t \eta$. But since we are dealing with moving domains we have to slightly modify these test functions in order to obtain admissible test functions. For $\sigma > 1$ we define $\boldsymbol{v}_\sigma$ by

$$\boldsymbol{v}_\sigma(x, y) = (\sigma v_1(x, \sigma y), v_2(x, \sigma y)). \tag{1.57}$$

If $\boldsymbol{v}$ is divergence free, $\boldsymbol{v}_\sigma$ is also divergence free.

We set

$$\boldsymbol{\phi} = \int_{t-h}^t (\overline{\boldsymbol{u}})_\sigma(s) ds, \quad b = \int_{t-h}^t \partial_t \eta(s) ds.$$

The function $\boldsymbol{\phi}$ belongs to $H^1(0, T; H^1(B))$ and b belongs to $H^1(0, T; H_0^1(0, L))$. Remember that η has been extended by η_0 for $t < 0$ and $\overline{\boldsymbol{u}}$ and $\partial_t \eta$ extended by 0 for $t < 0$. The function $\boldsymbol{\phi}$ is divergence free.

Moreover since $\|\eta\|_{H^1(0,T;H_0^1(0,L))} \leq C$ and

$$H^1(0, T; H_0^1(0, L))) \hookrightarrow C^{0,1/2}([0, T]; C^0([0, L])),$$

we have

$$\|\eta - \eta^-\|_{L^\infty((0,T)\times(0,L))} \leq C\sqrt{h},$$

and

$$\|\eta^* - (\eta^*)^-\|_{L^\infty((0,T)\times(0,L))} \leq C\sqrt{h}.$$

Thus, if σ is such as $\sigma \geq 1 + \dfrac{2C}{\alpha}\sqrt{h}$, we have

$$\boldsymbol{\phi}(t, x, 1 + \eta^*(t, x)) = \left(0, \int_{t-h}^t \partial_t \eta(s, x) ds\right)^T \quad \text{on } (0, L).$$

Remark 1.6. We have used, in particular to prove Lemma 1.2, the additional damping on the structure equation. In the next subsection we will see that the dissipation coming from the fluid enables to have nevertheless this type of property for the coupled system.

Once we have obtained the desired bounds on the sequence of solutions we can pass to the limit as the regularization parameter tends to zero. Indeed, by denoting by ε regularization parameter, we have a sequence $(\boldsymbol{u}_\varepsilon, \partial\eta_\varepsilon)_\varepsilon$ solution of (1.52) that satisfies the following convergences. Let $T > 0$ such that $\inf_\varepsilon \min_{[0,T]\times[0,L]}(1+\eta_\varepsilon) \geq \alpha > 0$. Note that T does not depend on ε since (η_ε) is uniformly bounded in C^0. Let us denote by $(\eta, \tilde{\boldsymbol{u}})$ the limit of a subsequence of $(\eta_\varepsilon, \overline{\boldsymbol{u}}_\varepsilon)_{\varepsilon>0}$ (with the extension defined by (1.54)). We denote any subsequence of $(\eta_\varepsilon, \overline{\boldsymbol{u}}_\varepsilon)_{\varepsilon>0}$ by $(\eta_\varepsilon, \overline{\boldsymbol{u}}_\varepsilon)_{\varepsilon>0}$. We have the following convergences as ε goes to zero:

$$\begin{array}{ll} \eta_\varepsilon & \to \eta \text{ in } C^0([0,T]\times[0,L]) \\ \eta_\varepsilon & \rightharpoonup \eta \text{ in } H^1(0,T;H^1_0(0,L)) \\ \partial_t\eta_\varepsilon & \to \partial_t\eta \text{ in } L^2(0,T;L^2(0,L)) \\ \overline{\boldsymbol{u}}_\varepsilon & \to \tilde{\boldsymbol{u}} \text{ in } L^2(0,T;L^2(B)) \\ \rho_\varepsilon\boldsymbol{u}_\varepsilon & \to \rho\tilde{\boldsymbol{u}} \text{ in } L^2(0,T;L^2(B)) \\ \overline{\boldsymbol{u}}_\varepsilon & \rightharpoonup \tilde{\boldsymbol{u}} \text{ in } L^2(0,T;H^1_{0,\Gamma_0}(B)) \\ \eta^*_\varepsilon & \to \eta \text{ in } C^0([0,T]\times[0,L]) \\ \partial_t\eta^*_\varepsilon & \to \partial_t\eta \text{ in } L^2(0,T;L^2(\omega)) \\ \boldsymbol{u}^*_\varepsilon & \to \tilde{\boldsymbol{u}} \text{ in } L^2(0,T;L^2(B)) \\ \rho_\varepsilon\nabla\boldsymbol{u}_\varepsilon & \rightharpoonup \rho\nabla\tilde{\boldsymbol{u}} \text{ in } L^2(0,T;L^2(B)) \end{array} \tag{1.58}$$

First we take care of the equality

$$\boldsymbol{u}_\varepsilon(t,x,1+\eta^*_\varepsilon(t,x)) = (0,\partial_t\eta_\varepsilon(t,x))^T$$

on $(0,T)\times(0,L)$. The right hand side converges to $(0,\partial_t\eta)^T$ in $L^2(0,T;L^2(0,L))$. For the left hand side, we consider the function $\boldsymbol{w}_\varepsilon$ defined for a.e. t by

$$\boldsymbol{w}_\varepsilon = \left| \begin{array}{l} (0,\partial_t\eta_\varepsilon)^T \text{ in } B\setminus C_{\alpha/2} \\ \mathcal{R}(0,\partial_t\eta_\varepsilon)^T \text{ in } C_{\alpha/2}, \end{array} \right.$$

where $\mathcal{R}$ is a linear continuous lifting operator from $H^{\frac{1}{2}}((0,L)\times\{\alpha/2\})$ to $H^1_{0,\Gamma_f}(C_{\alpha/2})$ such that $\mathcal{R}(0,\partial_t\eta_\varepsilon)^T$ is divergence free and where $C_\beta = (0,L)\times(0,\beta)$. Then, the function $\overline{\boldsymbol{u}}_\varepsilon - \boldsymbol{w}_\varepsilon$ belongs to $L^2(0,T;H^1(B))$ and is bounded in this space independently of ε. Thus, a subsequence of $(\overline{\boldsymbol{u}}_\varepsilon - \boldsymbol{w}_\varepsilon)_{\varepsilon>0}$ converges weakly in $L^2(0,T;H^1(B))$ to $\boldsymbol{w}_0 = \tilde{\boldsymbol{u}} - \boldsymbol{w}$ (with an obvious definition of $\boldsymbol{w}$). We have $\boldsymbol{w}_0 = \boldsymbol{0}$ on Γ_f and in $\Omega^T_{\eta+\delta}$ for all $\delta > 0$, since η^*_ε converges uniformly to η. Thus $\tilde{\boldsymbol{u}} = (0,\partial_t\eta)^T$ in $((0,T)\times B)\setminus\Omega^T_\eta$ and $\boldsymbol{w}_0 \in L^2(0,T;H^1_0(\Omega_\eta(t)))$. Thus $\boldsymbol{w}_0(t,x,R+\eta(t,x)) = \boldsymbol{0}$ on $(0,L)$, but $\boldsymbol{w}(t,x,R+\eta(t,x)) = (0,\partial_t\eta)^T$, hence

$$\tilde{\boldsymbol{u}}(t,x,R+\eta(t,x)) = (0,\partial_t\eta(t,x))^T,$$

on $(0,T)\times(0,L)$. Next we pass to the limit in the weak formulation, the fluid test functions a priori depend on ε. However, it is sufficient to consider test functions that do not depend on ε and that are admissible for ε small enough.

We consider first test functions of the form $(\boldsymbol{\phi}^0, 0)$, with $\boldsymbol{\phi}^0 \in \mathcal{D}(\Omega^T\eta)$ and $\operatorname{div}\boldsymbol{\phi}^0 = 0$. These test functions satisfy the property that $\boldsymbol{\phi}^0(t,\cdot) \in \mathcal{D}(\Omega_\eta(t))$ for every t. Hence for ε small enough, $\boldsymbol{\phi}^0 \in \mathcal{V}_{\eta_\varepsilon^*}$, since η_ε^* converges uniformly to η as $\varepsilon \downarrow 0$.

The second pair of test functions we consider is $(\boldsymbol{\phi}^1, b)$ where b belongs to $C^1([0,T]; H_0^1(0,L))$, with $\int_\omega b = 0$ and for a.e. t:

$$\boldsymbol{\phi}^1 = \left| \begin{array}{l} (0,b)^T \text{ in } B \setminus C_{\alpha/2} \\ \mathcal{R}(0,b)^T \text{ in } C_{\alpha/2}. \end{array} \right.$$

Since $\min_{[0,T]\times[0,L]} (1+\eta_\varepsilon^*) \geq \alpha/2$, $(\boldsymbol{\phi}^1, b)$ is a pair of admissible test functions for all ε.

With both types of test functions, it is easy to pass to the limit in the weak formulation as ε goes to zero. We obtain the existence of a weak solution on $(0,T)$ satisfying energy estimates.

Eventually, we show that we can extend the solution as long as we have $\min_{[0,T]\times[0,L]} (1+\eta) > 0$. Let us build an increasing sequence of times $(T_k)_{k\geq 1}$ as follows. First we choose a time $T_1 > 0$ such that there exists a weak solution up to T_1, with $m_1 = \min_{[0,T_1]\times[0,L]} (1+\eta) > 0$. Possibly changing slightly T_1, we may assume that $\partial_t \eta(T_1) \in L^2(0,L)$ and $\boldsymbol{u}(T_1) \in L^2(\Omega_\eta(T_1))$ (since this is true for almost every time).

Now, let $k \geq 1$ and assume that we have built a solution up to some time T_k, with $m_k = \min_{[0,T_k]\times[0,L]} (1+\eta) > 0$. Our construction allows us to build an extension of our solution, on some time interval starting from T_k. Thanks to the a priori energy estimate, we have for $s \geq T_k$

$$1+\eta(s) \geq 1+\eta(T_k) - (s-T_k)^{\frac{1}{2}} C(T_k,s) \geq m_k - (s-T_k)^{\frac{1}{2}} C(T_k,s), \quad (1.59)$$

with

$$C(T_k,s) = \tilde{C}\Big(\|\boldsymbol{u}(T_k)\|_{L^2(\Omega_{\eta 0})}, \|\partial_t \eta(T_k)\|_{H_0^1(0,L)}, \int_{T_k}^s \exp.(s-u)(\|\boldsymbol{f}\|_{L^2(\Omega_\eta(u))}(u) + \|g\|_{L^2(0,L)}(u)) du \Big),$$

where $\tilde{C}$ is positive and nondecreasing with respect to its arguments, and $C(T_k,s) \leq C(0,s)$. This a priori estimate shows that if we let

$$\tau_k = \min\{1, (m_k/2C(T_k, T_k+1))^2\},$$

we can build a solution starting from $\boldsymbol{u}(T_k)$, $\eta(T_k)$ and $\partial_t \eta(T_k)$ up to the time $T_k + \tau_k$ (this corresponds to choosing $\alpha = m_k/2$ in the construction of the solution). The time T_{k+1} is chosen close to $T_k + \tau_k$ (in $[T_k + \tau_k/2, T_k + \tau_k]$), in order to have also $\partial_t \eta(T_{k+1}) \in L^2(\omega)$ and $\boldsymbol{u}(T_{k+1}) \in L^2(\Omega_\eta(T_{k+1}))$.

We let $T^* = \sup_k T_k$. If $T^* < +\infty$, then $m^* = \min_{[0,T^*]\times[0,L]} (1+\eta) = 0$. Otherwise, since $m_k \geq m^*$ for all k, $\tau_k \geq \min\{1, (m^*/2C(0,T^*))^2\} > 0$. But

$T_{k+1} - T_k \geq \tau_k/2$ and goes to zero, which is a contradiction. This achieves the proof of existence of a weak solution as long as no contact occurs between the elastic structure and the bottom of the fluid cavity.

Undamped Wave Equation: $\beta_2 \longrightarrow 0$

We explain in this subsection how one can pass to the limit in the coupled problem as the structure viscosity β_2 goes to zero. As we will see, the fluid dissipation enables to control the space of high frequencies of the structure velocity without any added viscosity on the wave equation.

The elastic displacement η in the case $\beta_2 = 0$ has only a hyperbolic regularity and belongs to $L^\infty(0, T; H_0^1(0, L)) \cap W^{1,\infty}(0, T; L^2(0, L))$. We easily verify that these regularities are sufficient to define all the functional spaces and to give a sense to the trace of the fluid velocity at the interface: $\boldsymbol{u}(t, x, R + \eta(t, x))$ on $(0, L)$. They are also sufficient to prove that the existence time of the weak solutions obtained for $\beta_2 > 0$ is bounded from below independently of β_2. Indeed η is uniformly continuous in space and time independently of β_2. Consequently this ensures that there exists a time $T > 0$ such that $R + \eta$ stays away from zero.

Yet the uniform energy estimates are not sufficient to obtain the compactness in $L^2(0, T; L^2(0, L))$ of the structure velocity. As noted at Remark 1.6, in the previous subsection this compactness relied on the fact that $\partial_t \eta \in L^2(0, T; H^1(0, L))$ thanks to the parabolic regularization of the elastic equation. Here one cannot hope to obtain (1.55) or (1.56) and in particular we cannot hope to have the convergence rate $\sqrt{h}$. But in order to obtain a uniform decay, as h goes to zero, it is sufficient to have some space regularity of $\partial_t \eta$. The idea is then to take advantage of the kinematic condition $(0, \partial_t \eta(t, x))^T = \boldsymbol{u}(t, x, R + \eta(t, x))$ on $(0, L)$, and of the fact that the fluid is viscous.

Indeed, if η had Lipschitzian regularity, since $\boldsymbol{u} \in L^2(0, T; H^1(\Omega_\eta(t))$, one would deduce that $\partial_t \eta \in L^2(0, T; H^{1/2}(0, L))$ as the trace of $\boldsymbol{u}$ and the space of high frequencies of the structure velocity would then be controlled. Here these regularities are not satisfied by any weak solution (unless $\gamma > 0$). But thanks to Sobolev injections $\eta \in C^0([0, T] \times [0, L]) \cap L^\infty(0, T; H_0^1(0, L))$ and this regularity of the fluid–structure interface enables to obtain that $\partial_t \eta \in L^2(0, T; H^s(0, L))$ for any $s < 1/4$.

The question is now: is this dissipation sufficient to derive the same kind of result enunciate at Lemma 1.2? The answer is yes and can be summarized by the following lemma:

Lemma 1.3. *Let* $T > 0$ *such that* $\min_{[0,T]\times(0,L)}(1 + \eta) \geq \alpha > 0$. *We have* $\forall \varepsilon > 0$, $\exists h_0 > 0$, *s. t.* $\forall \beta_2 > 0$, $\forall h \leq h_0$

$$\int_0^T \int_B \rho |\overline{\boldsymbol{u}} - (\overline{\boldsymbol{u}})^-|^2 + \int_0^T \int_0^L (\partial_t \eta - \partial_t \eta^-)^2 \leq \varepsilon \tag{1.60}$$

and

$$\int_0^T \int_B |\rho \overline{\boldsymbol{u}} - \rho^-(\overline{\boldsymbol{u}})^-|^2 \leq \varepsilon, \tag{1.61}$$

with η extended by η_0 for $t < 0$ and $\overline{\boldsymbol{u}}$ and $\partial_t \eta$ extended by 0 for $t < 0$, and where ρ denotes the characteristic function of $\hat{\Omega}_\eta^T$.

Remark 1.7. The main difference between the two lemmas is that in Lemma 1.2, a uniform rate is obtained, whereas in Lemma 1.3, we have only a convergence to zero, as h goes to zero, uniformly in β_2. Note that the dissipation of the fluid, which induces dissipation of the structure, is crucial.

In order to prove this lemma we may follow the same lines as for the proof of Lemma 1.2 and take advantage of the fact that $\partial_t \eta \in L^2(0,T;H^s(0,L))$ for any $s < 1/4$ thanks to the equality of the velocities at the interface. Note that it is necessary to split the space of high and low frequencies of the structure velocity. The space of high frequencies are controlled since $\partial_t \eta \in L^2(0,T;H^s(0,L))$, with $s < 1/4$ and the low frequencies will be controlled thanks to the variational formulation and a good choice of the test functions. Let us consider the eigenfunctions $\boldsymbol{\xi}_i$ associated with the Laplace operator on $(0,L)$ with homogeneous Dirichlet boundary conditions and satisfying the additional constraint $\int_0^L \boldsymbol{\xi}_i = 0$. They form a basis of $H_0^1(0,L) \cap L_0^2(0,L)$. Consequently the L^2-projection, denoted η^{N_0}, of η in the space generated by the N_0 first eigenvectors satisfies

$$\|\partial_t \eta - \partial_t \eta^{N_0}\|_{L^2(0,L)} \leq C\lambda_{N_0}^{-s/2} \|\partial_t \eta\|_{H^s(0,L)}, s < 1/4,$$

where λ_{N_0} is the eigenvalue associated with $\boldsymbol{\xi}_{N_0}$. Then the high frequencies of $\partial_t \eta$ are controlled in $L^2(0,T;L^2(0,L))$ uniformly in β_2 and we have

$$\int_0^T \int_0^L (\partial_t \eta^{hf,N_0} - (\partial_t \eta^{hf,N_0})(t-h))^2 \leq C\lambda_{N_0}^{-\frac{s}{2}}, s < 1/4,$$

with $\eta^{hf,N_0} = \eta - \eta^{N_0}$ and where C is independent of the structure viscosity β_2.

For the low frequency part $\partial_t \eta^{N_0}$, the variational formulation will be used to obtain a uniform convergence (with respect to β_2 when h goes to zero) of quantities $\|\boldsymbol{u}(t+h) - \boldsymbol{u}(t)\|_{L^2_{tx}}$ and $\|\partial_t \eta^{N_0}(t+h) - \partial_t \eta^{N_0}(t)\|_{L^2_{tx}}$. The structure test function is $b = \int_{t-h}^t \partial_t \eta_\varepsilon^{N_0}$. The fluid test function has to be chosen carefully. We set

$$\boldsymbol{\phi} = \int_{t-h}^t \left(\overline{(\boldsymbol{u} - \mathcal{R}_\alpha(\partial_t \eta))^\lambda} \right)_\sigma (s)ds + \int_{t-h}^t \overline{\mathcal{R}_\alpha(\partial_t \eta^{N_0})}(s)ds,$$

where the extension $\boldsymbol{v} \mapsto \overline{\boldsymbol{v}}$ is defined by (1.54) and where $\mathcal{R}_\alpha$ is a lifting operator defined by

$$\mathcal{R}_\alpha(b) = \left| \begin{array}{l} (0,0,b)^T \text{ for } z \geq \alpha, \\ (0,0,\frac{z}{\alpha}b)^T + \mathbf{w}_\alpha \text{ in } C_\alpha \ , \end{array} \right. \quad \text{for a.e. } t, \tag{1.62}$$

with $\mathbf{w}_\alpha$ such that div $(\mathbf{w}_\alpha) = b$ and $\mathbf{w}_\alpha \in H_0^1(C_\alpha)$, $\|\mathbf{w}_\alpha\|_{H_0^1(C_\alpha)} \leq C\|b\|_{L^2(0,L)}$, for a.e. t. Note that $\mathbf{w}_\alpha$ exists because b has a zero mean. The first term of $\boldsymbol{\phi}$ has a trace equal to zero on the interface, the second term matches the structure test function at the interface. Note moreover that a space regularization of $\boldsymbol{v} = \boldsymbol{u} - \mathcal{R}_\alpha(\partial_t \eta)$, denoted by $\boldsymbol{v}^\lambda$ has been introduced in order to have $\boldsymbol{\phi}$ bounded in $H^1(0,T;H^1(B))$ independently of β_2. It verifies div $(\boldsymbol{v}^\lambda) = 0$, $\boldsymbol{v}^\lambda \in L^2(0,T;H_0^1(\Omega_\eta(t)))$ and

$$\begin{array}{l} \|\boldsymbol{v} - \boldsymbol{v}^\lambda\|_{L^2(0,T;L^2(\Omega_\eta(t)))} \longrightarrow 0, \text{ uniformly in } \beta_2, \text{ as } \lambda \text{ goes to zero,} \\ \|\boldsymbol{v}^\lambda\|_{L^2(0,T;H^1(\Omega_\eta(t)))} \leq C_\lambda. \end{array} \tag{1.63}$$

The construction of $\boldsymbol{v}^\lambda$ relies on the fact that the elastic interface does not touch the bottom of the fluid cavity. Moreover, the uniform convergence of $(\boldsymbol{v}^\lambda)_\lambda$ as $\lambda \longrightarrow 0$ in $L^2(0,T;L^2(\Omega_\eta(t)))$ is made possible since $\overline{\boldsymbol{v}}$ is uniformly bounded in $L^2(0,T;H^s(B)), 0 < s < 1/4$. With this choice,

$$\|\partial_t \eta^{N_0}\|_{L^\infty(0,T;L^2(0,L))} \leq C, \quad \|\partial_t \eta^{N_0}\|_{L^2(0,T;H^s(0,L))} \leq C,$$

$$\|b\|_{W^{1,\infty}(0,T;L^2(0,L))} \leq C, \quad \|b\|_{H^1(0,T;H^s(0,L))} \leq C,$$

$$\|\overline{\boldsymbol{v}^\lambda}\|_{L^\infty(0,T;L^2(B))} \leq C, \|\overline{\boldsymbol{v}^\lambda}\|_{L^2(0,T;H^s((B))} \leq C, \forall s' < s < 1/4,$$

and

$$\|\partial_t \eta^{N_0}\|_{L^\infty(0,T;H^1(0,L))} \leq C_{N_0}, \|b\|_{W^{1,\infty}(0,T;H^1(0,L))} \leq C_{N_0}, \|\overline{\boldsymbol{v}^\lambda}\|_{L^2(0,T;H^1(B))} \leq C_\lambda,$$

where C denotes a strictly positive constant that depends only on the data and not on β_2 and N_0, and C_{N_0} (resp. C_λ) denotes and will denote a strictly positive constant that depends on the data and not on β_2 but may depend on N_0 (resp. λ). The integer N_0 (resp. the real λ) is chosen large enough (resp. small enough). Then for well-chosen σ, $(\boldsymbol{\phi}, b)$ are admissible test functions. Indeed, $\boldsymbol{\phi}$ is divergence free thanks to the definitions of the lifting operator $\mathcal{R}_\alpha$, the extension operator $\boldsymbol{v} \mapsto \overline{\boldsymbol{v}}$, the operator $\boldsymbol{v} \mapsto \boldsymbol{v}_\sigma$ and the definition of the regularization $\boldsymbol{v} \mapsto \boldsymbol{v}^\lambda$. Moreover $\boldsymbol{\phi}$ belongs to $H^1(0,T;H^1(B))$. The function b belongs to $H^1(0,T;H_0^1(0,L))$. Both of them are bounded in the previous spaces independently of β_2 but not of N_0 and λ.

With this choice of test functions we have for $\varepsilon > 0$

$$\int_0^T \int_{\Omega_\eta(t)} |\boldsymbol{u}(t)-\bar{\boldsymbol{u}}(t-h)|^2 + \int_0^T \int_0^L (\partial_t \eta^{N_0}(t)-\partial_t \eta^{N_0}(t-h))^2 \leq C_{N_0} h^\gamma + C\lambda_{N_0}^{-\frac{s}{2}} + C\varepsilon,$$
$$\forall \beta_2 \text{ small enough},$$

with $s < \frac{1}{4}$ and for some $0 < \gamma < 1/4$. It implies finally

$$\int_0^T \int_{\Omega_\eta(t)} |\boldsymbol{u}(t)-\bar{\boldsymbol{u}}(t-h)|^2 + \int_0^T \int_0^L (\partial_t \eta(t)-\partial_t \eta(t-h))^2 \leq C_{N_0} h^\gamma + C\lambda_{N_0}^{-\frac{s}{2}} + C\varepsilon,$$
$$\forall \beta_2 \text{ small enough}.$$

This result together with the energy estimates enable to obtain the desired convergences, compactness and to pass to the limit in the weak formulation. We then obtain the existence of at least one weak solution until the elastic structure touches the bottom of the fluid cavity. Note that recently Muha and Canic [129] prove the same kind of result for a $2D/1D$ coupled problem with $\gamma > 0$. The proof is based on a time discretization and a splitting of the structure equations, inspires from a numerical strategy, that enables to have stability for an explicit coupling independently of the added mass effect [56, 98].

In the case $\gamma = 0$ since we do not have a lot regularity of the structure displacement at the interface, the proof relies strongly on the only transverse motion. In the general case where we do not neglect the longitudinal displacement and if (ξ, η) satisfies only the energy estimates, the displacement of the structure is not regular enough to properly define the problem. Thus there is a need to work with smooth solutions.

In the next subsection we then explain how one can obtain existence of a unique strong solution and review some results of Lequeurre that considers also the case where $\xi = 0$. It is nevertheless a first step to further be able to consider the full coupled problem.

1.2.3 Existence of Strong Solutions

In this subsection we will give the general steps of the results obtained by Lequeurre [119, 120], which are a generalization of the one obtained in [10]. In these studies, two-dimensional as well as three-dimensional problems with $\gamma > 0$ or $\gamma = 0$ and with $\beta_2 > 0$ have been considered, but the general scheme of proof is the same in any case. The main difference comes from the regularity of the structure displacement η. We will focus on the general steps without detailing the proofs which are rather technical. The goal is to underline the possible links between existence of solutions and numerical schemes. In particular we will see in Sect. 1.3 a semi-implicit scheme based on the same kind of splitting ideas.

We consider the coupled problem (1.1), (1.2), (1.7), (1.11), (1.15), (1.17), with $\xi = 0$, with initial conditions (1.3), (1.12) and with no external forces applied to the coupled system. For the fluid boundary conditions one could assume that the fluid velocity satisfies Dirichlet homogeneous boundary conditions as in [119] or considers periodicity in the x variable as it is done in [120]. Note that this periodicity assumption seems to be necessary in the case where $\gamma = 0$. Remember that in the case of homogeneous Dirichlet boundary conditions for the fluid velocity on Γ_f (or periodic boundary conditions in x on $\Gamma_{\text{in}} \cup \Gamma_{\text{out}}$ together with homogeneous Dirichlet boundary conditions for the fluid velocity on Γ_0) the velocity of the structure satisfies the additional constraint: $\int_0^T \partial_t \eta = 0$. We assume moreover that $\beta_2 > 0$ so that we have a parabolic–parabolic coupling. Assuming that the elastic displacement is regular enough, the fluid equations can be rewritten in the reference configuration $\hat{\Omega}_f$, thanks to the change of variables $\boldsymbol{\phi}_f$ as follows:

$$\begin{aligned} J_f \rho_f \partial_t \boldsymbol{v} - \rho_f (\boldsymbol{w} B_f \nabla)\boldsymbol{v} + \rho_f(\boldsymbol{v} B_f \nabla)\boldsymbol{v} - \nu \operatorname{div}((A_f \nabla)\boldsymbol{v}) + (B_f \nabla) q = \mathbf{0}, \\ \operatorname{div}(B_f^T \boldsymbol{v}) = 0, \end{aligned} \tag{1.64}$$

where $\boldsymbol{v}(t,x) = \boldsymbol{u}(t, \boldsymbol{\phi}_f(t,x))$, $q = (t,x) = p(t, \boldsymbol{\phi}_f(t,x))$, $\boldsymbol{w} = \partial_t \boldsymbol{\phi}_f$, $J_f = \det \nabla \boldsymbol{\phi}_f$ and with the following notations for the matrices A_f and B_f:

$$A_f = \frac{1}{J_f} (\operatorname{cof} \nabla \boldsymbol{\phi}_f)^T \operatorname{cof} \nabla \boldsymbol{\phi}_f, \tag{1.65}$$

$$B_f = \operatorname{cof}(\nabla \boldsymbol{\phi}_f), \tag{1.66}$$

where cof A denotes the cofactor matrix of A. Note that $\boldsymbol{\phi}_f$ is a lifting of the interface deformation into the fluid domain. In the particular case of only transverse motion it can be chosen as:

$$\boldsymbol{\phi}_f(t,x,y) = \begin{pmatrix} x \\ y(1 + \eta(t,x)) \end{pmatrix}, \tag{1.67}$$

for $(x,y) \in \hat{\Omega}_f = (0,L) \times (0,R)$. In the case where the structure displacement is regular enough we could consider $B_f^T \boldsymbol{v}$ as a new unknown, so that one works with divergence free velocities. By doing so, when studying the linearized associated problem we keep this divergence free condition unchanged. That was done [153] in the case of a rigid solid moving a viscous flow. Here it is possible, for instance, in the two-dimensional setting for $\gamma > 0$. In the general case we cannot make such a change of unknown because of the lack of regularity of the solution and in particular of the elastic displacement.

Remark 1.8. The mapping $\boldsymbol{\phi}_f$ could also be the flow associated with the fluid velocity. In this case, no convection terms appears in (1.64). This change of variables has, for instance, been done in [89] or [18] in the context of fluid–solid interaction.

The compatibility conditions that the data have to satisfy are now

$$\begin{cases} \min\limits_{[0,L]}(R+\eta_0) > 0, \\ \operatorname{div} \boldsymbol{u}_0 = 0 \text{ in } \Omega_\eta(0), \\ \boldsymbol{u}_0 = 0 \text{ on } \Gamma_0, \text{ periodic in } x \\ \boldsymbol{u}_0(t,x,R+\eta_0(x)) = (0,\eta_1(x))^T \text{ on } (0,L), \\ \displaystyle\int_0^L \eta_1 = 0. \end{cases} \tag{1.68}$$

The result presented in [120] for $\gamma = 0$ and written in terms of $(\boldsymbol{v}, q, \eta)$ can be summarized as follows:

Theorem 1.2. *Let us consider* $(\boldsymbol{u}_0, \eta_0, \eta_1) \in H^1_\#(\Omega_\eta(0)) \times H^2_\#(0,L) \times H^1_\#(0,L)$ *satisfying* (1.68), *then there exists a time* $T^* > 0$ *such that* (1.64), (1.11), (1.7), (1.3), (1.12) *has got a unique solution* $(\boldsymbol{v}, q, \eta)$ *that belongs to* $\left(L^2(0,T^*; H^2_\#(\hat{\Omega}_f)) \cap H^1(0,T^*; L^2_\#(\hat{\Omega}_f))\right) \times L^2(0,T^*; H^1_\#(\hat{\Omega}_f)) \times \left(H^1(0,T^*; H^2_\#(0,L)) \cap H^2(0,T^*; L^2_\#(0,L))\right)$, *where the subscript # stands for the x-periodicity.*

Remark 1.9. For the initial conditions we will assume, for the sake of simplicity, that $\eta_0 = 0$. Consequently $\hat{\Omega}_f = \Omega_\eta(0)$. Remark that the results in [119, 120] seem to be only correct in this case (or for a η_0 more regular than $H^2_\#(0,L)$) and that some, surely feasible, adaptations should to be done when $\eta_0 \neq 0$. Nevertheless to suppose that $\eta_0 = 0$ is a quite reasonable assumption that is not so restrictive.

The steps of the proof are the following:

- Study the linear coupled problem, with, in particular, nonhomogeneous divergence condition;
- Study the nonlinear terms and prove that they are small for small times to obtain that there exists a unique fixed point thanks to Picard theorem for small enough time or for small enough data.

The first point consists then in studying the following nonhomogenous coupled problem:

$$\begin{aligned} \rho_f \partial_t \boldsymbol{v} - \nu \Delta \boldsymbol{v} + \nabla q &= \boldsymbol{f}, && \text{in } \hat{\Omega}_f, \\ \operatorname{div} \boldsymbol{v} &= \operatorname{div} \boldsymbol{g}, && \text{in } \hat{\Omega}_f, \\ \boldsymbol{v} &= (0, \partial_t \eta)^T, && \text{on } \hat{\Sigma}, \\ \boldsymbol{v} &= \boldsymbol{0}, && \text{on } \Gamma_0, \\ \rho_s e \partial_{tt}\eta - \alpha_2 \partial_{xx}\eta + \gamma \partial_x^4 \eta - \beta_2 \partial_{xx}\partial_t \eta &= -(\boldsymbol{\sigma}(\boldsymbol{v}, q)\cdot \hat{\boldsymbol{n}})_2 + h, && \text{on } (0,L), \end{aligned} \tag{1.69}$$

for given $\boldsymbol{f}, \boldsymbol{g}, h$ and with the additional constraint that $\partial_t \eta$ has a zero mean. This system has to be completed by initial conditions and we assume that periodic boundary conditions in x are satisfied. In the fixed point procedure the right hand side terms will typically contain nonlinear terms that write $\nu \operatorname{div}(((I - A_f)\nabla)\tilde{\boldsymbol{v}})$ for $\boldsymbol{f}$ or the second component of $((I - A_f)\nabla)\tilde{\boldsymbol{v}} \cdot \hat{\boldsymbol{n}}$ for h, with $\tilde{\boldsymbol{v}}$ a given velocity in $L^2(0,T; H^2_\#(\hat{\Omega}_f)) \cap H^1(0,T^*; L^2_\#(\hat{\Omega}_f))$. Under the assumption that $\eta_0 = 0$ these terms shall be small in suitable spaces for small time since at initial time $A_f = B_f = I$.

Note that, since $\partial_t \eta$ has a zero average, $\boldsymbol{g}$ has to satisfy the compatibility condition $\int_{\hat{\Sigma} \cup \Gamma_0} \boldsymbol{g} \cdot \boldsymbol{n} = 0$. This property enables to consider the following lifting of the nonhomogeneous divergence:

$$
\begin{aligned}
-\Delta z + \nabla \mu &= 0, && \text{in } \hat{\Omega}_f,\\
\operatorname{div} z &= \operatorname{div} \boldsymbol{g}, && \text{in } \hat{\Omega}_f,\\
z &= \boldsymbol{0}, && \text{on } \hat{\Sigma} \cup \Gamma_0,\\
x &- \text{periodic}.
\end{aligned}
$$

Let us underline that this lifting does not modify the kinematic coupling condition and consequently the new velocity $\boldsymbol{v} - z$ is still equal to $(0, \partial_t \eta)^T$ on the fluid–structure interface. It is strongly linked to the only transverse motion. Note moreover that, in the fixed point procedure, $\boldsymbol{g}$ will write $(I - B_f)^T \tilde{\boldsymbol{v}}$, with $\tilde{\boldsymbol{v}}$ satisfying $v_1 = 0$ on $\hat{\Sigma}$ and $\tilde{\boldsymbol{v}} = \boldsymbol{0}$ on Γ_0. Consequently $\boldsymbol{g} = \boldsymbol{0}$ on $\hat{\Sigma}$. It implies that if $\partial_t \boldsymbol{g} \in L^2(0,T; L^2_\#(\hat{\Omega}_f))$ then $\operatorname{div}(\partial_t \boldsymbol{g})$ belongs to $L^2(0,T; (H^1_\#(\hat{\Omega}_f))')$. Furthermore, $\operatorname{div} \boldsymbol{g} = (I - B_f) : \nabla \tilde{\boldsymbol{v}}$ since B_f is the cofactor matrix of the deformation gradient $\nabla \boldsymbol{\phi}_f$ and thanks to the Piola's identity [31]. Consequently $\operatorname{div} \boldsymbol{g}$ will have the regularity of $(I - B_f) : \nabla \tilde{\boldsymbol{v}}$.

With this lifting we have to study the following coupled problem, by changing the right hand side still denoted here $\boldsymbol{f}$ and h:

$$
\begin{aligned}
\rho_f \partial_t \boldsymbol{w} - \nu \Delta \boldsymbol{w} + \nabla \pi &= \boldsymbol{f}, && \text{in } \hat{\Omega}_f.\\
\operatorname{div} \boldsymbol{w} &= 0, && \text{in } \hat{\Omega}_f.\\
\boldsymbol{w} &= (0, \partial_t \eta)^T, && \text{on } \hat{\Sigma}.\\
\boldsymbol{w} &= \boldsymbol{0}, && \text{on } \Gamma_0.\\
\rho_s e \partial_{tt} \eta - \alpha_2 \partial_{xx} \eta - \beta_2 \partial_{xx} \partial_t \eta &= (\boldsymbol{\sigma}(\boldsymbol{w}, \pi) \cdot \hat{\boldsymbol{n}})_2 + h, && \text{on } (0, L),\\
x &- \text{periodic}.
\end{aligned}
\tag{1.70}
$$

As we have seen in Sect. 1.2.1, to naively decouple the fluid from the structure to prove existence of a solution of (1.70) leads to impose a condition on the density of structure (that has to be large enough with respect to the fluid one). The key idea is then to rewrite this coupled system thanks to an added mass operator and to spilt the fluid problem into two sub-problems. We decompose $\boldsymbol{w} = \boldsymbol{w}_e + \boldsymbol{w}_s$, with $\boldsymbol{w}_s = (I - P)\boldsymbol{w}$, $\boldsymbol{w}_e = P\boldsymbol{w}$, where P denotes the Leray operator on the subspace of $L^2(\hat{\Omega}_f)$ of divergence free vectors with zero normal trace on the boundary. In the same way $\pi = \pi_e + \pi_s$. Then $\boldsymbol{w}_s = \nabla \theta$, where θ is the solution (up to an additive constant) of

$$\begin{aligned} \Delta\theta &= 0, && \text{in } \hat{\Omega}_f, \\ \frac{\partial\theta}{\partial \boldsymbol{n}} &= \partial_t \eta, && \text{on } \hat{\Sigma}, \\ \frac{\partial\theta}{\partial \boldsymbol{n}} &= 0, && \text{on } \Gamma_0. \end{aligned}$$

Consequently $\boldsymbol{w}_s$ is divergence free (as $\boldsymbol{w}$ is) and satisfies $\boldsymbol{w}_s \cdot \boldsymbol{n} = \partial_t \eta$. Moreover $\Delta \boldsymbol{w}_s = \boldsymbol{0}$. We next define $\pi_s = \pi(\boldsymbol{f}) - \mu \mathrm{m}$ where μ is defined by

$$\begin{aligned} -\Delta\mu &= 0, && \text{in } \hat{\Omega}_f, \\ \frac{\partial\mu}{\partial \boldsymbol{n}} &= -\rho_f \partial_{tt} \eta, && \text{on } \hat{\Sigma}, \\ \frac{\partial\mu}{\partial \boldsymbol{n}} &= 0, && \text{on } \Gamma_0, \end{aligned}$$

and $\pi(\boldsymbol{f})$ is the solution of:

$$\begin{aligned} -\Delta\pi(\boldsymbol{f}) &= -\operatorname{div}(\boldsymbol{f}), && \text{in } \hat{\Omega}_f, \\ \frac{\partial\pi(\boldsymbol{f})}{\partial \boldsymbol{n}} &= \boldsymbol{f} \cdot \boldsymbol{n}, && \text{on } \Gamma_0 \cup \hat{\Sigma}. \end{aligned}$$

This Neumann problem has to be understood in a weak way, so that $\pi(\boldsymbol{f})$ is defined for $\boldsymbol{f}$ belonging only to $L^2_\#(\hat{\Omega}_f)$.

Next $(\boldsymbol{w}_e, p_e)$ is the solution of:

$$\begin{aligned} \rho_f \partial_t \boldsymbol{w}_e - \nu \Delta \boldsymbol{w}_e + \nabla \pi_e &= P(\boldsymbol{f}), && \text{in } \hat{\Omega}_f, \\ \operatorname{div} \boldsymbol{w}_e &= 0, && \text{in } \hat{\Omega}_f, \\ \boldsymbol{w}_e \cdot \boldsymbol{n} &= 0, && \text{on } \Gamma_0 \cup \hat{\Sigma}, \\ \boldsymbol{w}_e \cdot \boldsymbol{\tau} &= \boldsymbol{w}_s \cdot \boldsymbol{\tau}, && \text{on } \Gamma_0 \cup \hat{\Sigma}. \end{aligned} \tag{1.71}$$

Consequently the structure equation can be rewritten as follows:

$$(\rho_s e \mathcal{I} + \rho_f \mathcal{M}) \partial_{tt} \eta - \alpha_2 \partial_{xx} \eta - \beta_2 \partial_{xx} \partial_t \eta = (-\boldsymbol{\sigma}(\boldsymbol{w}_e, \pi_e + c) \cdot \boldsymbol{n})_2 + \pi(\boldsymbol{f}) + h, \tag{1.72}$$

where $\mathcal{M}$ is the Neumann to Dirichlet operator which associate to any v, the trace over $\hat{\Sigma}$ of p solution of

$$\begin{aligned} -\Delta p &= 0, && \text{in } \hat{\Omega}_f, \\ \frac{\partial p}{\partial \boldsymbol{n}} &= v, && \text{on } \hat{\Sigma}, \\ \frac{\partial p}{\partial \boldsymbol{n}} &= 0, && \text{on } \Gamma_0. \end{aligned}$$

With this notation we have: $\mu = -\rho_f \mathcal{M}(\partial_{tt}\eta)$. Note that every intermediate pressure is defined up to an additive constant that can be taken such that the average of each pressure is zero. Consequently we have added a constant c in the right hand side which is the Lagrange multiplier of the constraint $\int_0^L \partial_t \eta = 0$. Now we are in a position to decouple the fluid from the structure equations and more precisely we decouple (1.71) from (1.72). Then the fixed point procedure can be applied to prove the existence of a regular solution of the linear nonhomogenous problem since now the added mass is treated implicitly. Indeed the regularity in time of $(\boldsymbol{w}_e, \pi_e)$ depends only on the regularity of $\partial_t \eta$ and not on the one of $\partial_{tt}\eta$. Note that here one needs to have $\beta_2 > 0$ in order to obtain enough regularity of the structure velocity.

Remark 1.10. To decouple the fluid from the structure in this way does not allow to take advantage of the fluid dissipation that induces, on the full coupled problem, dissipation of the elastic structure.

The second step is the fixed point procedure. It is based on the Picard fixed point theorem. To a given $(\tilde{\eta}, \tilde{\boldsymbol{v}}, \tilde{p})$ we associated $(\eta, \boldsymbol{v}, p)$ solution of the linear coupled problem, with right hand sides $\boldsymbol{f}, \boldsymbol{g}, h$ depending on $(\tilde{\eta}, \tilde{\boldsymbol{v}}, \tilde{p})$. Typically $\boldsymbol{f}$ contains terms of the form $\mathrm{div}((I - \tilde{A}_f)\nabla\tilde{\boldsymbol{v}})$ or $(I - \tilde{B}_f)\nabla\tilde{q}$ with $\tilde{A}_f$ and $\tilde{B}_f$ defined by (1.65), (1.66), (1.67). The key argument of this step is that for a given $\tilde{\eta} \in L^2(0,T; H^2_\#(\hat{\Omega}_f)) \cap H^1(0,T^*; L^2_\#(\hat{\Omega}_f))$, with $\tilde{\eta}(0) = 0$ then $\tilde{A}_f$, $\tilde{B}_f$ will stay, for small enough time, in a neighborhood of the identity matrix in spaces that are multiplayer of $L^2(0,T; H^1_\#(\hat{\Omega}_f))$, which is the space to which $\nabla\tilde{\boldsymbol{v}}$ belongs to if $\tilde{\boldsymbol{v}} \in L^2(0,T; H^2_\#(\hat{\Omega}_f)) \cap H^1(0,T^*; L^2_\#(\hat{\Omega}_f))$. Note that, at this step, one has to pay a particular attention on the dependency of the various constants with respect to the time since one wants to prove existence for small time.

The open questions raised by this study are numerous. First could we prove existence of strong solutions with no additional viscosity of the elastic part (i.e., $\beta_2 = 0$) and do these solutions exist till the elastic boundary touches the bottom of the fluid cavity? Moreover could we consider other type of boundary conditions for the fluid such as Neumann boundary conditions? One of the key points is then the regularity of the fluid velocity and pressure that satisfy a Stokes like system with mixed Neumann–Dirichlet boundary conditions. Note that considering (1.5), (1.6) or (1.8), (1.9) will not give the same type of regularities. Another question is: could we include also longitudinal motion of the elastic part of the boundary as for the steady state case (see [86])?

1.2.4 Non-Newtonian Shear-Dependent Fluid

The aim of this section is to present analysis of the fluid–structure interaction problem for some non-Newtonian fluids. Moreover, we will also present a complementary proof of the existence of the weak solution that is based on the so-called global iterative method with respect to the domain deformation and special $\varepsilon-, \kappa-$

approximation of the original problem. Let us start with the description of our fluid–structure interaction problem. We consider a two-dimensional fluid motion governed by the same momentum and the continuity equations, as previously

$$\rho_f \partial_t \boldsymbol{u} + \rho_f (\boldsymbol{u} \cdot \nabla) \boldsymbol{u} - \operatorname{div} \boldsymbol{\tau} + \nabla p = 0, \qquad (1.73)$$
$$\operatorname{div} \boldsymbol{u} = 0,$$

but with a fluid stress tensor given by

$$\boldsymbol{\tau} = \boldsymbol{\tau}(\mathbb{D}(\boldsymbol{u})) = 2\mu(|\mathbb{D}(\boldsymbol{u})|)\mathbb{D}(\boldsymbol{u}),$$

with the notations introduced in the introduction

$$\boldsymbol{\sigma}_f(\boldsymbol{u}, p) = \boldsymbol{\tau} - p\boldsymbol{I}. \qquad (1.74)$$

Moreover we assume that there exists a potential $\mathcal{U} \in C^2(\mathbb{R}^{2\times 2})$ of the stress tensor $\boldsymbol{\tau}$, such that for some $1 < q < \infty$, $C_1, C_2 > 0$ we have for all $\chi, \xi \in \mathbb{R}^{2\times 2}_{\text{sym}}$ and $i, j, k, l \in \{1,2\}$, cf. [123]

$$\frac{\partial \mathcal{U}(\chi)}{\partial \chi_{ij}} = \boldsymbol{\tau}_{ij}(\chi), \qquad \mathcal{U}(\mathbf{0}) = \frac{\partial \mathcal{U}(\mathbf{0})}{\partial \chi_{ij}} = 0, \qquad (1.75)$$

$$\frac{\partial^2 \mathcal{U}(\chi)}{\partial \chi_{mn} \partial \chi_{rs}} \xi_{mn} \xi_{rs} \geq C_1 (1 + |\chi|)^{q-2} |\xi|^2, \qquad (1.76)$$

$$\left| \frac{\partial^2 \mathcal{U}(\chi)}{\partial \chi_{ij} \partial \chi_{kl}} \right| \leq C_2 (1 + |\chi|)^{q-2}. \qquad (1.77)$$

A typical example satisfying the above properties is

$$\mu(|\mathbb{D}(\boldsymbol{u})|) = \mu(1 + |\mathbb{D}(\boldsymbol{u})|^2)^{\frac{q-2}{2}} \quad q > 1, \qquad (1.78)$$

cf. [109, 122, 123, 159]. Note that according to the parameter q, the non-Newtonian fluid is either shear-thinning ($q < 2$) or shear-thickening ($q \geq 2$). Models for fluids with the shear-dependent viscosity are used in many areas such as geophysics, glaciology, polymer mechanics, blood or food rheology. For $q > 2$ this model is an analogy of the so-called Ladyzhenskaja's fluid, for $q = 3$ it yields the Smagorinskij model of turbulence. In numerical simulations presented in recent papers [109, 121] the shear-thinning model of Carreau has been used in order to model blood flow in compliant vessels. For the simplicity of presentation we will consider here only the case of shear-thickening fluids, i.e., $q \geq 2$. The generalization for shear-thinning fluids may be done in an analogous way as here, using an appropriate techniques for shear-thinning fluids, see [42, 73, 158].

The two-dimensional deformable fluid domain

$$\Omega_\eta(t) \equiv \{(x, y);\ 0 < x < L,\ 0 < y < R_0(x) + \eta(t, x)\},\ 0 < t < T$$

is given by a reference radius function $R_0(x_1)$ and the unknown free boundary function $\eta(t, x)$ describing the domain of deformation. The fluid and the geometry of the computational domain are coupled through the following Dirichlet boundary condition on the deformable part of the boundary $\Sigma(t)$

$$\boldsymbol{u}(t, x, R_0(x) + \eta(t, x)) = (0, \partial_t \eta(t, x))^T, \tag{1.79}$$

where $\Sigma(t) = \{(x, y);\ y = R_0(x) + \eta(t, x),\ x \in (0, L)\}$. Note that compared to the previous subsections we have introduced a nonconstant radius $R_0(x)$ to model, for example, stenotic vessels. Nevertheless it adds no further difficulties.

Our structural model is similar to (1.11) but we set $\gamma = 0$ and use instead of a higher order friction term (term with the coefficient β_2)

$$\rho_s e \partial_{tt}\eta - \alpha_2 \partial_{xx}\eta + \tilde{\delta}\eta + \beta_2 \partial_x^4 \partial_t \eta - \alpha_2 \partial_{xx} R_0 = H(\boldsymbol{u}, p) \quad \text{on} \quad \hat{\Sigma}, \tag{1.80}$$

where

$$H(\boldsymbol{u}, p) = -(\boldsymbol{T}_f)_2. \tag{1.81}$$

Here $\boldsymbol{T}_f$ is defined in the same way as (1.16) or (1.17) but with $\boldsymbol{\sigma}_f$ defined by (1.74) (see also (1.129)) and

$$\hat{\Sigma} := \left\{(x, y) \in \mathbb{R}^2 : 0 < x < L,\ y = R_0(x)\right\}, \tag{1.82}$$

see also Sect. 1.3.3, for more details. Equation (1.80) is equipped with the following boundary and initial conditions

$$\begin{aligned} &\eta(t, 0) = \eta(t, L) = 0 \quad \text{and} \quad \eta(0, x) = \partial_t \eta(0, x) = 0, \\ &\partial_x \eta(t, 0) = \partial_x \eta(t, L) = 0, \end{aligned} \tag{1.83}$$

We complete the system (1.73) with the boundary and initial conditions. We assume that the flow motion is driven by a pressure drop: on the inflow part of the boundary (same as (1.8)), which we denote Γ_{in}, we set

$$u_2(t, 0, y) = 0, \tag{1.84}$$

$$\left(2\mu(|\mathbb{D}(\boldsymbol{u})|)\frac{\partial u_1}{\partial x} - p + p_{in} - \frac{\rho_f}{2}|u_1|^2\right)(t, 0, y) = 0 \tag{1.85}$$

for any $0 < y < R_0(0)$, $0 < t < T$ and for a given function $p_{in} = p_{in}(y, t)$. On the opposite, outflow part of the boundary Γ_{out}, (see (1.9)) we set

$$u_2(t, L, y) = 0 \,, \tag{1.86}$$

$$\left(2\mu(|\mathbb{D}(\boldsymbol{v})|)\frac{\partial u_1}{\partial x} - p + p_{out} - \frac{\rho_f}{2}|u_1|^2\right)(t, L, y) = 0 \tag{1.87}$$

for any $0 < y < R_0(L)$, $0 < t < T$ and for a given function $p_{out} = p_{out}(t, y)$. Note that we require here that the so-called kinematic pressure (or total pressure) is prescribed on the inflow and outflow boundary. This implies that the fluxes of kinetic energy on inflow and outflow boundary will disappear in the weak formulation as already stated in the introduction. Finally, on the remaining part of the boundary, Γ_0, we set the flow symmetry condition

$$u_2(t, x, 0) = 0 \,, \qquad \mu(|\mathbb{D}(\boldsymbol{u})|)\frac{\partial u_1}{\partial y}(t, x, 0) = 0 \tag{1.88}$$

for any $0 < x < L$, $0 < t < T$. The initial conditions read

$$\boldsymbol{u}(0, x, y) = \boldsymbol{0} \qquad \text{for any } 0 < x < L,\ 0 < y < R_0(x). \tag{1.89}$$

The proof of the main result formulated in Theorem 1.1 will be realized in several steps:

- approximation of the solenoidal spaces on a moving domain by the artificial compressibility approach: ε-approximation
- splitting of the boundary conditions (1.79)–(1.80) by introducing the semi-pervious boundary: κ-approximation
- transformation of the weak formulation on a time-dependent domain $\Omega_\eta(t)$ to a fixed reference domain $\hat{\Omega}_f = (0, L) \times (0, 1)$ using a given domain deformation $\eta = \eta^1$. Note that we scale in the y-direction to $(0, 1)$ interval. This step requires that the domain motion is regular enough (here η (or η^1) should be at least C^1 in space, which was not the case in the previous subsections for weak solutions).
- limiting process for $\varepsilon \to 0$, $\kappa \to \infty$ and a fixed point on the geometry domain, respectively.

Weak Formulation

In this subsection our aim is to present the weak formulation of the problem (1.73)–(1.89). Assuming that η is enough regular (see below) and taking into account the results from [26] we can define the functional spaces that gives sense to the trace of velocity from $W^{1,p}(\Omega_\eta(t))$ and thus to define the weak solution of the problem. We assume that $R_0 \in C_0^2(0, L)$.

Definition 1.1 (Weak Formulation). We say that $(\boldsymbol{u}, \eta)$ is a weak solution of (1.73)–(1.89) on $[0, T)$ if the following conditions hold

- $\boldsymbol{u} \in L^p(0, T; W^{1,q}(\Omega_\eta(t))) \cap L^\infty(0, T; L^2(\Omega_\eta(t)))$,
- $\eta \in W^{1,\infty}(0, T; L^2(0, L)) \cap H^1(0, T; H_0^2(0, L))$,
- $\operatorname{div} \boldsymbol{u} = 0$ a.e. on $\Omega_\eta(t)$,
- $\boldsymbol{u}\big|_{\Sigma(t)} = (0, \partial_t \eta)^T$ for a.e. $x \in \Sigma(t)$, $t \in (0, T)$, $u_2\big|_{\Gamma_{\text{in}} \cup \Gamma_{\text{out}} \cup \Gamma_0} = 0$,

$$
\begin{aligned}
&\int_0^T \int_{\Omega_\eta(t)} \left\{ -\rho_f \boldsymbol{u} \cdot \partial_t \boldsymbol{\varphi} + 2\mu(|\mathbb{D}(\boldsymbol{u})|)\mathbb{D}(\boldsymbol{u})\mathbb{D}(\boldsymbol{\varphi}) + \rho_f \sum_{i,j=1}^{2} u_i \partial_{x_i} u_j \boldsymbol{\varphi}_j \right\} \\
&+ \int_0^T \int_0^{R_0(L)} \left(p_{\text{out}} - \frac{\rho_f}{2} |u_1|^2 \right) \varphi_1|_{x=L} \\
&- \int_0^T \int_0^{R_0(0)} \left(p_{\text{in}} - \frac{\rho_f}{2} |u_1|^2 \right) \varphi_1|_{x=0} \\
&+ \int_0^T \int_0^L \alpha_2 \partial_{xx} R_0 \xi \\
&+ \int_0^T \int_0^L -\rho_s e \partial_t \eta \partial_t \xi + \beta_2\, \partial_{xx} \partial_t \eta \partial_{xx} \xi + \alpha_2 \partial_x \eta \partial_x \xi + \tilde{\delta} \eta\, \xi \;=\; 0
\end{aligned}
\tag{1.90}
$$

for every test functions

$$
\begin{aligned}
&\boldsymbol{\varphi}(t, x, y) \in H^1(0, T; W^{1,q}(\Omega_\eta(t))) \text{ such that} \\
&\operatorname{div} \boldsymbol{\varphi} = 0 \text{ a.e on } \Omega_\eta(t), \\
&\varphi_2\big|_{\Sigma(t)} \in H^1(0, T; H_0^2(\Sigma(t))), \quad \varphi_2\big|_{\Gamma_{\text{in}} \cup \Gamma_{\text{out}} \cup \Gamma_0} = \varphi_1\big|_{\Sigma(t)} = 0 \quad \text{and} \\
&\xi(t, x) = \varphi_2(t, x, R_0(x) + \eta(t, x)) \text{ on } (0, T) \times (0, L).
\end{aligned}
\tag{1.91}
$$

Theorem 1.3 (Existence of a Weak Solution). *Let $q \geq 2$. Assume that the boundary data fulfill $p_{\text{in}} \in L^{q'}(0, T; L^2(0, R_0(0)))$, $p_{\text{out}} \in L^{q'}(0, T; L^2(0, R_0(L)))$, $p_w \in L^{q'}(0, T; L^2(0, L))$, $\frac{1}{q} + \frac{1}{q'} = 1$. Furthermore, assume that the properties (1.75)–(1.77) for the viscous stress tensor hold. Then there exists a weak solution $(\boldsymbol{u}, \eta)$ of the problem (1.73)–(1.89) such that*

i) $\boldsymbol{u} \in L^q(0, T; W^{1,q}(\Omega_\eta(t))) \cap L^\infty(0, T; L^2(\Omega_\eta(t)))$,
$\eta \in W^{1,\infty}(0, T; L^2(0, L)) \cap H^1(0, T; H_0^2(0, L))$,
ii) $\boldsymbol{u}\big|_{\Sigma(t)} = (0, \partial_t \eta)^T$ *for a.e.* $x \in \Sigma(t)$, $t \in (0, T)$, $u_2\big|_{\Gamma_{\text{in}} \cup \Gamma_{\text{out}} \cup \Gamma_0} = 0$,
iii) $\boldsymbol{u}$ *satisfies the condition* $\operatorname{div} \boldsymbol{u} = 0$ *a.e on* $\Omega_\eta(t)$ *and (1.90) holds.*

Auxiliary Problem: $(\kappa, \varepsilon$, given domain)-Approximation

In what follows we will formulate a suitable approximation of the original problem (1.73)–(1.89).

First of all we approximate the deformable boundary $\Sigma(t)$ by a given function $h = R_0 + \delta$, $\delta \in H^1(0,T;H_0^2(0,L)) \cap W^{1,\infty}(0,T;L^2(0,L))$, $R_0(x) \in C^2[0,L]$ satisfying for all $x \in [0,L]$

$$0 < \alpha \leq h(t,x) \leq \alpha^{-1}, \quad |\partial_x h(t,x)| + \int_0^T |\partial_t h(t,x)|^2 \leq K < \infty \tag{1.92}$$

$h(t,0) = R_0(0)$, $h(t,L) = R_0(L)$.

Consider now one step of the iterative process with respect to the domain deformation: we look for a solution $(\boldsymbol{u}, p, \eta)$ of the following problem

$$\rho_f \partial_t \boldsymbol{u} + \rho_f (\boldsymbol{u} \cdot \nabla)\boldsymbol{u} = \operatorname{div} \boldsymbol{\tau} - \nabla p \quad \text{in} \quad \Omega_\delta(t) =: \Omega_h(t), \tag{1.93}$$

and for all $x \in (0,L)$, $0 < t < T$

$$\Big[\rho_s e \partial_{tt}\eta - \alpha_2 \partial_{xx}\eta + \tilde{\delta}\eta - \beta_2 \partial_x^4 \partial_t \eta - \alpha_2 \partial_{xx} R_0\Big](t,x) = \hat{H}(\boldsymbol{u}, p)(t,\bar{x}) \tag{1.94}$$

$$\hat{H}(t,\bar{x}) = \big[\mu(|\mathbb{D}(\boldsymbol{u})|)\left\{-\left(\partial_x u_2 + \partial_y u_1\right)\partial_x h + 2\partial_y u_2\right\} - p\big](t,\bar{x}), \tag{1.95}$$

$$\boldsymbol{u}(t,\bar{x}) = (0, \partial_t \eta(t,x))^T, \tag{1.96}$$

$\bar{x} = (x, h(t,x))$. Note here that we have written the forcing term thanks to a change of variables depending on the given h.

Furthermore, in the analysis of problem (1.73)–(1.89) the boundary conditions (1.79) and the structure equation (1.80), cf. (1.94)–(1.96) are splitted in the following way, see [66]

$$\begin{aligned} &\hat{H}(\boldsymbol{u}, p)(t,\bar{x}) - \frac{\rho_f}{2} u_2\Big(u_2(t,\bar{x}) - \partial_t \delta(t,x)\Big) \\ &= \big[\mu(|\mathbb{D}(\boldsymbol{u})|)\left\{-\left(\partial_x u_2 + \partial_y u_1\right)\partial_x h + 2\partial_y u_2\right\} - p\big](t,\bar{x}) \\ &-\frac{\rho_f}{2} u_2\Big(u_2(t,\bar{x}) - \partial_t \delta(t,x)\Big) = \kappa\Big[\partial_t \eta(t,x) - u_2(t,\bar{x})\Big] \end{aligned} \tag{1.97}$$

and

$$-\Big[\rho_s e \partial_{tt}\eta - \alpha_2 \partial_{xx}\eta + \tilde{\delta}\eta + \beta_2 \partial_x^4 \partial_t \eta - \alpha_2 \partial_{xx} R_0\Big](t,x) = \kappa\Big[\partial_t \eta(t,x) - u_2(t,\bar{x})\Big]$$

with $\kappa \gg 1$. (1.98)

We will show later that the approximation with κ is reasonable. One of the possible physical interpretations for introducing finite κ comes from the mathematical modeling of semi-pervious boundary, where this type of boundary condition occurs. In our case, the boundary Σ seems to be partly permeable for finite κ, but letting $\kappa \to \infty$ it becomes impervious. In fact, we prove the existence of solution if $\kappa \to \infty$ and thus we get the original boundary conditions and structure equation (1.94)–(1.96). Note that we have added the term $-\frac{\rho_f}{2} u_2\Big(u_2(t,\bar{x}) - \partial_t \delta(t,x)\Big)$ to the fluid forces $\hat{H}$ to conserve an energy estimate in the approximation process. This term will vanish in the limit.

Furthermore, we overcome the difficulties of solenoidal spaces by means of the artificial compressibility. We approximate the continuity equation similarly as in [66] with

$$\begin{aligned} \boldsymbol{u}_\varepsilon\left(\frac{\partial p_\varepsilon}{\partial t}-\Delta p_\varepsilon\right)+\operatorname{div}\boldsymbol{u}_\varepsilon=0 \quad &\text{in } \Omega_\delta(t), \\ \frac{\partial p_\varepsilon}{\partial \boldsymbol{n}}=0, \quad \text{on } \partial\Omega_\delta(t), \quad &\varepsilon>0. \end{aligned} \tag{1.99}$$

By letting $\varepsilon \to 0$ we show that $\boldsymbol{u}_\varepsilon \to \boldsymbol{u}$, where $\boldsymbol{u}$ is the weak solution of (1.73). For fixed ε, due to the lack of solenoidal property for velocity, we have the additional term $\frac{\rho_f}{2}u_i \operatorname{div}\boldsymbol{u}$ in momentum equation, which can be included into the convective term, see (1.106). With both these approximation strategies we will avoid the added mass effect for this approximation step.

The approximated problem is defined on a moving domain depending on function $h = R_0+\delta$. Now we reformulate it to a fixed rectangular domain. This step requires that the deformation h is regular enough. Let us set

$$\begin{aligned} \hat{\boldsymbol{u}}(t,\hat{x},\hat{y}) &\overset{\text{def}}{=} \boldsymbol{u}(t,\hat{x},h(t,\hat{x})\hat{y}) \\ \hat{p}(t,\hat{x},\hat{y}) &\overset{\text{def}}{=} p(t,\hat{x},h(t,\hat{x})\hat{y}) \\ \sigma(t,\hat{x}) &\overset{\text{def}}{=} \partial_t\eta(t,\hat{x}) \end{aligned} \tag{1.100}$$

for $\mathbf{y} \in D = \{(\hat{x},\hat{y});\ 0<\hat{x}<L,\ 0<\hat{y}<1\}, 0<t<T$.

We define the following space

$$\begin{aligned} V &\equiv \left\{\boldsymbol{w}\in W^{1,p}(D):\ w_1=0 \ \text{ on } S_w,\ w_2=0 \ \text{ on } S_{\text{in}}\cup S_{\text{out}}\cup S_c\right\}, \\ S_w &= \{(\hat{x},1):0<\hat{x}<L\}, \qquad S_{\text{in}}=\{(0,\hat{y}):0<\hat{y}<1\}, \\ S_{\text{out}} &= \{(L,\hat{y}):0<\hat{y}<1\}, \qquad S_c=\{(\hat{x},0):0<\hat{x}<L\}. \end{aligned} \tag{1.101}$$

Let us introduce the following notations that correspond to the change of variables in the problem: for the divergence equation

$$\operatorname{div}_h\boldsymbol{u} \overset{\text{def}}{=} \partial_{\hat{x}}\,\widehat{u_1}-\frac{\hat{y}}{h}\partial_{\hat{x}}\,h\,\partial_{\hat{y}}\,\widehat{u_1}+\frac{1}{h}\partial_{\hat{y}}\,\widehat{u_2},$$

for the bilinear form involving the pressure

$$a_h(q,\phi)=\int_D h\,\hat{\partial}_1\hat{q}\,\partial_{\hat{x}}\phi+\left(\hat{\partial}_2\hat{q}-\hat{y}\,\partial_{\hat{x}}h\,\hat{\partial}_1\hat{q}\right)\partial_{\hat{y}}\phi, \tag{1.102}$$

with

$$\hat{\partial}_1 = \left(\frac{\partial}{\partial \hat{x}} - \frac{\hat{y}}{h}\frac{\partial h}{\partial \hat{x}}\frac{\partial}{\partial \hat{y}}\right), \quad \hat{\partial}_2 = \frac{1}{h}\frac{\partial}{\partial \hat{y}}.$$

With the same notations we define for the viscous term

$$((\hat{\boldsymbol{u}}, \boldsymbol{\psi}))_h = \int_D h\boldsymbol{\tau}_{ij}(\hat{\mathbb{D}}(\hat{\boldsymbol{u}}))\hat{\mathbb{D}}_{ij}(\boldsymbol{\psi}), \tag{1.103}$$

$$\boldsymbol{\tau}_{ij}(\hat{\mathbb{D}}(\hat{\boldsymbol{u}})) = 2\mu(|\hat{\mathbb{D}}(\hat{\boldsymbol{u}})|)\hat{\mathbb{D}}_{ij}(\hat{\boldsymbol{u}}), \tag{1.104}$$

$$\hat{\mathbb{D}}_{ij}(\hat{\boldsymbol{u}}) = \frac{1}{2}(\hat{\partial}_i(\hat{u}_j) + \hat{\partial}_j(\hat{u}_i)), \tag{1.105}$$

and for the convective term

$$\begin{aligned} b_h(\hat{\boldsymbol{u}}, z, \boldsymbol{\psi}) = \rho_f \Bigg(\int_D \left(h\widehat{u_1}\hat{\partial}_1 z + u_2\partial_{\hat{y}} z\right) \cdot \boldsymbol{\psi} + \frac{h}{2} z \cdot \boldsymbol{\psi}\, \mathrm{div}_h \hat{\boldsymbol{u}} \\ -\frac{1}{2}\int_0^1 R_0\widehat{u_1} z_1 \psi_1|_{\hat{x}=L} + \frac{1}{2}\int_0^1 R_0\widehat{u_1} z_1\psi_1|_{\hat{x}=0} \\ -\frac{1}{2}\int_0^L \widehat{u_2} z_2 \psi_2|_{\hat{y}=1} \Bigg). \end{aligned} \tag{1.106}$$

Remark 1.11. Note that the transformed stress tensor $\boldsymbol{\tau}_{ij} = 2\mu(|\hat{\mathbb{D}}(\hat{\boldsymbol{u}})|)\hat{\mathbb{D}}_{ij}(\hat{\boldsymbol{u}})$ from (1.103) with $\mu(|\hat{\mathbb{D}}(\hat{\boldsymbol{u}})|)$ defined in (1.78) also satisfies (1.75)–(1.77).

Remark 1.12. Note that all these terms could be written by means of the notations introduced in the study of the existence of strong solutions. For instance,

$$\mathrm{div}_h \boldsymbol{u} = B_h : \nabla\hat{\boldsymbol{u}}(J_h)^{-1},$$

with $B_h = \mathrm{cof}\, \nabla\boldsymbol{\phi}_h$, $J_h = \det \nabla\phi_h$ and $\phi_h(\hat{x}, \hat{y}) = (\hat{x}, \hat{y}h(\hat{x}))$.

Definition 1.1 (Weak Solution of the Approximated Linearized Problem). Let $\hat{\boldsymbol{u}} \in L^p(0, T; V) \cap L^\infty(0, T; L^2(D))$, $\hat{p} \in L^2(0, T; H^1(D)) \cap L^\infty(0, T; L^2(D))$ and $\sigma \in L^\infty(0, T; L^2(0, L)) \cap L^2(0, T; H_0^2(0, L))$. A triple $\boldsymbol{w} = (\hat{\boldsymbol{u}}, \hat{p}, \sigma)$ is called a weak solution of the regularized problem (1.73)–(1.89) if the following equation holds (for simplicity in what follows all the physical constants are assumed to be equal to one)

$$\begin{aligned} &-\int_0^T \langle \partial_t(h\hat{\boldsymbol{u}}), \boldsymbol{\psi}\rangle \\ &= \int_0^T \left(\int_D -\partial_t h\, \partial_{\hat{y}}(\hat{y}\hat{\boldsymbol{u}}) \cdot \boldsymbol{\psi} - h\,\hat{p}\, \mathrm{div}_h \boldsymbol{\psi} \right) + b_h(\hat{\boldsymbol{u}}, \hat{\boldsymbol{u}}, \boldsymbol{\psi}) + ((\hat{\boldsymbol{u}}, \boldsymbol{\psi}))_h \end{aligned}$$

$$
\begin{aligned}
&+\int_0^T\int_0^1 \hat{p}_{\text{out}}\, h|_{\hat{x}=L}\, \psi_1|_{\hat{x}=L} - \hat{p}_{\text{in}}\, h|_{\hat{x}=0}\, \psi_1|_{\hat{x}=0}\\
&+\int_0^T\int_0^L \left(\frac{1}{2}\widehat{u_2}\,\partial_t h + \kappa\,(\widehat{u_2}-\sigma)\right)\psi_2|_{\hat{y}=1}\\
&+\varepsilon\int_0^T \langle \partial_t(h\hat{p}), \phi\rangle \qquad (1.107)\\
&+\int_0^T\int_D \left(-\varepsilon\partial_t h\,\partial_{\hat{y}}(\hat{y}\hat{p})\,\phi + \varepsilon a_h(\hat{p},\phi) + h\,\mathrm{div}_h\hat{\boldsymbol{u}}\,\phi\right)\\
&+\frac{\varepsilon}{2}\int_0^T\int_0^L \partial_t h\,\hat{p}\,\phi|_{\hat{y}=1}\\
&+\int_0^T\int_0^L \Bigg(\partial_t\sigma\xi + \partial_{\hat{x}\hat{x}}\sigma\,\partial_{\hat{x}\hat{x}}\xi + \partial_{\hat{x}}\left(\int_0^t \sigma(s,\hat{x})ds\right)\partial_{\hat{x}}\xi\\
&\qquad + \left(\int_0^t \sigma(s,\hat{x})ds\right)\xi - \partial_{\hat{x}\hat{x}}R_0\xi + \kappa\,(\sigma-\widehat{u_2})\,\xi\Bigg)
\end{aligned}
$$

for every $(\boldsymbol{\psi},\phi,\xi)\in H_0^1(0,T;\boldsymbol{V})\times L^2(0,T;H^1(D))\times L^2(0,T;H_0^2(0,L))$.

Theorem 1.4 (Existence of the Approximated Linearized Weak Solution). *Let $\varepsilon,\ \kappa,\ \delta$ be fixed. Assume (1.75)–(1.77), (1.92), $\hat{p}_{\text{in}},\ \hat{p}_{\text{out}}\in L^{q'}(0,T;L^2(0,1))$. Then there exists a weak solution of the $(\kappa,\varepsilon,\delta)$-approximated problem transformed to the fixed domain, in the sense of integral identity (1.107). Moreover,*

$$
\partial_t(h\hat{\boldsymbol{u}})\in\begin{cases} L^{q'}(0,T;\boldsymbol{V}^*)\ \textit{for}\ 2<p<\infty,\\ L^{q'}(0,T;\boldsymbol{V}^*)\oplus L^{4/3}((0,T)\times D),\quad \partial_t(h\hat{p})\in L^2(0,T;H^{-1}(D)),\\ \qquad\qquad \textit{for}\ p=2,\end{cases}
$$

such that

$$
\int_0^T \left\langle \partial_t(h\hat{\boldsymbol{u}}),\boldsymbol{\psi}\right\rangle = -\int_0^T\int_D h\hat{\boldsymbol{u}}\cdot\partial_t\boldsymbol{\psi}.
$$

Solution of this approximated linearized problem has been obtained in the recent work [111] by means of the standard energy method using the Galerkin approximate solutions and showing the a priori estimates. To guarantee the coercivity of the nonlinear viscous term we need to apply the generalized Korn inequality with variable coefficients $\int_D|\hat{\mathbb{D}}(\boldsymbol{u})|^p \geq c(K,\alpha)\int_D|\nabla\boldsymbol{u}|^p$. This requires that at least $h\in L^\infty(0,T;W^{1,\infty}(0,L))$, cf. [111] for more details. We would like to point out that this is a crucial step forcing us to require also higher order friction term in modeling the string equation.

Since the problem for a given δ, and fixed ε, κ is still nonlinear because of the convection and nonlinear viscous terms, we have to obtain some compactness in order to prove that the Galerkin approximation converges. Due to the additional compressibility we are able to prove the following a priori estimates: the fluid velocity is bounded in $L^\infty(0,T;L^2(D)) \cap L^q(0,T;W^{1,q}(D))$ and its time derivative is bounded in $L^{q'}(0,T;(W^{1,q}(D))')$, whereas the structure velocity is bounded in $L^\infty(0,T;L^2(0,L)) \cap L^2((0,T);H^2(0,L))$, and its time derivative is bounded in $L^2((0,T);L^2(0,L))$. Note that the latter estimate arises due to the κ-approximation that implies that the right hand side of the structure equation can be represented in terms of velocities and not of the fluid stress.

Now applying the Lions–Aubin lemma the compactness results and the corresponding strong convergence in $L^q((0,T)\times D)$ for $\hat{\boldsymbol{u}}$ and in $L^2(0,T;H^1(0,L))$ for σ are obtained. The limiting process in the nonlinear viscous term is realized by applying the Minty-trick and theory of monotone operators.

The weak solution from Theorem 1.4 depends on the parameters ε, κ and on the given geometry h. Keeping geometry fixed but passing to the limit with $\varepsilon \to 0$, $\kappa \to \infty$ we can obtain the weak solution of the original problem (1.73)–(1.89) defined on Ω_δ. Note that now when passing to the limit with ϵ and κ we cannot apply the Lions–Aubin lemma anymore and to get the strong convergence for $\hat{\boldsymbol{u}}$ and σ the equicontinuity in time is applied (see Lemma 1.4). Again the limiting process in the nonlinear viscous term is realized by applying the Minty-trick and theory of monotone operators.

Fixed Point Iterations

We have proved the existence of weak solution of the original problem in a domain given by a known deformation function $h = R_0 + \delta$. The aim of this subsection is to prove that the mapping that associates η with h (or equivalently to δ) has at least one fixed point. To prove it we apply the Schauder fixed point theorem. The compactness argument will be based on the equicontinuity in time, see Lemma 1.4. Consequently we obtain the final result: existence of weak solution for a fully coupled fluid–structure interaction problem (1.73)–(1.89).

We denote the space $Y = H^1(0,T;L^2(0,L))$. Let us assume that δ belongs to the ball $B_{\alpha,K}$ defined by

$$
\begin{aligned}
B_{\alpha,K} = \Big\{ & \delta \in Y;\ \|\delta\|_Y \le C_{\alpha,K},\ 0 < \alpha \le R_0(\hat{x}) + \delta(t,\hat{x}) \le \alpha^{-1}, \\
& \Big|\partial_{\hat{x}}\delta(t,\hat{x})\Big| \le K,\ \ \eta(\hat{x},0) = 0,\ \forall \hat{x} \in [0,L],\ \forall t \in [0,T], \\
& \int_0^T |\partial_t \eta(t,\hat{x})|^2 \le K,\ \forall \hat{x} \in [0,L] \Big\},
\end{aligned}
$$

where $C_{\alpha,K}$ is a suitable constant large enough with respect to K, α and the data.

Note that here we ask the domain interface to be at least $W^{1,\infty}$ in space so that all the nonlinear geometric terms of the formulation coming from the change of variables are well defined.

For a given $h = R_0 + \delta, \delta \in B_{\alpha,K}$ we consider

$$X = \{\boldsymbol{\psi}(t) \in V_{\text{div}}\,;\ \boldsymbol{\psi}_2(t)\big|_{S_w} \in H_0^2(0,L),\ \xi(t) = \boldsymbol{\psi}_2(t)\big|_{S_w}\}, \tag{1.108}$$
$$V_{\text{div}} := \{f \in \boldsymbol{V},\ \text{div}_h f = 0\ a.e.\text{ on } D\},\ \text{cf. (1.101).}$$

For each test function $\boldsymbol{\psi} \in L^q(0,T;X)$, $\boldsymbol{\psi}(T) = 0$, and for any $h \in Y$, such that (1.92) holds, we construct solutions $(\hat{\boldsymbol{u}},\ \eta)$ of the following problem defined on the reference domain D, $\sigma = \partial_t \eta$

$$\begin{aligned}
&-\int_0^T \langle \partial_t(h\hat{\boldsymbol{u}}), \boldsymbol{\psi}\rangle \\
&= \int_0^T \left(\int_D -\partial_t h\, \partial_{\hat{y}}(\hat{y}\hat{\boldsymbol{u}}) \cdot \boldsymbol{\psi} - h\,\hat{p}\,\text{div}_h \boldsymbol{\psi} \right) + b_h(\hat{\boldsymbol{u}}, \hat{\boldsymbol{u}}, \boldsymbol{\psi}) + ((\hat{\boldsymbol{u}}, \boldsymbol{\psi}))_h \\
&+ \int_0^T \int_0^1 \hat{p}_{\text{out}}\, h|_{\hat{x}=L}\, \psi_1|_{\hat{x}=L} - \hat{p}_{\text{in}}\, h|_{\hat{x}=0}\, \psi_1|_{\hat{x}=0} \\
&+ \int_0^T \int_0^L \frac{1}{2}\sigma\, \partial_t h \psi_2|_{\hat{y}=1} \\
&+ \int_0^T \int_0^L \left(\partial_t \sigma \xi + \partial_{\hat{x}\hat{x}}\sigma\, \partial_{\hat{x}\hat{x}}\xi + \partial_{\hat{x}} \left(\int_0^t \sigma(s,\hat{x})ds \right) \partial_{\hat{x}}\xi \right. \\
&\qquad \left. + \left(\int_0^t \sigma(s,\hat{x})ds \right) \xi - \partial_{\hat{x}\hat{x}} R_0 \xi \right).
\end{aligned}$$

By choosing $\delta \in B_{\alpha,K}$ and applying the Korn's inequality we ensure that the viscous bilinear form is coercive.

The following energy estimate holds for all $\infty > q \geq 2$ uniformly in δ

$$\begin{aligned}
&\|\hat{\boldsymbol{u}}\|^2_{L^\infty(0,T;L^2(D))} + \|\hat{\boldsymbol{u}}\|^q_{L^q(0,T;W^{1,q}(D))} \\
&\quad + \|\partial_t \eta\|^2_{L^\infty(0,T;L^2(0,L))} + \|\partial_t \eta\|^2_{L^2(0,T;H^2(0,L))} + \|\eta\|^2_{L^\infty(0,T;H^1(0,L))} \\
&\leq c(T,p,K,\alpha) \left(\|\hat{p}_{\partial D}\|_{L^{q'}(),T,L^2(\partial\Omega)} + \|R_0\|^2_{C^2[0,L]} \right).
\end{aligned} \tag{1.109}$$

Note that this energy estimate takes the same form as the one obtained in the Newtonian case except that the space $L^2(0,T;H^1(D))$ is replaced by $L^q(0,T;W^{1,q}(D))$.

Now, let us define the following mapping,

$$\mathcal{F} : B_{\alpha,K} \to Y;$$
$$\mathcal{F}(h) \equiv \mathcal{F}(R_0 + \delta) = \eta.$$

Our aim is to apply the Schauder fixed point theorem and prove that the mapping $\mathcal{F}$ has at least one fixed point. This implies the existence of the weak solution to our fully coupled problem. It is necessary to check the following properties:

- $\mathcal{F}(B_{\alpha,K}) \subset B_{\alpha,K}$
- $(\mathcal{F}(R_0 + \delta^{(k)}))_k$ is relatively compact in Y, for any sequence in $(\delta^{(k)})_k \in B_{\alpha,K}$ (We show the equicontinuity in time, Lemma 1.4, which implies, together with the energy estimate, the strong convergence in Y.)
- The continuity of the mapping $\mathcal{F}$ with respect to the strong topology in Y must be proven.

Relative Compactness of the Fixed Point Mapping $\mathcal{F}$

We focus on the integral equicontinuity in time and the relative compactness in Y. Lemma 1.4 provides the equicontinuity result that holds independently on k. The lemma is the analogue of the two Lemmas 1.2 and 1.3 and writes (with the same notations as in the previous subsections, see (1.54))

Lemma 1.4. *For the weak solution* $(\boldsymbol{u}^{(k)}, \partial_t \eta^{(k)}) = (\boldsymbol{u}^{(k)}, \sigma^{(k)})$ *of the problem associated with a given* $\delta^{(k)} \in B_{\alpha,K}$ *it holds*

$$\int_0^{T-\tau}\int_B \rho_t^{(k)} |\bar{\boldsymbol{u}}^{(k)}(t+\tau) - \bar{\boldsymbol{u}}^{(k)}(t)|^2 + \int_0^{T-\tau}\int_0^L |\partial_t \eta^{(k)}(t+\tau) - \partial_t \eta^{(k)}(t)|^2$$
$$\leq C(\tau^{1/q} + \tau^{1/2}). \tag{1.110}$$

Here $\rho_t^{(k)}$ *denotes the characteristic function of* $\Omega_{h^{(k)}}(t)$. *The constant* $C = C(K, \alpha)$ *does not depend on* k.

From (1.110) we can get that

$$\int_0^{T-\tau}\int_B |\rho_{t+\tau}^{(k)} \bar{\boldsymbol{u}}^{(k)}(t+\tau) - \rho_t^{(k)} \bar{\boldsymbol{u}}^{(k)}(t)|^2 + \int_0^{T-\tau}\int_0^L |\partial_t \eta^{(k)}(t+\tau) - \partial_t \eta^{(k)}(t)|^2$$
$$\leq C(\tau^{1/q} + \tau^{1/2}). \tag{1.111}$$

It implies that $\rho_t^{(k)} \bar{\boldsymbol{u}}^{(k)}(t)$, and thus $\bar{\boldsymbol{u}}^{(k)}(t)$ is relatively compact in $L^2((0,T) \times B)$.

Consequently, the Riesz–Fréchet–Kolmogorov compactness argument [20, Theorem IV.26] (or see Lemma 1.1) based on (1.111) implies the relative compactness of $\partial_t \eta^{(k)}, \bar{\boldsymbol{u}}^{(k)}$ in $L^2(0,T; L^2(0,L))$, $L^2(0,T; L^2(B))$, respectively. Additionally, the standard interpolations give us the compactness of $\bar{\boldsymbol{u}}^{(k)}$ in $L^r((0,T) \times B)$, $1 \leq r < 4$ and $\partial_t \eta^{(k)}$ in $L^s((0,T) \times (0,L))$, $1 \leq s < 6$.

Continuity of the Mapping $\mathcal{F}$

We can show by limiting process for $k \to \infty$ in the weak formulation satisfied by $(\boldsymbol{u}^{(k)}, \eta^{(k)})$ associated with $h^{(k)}$ that

$$\mathcal{F}(h^{(k)}) \to \mathcal{F}(h)$$

whenever $h^{(k)}$ converges to h in $B_{\alpha,K}$. As already shown above $(\hat{\boldsymbol{u}}^{(k)}, \eta^{(k)})$ converges strongly to some $(\hat{\boldsymbol{u}}, \eta)$ in Y, i.e., we have $\eta^{(k)} \to \eta$ in $H^1(0,T;L^2(0,L))$ as $k \to \infty$. Due to the boundedness of η from a priori estimate (1.109) and the imbeddings in one dimension we have even stronger result—the uniform convergence of $\partial_{\hat{x}}\eta^{(k)}$ in $C([0,T] \times [0,L])$. Indeed,

$$\eta^{(k)} \in L^\infty(0,T;H^2(0,L)) \cap W^{1,\infty}(0,T;L^2(0,L)) \hookrightarrow C^{0,1-\beta}(0,T;H^{2\beta}(0,L)) \tag{1.112}$$

for $0 < \beta < 1$. From the continuous imbedding of $H^{2\beta}(0,L)$ into $H^{2\beta-\epsilon}(0,L)$ and the Arzelá–Ascoli Lemma we conclude that a subsequence of $\eta^{(k)}$ converges strongly in $C([0,T];H^s(0,L))$, $0 < s < 2$. Since for $s > 3/2$ we also have continuous imbedding $H^s(0,L) \hookrightarrow C^1[0,L]$, we can conclude, that $\eta^{(k)} \to \eta$ strongly in $C(0,T;C^1[0,L])$. Let us summarize available convergences

$$\begin{aligned}
\hat{\boldsymbol{u}}^{(k)} &\rightharpoonup \boldsymbol{u} && \text{weakly in } L^q(0,T;W^{1,q}(D)),\\
\bar{\hat{\boldsymbol{u}}}^{(k)} &\to \bar{\boldsymbol{u}} && \text{strongly in } L^r((0,T)\times B),\ 1 \le r < 4,\\
\hat{\boldsymbol{u}}^{(k)} &\to \hat{\boldsymbol{u}} && \text{strongly in } L^r((0,T)\times D),\ 1 \le r < 4,\\
\eta^{(k)} &\rightharpoonup \eta && \text{weakly in } H^1(0,T;H^2(0,L)),\\
\eta^{(k)} &\rightharpoonup^* \eta && \text{weakly* in } L^\infty(0,T;L^2(0,L)),\\
\eta^{(k)} &\to \eta && \text{uniformly in } C(0,T;C^1[0,L]),\\
\partial_t\eta^{(k)} &\to \partial_t\eta && \text{strongly in } L^s((0,T)\times(0,L)),\ 1 \le s < 6.
\end{aligned} \tag{1.113}$$

Limiting Process

Now we let $k \to \infty$ in the weak formulation satisfied by $(\boldsymbol{u}^{(k)}, \eta^{(k)})$. The previous convergences allow to pass to the limit. As already mentioned the main difficulty at this step is to deal with test functions that depend on the solution. Nevertheless as mentioned in Sect. 1.2.2, one can choose test functions that do not depend on k. Let us mention two important limits for $k \to \infty$:

$$\int_0^T \Big(b_{h^{(k)}}(\hat{\boldsymbol{u}}^{(k)}, \hat{\boldsymbol{u}}^{(k)}, \boldsymbol{\psi}) - b_h(\hat{\boldsymbol{u}}, \hat{\boldsymbol{u}}, \boldsymbol{\psi})\Big) \to 0,$$

$$\int_0^T ((\hat{\boldsymbol{u}}^{(k)}, \boldsymbol{\psi}))_{h^{(k)}} - ((\hat{\boldsymbol{u}}, \boldsymbol{\psi}))_h \to 0. \tag{1.114}$$

This concludes the limiting process in (1.109). We found out that $\mathcal{F}(h^{(k)}) \to \mathcal{F}(h)$, as $h^{(k)} = R_0 + \delta^{(k)}$ converges to $h = R_0 + \delta$ as $k \to \infty$.

Finally, using the continuity of the mapping $\mathcal{F}$, its relative compactness in Y and the property $\mathcal{F}(B_{\alpha,K}) \subset B_{\alpha,K}$ we deduce from the Schauder fixed point theorem, that there exists at least one fixed point of the mapping $\mathcal{F}$ defined by the weak formulation (1.109), $\mathcal{F}(R_0 + \eta) = \eta$. Thus, we obtain the existence of at least one weak solution (1.90) of the original unsteady fluid–structure interaction problem (1.73)–(1.89). The proof of the Theorem 1.3 is now completed.

Let us point out that we have obtained the existence of weak solution until some time T^*. We remind that this time is obtained in order to achieve the fixed point of the mapping $\mathcal{F}$ and to avoid the contact of the elastic boundary $\hat{\Sigma}$ with the fixed boundary for given data $\hat{p}_{\partial D}$, R_0 and α, K. Similarly as in [26] and as explained in the previous subsection for the Newtonian case, we can prolongate the solution in time and even obtain the global existence until the contact with the solid bottom.

Remark 1.13. The result on the existence of weak solution for the coupled fluid–structure interaction problem for shear-thickening power-law fluids is shown for the generalized string equation with a regularizing term of type $-\partial_x^4 \partial_t \eta$. The same existence result can be obtained for other regularizing terms in the structure equation. Instead of $-\partial_{xx}\eta + \partial_x^4 \partial_t \eta$ we can consider $\partial_x^4 \eta - \partial_{xx}\partial_t \eta$. The regularity of the domain deformation coming from the term $\partial_x^4 \eta$ is essential to obtain that $\eta \in L^\infty(0, T; W^{1,\infty}(0, L))$. This is a sufficient condition for generalized Korn's inequality for $q \neq 2$. As seen in the Newtonian case (see also [26, 88]) such a condition for η is not required for Korn's equality in the moving domain $\Omega_\eta(t)$ to hold. Consequently, in three-dimensional case a plate with no additional viscosity may be used and in the two-dimensional case a string model with no additional viscosity may be used.

In this section we have reviewed some existence results. In the unsteady state case for the existence of weak solutions, we have seen that the nonlinear geometrical term could be decoupled from the fluid–structure problem. In the case, when some decoupling of the fluid and the structure is used (e.g., to take advantage of the already known properties/regularity of solutions of the Stokes system or of the elastic one [119, 120] or because of the chosen strategy used to obtain the desired result [111, 129]), it is still necessary to decouple the fluid and the structure in an appropriate way, because of the added mass effect. This is in particular true for numerical schemes as we will detail it hereafter.

1.3 Numerical Analysis

As we have seen previously the coupling conditions at the interface are of two types: kinematic (equality of the velocities in the case of a viscous flow) and dynamic (the action–reaction principle). From a numerical point of view the most direct way to satisfy both these two conditions is to solve the coupled fluid–structure problem thanks to a unique solver based on a global weak formulation. One obtains then schemes that are called *monolithic*. Examples of such approach are numerous.

Without being exhaustive, one can refer to [101, 144] based on an ALE formulation of the fluid equations, [4, 36] based on fictitious domain method or to [65].

By construction both the kinematic and the dynamic conditions are satisfied and it leads to *strongly coupled* schemes. These methods conserve the energy at the fluid–structure interface and are consequently usually stable. Nevertheless they are not really modular and do not allow to take easily advantage of the specificities of each sub-problem. From the numerical analysis point of view, we refer to [44, 92, 115, 148], for stability and convergences studies on fully coupled schemes.

With the *stagerred* schemes the fluid and the structure are computed by two specific solvers and one question is then: *how to couple these two solvers efficiently?* A first strategy is to use a *weakly coupled* or *explicit* schemes where the fluid and the structure equations are solved once per time step. The advantage of such methods is that they are cheap since they require the resolution of the fluid and the structure only once per time step. These methods have been wildly used for the simulation of compressible flows (see, for instance, [134–136]). We refer to [51] for a state of the art of such methods.

Nevertheless the kinematic condition and the dynamic condition are not both satisfied and it may lead to numerical instabilities in particular in the case of strong added mass effect, as already mentioned in the previous section and as we will see in Sect. 1.3.1. These instabilities were seen in the case of blood flows in arteries, where the incompressible flow interacts with a structure whose density is closed to the fluid density. The role played by the added mass was underlined in [116] and formalized in [25] (and later on in [72] for more general time discretization schemes). We will see, on the toy model introduced in Sect. 1.2.1 that an explicit scheme, based on a Dirichlet to Neumann splitting, is unconditionally unstable if the fluid–structure density ratio is less than a given quantity (that may depend on the geometrical characteristics of the considered problem). It is strongly linked to the fluid incompressibility.

Consequently, *strongly* coupled or *implicit* schemes have been further developed and used to obtain stable numerical methods. These schemes are naturally stable since they conserve the energy balance at the interface. Nevertheless they are costly since they require many requests to each solver at each time step. Thus, in the early 2000 year efforts have been made to accelerate the convergence and the efficiency of fixed point procedures [37, 116, 127, 128, 131] or of inexact Newton methods [101, 124, 125, 156] or exact Newton methods [58–61]. Moreover more recently methods have been introduced based on Robin–Robin decomposition [6] to optimize the partitioned procedure, see also [8] for further results on Robin transmission conditions. In [37] further results on acceleration using the so-called transpiration conditions were presented, see also [38] for a nice review on different coupling approaches based on the domain decomposition techniques and approximation of the Steklov–Poincaré operator.

Nevertheless, these schemes stay expensive and there was a need to find new efficient, cheap and stable strategies. As already mentioned, the most natural splitting that can be made for the coupled problem we are considering in this chapter is based on a *Dirichlet to Neumann* decomposition, where the fluid is solved by imposing Dirichlet boundary conditions at the interface whereas the structure depends on the stress applied by the fluid. Nevertheless, in this case one has to deal with the added mass effect. Consequently, semi-implicit [63] has been introduced based on the implicit treatment of the added mass effect and the explicit treatment of the viscous stress. This splitting can be obtained by applying a Chorin–Temam scheme for the resolution of the fluid part [29, 30, 155] (see also [97] for a state of the art on projection schemes) and has a lot of similarities with the strategy of proof developed in [119, 120] and summarized in the previous section. We refer to [56] for a nice review on these semi-implicit schemes. In [7, 137] algebraic splitting methods are used. Moreover, in the case where the structure is a thin structure then it can be viewed as generalized Robin condition for the fluid [130].

More recently, in the case of a thin structure, a splitting of the structure equation has been proposed in [98], see also [56, 57, 62, 110, 121] for further results and generalizations. Note that the kinematic splitting method introduced in [98] applies only for a thin viscoelastic structure. If the structure is viscoelastic and we treat the friction term from the structure within the fluid solver, then the fluid problem conserves the complexity of the full coupled problem since one has a differential and nonlocal operator at the fluid interface. On the other hand, we can left the friction terms in the structure equation which yields only the local structural operator within the fluid solver, see, e.g., [55]. In the recent work of Čanič [22] a generalization of the kinematic splitting algorithm for coupling a thick structure with the fluid but with an intermediate thin layer was presented. These well-adapted methods in the case of a thin structure are stable and do not depend on the added mass effect. But they may dissipate the energy and need precise time integration scheme and prediction to be accurate [55]. The major question now in the fluid–structure interaction discretization is: *how can we improve these cheap and stable kinematically coupled schemes so that they become more precise*? In the recent work of Čanič [23] the so-called β-schemes have been derived that point out the implicit inclusion of the structure inertia into the fluid sub-problem to avoid the added mass effects and improve scheme's accuracy. The questions of stability and accuracy are discussed here as well. One way to increase the accuracy of the kinematic coupling schemes is to apply the higher order splitting schemes, such as the Strang splitting. It has been shown in the recent works of Lukáčová et al. in [110, 121] that in such a way a new kinematic splitting scheme can be obtained that is stable and second order accurate. Another way, presented in [56] is based on the so-called incremental displacement-correction scheme, i.e., interpolation of the structure velocity used in the hydrodynamic operator to improve accuracy of the coupling. As mentioned above, these schemes were termed kinematically coupled, since they treat implicitly the hydrodynamic fluid–structure coupling (the so-called added mass effect) and explicitly the contribution of elastic structure.

In the next subsections we will underline, once again, why, in the case of strong added mass effect a naive explicit scheme may be unconditionally unstable. Next we will present a semi-implicit scheme based on the Chorin–Temam projection strategy for the fluid part that is conditionally stable and convergent. Further, we also present main ideas of the kinematically coupled schemes and present their stability analysis.

1.3.1 Explicit Scheme and Added Mass Effect

In this subsection we recall some results obtained in [25] to illustrate the so-called added mass effect on an explicit Dirichlet to Neumann time marching scheme. We consider the same kind of linear toy problem described in Sect. 1.2.1. The fluid satisfies (1.30), (1.31), (1.32), (1.33), (1.34), with the coupling condition (1.36), whereas we further simplify the structure equation and consider

$$\rho_s e \partial_{tt} \eta + a\eta = p, \text{ on } (0, L). \tag{1.115}$$

Let $\Delta t > 0$ denote the time step. Let us denote by $\boldsymbol{u}^n$, η^n the approximations of $\boldsymbol{u}$, η at time $n\Delta t$, $n \in \mathbb{N}$. The fluid equations (1.30) are discretized with respect to the time by an implicit Euler scheme

$$\rho_f \frac{\boldsymbol{u}^{n+1} - \boldsymbol{u}^n}{\Delta t} + \nabla p^n = \mathbf{0} \quad \text{in } \hat{\Omega}_f. \tag{1.116}$$

Moreover we impose a Dirichlet boundary condition on the fluid–structure interface, that is an explicit discretization of (1.36):

$$\boldsymbol{u}^{n+1} \cdot \hat{\boldsymbol{n}} = \frac{\eta^n - \eta^{n-1}}{\Delta t}, \text{ on } \hat{\Sigma}. \tag{1.117}$$

The structure equation is discretized with an explicit leap frog scheme:

$$\rho_s e \frac{\eta^{n+1} - 2\eta^n + \eta^{n-1}}{\Delta t^2} + a\eta^n = p^{n+1}, \text{ on } (0, L). \tag{1.118}$$

With the same formalism as previously by introducing the added mass operator $\mathcal{M}$ defined by the Poisson problem (1.45), (1.46), (1.47), (1.48), it leads to

$$\rho_s e \frac{\eta^{n+1} - 2\eta^n + \eta^{n-1}}{\Delta t^2} + \rho_f \mathcal{M}\left(\frac{\eta^n - 2\eta^{n-1} + \eta^n}{\Delta t^2}\right) + a\eta^n = q^n, \text{ on } (0, L), \tag{1.119}$$

where the pressure q is defined by (1.41), (1.42), (1.43) (1.44). Then we have the following proposition [25]:

Proposition 1.1. *Let $\mu_{\max}$ be the largest eigenvalue of the compact, self-adjoint operator $\mathcal{M}$ in $\mathcal{L}(L^2(\hat{\Sigma}))$, then the scheme* (1.119) *is unconditionally unstable if*

$$\frac{\rho_s e}{\rho_f \mu_{\max}} < 1. \tag{1.120}$$

The proof of this proposition is elementary: by expanding η^n on the orthonormal $L^2(\hat{\Sigma})$ basis of eigenvectors (ξ_i) of $\mathcal{M}$, we obtain

$$\rho_s e \frac{\eta_i^{n+1} - 2\eta_i^n + \eta_i^{n-1}}{\Delta t^2} + \rho_f \mu_i \frac{\eta_i^n - 2\eta_i^{n-1} + \eta_i^{n-2}}{\Delta t^2} + a\eta_i^n = q_i^n, \text{ on } (0, L),$$

where $\mu_i > 0$ is the ith eigenvalue of $\mathcal{M}$ and $\eta^n = \sum_i \eta_i^n \xi_i$ and $q^n = \sum_i q_i^n \xi_i$. By a direct calculation of the characteristic polynomial χ of the previous difference equation, we obtain that $\chi(-\infty) = -\infty$ and $\chi(-1) > 0$ as soon as there exists i such that $\rho_f \mu_i > \rho_s e$. Consequently, there exists a root of the polynomial χ that is strictly less than -1. Thus, if $\rho_f \mu_{\max} > \rho_s e$, the scheme is unstable unconditionally with respect to the time step Δt.

1.3.2 A Semi-Implicit Scheme

Here we present a semi-implicit scheme but only on the linear problem associated with (1.1), (1.2), (1.10), (1.11), (1.15), (1.16), (1.3), (1.12). It means that we neglect the variations of the fluid domain and consequently the nonlinear geometrical terms. As we have seen in the previous section these nonlinearities may be treated explicitly in the time advancing scheme. We refer to [132] for a study of the different possibility of choice to advance in time the fluid domain in the case where the ALE method is used. We also consider only the Stokes system. The fluid equations are discretized with a Chorin–Temam projection method. The time marching scheme for the structure is done by a first order leap frog scheme. Let us denote by Δt the time step, and $t^n = n\Delta t$. We assume that approximations $(\boldsymbol{u}^n, p^n, \xi^n, \eta^n)$ of the fluid velocity, fluid pressure, and structure displacement at time t^n are known. The algorithm writes:

- **Step 1:** Advection–diffusion step

$$\begin{cases} \rho_f \dfrac{\tilde{\boldsymbol{u}}^{n+1} - \boldsymbol{u}^n}{\Delta t} - \nu \Delta \tilde{\boldsymbol{u}}^{n+1} = \boldsymbol{0}, & \text{in} \quad \hat{\Omega}_f \\ \tilde{\boldsymbol{u}}^{n+1} = \left(\dfrac{\xi^n - \xi^{n-1}}{\Delta t}, \dfrac{\eta^n - \eta^{n-1}}{\Delta t} \right)^T = \dfrac{\boldsymbol{d}^n - \boldsymbol{d}^{n-1}}{\Delta t}, & \text{on} \quad \hat{\Sigma}. \end{cases} \tag{1.121}$$

- **Step 2** : Projection step (velocity correction)
  - **Step 2.1:**

$$\begin{cases} \rho_f \dfrac{\boldsymbol{u}^{n+1} - \tilde{\boldsymbol{u}}^{n+1}}{\Delta t} + \nabla p^{n+1} = \boldsymbol{0}, & \text{in} \quad \hat{\Omega}_f, \\ \operatorname{div} \boldsymbol{u}^{n+1} = 0, & \text{in} \quad \hat{\Omega}_f, \\ \boldsymbol{u}^{n+1} \cdot \hat{\boldsymbol{n}} = \dfrac{\eta^{n+1} - \eta^n}{\Delta t} \cdot \hat{\boldsymbol{n}}, & \text{on} \quad \hat{\Sigma}. \end{cases} \tag{1.122}$$

  - **Step 2.2: Structure equation on** $(0, L)$

$$\begin{cases} \rho_s e \dfrac{\xi^{n+1} - 2\xi^n + \xi^{n-1}}{\Delta t^2} - \alpha_1 \partial_{xx} \xi^{n+1} = (\boldsymbol{\sigma}_f(\tilde{\boldsymbol{u}}^{n+1}, p^{n+1}) \cdot \hat{\boldsymbol{n}})_1, \\ \rho_s e \dfrac{\eta^{n+1} - 2\eta^n + \eta^{n-1}}{\Delta t^2} + \gamma \partial_x^4 \eta^{n+1} - \alpha_2 \partial_{xx} \eta^{n+1} = (\boldsymbol{\sigma}_f(\tilde{\boldsymbol{u}}^{n+1}, p^{n+1}) \cdot \hat{\boldsymbol{n}})_2. \end{cases} \tag{1.123}$$

Here only the projection step is implicitly coupled with the structure equation.

Remark 1.14. When dealing with the full nonlinear problem all the nonlinearity (convection terms, geometrical nonlinearities) are treated at the explicit advection–diffusion step.

Remark 1.15. The projection step is written as a Darcy system but we could also have written it as a Poisson problem on the pressure as we did it for the toy problem (see (1.37)). This scheme is based on the same kind of splitting idea summarized in Sect. 1.2.3.

Remark 1.16. Note that if one computes the residual of the fluid equation on the interface it is not equal to zero but to

$$\int_{\hat{\Sigma}} \mathbb{D}(\tilde{\boldsymbol{u}}^{n+1}) \cdot \left(\tilde{\boldsymbol{u}}^{n+1} - \frac{\boldsymbol{d}^{n+1} - \boldsymbol{d}^n}{\Delta t} \right).$$

It involves only the viscous stress and does not involve the pressure so one could hope to control it.

Stability and Convergence

Now we are going to see why this scheme is stable and convergent for the coupled system. We are considering where a $2D$ fluid is interacting with a $1D$ structure. Note that in [64] and [1] more general coupled systems are considered in particular $3D/3D$ coupling. For the purpose of the analysis we assume that the coupled system is also discretized in space with a finite element discretization. We denote by h the space mesh size. We assume that the finite element spaces for the fluid velocity and the fluid pressure are, respectively, conform in $H^1(\hat{\Omega}_f)$ and $L^2(\hat{\Omega}_f)$ (we assume that these spaces satisfy the inf–sup or LBB condition even if it plays no role in our analysis). We assume moreover that the structure space approximation V_s^h is a conformal approximation of $V_s = H^1(0,L) \times H^2(0,L)$ if $\gamma > 0$ or $V_s = H^1(0,L) \times H^1(0,L)$ if $\gamma = 0$. Even if the approximations are conform for each sub-problem, the global approximation may be nonconform, in particular in the case $\gamma > 0$, where one could have, for instance, Lagrange finite element for the fluid part and Hermite finite element for the discretization of the transverse displacement. In this particular case, there is a nonconforming matching at the interface and it is necessary to introduce a matching operator π_h that could be an interpolation operator or an integral matching operator of mortar type [11–14]. The kinematic conditions satisfied by the fluid and the structure at the interface and by the test functions that appear in the variational formulation of the coupled problem discretized in space and time writes (by obvious notations)

$$\tilde{\boldsymbol{u}}_h^{n+1} = \pi_h\left(\frac{\boldsymbol{d}_h^n - \boldsymbol{d}_h^{n-1}}{\Delta t}\right), \qquad \text{on } \hat{\Sigma}$$

$$\boldsymbol{u}_h^{n+1}\cdot\hat{\boldsymbol{n}} = \pi_h\left(\frac{\eta_h^{n+1} - \eta_h^n}{\Delta t}\right), \qquad \text{on } \hat{\Sigma},$$

and for the test functions

$$\boldsymbol{v}_h = \pi_h(\boldsymbol{b}_h), \qquad \text{on } \hat{\Sigma}.$$

The following stability result is then satisfied [64].

Theorem 1.5. *Assume that the restriction on the interface of the matching operator π_h is stable in L^2. There exists a constant $C > 0$ that do not depend on the physical parameters of the problem such that, if*

$$\rho_s \geq C\left(\rho_f h + 2\frac{\nu\Delta t}{h}\right), \tag{1.124}$$

the discrete energy bounds are satisfied:

$$\frac{1}{\Delta t}\left[\frac{\rho_f}{2}\|\boldsymbol{u}_h^{n+1}\|^2_{L^2(\hat{\Omega}_f)} - \frac{\rho_f}{2}\|\boldsymbol{u}_h^n\|^2_{L^2(\hat{\Omega}_f)}\right] + \frac{1}{\Delta t}\left[\frac{\rho_s}{2}\left\|\frac{\boldsymbol{d}_h^{n+1} - \boldsymbol{d}_h^n}{\Delta t}\right\|^2_{L^2(\hat{\Sigma})}\right.$$

$$-\frac{\rho_s}{2}\left\|\frac{\boldsymbol{d}_h^n-\boldsymbol{d}_h^{n-1}}{\Delta t}\right\|^2_{L^2(\hat{\Sigma})}\Bigg]+\frac{1}{2\Delta t}\left[a_s(\boldsymbol{d}_h^{n+1},\boldsymbol{d}_h^{n+1})-a_s(\boldsymbol{d}_h^n,\boldsymbol{d}_h^n)\right]$$
$$+\nu\|\nabla(\tilde{\boldsymbol{u}}_h^{n+1})\|^2_{L^2(\hat{\Omega}_f)}\leq 0, \quad (1.125)$$

where

$$a_s(\boldsymbol{d},\boldsymbol{b})=\alpha_1\int_0^L \partial_x d_1\partial_x b_1+\alpha_2\int_0^L \partial_x d_2\partial_x b_2+\gamma\int_0^L \partial_{xx} d_2\partial_{xx} b_2$$

with $\boldsymbol{d}=(d_1,d_2)^T$ *and* $\boldsymbol{b}=(b_1,b_2)^T$.

Remark 1.17. In the really special case, when $\xi = 0$, we can prove the same stability result on the semi-discrete scheme in time (i.e., without space discretization) provided the time step is small enough.

Furthermore, the following convergence theorem holds true.

Theorem 1.6. *Assuming that the finite element spaces have good approximation properties, under the stability conditions (1.124) and assuming that the exact solution of the linear coupled problem is regular enough then the sequence of discrete solutions converge through the continuous solution and satisfy the error estimates:*

$$\frac{\rho_f}{2}\|\boldsymbol{u}_h^{n+1}-\boldsymbol{u}(t^{n+1})\|^2_{L^2(\hat{\Omega}_f)}+\frac{\rho_s}{2}\left\|\frac{\boldsymbol{d}_h^{n+1}-\boldsymbol{d}(t^{n+1})-\boldsymbol{d}_h^n-\boldsymbol{d}(t^n)}{\Delta t}\right\|^2_{L^2(\hat{\Sigma})}$$
$$+\frac{1}{4}(a_s(\boldsymbol{d}_h^{n+1}-\boldsymbol{d}(t^{n+1}),\boldsymbol{d}_h^{n+1}-\boldsymbol{d}(t^{n+1})))+\sum_{n=0}^{N}\nu\Delta t\|\nabla(\tilde{\boldsymbol{u}}_h^{n+1}-\boldsymbol{u}(t^{n+1}))\|^2_{L^2(\hat{\Omega}_f)}$$
$$\leq C\Delta t+Ch^{2k}+Ch^{2m}+Ch^{2l}, \quad (1.126)$$

k denotes the degree of the Lagrange polynomial associated with the fluid finite element discretization and m depends on the choice of space discretization for the structure is such that, for $\boldsymbol{v}\in V_s$,

$$\inf_{\boldsymbol{v}_h\in V_s^h} a_s(\boldsymbol{v}-\boldsymbol{v}_h,\boldsymbol{v}-\boldsymbol{v}_h)\leq Ch^{2m},$$

and whereas l depends on the choice of the matching operator at the interface π_h*:*

- $l\leq\min(1,k+1)$ *if* $\gamma=0$, $l\leq\min(2,k+1)$ *if* $\gamma=0$ *for the interpolation operator;*
- $\frac{1}{2}\leq l\leq k$ *for the mortar operator.*

These results, which are the adaptation of [1, 64], to our setting are to be commented. First of all, concerning the stability, it is obtained under a sufficient condition that requires that the time steps is small enough with respect to the mesh size. In the case where one considers a more general setting, for instance a $3D$ fluid coupled with a $3D$ structure then, to obtain the same kind of stability property, one has to consider different space discretizations for the fluid and for the structure. Nevertheless this restriction is not seen in practice. Note moreover that the sufficient condition depends on the physical parameters of the problem and on the ratio of the densities. Once again this restriction has not been observed for practical applications such as blood flow in arteries.

This sufficient condition comes from the fact that we have to control the viscous residual $\int_{\hat{\Sigma}} \mu \mathbb{D}(\tilde{\boldsymbol{u}}^{n+1})(\tilde{\boldsymbol{u}}_h^{n+1} - \partial_t \boldsymbol{d}_h^{n+1})$ since the viscous effect is treated explicitly (see Remark 1.16). It will be the same sufficient condition if one considers a heat-wave (parabolic–hyperbolic) explicit coupling. To prove these stability result, one would like to choose as test functions

$$\tilde{\boldsymbol{u}}_h^{n+1}, \quad \boldsymbol{u}_h^{n+1}, \quad \frac{\boldsymbol{d}_h^{n+1} - \boldsymbol{d}_h^n}{\Delta t}.$$

But at the interface we have $\tilde{\boldsymbol{u}}_h^{n+1} = \pi_h(\frac{\boldsymbol{d}_h^n - \boldsymbol{d}_h^{n-1}}{\Delta t}) \neq \pi_h(\frac{\boldsymbol{d}_h^{n+1} - \boldsymbol{d}_h^n}{\Delta t})$. Thus we choose

$$\tilde{\boldsymbol{u}}_h^{n+1} + \mathrm{Ext}_h\left(\pi_h(\frac{\boldsymbol{d}_h^{n+1} - 2\boldsymbol{d}_h^n + \boldsymbol{d}_h^{n-1}}{\Delta t})\right), \quad \boldsymbol{u}_h^{n+1}, \quad \frac{\boldsymbol{d}_h^{n+1} - \boldsymbol{d}_h^n}{\Delta t},$$

as test functions, where Ext_h is an extension of the structure test function in the fluid domain such that the L^2 norm is of order h and the H^1 norm blows up as $\frac{1}{h}$. The difficulty here is to control the extra terms coming from $\mathrm{Ext}_h\left(\pi_h(\frac{\boldsymbol{d}_h^{n+1} - 2\boldsymbol{d}_h^n + \boldsymbol{d}_h^{n-1}}{\Delta t})\right)$. Note that these terms are easily controlled by the dissipation coming from the Leap–Frog chosen scheme for the time discretization of the structure. In the case of Newmark scheme, which is not dissipative, the proof does not hold any more. Nevertheless the same stability estimate can be obtained provided that the kinematic boundary conditions are imposed in a weak way thanks to Nitsche's method [3], since Nitsche's method introduces some dissipation in the coupled scheme.

Concerning the convergence result, the time error is at least of order $\sqrt{\Delta t}$. It is well known that a non-incremental Chorin–Temam scheme has time rate of convergence that is less than one, for the pressure in norm $L^\infty(0,T; L^2(\hat{\Omega}_f))$, and for the velocity in norm $L^\infty(0,T; H^1(\hat{\Omega}_f))$ (see [5,47,95,96,143] for details). Here it is also the case for the fluid velocity in $L^\infty(0,T; L^2(\hat{\Omega}_f))$ norm. The reasons for this may be due to our proof technique, that may not be optimal, or due to the fact that the error for the fluid pressure propagates through the structure to the whole coupled system.

Concerning the non-matching operator at the interface the error estimate is optimal in any case when considering the Mortar matching. This is conform with what can be found in [116]. Yet, in these case of a pointwise matching, then the optimality depends on the structure type and of the polynomial degree of the fluid Lagrange finite element.

For a discussion on non-matching grids for fluid–structure interaction problems, we also refer to [48].

1.3.3 Kinematically Coupled Schemes

The idea of kinematically coupled schemes has been firstly introduced in [98] and later generalized to the second order schemes in [110, 121] using the higher order splitting methods and in [56] applying the interpolation of the structure velocity used in the hydrodynamic equations, see also [23, 57, 62] for related works. Let us consider the following generalized string model introduced in [109]

$$\rho_s e \partial_{tt}\eta - \alpha_2 \partial_{xx}\eta + \tilde{\delta}\eta - \beta_2 \partial_{xx}\partial_t \eta = H(\boldsymbol{u}, p) \quad \text{on} \quad \hat{\Sigma}, \tag{1.127}$$

where

$$\hat{\Sigma} := \left\{(x, y) \in \mathbb{R}^2 : 0 < x < L,\ y = R_0(x)\right\}, \tag{1.128}$$

$R_0(x)$ is a given reference domain radius and $H(\boldsymbol{u}, p) = (\boldsymbol{T}_f)_2$ represents the forces exhibited by the normal fluid stress acting on the elastic vessel wall and transformed to $\hat{\Sigma}$. In detail we have

$$H(\boldsymbol{u}, p) := -\left(\left((\boldsymbol{\sigma}_f + p_{\text{ext}}\,\mathbf{I})\boldsymbol{n}\right)\Big|_{\hat{\Sigma}} \cdot \boldsymbol{e}_r\right) \frac{R}{R_0} \frac{\sqrt{1 + (\partial_x R)^2}}{\sqrt{1 + (\partial_x R_0)^2}}, \qquad R = R_0 + \eta. \tag{1.129}$$

The term with square roots corresponds to the Jacobian of the transformation between the Eulerian framework used for the description of fluid and the Lagrangian framework used for the structure, $\boldsymbol{e}_r$ denotes a unit vector in the y-direction (radial direction) and p_{ext} is a given external pressure.

For the structure we assume analogously as in (1.11) that we have the third order viscoelastic term $-\partial_t \partial_x^2 \eta$ and moreover that $\gamma = 0$. The parameters are defined as follows:

$$\alpha_2 = |\sigma_x| e \left[1 + \left(\frac{\partial R_0}{\partial x}\right)^2\right]^{-2}, \quad \tilde{\delta} = \frac{E e}{R_0^2(1 - \tilde{\xi}^2)}. \tag{1.130}$$

Recall that E is the Young modulus, e the thickness of the vessel wall, ρ_s its density, $|\sigma_x| = G\kappa$ is the longitudinal stress, $G = E/(2(1 + \tilde{\xi}))$ denotes the shear modulus,

$\tilde{\xi}$ is the Poisson ratio, κ is the Timoshenko shear correction factor, and β_2 a positive viscoelastic constant.

The kinematic splitting algorithm is based on the kinematic coupling condition

$$\boldsymbol{u} = \boldsymbol{w} := \left(0, \partial_t \eta\right)^T \quad \text{on} \quad \Sigma(t) \tag{1.131}$$

and special splitting of the structure equation into the hyperbolic and parabolic part. We define the operator A that includes the fluid solver and viscoelastic part of the structure equation

$$\textbf{A operator} \text{ (hydrodynamic)} \begin{cases} \text{fluid solver } (\boldsymbol{u}, p), \\ \sigma := u_2|_\Sigma, \\ \rho_s e \dfrac{\partial \sigma}{\partial t} = \beta_2 \dfrac{\partial^2 \sigma}{\partial x^2} + H(\boldsymbol{u}, p), \end{cases} \tag{1.132}$$

and the operator B for purely elastic load of the structure

$$\textbf{B operator} \text{ (elastic)} \begin{cases} \dfrac{\partial \eta}{\partial t} = \sigma, \\ \dfrac{\partial \sigma}{\partial t} = \dfrac{\alpha_2}{\rho_s e} \dfrac{\partial^2 \eta}{\partial x^2} - \dfrac{\tilde{\delta}}{\rho_s e} \eta + G(R_0), \end{cases} \tag{1.133}$$

where $G(R_0) := \frac{\alpha_2}{\rho_s e} \frac{\partial^2 R_0}{\partial x^2}$. Here we note that the coupling condition allowed us to rewrite the hydrodynamic part of structure equation in terms of wall velocity σ. The splitting works well also in the case that the structure is just elastic, i.e., without any friction term, $\beta_2 = 0$. Time discretization of our problem is done in the following way: from the fluid equation we compute new velocities $\boldsymbol{u}^{n+1}$ and pressures p^{n+1} for $x^n \in \Omega^n$ (i.e., $\Omega^n := \Omega_\eta(t^n)$). Note that $\tilde{\boldsymbol{u}}^{n+1} = \boldsymbol{u}^{n+1} \circ \boldsymbol{\phi}_f(t^{n+1}) \circ \boldsymbol{\phi}_f(t^n)^{-1}$ and $\tilde{p}^{n+1} = p^{n+1} \circ \boldsymbol{\phi}_f(t^{n+1}) \circ \boldsymbol{\phi}_f(t^n)^{-1}$, where $\boldsymbol{\phi}_f(t^n)$ is the ALE-mapping from a reference domain $\hat{\Omega}_f$ onto Ω^n. Then we continue with computing of the wall velocity $\sigma^{n+\frac{1}{2}}$ from the hydrodynamic part of structure equation (1.132). Further we proceed with the operator B and compute new wall displacement η^{n+1} and new wall velocity σ^{n+1}. Finally, knowing η^{n+1} the geometry is updated from Ω^n to Ω^{n+1} and new values of fluid velocity $\boldsymbol{u}^{n+1}$ and pressure p^{n+1} are transformed onto Ω^{n+1}. In order to update the domain Ω^n we need to define the grid velocity $\boldsymbol{w}$. First, we set $\boldsymbol{w}|_\Sigma = \sigma^{n+1}$. In order to prescribe the grid velocity also inside the fluid domain we can solve an auxiliary problem, e.g., harmonic extension of $\boldsymbol{x}^{n+1}\Big|_{\Sigma^{n+1}} \equiv \eta^{n+1}$ from Σ^{n+1} inside Ω^{n+1}. Consequently, we get $\boldsymbol{w}^{n+1} := \dfrac{\boldsymbol{x}^{n+1} - \boldsymbol{x}^n}{t^{n+1} - t^n}$, $\boldsymbol{x}^{n+1} \in \Omega^{n+1}$.

Stability Analysis

In what follows we will briefly present the stability analysis of the semi-discrete scheme for the kinematic coupling approach. More details on the derivation can be found in [121]. Now, let us consider the weak formulation of the fluid equation. Integrating over Ω^n and approximating the time derivative by the backward Euler method the operator A yields the following equation for new intermediate velocities $\tilde{\boldsymbol{u}}^{n+1}, \sigma^{n+\frac{1}{2}}$

$$\int_{\Omega^n} \tilde{\boldsymbol{u}}^{n+1} \cdot \frac{\tilde{\boldsymbol{u}}^{n+1} - \boldsymbol{u}^n}{\Delta t} + \frac{2}{\rho_f} \int_{\Omega^n} \mu(|\mathbb{D}(\tilde{\boldsymbol{u}}^{n+1})|)\, \mathbb{D}(\tilde{\boldsymbol{u}}^{n+1}) : \mathbb{D}(\tilde{\boldsymbol{u}}^{n+1})$$

$$+\frac{1}{2} \int_{\Omega^{n+1/2}} |\hat{\boldsymbol{u}}^{n+1}|^2 \mathrm{div}\, \boldsymbol{w}^{n+1/2} \quad = \quad -\frac{\rho_s e}{\rho_f} \int_{\hat{\Sigma}} \left[\frac{\sigma^{n+\frac{1}{2}} - \sigma^n}{\Delta t} \right] \sigma^{n+\frac{1}{2}} \tag{1.134}$$

$$-\frac{\beta_2}{\rho_f} \int_{\hat{\Sigma}} \left[\frac{\partial \sigma^{n+\frac{1}{2}}}{\partial x} \right]^2 \mathrm{d}l_0 - \frac{1}{\rho_f} \int_{\Sigma^n} \frac{p_{\mathrm{ext}}(t^{n+1})\, \tilde{u}_2^{n+1}}{\sqrt{1 + (\partial_x R_0)^2}} \,\mathrm{d}l \; + \frac{1}{\rho_f} \int_{\Omega^n} \tilde{\boldsymbol{u}}^{n+1} \cdot \boldsymbol{f}^{n+1}$$

$$+\frac{1}{\rho_f} \int_0^{R_0(0)} p_{\mathrm{in}}(t^{n+1}) \tilde{u}_1^{n+1}|_{x=0} - \frac{1}{\rho_f} \int_0^{R_0(L)} p_{\mathrm{out}}(t^{n+1}) \tilde{u}_1^{n+1}|_{x=L}.$$

where $\hat{\boldsymbol{u}}^{n+1} = \boldsymbol{u}^{n+1} \circ \boldsymbol{\phi}_f(t^{n+1}) \circ \boldsymbol{\phi}_f^{-1}(t^{n+1/2})$.

Moreover, we have $\mathrm{div}\, \tilde{\boldsymbol{u}}^{n+1} = 0$ in Ω^n. The operator B is discretized in time via the Crank–Nicolson scheme, i.e.,

$$\frac{\eta^{n+1} - \eta^n}{\Delta t} = \frac{1}{2}\left(\sigma^{n+1} + \sigma^{n+\frac{1}{2}}\right), \tag{1.135}$$

$$\frac{\sigma^{n+1} - \sigma^{n+\frac{1}{2}}}{\Delta t} = \frac{\alpha_2}{2\rho_s e}\left(\partial_{xx}\eta^{n+1} + \partial_{xx}\eta^n\right) - \frac{\tilde{\delta}}{2\rho_s e}\left(\eta^{n+1} + \eta^n\right) + G(R_0). \tag{1.136}$$

The discrete scheme (1.135)–(1.136) is also reported in literature as the Newmark scheme.

First we look for an energy estimate of the semi-discrete weak formulation of the momentum equation (1.135). In order to control the energy of the operator A we apply the Young and the trace inequalities. Further using the generalized Korn inequality for nonlinear viscous term we can show the coercivity of the nonlinear viscous fluid term. After some manipulations, cf. [121], we obtain

$$||\tilde{\boldsymbol{u}}^{n+1}||^2_{L^2(\Omega^n)} + \frac{\Delta t}{2} \int_{\Omega^{n+1/2}} |\hat{\boldsymbol{u}}^{n+1}|^2 \mathrm{div}\, \boldsymbol{w}^{n+1/2}$$

$$
\begin{aligned}
&+C^*\Delta t||\tilde{\boldsymbol{u}}^{n+1}||^q_{W^{1,q}(\Omega^n)} \qquad (1.137)\\
&+\frac{\rho_s e}{\rho_f}\left[||\sigma^{n+\frac{1}{2}}||^2_{L^2(\hat{\Sigma})}-||\sigma^n||^2_{L^2(\hat{\Sigma})}\right]+2\Delta t\frac{\beta_2}{\rho_f}\,||\partial_x\sigma^{n+\frac{1}{2}}||^2_{L^2(\hat{\Sigma})}\Big]\\
&\le ||\boldsymbol{u}^n||^2_{L^2(\Omega^n)}+\frac{2\Delta t C_2}{\varepsilon^{q'/q}}\,\mathrm{RHS}^{\mathrm{n+1}}+2C^*\kappa\,\Delta t,
\end{aligned}
$$

where $\kappa = 0$ for $q \geq 2$ and $\kappa = 1$ for $1 \leq q < 2$,

$$
\begin{aligned}
\mathrm{RHS}^{\mathrm{n+1}} :=\, &||p_{\mathrm{in}}(t^{n+1})||^{p'}_{L^{q'}(\Gamma_{\mathrm{in}})}+||p_{\mathrm{out}}(t^{n+1})||^{q'}_{L^{q'}(\Gamma_{\mathrm{out}})}+||p_{\mathrm{ext}}(t^{n+1})||^{q'}_{L^{q'}(\Sigma^n)}\\
&+||\boldsymbol{f}^{n+1}||^{q'}_{L^{q'}(\Omega^{n+1})}
\end{aligned}
$$

and C^*, C^{tr}, ε are positive constants. The dual argument $q' \geq 1$ satisfies $1/q + 1/q' = 1$.

In order to rewrite the term containing the norm $||\tilde{\boldsymbol{u}}^{n+1}||_{L^2(\Omega^n)}$ by means of $||\boldsymbol{u}^{n+1}||_{L^2(\Omega^{n+1})}$ we use the well-known Reynolds transport theorem:

$$
||\boldsymbol{u}^{n+1}||^2_{L^2(\Omega^{n+1})}-||\tilde{\boldsymbol{u}}^{n+1}||^2_{L^2(\Omega^n)}=\int_{t^n}^{t^{n+1}}\int_{\boldsymbol{\phi}_f(t)(\hat{\Omega}_f)}|\boldsymbol{u}^{n+1}\circ\boldsymbol{\phi}_f(t^{n+1})\circ\boldsymbol{\phi}_f^{-1}(t)|^2\,\mathrm{div}\,\boldsymbol{w} \tag{1.138}
$$

Applying the midpoint rule for approximation of the convective ALE-term in two-dimensional case the integrand on the left hand side of (1.138) can be exactly computed, cf. [67], or [50] i.e.,

$$
\int_{t^n}^{t^{n+1}}\int_{\boldsymbol{\phi}_f(t)(\hat{\Omega}_f)}|\boldsymbol{u}^{n+1}\circ\boldsymbol{\phi}_f(t^{n+1})\circ\boldsymbol{\phi}_f^{-1}(t)|^2\,\mathrm{div}\,\boldsymbol{w}=\Delta t\int_{\Omega^{n+1/2}}|\hat{\boldsymbol{u}}^{n+1}|^2\,\mathrm{div}\,\boldsymbol{w}^{n+1/2}. \tag{1.139}
$$

The latter equation is called the Geometric Conservation Law. Thus we obtain an estimate that does not depend on the mesh velocity.

$$
\begin{aligned}
&||\boldsymbol{u}^{n+1}||^2_{L^2(\Omega^n)}+C^*\Delta t||\tilde{\boldsymbol{u}}^{n+1}||^q_{W^{1,q}(\Omega^n)} \qquad (1.140)\\
&+\frac{\rho_s e}{\rho_f}\left[||\sigma^{n+\frac{1}{2}}||^2_{L^2(\hat{\Sigma})}-||\sigma^n||^2_{L^2(\hat{\Sigma})}\right]+2\Delta t\frac{\beta_2}{\rho_f}\,||\partial_x\sigma^{n+\frac{1}{2}}||^2_{L^2(\hat{\Sigma})}\Big]\\
&\le ||\boldsymbol{u}^n||^2_{L^2(\Omega^n)}+\frac{2\Delta t C_2}{\varepsilon^{q'/q}}\,\mathrm{RHS}^{\mathrm{n+1}}+2C^*\kappa\,\Delta t,
\end{aligned}
$$

Remark 1.18. Note that if we do not assume that the Geometric Conservation Law (1.139) is satisfied exactly, but we only approximate time integral on the left hand side by, e.g., the rectangle rule, then the estimate will depend on the mesh velocity

$\boldsymbol{w}$ and a condition on the time step involving $\|\mathrm{div}\ \boldsymbol{w}\|_{L^\infty}$ appears (see [121]). This condition requires that the time step is small enough and that the mesh velocity that depends on the structure velocity is regular enough. Note however that we do not always have the required regularity at the continuous level.

Remark 1.19. We have considered here a nonconservative form of the fluid equations. In [67] they obtain an energy estimate depending on the mesh velocity even in the case where the Geometric Conservation Law is satisfied. This is not in contradiction with the result presented here since we solve the fluid equations on Ω^n whereas in [67] it was solved on Ω^{n+1}.

The next step is to estimate the operator B. To this end we firstly multiply Eq. (1.135) by $\delta(\eta^{n+1}+\eta^n)$ and Eq. (1.136) by $(\sigma^{n+1}+\sigma^{n+\frac{1}{2}})$, sum up the multiplied equations, and then integrate them over $\hat{\Sigma}$. After some manipulations this yields, cf. [121],

$$\begin{aligned}&\alpha_2||\partial_x\eta^{n+1}||^2_{L^2(\hat{\Sigma})} + \tilde{\delta}\ ||\eta^{n+1}||^2_{L^2(\hat{\Sigma})} + \rho_s e||\sigma^{n+1}||^2_{L^2(\hat{\Sigma})}\\ &\le \alpha_2||\partial_x\eta^0||^2_{L^2(\hat{\Sigma})} + \tilde{\delta}\ ||\eta^0||^2_{L^2(\hat{\Sigma})} + \rho_s e||\sigma^0||^2_{L^2(\hat{\Sigma})}\\ &+\rho_s e\sum_{i=0}^{n}\Big(||\sigma^{i+\frac{1}{2}}||^2_{L^2(\hat{\Sigma})} - ||\sigma^i||^2_{L^2(\hat{\Sigma})}\Big) + 2\rho_s e\int_{\hat{\Sigma}} G(R_0)(\eta^{n+1}-\eta^0).\end{aligned} \tag{1.141}$$

It can be shown easily that for some sufficiently small ε we have

$$\begin{aligned}2\rho_s e\int_{\hat{\Sigma}} G(R_0)\ (\eta^{n+1}-\eta^0)\ \mathrm{d}l_0 \le{}& \frac{\alpha_2\rho_s e L\ |\hat{\Sigma}|}{\varepsilon}\\ &+\frac{\tilde{\delta}}{2}\Big[||\eta^{n+1}||^2_{L^2(\hat{\Sigma})} + ||\eta^0||^2_{L^2(\hat{\Sigma})}\Big].\end{aligned} \tag{1.142}$$

Inserting (1.143) into (1.142) we obtain an estimate of the operator B

$$\begin{aligned}&\alpha_2||\partial_x\eta^{n+1}||^2_{L^2(\hat{\Sigma})} + \frac{\tilde{\delta}}{2}\ ||\eta^{n+1}||^2_{L^2(\hat{\Sigma})} + \rho_s e||\sigma^{n+1}||^2_{L^2(\hat{\Sigma})}\\ &\le \alpha_2||\partial_x\eta^0||^2_{L^2(\hat{\Sigma})} + \frac{3\tilde{\delta}}{2}\ ||\eta^0||^2_{L^2(\hat{\Sigma})} + \rho_s e||\sigma^0||^2_{L^2(\hat{\Sigma})}\\ &+\rho_s e\sum_{i=0}^{n}\Big(||\sigma^{i+\frac{1}{2}}||^2_{L^2(\hat{\Sigma})} - ||\sigma^i||^2_{L^2(\hat{\Sigma})}\Big) + \frac{\alpha_2\rho_s e L|\hat{\Sigma}|}{\varepsilon}.\end{aligned} \tag{1.143}$$

Note that in our model we have $\eta^0 = 0$ and $\sigma^0 = u_2^0|_\Sigma$.

Taking into account (1.141) and (1.143) and applying discrete Gronwall Lemma, the total energy at the new time step t^{n+1} will be bounded only with the initial energy and the boundary data

$$E^{n+1} + \Delta t \sum_{i=1}^{n+1} G^i \leq E^0 + \frac{\rho_s e}{\rho_f}\tilde{\delta}||\eta^0||^2_{L^2(\hat{\Sigma})} + \frac{\rho_s e \alpha_2 L}{\rho_f}\frac{|\hat{\Sigma}|}{\varepsilon} \qquad (1.144)$$

$$+2C^*\kappa T + \frac{2\Delta t\ C_2}{\varepsilon^{q'/q}} \sum_{i=1}^{n+1} P^i,$$

where

$$E^i := ||\boldsymbol{u}^i||^2_{L^2(\Omega^i)} + \frac{1}{\rho_f}\left[\alpha_2||\eta^i_x||^2_{L^2(\hat{\Sigma})} + \frac{\tilde{\delta}}{2}||\eta^i||^2_{L^2(\hat{\Sigma})} + \rho_s e||\sigma^i||^2_{L^2(\hat{\Sigma})}\right],$$

$$G^i := C^*||\tilde{\boldsymbol{u}}^i||^q_{W^{1,q}(\Omega^{i-1})} + \frac{2\beta_2}{\rho_f}||\partial_x \sigma^{i-\frac{1}{2}}||^2_{L^2(\hat{\Sigma})},$$

$$P^i := \frac{2C_2}{\varepsilon^{q'/q}}\,\mathrm{RHS}^{\mathrm{i}},$$

and $i = 0, \ldots, n+1$.

Numerical Experiments

The aim of this section is to illustrate performance of numerical methods presented above, the semi-implicit schemes presented in Sect. 1.3.2 as well as the kinematically coupled schemes presented in Sect. 1.3.3.

We start with the semi-implicit schemes and illustrate by numerical experiments performance of the semi-implicit scheme for strong added mass effect. It is worth mentioning that in the Newmark scheme is used in every numerical test without any stability issue. In every test case the projection step is solved by a Poisson equation on the pressure (see (1.37), (1.38)), for which one has really performant algorithms, and not the mixed Darcy problem (see (1.30), (1.31)).

First of all we have compared for $\sigma = 0$ and $\gamma = 0$ three difference schemes relying on *explicit*, *semi-implicit*, and *implicit* strategies (Fig. 1.2.)

The second test case deals with a three-dimensional problem, where the structure satisfies a shell equation. It is the benchmark case proposed in [69] (for the used parameters we refer to [64]). The fluid is discretized by $\mathbb{Q}_1/\mathbb{Q}_1$ finite elements and the structure by shell element MITC4 [27]. The implicit step is solved by a Newton algorithm (Fig. 1.3).

These results are in agreement with [61, 69, 81]. Note moreover that this scheme has the same precision as an implicit scheme with a computational cost which is much lower. Moreover, the theoretical convergence rate in time has not been recovered easily in practice and only on well-chosen cases [1].

In the following experiment we compare the accuracy of the first and second order kinematically coupled fluid–structure interaction schemes. To derive the first or second order methods we can apply, respectively, the first order Marchuk–

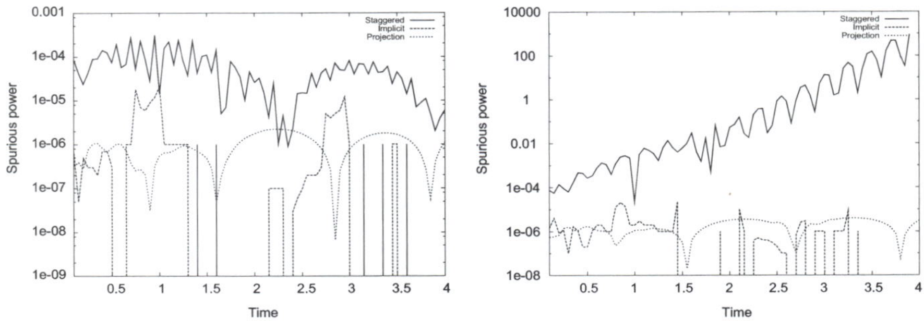

Fig. 1.2 *Left*: Residual power at the interface for $\rho_s = 30$. Added has effect are small—all the schemes are stable. *Right*: Residual power at the interface for $\rho_s = 20$. The explicit scheme is unstable whereas the two other are stable

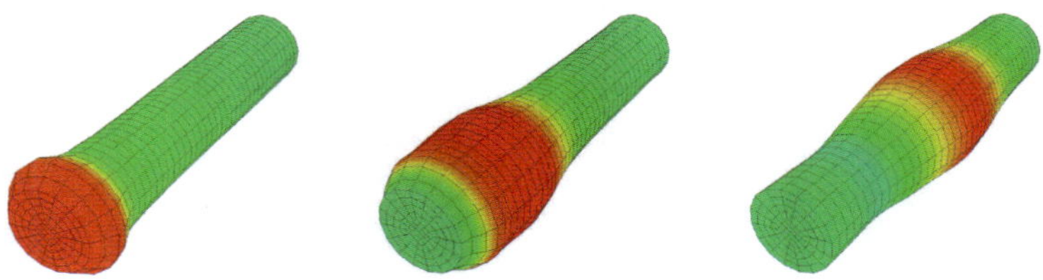

Fig. 1.3 Wave propagation in a cylinder, $t = 0.0018, 0.0058$ and 0.0098 s

Yanenko splitting scheme

$$\mathcal{U}^{n+1} = B_{\triangle t} A_{\triangle t}\, \mathcal{U}^n, \tag{1.145}$$

or the second order Strang splitting scheme

$$\mathcal{U}^{n+1} = B_{\triangle t/2} A_{\triangle t} B_{\triangle t/2}\, \mathcal{U}^n. \tag{1.146}$$

Here we have denoted by $\mathcal{U}^n$ the approximate solution of coupled problem at the time level t^n and by $A_{\triangle t}$ and $B_{\triangle t}$ the operator A, cf. (1.132), and the operator B, cf. (1.133), acting on interval $(t^n, t^{n+1}]$, $\triangle t = t^{n+1} - t^n$, respectively. Further, $B_{\triangle t/2}$ denotes the operator B acting on the interval of length $\Delta t/2$. The operator A is discretized by the finite volume method using the so-called dual grids of the basic triangular mesh. Velocity as well as pressures is approximated by the piecewise linear functions over the underlying triangular grid. Artificial compressibility is used to relax the solenoidal condition for velocities, more precisely we have

$$-\varepsilon \Delta \tilde{p}^{n+1} + \nabla \cdot \tilde{\boldsymbol{u}}^{n+1} = 0 \text{ on } \Omega^n.$$

Here ε is a small positive constant, $\varepsilon \approx \Delta h^2$, where Δh denotes the grid size. In order to stabilize the convective term the upwinding technique is used, nonlinearity

Table 1.1 Convergence rates in space; Marchuk–Yanenko splitting scheme, Carreau viscosity

	L^q-norm				L^2-norm	
# refin (Δh)	Err ($\boldsymbol{u}$)	EOC ($\boldsymbol{u}$)	Err ($\nabla \boldsymbol{u}$)	EOC ($\nabla \boldsymbol{u}$)	Err (η)	EOC (η)
1	0.24542		0.0219		0.2942	
2	0.07939	1.63	0.0637	1.78	0.1087	1.44
3	0.02861	1.46	0.0303	1.07	0.0425	1.35
4	0.00729	1.99	0.0110	1.46	0.0150	1.50

in the convective term and in the viscosity is approximated by the fixed point iterations. The operator B is approximated by the Newmark finite difference scheme, we refer the reader to [121] for more details of the discretization schemes.

In hemodynamic applications the nonlinear viscosity function is typically used to describe non-Newtonian effects due to the aggregation and deformability of the red blood cells. In [121] Lukáčová et al. considered the so-called Carreu viscosity model that describes shear-thinning behavior of blood

$$\mu(|\mathbb{D}(\boldsymbol{u})|) = \mu_\infty + (\mu_0 - \mu_\infty)\left[1 + 2(\lambda|\mathbb{D}(\boldsymbol{u})|)^2\right]^{\frac{q-2}{2}}, \tag{1.147}$$

$\mathbb{D} = 1/2(\nabla \boldsymbol{u} + \nabla \boldsymbol{u}^T)$ denotes the rate of the deformation tensor and $\mu_\infty := \lim_{|\mathbb{D}(\boldsymbol{u})|\to\infty} \mu(|\mathbb{D}(\boldsymbol{u})|)$, $\mu_0 := \lim_{|\mathbb{D}(\boldsymbol{u})|\to 0} \mu(|\mathbb{D}(\boldsymbol{u})|)$. In the experiments presented below the following parameters are used $\mu_0 = 1.26P$, $\mu_\infty = 0.63P$, $q = 1.6$, $\lambda = 1$. We compute a reference solution on a fine mesh (consisting of 32,768 elements) and compare the L^2 or L^q norms of the difference between the reference $\boldsymbol{u}_{\text{ref}}$ and the approximate solutions, respectively. Thus, the normalized L^q error is given as

$$\text{Err}(\boldsymbol{u}) = \frac{||\boldsymbol{u}_{\Delta h,\Delta t} - \boldsymbol{u}_{\text{ref}}||_{L^q}}{|\Omega_{\Delta h}|^{1/q}} \tag{1.148}$$

and the experimental order of convergence (EOC) is computed in the following way

$$\text{EOC}(\boldsymbol{u}) = \log_2 \frac{||\boldsymbol{u}_{\Delta h,\Delta t} - \boldsymbol{u}_{\text{ref}}||_{L^q}/|\Omega_{\Delta h}|^{1/q}}{||\boldsymbol{u}_{\Delta h/2,\Delta t} - \boldsymbol{u}_{\text{ref}}||_{L^q}/|\Omega_{\Delta h/2}|^{1/q}}. \tag{1.149}$$

Tables 1.1 and 1.2 present the EOC results in space obtained by the kinematic splitting scheme and the Strang splitting approach. To this end we compute solution at different grid refinements having 32, 128, 512, and 2,048 elements, respectively. The final time was taken $T = 0.004$ s and a fixed time step $\Delta t = 10^{-4}$ s have been used. We can see that both schemes have very similar error behavior, in particular we get the second order convergence for velocities and pressures. As it is expected, the Strang splitting technique has not visibly influenced the convergence rates in space.

Table 1.2 Convergence rates in space; Strang splitting scheme, Carreau viscosity

	L^q-norm				L^2-norm	
# refin (Δh)	Err ($\boldsymbol{u}$)	EOC ($\boldsymbol{u}$)	Err ($\nabla\boldsymbol{u}$)	EOC ($\nabla\boldsymbol{u}$)	Err (η)	EOC (η)
1	0.24469		0.0218		0.3068	
2	0.07891	1.63	0.0796	1.45	0.1186	1.37
3	0.02867	1.46	0.0351	1.18	0.0514	1.21
4	0.00709	2.02	0.0117	1.58	0.0190	1.43

Table 1.3 Convergence rates in time; Marchuk–Yanenko splitting, Carreau viscosity

	$L^q(L^q)$-norm				$L^2(L^2)$-norm	
# refin (Δt)	Err ($\boldsymbol{u}$)	EOC ($\boldsymbol{u}$)	Err ($\nabla\boldsymbol{u}$)	EOC ($\nabla\boldsymbol{u}$)	Err (η)	EOC (η)
2/1	0.1532		0.1600		0.2706	
3/2	0.0705	1.12	0.0747	1.10	0.2000	0.44
4/3	0.0218	1.69	0.0234	1.67	0.0915	1.13

Setting the final computational time $T = \sum_{j=1}^{N} \Delta t = \Delta t\ N$, the experimental order of convergence in time is computed by the following formula

$$\mathrm{EOC}(\boldsymbol{u}) = \log_2 \frac{\left(\sum_{j=1}^{N} ||\boldsymbol{u}^j_{\Delta h,\Delta t} - \boldsymbol{u}^j_{\Delta h,\Delta t/2}||^q_{L^q} / |\Omega^j_{\Delta h,\Delta t}|^q\right)^{1/q}}{\left(1/2 \sum_{j=1}^{2N} ||\boldsymbol{u}^j_{\Delta h,\Delta t/2} - \boldsymbol{u}^j_{\Delta h,\Delta t/4}||^q_{L^q} / |\Omega^j_{\Delta h,\Delta t/2}|^q\right)^{1/q}}. \tag{1.150}$$

Moreover, the normalized relative $L^q(0,T;L^q(\Omega))$ error in time is defined by

$$\mathrm{Err}(\boldsymbol{u}) = \frac{1}{T}\left(\sum_{j=1}^{N} \Delta t \left(\frac{||\boldsymbol{u}^j_{\Delta h,\Delta t} - \boldsymbol{u}^j_{\Delta h,\Delta t/2}||_{L^q}}{|\Omega^j_{\Delta h,\Delta t}|},\right)^q\right)^{1/q} \tag{1.151}$$

where $\boldsymbol{u}^j_{\Delta h,\Delta t}$ and $\Omega^j_{\Delta h,\Delta t}$ denotes the velocity and the computational domain associated with the time instant $j\ \Delta\ t$, respectively, using the fixed grid size Δh The EOC in time (1.150) was computed on a computational mesh consisting of 585 elements. Going from one time refinement to the finer one, the time step was halved. The time period for the computation was $t \in [0.2; 0.8]$ s and the initial time step was $\Delta t = 0.0125$ s. The initial computational domain is a rectangular reference domain $\hat{\Omega}_f$. In Tables 1.3 and 1.4 we present experimental order of convergence in time for the first order Marchuk–Yanenko and the second order Strang splitting schemes. We see that the global errors are significantly smaller for the Strang splitting scheme in comparison with the first order Marchuk–Yanenko splitting scheme, see also [121] for further experiments.

In the following experiment we have analyzed the effects of different viscosity models used in hemodynamics, the Carreu model (1.147), the logarithmic model

Table 1.4 Convergence rates in time; Strang splitting, Carreau viscosity

	$L^q(L^q)$-norm				$L^2(L^2)$-norm	
# refin (Δt)	Err ($\boldsymbol{u}$)	EOC ($\boldsymbol{u}$)	Err ($\nabla \boldsymbol{u}$)	EOC ($\nabla \boldsymbol{u}$)	Err (η)	EOC (η)
2/1	0.0578		0.0609		0.1140	
3/2	0.0241	1.26	0.0243	1.32	0.0441	1.37
4/3	0.0088	1.44	0.0078	1.64	0.0173	1.35

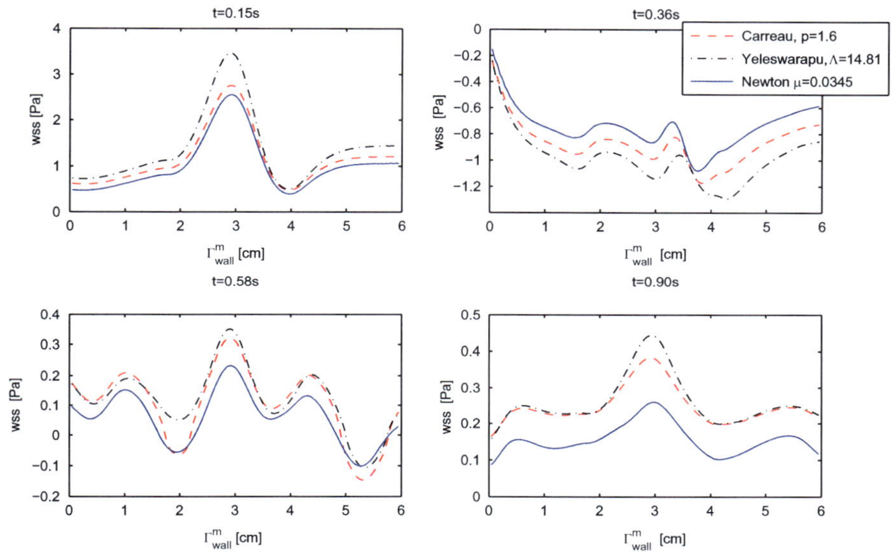

Fig. 1.4 WSS along Σ for stenotic vessel geometry at several time instants

of Yeleswarapu, see, e.g., [121, 159], and the Navier–Stokes model. In Fig. 1.4 we compare the so-called wall shear stress (WSS), i.e., the normal component of the fluid stress tensor evaluated at the moving wall $\Sigma(t)$, projected in the tangential direction. The computational domain has a reference radius

$$R_0(x_1) = \begin{cases} R_0(0)\left[1 - 0.15\left(1 + \cos\left(\dfrac{5\pi(x_1 - L/2)}{L}\right)\right)\right] & \text{if} \quad x_1 \in [0.3L; 0.7L], \\ R_0(0) & \text{if} \quad x_1 \in [0; 0.3L) \cup (0.7L; L], \end{cases} \tag{1.152}$$

which models a stenotic occlusion. We set $L = 6\,\text{cm}$ and $R(0) = 0.6\,\text{cm}$.

In Fig. 1.4 we see the distribution of WSS for different time instants during the cardiac cycle along the moving boundary of stenosed vessel. Negative values of WSS along the moving boundary are visible. In both cases we observe that the WSS

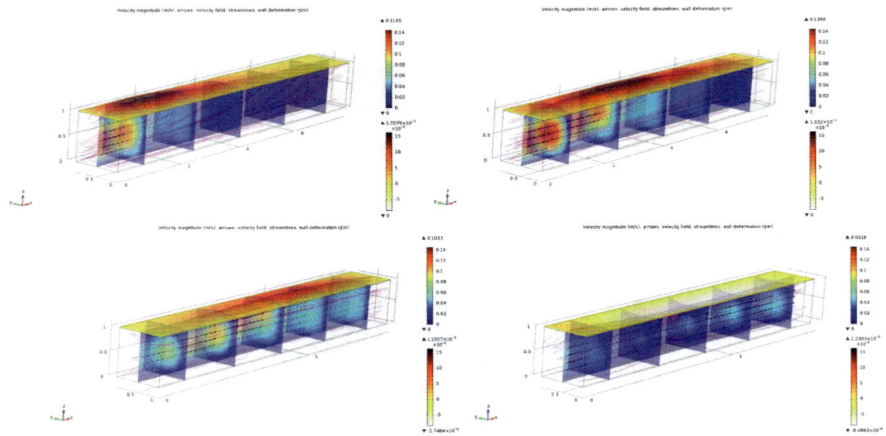

Fig. 1.5 Fluid–structure interaction of three-dimensional Newtonian fluid and a thin viscoelastic structure at time instants $t = 0.4, 0.6, 0.8, 1.2\,\mathrm{s}$ (*from left to right* and *top to down*); velocity magnitudes, velocity vectors and streamlines

corresponding to the non-Newtonian model considerably differs from that given by the Navier–Stokes equations, indeed the WSS gives higher extremal values.

In the last experiment we present a generalization of the kinematically coupled schemes to 3D/2D coupling problem, i.e., we consider three-dimensional (Newtonian) fluid and two-dimensional membrane. We take structure to be clamped. The equation describing time evolution of the transversal displacement η; $\eta(x_1, x_2, t) \in \mathbb{R}$ reads

$$\rho_s e \frac{\partial^2 \eta}{\partial t^2} - \alpha_2 \Delta \eta - \beta_2 \Delta \frac{\partial \eta}{\partial t} = H(\boldsymbol{u}, p) \quad \text{on } \hat{\Sigma} \subset \mathbb{R}^2. \tag{1.153}$$

We impose the Dirichlet boundary conditions for fluid at the inflow part. Inflow velocity has the maximum value 16 cm/s and is multiplied by $\sin(\pi t)$ function to model time-dependent inflow, $t \in (0, T)$, $T = 2.0$. On the outflow part the normal stress is set to 0. We use the no-slip boundary condition for the remaining parts of boundary Γ_0. For the computational domain the following parameters have been used: $L_1 = 8\,\mathrm{cm}$, $L_2 = 1\,\mathrm{cm}$, $R_0(x_1, x_2) = 1\,\mathrm{cm}$. Further, we set the fluid viscosity to 0.0345 P and density to $1\,\mathrm{g\,cm^{-3}}$. For structure we use the following parameters: wall thickness is $h = 0.1\,\mathrm{cm}$, wall density is $\rho_w = 1.1\,\mathrm{g\,cm^{-3}}$, Timoshenko's factor is $\kappa = 1$, Poisson's ratio $\tilde{\xi} = 0.5$, $E = 0.75 \times 10^6\,\mathrm{Pa}$, viscoelasticity coefficient $\gamma = 0.2 \times 10^6$, cf. [121].

Computational domain for fluid is discretized by 60,915 tetrahedral elements. In Fig. 1.5 we present results of the fluid–structure interaction simulations obtained by the first order kinematically coupling method at different time instants 0.4, 0.6, 0.8, 1.2 s. We depict velocity magnitudes, velocity vectors as well as streamlines.

1.4 Conclusions

Our main aim in this chapter was to discuss analytical and numerical difficulties arising in fluid–structure interaction problems. Due to the nonlinear geometrical coupling the so-called added mass effects appear and play a crucial role in both analysis and numerical simulations. The applications we have in mind are, for example, blood flow in elastic vessels, but more general applications of fluid flow in compliant domains are governed as well. More importantly, if the coupled problem has biological applications the fraction of fluid and structure densities is typically close to one and the added mass effects due to decoupling fluid and structure are more profound. We have presented several approaches to obtain existence of global weak solutions for both Newtonian and non-Newtonian shear-dependent fluids. Moreover, we present also main ideas and corresponding results for local existence of strong solutions. In this chapter we concentrate on the case when a two-dimensional fluid interacts with a viscoelastic membrane. However, many of the results can be generalized for three-dimensional situations and purely elastic structure and we present a broad literature overview. In the second part we concentrate on the numerical analysis and try to underline connections between analytical results and construction of efficient and stable schemes. We present in more detail two partitioned schemes, the semi-implicit scheme, cf. Sect. 1.3.2 and the kinematically coupled scheme based on a suitable operator splitting technique 1.3.3. We also study their stability and accuracy from both theoretical and experimental point of view. The chapter is closed with some numerical experiments that demonstrate reliability of the presented numerical approaches.

Acknowledgements Céline Grandmont was supported by the grant ANR-08-JCJC-013-01 (M3RS project) of the French Research National Agency and by REO Project, Inria Paris Rocquencourt, Inria, France & LJLL, UPMC Univeristy, Paris, France, Mária Lukáčová-Medvid'ová was supported by the German Science Foundation under the grant LU 1470/2-2,3, and Šárka Nečasová was supported by the Grant Agency of the Czech Republic n. P 201/11/1304 and by RVO 67985840.

References

1. M. Astorino, C. Grandmont, Convergence analysis of a projection semi-implicit coupling scheme for fluid-structure interaction problems. Numer. Math. **116**(4), 721–767 (2010)
2. M. Astorino, F. Chouly, M.A. Fernández, An added-mass free semi-implicit coupling scheme for fluid-structure interaction. C. R. Math. Acad. Sci. Paris **347**(1–2), 99–104 (2009)
3. M. Astorino, F. Chouly, M.A. Fernández, Robin based semi-implicit coupling in fluid-structure interaction: stability analysis and numerics. SIAM J. Sci. Comput. **31**(6), 4041–4065 (2009/2010)
4. F. Baaijens, A fictitious domain/mortar element method for fluid-structure interaction. Int. J. Numer. Methods Fluids **35**, 743–761 (2001)
5. S. Badia, R. Codina, Convergence analysis of the FEM approximation the first order projection method for incompressible flows with and without the inf-sup condition. Numer. Math. **107**, 533–557 (2007)

6. S. Badia, F. Nobile, C. Vergara, Fluid-structure partitioned procedures based on Robin transmission conditions. J. Comput. Phys. **227**, 7027–7051 (2008)
7. S. Badia, A. Quaini, A. Quarteroni, Splitting methods based on algebraic factorization for fluid-structure interaction. SIAM J. Sci. Comput. **30** , 1778–1805 (2008)
8. S. Badia, F. Nobile, Ch. Vergara, Robin-Robin preconditioned Krylov methods for fluid-structure interaction problems. Comput. Methods Appl. Mech. Eng. **198**, 2768–2784 (2009)
9. L. Baffico, C. Grandmont, B. Maury, Multiscale modeling of the respiratory tract. Math. Models Methods Appl. Sci. **20**(1), 59–93 (2010)
10. H. Beirão da Veiga, On the existence of strong solutions to a coupled fluid-structure evolution problem. J. Math. Fluid Mech. **6** (1), 21–52 (2004)
11. F.B. Belgacem, The mortar finite element method with Lagrange multipliers. Numer. Math. **84**, 173–197 (1999)
12. F.B. Belgacem, Y. Maday, The mortar element method for three-dimensional finite elements. RAIRO Modèl. Math. Anal. Numér. **31**, 289–302 (1997)
13. C. Bernardi, Y. Maday, A.T. Patera, Domain decomposition by the mortar element method, in *Asymptotic and Numerical Methods for Partial Differential Equations with Critical Parameters (Beaune, 1992)*. NATO Advanced Science Institutes Series C, Mathematical and Physical Sciences, vol. 384 (Kluwer Academic, Dordrecht, 1993), pp. 269–286
14. C. Bernardi, Y. Maday, A.T. Patera, A new nonconforming approach to domain decomposition: the mortar element method, in *Nonlinear Partial Differential Equations and Their Applications. Collège de France Seminar, vol. XI (Paris, 1989–1991)*. Pitman Research notes in mathematics Series, vol. 299 (Longmann House, Harlow, 1994), pp. 13–51
15. M. Boulakia, Existence of weak solutions for an interaction problem between an elastic structure and a compressible viscous fluid. J. Math. Pures Appl. (9) **84**(11), 1515–1554 (2005)
16. M. Boulakia, Existence of weak solutions for the three-dimensional motion of an elastic structure in an incompressible fluid. J. Math. Fluid Mech. **9**(2), 262–294 (2007)
17. M. Boulakia, S. Guerrero, A regularity result for a solid-fluid system associated to the compressible Navier–Stokes equations. Ann. Inst. H. PoincarÁnal. Non Lináire **26**(3), 777–813 (2009)
18. M. Boulakia, S. Guerrero, Regular solutions of a problem coupling a compressible fluid an elastic structure. J. Math. Pures Appl. (9) **94**(4), 341–365 (2010)
19. M. Boulakia, E. Schwindt, T. Takahashi, Existence of strong solutions for the motion of an elastic structure in an incompressible viscous fluid. Interfaces Free Boundaries **14**(3), 273–306 (2012)
20. H. Brezis, *Analyse Fonctionelle- Théorie et Applications* (Masson, Paris, 1983)
21. M. Bukač, S. Čanić, R. Glowinski, J. Tambaca, A. Quaini, Fluid-structure interaction in blood flow capturing non-zero longitudinal structure displacement. J. Comput. Phys. **235**, 515–541 (2013)
22. S. Čaniċ, B. Muha, *A Nonlinear Moving-Boundary Problem of Parabolic-Hyperbolic-Hyperbolic Type Arising in Fluid-Multi-Layered Structure Interaction Problems*, ed. by F. Ancona, A. Bressan, P. Marcati, A. Marson (American Institute of Mathematical Sciences (AIMS) Publications, 2014), pp. 389–398
23. S. Čaniċ, B. Muha, M. Bukač, *Stability of the Kinematically Coupled Beta-Scheme for Fluid-Structure Interaction Problems in Hemodynamics* (Accepted)
24. S. Čanić, D. Lamponi, A. Mikelič, J. Tambača, Self-consistent effective equations modeling blood ow in medium-to-large compliant arteries. SIAM J. Multiscale Model. Simul. **3**, 559–596 (2005)
25. P. Causin, J.F. Gerbeau, F. Nobile, Added-mass effect in the design of partitioned algorithms for fluid-structure problems. Comput. Methods Appl. Mech. Eng. **194**, 42–44 (2005)
26. A. Chambolle, B. Desjardins, M.J. Esteban, C.Grandmont, Existence of weak solutions for the unsteady interaction of a viscous fluid with an elastic plate. J. Math. Fluid Mech. **7**(3), 368–404 (2005)
27. D. Chapelle, K.J. Bathe, *The Finite Element Analysis of Shells: Fundamentals* (Springer, New York, 2003)

28. N. Chemetov, Š. Nečasová, *The Motion of the Rigid Body in Viscous Fluid Including Collisions. Global Solvability Result* (Submitted)
29. A.J. Chorin, Numerical solution of the Navier-Stokes equations. Math. Comput. **22**, 745–762 (1968)
30. A.J. Chorin, On the convergence of discrete approximations to the Navier-Stokes equations. Math. Comput. **23**, 341–353 (1969)
31. P.G. Ciarlet, *Mathematical Elasticity*, vol. I (North-Holland, Amsterdam, 1988)
32. C. Conca, F. Murat, O. Pironneau, The Stokes and Navier-Stokes equations with boundary conditions involving the pressure. Jpn. J. Math. (N.S.) **20**(2), 279–318 (1994)
33. C. Conca, J.H. San Martin, M. Tucsnak, Existence of solutions for the equations modelling the motion of a rigid body in a viscous fluid. Commun. Partial Differ. Equ. **25**(5–6), 1019–1042 (2000)
34. D. Coutand, S. Shkoller, Motion of an elastic solid inside an incompressible viscous fluid. Arch. Ration. Mech. Anal. **176**(1), 25–102 (2005)
35. D. Coutand, S. Shkoller, The interaction between quasilinear elastodynamics and the Navier-Stokes equations. Arch. Ration. Mech. Anal. **179**(3), 303–352 (2006)
36. J. De Hart, G.W.M. Peters, P.J.G. Schreurs, F.P.T. Baaijens, A three-dimensional computational analysis of fluid-structure interaction in the aortic valve. J. Biomech. **36**, 103–112 (2003)
37. S. Deparis, M.A. Fernández, L. Formaggia, Acceleration of a fixed point algorithm for fluid-structure interaction using transpiration conditions. M2AN Math. Model. Numer. Anal. **37**, 601–616 (2003)
38. S. Deparis, M. Discacciati, G. Fourestey, A. Quarteroni, Fluid-structure algorithms based on Steklov-Poincar operators. Comput. Methods Appl. Mech. Eng. **195**(41–43), 5797–5812 (2006)
39. B. Desjardins, M.J. Esteban, Existence of weak solutions for the motion of rigid bodies in a viscous fluid. Arch. Ration. Mech. Anal. **146**(1), 59–71 (1999)
40. B. Desjardins, M.J. Esteban, On weak solutions for fluid-rigid structure interaction: compressible and incompressible models. Commun. Partial Differ. Equ. **25**(7–8), 1399–1413 (2000)
41. B. Desjardins, M.J. Esteban, C. Grandmont, P. Le Tallec, Weak solutions for a fluid-elastic structure interaction model. Rev. Mat. Complut. **14**(2), 523–538 (2001)
42. L. Diening, M. Růžička, J. Wolf, Existence of weak solutions for unsteady motions of generalized Newtonian fluids. Ann. Sc. Norm. Super. Pisa Cl. Sci. **9**(1), 1–46 (2010)
43. J. Donea, S. Giuliani, J.P. Halleux, An arbitrary Lagrangian-Eulerian finite element method for transient dynamic fluid-structure interactions. Comput. Methods Appl. Mech. Eng. **33**, 689–723 (1982)
44. Q. Du, M.D. Gunzburger, L.S. Hou, J. Lee., Semidiscrete finite element approximations of a linear fluid-structure interaction problem. SIAM J. Numer. Anal. **42**, 1–29 (2004)
45. B. Ducomet, Š. Nečasová, On the motion of several rigid bodies in an incompressible viscous fluid under the influence of selfgravitating forces. Prog. Nonlinear Differ. Equ. Appl. **80**, 167–192 (2011)
46. B. Ducomet, Š. Nečasová, On the motion of rigid bodies in an incompressible or compressible viscous fluid under the action of gravitational force. Discrete Continuous Dyn. Syst. Ser. S **6**(5), 1193–1213 (2013)
47. A. Ern, J.-L. Guermond, in *Theory and Practice of Finite Elements*. Applied Mathematical Sciences, vol. 159 (Springer, New York, 2004)
48. C. Farhat, M. Lesoinne, P. Le Tallec, Load and motion transfer algorithms for fluid/structure interaction problems with non-matching discrete interfaces: momentum and energy conservation, optimal discretization and application to aeroelasticity. Comput. Methods Appl. Mech. Eng. **157**, 95–114 (1998)
49. C. Farhat, K. Pierson, M. Lesoinne, The second generation of FETI methods and their application to the parallel solution of large-scale linear and geometrically nonlinear structural analysis problems. Comput. Methods Appl. Mech. Eng. **184** , 333–374 (2000)

50. C. Farhat, P. Geuzaine, C. Grandmont, The discrete geometric conservation law and the nonlinear stability of ALE schemes for the solution of flow problems on moving grids. J. Comput. Phys. **174**(2), 669–694 (2001)
51. C. Farhat, K. van der Zee, Ph. Geuzaine, Provably second-order time-accurate loosely-coupled solution algorithms for transient nonlinear computational aeroelasticity. Comput. Methods Appl. Mech. Eng. **195**, 1973–2001 (2006)
52. E. Feireisl, On the motion of rigid bodies in a viscous compressible fluid. Arch. Ration. Mech. Anal. **167**, 281–308 (2003)
53. E. Feireisl, Š. Nečasová, On the long-time behavior of a rigid body immersed in a viscous fluid. Appl. Anal. **90**(1), 59–66 (2011)
54. E. Feireisl, M. Hillairet, Š. Nečasová, On the motion of several rigid bodies in an incompressible non-Newtonian fluid. Nonlinearity **21**, 1349–1366 (2008)
55. M.A. Fernández, Coupling schemes for incompressible fluid-structure interaction: implicit, semi-implicit and explicit. SeMA J. **55**, 59–108 (2011)
56. M.A. Fernández, Incremental displacement-correction schemes for incompressible fluid-structure interaction. Numer. Math. **123**(1), 21–65 (2013)
57. M.A. Fernández, M. Landajuela, A fully decoupled scheme for the interaction of a thin-walled structure with an incompressible fluid. C. R. Math. Acad. Sci. Paris **351**(3–4), 161–164 (2013)
58. M.A. Fernández, M. Moubachir, An exact block-Newton algorithm for solving fluid-structure interaction problems. C. R. Math. Acad. Sci. Paris **336**, 681–686 (2003)
59. M.A. Fernández, M. Moubachir, An exact Block-Newton algorithm for the solution of implicit time discretized coupled systems involved in fluid-structure interaction problems, in *Second M.I.T. conference on computational fluid and solid mechanics*, ed. by K.J. Bathe (Elsevier, Amsterdam, 2003), pp. 1337–1341
60. M.A. Fernández, M. Moubachir, Numerical simulation of fluid-structure systems via Newtons method with exact Jacobians, in *4th European Congress on Computational Methods in Applied Sciences and Engineering*, Jyvaskyla, vol. 1, and ECCOMAS 2004, ed. by P. Neittaanmäki, T. Rossi, S. Korotov, E. Oñate, J. Périaux, D. Knörzer, Jyväskylä, 24–28 July 2004
61. M.A. Fernandez, M. Moubachir, A Newton method using exact Jacobians for solving fluid-structure coupling. Comput. Struct. **83**, 127–142 (2005)
62. M.A. Fernández, J. Mullaert, Displacement-velocity correction schemes for incompressible fluid-structure interaction. C. R. Math. Acad. Sci. Paris **349**(17–18), 1011–1015 (2011)
63. M.A. Fernández, J.-F. Gerbeau, C. Grandmont, A projection algorithm for fluid-structure interaction problems with strong added-mass effect. C. R. Math. Acad. Sci. Paris **342**(4), 279–284 (2006)
64. M.A. Fernández, J.-F. Gerbeau, C. Grandmont, A projection semi-implicit scheme for the coupling of an elastic structure with an incompressible fluid. Int. J. Numer. Methods Eng. **69**(4), 794–821 (2007)
65. C. Figueroa, I. Vignon-Clementel, K. Jansen, T.J.R. Hughes, C. Taylor, A coupled momentum method for modeling blood flow in three-dimensional deformable arteries. Comput. Methods Appl. Mech. Eng. **195**(41–43), 5685–5706 (2006)
66. J. Filo, A. Zaušková, 2D Navier-Stokes equations in a time dependent domain with Neumann type boundary conditions. J. Math. Fluid Mech. **12**(1) 1–46 (2010)
67. L. Formaggia, F. Nobile, A stability analysis for the arbitrary Lagrangian Eulerian formulation with finite elements. East-West J. Numer. Math. **7**, 105–131 (1999)
68. L. Formaggia, F. Nobile, Stability analysis of second-order time accurate schemes for ALE-FEM. Comput. Methods Appl. Mech. Eng. **193**, 4097–4116 (2004)
69. L. Formaggia, J.F. Gerbeau, F. Nobile, A. Quarteroni, On the coupling of 3D and 1D Navier-Stokes equations for flow problems in compliant vessels. Comput. Methods Appl. Mech. Eng. **191**(6–7), 561–582 (2001)
70. L. Formaggia, J.-F. Gerbeau, F. Nobile, A. Quarteroni, Numerical treatment of defective boundary conditions for the Navier-Stokes equations. SIAM J. Numer. Anal. **40**(1), 376–401 (2002)

71. L. Formaggia, A. Moura, F. Nobile, On the stability of the coupling of 3D and 1D fluid-structure interaction models for blood flow simulations. M2AN Math. Model. Numer. Anal. **41**(4), 743–769 (2007)
72. C. Forster, W.A. Wall, E. Ramm, Artificial added mass instabilities in sequential staggered coupling of nonlinear structures and incompressible viscous flows. Comput. Methods Appl. Mech. Eng. **7**, 1278–1293 (2007)
73. J. Frehse, J. Málek, M. Steinhauer, An existence result for fluids with shear dependent viscosity, steady flows. Nonlinear Anal. **30**, 3041–3049 (1997)
74. H. Fujita, N. Sauer, On existence of weak solutions of the Navier-Stokes equations in regions with moving boundaries. J. Fac. Sci. Univ. Tokyo Sect. I **17**, 403–420 (1970)
75. G.P. Galdi, On the motion of a rigid body in a viscous fluid: a mathematical analysis with applications, in *Handbook of Mathematical Fluid Dynamics*, vol. I (Elsevier Science, Amsterdam, 2002)
76. G.P. Galdi, M. Kyed, Steady flow of a Navier-Stokes liquid past an elastic body. Arch. Ration. Mech. Anal. **194**(3), 849–875 (2009)
77. D. Gérard-Varet, M. Hillairet, Regularity issues in the problem of fluid structure. Arch. Ration. Mech. Anal. **195**(2), 375–407 (2010)
78. M. Gérard-Varet, M. Hillairet, Computation of the drag force on a sphere close to a wall: the roughness issue. ESAIM Math. Model. Numer. Anal. **46**(5), 1201–1224 (2012)
79. D. Gérard-Varet, M. Hillairet, Existence of Weak Solutions Up To Collision for Viscous Fluid-Solid Systems with Slip. Comm. Pure Appl. Math. (Submitted)
80. D. Gérard-Varet, M. Hillairet, C. Wang, The Influence of Boundary Conditions on the Contact Problem in a 3D Navier-Stokes Flow. J. Math. Pures Appl. (Appear in)
81. J.-F. Gerbeau, M. Vidrascu, A quasi–Newton algorithm based on a reduced model for fluid-structure interactions problems in blood flows. Math. Model. Numer. Anal. **37**, 631–648 (2003)
82. P. Geuzaine, C. Grandmont, C. Farhat, Design and analysis of ALE schemes with provable second-order time-accuracy for inviscid and viscous flow simulations. J. Comput. Phys. **191**(1), 206–227 (2003)
83. O. Glass, F. Sueur, The movement of a solid in an incompressible perfect fluid as a geodesic flow. Proc. Am. Math. Soc. **140**(6), 2155–2168 (2012)
84. O. Glass, F. Sueur, T. Takahashi, Smoothness of the motion of a rigid body immersed in an incompressible perfect fluid. Ann. Sci. Ec. Norm. Supér. (4) **45**(1), 1–51 (2012)
85. R. Glowinski, T.W. Pan, J. Périaux, A fictitious domain method for Dirichlet problem and applications. Comput. Methods Appl. Mech. Eng. **111**, 283–303 (1994)
86. C. Grandmont, Existence et unicit de solutions d'un problme de couplage fluide-structure bidimensionnel stationnaire (French) [Existence and uniqueness for a two-dimensional steady-state fluid-structure interaction problem]. C. R. Acad. Sci. Paris Sér. I Math. **326**(5), 651–656 (1998)
87. C. Grandmont, Existence for a three-dimensional steady state fluid–structure interaction problem. J. Math. Fluid Mech. **4** (1), 76–94 (2002)
88. C. Grandmont, Existence of weak solutions for the unsteady interaction of a viscous fluid with an elastic plate. SIAM J. Math. Anal. **40**(2), 716–737 (2008)
89. C. Grandmont, Y. Maday, Existence for an unsteady fluid-structure interaction problem. M2AN Math. Model. Numer. Anal. **34**(3), 609–636 (2000)
90. C. Grandmont, Y. Maday, Some remarks on fluid-structure interaction problems in case of rigid body plus small perturbations, in *Coupling of Fluids, Structures and Waves in Aeronautics*. Notes on Numerical Fluid Mechanics and Multidisciplinary Design, vol. 85 (Springer, Berlin, 2003), pp. 239–250
91. C. Grandmont, A. Soualah, Solutions fortes des quations de Navier-Stokes avec conditions dissipatives naturelles (French) [Strong solutions of Navier-Stokes equations with natural dissipative conditions], in *Paris-Sud Working Group on Modelling and Scientific Computing 2007–2008*. ESAIM Proceedings, vol. 25 (EDP Science, Les Ulis, 2008), pp. 1–18

92. C. Grandmont, V. Guimet, Y. Maday, Numerical analysis of some decoupling techniques for the approximation of the unsteady fluid-structure interaction. Math. Models Methods Appl. Sci. **11**(8), 1349–1377 (2001)
93. C. Grandmont, B. Maury, A. Soualah, Multiscale modelling of the respiratory track: a theoretical framework, in *Mathematical and Numerical Modelling of the Human Lung*. ESAIM Proceedings, vol. 23 (EDP Science, Les Ulis, 2008), pp. 10–29
94. V. Gravemeier, A. Comerford, L. Yoshihara, M. Ismail, W.A. Wall, A novel formulation for neumann inflow boundary conditions in biomechanics. Int. J. Numer. Methods Biomed. Eng. **28**, 560–573 (2012)
95. J.-L. Guermond, L. Quartapelle, On stability and convergence of projection methods based on pressure Poisson equation. Int. J. Numer. Methods Fluids **26**, 1039–1053 (1998)
96. J.-L. Guermond, L. Quartapelle, On the approximation of the unsteady Navier–Stokes equation by finite element projection methods. Numer. Math. **80**, 207–238 (1998)
97. J.-L. Guermond, P. Minev, J. Shen, An overview of projection methods for incompressible flows. Comput. Methods Appl. Mech. Eng. **195**, 6011–6045 (2006)
98. G. Guidoboni, R. Glowinski, N. Cavallini, S. Canic, Stable loosely-coupled-type algorithm for fluid-structure interaction in blood flow. J. Comput. Phys. **228**(18), 6916–6937 (2009)
99. G. Guidoboni, M. Guidorzi, M. Padula, Continuous dependence on initial data in fluid-structure motions. J. Math. Fluid Mech. **14**(1), 1–32 (2012)
100. M. Guidorzi, M. Padula, P.I. Plotnikov, Hopf solutions to a fluid-elastic interaction model. Math. Models Methods Appl. Sci. **18**(2), 215–269 (2008)
101. M. Heil, An efficient solver for the fully coupled solution of large-displacement fluid-structure interaction problems. Comput. Methods Appl. Mech. Eng. **193**, 1–2 (2004)
102. T.I. Hesla, Collision of smooth bodies in a viscous fluid: a mathematical investigation. Ph.D. thesis, Minnesota, 2005
103. J. G. Heywood, R. Rannacher, S. Turek, Artificial boundaries and flux and pressure conditions for the incompressible Navier–Stokes equations. Int. J. Numer. Methods Fluids **22**(5), 325–352 (1996)
104. M. Hillairet, Lack of collision between solid bodies in a 2D incompressible viscous flow. Commun. Partial Differ. Equ. **32**, 1345–1371 (2007)
105. M. Hillairet, T. Takahashi, Collisions in three-dimensional fluid structure interaction problems. SIAM J. Math. Anal. **40**(6), 2451–2477 (2009)
106. M. Hillairet, T. Takahashi, Blow up and grazing collision in viscous fluid solid interaction systems. Ann. Inst. H. Poincaré Anal. Non Linéaire **27**(1), 291–313 (2010)
107. K.H. Hoffmann, V.N. Starovoitov, On a motion of a solid body in a viscous fluid: two-dimensional case. Adv. Math. Sci. Appl. **9**(2), 633–648 (1999)
108. K.H. Hoffmann, V.N. Starovoitov, Zur Bewegung einer Kugel in einer zhen Flssigkeit (German) [On the motion of a sphere in a viscous fluid]. Doc. Math. **5**, 15–21 (2000)
109. A. Hundertmark-Zaušková, M. Lukáčová-Medvid'ová, Numerical study of shear-dependent non-Newtonian fluids in compliant vessels. Comput. Math. Appl. **60**, 572–590 (2010)
110. A. Hundertmark-Zaušková, M. Lukáčová-Medvid'ová, G. Rusnáková, Fluid-structure interaction for shear-dependent non-Newtonian fluids, in *Topics in Mathematical Modeling and Analysis*, Nečas Center for Mathematical Modeling, Lecture Notes, vol. 7 (Matfyzpress, Praha, 2012), pp. 109–158
111. A. Hundertmark-Zaušková, M. Lukáčová-Medvid'ová, Š. Nečasová, *On the Existence of Weak Solution to the Coupled Fluid-Structure Interaction Problem for Non-Newtonian Shear-Dependent Fluid* (submitted)
112. N.V. Judakov, The solvability of the problem of the motion of a rigid body in a viscous incompressible fluid (Russian). Dinamika Splošn. Sredy Vyp. **18**, 249–253 (1974)
113. H.J. Kim, C.A. Figueroa, T.J.R. Hughes, K.E. Jansen, C.A. Taylor, Augmented Lagrangian method for constraining the shape of velocity profiles at outlet boundaries for three-dimensional finite element simulations of blood flow. Comput. Methods Appl. Mech. **198**, 3551–3566 (2009)

114. I. Kukavica, A. Tuffaha, Well-posedness for the compressible Navier-Stokes-Lamé system with a free interface. Nonlinearity **25**(11), 3111–3137 (2012)
115. P. Le Tallec, S. Mani, Numerical analysis of a linearized fluid-structure interaction problem. Numer. Math. **87**, 317–354 (2000)
116. P. Le Tallec, J. Mouro, Fluid structure interaction with large structural displacements. Comput. Methods Appl. Mech. Eng. **190**, 3039–3067 (2001)
117. D. Lengeler, *Global Weak Solutions for An Incompressible, Generalized Newtonian Fluid Interacting with a Linearly Elastic Koiter Shell* (Preprint)
118. D. Lengeler, M. Růžička, Global Weak Solutions for an Incompressible Newtonian Fluid Interacting with a Linearly Elastic Koiter Shell. Arch. Ration. Mech. Anal. **211**(1), 205–255 (2014)
119. J. Lequeurre, Existence of strong solutions to a fluid-structure system. SIAM J. Math. Anal. **43**(1), 389–410 (2011)
120. J. Lequeurre, Existence of strong solutions for a system coupling the Navier-Stokes equations and a damped wave equation. J. Math. Fluid Mech. **15**(2), 249–271 (2013)
121. M. Lukáčová-Medviďová, G. Rusnáková, A. Hundertmark-Zaušková, Kinematic splitting algorithm for fluid-structure interaction in hemodynamics. Comput. Methods Appl. Mech. Eng. **265**, 83–106 (2013)
122. J. Málek, K.R. Rajagopal, Mathematical issues concerning the Navier-Stokes equations and some of its generalizations, in *Handbook of Differential Equations*, ed. by C.M. Dafermos, E. Feireisl (North-Holland, Boston, 2005)
123. J. Málek, J. Nečas, M. Rokyta, M. Ružička, *Weak and Measure-Valued Solutions to Evolutionary PDE's* (Chapman and Hall, London, 1996)
124. H.G. Matthies, J. Steindorf, Partitioned but strongly coupled iteration schemes for nonlinear fluid-structure interaction. Comput. Struct. **80**, 1991–1999 (2002)
125. H.G. Matthies, J. Steindorf, Partitioned strong coupling algorithms for fluid-structure interaction. Comput. Struct. **81**, 805–812 (2003)
126. M.E. Moghadam, Y. Bazilevs, T.-Y. Hsia, I.E. Vignon-Clementel, A.L. Marsden, Modeling of Congenital Hearts Alliance (MOCHA), a comparison of outlet boundary treatments for prevention of backflow divergence with relevance to blood flow simulations. Comput. Mech. **48**, 277–291 (2011)
127. D.P. Mok, W.A. Wall, Partitioned analysis schemes for the transient interaction of incompressible flows and nonlinear flexible structures, in *Trends in Computational Structural Mechanics*, ed. by K. Schweizerhof, W.A. Wall, K.U. Bletzinger (International Center for Numerical Methods in Engineering (CIMNE), Barcelona, 2001)
128. D.P. Mok, W.A. Wall, E. Ramm, Accelerated iterative substructuring schemes for instationary fluid-structure interaction, in *Computational Fluid and Solid Mechanics*, ed. by K.J. Bathe (Elsevier, Amsterdam, 2001), pp. 1325–1328
129. B. Muha, S. Canić, Existence of a weak solution to a nonlinear fluid-structure interaction problem modeling the flow of an incompressible, viscous fluid in a cylinder with deformable walls. Arch. Ration. Mech. Anal. **207**(3), 919–968 (2013)
130. F. Nobile, C. Vergara, An effective fluid-structure interaction formulation for vascular dynamics by generalized Robin conditions. SIAM J. Sci. Comput. **30**(2), 731–763 (2008)
131. F. Nobile, C. Vergara, Partitioned algorithms for fluid-structure interaction problems in haemodynamics. Milan J. Math. **80**, 443–467 (2012)
132. F. Nobile, M. Pozzoli, C. Vergara, Time accurate partitioned algorithms for the solution of fluid-structure interaction problems in haemodynamics. Comput. Fluids **86**, 470–482 (2013)
133. C.S. Peskin, D.M. McQueen, A three-dimensional computational method for blood flow in the heart. I. Immersed elastic fibers in a viscous incompressible fluid. J. Comput. Phys. **81**, 372–405 (1989)
134. S. Piperno, Explicit/implicit fluid/structure staggered procedures with a structural predictor and fluid subcycling for 2D inviscid aeroelastic simulations. Int. J. Numer. Methods Fluids **25**, 1207–1226 (1997)

135. S. Piperno, C. Farhat, Design of efficient partitioned procedures for the transient solution of aeroelastic problems, in *Fluid-Structure Interaction*. Innovative Technology Series (Kogan Page Science, London, 2003), pp. 23–49
136. S. Piperno, C. Farhat, B. Larrouturou, Partitioned procedures for the transient solution of coupled aeroelastic problems. Part I: model problem, theory and two-dimensional application. Comput. Methods Appl. Mech. Eng. **124**, 79–112 (1995)
137. A. Quaini, A. Quarteroni, A semi-implicit approach for fluid-structure interaction based on an algebraic fractional step method. Math. Models Methods Appl. Sci. **17**, 957–983 (2007)
138. A. Quarteroni, A. Veneziani, Analysis of a geometrical multiscale model based on the coupling of ODEs and PDEs for blood flow simulations. Multiscale Model. Simul. **1**(2), 173–195 (2003)
139. A. Quarteroni, M. Tuveri, A. Veneziani, Computational vascular fluid dynamics: problems, models and methods. Comput. Vis. Sci. **2**, 163–197 (2000)
140. A. Quarteroni, S. Ragni, A. Veneziani, Coupling between lumped and distributed models for blood flow problems. Second AMIF International Conference (Il Ciocco, 2000). Computing and Visualization in Science **4**(2), 111–124 (2001)
141. P. Raback, J. Ruokolainen, M. Lyly, E. Jarvinen, in *Fluid–structure interaction boundary conditions by artificial compressibility*, ECCOMAS 2008, Venice, Italy, June 30–4 July 2008
142. R. Raghu, I. Vignon-Clementel, C. Figueroa, C. Taylor, Comparative study of viscoelastic arterial wall models in nonlinear one-dimensional finite element simulations of blood flow. J. Biomech. Eng. **133**(8), 081003 (2011)
143. R. Rannacher, in *On Chorin's projection method for the incompressible Navier-Stokes equations*. The Navier-Stokes equations II–theory and numerical methods (Oberwolfach, 1991). Lecture Notes in Mathematics, vol. 1530 (Springer, Berlin, 1992), pp. 167–183
144. S. Rugonyi, K.J. Bathe, On finite element analysis of fluid flows coupled with structural interaction. CMES-Comput. Model. Eng. Sci. **2**, 195–212 (2001)
145. R. Salvi, On the existence of free surface problem for viscous incompressible flow. Navier-Stokes equations and related nonlinear problems (Ferrara, 1999). Ann. Univ. Ferrara Sez. VII (N.S.) **46**, 251–266 (2000)
146. J. San Martin, V. Starovoitov, M. Tucsnak, Global weak solutions for the two-dimensional motion of several rigid bodies in an incompressible viscous fluid. Arch. Ration. Mech. Anal. **161**(2), 113–147 (2002)
147. J.A. San Martin, V. Starovoitov, M. Tucsnak, Global weak solutions for the two dimensional motion of several rigid bodies in an incompressible viscous fluid. Arch. Ration. Mech. Anal. **161**, 93–112 (2002)
148. J. San Martin, J.F. Scheid, T. Takahashi, M. Tucsnak, Convergence of the Lagrange-Galerkin method for the equations modelling the motion of a fluid-rigid system. SIAM J. Numer. Anal. **43**, 1536–1571 (2005)
149. D. Serre, Chute libre d'un solide dans un fluide visqueux incompressible. Existence (French) [Free fall of a rigid body in an incompressible viscous fluid. Existence]. Jpn J. Appl. Math. **4**(1), 99–110 (1987)
150. J. Simon, Compact sets in the space $L^p(0,T;B)$. Ann. Mat. Pura Appl. (4), **146**, 65–96 (1987)
151. V.N. Starovoitov, Behavior of a rigid body in an incompressible viscous fluid near a boundary, in *Free Boundary Problems (Trento, 2002)*. International Series of Numerical Mathematics, vol. 147 (Birkhauser, Basel, 2004), pp. 313–327
152. C. Surulescu, On the stationary interaction of a Navier–Stokes fluid with an elastic tube wall. Appl. Anal. **86**, 149–165 (2007)
153. T. Takahashi, Analysis of strong solutions for the equations modeling the motion of a rigid-fluid system in a bounded domain. Adv. Differ. Equ. **8**(12), 1499–1532 (2003)
154. T. Takahashi, M. Tucsnak, Global strong solutions for the two-dimensional motion of an infinite cylinder in a viscous fluid. J. Math. Fluid Mech. **6**(1), 53–77 (2004)
155. R. Temam, Une méthode d'approximation de la solution des équations de Navier-Stokes. Bull. Soc. Math. France **96**, 115–152 (1968)

156. T.E. Tezduyar, Finite element methods for fluid dynamics with moving boundaries and interfaces. Arch. Comput. Methods Eng. **8**, 83–130 (2001)
157. I. Vignon, C.A. Taylor, Outflow boundary conditions for one-dimensional finite element modeling of blood flow and pressure waves in arteries: new computational methods for wave propagation. Wave Motion **39**(4), 361–374 (2004)
158. J. Wolf, Existence of weak solution to the equations of non-stationary motion of non-Newtonian fluids with shear rate dependent viscosity. J. Math. Fluid. Mech. **9**(1), 104–138 (2007)
159. K.K. Yeleswarapu, Evaluation of continuum models for characterizing the constitutive behavior of blood. Ph.D. Thesis, University of Pittsburgh, Pittsburgh, 1996

C. Grandmont (✉)
INRIA Paris-Rocquencourt, Paris, France
e-mail: celine.grandmont@inria.fr

M. Lukáčová-Medvid'ová
Institut für Mathematik, Johannes Gutenberg Universität Mainz, Mainz, Germany
e-mail: lukacova@mathematik.uni-mainz.de

Š. Nečasová
Institute of Mathematics, Academy of Sciences of the Czech Republic, Prague, Czech Republic
e-mail: matus@math.cas.cz

Chapter 2
Fluid–Structure Interaction in Hemodynamics: Modeling, Analysis, and Numerical Simulation

Sunčica Čanić, Boris Muha, and Martina Bukač

Abstract Fluid–structure interaction (FSI) problems arise in many applications. They include multi-physics problems in engineering such as aeroelasticity and propeller turbines, as well as biofluidic application such as self-propulsion organisms, fluid–cell interactions, and the interaction between blood flow and cardiovascular tissue. A comprehensive study of these problems remains to be a challenge due to their strong nonlinearity and multi-physics nature. To make things worse, in many biological applications the structure is composed of several layers, each with different mechanical characteristics. This is, for example, the case with arterial walls, which are composed of three main layers: the intima, media, and adventitia, separated by thin elastic laminae. A stable and efficient FSI solver that simulates the interaction between an incompressible, viscous fluid and a multi-layered structure would be an indispensable tool for the computational studies of solutions.

The multi-physics nature of this class of problems suggests the use of partitioned, modular algorithms based on an operator splitting approach that would separate the different physics in the problem. This chapter presents such a scheme, which can be used not only in computations, but also to prove existence of weak solutions to this class of problems. Particular attention will be payed to multi-physics FSI problems involving structures consisting of multiple layers.

Keywords Aeroelasticity • ALE method • Biomechanics of voice • Compressible flow • Coupling algorithm • Discontinuous Galerkin method • Dynamic elasticity problem • Fluid–structure interaction • Navier–Stokes equations • Stabilized finite element method • Time and space discretization • Two degrees of freedom model • Vocal folds

MSC2010: 74F10, 76Z05, 76M10, 74S05, 74H15

T. Bodnár et al. (eds.), *Fluid-Structure Interaction and Biomedical Applications*, Advances in Mathematical Fluid Mechanics, DOI 10.1007/978-3-0348-0822-4_2,

2.1 Introduction

Fluid–structure interaction (FSI) problems arise in many applications. The widely known examples are aeroelasticity and biofluids. In aeroelasticity, where the structure (wing of an airplane) is much heavier than the fluid (air), it is sometimes of interest to study small vibrations of the structure in which case linear coupling between the fluid and the structure may be sufficient to capture the main features of the solutions. In that case the fluid domain remains fixed in the FSI model, and only the location of the structure is computed based on the fluid loading (one-way coupling). In biofluidic applications, such as the interaction between blood flow and cardiovascular tissue where the density of the structure (arterial walls) is roughly equal to the density of the fluid (blood), the coupling between the fluid and the relatively light structure is *highly nonlinear*. In that case the fluid domain is not fixed in the FSI model, and its location is determined by the location of the structure. The elastodynamics of the structure influences the motion of the fluid through the contact force exerted by the structure onto the fluid, while the structure location is computed based on the fluid loading expressed through the contact force exerted by the fluid onto the structure (two-way coupling). It has recently been shown that classical "partitioned" time-marching numerical algorithms, which are based on subsequent solutions of the fluid and structure sub-problems, are unconditionally unstable in problems in which the density of the structure and of the fluid are comparable [30]. The exchange of energy between the moving fluid and structure is so significant, that a mismatch between the energy of the discretized problem and the energy of the continuous problem causes instabilities in classical "loosely coupled" partitioned schemes. The difficulties associated with the significant energy exchange and the high geometric nonlinearity of the fluid–structure interface are reflected not only in the design of numerical schemes but also in the theoretical studies of existence and stability of solutions to this class of problems. A comprehensive study of these problems remains to be a challenge due to their strong nonlinearity and multi-physics nature.

In the blood flow application, the problems are further exacerbated by the fact that arterial walls of major arteries are composed of several layers, each with different mechanical characteristics. The main layers are the tunica intima, media, and adventitia. They are separated by the thin elastic laminae, see Fig. 2.1. Recent developments in ultrasound speckle tracking methods revealed significant shear strain between the different layers in high adrenaline situations [2, 39, 40]. It was noted that the consequences of this phenomenon on cardiovascular disease are yet to be explored! An example of a disease which is associated with a pathophysiology of the aortic wall layers is aortic dissection: tears in the intimal layer result in separation of the aortic wall layers causing blood to flow within the aortic wall.

Until recently, there have been no FSI models or computational solvers of arterial flow that take into account the multi-layered structure of arterial walls. In this chapter we take a step in this direction by studying a benchmark problem in fluid-multi-layered-structure interaction in which the structure consists of two layers, a

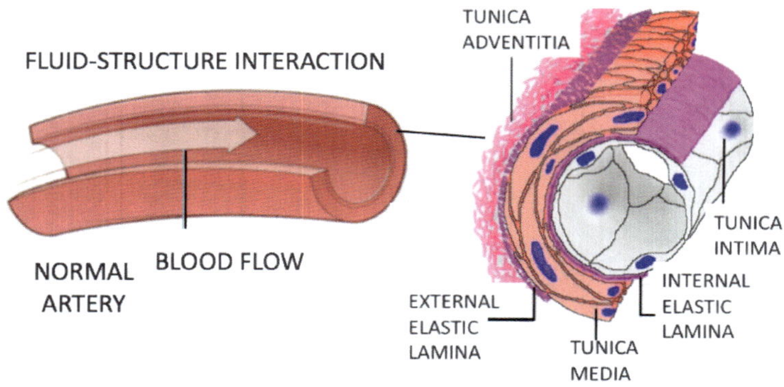

Fig. 2.1 FSI between blood flow and arterial walls, which are composed of multiple structural layers

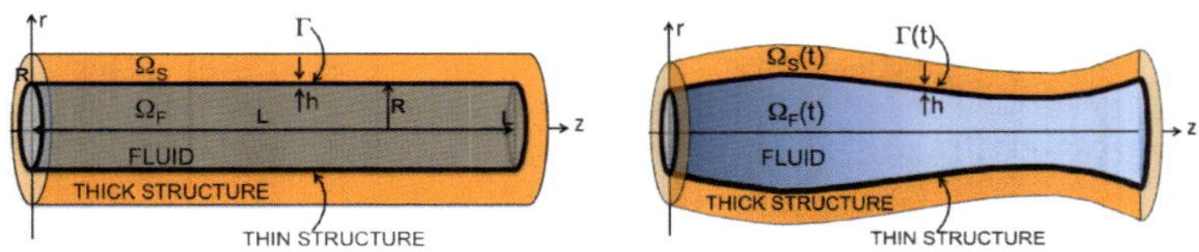

Fig. 2.2 *Left*: Reference domain. *Right*: Deformed domain

thin and a thick layer. See Fig. 2.2. The fluid flow will be modeled by the Navier–Stokes equations for an incompressible, viscous, Newtonian fluid. This is a good approximation for blood flow in major arteries, such as the aorta or coronary arteries. The thin structural layer will be modeled by the cylindrical Koiter shell equations, and the thick structural layer will be modeled by the equations of linear elasticity. The thin structural layer located between the fluid and the thick structure serves as a fluid–structure interface with mass. The proposed problem is a nonlinear moving-boundary problem.

THE BENCHMARK PROBLEM

Fluid: Navier–Stokes equations for an incompressible, viscous fluid;

Thin Structure: Cylindrical Koiter shell equations;

Thick Structure: Classical equations of linear elasticity.

This is a *multi-physics problem* which consists of three different physical models: a model for fluid flow, a model describing the elastodynamics of the thin structure, and a model describing the elastodynamics of the thick structural layer. The multi-physics nature of the problem strongly suggests the use of a partitioned algorithm that would solve the underlying coupled problem by splitting the problem into sub-problems determined by the different physics. This has the advantage of allowing modular implementations in both the numerical method development and in constructing the proof of existence of solutions for this class of problems. In this chapter we present a stable, convergent, modular scheme with precisely these properties, called the Kinematically Coupled β-Scheme. This scheme was originally constructed to study FSI problems with a single structural layer, modeled by the cylindrical Koiter shell equations in [85, 86], and then recently improved for higher accuracy in [21] (Kinematically Coupled β-Scheme). Modifications of this scheme can be applied to a much larger class of multi-physics problems associated with FSI, such as FSI involving stent–artery–fluid interaction [122], FSI involving a multi-layered elastic porous medium [23], and FSI involving a non-Newtonian fluid [95, 96].

In this chapter we present a *general "recipe"* describing the construction of the main steps of such a scheme that can be used to:

- prove existence of weak solutions, and/or
- construct a numerical solver

to study a class of FSI problems that include:

- problems with viscoelastic and/or purely elastic structural models,
- problems with different coupling conditions (no-slip, slip),
- problems with nonlinear thin structure models,
- 2D and 3D scenarios.

An interesting new feature of the class of problems studied in this chapter is the fact that the presence of a thin fluid–structure interface with mass regularizes solutions of this class of FSI problems. More precisely, the energy estimates presented in this chapter will show that the thin structure inertia regularizes evolution of the thin structure, which affects the solution of the entire coupled FSI problem. Namely, if we were considering a problem in which the structure consisted of only one layer, modeled by the equations of linear elasticity, from the energy estimates we would not be able to conclude that the fluid–structure interface is even continuous. With the presence of a thin elastic fluid–structure interface with mass (modeled, e.g., by the linear wave equation), the energy estimates imply that the displacement of the thin interface is in $H^1(\Gamma)$, which, due to the Sobolev embeddings, implies that the interface is Hölder continuous $C^{0,1/2}(\Gamma)$. The inertia of the fluid–structure interface with mass serves as a regularizing mechanism for the entire FSI problem. It will be shown in Sect. 2.7 that numerical simulations confirm this behavior.

This is reminiscent of the results by Hansen and Zuazua [90] in which the presence of a point mass at the interface between two linearly elastic strings with solutions in asymmetric spaces (different regularity on each side) allowed the proof of well-posedness due to the regularizing effects by the point mass. More precisely, they considered two elastic strings modeled by the linear wave equations, connected by a point mass, with initial data of different regularity on the left or right side of the point mass. They showed that the rough waves traveling through the point mass, which served as an interface with mass between the two elastic strings, were regularized due to the inertia effects of the point mass. See Sect. 2.7.7 for more details. For a reader with further interest in this area we also mention [91, 134, 145]. Further research in this direction, directly relevant to the FSI with multiple layers, is under way by the authors.

We begin by a review of models used in FSI studies to describe mechanical properties of arterial walls.

2.2 Mathematical Models of Arterial Walls

The walls of blood vessels are composed of three layers: the intima, media, and adventitia. They are separated by thin elastic laminae. See Fig. 2.1. The intima is the innermost layer and it is mainly composed of endothelial cells. The media is the middle layer and it is mainly composed of elongated smooth muscle cells, and also elastin and collagen. Most blood vessels contain smooth muscle arranged in either circular or spiral layers. The media gives rise to the majority of the vessel's viscoelastic behavior. The adventitia is the outermost layer, and it is manly composed of collagen fibrils, elastic sheets, and elastic fibrils. The layers of smooth muscle and connective tissue surrounding the intima vary in thickness in different vessels.

The *aorta and major arteries* are characterized by walls that have a thick smooth muscle layer and large amounts of elastic and fibrous tissue. Because of the stiffness of the fibrous tissue, substantial amounts of energy are required to stretch the walls of an artery outward. This energy comes from the high blood pressure exerted onto the arterial walls during the systolic part of cardiac cycle, when the left ventricle of the heart contracts, and squeezes blood through the aortic valve on to the aorta. Once the artery is distended with blood, energy stored by stretching elastic fibers is released through elastic recoil. Elastic recoil takes place during the diastolic part of cardiac cycle, when the left ventricle relaxes and gets refilled by blood. During that time the elastic recoil of arteries helps propel blood to the far most parts of the cardiovascular system.

Downstream from the arteries, small vessels called *arterioles* create a high-resistance outlet for arterial blood flow. Arterioles direct distribution of blood flow to individual tissues by selectively constricting and dilating. Arteriolar diameter is regulated by both local factors, such as tissue oxygen, and homeostatic control.

Downstream from the arterioles are *capillaries*. A leaky epithelium in the capillaries allows exchange of materials between the blood plasma, the interstitial fluid, and the cells of the body. At the distal end of the capillaries, blood flows into the venous side of the circulation and from there back to the right heart.

Depending on the types of questions one is trying to answer, and depending on the thickness of the vessel wall with respect to the diameter of the corresponding vessel, different arterial wall models can be used to describe the mechanical properties of arterial walls [68, 69, 88, 133]. Table 2.1 summarizes the mean diameter and wall thickness for arteries, arterioles, and capillaries [136].

In FSI studies, the coupling between blood flow and vascular tissue is so complicated that several simplifying assumptions have to be taken into account to

Table 2.1 Mean diameter and wall thickness for human vasculature

	Artery (mm)	Arteriole (μm)	Capillary (μm)
Mean diameter	4.0	30.0	8.0
Mean wall thickness	1.0	6.0	0.5

make the computer simulations feasible. A common set of simplifying assumptions that captures only the most important physics in the description of the mechanical properties of arterial walls includes homogeneity and isotropy, capturing the average mechanical properties of arterial walls. Further simplifying assumptions that are often used in hemodynamics FSI literature are "small" displacements and "small" deformation gradients leading to the hypothesis of linear elasticity.

Depending on the relative thickness of the structure (arterial walls) with respect to the diameter of the cylindrical fluid domain (arterial lumen), different modes have been used to approximate the overall (average) mechanical behavior of arterial walls. Three-dimensional equations of elasticity have been used under the assumption that the thickness of arterial walls is comparable to the diameter of the vessel lumen, while reduced shell or membrane models have been used under the assumption that the ratio between the thickness of the vessel wall and the vessel radius is small ($\epsilon \ll 1$). In the latter case, most FSI hemodynamics literature assumes that only the radial component of displacement of the thin structural wall is non-negligible. Recent developments in ultrasound speckle tracking methods revealed, however, that the axial (longitudinal) component of displacement of arterial walls may be significant in certain situations. Moreover, it was revealed that there is significant axial shear strain between the different layers (the intima–media complex and the adventitia) in high adrenaline situations [39, 40, 129, 138]. It was noted that the consequences of this phenomenon on cardiovascular disease are yet to be explored. Motivated by these experimental findings, recent progress in designing an FSI solver capturing both longitudinal and radial displacement of a thin Koiter shell modeling arterial walls was reported in [20, 21].

Finally, a further simplification that can be utilized in certain situations is axial symmetry of the loading exerted by the blood flow to the vessel walls in the approximately straight cylindrical sections, leading to the axially symmetric models with a potential of further reduction to 1D FSI models.

We give a brief review of these models next.

2.2.1 *Elastodynamics of Thin Structures*

The equations of shell theory have been derived by many authors, see [51] and the references therein. Due to variations in approach and rigor the variety of equations occurring in literature is overwhelming. Among all the equations of shell theory the Koiter shell equations appear to be the simplest consistent first approximation in the general theory of thin elastic shells [97, 98]. In addition, they have been

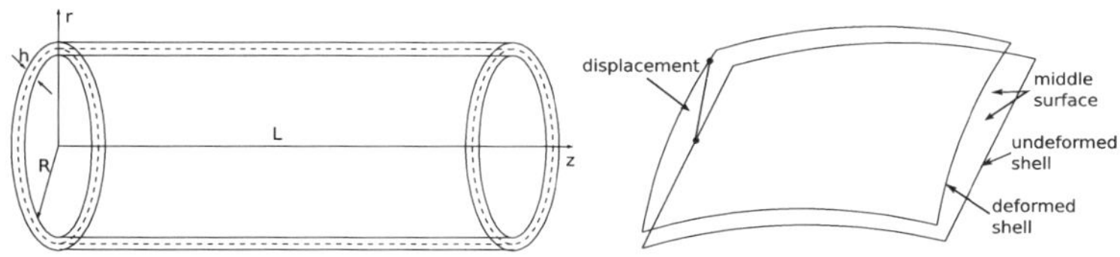

Fig. 2.3 *Left*: Cylindrical shell in reference configuration with middle surface radius R and shell thickness h. *Right*: Deformed shell

mathematically justified using asymptotic methods to be consistent with three-dimensional elasticity [37]. Ciarlet and Lods showed in [37] that the Koiter shell model has the same asymptotic behavior as the three-dimensional membrane model, the bending model and the generalized membrane model in the respective regimes in which each of them holds. Motivated by these remarkable properties of the Koiter shell model, in [27, 28] Čanić et al. derived the Koiter shell equations for the cylindrical geometry with the purpose of using the equations as a model to study the mechanical behavior of arterial walls. The models in [27, 28], and a portion of the text presented in this section, were based on the derivations of the cylindrical Koiter shell equations, obtained by Tambača in [139]. The cylindrical Koiter shell equations are a generalization of several classes of models that have been used in modeling of arterial walls. They include the linear string model proposed by Quarteroni et al. in [26, 133] as a benchmark problem for testing numerical schemes for FSI in blood flow, the independent ring model [133], and the cylindrical membrane model.

In [27,28] Čanić et al. have extended the linearly elastic cylindrical Koiter model to include the viscous effects of Kelvin–Voigt type, observed in the measurements of the mechanical properties of vessel walls [3,4,14]. It was shown in [3,4,14] that the Kelvin–Voigt model approximates well the experimentally measured viscoelastic properties of the canine aorta and of the human femoral and carotid arteries. In [27, 28] it was shown that a reduced FSI model between the linearly elastic cylindrical Koiter shell and the flow of an incompressible, viscous fluid, approximates well the experimentally measured data presented in [3,4,14]. The Kelvin–Voight model was also used in [130] to model the arterial walls as a linearly viscoelastic membrane. We summarize the derivation of the *Koiter shell model* next.

The Cylindrical Koiter Shell Equations: General Framework

Consider a clamped cylindrical shell of thickness h, length L, and reference radius of the middle surface equal to R. See Fig. 2.3. This reference configuration, which we denote by Γ, can be defined via the parameterization

$$\varphi : \omega \to \mathbb{R}^3, \quad \varphi(z, \theta) = (R\cos\theta, R\sin\theta, z)^t,$$

where $\omega = (0, L) \times (0, 2\pi)$ and $R > 0$. Therefore, the reference configuration is

$$\Gamma = \{\mathbf{x} = (R\cos\theta, R\sin\theta, z) \in \mathbb{R}^3 \; : \; \theta \in (0, 2\pi), z \in (0, L)\}. \tag{2.1}$$

The associated covariant A_c and contravariant A^c metric tensors of this (non-deformed) cylinder are given by:

$$\mathbf{A}_c = \begin{pmatrix} 1 & 0 \\ 0 & R^2 \end{pmatrix}, \; \mathbf{A}^c = \begin{pmatrix} 1 & 0 \\ 0 & \frac{1}{R^2} \end{pmatrix}, \tag{2.2}$$

and the area element along cylinder Γ is $dS = \sqrt{a}dy := \sqrt{\det A_c}dy = Rdy$. The corresponding curvature tensor in covariant components is given by

$$\mathbf{B}_c = \begin{pmatrix} 0 & 0 \\ 0 & R \end{pmatrix}.$$

We define the elasticity properties of this cylindrical shell by the following elasticity tensor $\mathcal{A}$:

$$\mathcal{A}\mathbf{E} = \frac{4\lambda\mu}{\lambda + 2\mu}(\mathbf{A}^c \cdot \mathbf{E})\mathbf{A}^c + 4\mu\mathbf{A}^c\mathbf{E}\mathbf{A}^c, \quad \mathbf{E} \in \mathrm{Sym}(\mathcal{M}_2), \tag{2.3}$$

where μ and λ are the Lamé coefficients.

Using the following relationships between the Lamé constants and the Young's modulus of elasticity E and Poisson ratio σ:

$$\frac{2\mu\lambda}{\lambda+2\mu} + 2\mu = 4\mu\frac{\lambda+\mu}{\lambda+2\mu} = \frac{E}{1-\sigma^2}, \quad \frac{2\mu\lambda}{\lambda+2\mu} = 4\mu\frac{\lambda+\mu}{\lambda+2\mu}\frac{1}{2}\frac{\lambda}{\lambda+\mu} = \frac{E}{1-\sigma^2}\sigma, \tag{2.4}$$

the elasticity tensor $\mathcal{A}$ can also be written as:

$$\mathcal{A}\mathbf{E} = \frac{2E\sigma}{1-\sigma^2}(\mathbf{A}^c \cdot \mathbf{E})\mathbf{A}^c + \frac{2E}{1+\sigma}\mathbf{A}^c\mathbf{E}\mathbf{A}^c, \quad \mathbf{E} \in \mathrm{Sym}\,(\mathbb{R}^2).$$

A Koiter shell can undergo stretching of the middle surface, and flexure (bending). Namely, the Koiter shell model accounts for both the membrane effects (stretching) and shell effects (flexure). Stretching of the middle surface is measured by the change of metric tensor, while flexure is measured by the change of curvature tensor. Consider an arbitrary displacement field $\boldsymbol{\eta} = (\eta_z, \eta_\theta, \eta_r)$ from the reference configuration Γ. Then, the corresponding change of metric, and change of curvature tensors for the deformed shell, in covariant components, are defined by:

$$\underbrace{\mathbf{G}(\boldsymbol{\eta}) = \frac{1}{2}\left(\mathbf{A}_c(\boldsymbol{\eta}) - \mathbf{A}_c\right)}_{\text{The Change of Metric Tensor}}, \text{ and } \underbrace{\mathbf{R}(\boldsymbol{\eta}) = \frac{1}{2}\left(\mathbf{B}_c(\boldsymbol{\eta}) - \mathbf{B}_c\right)}_{\text{The Change of Curvature Tensor}}, \tag{2.5}$$

where $\mathbf{A}_c(\boldsymbol{\eta})$ and $\mathbf{B}_c(\boldsymbol{\eta})$ are the covariant metric and curvature tensors, respectively, of the deformed shell. These will be specified below for the problem we consider in this chapter.

With the corresponding change of metric and change of curvature tensors we can now write formally the corresponding elastic energy of the deformed shell. The elastic energy of the cylindrical Koiter shell is given by [35, 36, 38, 98]:

$$E_{\mathrm{el}}(\boldsymbol{\eta}) = \frac{h}{4}\int_\omega \mathcal{A}\mathbf{G}(\boldsymbol{\eta}) : \mathbf{G}(\boldsymbol{\eta})\sqrt{a} + \frac{h^3}{48}\int_\omega \mathcal{A}\mathbf{R}(\boldsymbol{\eta}) : \mathbf{R}(\boldsymbol{\eta})\sqrt{a}, \tag{2.6}$$

where : denotes the scalar product

$$\mathbf{A} : \mathbf{B} := \mathrm{Tr}\left(\mathbf{A}\mathbf{B}^T\right) \; \mathbf{A}, \mathbf{B} \in \mathbf{M}_2(\mathbb{R}) \cong \mathbb{R}^4. \tag{2.7}$$

Given a force with surface force density $\mathbf{f}$, the loaded shell deforms and the corresponding displacement $\boldsymbol{\eta}$ of the deformed shell is a minimizer of the energy functional [35, 36, 38, 98]:

$$\mathbf{J}(\boldsymbol{\eta}) = \frac{h}{4}\int_\omega \mathcal{A}\mathbf{G}(\boldsymbol{\eta}) : \mathbf{G}(\boldsymbol{\eta})\sqrt{a} + \frac{h^3}{48}\int_\omega \mathcal{A}\mathbf{R}(\boldsymbol{\eta}) : \mathbf{R}(\boldsymbol{\eta})\sqrt{a} - \int_\omega \mathbf{f}\cdot\boldsymbol{\eta}\sqrt{a}. \tag{2.8}$$

The corresponding weak formulation can be written as:

$$\begin{aligned} &\frac{h}{2}\int_\omega \mathcal{A}\mathbf{G}(\boldsymbol{\eta}) : \mathbf{G}'(\boldsymbol{\eta})\boldsymbol{\psi}\sqrt{a} + \frac{h^3}{48}\int_\omega \mathcal{A}\mathbf{R}(\boldsymbol{\eta}) : \mathbf{R}'(\boldsymbol{\eta})\boldsymbol{\psi}\sqrt{a} \\ &\quad = \int_\omega \mathbf{f}\cdot\boldsymbol{\psi}\sqrt{a}, \; \forall \boldsymbol{\psi} \in C_c^\infty, \end{aligned} \tag{2.9}$$

where $\mathbf{G}'$ is the Gateux derivative of $\mathbf{G}$.

The weak formulation of the corresponding *elastodynamics problem* is given by the following:

$$\begin{aligned} &\rho_K h \int_\omega \partial_t^2 \boldsymbol{\eta}\boldsymbol{\psi}\sqrt{a} + \frac{h}{2}\int_\omega \mathcal{A}\mathbf{G}(\boldsymbol{\eta}) : \mathbf{G}'(\boldsymbol{\eta})\boldsymbol{\psi}\sqrt{a} + \frac{h^3}{48}\int_\omega \mathcal{A}\mathbf{R}(\boldsymbol{\eta}) : \mathbf{R}'(\boldsymbol{\eta})\boldsymbol{\psi}\sqrt{a} \\ &\quad = \int_\omega \mathbf{f}\cdot\boldsymbol{\psi}\sqrt{a}, \; \forall \boldsymbol{\psi} \in C_c^\infty, \end{aligned} \tag{2.10}$$

where ρ_K and h are the Koiter shell density and thickness.

Associated with this problem are the following physical quantities:

- *Stress Resultant (Internal Force)*, which relates the internal force with the change of metric tensor, and is defined by

$$N := \frac{h}{2}\mathcal{A}\mathbf{G}(\boldsymbol{\eta}), \tag{2.11}$$

and

- *Stress Couples (Bending Moment)*, which describe the bending moments in terms of the change of curvature tensor, and are defined by

$$\boldsymbol{M} := \frac{h^3}{24}\mathcal{A}\mathbf{R}(\boldsymbol{\eta}).$$

At this point we also introduce the effects of *prestress* by defining the stress resultant $\boldsymbol{N}_{\text{ref}}$ that relates the reference pressure p_{ref} with circumferential strain [46, 113, 114]

$$\frac{h}{2}\boldsymbol{N}_{\text{ref}} = hR\mathbf{A}^c \begin{bmatrix} 0 & 0 \\ 0 & p_{\text{ref}}\frac{R}{h}\eta_r \end{bmatrix} \mathbf{A}^c \tag{2.12}$$

so that the total stress resultant, including the effects of prestress, reads

- *Stress Resultant for a prestressed elastic Koiter shell*

$$\boldsymbol{N} = \frac{h}{2}\mathcal{A}\mathbf{G}(\boldsymbol{\eta}) + \frac{h}{2}\boldsymbol{N}_{\text{ref}}. \tag{2.13}$$

In what follows, we will be providing more specific details on a few concrete examples of the general framework described above.

Example 1: The Linearly Elastic Cylindrical Koiter Shell with Radial Displacement

We present the cylindrical Koiter shell equations *without the assumption of axial symmetry*. This means that the displacement $\boldsymbol{\eta}$ can be written as:

$$\boldsymbol{\eta}(t,z,\theta) = (\eta_z(t,z,\theta), \eta_\theta(t,z,\theta), \eta_r(t,z,\theta)).$$

However, as is common in the blood flow literature, we will be assuming that the azimuthal and longitudinal components of the displacement are negligible $\eta_\theta \approx 0, \eta_z \approx 0$, i.e., only the radial component of the displacement is different from zero, so that:

$$\boldsymbol{\eta}(t,z,\theta) = (0,0,\eta_r(t,z,\theta),) = \eta(t,z,\theta)\mathbf{e}_r(\theta),$$

where $\mathbf{e}_r(\theta)$ is the unit vector pointing in the radial direction. Notice that this does not mean that the flow is axially symmetric, since the radial displacement is a function of both θ and z.

In this case, the corresponding *linearized* change of metric, and change of curvature tensors (2.5) take the following form:

$$\boldsymbol{G}(\boldsymbol{\eta}) = \begin{bmatrix} 0 & 0 \\ 0 & R\eta_r \end{bmatrix}, \quad \boldsymbol{R}(\boldsymbol{\eta}) = \begin{bmatrix} -\partial_z^2\eta & -\partial_{z\theta}^2\eta \\ -\partial_{z\theta}^2\eta & -\partial_\theta^2\eta + \eta \end{bmatrix}. \tag{2.14}$$

The elastic energy of the shell is defined by:

$$E_{\mathrm{el}}(\eta) = \frac{h}{4}\int_\omega \mathcal{A}\boldsymbol{G}(\eta) : \boldsymbol{G}(\eta)\sqrt{a} + \frac{h^3}{48}\int_\omega \mathcal{A}\boldsymbol{R}(\eta) : \boldsymbol{R}(\eta)\sqrt{a}, \tag{2.15}$$

where η is the scalar displacement function. We will be assuming that the shell is clamped at the end points, satisfying the following boundary conditions:

$$\eta = \frac{\partial \eta}{\partial \mathbf{n}} = 0 \text{ on } \partial\omega.$$

The dynamics of the linearly elastic cylindrical Koiter shell is given by the following weak formulation: find $\eta \in H_0^2(\omega)$ such that

$$\begin{aligned} \rho_K h \int_\omega \partial_t^2 \eta \psi \sqrt{a} + \frac{h}{2}\int_\omega \mathcal{A}\boldsymbol{G}(\eta) : \boldsymbol{G}(\psi)\sqrt{a} + \frac{h^3}{24}\int_\omega \mathcal{A}\boldsymbol{R}(\eta) : \boldsymbol{R}(\psi)\sqrt{a} \\ = \int_\omega f\psi\sqrt{a} \end{aligned} \tag{2.16}$$

for all $\psi \in H_0^2(\omega)$, where f is the radial component of the surface force density applied to the shell. Here, we have used the fact that for *linear problems*:

$$\mathbf{G}'(\eta)\psi = \mathbf{G}(\psi).$$

We define the corresponding linear elasticity operator $\mathcal{L}_{\mathrm{el}}$:

$$\langle \mathcal{L}_{\mathrm{el}}\eta, \psi\rangle = \frac{h}{2}\int_\omega \mathcal{A}\boldsymbol{G}(\eta) : \boldsymbol{G}(\psi)\sqrt{a} + \frac{h^3}{24}\int_\omega \mathcal{A}\boldsymbol{R}(\eta) : \boldsymbol{R}(\psi)\sqrt{a}, \ \forall\psi \in H_0^2(\omega).$$

A calculation shows that the operator $\mathcal{L}_{\mathrm{el}}$ in differential form reads:

$$\begin{aligned} \mathcal{L}_{\mathrm{el}}\eta = & \frac{h^3\mu}{3R^4(\lambda+2\mu)}\Big((\lambda+\mu)\partial_\theta^4\eta + R^4(\lambda+\mu)\partial_z^4\eta + 2R^2(\lambda+\mu)\partial_z^2\partial_\theta^2\eta \\ & -R^2\lambda\partial_z^2\eta - 2(\lambda+\mu)\partial_\theta^2\eta + (\lambda+\mu)\eta\Big) + \frac{4h}{R^2}\frac{(\lambda+\mu)\mu}{\lambda+2\mu}\eta. \end{aligned}$$

By using the relationships between the Lamé constants and Young's modulus of elasticity E and Poisson ratio σ, given by (2.4), operator $\mathcal{L}_{\text{el}}$ can be written as:

$$\begin{aligned}\mathcal{L}_{\text{el}}\eta = &\frac{h^3 E}{12R^4(1-\sigma^2)}\left(\partial_\theta^4\eta + R^4\partial_z^4\eta + 2R^2\partial_z^2\partial_\theta^2\eta - 2\partial_\theta^2\eta + \eta\right)\\ &+\frac{h^3 E\sigma}{6R^2(1-\sigma^2)}\partial_z^2\eta + \frac{hE}{R^2(1-\sigma^2)}\eta.\end{aligned} \tag{2.17}$$

Example 2: The Axially Symmetric Koiter Shell Allowing Both Radial and Longitudinal Displacement

Here, we assume that nothing in the problem depends on θ. The problem is axially symmetric, and the displacement $\boldsymbol{\eta}$ is given by

$$\boldsymbol{\eta}(t,z) = (\eta_z(r,z), \eta_r(t,z)).$$

The linearized change of metric tensor and the linearized change of curvature tensor are given, respectively, by:

$$\boldsymbol{G}(\boldsymbol{\eta}) = \begin{bmatrix} \partial_z\eta_z & 0 \\ 0 & R\eta_r \end{bmatrix}, \quad \boldsymbol{R}(\boldsymbol{\eta}) = \begin{bmatrix} -\partial_{zz}\eta_r & 0 \\ 0 & \eta_r \end{bmatrix}. \tag{2.18}$$

The elastic energy of the problem is given by:

$$E_{\text{el}}(\boldsymbol{\eta}) = \frac{h}{2}\int_0^L \mathcal{A}\boldsymbol{G}(\boldsymbol{\eta}) : \boldsymbol{G}(\boldsymbol{\eta})Rdz + \frac{h^3}{24}\int_0^L \mathcal{A}\boldsymbol{R}(\boldsymbol{\eta}) : \boldsymbol{R}(\boldsymbol{\eta})Rdz. \tag{2.19}$$

To define a weak formulation of the problem, introduce the following function space:

$$\begin{aligned}V_c = H_0^1(0,L)\times H_0^2(0,L) = \{&(\psi_z,\psi_r)\in H^1(0,L)\times H^2(0,L) :\\ &\psi_z(0)=\psi_z(L)=\psi_r(0)=\psi_r(L)=0, \partial_z\psi_r(0)=\partial_z\psi_r(L)=0\}.\end{aligned}$$

Then the weak formulation of the linearly elastic cylindrical Koiter shell is given by the following: find $\boldsymbol{\eta} = (\eta_z,\eta_r)\in V_c$ such that

$$\begin{aligned}&\frac{h}{2}\int_0^L \mathcal{A}\boldsymbol{G}(\boldsymbol{\eta}) : \boldsymbol{G}(\boldsymbol{\psi})Rdz + \frac{h^3}{24}\int_0^L \mathcal{A}\boldsymbol{R}(\boldsymbol{\eta}) : \boldsymbol{R}(\boldsymbol{\psi})Rdz\\ &= \int_0^L \boldsymbol{f}\cdot\boldsymbol{\psi}Rdz, \quad \forall\boldsymbol{\psi}\in V_c,\end{aligned} \tag{2.20}$$

Here $\boldsymbol{f}$ is the surface density of the force applied to the shell, and $\mathcal{A}$ is the elasticity tensor given by (2.3).

The weak formulation of the associated *elastodynamics problem* is given by:

$$\rho_K h \int_0^L \partial_t^2 \boldsymbol{\eta} \boldsymbol{\psi} R dz + \frac{h}{2} \int_0^L \mathcal{A} \boldsymbol{G}(\boldsymbol{\eta}) : \boldsymbol{G}(\boldsymbol{\psi}) R dz + \frac{h^3}{24} \int_0^L \mathcal{A} \boldsymbol{R}(\boldsymbol{\eta}) : \boldsymbol{R}(\boldsymbol{\psi}) R dz$$

$$= \int_0^L \boldsymbol{f} \cdot \boldsymbol{\psi} R dz, \quad \forall \boldsymbol{\psi} \in V_c, \tag{2.21}$$

To write the weak form explicitly in terms of displacement, we introduce a simpler notation for the spatial derivative with respect to z, and for the time derivative. Namely, in this section we will be using $'$ to denote the partial derivative with respect to z, and $\dot{}$ to denote the partial derivative with respect to time. Namely, for an arbitrary function f:

$$f' := \frac{\partial f}{\partial z}, \quad \dot{f} := \frac{\partial f}{\partial t}.$$

Using this notation, the weak formulation written explicitly in terms of the displacement now reads:

$$\rho_K h \int_0^L \ddot{\eta}_z \psi_z + \ddot{\eta}_r \psi_r$$

$$+ \frac{h}{2} \int_0^L \left(\frac{4\mu\lambda}{\lambda + 2\mu} \left(\eta_z' + \frac{1}{R} \eta_r \right) \cdot \left(\xi_z' + \frac{1}{R} \xi_r \right) + 4\mu \left(\eta_z' \xi_z' + \frac{1}{R^2} \eta_r \xi_r \right) \right) dz$$

$$+ \frac{h^3}{24} \int_0^L \left(\frac{4\mu\lambda}{\lambda + 2\mu} \left(-\eta_r'' + \frac{1}{R^2} \eta_r \right) \cdot \left(-\xi_r'' + \frac{1}{R^2} \xi_r \right) + 4\mu \left(\eta_r'' \xi_r'' + \frac{1}{R^4} \eta_r \xi_r \right) \right) dz$$

$$= \int_0^L (f_z \xi_z + f_r \xi_r) dz, \quad \forall (\xi_z, \xi_r) \in V_c.$$

By using the relationships between λ, μ and E, σ, given by (2.4), the weak formulation in terms of E and σ reads:

$$\rho_K h \int_0^L \ddot{\eta}_z \psi_z + \ddot{\eta}_r \psi_r$$

$$+ h \int_0^L \left(\frac{E\sigma}{1 - \sigma^2} \left(\eta_z' + \frac{1}{R} \eta_r \right) \left(\xi_z' + \frac{1}{R} \xi_r \right) + \frac{E}{1 + \sigma} \left(\eta_z' \xi_z' + \frac{1}{R^2} \eta_r \xi_r \right) \right) dz \tag{2.22}$$

$$+ \frac{h^3}{12} \int_0^L \left(\frac{E\sigma}{1 - \sigma^2} \left(-\eta_r'' + \frac{1}{R^2} \eta_r \right) \left(-\xi_r'' + \frac{1}{R^2} \xi_r \right) + \frac{E}{1 + \sigma} \left(\eta_r'' \xi_r'' + \frac{1}{R^4} \eta_r \xi_r \right) \right) dz$$

$$= \int_0^L (f_z \xi_z + f_r \xi_r) dz, \quad (\xi_z, \xi_r) \in V_c. \tag{2.23}$$

The terms multiplying $h/2$ account for the stored energy density due to stretching (membrane effects) and the terms multiplying $h^3/12$ account for the stored energy density due to bending (flexural shell effects). Integration by parts gives rise to the following dynamics equilibrium equations in differential form:

LINEARLY ELASTIC, AXIALLY SYMMETRIC CYLINDRICAL KOITER SHELL

$$\rho_K h\, \ddot{\eta}_z - \frac{hE}{1-\sigma^2}\left(\eta_z'' + \sigma\frac{1}{R}\eta_r'\right) = f_z,$$

$$\rho_K h\, \ddot{\eta}_r + \frac{hE}{R(1-\sigma^2)}\left(\sigma\eta_z' + \frac{\eta_r}{R}\right) + \frac{h^3 E}{12(1-\sigma^2)}\left(\eta_r'''' - 2\sigma\frac{1}{R^2}\eta_r' + \frac{1}{R^4}\eta_r\right) = f_r. \quad (2.24)$$

By ignoring the terms accounting for the bending energy (shell effects), the resulting equations representing a model for the linearly elastic, axially symmetric cylindrical Koiter membrane take the following form:

LINEARLY ELASTIC, AXIALLY SYMMETRIC CYLINDRICAL KOITER MEMBRANE

$$\begin{aligned} \rho_K h\, \ddot{\eta}_z - \frac{hE}{1-\sigma^2}\left(\eta_z' + \sigma\frac{1}{R}\eta_r'\right) &= f_z, \\ \rho_K h\, \ddot{\eta}_r + \frac{hE}{R(1-\sigma^2)}\left(\sigma\eta_z' + \frac{\eta_r}{R}\right) &= f_r. \end{aligned} \quad (2.25)$$

Example 3: A Nonlinearly Elastic, Axially Symmetric Koiter Membrane with Only Radial Displacement

As in the previous example, we assume that nothing in the problem depends on θ. Also, for simplicity, we will be assuming that only the radial component of the displacement is different from zero, so that

$$\boldsymbol{\eta} = \boldsymbol{\eta}(t,z) = \eta \mathbf{e}_r.$$

Therefore, we consider axially symmetric deformations of a nonlinearly elastic Koiter membrane from the reference configuration Γ given by (2.1), with only the radial component of displacement different from zero. See Fig. 2.4. The corresponding change of metric tensor capturing membrane effects is given by

$$\mathbf{G}(\eta) = \frac{1}{2}\begin{pmatrix} (\partial_z \eta)^2 & 0 \\ 0 & 2R\eta + \eta^2 \end{pmatrix}. \quad (2.26)$$

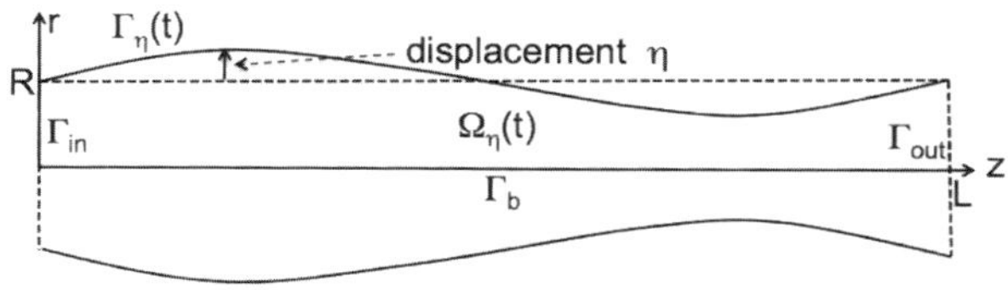

Fig. 2.4 A sketch of an axially symmetric fluid domain with radial displacement η

The elastic energy of the Koiter membrane is given by the following:

$$E_{\rm el}^{\rm mem}(\eta) = \frac{h}{2}\int_0^L \mathcal{A}\mathbf{G}(\eta)\cdot\mathbf{G}(\eta)\ R dz \tag{2.27}$$

We consider the dynamics of the Koiter membrane with fixed end points, modeled by the boundary conditions

$$\eta(0) = \eta(L) = 0.$$

Following (2.10), the variational formulation for the nonlinearly elastic Koiter membrane problem is given by the following:

$$\int_0^L \rho_K h \partial_t^2 \eta \xi\ Rdz + \frac{h}{2}\int_0^L \mathcal{A}\mathbf{G}(\eta)\cdot\mathbf{G}'(\eta)\xi\ Rdz = \int_0^L f\xi\ Rdz,\ \forall \xi \in H_0^2(0,L), \tag{2.28}$$

where $\mathbf{G}'$ is Gateux derivative of $\mathbf{G}$ given by:

$$\mathbf{G}'(\eta)\xi = \begin{pmatrix} \partial_z\eta\partial_z\xi & 0 \\ 0 & (R+\eta)\xi \end{pmatrix}.$$

This defines the following (nonlinear) differential operator $\mathcal{L}_{\rm el}^{\rm mem}$:

$$\langle \mathcal{L}_{\rm el}^{\rm mem}(\eta), \xi\rangle := \frac{h}{2}\int_0^L \mathcal{A}\mathbf{G}(\eta)\cdot\mathbf{G}'(\eta)\xi\ Rdz, \quad \forall \xi \in C_c^\infty(0,L).$$

Integration by parts yields the following formula:

$$\begin{aligned}\mathcal{L}_{\rm el}^{\rm mem}(\eta) &= -\partial_z\Big[\Big(\frac{hE}{2(1-\nu^2)}(\partial_z\eta)^2 + \frac{hE\nu}{1-\nu^2}\big(\frac{1}{R}\eta + \frac{1}{2R^2}\eta^2\big)\Big)\partial_z\eta\Big] \\ &+\Big(\frac{hE}{1-\nu^2}\big(\frac{1}{R}\eta + \frac{1}{2R^2}\eta^2\big) + \frac{hE\nu}{2(1-\nu^2)}(\partial_z\eta)^2\Big)\big(\frac{1}{R} + \frac{1}{R^2}\eta\big), \quad \eta \in W_0^{2,4}(0,L).\end{aligned} \tag{2.29}$$

With this notation, the corresponding differential formulation of (2.28) can be written as:

$$\rho_K h\ \partial_t^2\eta + \mathcal{L}_{\rm el}^{\rm mem}(\eta) = f. \tag{2.30}$$

Here, ρ_K is the structure density, h is the structure thickness, and f is the force density in the radial (vertical) $\mathbf{e}_r$ direction acting on the structure.

Example 4: A Linearly Viscoelastic Koiter Shell

We introduce the viscoelastic effects to the linearly elastic Koiter shell by considering viscoelasticity of Kelvin–Voigt type in which the stress is linearly proportional to strain plus the time derivative of strain. For this purpose we introduce the following equivalent of the elasticity tensor $\mathcal{A}$ given by (2.3), which we denote by $\mathcal{B}$:

$$\mathcal{B}\mathbf{E} = \frac{4\lambda_v\mu_v}{\lambda_v + 2\mu_v}(\mathbf{A}^c \cdot \mathbf{E})\mathbf{A}^c + 4\mu_v\mathbf{A}^c\mathbf{E}\mathbf{A}^c, \quad \mathbf{E} \in \mathrm{Sym}(\mathcal{M}_2), \tag{2.31}$$

where λ_v and μ_v are the viscoelastic counterparts of the Lamé constants of elasticity. Here $\mathbf{A}^c$ is the contra variant metric tensor of the reference configuration Γ, given in (2.2).

Given the force density $\mathbf{f}$, the displacement of the deformed *linearly viscoelastic* Koiter shell can be found by solving the following variational formulation for $\boldsymbol{\eta}$:

$$\begin{aligned}
&\frac{h}{2}\int_\omega (\mathcal{A}\mathbf{G}(\boldsymbol{\eta}) + \mathcal{B}\mathbf{G}(\dot{\boldsymbol{\eta}})) : \mathbf{G}(\boldsymbol{\psi})\sqrt{a} + \frac{h^3}{48}\int_\omega (\mathcal{A}\mathbf{R}(\boldsymbol{\eta}) + \mathcal{B}\mathbf{R}(\dot{\boldsymbol{\eta}})) : \mathbf{R}(\boldsymbol{\psi})\sqrt{a} \\
&= \int_\omega \mathbf{f}\cdot\boldsymbol{\psi}\sqrt{a}, \ \forall\boldsymbol{\psi} \in C_c^\infty,
\end{aligned} \tag{2.32}$$

where ρ_K and h are the Koiter shell density and thickness, respectively.

The energy of this problem is given by:

$$\begin{aligned}
E(\boldsymbol{\eta}) &= \frac{h}{2}\int_\omega \mathcal{A}\mathbf{G}(\boldsymbol{\eta}) : \mathbf{G}(\boldsymbol{\eta})\sqrt{a} + \frac{h^3}{48}\int_\omega \mathcal{A}\mathbf{R}(\boldsymbol{\eta}) : \mathbf{R}(\boldsymbol{\eta})\sqrt{a} \\
&\quad + \frac{h}{4}\frac{d}{dt}\int_\omega \mathcal{B}\mathbf{G}(\boldsymbol{\eta}) : \mathbf{G}(\boldsymbol{\eta})\sqrt{a} + \frac{h^3}{96}\frac{d}{dt}\int_\omega \mathcal{B}\mathbf{R}(\boldsymbol{\eta}) : \mathbf{R}(\boldsymbol{\eta})\sqrt{a} \\
&= E_{\mathrm{el}}(\boldsymbol{\eta}) + \frac{1}{2}\frac{d}{dt}E_{\mathrm{vis}}(\boldsymbol{\eta}),
\end{aligned} \tag{2.33}$$

where

$$E_{\mathrm{el}}(\boldsymbol{\eta}) = \frac{h}{2}\int_\omega \mathcal{A}\mathbf{G}(\boldsymbol{\eta}) : \mathbf{G}(\boldsymbol{\eta})\sqrt{a} + \frac{h^3}{48}\int_\omega \mathcal{A}\mathbf{R}(\boldsymbol{\eta}) : \mathbf{R}(\boldsymbol{\eta})\sqrt{a} \tag{2.34}$$

$$E_{\mathrm{vis}}(\boldsymbol{\eta}) = \frac{h}{2}\int_\omega \mathcal{B}\mathbf{G}(\boldsymbol{\eta}) : \mathbf{G}(\boldsymbol{\eta})\sqrt{a} + \frac{h^3}{48}\int_\omega \mathcal{B}\mathbf{R}(\boldsymbol{\eta}) : \mathbf{R}(\boldsymbol{\eta})\sqrt{a} \tag{2.35}$$

The corresponding *elastodynamics problem* is given by:

$$\rho_K h \int_\omega \partial_t^2 \boldsymbol{\eta} \boldsymbol{\psi} \sqrt{a} + \frac{h}{2} \int_\omega (\mathcal{A}\mathbf{G}(\boldsymbol{\eta}) + \mathcal{B}\mathbf{G}(\dot{\boldsymbol{\eta}})) : \mathbf{G}(\boldsymbol{\psi}) \sqrt{a}$$
$$+ \frac{h^3}{48} \int_\omega (\mathcal{A}\mathbf{R}(\boldsymbol{\eta}) + \mathcal{B}\mathbf{R}(\dot{\boldsymbol{\eta}})) : \mathbf{R}(\boldsymbol{\psi}) \sqrt{a} = \int_\omega \mathbf{f} \cdot \boldsymbol{\psi} \sqrt{a}, \ \forall \boldsymbol{\psi} \in C_c^\infty, \quad (2.36)$$

Introduce the following notation for the corresponding elastic and viscoelastic operators:

$$\langle \mathcal{L}_{\mathrm{el}} \boldsymbol{\eta}, \boldsymbol{\psi} \rangle := \frac{h}{2} \int_\omega \mathcal{A}\mathbf{G}(\boldsymbol{\eta}) : \mathbf{G}(\boldsymbol{\psi}) \sqrt{a} + \frac{h^3}{48} \int_\omega \mathcal{A}\mathbf{R}(\boldsymbol{\eta}) : \mathbf{R}(\boldsymbol{\psi}) \sqrt{a}. \quad (2.37)$$

$$\langle \mathcal{L}_{\mathrm{vis}} \dot{\boldsymbol{\eta}}, \boldsymbol{\psi} \rangle := \frac{h}{2} \int_\omega \mathcal{B}\mathbf{G}(\dot{\boldsymbol{\eta}}) : \mathbf{G}(\boldsymbol{\psi}) \sqrt{a} + \frac{h^3}{48} \int_\omega \mathcal{B}\mathbf{R}(\dot{\boldsymbol{\eta}}) : \mathbf{R}(\boldsymbol{\psi}) \sqrt{a}. \quad (2.38)$$

Then, we can write (2.36) as

$$\rho_K h \int_\omega \partial_t^2 \boldsymbol{\eta} \boldsymbol{\psi} \sqrt{a} + \langle \mathcal{L}_{\mathrm{el}} \boldsymbol{\eta}, \boldsymbol{\psi} \rangle + \langle \mathcal{L}_{\mathrm{vis}} \dot{\boldsymbol{\eta}}, \boldsymbol{\psi} \rangle = \int_\omega \mathbf{f} \cdot \boldsymbol{\psi} \sqrt{a}, \ \forall \boldsymbol{\psi} \in C_c^\infty.$$

We now write the explicit form of these equations for the case when the structure displacement is independent of θ so that:

$$\boldsymbol{\eta}(t,z) = (\eta_z(t,z), \eta_r(t,z)),$$

and with the boundary conditions corresponding to a clamped shell

$$\eta(0) = \partial_z \eta(0) = \eta(L) = \partial_z \eta(L) = 0.$$

Therefore, to simplify the form of the explicit equations, we assume axial symmetry of the problem. In this case, we look for a weak solution which is in the space

$$V_c = H_0^1(0,L) \times H_0^2(0,L) = \{(\psi_z, \psi_r) \in H^1(0,L) \times H^2(0,L) :$$
$$\psi_z(0) = \psi_z(L) = \psi_r(0) = \psi_r(L) = 0, \partial_z \psi_r(0) = \partial_z \psi_r(L) = 0\}.$$

The corresponding weak formulation is given by (2.36), where we can replace the test space C_c^∞ by the space V_c. We write the differential form of the elastodynamics equations in terms of the Young's modulus of elasticity E and Poisson ratio σ, and the corresponding viscoelastic equivalents, which we denote be E_v and σ_v. The relationship between λ, μ and E, σ is given by (2.4). The same relationship holds between the corresponding viscoelastic constants λ_v, μ_v and E_v, σ_v.

After writing out the weak form (2.36), and after performing integration by parts, the corresponding dynamic equilibrium equations for the linearly viscoelastic Koiter shell in differential form are given by:

$$\rho_K h \frac{\partial^2 \eta_z}{\partial t^2} - C_2 \frac{\partial \eta_r}{\partial z} - C_3 \frac{\partial^2 \eta_z}{\partial z^2} - D_2 \frac{\partial^2 \eta_r}{\partial t \partial z} - D_3 \frac{\partial^3 \eta_z}{\partial t \partial z^2} = f_z, \tag{2.39}$$

$$\rho_K h \frac{\partial^2 \eta_r}{\partial t^2} + C_0 \eta_r - C_1 \frac{\partial^2 \eta_r}{\partial z^2} + C_2 \frac{\partial \eta_z}{\partial z} + C_4 \frac{\partial^4 \eta_r}{\partial z^4} + D_0 \frac{\partial \eta_r}{\partial t} - D_1 \frac{\partial^3 \eta_r}{\partial t \partial z^2}$$
$$+ D_2 \frac{\partial^2 \eta_z}{\partial t \partial z} + D_4 \frac{\partial^5 \eta_r}{\partial t \partial z^4} = f_r, \tag{2.40}$$

where

$$\begin{array}{lll} C_0 = \frac{hE}{R^2(1-\sigma^2)}(1 + \frac{h^2}{12R^2}), & C_1 = \frac{h^3}{6}\frac{E\sigma}{R^2(1-\sigma^2)}, & C_2 = \frac{h}{R}\frac{E\sigma}{1-\sigma^2}, \\ C_3 = \frac{hE}{1-\sigma^2}, & C_4 = \frac{h^3}{12}\frac{E}{1-\sigma^2}, & \\ D_0 = \frac{h}{R^2}C_v(1 + \frac{h^2}{12R^2}), & D_1 = \frac{h^3}{6}\frac{D_v}{R^2}, & D_2 = \frac{hD_v}{R}, \\ D_3 = hC_v, & D_4 = \frac{h^3}{12}C_v, & \end{array} \tag{2.41}$$

and

$$C_v := \frac{E_v}{1-\sigma_v^2}, \quad D_v := \frac{E_v \sigma_v}{1-\sigma_v^2}.$$

We can write this problem using the operators $\mathcal{L}_{el}$ and $\mathcal{L}_{vis}$ as:

$$\rho_K h \frac{\partial^2 \boldsymbol{\eta}}{\partial t^2} + \mathcal{L}_{el}\boldsymbol{\eta} + \mathcal{L}_{vis}\frac{\partial \boldsymbol{\eta}}{\partial t} = \mathbf{f}, \tag{2.42}$$

where

$$\mathcal{L}_{el}\boldsymbol{\eta} = \begin{pmatrix} -C_2 \frac{\partial \eta_r}{\partial z} - C_3 \frac{\partial^2 \eta_z}{\partial z^2} \\ C_0 \eta_r - C_1 \frac{\partial^2 \eta_r}{\partial z^2} + C_2 \frac{\partial \eta_z}{\partial z} + C_4 \frac{\partial^4 \eta_r}{\partial z^4} \end{pmatrix}, \tag{2.43}$$

and

$$\mathcal{L}_{vis}\frac{\partial \boldsymbol{\eta}}{\partial t} = \begin{pmatrix} -D_2 \frac{\partial^2 \eta_r}{\partial t \partial z} - D_3 \frac{\partial^3 \eta_z}{\partial t \partial z^2} \\ D_0 \frac{\partial \eta_r}{\partial t} - D_1 \frac{\partial^3 \eta_r}{\partial t \partial z^2} + D_2 \frac{\partial^2 \eta_z}{\partial t \partial z} + D_4 \frac{\partial^5 \eta_r}{\partial t \partial z^4} \end{pmatrix}. \tag{2.44}$$

The typical values of the model parameters for the aorta and iliac arteries are given in Table 2.2

Table 2.2 Table with typical wall parameter values for the aorta and iliac arteries

Parameters	Aorta/iliacs
Char. radius R (m)	0.006–0.012 [133]
Wall thickness h (m)	$1–2 \times 10^{-3}$[133]
Wall density ρ_K (kg/m^3)	1.1×10^3[133]
Young's modulus E (Pa)	$10^5–10^6$ [3, 14, 133]
Wall viscosity coef. hC_v/R (Pa s)	$10^3–8 \times 10^3$[3, 4, 14]
Poisson's ratio σ	0.5

Table 2.3 Structure parameters for Example 5

Parameters	Values for model problem
Shear mod. G (dynes/cm^2)	0.25×10^6
Timoshenko factor k	1
Viscoelasticity γ (poise cm)	0.01
Radius R (cm)	0.5
Wall density ρ_s (g/cm^3)	1.1
Wall thickness h_s (cm)	0.1
Young's mod. E(dynes/cm^2)	0.75×10^6
Poisson's ratio σ	0.5

Example 5: The Linearly Elastic String Model

We present here a model which has been used by several authors to test numerical solvers for FSI in blood flow [8, 9, 21, 85, 125, 131]. This model problem was first introduced by Formaggia et al. in [67]. The structure model for this benchmark problem is of the form

$$\rho_s h \frac{\partial^2 \eta_r}{\partial t^2} - kGh \frac{\partial^2 \eta_r}{\partial z^2} + \frac{Eh}{1-\sigma^2} \frac{\eta_r}{R^2} - \gamma \frac{\partial^3 \eta_r}{\partial z^2 \partial t} = f. \tag{2.45}$$

Here $G = \frac{E}{2(1+\sigma)}$ is the *shear modulus* and k is the *Timoshenko shear correction factor*. The values of the model parameters used in [67] are given in Table 2.3. Notice that this model can be recovered from the linearly viscoelastic Koiter shell model (2.42) by taking the longitudinal component of displacement to be equal to zero, and by choosing the following values for the coefficients in (2.43), (2.44):

$$C_0 = \frac{Eh}{R^2(1-\sigma^2)}, \; C_1 = -kGh, \; D_2 = -\gamma,$$

with all the other coefficients equal to zero. The typical values of the parameters in this model are given in Table 2.3 [67]. The Young's modulus E and viscoelasticity γ are smaller than the physiological values. This means that the arterial wall in this example is rather elastic. The relatively large value of the coefficient in front of the second-order derivative with respect to z (describing bending rigidity) minimizes the oscillations that would normally appear in such structures. For the typical physiological values of these parameters see Table 2.2.

Example 6: The Independent Ring Model

The independent ring model has been extensively used in modeling elastic properties of arterial walls. See, e.g., [26, 133], and the references therein. The model is particularly suitable to study blood flow in compliant arteries using a reduced, 1D model, studied in, e.g., [26].

The independent ring model reads

$$p - p_{\mathrm{ref}} = \frac{hE}{R^2(1-\sigma^2)}\eta, \tag{2.46}$$

where p_{ref} is the reference pressure, i.e., the pressure at which the displacement from the reference configuration is equal to zero.

Notice that this model is included in the Koiter shell equations (2.39), (2.40). Indeed, if we ignore the longitudinal displacement and take only the terms that follow from the *membrane effects* ($\frac{h}{2}\int \mathcal{A}\mathbf{G}(\eta) : \mathbf{G}(\eta)$) we obtain exactly the Independent Ring Model:

$$f_r = C_0\eta_r = \frac{hE}{R^2(1-\sigma^2)}\eta. \tag{2.47}$$

We conclude this example by proposing a *Nonlinear Independent Ring model* consistent with the Koiter membrane theory. To obtain this model, consider the nonlinearly elastic Koiter membrane model (2.29), (2.30), which assumes axial symmetry, and only the radial component of displacement to be different from zero. By assuming, additionally, that the gradient of the radial displacement $\partial_z\eta$ is negligible, one obtains the following Nonlinear Independent Ring Model:

$$p - p_{\mathrm{ref}} = \frac{hE}{(1-\sigma^2)R}\left(\frac{\eta}{R} + \frac{3}{2}\frac{\eta^2}{R^2} + \frac{1}{2}\frac{\eta^3}{R^3}\right). \tag{2.48}$$

For the parameter values given in Table 2.2, we calculated the pressure–displacement relationship for this model, which is depicted in Fig. 2.5. This figure also shows the pressure–displacement relationship for the Linear Independent Ring model.

We conclude this section by a remark on the nonlinearly elastic independent ring model of the form

$$p - p_{\mathrm{ref}} = \frac{hE}{R(1-\sigma^2)}\left(\left(\frac{R+\eta}{R}\right)^{\beta} - 1\right), \tag{2.49}$$

which was used by certain authors to model the nonlinearly elastic properties of arterial walls. For $\beta > 1$, this model is *not* consistent with the linearly elastic Independent Ring model, since its linearization does not coincide with the linearly elastic Independent Ring model. Figure 2.6 shows the plot of the

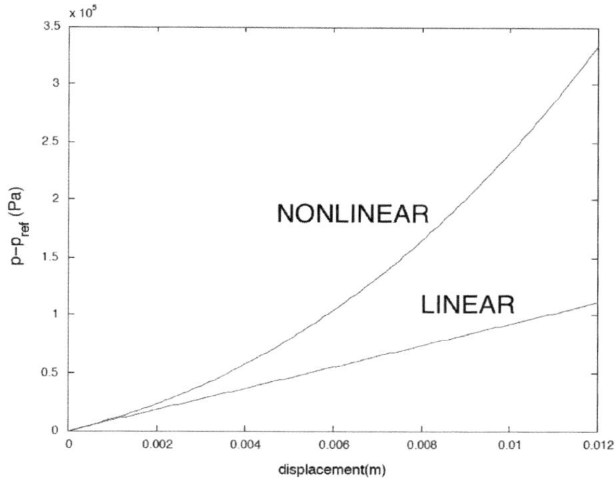

Fig. 2.5 The pressure–displacement relationship for the nonlinear Independent Ring Model (2.48), and the linear Independent Ring Model (2.46)

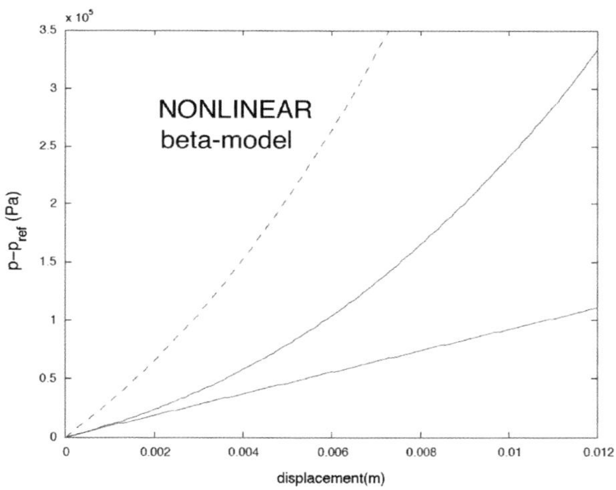

Fig. 2.6 The pressure–displacement relationship for the nonlinear Independent Ring Model (2.49), shown in *dashed line*, superimposed over the plots of the linearly elastic and nonlinearly elastic Independent Ring models (2.46) and (2.48). Notice how the slope at zero for the nonlinear Independent beta-model (2.49) does not coincide with that of (2.46) and (2.48)

pressure–displacement relationship given by (2.49) with $\beta = 3$, for the same values of the parameter, given in Table 2.2 as the plot shown in Fig. 2.5. Notice how the slope of the pressure–displacement curve for model (2.49), evaluated at $\eta = 0$, differs from the slope of the linearly elastic and nonlinearly elastic Independent Ring models given by (2.46) and (2.48). This means, in particular, that the leading-order coefficient modeling the stiffness of arterial walls for small displacements is different for the β-model (2.49), and cannot be approximated for small displacements by the physically reasonable one stated in the linearly elastic Independent Ring model.

Example 7: A Koiter Shell Model with Prestress

We follow the general description provided in (2.13) and calculate the differential form of the linearly elastic Koiter shell. The only difference with the examples presented above is in the coefficient multiplying the non-differentiated term, which will now have an extra term p_{ref}/R. Therefore, in Example 1, the linear operator $\mathcal{L}_{\text{el}}$ given by Eq. (2.17) now becomes

$$\begin{aligned}\mathcal{L}_{\text{el}}\eta = &\frac{h^3 E}{12R^4(1-\sigma^2)}\Big(\partial_\theta^4\eta + R^4\partial_z^4\eta + 2R^2\partial_z^2\partial_\theta^2\eta - 2\partial_\theta^2\eta + \eta\Big)\\ &+\frac{h^3 E\sigma}{6R^2(1-\sigma^2)}\partial_z^2\eta + \left(\frac{hE}{R^2(1-\sigma^2)} + \frac{p_{\text{ref}}}{R}\right)\eta.\end{aligned}$$

In Example 2, this gives rise to the following linearly elastic Koiter membrane equations with prestress:

$$\begin{aligned}\rho_K h\,\ddot{\eta}_z - \frac{hE}{1-\sigma^2}\left(\eta_z' + \sigma\frac{1}{R}\eta_r'\right) &= f_z,\\ \rho_K h\,\ddot{\eta}_z + \frac{hE\sigma}{R(1-\sigma^2)}\eta_z' + \left(\frac{hE}{R(1-\sigma^2)} + p_{\text{ref}}\right)\frac{\eta_r}{R} &= f_r.\end{aligned} \tag{2.50}$$

In Example 4, the prestress changes the constant C_0 in (2.41), which now becomes

$$C_0 = \frac{hE}{R^2(1-\sigma^2)}(1 + \frac{h^2}{12R^2}) + \frac{p_{\text{ref}}}{R}.$$

2.2.2 *Elastodynamics of Structures with Finite Thickness ("Thick Structures")*

The equations modeling elastodynamics of a structure are typically given in terms of the displacement vector field $\boldsymbol{d} = \boldsymbol{d}(t, \boldsymbol{x})$. Vector field $\boldsymbol{d}$ denotes the displacement from a given reference configuration Ω_S. We will be assuming that the reference configuration of the thick structure is given by a straight cylinder of radius R, length L and thickness H. See Fig. 2.2. The elastodynamics equations describe the second Newton's law of motion

$$\rho_s\,\partial_{tt}\mathbf{d} = \nabla\cdot\mathbf{S} \quad \text{in } \Omega_S,\ t\in(0,T), \tag{2.51}$$

where ρ_s denotes density of the thick structure, and $\mathbf{S}$ is the *first Piola–Kirchhoff stress tensor.*

To close the system, we need to specify the dependence of $\mathbf{S}$ on $\boldsymbol{d}$. The relationship between $\mathbf{S}$ and $\boldsymbol{d}$ depends on the material under consideration. In this chapter we will be assuming that our thick elastic structure is

- *homogeneous*, i.e., the material properties do not depend on $\boldsymbol{x}$, and
- *isotropic*, i.e., the response of the material deformation is the same in all directions.

Additionally, we will be assuming that

- *the displacement gradient* is small (i.e., $\nabla \boldsymbol{d} \ll 1$).

Under these assumptions, one of the simplest constitutive models for the mechanical behavior of linearly elastic structures, called the *linearized Saint Venant–Kirchhoff model*, takes the following form:

$$\mathbf{S} = \mu\,(\nabla \mathbf{d} + (\nabla \mathbf{d})^T) + \lambda(\nabla \cdot \mathbf{d})\mathbf{I}, \tag{2.52}$$

Here, λ and μ are the Lamé constants, accounting the compression and distortion of the structure, respectively.

Writing a constitutive model for the behavior or elastic structures in general is a bit more involving. Arterial walls are, in fact, nonlinear. The linear approximation written above is good as long as the displacement gradient and displacement are not too large, which in the blood flow application means displacement not larger than roughly 5 % of the reference radius of an artery. A typical displacement in a healthy artery under normal physiological conditions is between 5 and 10 %. Thus, many physiological and pathophysiological situations can exceed the linearly elastic regime. Depending on what types of questions is one trying to answer, linear or nonlinear models may be appropriate.

A typical assumption in biomedical literature on soft tissue mechanics is that arterial walls behave as a *hyperelastic material*. This means that the relationship between stress and strain in the structure can be written as the derivative of the energy density function with respect to strain. More precisely, if we denote by

- $\mathbf{\Pi}$—*the second Piola–Kirchhoff stress tensor*,
- $\boldsymbol{E}$—*the Green–Lagrange strain tensor*, and
- W—*the energy density function*,

then, for a *hyperelastic material*

$$\mathbf{\Pi}(\boldsymbol{E}) = \frac{\partial W}{\partial \boldsymbol{E}}(\boldsymbol{E}).$$

What is the relationship between the first and second Piola–Kirchhoff stress tensors $\mathbf{S}$ and Π, and between the Green–Lagrange strain tensor $\boldsymbol{E}$ and displacement $\boldsymbol{d}$? To explain these relationships we need to recall the notion of *deformation*. For each point $\boldsymbol{x} \in \Omega_S$ belonging to an undeformed, reference configuration Ω_S, *deformation* is a mapping $\boldsymbol{\varphi}$ which to each point $\boldsymbol{x} \in \Omega_S$ associates a point $\boldsymbol{\varphi}(\boldsymbol{x}) = \boldsymbol{x} + \boldsymbol{d}(\boldsymbol{x})$, where $\boldsymbol{d}$ denotes the displacement of $\boldsymbol{x}$. Deformation gradient will be denoted by $\boldsymbol{F} = \nabla\boldsymbol{\varphi}$. Namely,

$$\boldsymbol{F} = \nabla\boldsymbol{\varphi} = \frac{\partial \boldsymbol{\varphi}_i}{\partial \boldsymbol{x}_j} = \boldsymbol{I} + \nabla \boldsymbol{d} = \boldsymbol{I} + \frac{\partial \boldsymbol{d}_i}{\partial \boldsymbol{x}_j}. \tag{2.53}$$

$\boldsymbol{F}$ plays a key role in specifying the relationship between the first and second Piola–Kirchhoff stress tensors, and in the relationship between strain and displacement. The first and second Piola–Kirchhoff stress tensors are related through the gradient of deformation as follows:

$$\boldsymbol{S} = \boldsymbol{F}\boldsymbol{\Pi}. \tag{2.54}$$

While the first Piola–Kirchhoff stress tensor is not generally symmetric, the second Piola–Kirchoff stress tensor is, and is, therefore, more suited for the description of physical properties of materials in terms of constitutive relations.

Constitutive relations, which specify the material properties of a structure, typically express a relationship between stress and strain, more precisely, between the second Piola–Kirchhoff stress tensor $\boldsymbol{\Pi}$ and the Green–Lagrange strain tensor $\boldsymbol{E}$:

$$\boldsymbol{\Pi} = \boldsymbol{\Pi}(\boldsymbol{E}),$$

where the Green–Lagrange strain tensor is defined via deformation gradient as

$$\boldsymbol{E} := \frac{1}{2}\left(\boldsymbol{F}^T\boldsymbol{F} - I\right). \tag{2.55}$$

A calculation shows that in terms of the displacement gradient, $\boldsymbol{E}$ is given by:

$$\boldsymbol{E} := \frac{1}{2}\left(\nabla\boldsymbol{d} + \nabla\boldsymbol{d}^T + \nabla\boldsymbol{d}\nabla\boldsymbol{d}^T\right). \tag{2.56}$$

Therefore, a general relationship between strain and displacement gradient is quadratic. For small displacement gradients, the quadratic term can be neglected, and the relationship becomes linear:

$$\boldsymbol{E} \approx \boldsymbol{\varepsilon} := \frac{1}{2}\left(\nabla\boldsymbol{d} + \nabla\boldsymbol{d}^T\right) = \boldsymbol{D}(\boldsymbol{d}), \tag{2.57}$$

where $\boldsymbol{D}$ is known as the symmetrized gradient of displacement.

Therefore, in summary, the *elastodynamics* of elastic structures is described by the second Newton's law of motion

$$\boxed{\rho_s\, \partial_{tt}\mathbf{d} = \nabla\cdot\mathbf{S}} \quad \text{in } \Omega_S \times (0,T),$$

where

- $\boldsymbol{S} = \boldsymbol{F}\boldsymbol{\Pi}$ is the first Piola–Kirchhoff stress tensor,
- $\boldsymbol{\Pi}$ is the second Piola–Kirchhoff stress tensor,

- $\boldsymbol{F} = \nabla\boldsymbol{\varphi} = I + \nabla\boldsymbol{d}$ is the deformation gradient,
- $\boldsymbol{\varphi}(\boldsymbol{x}) = \boldsymbol{x} + \boldsymbol{d}(\boldsymbol{x}), \boldsymbol{x} \in \Omega_S$ is deformation of Ω_S, and
- $\boldsymbol{d}$ is displacement from the reference configuration.

To close the system, a *constitutive relation* needs to be specified:

$$\boxed{\boldsymbol{\Pi} = \boldsymbol{\Pi}(\boldsymbol{E})},$$

where

- $\boldsymbol{E} = \frac{1}{2}\left(\boldsymbol{F}^T\boldsymbol{F} - I\right)$ is the Green–Lagrange strain tensor, also expressed as
- $\boldsymbol{E} = \frac{1}{2}\left(\nabla\boldsymbol{d} + \nabla\boldsymbol{d}^T + \nabla\boldsymbol{d}\nabla\boldsymbol{d}^T\right)$ in terms of displacement gradient.

Therefore, the elastodynamics equations in closed form can be written as

$$\rho_s\, \partial_{tt}\mathbf{d} = \nabla \cdot [(I + \nabla\boldsymbol{d})\boldsymbol{\Pi}(\underbrace{(\nabla\boldsymbol{d} + \nabla\boldsymbol{d}^T)/2 + \nabla\boldsymbol{d}\nabla\boldsymbol{d}^T/2}_{\boldsymbol{E}})],$$

where $\boldsymbol{\Pi}$ is a given function via a constitutive relation.

For *hyperelastic materials* we have

$$\boldsymbol{\Pi}(\boldsymbol{E}) = \partial W/\partial \boldsymbol{E}.$$

Examples of hyperelastic materials include:

- *The Saint Venant–Kirchhoff model* for which

$$W(\boldsymbol{E}) = \frac{\lambda}{2}[\mathrm{tr}\boldsymbol{E}]^2 + \mu[\mathrm{tr}(\boldsymbol{E}^2)], \quad \text{and so} \quad \boldsymbol{\Pi}(\boldsymbol{E}) = \lambda[\mathrm{tr}\boldsymbol{E}]I + 2\mu\boldsymbol{E}.$$

- *The linearized Saint Venant–Kirchhoff model* for which

$$\boldsymbol{E} \approx \boldsymbol{D}(\boldsymbol{d}) = (\nabla\boldsymbol{d} + \nabla\boldsymbol{d}^T)/2$$

  and so

$$\boldsymbol{\Pi} \approx \boldsymbol{\Pi}(\boldsymbol{D}(\boldsymbol{d})) = \lambda[\mathrm{tr}\boldsymbol{D}(\boldsymbol{d})]I + 2\mu\boldsymbol{D}(\boldsymbol{d}), \text{ and } \mathbf{S} \approx \mu\,(\nabla\mathbf{d} + (\nabla\mathbf{d})^T) + \lambda(\nabla\cdot\mathbf{d})\mathbf{I},$$

  where $\boldsymbol{D}(\boldsymbol{d})$ is the symmetrized gradient of displacement.
- *The exponential stiffening stress–strain law* of Fung [68, 69], providing a more realistic model of the mechanical properties of arterial walls, for which

$$W(\boldsymbol{E}) = C\exp(a_1 E_{\theta\theta}^2 + a_2 E_{zz}^2 + a_3 E_{\theta\theta} E_{zz}),$$

  where $E_{\theta\theta}$ and E_{zz} are strains in the circumferential direction (θ) and longitudinal direction (z), respectively, and C, a_1, a_2, a_3 are constants.

In the rest of this chapter we will be working with the linearized Saint Venant–Kirchhoff model.

2.3 A Benchmark Problem

In this section we focus on a benchmark problem in fluid-multi-layered structure interaction. The problem consists of studying FSI between an incompressible, viscous fluid, and a structure consisting of two layers: a thin layer modeled by the Koiter shell equations, and a thick layer modeled by the equations of linear elasticity. The methods presented in this chapter work for an entire class of problems in which the thin structural layer can be described by either the full cylindrical linearly elastic Koiter shell model, described in Examples 1 and 2 of Sect. 2.2.1, the linearly elastic membrane equations, presented in Example 2 of Sect. 2.2.1, the *nonlinearly* elastic Koiter membrane/shell model, described in Example 3 of Sect. 2.2.1, the cylindrical linearly *viscoelastic* Koiter shell model, presented in Example 4 of Sect. 2.2.1, or the elastic string model described in Example 5 of Sect. 2.2.1.

2.3.1 *The Model Equations*

The thin structural layer is modeled by the reduced equations of linear (visco)elasticity, discussed in Sect. 2.2.1, which take the general form:

$$\textbf{THIN STRUCTURE}: \quad \rho_K h \frac{\partial^2 \boldsymbol{\eta}}{\partial t^2} + \mathcal{L}_{\text{el}}(\boldsymbol{\eta}) + \mathcal{L}_{\text{vis}} \frac{\partial \boldsymbol{\eta}}{\partial t} = \boldsymbol{f}, \quad \text{on } \Gamma \times (0,T), \tag{2.58}$$

These equations are defined on the reference domain which is a cylinder of radius R:

$$\Gamma = \{(R\cos\theta, R\sin\theta, z) \in \mathbb{R}^3 : z \in (0,L),\ \theta \in (0,2\pi)\}.$$

As discussed in Sect. 2.2.1, $\mathcal{L}_{\text{el}}$ may be a linear or a nonlinear operator modeling the elastic properties of shells or membranes, and $\mathcal{L}_{visc}$ denotes a linear operator modeling their viscoelastic properties. Operator $\mathcal{L}_{\text{vis}}$ may be equal to the zero operator. The methodology presented in this chapter is robust in the sense that it can be applied to solving both the viscoelastic and purely elastic thin structure models.

The thick structural layer is modeled by the equations of linear elasticity, discussed in Sect. 2.2.2:

$$\textbf{THICK STRUCTURE}: \begin{cases} \rho_s\, \partial_{tt}\mathbf{d} = \nabla \cdot \mathbf{S}, \quad \text{on } \Omega_S \times (0,T), \text{ where} \\ \mathbf{S} = \mu\,(\nabla \mathbf{d} + (\nabla \mathbf{d})^T) + \lambda(\nabla \cdot \mathbf{d})\mathbf{I}. \end{cases} \tag{2.59}$$

Table 2.4 Blood density and dynamics viscosity coefficients

Blood density ρ_F (kg/m^3)	1.1×10^3[133]
Blood dynamic viscosity μ_F (kg/ms)	1,050 [133]

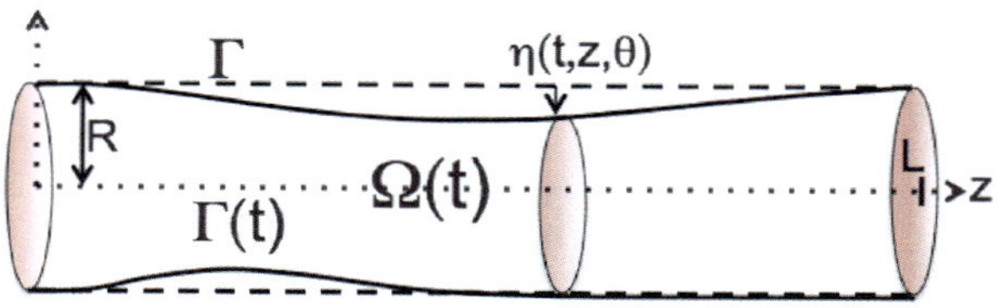

Fig. 2.7 Domain sketch and notation

These equations are defined on the reference domain

$$\Omega_S = \{(x, y, z) \in \mathbb{R}^3 : z \in (0, L),\ R < \sqrt{x^2 + y^2} < R + H)\}.$$

The flow of an incompressible, viscous fluid is modeled by the Navier–Stokes equations. They are defined on a time-dependent cylindrical fluid domain $\Omega_F(t)$, which is not known a priori:

$$\textbf{FLUID}: \quad \left.\begin{array}{r} \rho_F(\partial_t \mathbf{u} + \mathbf{u} \cdot \nabla \mathbf{u}) = \nabla \cdot \boldsymbol{\sigma}, \\ \nabla \cdot \mathbf{u} = 0, \end{array}\right\} \text{ in } \Omega_F(t),\ t \in (0, T), \tag{2.60}$$

where ρ_F denotes the fluid density; $\mathbf{u}$ the fluid velocity; $\boldsymbol{\sigma} = -p\mathbf{I} + 2\mu_F \mathbf{D}(\mathbf{u})$ is the fluid Cauchy stress tensor; p is the fluid pressure; μ_F is the dynamic viscosity coefficient; and $\mathbf{D}(\mathbf{u}) = \frac{1}{2}(\nabla\mathbf{u} + \nabla^\tau \mathbf{u})$ is the symmetrized gradient of $\mathbf{u}$. The typical values of the parameters ρ_F and μ_F for blood are given in Table 2.4.

We will be working with the fluid equations written in Cartesian coordinates (x, y, z), while the structure equations will be written in cylindrical coordinates (r, θ, z). For any function f given in Cartesian coordinates, we define $\tilde{f}$ to be the corresponding function given in cylindrical coordinates:

$$\tilde{f}(r, \theta, z) := f(x, y, z).$$

For simplicity, in the rest of this chapter, we drop the tilde notation.

The cylindrical fluid domain is of length L, with reference radius $r = R$. See Fig. 2.7. The thin structure, described by Eq. (2.58), serves as a fluid–structure interface. The nonzero inertia term $\rho_K h \partial^2 \boldsymbol{\eta}/\partial t^2$ indicates that our fluid–structure interface has mass. This has important implications for the analysis and numerical simulation of FSI problems, discussed in Sect. 2.7.7.

For simplicity, in the rest of this chapter, we will be assuming that only the radial component of the displacement of the thin structure is different from zero, i.e., we will be assuming

$$\textbf{ASSUMPTION}: \quad \boldsymbol{\eta} = (\eta_r, \eta_\theta, \eta_z) = (\eta_r, 0, 0) =: \eta \mathbf{e}_r, \tag{2.61}$$

where $\mathbf{e}_r = \mathbf{e}_r(\theta, z)$ is the unit vector in the r-direction. This is a common assumption in the literature on FSI in blood flow. For problems with nonzero radial and longitudinal displacement $\eta_r, \eta_z \neq 0$, please see [20, 21].

The radius of the deformed domain is equal to $R + \eta(t, \theta, z)$. Thus, the fluid domain, sketched in Fig. 2.7, is given by

$$\Omega_F(t) = \{(x, y, z) \in \mathbb{R}^3 : z \in (0, L), \sqrt{x^2 + y^2} < (0, R + \eta(t, \theta, z))\},$$

where the lateral boundary of the cylinder corresponds to fluid–structure interface, denoted by

$$\Gamma(t) = \{(x, y, z) \in \mathbb{R}^3 : z \in (0, L), \sqrt{x^2 + y^2} < (0, R + \eta(t, \theta, z))\}.$$

The inlet and outlet boundary of the fluid domain will be denoted by Γ_{in} and Γ_{out}, respectively.

2.3.2 *The Coupling Conditions*

Since we have three different physical models describing three different physical processes which are coupled, we need to describe the physics of the coupling between all of them. This includes prescribing coupling conditions between the fluid and structure, and prescribing coupling conditions between the thin and thick structure.

The coupling between the fluid, the thin structural layer, and the thick structural layer is achieved via two sets of coupling conditions: the kinematic coupling condition and the dynamic coupling condition. The kinematic coupling condition addresses the coupling of kinematic quantities, such as velocity. The dynamic coupling condition describes balance of forces that occurs at the interface between different physical models. These two sets of conditions give rise to a well-defined mathematical problem, while, at the same time, they capture the basic physical laws of the coupling.

In our problem, the thin structure serves both as a fluid–structure interface, and as a structure–structure interface. In this chapter we will be assuming that the kinematic coupling condition is the *no-slip* boundary condition between both the fluid and thin structure, as well as between the thin and thick structural layers.

Concerning the dynamic coupling condition, since $\Gamma(t)$ is a fluid–structure interface with mass, the dynamic coupling condition is simply the second Newton's Law of motion. It states that mass times acceleration of the interface is balanced by the sum of total forces acting on, or within, $\Gamma(t)$. This includes the contribution due to the elastic energy of the structure, and the balance of contact forces exerted by the fluid and the thick structure onto $\Gamma(t)$. More precisely, we have the following set of coupling conditions written in Lagrangian framework:

- **Kinematic Coupling Condition:**

$$\begin{aligned} \partial_t \eta(t,\theta,z)\mathbf{e}_r(\theta,z) &= \boldsymbol{u}(t, R+\eta(t,\theta,z),\theta,z), && \text{(continuity of velocity)} \\ \eta(t,\theta,z)\mathbf{e}_r(\theta,z) &= \mathbf{d}(t,R,\theta,z), && \text{(continuity of displacement)} \end{aligned} \tag{2.62}$$

  where $\mathbf{e}_r(\theta,z)$ is the unit vector in the r-direction.
- **Dynamic Coupling Condition:**

$$\rho_K h \partial_{tt}\eta + \mathcal{L}_{\mathrm{el}}(\eta) + \mathcal{L}_{\mathrm{vis}}\frac{\partial \eta}{\partial t} = -J(\sigma\mathbf{n})|_{(t,R+\eta,\theta,z)} \cdot \mathbf{e}_r + R\mathbf{S}|_{(t,R,\theta,z)}\mathbf{e}_r \cdot \mathbf{e}_r. \tag{2.63}$$

  Here $J = J(t,\theta,z) = \sqrt{[1+(\partial_z\eta)^2][R+\eta]^2 + \partial_\theta\eta^2}$ denotes the Jacobian of the composite function which includes the transformation from Eulerian to Lagrangian coordinates, and the transformation from cylindrical to Cartesian coordinates; the R in front of $\mathbf{S}$ in (2.63) denotes the Jacobian of the transformation from cylindrical to Cartesian coordinates evaluated at $r = R$, and $\mathbf{n}$ evaluated at $(t, R+\eta, \theta, z)$ is the outward unit normal vector to the deformed fluid–structure interface $\Gamma(t)$. As before, $\eta = \eta(t,\theta,z)$, and $\mathbf{e}_r = \mathbf{e}_r(\theta,z)$ is the unit vector in the r-direction.

If we did not have the thin structure with mass present, i.e., if we only had the fluid and thick structure interacting with each other, the dynamic coupling condition would look slightly different. The balance of contact forces at the fluid–structure interface would be given by the following:

$$-J(\sigma\mathbf{n})|_{(t,R+\eta,\theta,z)} \cdot \mathbf{e}_r + R\mathbf{S}|_{(t,R,\theta,z)}\mathbf{e}_r \cdot \mathbf{e}_r = 0, \ \text{on } \Gamma \times (0,T).$$

Namely, the dynamic coupling condition in this case reads that the normal stress exerted by the fluid onto the thick structure is balanced the normal stress exerted by the thick structure onto the fluid.

2.3.3 *The Boundary and Initial Conditions*

To get to a well-defined mathematical problem, Eqs. (2.58)–(2.63) need to be supplemented with initial and boundary conditions.

Fluid Inlet and Outlet Boundary Conditions. Examples of the inlet and outlet boundary conditions for the *fluid* include the following:

- *Dynamic pressure data:* In the existence proof presented later in the chapter, we will be working with the following inlet/outlet data:

$$\left.\begin{aligned} p + \frac{\rho_F}{2}|\mathbf{u}|^2 &= P_{\mathrm{in/out}}(t), \\ \mathbf{u} &= u_z\mathbf{e}_z, \end{aligned}\right\} \quad \text{on } \Gamma_{\mathrm{in/out}}, \tag{2.64}$$

where $P_{\text{in/out}} \in L^2_{loc}(0, \infty)$ are given, and $\mathbf{e}_z$ is the outer unit normal to $\Gamma_{\text{in/out}}$. Therefore, the fluid flow is driven by a prescribed dynamic pressure drop, and the flow enters and leaves the fluid domain orthogonally to the inlet and outlet boundary.

- *Normal stress data:* In the section on numerical simulations, presented later in this chapter, we will be using the following inlet/outlet data:

$$\begin{aligned} \boldsymbol{\sigma}\mathbf{n}|_{\Gamma_{\text{in}}} &= P_{\text{in}}(t), \\ \boldsymbol{\sigma}\mathbf{n}|_{\Gamma_{\text{out}}} &= P_{\text{out}}(T). \end{aligned} \tag{2.65}$$

Structure Inlet and Outlet Boundary Conditions. Examples of the boundary conditions for the *structure* at the inlet and outlet ends of the tube include the following:

- *The thin structure data.* At the end points of the thin structure we prescribe zero displacement:

$$\eta(t, r, \theta, 0) = \eta(t, r, \theta, L) = 0, \;\; r \in (0, R), \;\; \theta \in (0, 2\pi), t \in (0, T). \tag{2.66}$$

  If 4-th order derivative terms with respect to z appear in the model (i.e., if bending rigidity is included in the model), we consider a clamped Kotier shell with the additional boundary conditions

$$\eta_z(t, r, \theta, 0) = \eta_z(t, r, \theta, L) = 0, \;\; r \in (0, R), \;\; \theta \in (0, 2\pi), t \in (0, T). \tag{2.67}$$

- *The thick structure data.* At the end points of the annular sections of the thick structure we prescribe zero displacement:

$$\mathbf{d}(t, r, \theta, 0) = \mathbf{d}(t, r, \theta, L) = 0, \quad \text{for } r \in (R, R + H), \theta \in (0, 2\pi).$$

The External Boundary Condition. We will be assuming that the external boundary of the thick structure

$$\Gamma_{\text{ext}} = \{(x, y, z) \in \mathbb{R}^3 : z \in (0, L), x^2 + y^2 = (R + H)^2\}$$

is exposed to an external ambient pressure P_e:

$$\mathbf{S}\mathbf{e}_r = -P_e \mathbf{e}_r, \quad \text{on } \Gamma_{\text{ext}}. \tag{2.68}$$

Initial Data. The initial fluid and structure velocities, and the initial displacements, are given by

$$\mathbf{u}(0, .) = \mathbf{u}_0, \;\; \eta(0, .) = \eta_0, \;\; \partial_t \eta(0, .) = v_0, \;\; \mathbf{d}(0, .) = \mathbf{d}_0, \;\; \partial_t \mathbf{d}(0, .) = \mathbf{V}_0, \tag{2.69}$$

and are assumed to belong to the following spaces: $\mathbf{u}_0 \in L^2(\Omega_F(0))$, $\eta_0 \in H^1_0(0, 1)$, $v_0 \in L^2(0, 1)$, $\boldsymbol{V}_0 \in L^2(\Omega_S)$, $\boldsymbol{d}_0 \in H^1(\Omega_S)$.

A Summary of the Benchmark Problem. The benchmark problem in fluid-multi-layered-structure interaction that we are interested in studying is given by the following:

Find $\mathbf{u}$, p, η and $\mathbf{d}$ such that:

$$\left.\begin{aligned} \rho_F\,(\partial_t \mathbf{u} + (\mathbf{u}\cdot\nabla)\mathbf{u}) &= \nabla\cdot\boldsymbol{\sigma} \\ \nabla\cdot\mathbf{u} &= 0 \end{aligned}\right\}\ \text{in } \Omega_F(t),\ t\in(0,T),$$

$$\rho_S\,\partial_{tt}\mathbf{d} = \nabla\cdot\mathbf{S} \quad \text{in } \Omega_S\times(0,T),$$

$$\left.\begin{aligned} &\partial_t\eta\mathbf{e}_r = \boldsymbol{u}|_{R+\eta}, \\ &\eta\mathbf{e}_r = \mathbf{d}|_R, \\ &\rho_K h\partial_{tt}\eta + \mathcal{L}_{\text{el}}(\eta) + \mathcal{L}_{\text{vis}}\tfrac{\partial\eta}{\partial t} = -J(\sigma\mathbf{n})|_{R+\eta}\cdot\mathbf{e}_r + R\mathbf{S}|_R\mathbf{e}_r\cdot\mathbf{e}_r \end{aligned}\right\}\ \text{on } \Gamma\times(0,T).$$

where $\boldsymbol{\sigma}$, $\mathbf{S}$, $\mathcal{L}_{\text{el}}$, and $\mathcal{L}_{\text{vis}}$ are defined above. In this formulation, the fluid and thick structure equations are defined in Cartesian coordinates, while the thin structure model is given in cylindrical coordinates. Furthermore, the fluid equations are given in Eulerian framework, while the structure equations are given in Lagrangian framework. To account for the different coordinates, J in the dynamic coupling condition denotes the Jacobian of the transformation from the Eulerian to the Lagrangian framework, and from Cartesian to cylindrical coordinates. Similarly, the factor R in front of the first Piola–Kirchhoff stress tensor $\mathbf{S}$ is the Jacobian of the transformation between the Cartesian and cylindrical coordinates. While the coupling conditions are calculated at the deformed interface $\Gamma(t)$, they are written in terms of the reference configuration of the fluid–structure interface, namely, they are written in terms of points on Γ.

Supplemented with initial and boundary conditions, this problem defines a nonlinear, moving-boundary problem of mixed, parabolic–hyperbolic type. Hyperbolicity is associated with the thick structure problem and with the thin structure problem when no viscoelastic effects are taken into account, i.e., when $\mathcal{L}_{\text{vis}} = 0$. Parabolicity describes the properties of the fluid problem.

We will be studying this class of problems from both numerical, and theoretical point of view. Numerical method development for this class of problems will be presented in Sect. 2.7, while existence of solutions will be studied in Sect. 2.6. In those sections, concrete examples of this class of problems will be presented and studied. A simplified version of these equations in 2D will be presented.

2.4 FSI Literature Review

Fluid–structure interaction problems have been extensively studied for the past 20 years by many authors. The focus has been exclusively on FSI problems with structures consisting of a single material, except for the numerical simulations using the Immersed Boundary Method which is particularly suitable to deal with structures composed of fibers. The field has evolved from first studying FSI between an incompressible, viscous fluid and a rigid structure immersed in a fluid, to considering compliant (elastic/viscoelastic) structures interacting with a fluid. Concerning compliant structures, the coupling between the structure and fluid was first assumed to take place along a fixed fluid domain boundary (linear coupling). This was then extended to FSI problems in which the coupling was evaluated at a deformed fluid–structure interface, giving rise to an additional nonlinearity in the problem (nonlinear coupling).

2.4.1 Literature on Analysis of FSI Problems

Well-posedness results in which the structure was assumed to be a rigid body immersed in a fluid, or described by a finite number of modal functions, were studied in [19,41,45,49,50,58,71,72,74–76,135]. FSI problems coupling the Navier–Stokes equations with linear elasticity where the coupling was calculated at a fixed fluid domain boundary were considered in [53], and in [11, 12, 101] where an additional nonlinear coupling term was added at the interface. A study of well-posedness for FSI problems between an incompressible, viscous fluid and an elastic/viscoelastic structure with nonlinear coupling evaluated at a moving interface started with the result by daVeiga [17], where existence of a strong solution was obtained locally in time for an interaction between a $2D$ fluid and a $1D$ viscoelastic string, assuming periodic boundary conditions. This result was extended by Lequeurre in [108, 109], where the existence of a unique, local in time, strong solution for any data, and the existence of a global strong solution for small data, was proved in the case when the structure was modeled as a clamped viscoelastic beam. D. Coutand and S. Shkoller proved existence, locally in time, of a unique, regular solution for an interaction between a viscous, incompressible fluid in $3D$ and a $3D$ structure, immersed in the fluid, where the structure was modeled by the equations of linear [43], or quasi-linear [44] elasticity. In the case when the structure (solid) is modeled by a linear wave equation, I. Kukavica and A. Tufahha proved the existence, locally in time, of a strong solution, assuming lower regularity for the initial data [102]. A similar result for compressible flows can be found in [103]. A FSI between a viscous, incompressible fluid in $3D$, and $2D$ elastic shells was considered in [33, 34] where existence, locally in time, of a unique regular solution was proved. All the above-mentioned existence results for strong solutions are local in time. We also mention that the works of Shkoller et al., and Kukavica et al. were obtained in the context of Lagrangian coordinates, which were used for both the structure and fluid problems.

In the context of weak solutions, the following results have been obtained. Continuous dependence of weak solutions on initial data for an FSI problem with a free-boundary type coupling condition was studied in [87]. Existence of a weak solution for an FSI problem between a $3D$ incompressible, viscous fluid and a $2D$ viscoelastic plate was considered by Chambolle et al. in [32], while Grandmont improved this result in [79] to hold for a $2D$ elastic plate. These results were extended to a more general geometry in [106], and then to the case of generalized Newtonian fluids in [105], and to a non-Newtonian shear-dependent fluid in [94,95]. In these works existence of a weak solution was proved for as long as the elastic boundary does not touch "the bottom" (rigid) portion of the fluid domain boundary.

Muha and Čanić recently proved existence of weak solutions to a class of FSI problems modeling the flow of an incompressible, viscous, Newtonian fluid flowing through a cylinder whose lateral wall was modeled either by the linearly viscoelastic or by the linearly elastic Koiter shell equations [119], assuming nonlinear coupling at the deformed fluid–structure interface. The fluid flow boundary conditions were not periodic, but rather, the flow was driven by the dynamic pressure drop data. The methodology of proof in [119] was based on a semi-discrete, operator splitting Lie scheme which we discuss later in this chapter, and which was also used in [85] to design a stable, loosely coupled partitioned numerical scheme, called the kinematically coupled scheme (see also [21]). Ideas based on the Lie operator splitting scheme were also used by Temam in [140] to prove the existence of a solution to the nonlinear Carleman equation.

Finally, we also mention the results in [73] where a free-boundary problem for a steady flow of the incompressible, viscous fluid past a three-dimensional elastic body was studied, and the results in [18] where the authors consider a rigid body floating on the free surface of the fluid.

2.4.2 Literature on Numerical Simulation of FSI Problems

The development of numerical solvers for FSI problems has become particularly active since the 1980s. Among the most popular techniques are the Immersed Boundary Method [57,66,80–84,111,117,127,128] and the Arbitrary Lagrangian–Eulerian (ALE) method [52, 92, 93, 104, 110, 132, 133]. We further mention the Fictitious Domain Method in combination with the mortar element method or ALE approach [7, 142], and the methods recently proposed for the use in the blood flow application such as the Lattice Boltzmann method [56, 59, 99, 100], the Level Set Method [42], and the Coupled Momentum Method [65].

Until recently, only monolithic algorithms seemed applicable to blood flow simulations [15, 16, 47, 65, 77, 126, 144]. These algorithms are based on solving the entire nonlinear coupled problem as one monolithic system. They are, however, generally quite expensive in terms of the computational time, programming time, and memory requirements, since they require solving a sequence of strongly coupled problems using, e.g., the fixed point and Newton's methods [31,47,62,92,115,126], or the Steklov–Poincaré-based domain decomposition methods [48].

The multi-physics features of the blood flow problem strongly suggest to employ partitioned (or staggered) numerical algorithms, where the coupled fluid–structure problem is separated into a pure fluid sub-problem and a pure structure sub-problem. The fluid and structure sub-problems are integrated in time in an alternating way, and the coupling conditions are enforced asynchronously. When the density of the structure is much larger than the density of the fluid, as is the case in aeroelasticity, it is sufficient to solve, at every time step, just one fluid sub-problem and one structure sub-problem to obtain a solution. The classical *loosely coupled* partitioned schemes of this kind typically use the structure velocity in the *fluid sub-problem* as *Dirichlet* data for the fluid velocity (enforcing the no-slip boundary condition at the fluid–structure interface), while in the *structure sub-problem* the structure is loaded by the fluid *normal stress* calculated in the fluid sub-problem. These *Dirichlet–Neumann* loosely coupled schemes work well for problems in which the structure is much heavier than the fluid. Unfortunately, when fluid and structure have comparable densities, which is the case in the blood flow application, the simple strategy of separating the fluid from the structure suffers from severe stability issues [30, 116]. This is because the energy of the discretized problem in Dirichlet–Neumann loosely coupled schemes does not approximate well the energy of the continuous problem. A partial solution to this problem is to iterate several times between the fluid and structure sub-solvers at every time step until the energy of the continuous problem is well approximated. These *strongly coupled* partitioned schemes, however, are computationally expensive and may suffer from convergence issues for certain parameter values [30].

To get around these difficulties, and to retain the main advantages of loosely coupled partitioned schemes such as modularity, simple implementation, and low computational costs, several new loosely coupled algorithms have been proposed recently. The method proposed in [9] uses a simple membrane model for the structure which can be easily embedded into the fluid problem where it appears as a generalized Robin boundary condition. In this way the original problem reduces to a sequence of fluid problems. A similar approach was proposed in [126] where the fluid and structure were split in the classical way, but the fluid and structure sub-problems were linked via novel transmission (coupling) conditions that improve the convergence rate. A different approach to stabilization of loosely coupled schemes was proposed in [25] where a stabilization based on Nitsche's method [89] was used. We further mention the scheme proposed in [10] where a Robin–Robin type preconditioner was combined with Krylov iterations for a solution of an interface system. For completeness, we also mention several semi-implicit FSI schemes. The schemes proposed in [5, 6, 64] separate the computation of fluid velocity from the coupled pressure-structure velocity system, thereby reducing the computational costs. Similar schemes, derived from algebraic splitting, were proposed in [8, 132]. We also mention [124] where an optimization problem was solved at each time step to enforce the coupling conditions.

Recently, a novel loosely coupled partitioned scheme, called the "kinematically coupled β-scheme," was introduced by Bukač et al. in [21], and applied to FSI problem with thin elastic and viscoelastic structures, modeled by the membrane or shell equations. This scheme successfully deals with stability problems associated with loosely coupled schemes in a way different from those reported above. Stability is achieved by combining the structure inertia with the fluid sub-problem to mimic the energy balance of the continuous problem. It was shown in [29] that the scheme is unconditionally stable even for the parameters associated with the blood flow applications. Additionally, Čanić and Muha showed that a version of this scheme with $\beta = 0$ converges to a weak solution of the fully nonlinear FSI problem [119]. This result uses energy estimates combined with compactness arguments to show that the approximate solutions converge to a weak solution of the problem as the time discretization tends to zero. This is a significant result since it proves the existence of a (weak) solution to a nonlinear FSI problem in a constructive way, by using a computational scheme to construct a solution. See [54,55] for the related results concerning linear FSI problems.

The case $\beta = 0$ considered in [119] corresponds to the classical kinematically coupled scheme, first introduced in [85]. Parameter β was introduced in [21] to increase the accuracy of the scheme. It was shown in [21] that the accuracy of the kinematically coupled β-scheme with $\beta = 1$ was comparable to that of monolithic scheme by Badia et al. in [8] when applied to the nonlinear benchmark FSI problem in hemodynamics, introduced by Formaggia et al. in [67]. A different approach to increasing the accuracy of the classical kinematically coupled scheme was recently proposed by Fernández et al. [60,61,63]. Their modified kinematically coupled scheme, called "the incremental displacement-correction scheme" threats the structure displacement explicitly in the fluid sub-step and then corrects it in the structure sub-step. Fernández et al. showed that the accuracy of the incremental displacement-correction scheme is first-order in time. The results were obtained for an FSI problem involving a thin elastic structure.

These recent results indicate that the kinematically coupled scheme and its modifications provide an appealing way to solve FSI problems using partitioned approach. This scheme is particularly suitable for problems in which the structure consists of several layers, since modeling each additional layer can be accomplished by adding a new module to the partitioned scheme. Indeed, in the sections that follow, we present the kinematically coupled scheme, discuss the numerical results, and show the main steps in the proof of the existence of a weak solution to the class of fluid-multi-layered structure interaction problems discussed in Sect. 2.3.

2.5 Solution Framework

To study numerical simulation and existence of solutions to the class of problems (2.58)–(2.63) we present here a stable, partitioned approach that splits the fluid from the structure problem by using the Lie splitting, also known as the Marchuk–Yanenko scheme. The Lie splitting scheme has been widely used in numerical computations, see [78] and the references therein. Here we discuss an extension of this approach to study fluid-multi-layered structure interaction problems via the Kinematically Coupled Scheme, which is based on the time-discretization via Lie splitting. The time-dependent coupled problem is discretized in time (semi-discretization) in such a way that at each time step the coupled problem is split into a fluid and a structure sub-problem in a particular way so that the energy of the discretized problem mimics the energy of the continuous problem. As we shall see later, this guarantees stability of the scheme.

2.5.1 The Energy of the Coupled Problem

We present here a general approach to deriving an energy estimate of the coupled FSI benchmark problem, described in Sect. 2.3, for the class of problems in which the Koiter shell is linear. Thus, we consider a *clamped linearly (visco)elastic Koiter shell* (2.58), coupled with the equations of linear elasticity (2.59), and the flow of an incompressible, viscous fluid modeled by the Navier–Stokes equations (2.60). The inlet and outlet data are given by the *dynamic pressure data*, specified in (2.64).

We first recall from Sect. 2.2.1 that the linear operators $\mathcal{L}_{\text{el}}$ and $\mathcal{L}_{\text{vis}}$ are defined as follows:

$$\langle \mathcal{L}_{\text{el}} \boldsymbol{\eta}, \boldsymbol{\psi} \rangle := \frac{h}{2} \int_\omega \mathcal{A}\mathbf{G}(\eta) : \mathbf{G}(\boldsymbol{\psi})\sqrt{a} + \frac{h^3}{48} \int_\omega \mathcal{A}\mathbf{R}(\eta) : \mathbf{R}(\boldsymbol{\psi})\sqrt{a}, \ \forall \boldsymbol{\psi} \in C_c^\infty. \tag{2.70}$$

$$\langle \mathcal{L}_{\text{el}} \dot{\boldsymbol{\eta}}, \boldsymbol{\psi} \rangle := \frac{h}{2} \int_\omega \mathcal{B}\mathbf{G}(\dot{\eta}) : \mathbf{G}(\boldsymbol{\psi})\sqrt{a} + \frac{h^3}{48} \int_\omega \mathcal{B}\mathbf{R}(\dot{\eta}) : \mathbf{R}(\boldsymbol{\psi})\sqrt{a}, \ \forall \boldsymbol{\psi} \in C_c^\infty, \tag{2.71}$$

where $\mathbf{G}$ and $\mathbf{R}$ are the change of metric, and change of curvature tensors, respectively, and $\mathcal{A}$ and $\mathcal{B}$ are the elasticity tensor and the viscoelasticity tensor, respectively, defined in Example 4 of Sect. 2.2.1. This will be used to obtain the following energy estimate for the coupled problem:

Proposition 2.1. *The coupled FSI benchmark problem* (2.58)–(2.69) *with dynamic inlet and outlet pressure data satisfies the following energy estimate:*

$$\frac{d}{dt}(E_{\text{kin}}(t) + E_{\text{el}}(t)) + D(t) \leq C(P_{\text{in}}(t), P_{\text{out}}(t)), \tag{2.72}$$

where

$$\begin{aligned} E_{\text{kin}}(t) &:= \tfrac{1}{2}\left(\rho_F \|\mathbf{u}\|^2_{L^2(\Omega_F(t))} + \rho_K h \|\partial_t \boldsymbol{\eta}\|^2_{L^2(\Gamma)} + \rho_S \|\partial_t \mathbf{d}\|^2_{L^2(\Omega_S)}\right), \\ E_{\text{el}}(t) &:= \tfrac{1}{2}\left(E_{\text{el}}(\boldsymbol{\eta}) + 2\mu \|\mathbf{D}(\mathbf{d})\|^2_{L^2(\Omega_S)} + \lambda \|\nabla \cdot \mathbf{d}\|^2_{L^2(\Omega_S)}\right), \end{aligned} \tag{2.73}$$

denote the kinetic and elastic (internal) energy of the coupled problem, respectively, and the term $D(t)$ captures viscous dissipation:

$$D(t) := E_{\text{vis}}(\partial_t \boldsymbol{\eta}) + \mu_F \|\mathbf{D}(\mathbf{u})\|^2_{L^2(\Omega_F(t))}. \tag{2.74}$$

The constant $C(P_{\text{in}}(t), P_{\text{out}}(t)))$ depends only on the inlet and outlet pressure data, which are both functions of time.

The expressions for the energy associated with the Koiter shell are given by:

$$E_{\text{el}}(\eta) = \frac{h}{2}\int_{\omega} \mathcal{A}\mathbf{G}(\eta) : \mathbf{G}(\eta)R + \frac{h^3}{48}\int_{\omega} \mathcal{A}\mathbf{R}(\eta) : \mathbf{R}(\eta)R,$$

$$E_{\text{vis}}(\partial_t \eta) = \frac{h}{2}\int_{\omega} \mathcal{B}\mathbf{G}(\partial_t \eta) : \mathbf{G}(\partial_t \eta)R + \frac{h^3}{48}\int_{\omega} \mathcal{B}\mathbf{R}(\partial_t \eta) : \mathbf{R}(\partial_t \eta)R.$$

Notice that, due to the presence of an elastic fluid–structure interface with mass, the kinetic energy term $E_{\text{kin}}(t)$ contains a contribution from the kinetic energy of the fluid–structure interface $\|\partial_t \boldsymbol{\eta}\|^2_{L^2(\Gamma)}$ incorporating the interface inertia. Furthermore, the elastic energy $E_{\text{el}}(t)$ of the FSI problem accounts for the elastic energy $\|\partial_z \boldsymbol{\eta}\|^2_{L^2(\Gamma)}$ of the interface. If an FSI problem between the fluid and a thick structure was considered without the thin FSI interface with mass, these terms would not be present. In fact, the traces of the displacement and velocity at the fluid–structure interface of that FSI problem would not have been even defined for weak solutions.

Proof. A formal calculation of the energy estimate for this class of problems typically entails multiplying the fluid and structure equations in differential form by the fluid and structure velocities, respectively, and performing integration by parts. Integration by parts of the fluid equations takes into account the boundary conditions, which are the conditions at the inlet and outlet boundary of the fluid domain, and the conditions at the lateral boundary of the fluid domain. At the lateral boundary of the fluid domain, the normal fluid stress is coupled with the structure equations, and here is where the dynamic and kinematic coupling conditions come into play. By taking these coupling conditions into account, the energy of the fluid and the energy of the structure are coupled together into the total energy of the coupled FSI problem.

More precisely, we first multiply equation (2.60) by $\mathbf{u}$, integrate over $\Omega_F(t)$, and formally integrate by parts to obtain:

$$\int_{\Omega_F(t)} \rho_F \big(\partial_t \mathbf{u}\cdot\mathbf{u} + (\mathbf{u}\cdot\nabla)\mathbf{u}\cdot\mathbf{u}\big) + 2\mu_F \int_{\Omega_F(t)} |\mathbf{D}(\mathbf{u})|^2 - \int_{\partial\Omega_F(t)} (-p\mathbf{I} + 2\mu_F \mathbf{D}(\mathbf{u}))\mathbf{n}(t)\cdot\mathbf{u} = 0. \tag{2.75}$$

To deal with the inertia term we first recall that $\Omega_F(t)$ is moving in time and that the velocity of the lateral boundary is given by $\mathbf{u}|_{\Gamma(t)}$. The transport theorem applied to the first term on the left-hand side of the above equation then gives:

$$\int_{\Omega_F(t)} \partial_t \mathbf{u}\cdot\mathbf{u} = \frac{1}{2}\frac{d}{dt}\int_{\Omega_F(t)} |\mathbf{u}|^2 - \frac{1}{2}\int_{\Gamma(t)} |\mathbf{u}|^2 \mathbf{u}\cdot\mathbf{n}(t).$$

The second term on the left-hand side can be rewritten by using integration by parts, and the divergence-free condition, to obtain:

$$\int_{\Omega_F(t)} (\mathbf{u}\cdot\nabla)\mathbf{u}\cdot\mathbf{u} = \frac{1}{2}\int_{\partial\Omega_F(t)} |\mathbf{u}|^2\mathbf{u}\cdot\mathbf{n}(t) = \frac{1}{2}\Big(\int_{\Gamma(t)} |\mathbf{u}|^2\mathbf{u}\cdot\mathbf{n}(t) - \int_{\Gamma_{\text{in}}} |\mathbf{u}|^2 u_z + \int_{\Gamma_{\text{out}}} |\mathbf{u}|^2 u_z.\Big)$$

These two terms added together give

$$\rho_F \int_{\Omega_F(t)} (\partial_t\mathbf{u}\cdot\mathbf{u} + (\mathbf{u}\cdot\nabla)\mathbf{u}\cdot\mathbf{u}) = \frac{\rho_F}{2}\frac{d}{dt}\int_{\Omega_F(t)} |\mathbf{u}|^2 - \frac{\rho_F}{2}\int_{\Gamma_{\text{in}}} |\mathbf{u}|^2 u_z + \frac{\rho_F}{2}\int_{\Gamma_{\text{out}}} |\mathbf{u}|^2 u_z. \tag{2.76}$$

Notice the importance of nonlinear advection in canceling the cubic term $\int_{\Gamma(t)} |\mathbf{u}|^2\mathbf{u}\cdot \mathbf{n}(t)$!

To deal with the boundary integral over $\partial\Omega_F(t)$, first notice

$$\int_{\partial\Omega_F(t)} (-p\mathbf{I} + 2\mu_F\mathbf{D}(\mathbf{u}))\mathbf{n}\cdot\mathbf{u} = \int_{\Gamma_{\text{in/out}}\cup\Gamma(t)} (-p\mathbf{I} + 2\mu_F\mathbf{D}(\mathbf{u}))\mathbf{n}\cdot\mathbf{u}. \tag{2.77}$$

To calculate the contribution of the integral over $\Gamma_{\text{in/out}}$, notice that on $\Gamma_{\text{in/out}}$ the outward unit normal is given by $\pm\mathbf{e}_z$. Furthermore, the boundary condition (2.64) implies $u_r = u_\theta = 0$, or, in Cartesian coordinates $u_x = u_y = 0$. Combined with the divergence-free condition one obtains $\partial_z u_z = 0$. This implies $\mathbf{D}(\mathbf{u}) = 0$ on $\Gamma_{\text{in/out}}$. Therefore,

$$\int_{\Gamma_{\text{in/out}}} (-p\mathbf{I} + 2\mu_F\mathbf{D}(\mathbf{u}))\mathbf{n}\cdot\mathbf{u} = \int_{\Gamma_{\text{in}}} p\, u_z - \int_{\Gamma_{\text{out}}} p\, u_z. \tag{2.78}$$

What is left is to calculate the remaining boundary integral over $\Gamma(t)$, namely

$$-\int_{\Gamma(t)} (-p\mathbf{I} + 2\mu_F\mathbf{D}(\mathbf{u}))\mathbf{n}(t)\cdot\mathbf{u} = -\int_{\Gamma(t)} \boldsymbol{\sigma}\mathbf{n}(t)\cdot\mathbf{u}.$$

By enforcing the dynamic and kinematic coupling conditions (2.62), (2.63), we obtain

$$-\int_{\Gamma(t)} \sigma\mathbf{n}(t)\cdot\mathbf{u} = -\int_{\Gamma} J\sigma\mathbf{n}\cdot\mathbf{u} = \int_{\omega} R(f - \mathbf{S}\mathbf{e}_r\cdot\mathbf{e}_r)\partial_t\eta, \tag{2.79}$$

where f is the function appearing on the right-hand side of the Koiter shell equation (2.58).

The rest of the proof entails calculating the right-hand side of (2.79) in terms of the energy of the thin and thick structure problems. We begin with the Koiter shell problem:

$$\rho_K h \partial_{tt}\boldsymbol{\eta} + \mathcal{L}_{\mathrm{el}}\boldsymbol{\eta} + \mathcal{L}_{\mathrm{vis}}\partial_t\boldsymbol{\eta} = \boldsymbol{f}.$$

Multiply this equation by $\partial_t\boldsymbol{\eta}$ and formally integrate by parts, using, on the way, the definition of operators $\mathcal{L}_{\mathrm{el}}$ and $\mathcal{L}_{\mathrm{vis}}$, given in (2.70) and (2.71), respectively. The resulting equation is given by the following:

$$\frac{1}{2}\frac{d}{dt}\int_\omega \rho_K h\,(\boldsymbol{\eta}_t)^2\,R + < \mathcal{L}_{\mathrm{el}}\boldsymbol{\eta}, \partial_t\boldsymbol{\eta} > + < \mathcal{L}_{\mathrm{vis}}\partial_t\boldsymbol{\eta}, \partial_t\boldsymbol{\eta} > = \int_\omega \boldsymbol{f}\,\partial_t\boldsymbol{\eta}R,$$

or, by expanding the operators $\mathcal{L}_{\mathrm{el}}$ and $\mathcal{L}_{\mathrm{vis}}$, and using $\dot{\boldsymbol{\eta}}$ to denote $\partial_t\boldsymbol{\eta}$:

$$\begin{aligned}&\frac{1}{2}\frac{d}{dt}\int_\omega \rho_K h\,(\boldsymbol{\eta}_t)^2\,R + \frac{h}{2}\int_\omega \mathcal{A}\mathbf{G}(\boldsymbol{\eta}) : \mathbf{G}(\dot{\boldsymbol{\eta}})R + \frac{h^3}{48}\int_\omega \mathcal{A}\mathbf{R}(\dot{\boldsymbol{\eta}}) : \mathbf{R}(\dot{\boldsymbol{\eta}})R\\ &\frac{h}{2}\int_\omega \mathcal{B}\mathbf{G}(\dot{\boldsymbol{\eta}}) : \mathbf{G}(\dot{\boldsymbol{\eta}})R + \frac{h^3}{48}\int_\omega \mathcal{B}\mathbf{R}(\dot{\boldsymbol{\eta}}) : \mathbf{R}(\dot{\boldsymbol{\eta}})R = \int_\omega \boldsymbol{f}\,\partial_t\boldsymbol{\eta}R.\end{aligned} \tag{2.80}$$

By recalling the definitions of the elastic and viscous energy of the Koiter shell:

$$E_{\mathrm{el}}(\boldsymbol{\eta}) = \frac{h}{2}\int_\omega \mathcal{A}\mathbf{G}(\boldsymbol{\eta}) : \mathbf{G}(\boldsymbol{\eta})\sqrt{a} + \frac{h^3}{48}\int_\omega \mathcal{A}\mathbf{R}(\boldsymbol{\eta}) : \mathbf{R}(\boldsymbol{\eta})\sqrt{a},$$

$$E_{\mathrm{vis}}(\boldsymbol{\eta}) = \frac{h}{2}\int_\omega \mathcal{B}\mathbf{G}(\boldsymbol{\eta}) : \mathbf{G}(\boldsymbol{\eta})\sqrt{a} + \frac{h^3}{48}\int_\omega \mathcal{B}\mathbf{R}(\boldsymbol{\eta}) : \mathbf{R}(\boldsymbol{\eta})\sqrt{a},$$

Equation (2.80) can be written as

$$\frac{\rho_K h}{2}\frac{d}{dt}\|\partial_t\boldsymbol{\eta}\|^2_{L^2(\omega)} + \frac{1}{2}\frac{d}{dt}E_{\mathrm{el}}(\boldsymbol{\eta}) + E_{\mathrm{vis}}(\partial_t\boldsymbol{\eta}) = \int_\omega R\,\boldsymbol{f}\,\partial_t\boldsymbol{\eta}. \tag{2.81}$$

Next, consider the elasticity equation (2.59), multiply it by $\partial_t\mathbf{d}$ and integrate by parts to obtain:

$$\begin{aligned}&\frac{1}{2}\frac{d}{dt}\big(\rho_S\|\partial_t\mathbf{d}\|^2_{L^2(\Omega_S)} + 2\mu\|\mathbf{D}(\mathbf{d})\|^2_{L^2(\Omega_S)} + \lambda\|\nabla\cdot\mathbf{d}\|^2_{L^2(\Omega_S)}\big) = -\int_\Gamma \mathbf{S}\mathbf{e}_r\cdot\partial_t\mathbf{d}\\ &= -\int_\omega R\,\mathbf{S}\mathbf{e}_r\cdot\partial_t\mathbf{d}.\end{aligned} \tag{2.82}$$

Now, the right-hand side of Eq. (2.79) can be calculated by combining (2.81) and (2.82) to obtain

$$-\int_{\Gamma(t)} \sigma \mathbf{n}(t) \cdot \mathbf{u} = \frac{\rho_K h}{2} \frac{d}{dt} \|\partial_t \boldsymbol{\eta}\|^2_{L^2(\omega)} + \frac{1}{2}\frac{d}{dt} E_{\text{el}}(\boldsymbol{\eta}) + E_{\text{vis}}(\partial_t \boldsymbol{\eta})$$

$$\frac{1}{2}\frac{d}{dt}\left(\rho_s \|\partial_t \mathbf{d}\|^2_{L^2(\Omega_S)} + 2\mu \|\mathbf{D}(\mathbf{d})\|^2_{L^2(\Omega_S)} + \lambda \|\nabla \cdot \mathbf{d}\|^2_{L^2(\Omega_S)}\right). \quad (2.83)$$

By combining (2.75) with (2.76), (2.77), (2.78), and (2.83), one obtains the following energy equality:

$$\frac{1}{2}\frac{d}{dt}\Big\{\rho_F \|\mathbf{u}\|^2_{\Omega_F(t)} + \rho_K h \|\partial_t \boldsymbol{\eta}\|^2_{L^2(\Gamma)} + \rho_S \|\partial_t \mathbf{d}\|^2_{L^2(\Omega_S)} + E_{\text{el}}(\boldsymbol{\eta}) + 2\mu \|\mathbf{D}(\mathbf{d})\|^2_{L^2(\Omega_S)}$$

$$+ \lambda \|\nabla \cdot \mathbf{d}\|^2_{L^2(\Omega_S)}\Big\} + 2\mu_F \|\mathbf{D}(\mathbf{u})\|^2_{\Omega_F(t)} + E_{\text{vis}}(\partial_t \boldsymbol{\eta}) = \pm\, P_{\text{in/out}}(t) \int_{\Gamma_{\text{in/out}}} u_z$$

Finally, by using the trace inequality and Korn inequality one can estimate:

$$|P_{\text{in/out}}(t) \int_{\Gamma_{\text{in/out}}} u_z| \leq C |P_{\text{in/out}}| \|\mathbf{u}\|_{H^1(\Omega_F(t))} \leq \frac{C}{2\epsilon} |P_{\text{in/out}}|^2 + \frac{\epsilon C}{2} \|\mathbf{D}(\mathbf{u})\|^2_{L^2(\Omega_F(t))}.$$

By choosing ϵ such that $\frac{\epsilon C}{2} \leq \mu_F$ we get the energy inequality

$$\frac{1}{2}\frac{d}{dt}\Big\{\rho_F \|\mathbf{u}\|^2_{\Omega_F(t)} + \rho_K h \|\partial_t \boldsymbol{\eta}\|^2_{L^2(\Gamma)} + \rho_S \|\partial_t \mathbf{d}\|^2_{L^2(\Omega_S)} + E_{\text{el}}(\boldsymbol{\eta}) + 2\mu \|\mathbf{D}(\mathbf{d})\|^2_{L^2(\Omega_S)}$$

$$+ \lambda \|\nabla \cdot \mathbf{d}\|^2_{L^2(\Omega_S)}\Big\} + \mu_F \|\mathbf{D}(\mathbf{u})\|^2_{\Omega_F(t)} + E_{\text{vis}}(\partial_t \boldsymbol{\eta}) \leq C(P_{\text{in}}(t), P_{\text{out}}(t)).$$

□

2.5.2 ALE Formulation

Since the fluid–structure coupling studied in this chapter is performed along the moving fluid–structure interface, the fluid domain $\Omega(t)$ is not fixed. This is a problem from many points of view. In particular, defining the time discretization of the time derivative $\partial \mathbf{u}/\partial t$, for example $\partial \mathbf{u}/\partial t \approx (\mathbf{u}(t^{n+1}, .) - \mathbf{u}(t^n, .))/(t^{n+1} - t^n)$, is not well defined since $\mathbf{u}(t^{n+1}, .)$ and $\mathbf{u}(t^n, .)$ are not defined on the same domain at two different time steps. To resolve this difficulty, often times the fluid domain is mapped onto a fixed, reference domain via a smooth, invertible ALE mapping [52]:

$$\mathcal{A} : \Omega_F \to \Omega_F(t).$$

An example of such a mapping is the harmonic extension of the boundary $\partial\Omega_F(t)$ onto the fluid domain. See Sect. 2.7. Another example is a mapping particularly convenient for the existence proof, presented in Sect. 2.6. This introduces additional nonlinearities into the equations, reflecting the geometric nonlinearities of the moving interface. The transformed gradient, which we denote by $\nabla^{\boldsymbol{\eta}}$, will depend on the fluid–structure interface $\boldsymbol{\eta}$. Furthermore, by using the chain rule, one can see that the time derivative of the transformed fluid velocity will have an additional advection term with the coefficient given by the domain velocity $\mathbf{w} := \mathcal{A}_t \circ \mathcal{A}^{-1}$, where $\mathcal{A}_t$ denotes the time derivative of $\mathcal{A}$. Finally, the mapped fluid equations in Ω_F read:

$$\left.\begin{aligned} \rho_F\,(\partial_t \mathbf{u} + ((\mathbf{u}-\mathbf{w})\cdot\nabla^{\eta})\mathbf{u}) &= \nabla^{\eta}\cdot\boldsymbol{\sigma}^{\eta} \\ \nabla^{\eta}\cdot\mathbf{u} &= 0 \end{aligned}\right\} \text{ in } \Omega_F \times (0,T). \tag{2.84}$$

Here, the notation $\boldsymbol{\sigma}^{\eta}$ reflects the dependence of $\mathbf{D}^{\boldsymbol{\eta}}(\mathbf{u}) = \frac{1}{2}(\nabla^{\eta}\mathbf{u} + \nabla^{\eta^T}\mathbf{u})$ on η. Therefore, our problem in ALE formulation reads as follows:

The Coupled Problem in ALE Form defined on Ω_F Find $\mathbf{u}$, p, η and $\mathbf{d}$ such that:

$$\left.\begin{aligned} \rho_F\,(\partial_t \mathbf{u} + ((\mathbf{u}-\mathbf{w})\cdot\nabla^{\eta})\mathbf{u}) &= \nabla^{\eta}\cdot\boldsymbol{\sigma}^{\eta} \\ \nabla^{\eta}\cdot\mathbf{u} &= 0 \end{aligned}\right\} \text{ in } \Omega_F \times (0,T),$$

$$\rho_S \partial_{tt}\mathbf{d} = \nabla\cdot\mathbf{S} \quad \text{in } \Omega_S \times (0,T),$$

$$\left.\begin{aligned} &\partial_t \eta \mathbf{e}_r = \boldsymbol{u}|_{R+\eta}, \\ &\eta \mathbf{e}_r = \mathbf{d}, \\ &\rho_K h \partial_{tt}\eta + \mathcal{L}_{\text{el}}(\eta) + \mathcal{L}_{\text{vis}}\partial_t \eta = -J(\sigma\mathbf{n})|_{R+\eta}\cdot\mathbf{e}_r + R\,\mathbf{S}\mathbf{e}_r\cdot\mathbf{e}_r \end{aligned}\right\} \text{ on } \Gamma\times(0,T).$$

As we shall see in Sect. 2.7, the actual numerical simulations at each time step are typically performed on the current (fixed) domain $\Omega_F^n(t)$, with only the time-derivative calculated on Ω_F, thereby avoiding the need to calculate the transformed gradients $\nabla^{\boldsymbol{\eta}}$. The corresponding continuous problem in ALE form can be written as follows:

The Coupled Problem in ALE Form defined on $\Omega_F(t)$

Find $\mathbf{u}$, p, η and $\mathbf{d}$ such that:

$$\left.\begin{aligned} \rho_F\,(\partial_t \mathbf{u}|_{\Omega_F} + ((\mathbf{u}-\mathbf{w})\cdot\nabla)\mathbf{u}) &= \nabla\cdot\boldsymbol{\sigma} \\ \nabla\cdot\mathbf{u} &= 0 \end{aligned}\right\} \text{ in } \Omega_F(t) \times (0,T),$$

$$\rho_S \partial_{tt}\mathbf{d} = \nabla\cdot\mathbf{S} \quad \text{in } \Omega_S \times (0,T),$$

$$\left.\begin{aligned} &\partial_t \eta \mathbf{e}_r = \boldsymbol{u}|_{R+\eta}, \\ &\eta \mathbf{e}_r = \mathbf{d}, \\ &\rho_K h \partial_{tt}\eta + \mathcal{L}_{\text{el}}(\eta) + \mathcal{L}_{\text{vis}}\partial_t \eta = -J(\sigma\mathbf{n})|_{R+\eta}\cdot\mathbf{e}_r + R\,\mathbf{S}\mathbf{e}_r\cdot\mathbf{e}_r \end{aligned}\right\} \text{ on } \Gamma\times(0,T).$$

Here, $\partial_t \mathbf{u}|_{\Omega_F}$ denotes the time derivative calculated on Ω_F. This approach is standard in ALE methods applied to partitioned schemes. In our existence proof, and in our definition of the splitting scheme, however, it will be convenient to use the fully mapped problem onto the fixed reference domain Ω_F.

2.5.3 The Splitting Scheme: General Framework

To apply the Lie splitting scheme the problem must first be written as a first-order system in time:

$$\frac{\partial \phi}{\partial t} + A(\phi) = 0, \quad \text{in } (0, T), \tag{2.85}$$

$$\phi(0) = \phi_0, \tag{2.86}$$

where A is an operator from a Hilbert space into itself. Operator A is then split, in a nontrivial decomposition, as

$$A = \sum_{i=1}^{I} A_i. \tag{2.87}$$

The Lie scheme consists of the following. Let $\Delta t > 0$ be a time discretization step. Denote $t^n = n\Delta t$ and let ϕ^n be an approximation of $\phi(t^n)$. Set $\phi^0 = \phi_0$. Then, for $n \geq 0$ compute ϕ^{n+1} by solving

$$\frac{\partial \phi_i}{\partial t} + A_i(\phi_i) = 0 \quad \text{in } (t^n, t^{n+1}), \tag{2.88}$$

$$\phi_i(t^n) = \phi^{n+(i-1)/I}, \tag{2.89}$$

and then set $\phi^{n+i/I} = \phi_i(t^{n+1})$, for $i = 1, \ldots, I$. Thus, the value at $t = t^{n+1}$ of the solution of the i-th problem is taken as the initial data for the $(i+1)$-st problem on (t^n, t^{n+1}).

This method is first-order accurate in time. More precisely, if (2.85) is defined on a finite-dimensional space, and if operators A_i are smooth enough, then $\|\phi(t^n) - \phi^n\| = O(\Delta t)$ [78].

To solve the class of problems (2.58)–(2.63), we split the fluid from the structure sub-problem to separate the different physics in the coupled problem. Thus, the coupled problem, which defines operator A, is split into a sum of two operators:

1. An elastodynamics problem for the thick structure, and
2. A fluid problem with suitable boundary conditions involving structure velocity and fluid stress at the boundary.

The thin-structure problem will enter through the boundary conditions, enforcing the dynamic coupling condition between the fluid and thick structure.

Thus, this scheme works as follows: first the structure problem is solved on the time-interval (t^n, t^{n+1}) with the initial data obtained from the previous time step. Then, the fluid problem is solved on the same-time interval (t^n, t^{n+1}), but with the initial data obtained from the just calculated solution in the first step.

Not every splitting of this kind would lead to a stable, convergent scheme. Our strategy is to split the fluid from the structure sub-problem in such a way that the energy of the discretized problem approximates well the energy of the continuous problem. To achieve this goal, a key role is played by the kinematic coupling condition, which will be enforced implicitly in both steps of the splitting scheme, keeping the two sub-problems tightly coupled at all times.

Before we apply the Lie splitting, we rewrite our coupled problem in first-order form with respect to time. For this purpose we introduce the following notation:

- the trace of the fluid velocity at the moving interface $\Gamma(t)$ will be denoted by v, i.e.,

$$v\mathbf{e_r} := \mathbf{u}|_{\Gamma(t)}.$$

  Namely, v, which is defined on Γ, is equal to the trace of $\mathbf{u}$ evaluated at $R + \eta$. The kinematic coupling condition (no-slip) then reads $\partial_t \eta = v$.
- the thick structure velocity will be denoted by:

$$\mathbf{V} := \frac{\partial \mathbf{d}}{\partial t}.$$

The system in ALE form is now rewritten by using the abovementioned notation, and by employing the *kinematic coupling condition* in the thin structure model. This way the kinematic coupling condition will be enforced everywhere, in all the steps of the splitting scheme. The resulting coupled problem in first-order ALE form is given by the following:

The Coupled Problem in First-Order ALE Form Find $\mathbf{u}$, p, η, $\mathbf{d}$, v, and $\mathbf{V}$ such that:

$$\left.\begin{aligned} \rho_F\left(\partial_t \mathbf{u} + ((\mathbf{u}-\mathbf{w})\cdot\nabla^\eta)\mathbf{u}\right) &= \nabla^\eta\cdot\boldsymbol{\sigma}^\eta \\ \nabla^\eta\cdot\mathbf{u} &= 0 \end{aligned}\right\} \text{ in } \Omega_F\times(0,T),$$

$$\left.\begin{aligned} \rho_S \partial_t \mathbf{V} &= \nabla\cdot\mathbf{S} \\ \mathbf{d}_t &= \mathbf{V}, \end{aligned}\right\} \text{ in } \Omega_S\times(0,T),$$

$$\left.\begin{aligned} &\partial_t\eta = v, \\ &v\mathbf{e}_r = \mathbf{u}, \\ &\eta\mathbf{e}_r = \mathbf{d}, \\ &\rho_K h\partial_t v + \mathcal{L}_{\text{el}}(\eta) + \mathcal{L}_{\text{vis}} v = -J\,\sigma^\eta\mathbf{n}\cdot\mathbf{e}_r + R\,\mathbf{S}\mathbf{e}_r\cdot\mathbf{e}_r \end{aligned}\right\} \text{ on } \Gamma\times(0,T).$$

Notice that we have enforced the kinematic coupling condition both in the thin structure acceleration term and in the viscous part of the thin structure equation.

We are now ready to split the problem. For this purpose, observe that the portion $\rho_K h\ \partial_t v + \mathcal{L}_{\text{vis}} v = -J\boldsymbol{\sigma}\mathbf{n}\cdot\mathbf{e}_r$ of the dynamic coupling condition is all given in terms of the trace v of the fluid velocity on Γ (recall, $\boldsymbol{\sigma}$ depends on v). We can, therefore, use this as a lateral boundary condition on Γ for the fluid sub-problem. This observation is crucial because keeping the structure inertia term $\rho_K h\partial_t v$ together with the inertia of the fluid in the fluid sub-problem is of paramount importance for designing a stable and convergence scheme.

This is different from the classical loosely coupled schemes. In classical Dirichlet–Neumann loosely coupled scheme, the boundary condition for the fluid sub-problem is the Dirichlet condition for the fluid velocity v on Γ given in terms of the structure velocity $\partial\eta/\partial t$, namely $v = \partial\eta/\partial t$, where $\partial\eta/\partial t$ is *calculated at the previous time step*! This inclusion of the structure inertia from the previous time step (explicitly) makes the fluid sub-problem unstable for certain parameters values [30]. The main reason for this is that the kinetic energy at this time step includes only the fluid kinetic energy from the current time step, and not the thin structure kinetic energy, since the thin structure velocity enters from the previously calculated time step. For strong geometric nonlinearities, which often happen when the fluid and structure densities are comparable, this mismatch between the kinetic energy of the discretized problem (where only the fluid kinetic energy appears in the current time step) and the kinetic energy of the continuous problem (where both the fluid and structure kinetic energy are tied together in a strongly coupled FSI problem) gives rise to an unstable numerical scheme [30].

Therefore, the strategy of our splitting, mentioned above, to keep the thin structure inertia together with the fluid inertia in the fluid sub-step will give rise to the kinetic energy of the discretized problem that approximates well the kinetic energy of the continuous problem, giving rise to a scheme that is unconditionally stable for all the parameters in the problem [29]. In Sect. 2.6 we prove that the scheme converges to a weak solution to the underlying FSI problem.

We therefore define the operators A_1 and A_2 as follows:

Problem A1 : STRUCTURE

$$
\begin{aligned}
\rho_S \partial_t \mathbf{V} &= \nabla\cdot\mathbf{S}, && \text{in } \Omega_S\\
\partial_t \mathbf{d} &= \mathbf{V}, && \text{in } \Omega_S\\
\mathbf{d} &= \eta\mathbf{e}_r, && \text{on } \Gamma\\
\partial_t \eta &= v, && \text{on } \Gamma\\
\rho_K h \partial_t v &= \mathcal{L}_{\text{el}}(\eta) + R\,\mathbf{S}\mathbf{e}_r\cdot\mathbf{e}_r && \text{on } \Gamma
\end{aligned}
$$

Here, of course, the PDE system in Ω_S can be solved just as a single second-order PDE for $\mathbf{d}$: $\rho_S\mathbf{d}_{tt} = \nabla\cdot\mathbf{S}$. Problem A1 is solved with the initial data $(\mathbf{d}, \mathbf{V}, \eta, v)$ given by the solution from the previous time step. This means, in particular, that the thin structure velocity v is set to be equal to the trace of the fluid velocity on Γ, calculated in the previous time step. Thus, we are solving the elastodynamics problem for the linearly elastic thick structure in Ω_S, with the boundary condition at the lateral boundary Γ given by a PDE that determines the motion of the lateral boundary. The motion of the lateral boundary in this sub-problem is driven by the normal component of the first Piola–Kirchhoff stress tensor $\mathbf{S}$, and by the initial data for the velocity of the thin structure, which is given by the trace of the fluid velocity, just calculated in the previous time step. In this step we also calculate the domain velocity $\mathbf{w}$, which is given by the time-derivative of the ALE mapping, associated with Problem A1.

Problem A2 : FLUID

$$\begin{aligned} \partial_t \mathbf{u} + ((\hat{\mathbf{u}} - \mathbf{w}) \cdot \nabla^\eta)\mathbf{u} &= \nabla^\eta \cdot \boldsymbol{\sigma}^\eta, && \text{in } \Omega_F \\ \nabla^\eta \cdot \mathbf{u} &= 0, && \text{in } \Omega_F \\ \mathbf{u} &= v\mathbf{e}_r, && \text{on } \Gamma \\ \rho_K h \partial_t v + \mathcal{L}_{\text{vis}} v &= -J\boldsymbol{\sigma}^\eta \mathbf{n} \cdot \mathbf{e}_r && \text{on } \Gamma \end{aligned}$$

Here $\hat{\mathbf{u}}$ is the value of $\mathbf{u}$ from the previous time step, and $\mathbf{w}$, which is the domain velocity (the time derivative of the ALE mapping), is obtained from the just calculated Problem A1. Furthermore, $\nabla^{\boldsymbol{\eta}}$ is the transformed gradient, which is based on the value of $\boldsymbol{\eta}$ from the previous time step. The initial data for $\mathbf{u}$ is given from the previous time step, while the initial data for the trace of the fluid velocity v is given by the just calculated velocity of the thin structure $\partial_t \eta$.

This concludes our description of the general framework based on the Lie splitting scheme for solving the class of FSI problems (2.58)–(2.63) with multiple structural layers.

Before we continue, several remarks are in order:

- The splitting works as well when the thin structure is purely elastic, i.e., when $\mathcal{L}_{\text{vis}} = 0$.
- Switching the order of solution (fluid step first, structure second) works as well. The corresponding algorithm is explicitly shown below in the corresponding block-diagram.
- The symmetrized Lie splitting obtained by solving the structure problem, followed by the fluid problem, and then the structure problem, increases the accuracy of the scheme to second-order in time.
- A version of Strang splitting for this problem was performed by Lukačova et al. in [95, 96] achieving second-order accuracy in time.
- Adding additional modules to capture different physics in a given multi-physics problem can be accomplished in a similar way. See [23] for an application of this scheme to an FSI problem with multiple *poroelastic* structural layers. Also, see [122] for an application of this scheme to an FSI between a vascular device called *stent*, elastic arterial wall, and the flow of an incompressible, viscous fluid.
- A modification of this scheme to achieve higher accuracy within the class of first-order schemes was introduced in [20, 21]. Details of this modified scheme, called the Kinematically Coupled β-scheme, are presented next.

2.5.4 A Modified Splitting Scheme Achieving Higher Accuracy

To increase the accuracy, the kinematically coupled β-scheme is based on additionally splitting the normal fluid stress as follows:

$$\sigma\mathbf{n} = \underbrace{\sigma\mathbf{n} + \beta p\mathbf{n}}_{(I)} \underbrace{-\beta p\mathbf{n}}_{(II)},$$

where $\beta \in [0, 1]$. Part I of the fluid stress is treated with the fluid sub-problem, while Part II with the structure sub-problem. The new boundary condition for the fluid sub-problem, written in the framework in which the entire fluid sub-problem is considered on the fixed, reference domain Ω_F, becomes

$$\rho_S h\partial_t v + \mathcal{L}_{\text{vis}} v = -J\,(\sigma^\eta\mathbf{n} + \beta\hat{p}\mathbf{n})\cdot\mathbf{e}_r \quad \text{on } \Gamma,$$

where $\hat{p}$ denotes the explicit use of the pressure calculated from the previous step. Part II of the fluid stress is then used to load the structure so that the new boundary condition for the structure sub-problem becomes

$$\rho_K h\partial_t v = \mathcal{L}_{\text{el}}(\eta) + (R\,\mathbf{S}\mathbf{e}_r - \beta Jp\mathbf{n})\cdot\mathbf{e}_r \quad \text{on } \Gamma. \tag{2.90}$$

A block diagram shown in Fig. 2.8 below summarizes the splitting in each iteration.

The main reason for the increase in accuracy of the β-scheme is the inclusion of the pressure loading $\beta Jp\mathbf{n}$ by the fluid onto the structure in the structure sub-problem. This way the structure "feels" the fluid not only through the kinematic coupling condition enforced via the initial condition for the structure sub-problem ($\partial_t\eta = v$), but also through the leading contribution of the normal stress, i.e., the pressure, exerted by the fluid onto the structure. Typically, the highest accuracy is achieved for $\beta = 1$. The accuracy of this modified scheme is still first-order, but the error is closer to the error of a monolithic scheme, as we shall see in Sect. 2.7.

Numerical Implementation. Typically, numerical implementation of the fluid sub-problem entails solving the fluid equations on the "current domain." Namely, numerical implementation is performed for the ALE problem written on $\Omega_F(t)$, and not Ω_F, as was discussed at the end of Sect. 2.5.2. In this case, the fluid sub-problem written above takes the following form:

Problem A2 (FLUID)

$$\begin{aligned}
\partial_t\mathbf{u}|_{\Omega_F} + ((\mathbf{u}^{\mathrm{n}} - \mathbf{w}^{\mathbf{n+1}})\cdot\nabla)\mathbf{u} &= \nabla\cdot\sigma, && \text{in } \Omega_F^{\mathrm{n}}\\
\nabla\cdot\mathbf{u} &= 0, && \text{in } \Omega_F^{\mathrm{n}}\\
(\rho_K h\partial_t\mathbf{u}|_{R+\eta} + \mathcal{L}_{\text{vis}}\mathbf{u}|_{R+\eta})\cdot\mathbf{e}_r &= -J\,(\sigma\mathbf{n} + \beta p^{\mathrm{n}}\mathbf{n})\,|_{R+\eta}\cdot\mathbf{e}_r && \text{on } \Gamma_F^{\mathrm{n}}
\end{aligned}$$

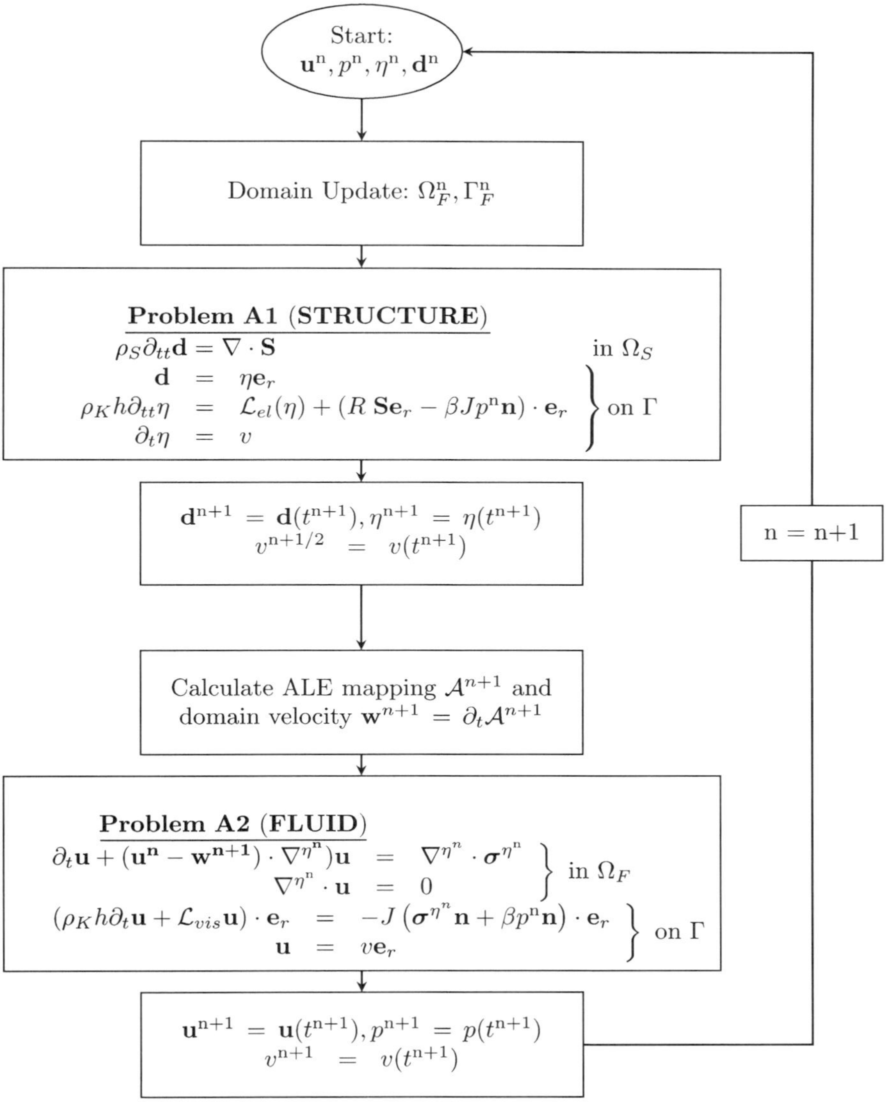

Fig. 2.8 A block diagram showing the main steps of the Kinematically Coupled β-Scheme

2.6 Existence of a Weak Solution

In this section we present details of the analysis of the existence of a weak solution to a previously discussed FSI problem with a multi-layered structure in two space dimensions. For related results in 3D, please see [121]. The most difficult case from the analysis point of view is the case when the viscoelastic effects in the thin structure model are ignored, and the highest order spatial derivatives (the 4-th order spatial derivatives) in the Koiter shell have the coefficients equal to zero. The resulting thin-structure model is the linear wave equation. Thus, in this section we study FSI between an incompressible, viscous fluid flowing through a 2D cylinder with compliant walls, consisting of a thin and a thick layer, modeled by the linear wave equation, and the equations of linear elasticity, respectively. As before, we will also be assuming that only the radial displacement in the thin structure is different from zero.

The existence result for the FSI problem in which the thin structure is modeled by the full linearly elastic/viscoelastic Koiter shell equations, and the thick structure is modeled by the equations of linear elasticity, can be obtained by combining the results of [119] and the results of [120].

2.6.1 Problem Definition

We consider the flow of an incompressible, viscous fluid modeled by the Navier–Stokes equations in a 2D, time-dependent cylindrical fluid domain $\Omega_F(t)$, which is not known a priori:

$$\textbf{FLUID}: \quad \left.\begin{array}{r} \rho_F(\partial_t \mathbf{u} + \mathbf{u}\cdot\nabla\mathbf{u}) = \nabla\cdot\boldsymbol{\sigma}, \\ \nabla\cdot\mathbf{u} = 0, \end{array}\right\} \text{ in } \Omega_F(t),\ t\in(0,T), \tag{2.91}$$

where ρ_F denotes the fluid density; $\mathbf{u}$ the fluid velocity; $\boldsymbol{\sigma} = -p\mathbf{I} + 2\mu_F\mathbf{D}(\mathbf{u})$ is the fluid Cauchy stress tensor; p is the fluid pressure; μ is the kinematic viscosity coefficient; and $\mathbf{D}(\mathbf{u}) = \frac{1}{2}(\nabla\mathbf{u} + \nabla^\tau\mathbf{u})$ is the symmetrized gradient of $\mathbf{u}$.

The cylindrical fluid domain is of length L, with reference radius $r = R$. The radial (vertical) displacement of the cylinder radius at time t and position $z\in(0,L)$ will be denoted by $\eta(t,z)$, giving rise to a deformed domain with radius $R+\eta(t,z)$. Thus, the fluid domain, sketched in Fig. 2.9, is given by

$$\Omega_F(t) = \{(z,r)\in\mathbb{R}^2 : z\in(0,L),\ r\in(0,R+\eta(t,z)\},$$

where the lateral boundary of the cylinder corresponds to the fluid–structure interface, denoted by

$$\Gamma(t) = \{(z,r)\in\mathbb{R}^2 : z\in(0,L),\ r = R+\eta(t,z)\}.$$

Without loss of generality we only consider the upper half of the fluid cylinder, with a symmetry boundary condition prescribed at the axis of symmetry, denoted by Γ_b in Fig. 2.9.

Fig. 2.9 2D domain sketch

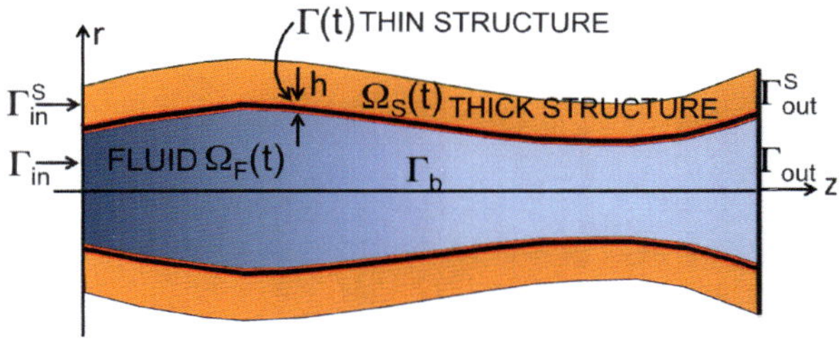

The fluid is in contact with a thin elastic structure, which is located between the fluid and the thick structural layer. The thin structure thereby serves as a fluid–structure interface with mass. We will be assuming that the elastodynamics of the thin elastic structure is governed by the 1D wave equation

$$\textbf{THIN STRUCTURE}: \quad \rho_K h\, \partial_{tt}\eta = c^2 \partial_{zz}\eta + f, \quad z \in (0, L),\ t \in (0, T), \tag{2.92}$$

where η denotes radial (vertical) displacement. Here, ρ_K is the structure density, h denotes structure thickness, and f denotes force density in the radial (vertical) direction acting on the structure. The wave equation can be viewed as a special case of the linearly (visco)elastic cylindrical Koiter shell model

$$\rho_K h \partial_t^2 \eta + C_0 \eta - C_1 \partial_z^2 \eta + C_2 \partial_z^4 \eta + D_0 \partial_t \eta - D_1 \partial_t \partial_z^2 \eta + D_2 \partial_t \partial_z^4 \eta = f, \tag{2.93}$$

with $C_0 = C_2 = D_0 = D_1 = D_2 = 0$. See Sect. 2.2.1.

The thick structural layer will be modeled by the equations of linear elasticity

$$\textbf{THICK STRUCTURE}: \quad \rho_S\, \partial_{tt}\mathbf{d} = \nabla \cdot \mathbf{S} \quad \text{in } \Omega_S,\ t \in (0, T), \tag{2.94}$$

where $\mathbf{d}(t, z, r) = (d_z(t, z, r), d_r(t, z, r))$ denotes structural displacement of the thick elastic wall at point $(z, r) \in \Omega_S$ and time t, $\boldsymbol{S}$ is the first Piola–Kirchhoff stress tensor given by $\mathbf{S} = \mu\, (\nabla\mathbf{d} + (\nabla\mathbf{d})^T) + \lambda(\nabla \cdot \mathbf{d})\mathbf{I}$, where λ and μ are the Lamé constants, and ρ_S is density of the thick structure. Domain Ω_S corresponds to a fixed, reference domain which is independent of time, and is given by

$$\Omega_S = (0, L) \times (R, R + H).$$

A deformation of Ω_S at time t is denoted by $\Omega_S(t)$ in Fig. 2.9.

The Coupling between the fluid, the thin structural layer, and the thick structural layer is achieved via

- the kinematic coupling condition:

$$\begin{aligned} (\partial_t \eta(t, z), 0) &= \boldsymbol{u}(t, z, R + \eta(t, z)), && \text{(continuity of velocity)} \\ (\eta(t, z), 0) &= \mathbf{d}(t, z, R), && \text{(continuity of displacement)} \end{aligned} \tag{2.95}$$

- the dynamic coupling condition:

$$\rho_{s_1} h \partial_{tt}\eta = c^2 \partial_{zz}\eta - J(t,z)(\sigma \mathbf{n})|_{(t,z,R+\eta(t,z))} \cdot \mathbf{e}_r + \mathbf{S}(t,z,R)\mathbf{e}_r \cdot \mathbf{e}_r. \tag{2.96}$$

  Here $J(t,z) = \sqrt{1 + (\partial_z \eta(t,z))^2}$ denotes the Jacobian of the transformation from Eulerian to Lagrangian coordinates, and $\mathbf{e}_r$ is the unit vector associated with the vertical, r-direction.

Notice that in this 2D problem both the structure and fluid equations are written in Cartesian coordinates, and so the Jacobian of the transformation between the cylindrical and Cartesian coordinates does not appear in these equations. This means, in particular, that the factor R that appears in equation (2.63) does not appear in (2.96).

The Boundary and Initial Conditions:

At the *inlet and outlet boundaries* to the fluid domain we prescribe zero tangential velocity and a given dynamic pressure $p + \frac{\rho_f}{2}|u|^2$:

$$\left.\begin{aligned} p + \frac{\rho_f}{2}|u|^2 &= P_{\text{in/out}}(t), \\ u_r &= 0, \end{aligned}\right\} \quad \text{on } \Gamma_{\text{in/out}}, \tag{2.97}$$

where $P_{\text{in/out}} \in L^2_{loc}(0,\infty)$ are given. Therefore, the fluid flow is driven by a prescribed dynamic pressure drop, and the flow enters and leaves the fluid domain orthogonally to the inlet and outlet boundary.

At the *bottom boundary* we prescribe the symmetry boundary condition:

$$u_r = \partial_r u_z = 0, \quad \text{on } \Gamma_b. \tag{2.98}$$

At the *end points of the thin structure* we prescribe zero displacement:

$$\eta(t,0) = \eta(t,L) = 0. \tag{2.99}$$

For the *thick structure*, we assume that the *external (top) boundary* $r = H$ is exposed to an external ambient pressure P_e:

$$\mathbf{S}\mathbf{e}_r = -P_e \mathbf{e}_r, \quad \text{on } \Gamma_{\text{ext}}, \tag{2.100}$$

while at the *end points* of the annular sections of the thick structure, $\Gamma^s_{\text{in/out}}$, we assume that the displacement is zero

$$\mathbf{d}(t,0,r) = \mathbf{d}(t,L,r) = 0, \quad \text{for } r \in (R,H).$$

The *initial* fluid and structural velocities, and the initial displacements are given by

$$\mathbf{u}(0,.) = \mathbf{u}_0,\ \eta(0,.) = \eta_0,\ \partial_t \eta(0,.) = v_0,\ \mathbf{d}(0,.) = \mathbf{d}_0,\ \partial_t \mathbf{d}(0,.) = \mathbf{V}_0, \tag{2.101}$$

and are assumed to belong to the following spaces: $\mathbf{u}_0 \in L^2(\Omega_F(0))$, $\eta_0 \in H_0^1(0,1)$, $v_0 \in L^2(0,1)$, $\boldsymbol{V}_0 \in L^2(\Omega_S)$, $\boldsymbol{d}_0 \in H^1(\Omega_S)$, satisfying the following compatibility conditions:

$$\begin{gathered}(\eta_0(z), 0) = \mathbf{d}_0(z, R),\\ \eta_0(0) = \eta_0(L) = v_0(0) = v_0(L) = 0 = \mathbf{d}_0(0,.) = \mathbf{d}_0(L,.) = \mathbf{V}_0(0,.) = \mathbf{V}_0(L,.),\\ R + \eta_0(z) > 0, \quad z \in [0, L].\end{gathered} \tag{2.102}$$

We study the existence of a weak solution to the nonlinear FSI problem (2.91)–(2.102), in which the flow is driven by the time-dependent inlet and outlet dynamic pressure data.

For simplicity, in the rest of this section, we will be setting all the parameters in the problem to be equal to 1. This includes the domain parameters R and L, the Lamé constants λ and μ, and the structure parameters ρ_K, ρ_S and h. Furthermore, we will be assuming that the external pressure, given in (2.68), is equal to zero. Alternatively, we subtract the constant external pressure data from the inlet and outlet dynamic pressure data to obtain an equivalent problem.

2.6.2 *The Energy of the Coupled Problem*

By using the same approach as described in Sect. 2.5.1, one can now show that the following energy estimate holds:

$$\frac{d}{dt}(E_{\text{kin}}(t) + E_{\text{el}}(t)) + D(t) \leq C(P_{\text{in}}(t), P_{\text{out}}(t)), \tag{2.103}$$

where

$$\begin{aligned} E_{\text{kin}}(t) &:= \frac{1}{2}\left(\|\mathbf{u}\|^2_{L^2(\Omega_F(t))} + \|\partial_t \eta\|^2_{L^2(\Gamma)} + \|\partial_t \mathbf{d}\|^2_{L^2(\Omega_S)}\right),\\ E_{\text{el}}(t) &:= \frac{1}{2}\left(\|\partial_z \eta\|^2_{L^2(\Gamma)} + 2\|\mathbf{D}(\mathbf{d})\|^2_{L^2(\Omega_S)} + \|\nabla \cdot \mathbf{d}\|^2_{L^2(\Omega_S)}\right), \end{aligned} \tag{2.104}$$

denote the kinetic and elastic energy of the coupled problem, respectively, and the term $D(t)$ captures viscous dissipation in the fluid:

$$D(t) := \|\mathbf{D}(\mathbf{u})\|^2_{L^2(\Omega_F(t))}. \tag{2.105}$$

The constant $C(P_{\text{in}}(t), P_{\text{out}}(t)))$ depends only on the inlet and outlet pressure data, which are both functions of time.

2.6.3 The ALE Formulation and Lie Splitting

First-Order ALE Formulation

As mentioned earlier, since we consider nonlinear coupling between the fluid and structure, the fluid domain changes in time. To prove the existence of a weak solution to (2.91)–(2.102) it is convenient to map the fluid domain onto a fixed domain Ω_F. The structural problems are already defined on fixed domains since they are formulated in the Lagrangian framework. We map our fluid domain $\Omega_F(t)$ onto Ω_F by using an Arbitrary Lagrangian–Eulerian (ALE) mapping [21,52,85,132,133]. We remark here that in our problem it is not convenient to use Lagrangian formulation for the fluid sub-problem, as is done in, e.g., [34,44,102], since, in our problem, the fluid domain consists of a fixed, control volume of a cylinder, with prescribed inlet and outlet pressure data, which does not follow Largangian flow.

We begin by defining a family of ALE mappings A_η parameterized by η:

$$A_\eta(t) : \Omega_F \to \Omega_F(t), \quad A_\eta(t)(\tilde{z}, \tilde{r}) := \begin{pmatrix} \tilde{z} \\ (1 + \eta(t, \tilde{z}))\tilde{r} \end{pmatrix}, \quad (\tilde{z}, \tilde{r}) \in \Omega_F, \tag{2.106}$$

where $(\tilde{z}, \tilde{r})$ denote the coordinates in the reference domain $\Omega_F = (0,1) \times (0,1)$. The mapping $A_\eta(t)$ is a bijection, and its Jacobian is given by

$$|\det \nabla A_\eta(t)| = |1 + \eta(t, \tilde{z})|. \tag{2.107}$$

Composite functions with the ALE mapping will be denoted by

$$\mathbf{u}^\eta(t,.) = \mathbf{u}(t,.) \circ A_\eta(t) \quad \text{and} \quad p^\eta(t,.) = p(t,.) \circ A_\eta(t). \tag{2.108}$$

The derivatives of composite functions satisfy:

$$\partial_t \mathbf{u} = \partial_t \mathbf{u}^\eta - (\mathbf{w}^\eta \cdot \nabla^\eta)\mathbf{u}^\eta, \quad \nabla \mathbf{u} = \nabla^\eta \mathbf{u}^\eta, \tag{2.109}$$

where the ALE domain velocity, $\mathbf{w}^\eta$, and the transformed gradient, ∇^η, are given by:

$$\mathbf{w}^\eta = \partial_t \eta \tilde{r} \mathbf{e}_r, \quad \nabla^\eta = \begin{pmatrix} \partial_{\tilde{z}} - \tilde{r} \dfrac{\partial_z \eta}{1+\eta} \partial_{\tilde{r}} \\ \dfrac{1}{1+\eta} \partial_{\tilde{r}} \end{pmatrix}. \tag{2.110}$$

One can see that $\nabla^\eta \mathbf{v} = \nabla \mathbf{v} (\nabla A_\eta)^{-1}$. For the purposes of the existence proof we also introduce the following notation:

$$\sigma^\eta = -p^\eta \mathbf{I} + 2\mathbf{D}^\eta(\mathbf{u}^\eta), \quad \mathbf{D}^\eta(\mathbf{u}^\eta) = \frac{1}{2}(\nabla^\eta \mathbf{u}^\eta + (\nabla^\eta)^\tau \mathbf{u}^\eta).$$

We are now ready to rewrite problem (2.91)–(2.102) in ALE formulation. However, before we do that, we will make one more important step in our strategy to prove the existence of a weak solution to (2.91)–(2.102). Namely, as mentioned earlier, we would like to "solve" the coupled FSI problem by approximating the problem using the time-discretization via Lie operator splitting. Since Lie operator splitting is defined for systems that are first-order in time, see Sect. 2.5.3, we have to replace the second-order time-derivatives of η and $\mathbf{d}$, with the first-order time-derivatives of the thin and thick structure velocities, respectively. In Sect. 2.5.3 we use the kinematic coupling condition (2.62) to achieve this goal. The kinematic coupling condition states that the fluid–structure interface velocity is equal to the normal trace of the fluid velocity on $\Gamma_\eta(t)$, and so we will introduce a new variable, v, to denote this trace, and replace $\partial_t \eta$ by v *everywhere* in the structure equation. We also introduce another new variable $\mathbf{V} = \partial_t \mathbf{d}$ which denotes the thick structure velocity. This enables us to rewrite problem (2.91)–(2.102) as a first-order system in time.

Thus, the ALE formulation of problem (2.91)–(2.102), defined on the reference domain Ω_F, and written as a first-order system in time, is given by the following (we drop the superscript η in $\mathbf{u}^\eta$ to simplify notation):

Find $\mathbf{u}(t,\tilde{z},\tilde{r})$, $p(t,\tilde{z},\tilde{r})$, $\eta(t,\tilde{z})$, $v(t,\tilde{z})$, $\mathbf{d}(t,\tilde{z})$ *and* $\mathbf{V}(t,\tilde{z})$ *such that*

$$\left.\begin{aligned} \partial_t \mathbf{u} + ((\mathbf{u}-\mathbf{w}^\eta)\cdot\nabla^\eta)\mathbf{u} &= \nabla^\eta\cdot\sigma^\eta, \\ \nabla^\eta\cdot\mathbf{u} &= 0, \end{aligned}\right\} \text{ in } (0,T)\times\Omega_F, \tag{2.111}$$

$$\left.\begin{aligned} u_r &= 0, \\ \partial_r u_z &= 0 \end{aligned}\right\} \text{ on } (0,T)\times\Gamma_b, \tag{2.112}$$

$$\left.\begin{aligned} p + \tfrac{1}{2}|u|^2 &= P_{\text{in/out}}(t), \\ u_r &= 0, \end{aligned}\right\} \text{ on } (0,T)\times\Gamma_{\text{in/out}}, \tag{2.113}$$

$$\left.\begin{aligned} \mathbf{u} &= v\mathbf{e}_r, \\ \mathbf{d} &= \eta\mathbf{e}_r, \\ \partial_t\eta &= v, \\ \partial_t v - \partial_z^2\eta &= -J\sigma\mathbf{n}\cdot\mathbf{e}_r + \mathbf{S}\mathbf{e}_r\cdot\mathbf{e}_r \end{aligned}\right\} \text{ on } (0,T)\times(0,1), \tag{2.114}$$

$$\left.\begin{aligned} \partial_t\mathbf{d} &= \mathbf{V}, \\ \partial_t\mathbf{V} &= \nabla\cdot\mathbf{S}, \end{aligned}\right\} \text{ on } \Omega_S, \tag{2.115}$$

$$\begin{aligned} \eta &= 0 \text{ on } (0,T)\times\partial\Gamma, \\ \mathbf{d} &= 0 \text{ on } (0,T)\times\Gamma^s_{\text{in/out}} \end{aligned} \tag{2.116}$$

$$\mathbf{S}\mathbf{e}_r = 0 \text{ on } (0,T)\times\Gamma_{\text{ext}}. \tag{2.117}$$

$$\mathbf{u}(0,.) = \mathbf{u}_0, \eta(0,.) = \eta_0, v(0,.) = v_0, \mathbf{d}(0,.) = \mathbf{d}_0, \mathbf{V}(0,.) = \mathbf{V}_0 \quad \text{at} \quad t = 0. \tag{2.118}$$

This defines a parabolic–hyperbolic–hyperbolic nonlinear moving-boundary problem. The nonlinearity appears in the equations (2.111), and in the coupling conditions (2.114) where the fluid quantities are evaluated at the deformed fluid–structure interface $\eta(t,z)$. Parabolic features are associated with the fluid problem (2.111)–(2.113), while hyperbolic features come from the 2D equations of elasticity, and from the 1D wave equation modeling the fluid–structure interface, described by the last equation in (2.114).

The Operator Splitting Scheme

To prove the existence of a weak solution to (2.111)–(2.118) we use the time-discretization via operator splitting, see Sect. 2.5.3. We apply the splitting strategy, described in Sect. 2.5.3, to separate the fluid sub-problem from the structure sub-problem.

Problem A1: The Structure Elastodynamics Problem. In this step we solve an elastodynamics problem for the location of the multi-layered cylindrical wall. The problem is driven only by the initial data, i.e., the initial boundary velocity, taken from the previous time step as the trace of the fluid velocity at the fluid–structure interface. The fluid velocity $\mathbf{u}$ remains unchanged in this step. More precisely, the problem reads:

Given $(\mathbf{u}^n, \eta^n, v^n, \boldsymbol{d}^n, V^n)$ *from the previous time step, find* $(\mathbf{u}, v, \eta, V, \boldsymbol{d})$ *such that:*

$$
\begin{aligned}
&\partial_t \mathbf{u} = 0, && \text{in } (t_n, t_{n+1}) \times \Omega_F,\\
&\left.\begin{aligned}\partial_t \mathbf{V} &= \nabla \cdot \mathbf{S},\\ \partial_t \mathbf{d} &= \mathbf{V}\end{aligned}\right\} && \text{in } (t_n, t_{n+1}) \times \Omega_S,\\
&\mathbf{d} = 0 && \text{on } \Gamma^s_{\text{in/out}},\\
&\mathbf{S}\mathbf{e}_r = 0 && \text{on } (t_n, t_{n+1}) \times \Gamma_{\text{ext}},
\end{aligned}
\tag{2.119}
$$

$$
\begin{aligned}
&\mathbf{d} = \eta \mathbf{e}_r && \text{on } (t_n, t_{n+1}) \times (0,1),\\
&\left.\begin{aligned}\partial_t v - \partial_z^2 \eta &= \mathbf{S}\mathbf{e}_r \cdot \mathbf{e}_r,\\ \partial_t \eta &= v\end{aligned}\right\} && \text{on } (t_n, t_{n+1}) \times (0,1),\\
&\eta(0) = \eta(1) = 0,
\end{aligned}
\tag{2.120}
$$

with $\mathbf{u}(t_n) = \mathbf{u}^n,\ \eta(t_n) = \eta^n,\ v(t_n) = v^n,\ \mathbf{d}(t_n) = \mathbf{d}^n,\ \mathbf{V}(t_n) = \mathbf{V}^n$.
Then set $\mathbf{u}^{n+\frac{1}{2}} = \mathbf{u}(t_{n+1})$, $\eta^{n+\frac{1}{2}} = \eta(t_{n+1})$, $v^{n+\frac{1}{2}} = v(t_{n+1})$, $\mathbf{d}^{n+\frac{1}{2}} = \mathbf{d}(t_{n+1})$, $\mathbf{V}^{n+\frac{1}{2}} = \mathbf{V}(t_{n+1})$.

Problem A2: The Fluid Problem. In this step we solve the Navier–Stokes equations coupled with structure inertia through a "Robin-type" boundary condition on Γ (lines 5 and 6 in (2.121) below). The kinematic coupling condition is implicitly satisfied. The structure displacement remains unchanged. With a slight abuse of notation, the problem can be written as follows:

Find $(\mathbf{u}, v, \eta, \boldsymbol{V}, \boldsymbol{d})$ *such that:*

$$
\begin{aligned}
&\partial_t \eta = 0 && \text{on } (t_n, t_{n+1}) \times (0,1),\\
&\partial_t \mathbf{d} = 0 && \text{on } (t_n, t_{n+1}) \times \Omega_S,\\
&\left.\begin{aligned}\partial_t \mathbf{u} + ((\mathbf{u}^n - \mathbf{w}^{\eta^{n+\frac{1}{2}}}) \cdot \nabla^{\eta^n})\mathbf{u} &= \nabla^{\eta^n} \cdot \sigma^{\eta^n}\\ \nabla^{\eta^n} \cdot \mathbf{u} &= 0\end{aligned}\right\} && \text{in } (t_n, t_{n+1}) \times \Omega_F,\\
&\left.\begin{aligned}\partial_t v &= -J\sigma\mathbf{n}\cdot\mathbf{e}_r\\ \mathbf{u} &= v\mathbf{e}_r\end{aligned}\right\} && \text{on } (t_n, t_{n+1}) \times (0,1),\\
&\left.\begin{aligned}u_r &= 0\\ \partial_r u_z &= 0\end{aligned}\right\} && \text{on } (t_n, t_{n+1}) \times \Gamma_b,\\
&\left.\begin{aligned}p + \frac{\rho_f}{2}|u|^2 &= P_{\text{in/out}}(t)\\ u_r &= 0\end{aligned}\right\} && \text{on } (t_n, t_{n+1}) \times \Gamma_{\text{in/out}},
\end{aligned}
\tag{2.121}
$$

with $\mathbf{u}(t_n,.) = \mathbf{u}^{n+\frac{1}{2}}$, $\eta(t_n,.) = \eta^{n+\frac{1}{2}}$, $v(t_n,.) = v^{n+\frac{1}{2}}$, $\mathbf{d}(t_n,.) = \mathbf{d}^{n+\frac{1}{2}}$, $\mathbf{V}(t_n,.) = \mathbf{V}^{n+\frac{1}{2}}$.

Then set $\mathbf{u}^{n+1} = \mathbf{u}(t_{n+1})$, $\eta^{n+1} = \eta(t_{n+1})$, $v^{n+1} = v(t_{n+1})$, $\mathbf{d}^{n+1} = \eta(t_{n+1})$, $\mathbf{V}^{n+1} = \mathbf{V}(t_{n+1})$.

Notice that, since in this step η does not change, this problem is linear.

In numerical simulations, one can use the ALE mapping A_{η^n} to "transform" only the time derivative term $\partial_t \mathbf{u}$ onto the fixed domain Ω_F while the rest of the PDE is discretized on the current domain $\Omega_F(t^n)$. This gives rise to the domain velocity term $\mathbf{w}$ in the equations, but avoids the unnecessary calculation of the transformed gradient ∇^{η^n}. See equation (2.195) in Sect. 2.7.3, and *Problem A2(b)* in Sect. 2.7.5. For the purposes of our proof, we will, however, remain working on the fixed, reference domain Ω_F.

It is important to notice that in Problem A2, the problem is "linearized" around the previous location of the boundary, i.e., we work with the domain determined by η^n, and not by $\eta^{n+1/2}$. This is in direct relation with the implementation of the numerical scheme studied in [21, 29]. However, we also notice that ALE velocity, $w^{n+\frac{1}{2}}$, is taken from the just calculated Problem A1! This choice is *crucial* for obtaining a semi-discrete version of an energy inequality, which will be discussed in Sect. 2.6.5.

Next we use the splitting scheme described above to define approximate solutions of (2.111)–(2.118) (or equivalently of problem (2.91)–(2.102)) and show that the approximate solutions converge to a weak solution, as $\Delta t \to 0$.

2.6.4 Weak Solutions

Notation and Function Spaces

Notation. To define weak solutions of the moving-boundary problem (2.91)–(2.102) and of the moving-boundary problem (2.111)–(2.118) defined on a fixed domain, the following notation will be useful:

- a_S will denote the bilinear form associated with the elastic energy of the thick structure:

$$a_S(\mathbf{d}, \boldsymbol{\psi}) = \int_{\Omega_S} \big(2\mathbf{D}(\mathbf{d}) : \mathbf{D}(\boldsymbol{\psi}) + (\nabla \cdot \mathbf{d}) \cdot (\nabla \cdot \boldsymbol{\psi})\big). \tag{2.122}$$

  Here ":" denotes the scalar product defined in (2.7).
- b will denote the following trilinear form corresponding to the (symmetrized) nonlinear advection term in the Navier–Stokes equations:

$$b(t, \mathbf{u}, \mathbf{v}, \mathbf{w}) = \frac{1}{2}\int_{\Omega_F(t)} (\mathbf{u} \cdot \nabla)\mathbf{v} \cdot \mathbf{w} - \frac{1}{2}\int_{\Omega_F(t)} (\mathbf{u} \cdot \nabla)\mathbf{w} \cdot \mathbf{v}. \tag{2.123}$$

- The linear functional which associates the inlet and outlet dynamic pressure boundary data with a test function $\mathbf{v}$ will be denoted by:

$$\langle F(t), \mathbf{v}\rangle_{\Gamma_{\text{in/out}}} = P_{\text{in}}(t)\int_{\Gamma_{\text{in}}} v_z - P_{\text{out}}(t)\int_{\Gamma_{\text{out}}} v_z.$$

Function Spaces. For the fluid velocity we would like to work with the classical function space associated with weak solutions of the Navier–Stokes equations. This, however, requires some additional consideration. Namely, since our thin structure is governed by the linear wave equation, lacking the bending rigidity terms, weak solutions cannot be expected to be Lipschitz-continuous. Indeed, from the energy inequality (2.103) we only have $\eta \in H^1(0, 1)$, and from Sobolev embedding we get that $\eta \in C^{0,1/2}(0, 1)$, which means that $\Omega_F(t)$ is not necessarily a Lipshitz domain. However, $\Omega_F(t)$ is locally a sub-graph of a Hölder continuous function. In that case one can define "Lagrangian" trace

$$\begin{aligned} \gamma_{\Gamma(t)} &: C^1(\overline{\Omega_F(t)}) \to C(\Gamma), \\ \gamma_{\Gamma(t)} &: v \mapsto v(t, z, r + \eta(t, z)). \end{aligned} \tag{2.124}$$

Furthermore, it was shown in [32, 79, 118] that the trace operator $\gamma_{\Gamma(t)}$ can be extended by continuity to a linear operator from $H^1(\Omega_F(t))$ to $H^s(\Gamma)$, $0 \le s < \frac{1}{4}$. For a precise statement of the results about "Lagrangian" trace, see Theorem 2.2. Now, the velocity solution space can be defined in the following way:

$$\begin{aligned} V_F(t) &= \{\mathbf{u} = (u_z, u_r) \in C^1(\overline{\Omega_F(t)})^2 : \nabla \cdot \mathbf{u} = 0, \\ &\qquad u_z = 0 \text{ on } \Gamma(t),\ u_r = 0 \text{ on } \partial\Omega_F(t) \setminus \Gamma(t)\}, \\ \mathcal{V}_F(t) &= \overline{V_F(t)}^{H^1(\Omega_F(t))}. \end{aligned} \tag{2.125}$$

Using the fact that $\Omega_F(t)$ is locally a sub-graph of a Hölder continuous function we can get the following characterization of the velocity solution space $\mathcal{V}_F(t)$ (see [32, 79]):

$$\begin{aligned} \mathcal{V}_F(t) = \{\mathbf{u} = (u_z, u_r) \in H^1(\Omega_\eta(t))^2 : \nabla \cdot \mathbf{u} = 0, \\ u_z = 0 \text{ on } \Gamma(t),\ u_r = 0 \text{ on } \partial\Omega_\eta(t) \setminus \Gamma(t)\}. \end{aligned} \tag{2.126}$$

The function space associated with weak solutions of the 1D linear wave equation and the thick wall are given, respectively, by

$$\mathcal{V}_W = H_0^1(0, 1), \tag{2.127}$$

$$\mathcal{V}_S = \{\boldsymbol{\psi} = (\psi_z, \psi_r) \in H^1(\Omega_S)^2 : \psi_z = 0 \text{ on } \Gamma,\ \boldsymbol{\psi} = 0 \text{ on } \Gamma^s_{\text{in/out}}\}. \tag{2.128}$$

Motivated by the energy inequality we also define the corresponding evolution spaces for the fluid and structure sub-problems, respectively:

$$\mathcal{W}_F(0, T) = L^\infty(0, T; L^2(\Omega_F(t))) \cap L^2(0, T; \mathcal{V}_F(t)), \tag{2.129}$$

$$\mathcal{W}_W(0, T) = W^{1,\infty}(0, T; L^2(0, 1)) \cap L^2(0, T; \mathcal{V}_W), \tag{2.130}$$

$$\mathcal{W}_S(0, T) = W^{1,\infty}(0, T; L^2(\Omega_S)) \cap L^2(0, T; \mathcal{V}_S). \tag{2.131}$$

Finally, we are in a position to define the solution space for the coupled fluid-multi-layered-structure interaction problem. This space involves the kinematic coupling condition, which is enforced in strong sense. The dynamic coupling condition will be enforced in weak sense, through integration by parts in the weak formulation of the problem. Thus, we define

$$\begin{aligned} \mathcal{W}(0, T) = \{(\mathbf{u}, \eta, \boldsymbol{d}) \in \mathcal{W}_F(0, T) \times \mathcal{W}_W(0, T) \times \mathcal{W}_S(0, T) : \\ \mathbf{u}(t, z, 1 + \eta(t, z)) = \partial_t \eta(t, z)\mathbf{e}_r,\ \boldsymbol{d}(t, z, 1) = \eta(t, z)\mathbf{e}_r\}. \end{aligned} \tag{2.132}$$

Equality $\mathbf{u}(t, z, 1 + \eta(t, z)) = \partial_t \eta(t, z)\mathbf{e}_r$ is taken in the sense defined in [32, 118]. The corresponding test space will be denoted by

$$\begin{aligned} \mathcal{Q}(0, T) = \{(\mathbf{q}, \psi, \boldsymbol{\psi}) \in C_c^1([0, T); \mathcal{V}_F \times \mathcal{V}_W \times \mathcal{V}_S) : \\ \mathbf{q}(t, z, 1 + \eta(t, z)) = \psi(t, z)\mathbf{e}_r, \boldsymbol{\psi}(t, z, 1) = \psi(t, z)\mathbf{e}_r\}. \end{aligned} \tag{2.133}$$

Notice the coupling conditions in the test space that are enforced at the fluid–structure interface.

Weak Solutions for the Problem Defined on the Moving Domain

We are now in a position to define weak solutions of fluid-multi-layered structure interaction problem, defined on the moving domain $\Omega_F(t)$.

Definition 2.1. We say that $(\mathbf{u}, \eta, \boldsymbol{d}) \in \mathcal{W}(0,T)$ is a weak solution of problem (2.91)–(2.102) if for every $(\mathbf{q}, \psi, \boldsymbol{\psi}) \in \mathcal{Q}(0,T)$ the following equality holds:

$$
\begin{aligned}
&-\int_0^T\int_{\Omega_F(t)} \mathbf{u}\cdot\partial_t\mathbf{q} + \int_0^T b(t,\mathbf{u},\mathbf{u},\mathbf{q}) + 2\int_0^T\int_{\Omega_F(t)} \mathbf{D}(\mathbf{u}):\mathbf{D}(\mathbf{q}) - \frac{1}{2}\int_0^T\int_0^1 (\partial_t\eta)^2\psi \\
&\quad -\int_0^T\int_0^1 \partial_t\eta\partial_t\psi + \int_0^T\int_0^1 \partial_z\eta\partial_z\psi - \int_0^T\int_{\Omega_S}\partial_t\boldsymbol{d}\cdot\partial_t\boldsymbol{\psi} + \int_0^T a_S(\boldsymbol{d},\boldsymbol{\psi}) \\
&\quad = \int_0^T \langle F(t),\mathbf{q}\rangle_{\Gamma_{\text{in/out}}} + \int_{\Omega_{\eta_0}} \mathbf{u}_0\cdot\mathbf{q}(0) + \int_0^1 v_0\psi(0) + \int_{\Omega_S}\mathbf{V}_0\cdot\boldsymbol{\psi}(0).
\end{aligned}
\tag{2.134}
$$

In deriving the weak formulation we used integration by parts, and the following equalities which hold for smooth functions:

$$
\begin{aligned}
\int_{\Omega_F(t)} (\mathbf{u}\cdot\nabla)\mathbf{u}\cdot\mathbf{q} &= \frac{1}{2}\int_{\Omega_F(t)}(\mathbf{u}\cdot\nabla)\mathbf{u}\cdot\mathbf{q} - \frac{1}{2}\int_{\Omega_F(t)}(\mathbf{u}\cdot\nabla)\mathbf{q}\cdot\mathbf{u} \\
&\quad + \frac{1}{2}\int_0^1 (\partial_t\eta)^2\psi \pm \frac{1}{2}\int_{\Gamma_{\text{out/in}}} |u_r|^2 v_r,
\end{aligned}
$$

$$
\int_0^T\int_{\Omega_F(t)} \partial_t\mathbf{u}\cdot\mathbf{q} = -\int_0^T\int_{\Omega_F(t)}\mathbf{u}\cdot\partial_t\mathbf{q} - \int_{\Omega_{\eta_0}}\mathbf{u}_0\cdot\mathbf{q}(0) - \int_0^T\int_0^1(\partial_t\eta)^2\psi.
$$

Weak Solutions for the Problem Defined on a Fixed, Reference Domain

Since most of the analysis will be performed on the problem mapped to Ω_F, we rewrite the above definition in terms of Ω_F using the ALE mapping $A_\eta(t)$ defined in (2.106). For this purpose, the following notation will be useful. We define the transformed trilinear functional b^η:

$$
b^\eta(\mathbf{u},\mathbf{u},\mathbf{q}) := \frac{1}{2}\int_{\Omega_F}(1+\eta)((\mathbf{u}-\mathbf{w}^\eta)\cdot\nabla^\eta)\mathbf{u}\cdot\mathbf{q} - \frac{1}{2}\int_{\Omega_F}(1+\eta)((\mathbf{u}-\mathbf{w}^\eta)\cdot\nabla^\eta)\mathbf{q}\cdot\mathbf{u},
\tag{2.135}
$$

where $1+\eta$ is the Jacobian of the ALE mapping, calculated in (2.107). Notice that we have included the ALE domain velocity $\mathbf{w}^\eta$ into b^η.

It is important to point out that the transformed fluid velocity $\mathbf{u}^\eta$ is not divergence-free anymore. Rather, it satisfies the transformed divergence-free condition $\nabla^\eta \cdot \mathbf{u}^\eta = 0$. Furthermore, since η is not a Lipschitz function, the ALE mapping is not necessarily a Lipschitz function either, and, as a result, $\mathbf{u}^\eta$ is not necessarily an H^1 function on Ω_F. Therefore we need to redefine the function spaces for the fluid velocity by introducing

$$\mathcal{V}_F^\eta = \{\mathbf{u}^\eta : \mathbf{u} \in \mathcal{V}_F(t)\},$$

where $\mathbf{u}^\eta$ is defined in (2.108). Under the assumption $1 + \eta(z) > 0$, $z \in [0,1]$, the following defines a scalar product on $\mathcal{V}_F^\eta$:

$$(\mathbf{u}^\eta, \mathbf{v}^\eta)_{\mathcal{V}_F^\eta} = \int_{\Omega_F} (1+\eta)\big(\mathbf{u}^\eta \cdot \mathbf{v}^\eta + \nabla^\eta \mathbf{u}^\eta : \nabla^\eta \mathbf{v}^\eta\big) = (\mathbf{u}, \mathbf{v})_{H^1(\Omega_F(t))}.$$

Therefore, $\mathbf{u} \mapsto \mathbf{u}^\eta$ is an isometric isomorphism between $\mathcal{V}_F(t)$ and $\mathcal{V}_F^\eta$, so $\mathcal{V}_F^\eta$ is also a Hilbert space. The function spaces $\mathcal{W}_F^\eta(0,T)$ and $\mathcal{W}^\eta(0,T)$ are defined as before, but with $\mathcal{V}_F^\eta$ instead $\mathcal{V}_F(t)$. More precisely:

$$\mathcal{W}_F^\eta(0,T) = L^\infty(0,T; L^2(\Omega_F)) \cap L^2(0,T; \mathcal{V}_F^\eta(t)), \tag{2.136}$$

$$\begin{aligned} \mathcal{W}^\equiv(0,T) = \{(\mathbf{u}, \eta, \boldsymbol{d}) \in \mathcal{W}_F^\eta(0,T) \times \mathcal{W}_W(0,T) \times \mathcal{W}_S(0,T) :\\ \mathbf{u}(t,z,1) = \partial_t \eta(t,z)\mathbf{e}_r,\ \eta(t,z) = \boldsymbol{d}(t,z,1)\}. \end{aligned} \tag{2.137}$$

The corresponding test space is defined by

$$\begin{aligned} \mathcal{Q}^\equiv(0,T) = \{(\mathbf{q}, \psi, \boldsymbol{\psi}) \in C_c^1([0,T); \mathcal{V}_F^\eta \times \mathcal{V}_W \times \mathcal{V}_S) :\\ \mathbf{q}(t,z,1) = \psi(t,z)\mathbf{e}_r, \boldsymbol{\psi}(t,z,1) = \psi(t,z)\mathbf{e}_r\}. \end{aligned} \tag{2.138}$$

Definition 2.2. We say that $(\mathbf{u}, \eta, \boldsymbol{d}) \in \mathcal{W}^\eta(0,T)$ is a weak solution of problem (2.111)–(2.118) defined on the reference domain Ω_F, if for every $(\mathbf{q}, \psi, \boldsymbol{\psi}) \in \mathcal{Q}^\equiv(0,T)$ the following equality holds:

$$\begin{aligned} -\int_0^T \int_{\Omega_F} (1+\eta)\mathbf{u} \cdot \partial_t \mathbf{q} + \int_0^T b^\eta(\mathbf{u}, \mathbf{u}, \mathbf{q}) + 2\int_0^T \int_{\Omega_F} (1+\eta)\mathbf{D}^\eta(\mathbf{u}) : \mathbf{D}^\eta(\mathbf{q}) \\ -\frac{1}{2}\int_0^T \int_{\Omega_F} (\partial_t \eta)\mathbf{u} \cdot \mathbf{q} - \int_0^T \int_0^1 \partial_t \eta \partial_t \psi + \int_0^T \int_0^1 \partial_z \eta \partial_z \psi \\ -\int_0^T \int_{\Omega_S} \partial_t \boldsymbol{d} \cdot \partial_t \boldsymbol{\psi} + \int_0^T a_S(\boldsymbol{d}, \boldsymbol{\psi}) \\ = \int_0^T \langle F(t), \mathbf{q} \rangle_{\Gamma_{\text{in/out}}} + \int_{\Omega_{\eta 0}} \mathbf{u}_0 \cdot \mathbf{q}(0) + \int_0^1 v_0 \psi(0) + \int_{\Omega_S} \mathbf{V}_0 \cdot \boldsymbol{\psi}(0). \end{aligned} \tag{2.139}$$

To see that this is consistent with the weak solution defined in Definition 2.1, we present the main steps in the transformation of the first integral on the left-hand side in (2.134), responsible for the fluid kinetic energy. Namely, we formally calculate:

$$-\int_{\Omega_F(t)} \mathbf{u}\cdot\partial_t\mathbf{q} = -\int_{\Omega_F}(1+\eta)\mathbf{u}^\eta\cdot(\partial_t\mathbf{q}-(\mathbf{w}^\eta\cdot\nabla^\eta)\mathbf{q}) = -\int_{\Omega_F}(1+\eta)\mathbf{u}^\eta\cdot\partial_t\mathbf{q}$$

$$+\frac{1}{2}\int_{\Omega_F}(1+\eta)(\mathbf{w}^\eta\cdot\nabla^\eta)\mathbf{q}\cdot\mathbf{u}^\eta+\frac{1}{2}\int_{\Omega_F}(1+\eta)(\mathbf{w}^\eta\cdot\nabla^\eta)\mathbf{q}\cdot\mathbf{u}^\eta.$$

In the last integral on the right-hand side we use the definition of $\mathbf{w}^\eta$ and of ∇^η, given in (2.110), to obtain

$$\int_{\Omega_F}(1+\eta)(\mathbf{w}^\eta\cdot\nabla^\eta)\mathbf{q}\cdot\mathbf{u}^\eta = \int_{\Omega_F}\partial_t\eta\,\tilde{r}\,\partial_{\tilde{r}}\mathbf{q}\cdot\mathbf{u}^\eta.$$

Using integration by parts with respect to r, keeping in mind that η does not depend on r, we obtain

$$-\int_{\Omega_F(t)} \mathbf{u}\cdot\partial_t\mathbf{q} = -\int_{\Omega_F}(1+\eta)\mathbf{u}^\eta\cdot(\partial_t\mathbf{q}-(\mathbf{w}^\eta\cdot\nabla^\eta)\mathbf{q}) = -\int_{\Omega_F}(1+\eta)\mathbf{u}^\eta\cdot\partial_t\mathbf{q}$$

$$+\frac{1}{2}\int_{\Omega_F}(1+\eta)(\mathbf{w}^\eta\cdot\nabla^\eta)\mathbf{q}\cdot\mathbf{u}^\eta-\frac{1}{2}\int_{\Omega_F}(1+\eta)(\mathbf{w}^\eta\cdot\nabla^\eta)\mathbf{u}^\eta\cdot\mathbf{q}$$

$$-\frac{1}{2}\int_{\Omega_F}\partial_t\eta\mathbf{u}^\eta\cdot\mathbf{q}+\frac{1}{2}\int_0^1(\partial_t\eta)^2\psi,$$

By using this identity in (2.134), and by recalling the definitions for b and b^η, we obtain exactly the weak form (2.139).

In the remainder of this section we will be working on the fluid-multi-layered structure interaction problem defined on the fixed domain Ω_F, satisfying the weak formulation presented in Definition 2.2. For brevity of notation, since no confusion is possible, we omit the superscript "tilde" which is used to denote the coordinates of points in Ω_F.

2.6.5 Approximate Solutions

In this section we use the Lie operator splitting scheme and semi-discretization to define a sequence of approximate solutions of the FSI problem (2.111)–(2.118). Each of the sub-problems defined by the Lie splitting in Sect. 2.6.3 as Problem A1 and Problem A2 will be discretized in time using the Backward Euler scheme. This approach defines a time step, which will be denoted by Δt, and a number of time sub-intervals $N \in \mathbb{N}$, so that

$$(0,T) = \cup_{n=0}^{N-1}(t^n, t^{n+1}), \quad t^n = n\Delta t,\ n = 0,\dots,N-1.$$

For every subdivision containing $N \in \mathbb{N}$ sub-intervals, the vector of unknown approximate solutions will be denoted by

$$\mathbf{X}_N^{n+\frac{i}{2}} = \left(\mathbf{u}_N^{n+\frac{i}{2}}, v_N^{n+\frac{i}{2}}, \eta_N^{n+\frac{i}{2}}, \boldsymbol{V}_N^{n+\frac{i}{2}}, \boldsymbol{d}_N^{n+\frac{i}{2}}\right)^T, n = 0,1,\dots,N-1,\ i = 1,2, \tag{2.140}$$

where $i = 1,2$ denotes the solution of Problem A1 or A2, respectively. The initial condition will be denoted by $\mathbf{X}^0 = (\mathbf{u}_0, v_0, \eta_0, \boldsymbol{V}_0, \boldsymbol{d}_0)^T$.

The semi-discretization and the splitting of the problem will be performed in such a way that the semi-discrete version of the energy inequality (2.103) is preserved at every time step. This is a crucial ingredient for the existence proof.

The semi-discrete versions of the kinetic and elastic energy (2.104), and of dissipation (2.105) are defined by the following:

$$\begin{aligned} E_{\text{kin},N}^{n+\frac{i}{2}} &= \frac{1}{2}\Big(\int_{\Omega_F}(1+\eta^{n-1+i})|\mathbf{u}_N^{n+\frac{i}{2}}|^2 + \|v_N^{n+\frac{i}{2}}\|_{L^2(0,1)}^2 + \|\boldsymbol{V}_N^{n+\frac{i}{2}}\|_{L^2(\Omega_S)}^2\Big), \\ E_{\text{el},N}^{n+1} &= \frac{1}{2}\Big(\|\partial_z \eta_N^{n+\frac{1}{2}}\|_{L^2(0,1)}^2 + 2\|\mathbf{D}(\boldsymbol{d}_N^{n+\frac{1}{2}})\|_{L^2(\Omega_S)}^2 + \|\nabla\cdot\boldsymbol{d}_N^{n+\frac{1}{2}}\|_{L^2(\Omega_S)}^2\Big), \\ E_N^{n+\frac{i}{2}} &= E_{\text{kin},N}^{n+\frac{i}{2}} + E_{\text{el},N}^{n+1}, \end{aligned} \tag{2.141}$$

$$D_N^{n+1} = \Delta t\int_{\Omega_F}(1+\eta^n)|D^{\eta^n}(\mathbf{u}_N^{n+1})|^2,\ n = 0,\dots,N-1,\ i = 0,1. \tag{2.142}$$

Throughout the rest of this section we fix the time step Δt, i.e., we keep $N \in \mathbb{N}$ fixed, and study the semi-discretized sub-problems defined by the Lie splitting. To simplify notation, we omit the subscript N and write $(\mathbf{u}^{n+\frac{i}{2}}, v^{n+\frac{i}{2}}, \eta^{n+\frac{i}{2}}, \boldsymbol{V}^{n+\frac{i}{2}}, \boldsymbol{d}^{n+\frac{i}{2}})$ instead of $(\mathbf{u}_N^{n+\frac{i}{2}}, v_N^{n+\frac{i}{2}}, \eta_N^{n+\frac{i}{2}}, \boldsymbol{V}_N^{n+\frac{i}{2}}, \boldsymbol{d}_N^{n+\frac{i}{2}})$.

Semi-discretization of Problem A1

In this step $\mathbf{u}$ does not change, and so

$$\mathbf{u}^{n+\frac{1}{2}} = \mathbf{u}^n.$$

Functions $(v^{n+\frac{1}{2}}, \eta^{n+\frac{1}{2}}, \mathbf{V}^{n+\frac{1}{2}}, \mathbf{U}^{n+\frac{1}{2}}) \in \mathcal{V}_W^2 \times \mathcal{V}_S^2$ define a weak solution of the semi-discretized Problem A1 if the following holds:

$$\boldsymbol{d}^{n+\frac{1}{2}}(z,1) = \eta^{n+\frac{1}{2}}(z,1)\mathbf{e}_r,\ z \in (0,1),$$

$$\frac{\boldsymbol{d}^{n+\frac{1}{2}}-\boldsymbol{d}^n}{\Delta t}=\boldsymbol{V}^{n+\frac{1}{2}},\quad \frac{\eta^{n+\frac{1}{2}}-\eta^n}{\Delta t}=v^{n+\frac{1}{2}}, \tag{2.143}$$

$$\int_{\Omega_S}\frac{\mathbf{V}^{n+\frac{1}{2}}-\mathbf{V}^n}{\Delta t}\cdot\boldsymbol{\Psi}+\int_0^1\frac{v^{n+\frac{1}{2}}-v^n}{\Delta t}\psi+a_S(\boldsymbol{d}^{n+\frac{1}{2}},\boldsymbol{\Psi})+\int_0^1\partial_z\eta^{n+\frac{1}{2}}\partial_z\psi=0,$$

for all $(\psi,\boldsymbol{\Psi})\in\mathcal{V}_W\times\mathcal{V}_S$ such that $\boldsymbol{\Psi}(t,z,1)=\psi(t,z)$. The first equation enforces the kinematic coupling condition, the second row in (2.6.5) introduces the structure velocities, while the third equation corresponds to a weak form of the semi-discretized elastodynamics problem. Notice that we solve the thin and thick structure problems as one problem. The thin structure enters as a boundary condition for the thick structure problem.

Proposition 2.2. *For each fixed* $\Delta t>0$, *problem* (2.6.5) *has a unique solution* $(v^{n+\frac{1}{2}},\eta^{n+\frac{1}{2}},\boldsymbol{V}^{n+\frac{1}{2}},\boldsymbol{d}^{n+\frac{1}{2}})\in\mathcal{V}_W^2\times\mathcal{V}_S^2$.

Proof. First notice that Korn's inequality implies that the bilinear form a_S is coercive on $\mathcal{V}_S$. From here, the proof is a direct consequence of the Lax-Milgram Lemma applied to the weak form

$$\int_0^1\eta^{n+\frac{1}{2}}\psi+\int_{\Omega_S}\boldsymbol{d}^{n+1}\cdot\boldsymbol{\Psi}+(\Delta t)^2\Big(\int_0^1\partial_z\eta\partial_z\psi+a_S(\boldsymbol{d}^{n+\frac{1}{2}},\boldsymbol{\Psi})\Big)$$
$$=\int_0^L(\Delta t v^n+\eta^n)\psi+\int_{\Omega_S}(\Delta t\boldsymbol{V}^n+\boldsymbol{d}^n)\cdot\boldsymbol{\Psi},\ \forall(\psi,\boldsymbol{\Psi})\in\{\mathcal{V}_W\times\mathcal{V}_S|\boldsymbol{\Psi}(t,z,1)$$
$$=\psi(z,1)\},$$

which is obtained after a substitution of $v^{n+\frac{1}{2}}$ and $\boldsymbol{V}^{n+\frac{1}{2}}$ in the third equation in (2.6.5), by using Eq. $(2.6.5)_2$. □

Proposition 2.3. *For each fixed* $\Delta t>0$, *solution of problem* (2.6.5) *satisfies the following discrete energy equality:*

$$E_{\mathrm{kin},N}^{n+\frac{1}{2}}+E_{\mathrm{el},N}^{n+1}+\frac{1}{2}\big(\|v^{n+\frac{1}{2}}-v^n\|^2_{L^2(0,1)}+\|\boldsymbol{V}^{n+\frac{1}{2}}-\boldsymbol{V}^n\|^2_{L^2(\Omega_S)}$$
$$+\|\partial_z(\eta^{n+\frac{1}{2}}-\eta^n)\|^2_{L^2(0,1)}+a_S(\boldsymbol{d}^{n+\frac{1}{2}}-\boldsymbol{d}^n,\boldsymbol{d}^{n+\frac{1}{2}}-\boldsymbol{d}^n)\big)=E_{\mathrm{kin},N}^n+E_{\mathrm{el},N}^n, \tag{2.144}$$

where the kinetic and elastic energy, $E_{\mathrm{kin},N}^n$, $E_{\mathrm{el},N}^n$, *are defined in* (2.141).

Proof. From the second row in (2.6.5) we immediately get

$$v^{n+\frac{1}{2}}=\frac{\eta^{n+\frac{1}{2}}-\eta^n}{\Delta t}\in\mathcal{V}_W,\quad \boldsymbol{V}^{n+\frac{1}{2}}=\frac{\boldsymbol{d}^{n+\frac{1}{2}}-\boldsymbol{d}^n}{\Delta t}\in\mathcal{V}_S.$$

Therefore, we can proceed as usual, by substituting the test functions in (2.6.5) with structure velocities. More precisely, we replace the test function $(\psi, \boldsymbol{\psi})$ by $(v^{n+\frac{1}{2}}, \boldsymbol{V}^{n+\frac{1}{2}})$ in the first term on the left-hand side, and then replace $(\psi, \boldsymbol{\psi})$ by $((\eta^{n+\frac{1}{2}} - \eta^n)/\Delta t, (\boldsymbol{d}^{n+\frac{1}{2}} - \boldsymbol{d}^n)/\Delta t)$ in the bilinear forms that correspond to the elastic energy. To deal with the terms $(v^{n+1/2} - v^n)v^{n+1/2}$, $(\eta^{n+1/2} - \eta^n)\eta^{n+1/2}$, $(\boldsymbol{V}^{n+1/2} - \boldsymbol{V}^n) \cdot \boldsymbol{V}^{n+1/2}$, and $(\boldsymbol{d}^{n+1/2} - \boldsymbol{d}^n) \cdot \boldsymbol{d}^{n+1/2}$, we use the algebraic identity $(a-b)\cdot a = \frac{1}{2}(|a|^2 + |a-b|^2 - |b|^2)$. After multiplying the entire equation by Δt, the third equation in (2.6.5) can be written as:

$$
\begin{aligned}
&(\|v^{n+\frac{1}{2}}\|^2_{L^2(0,1)} + \|v^{n+\frac{1}{2}} - v^n\|^2_{L^2(0,1)}) + (\|\boldsymbol{V}^{n+\frac{1}{2}}\|^2_{L^2(\Omega_S)} + \|\boldsymbol{V}^{n+\frac{1}{2}} - \boldsymbol{V}^n\|^2_{L^2(\Omega_S)}) \\
&\|\partial_z \eta^{n+\frac{1}{2}}\|^2_{L^2(0,1)} + \|\partial_z(\eta^{n+\frac{1}{2}} - \partial_z \eta^n)\|^2_{L^2(0,1)} + a_S(\boldsymbol{d}^{n+\frac{1}{2}}, \boldsymbol{d}^{n+\frac{1}{2}}) \\
&+ a_S(\boldsymbol{d}^{n+\frac{1}{2}} - \boldsymbol{d}^n, \boldsymbol{d}^{n+\frac{1}{2}} - \boldsymbol{d}^n) = \|v^n\|^2_{L^2(0,1)} + \|\boldsymbol{V}^n\|^2_{L^2(\Omega_S)} + \|\partial_z \eta^n\|^2_{L^2(0,1)} \\
&+ a_S(\boldsymbol{d}^n, \boldsymbol{d}^n).
\end{aligned}
$$

Since in this sub-problem $\mathbf{u}^{n+\frac{1}{2}} = \mathbf{u}^n$, we can add $\rho_f \int_{\Omega_F}(1+\eta^n)\mathbf{u}^{n+1/2}$ on the left-hand side, and $\rho_f \int_{\Omega_F}(1+\eta^n)\mathbf{u}^n$ on the right-hand side of the equation. Furthermore, displacements $\mathbf{d}^{n+\frac{1}{2}}$ and $\eta^{n+\frac{1}{2}}$ do not change in Problem A2 (see (2.145)), and so we can replace $\boldsymbol{d}^n$ and η^n on the right-hand side of the equation with $\boldsymbol{d}^{n-\frac{1}{2}}$ and $\eta^{n-\frac{1}{2}}$, respectively, to obtain exactly the energy equality (2.144). □

Semi-discretization of Problem A2

In this step η, $\boldsymbol{d}$ and $\boldsymbol{V}$ do not change, and so

$$
\eta^{n+1} = \eta^{n+\frac{1}{2}}, \ \boldsymbol{d}^{n+1} = \boldsymbol{d}^{n+\frac{1}{2}}, \ \boldsymbol{V}^{n+1} = \boldsymbol{V}^{n+\frac{1}{2}}. \tag{2.145}
$$

Then, define $(\mathbf{u}^{n+1}, v^{n+1}) \in \mathcal{V}_F^{\eta^n} \times L^2(0,1)$ to be a weak solution of Problem $A2$ (2.121) if the following holds for each $(\mathbf{q}, \psi) \in \mathcal{V}_F^{\eta^n} \times L^2(0,1)$ such that $\mathbf{q}_{|\Gamma} = \psi \mathbf{e}_r$, velocities $(\mathbf{u}^{n+1}, v^{n+1})$:

$$
\begin{aligned}
\int_\Omega (1+\eta^n)\Bigg(&\frac{\mathbf{u}^{n+1} - \mathbf{u}^{n+\frac{1}{2}}}{\Delta t}\cdot \mathbf{q} + \frac{1}{2}\left[(\mathbf{u}^n - v^{n+\frac{1}{2}} r\mathbf{e}_r)\cdot\nabla^{\eta^n}\right]\mathbf{u}^{n+1}\cdot\mathbf{q} \\
&-\frac{1}{2}\left[(\mathbf{u}^n - v^{n+\frac{1}{2}} r\mathbf{e}_r)\cdot\nabla^{\eta^n}\right]\mathbf{q}\cdot\mathbf{u}^{n+1}\Bigg) + \frac{1}{2}\int_\Omega v^{n+\frac{1}{2}}\mathbf{u}^{n+1}\cdot\mathbf{q} \\
&+2\int_\Omega (1+\eta^n)\mathbf{D}^{\eta^n}(\mathbf{u}) : \mathbf{D}^{\eta^n}(\mathbf{q})
\end{aligned} \tag{2.146}
$$

$$+\rho_s h \int_0^1 \frac{v^{n+1} - v^{n+\frac{1}{2}}}{\Delta t}\psi = \big(P_{\text{in}}^n \int_0^1 (q_z)_{|z=0} - P_{\text{out}}^n \int_0^1 (q_z)_{|z=L}\big),$$

$$\text{with } \nabla^{\eta^n} \cdot \mathbf{u}^{n+1} = 0, \quad \mathbf{u}_{|\Gamma}^{n+1} = v^{n+1}\mathbf{e}_r,$$

where $P_{\text{in/out}}^n = \dfrac{1}{\Delta t}\displaystyle\int_{n\Delta t}^{(n+1)\Delta t} P_{\text{in/out}}(t)dt.$

The existence of a unique weak solution and energy estimate are given by the following proposition.

Proposition 2.4. *Let $\Delta t > 0$, and assume that η^n are such that $1 + \eta^n \geq R_{\min} > 0, n = 0, \ldots, N$. Then:*

1. *The fluid sub-problem defined by (2.146) has a unique weak solution $(\mathbf{u}^{n+1}, v^{n+1}) \in \mathcal{V}_F^{\eta^n} \times L^2(0,1)$;*
2. *Solution of problem (2.146) satisfies the following discrete energy inequality:*

$$\begin{aligned} E_{\text{kin},N}^{n+1} + \frac{1}{2}\int_{\Omega_F}(1+\eta^n)|\mathbf{u}^{n+1} - \mathbf{u}^n|^2 + \frac{1}{2}\|v^{n+1} - v^{n+\frac{1}{2}}\|_{L^2(0,1)}^2 \\ + D_N^{n+1} \leq E_{\text{kin},N}^{n+\frac{1}{2}} + C\Delta t((P_{in}^n)^2 + (P_{\text{out}}^n)^2), \end{aligned} \tag{2.147}$$

where the kinetic energy E_N^n and dissipation D_N^n are defined in (2.141) *and* (2.142)*, and the constant C depends only on the parameters in the problem, and not on Δt (or N).*

The proof of this proposition is identical to the proof presented in [119] which concerns an FSI problem between an incompressible, viscous fluid and a thin elastic structure modeled by a linearly elastic Koiter shell model. The fluid sub-problems presented in [119] and in the present manuscript (Problem A2) are the same, except for the fact that η in this manuscript satisfies the linear wave equation. Since $\eta^{n+1/2}$ satisfies an elliptic problem for the Laplace operator with the right-hand side given in terms of approximate velocities $v^n, v^{n+1/2} \in L^2(0,1)$ (see equation (2.6.5)), the approximation $\eta^{n+1/2}$ is $H^2(0,1)$, and so the fluid domain in the semi-discretized Problem A2 is, in fact, Lipschitz. Therefore, the proof of Proposition 2.4 is the same as the proof of Proposition 3 [119] (for statement 1) and the proof of Proposition 4 [119] (for statement 2).

We pause for a second, and summarize what we have accomplished so far. For a given $\Delta t > 0$, the time interval $(0,T)$ was divided into $N = T/\Delta t$ sub-intervals $(t^n, t^{n+1}), n = 0, \ldots, N-1$. On each sub-interval (t^n, t^{n+1}) we "solved" the coupled FSI problem by applying the Lie splitting scheme. First, Problem A1 was solved for the structure position and velocity, both thick and thin, and then Problem A2 was solved to update fluid velocity and fluid–structure interface velocity. We showed that each sub-problem has a unique solution, provided that $1 + \eta^n \geq R_{\min} > 0, n = 0, \ldots, N$, and that each sub-problem solution satisfies an energy estimate. When combined, the two energy estimates provide a discrete version of the energy estimate (2.103). Thus, for each Δt we have designed a time-marching, splitting scheme, which defines an approximate solution on $(0, T)$ of our main FSI problem (2.111)–(2.118). Furthermore, the scheme is designed in such a way that for each $\Delta t > 0$ the approximate FSI solution satisfies a discrete version of an energy estimate for the continuous problem.

We would like to ultimately show that, as $\Delta t \to 0$, the sequence of solutions parameterized by N (or Δt) converges to a weak solution of (2.111)–(2.118). Furthermore, we also need to show that $1 + \eta^n \geq R_{\min} > 0$ is satisfied for each $n = 0, \ldots, N-1$. In order to obtain this result, it is crucial to show that the discrete energy of the approximate FSI solutions defined for each Δt is *uniformly bounded*, independently of Δt (or N). This result is obtained by the following Lemma.

Lemma 2.1 (The uniform energy estimates). *Let $\Delta t > 0$ and $N = T/\Delta t > 0$. Furthermore, let $E_N^{n+\frac{1}{2}}$, E_N^{n+1}, and D_N^j be the total energy and dissipation given by* (2.141) *and* (2.142)*, respectively.*

There exists a constant $C > 0$ independent of Δt (and N) such that the following estimates hold:

1. $E_N^{n+\frac{1}{2}} \leq C,\ E_N^{n+1} \leq C$, *for all* $n = 0, \ldots, N-1$,
2. $\sum_{j=1}^{N} D_N^j \leq C$,
3. $$\sum_{n=0}^{N-1}\Big(\int_{\Omega_F}(1+\eta^n)|\mathbf{u}^{n+1}-\mathbf{u}^n|^2 + \|v^{n+1}-v^{n+\frac{1}{2}}\|^2_{L^2(0,1)} + \|v^{n+\frac{1}{2}}-v^n\|^2_{L^2(0,1)} + \|\boldsymbol{V}^{n+1}-\boldsymbol{V}^n\|^2_{L^2(\Omega_S)}\Big) \leq C,$$
4. $$\sum_{n=0}^{N-1}\Big((\|\partial_z(\eta^{n+1}-\eta^n)\|^2_{L^2(0,1)} + a_S\big(\boldsymbol{d}^{n+1}-\boldsymbol{d}^n, \boldsymbol{d}^{n+1}-\boldsymbol{d}^n\big)\Big) \leq C.$$

In fact, $C = E_0 + \tilde{C}\left(\|P_{\text{in}}\|^2_{L^2(0,T)} + \|P_{\text{out}}\|^2_{L^2(0,T)}\right)$, *where $\tilde{C}$ is the constant from* (2.147)*, which depends only on the parameters in the problem.*

Proof. We begin by adding the energy estimates (2.144) and (2.147) to obtain

$$E_N^{n+1} + D_N^{n+1} + \frac{1}{2}\Big(\int_{\Omega_F}(1+\eta^n)|\mathbf{u}^{n+1}-\mathbf{u}^n|^2 + \|v^{n+1}-v^{n+\frac{1}{2}}\|^2_{L^2(0,1)}$$
$$+\|v^{n+\frac{1}{2}}-v^n\|^2_{L^2(0,1)} + \|\boldsymbol{V}^{n+1}-\boldsymbol{V}^n\|^2_{L^2(\Omega_S)} + \|\partial_z(\eta^{n+\frac{1}{2}}-\eta^n)\|^2_{L^2(0,1)}$$
$$+a_S\big(\boldsymbol{d}^{n+1}-\boldsymbol{d}^n, \boldsymbol{d}^{n+1}-\boldsymbol{d}^n\big)\Big) \leq E_N^n + \tilde{C}\Delta t((P_{\text{in}}^n)^2 + (P_{\text{out}}^n)^2), \quad n = 0, \ldots, N-1.$$

Then, we calculate the sum, on both sides, and cancel out like terms in the kinetic energy that appear on both sides of the inequality to obtain

$$E_N^N + \sum_{n=0}^{N-1} D_N^{n+1} + \frac{1}{2}\sum_{n=0}^{N-1}\Big(\int_{\Omega_F}(1+\eta^n)|\mathbf{u}^{n+1}-\mathbf{u}^n|^2 + \|v^{n+1}-v^{n+\frac{1}{2}}\|^2_{L^2(0,1)}$$
$$+\|v^{n+\frac{1}{2}}-v^n\|^2_{L^2(0,1)} + \|\boldsymbol{V}^{n+1}-\boldsymbol{V}^n\|^2_{L^2(\Omega_S)} + \|\partial_z(\eta^{n+\frac{1}{2}}-\eta^n)\|^2_{L^2(0,1)}$$

$$+a_S\big(\boldsymbol{d}^{n+1}-\boldsymbol{d}^n,\boldsymbol{d}^{n+1}-\boldsymbol{d}^n\big)\big) \le E_0+\tilde{C}\,\Delta t\sum_{n=0}^{N-1}((P_{in}^n)^2+(P_{out}^n)^2).$$

To estimate the term involving the inlet and outlet pressure, we recall that on every sub-interval (t^n, t^{n+1}) the pressure data is approximated by a constant which is equal to the average value of the pressure over that time interval. Therefore, we have, after using Hölder's inequality:

$$\Delta t\sum_{n=0}^{N-1}(P_{in}^n)^2=\Delta t\sum_{n=0}^{N-1}\left(\frac{1}{\Delta t}\int_{n\Delta t}^{(n+1)\Delta t}P_{in}(t)dt\right)^2 \le \|P_{in}\|^2_{L^2(0,T)}.$$

By using the pressure estimate to bound the right-hand side in the above energy estimate, we have obtained all the statements in the Lemma, with the constant C given by $C = E_0 + \tilde{C}\,\|P_{in/out}\|^2_{L^2(0,T)}$.

Notice that Statement 1 can be obtained in the same way by summing from 0 to $n-1$, for each n, instead of from 0 to $N-1$. □

We will use this Lemma in the next section to show convergence of approximate solutions.

2.6.6 Convergence of Approximate Solutions

We define approximate solutions of problem (2.111)–(2.118) on $(0,T)$ to be the functions which are piece-wise constant on each sub-interval $((n-1)\Delta t, n\Delta t]$, $n = 1\dots N$ of $(0,T)$, such that for $t\in((n-1)\Delta t, n\Delta t]$, $n=1\dots N$,

$$\mathbf{u}_N(t,.)=\mathbf{u}_N^n,\ \eta_N(t,.)=\eta_N^n,\ v_N(t,.)=v_N^n,\ v_N^*(t,.)=v_N^{n-\frac{1}{2}},\ \boldsymbol{d}_N(t,.)=\boldsymbol{d}_N^n,$$
$$V_N(t,.)=V_N^n. \tag{2.148}$$

See Fig. 2.10. Notice that functions $v_N^* = v_N^{n-1/2}$ are determined by Problem A1 (the elastodynamics sub-problem), while functions $v_N = v_N^n$ are determined by Problem A2 (the fluid sub-problem). As a consequence, functions v_N are equal to the normal trace of the fluid velocity on Γ, i.e., $\mathbf{u_N}=v_N\mathbf{e}_r$, which may be different from v_N^*. However, we will show later that $\|v_N - v_N^*\|_{L^2(0,1)}\to 0$, as $N\to\infty$.

Using Lemma 2.1 we now show that these sequences are uniformly bounded in the appropriate solution spaces.

We begin by showing that $(\eta_N)_{N\in\mathbb{N}}$ is uniformly bounded in $L^\infty(0,T;H_0^1(0,1))$, and that there exists a $T>0$ for which $1+\eta_N^n>0$ holds independently of N and n.

Proposition 2.5. *The sequence $(\eta_N)_{N\in\mathbb{N}}$ is uniformly bounded in*

$$L^\infty(0,T;H_0^1(0,1)).$$

Fig. 2.10 A sketch of u_N

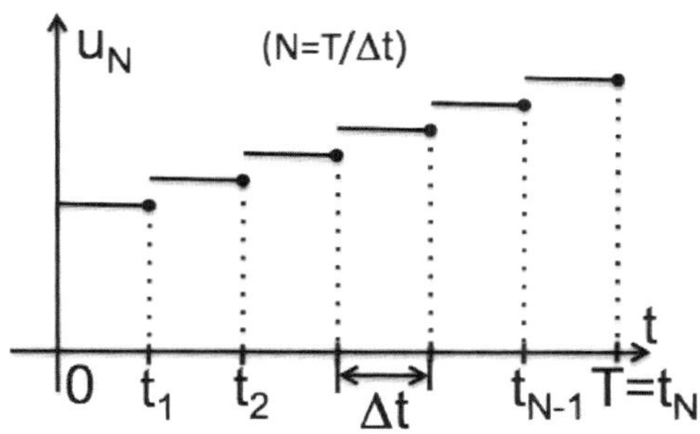

Moreover, for T small enough, we have

$$0 < R_{\min} \le 1 + \eta_N(t,z) \le R_{\max}, \ \forall N \in \mathbb{N}, z \in (0,1), t \in (0,T). \tag{2.149}$$

Proof. From the energy estimate in Lemma 2.1 we have

$$\|\eta_N(t)\|^2_{L^2(0,1)} + \|\partial_z \eta_N(t)\|^2_{L^2(0,1)}, \le C, \ \forall t \in [0,T],$$

which implies

$$\|\eta_N\|_{L^\infty(0,T;H_0^1(0,1))} \le C.$$

To show that the radius $1 + \eta_N$ is uniformly bounded away from zero for T small enough, we first notice that the above inequality implies

$$\|\eta_N^n - \eta_0\|_{H_0^1(0,1)} \le 2C, \ n = 1, \dots, N, \ N \in \mathbb{N}.$$

Furthermore,

$$\|\eta_N^n - \eta_0\|_{L^2(0,1)} \le \sum_{i=0}^{n-1} \|\eta_N^{i+1} - \eta_N^i\|_{L^2(0,1)} = \Delta t \sum_{i=0}^{n-1} \|v_N^{i+\frac{1}{2}}\|_{L^2(0,1)},$$

where we recall that $\eta_N^0 = \eta_0$. Lemma 2.1 implies that $E_N^{n+\frac{1}{2}} \le C$, where C is independent of N. Combined with the above inequality this implies

$$\|\eta_N^n - \eta_0\|_{L^2(0,1)} \le Cn\Delta t \le CT, \ n = 1, \dots, N, \ N \in \mathbb{N}.$$

Now, since $\|\eta_N^n - \eta_0\|_{L^2(0,1)}$ and $\|\eta_N^n - \eta_0\|_{H_0^1(0,1)}$ are uniformly bounded, we can use the interpolation inequality for Sobolev spaces, Theorem 4.17, p. 79 in [1], to get

$$\|\eta_N^n - \eta_0\|_{H^s(0,1)} \le 2CT^{1-s}, \ n = 1, \dots, N, \ N \in \mathbb{N}, \ \text{for } 0 < s < 1.$$

From Lemma 2.1 we see that C depends on T through the norms of the inlet and outlet data in such a way that C is an increasing function of T. Therefore, by choosing T small, we can make $\|\eta_N^n - \eta_0\|_{H^s(0,1)}$ arbitrarily small for $n = 1, \ldots, N$, $N \in \mathbb{N}$. Because of the Sobolev embedding of $H^s(0,1)$ into $C[0,1]$, for $s > 1/2$, we can also make $\|\eta_N^n - \eta_0\|_{C[0,1]}$ arbitrarily small. Since the initial data η_0 is such that $1 + \eta_0(z) > 0$ (due to the conditions listed in (2.102)), we see that for $T > 0$ small enough, there exist $R_{\min}, R_{\max} > 0$, such that

$$0 < R_{\min} \le 1 + \eta_N(t,z) \le R_{\max}, \quad \forall N \in \mathbb{N}, z \in (0,1), t \in (0,T). \qquad \square$$

We will show in the end that our existence result holds not only locally in time, i.e., for small $T > 0$, but rather, it can be extended all the way until either $T = \infty$, or until the lateral walls of the channel touch each other.

Proposition 2.5 implies, among other things, that the standard L^2-norm, and the following weighted L^2-norm are equivalent: for every $f \in L^2(\Omega_F)$, there exist constants $C_1, C_2 > 0$, which depend only on $R_{\min}, R_{\max}$, and not on f or N, such that

$$C_1 \int_{\Omega_F} (1 + \eta_N) f^2 \le \|f\|^2_{L^2(\Omega_F)} \le C_2 \int_{\Omega_F} (1 + \eta_N) f^2. \tag{2.150}$$

We will be using this property in the next section to prove strong convergence of approximate solutions.

Next we show that the sequences of approximate solutions for the velocity and its trace on the lateral boundary, as well as the displacement of the thick structure and the thick structure velocity, are uniformly bounded in the appropriate norms. To do that, we introduce the following notation which will be useful in the remainder of this section to prove compactness: denote by τ_h the translation in time by h of a function f

$$\tau_h f(t,.) = f(t-h,.), \quad h \in \mathbb{R}. \tag{2.151}$$

Proposition 2.6. *The following statements hold:*

1. $(v_N)_{N\in\mathbb{N}}, (v_N^*)_{N\in\mathbb{N}}$ *are uniformly bounded in* $L^\infty(0,T;L^2(0,1))$.
2. $(\mathbf{u}_N)_{N\in\mathbb{N}}$ *is uniformly bounded in* $L^\infty(0,T;L^2(\Omega_F))$.
3. $(\nabla^{\tau_{\Delta t}\eta_N}\mathbf{u}_N)_{N\in\mathbb{N}}$ *is uniformly bounded in* $L^2((0,T)\times\Omega_F)$.
4. $(\boldsymbol{d}_N)_{N\in\mathbb{N}}$ *is uniformly bounded in* $L^\infty(0,T;H^1(\Omega_S))$.
5. $(\boldsymbol{V}_N)_{N\in\mathbb{N}}$ *is uniformly bounded in* $L^\infty(0,T;L^2(\Omega_S))$.

Proof. The uniform boundedness of $(v_N)_{N\in\mathbb{N}}, (v_N^*)_{N\in\mathbb{N}}, (\boldsymbol{d}_N)_{N\in\mathbb{N}}, (\boldsymbol{V}_N)_{N\in\mathbb{N}}$, and the uniform boundedness of $(\mathbf{u}_N)_{N\in\mathbb{N}}$ in $L^\infty(0,T;L^2(\Omega_F))$ follow directly from Statements 1 and 2 of Lemma 2.1, and from the definition of $(v_N)_{n\in\mathbb{N}}, (v_N^*)_{N\in\mathbb{N}}$, $(\boldsymbol{d}_N)_{N\in\mathbb{N}}, (\boldsymbol{V}_N)_{N\in\mathbb{N}}$ and $(\mathbf{u}_N)_{N\in\mathbb{N}}$ as step-functions in t so that

$$\int_0^T \|v_N\|^2_{L^2(0,1)} dt = \sum_{n=0}^{N-1} \|v_N^n\|^2_{L^2(0,1)} \Delta t.$$

It remains to show uniform boundedness of $(\nabla^{\tau_{\Delta t}\eta_N}\mathbf{u}_N)_{N\in\mathbb{N}}$ in $L^2((0,T)\times\Omega_F)$. From Lemma 2.1 we only know that the *symmetrized gradient* is bounded in the following way:

$$\sum_{n=1}^{N}\int_{\Omega_F}(1+\eta_N^{n-1})|\mathbf{D}_N^{\eta^{n-1}}(\mathbf{u}_N^n)|^2\Delta t \le C. \tag{2.152}$$

We cannot immediately apply Korn's inequality since estimate (2.152) is given in terms of the transformed symmetrized gradient. Thus, there are some technical difficulties that need to be overcome due to the fact that our problem involves moving domains. To get around this difficulty we take the following approach. We first transform the problem back to the physical fluid domain $\Omega_F^{\eta_N^{n-1}}$ which is defined by the lateral boundary η_N^{n-1}, on which u_N is defined. There, instead of the transformed gradient, we have the standard gradient, and we can apply Korn's inequality in the usual way. However, since the Korn constant depends on the domain, we will need a result which provides a universal Korn constant, independent of the family of domains under consideration. Indeed, a result of this kind was obtained in [32, 119, 120, 143], assuming certain domain regularity. In particular, a calculation in [120] showed that the following Korn's equality holds for the space $\mathcal{V}_F(t)$:

$$\|\nabla\mathbf{u}^{N,n}\|^2_{L^2(\Omega_F^{\eta_N^{n-1}})} = 2\|\mathbf{D}(\mathbf{u}^{N,n})\|^2_{L^2(\Omega_F^{\eta_N^{n-1}})}. \tag{2.153}$$

Notice that the Korn constant (the number 2) is, indeed, domain independent. The proof of this Korn equality, presented in [120], is similar to the proof in Chambolle et al. [32, Lemma 6, p. 377], with the slightly different assumptions. By using (2.153) and by mapping everything back to the fixed domain Ω_F, one recovers the following Korn's equality on Ω_F:

$$2\int_{\Omega_F}(1+\eta_N^{n-1})|\mathbf{D}_N^{\eta^{n-1}}(\mathbf{u}_N^n)|^2 = \int_{\Omega_F}(1+\eta_N^{n-1})|\nabla_N^{\eta^{n-1}}(\mathbf{u}_N^n)|^2. \tag{2.154}$$

By summing equalities (2.154) for $n\ =, 1\ldots, N$, and by using (2.150), we get uniform boundedness of $(\nabla^{\tau_{\Delta t}\eta_N}\mathbf{u}_N)_{N\in\mathbb{N}}$ in $L^2((0,T)\times\Omega_F)$. □

From the uniform boundedness of approximate sequences, the following weak and weak* convergence results follow.

Lemma 2.2 (Weak and Weak* Convergence Results). *There exist subsequences* $(\eta_N)_{N\in\mathbb{N}}$, $(v_N)_{N\in\mathbb{N}}$, $(v_N^*)_{N\in\mathbb{N}}$, $(\boldsymbol{d}_N)_{N\in\mathbb{N}}$, $(\boldsymbol{V}_N)_{N\in\mathbb{N}}$ *and* $(\mathbf{u}_N)_{N\in\mathbb{N}}$, *and the functions* $\eta\in L^\infty(0,T;H_0^1(0,1))$, $v, v^*\in L^\infty(0,T;L^2(0,1))$, $\boldsymbol{d}\in L^\infty(0,T;\mathcal{V}_S)$, $\boldsymbol{V}\in L^\infty(0,T;L^2(\Omega_S))$, $\mathbf{u}\in L^\infty(0,T;L^2(\Omega_F))$ *and* $\mathbf{G}\in L^2((0,T)\times\Omega_F)$ *such that*

$$
\begin{aligned}
\eta_N &\rightharpoonup \eta \text{ weakly}* \text{ in } L^\infty(0,T;H_0^1(0,1)),\\
v_N &\rightharpoonup v \text{ weakly}* \text{ in } L^\infty(0,T;L^2(0,1)),\\
v_N^* &\rightharpoonup v^* \text{ weakly}* \text{ in } L^\infty(0,T;L^2(0,1)),\\
\\
\boldsymbol{d}_N &\rightharpoonup \boldsymbol{d} \text{ weakly}* \text{ in } L^\infty(0,T;H^1(\Omega_S)),\\
\boldsymbol{V}_N &\rightharpoonup \boldsymbol{V} \text{ weakly}* \text{ in } L^\infty(0,T;L^2(\Omega_S)),\\
\mathbf{u}_N &\rightharpoonup \mathbf{u} \text{ weakly}* \text{ in } L^\infty(0,T;L^2(\Omega_F)),\\
\nabla^{\tau_{\Delta t}\eta_N}\mathbf{u}_N &\rightharpoonup \mathbf{G} \text{ weakly in } L^2((0,T)\times\Omega_F).
\end{aligned}
\tag{2.155}
$$

Furthermore,

$$v = v^*. \tag{2.156}$$

Proof. The only thing left to show is that $v = v^*$. For this purpose, we multiply the second statement in Lemma 2.1 by Δt, and notice again that $\|v_N\|^2_{L^2((0,T)\times(0,1))} = \Delta t\sum_{n=1}^N \|v_N^n\|^2_{L^2(0,1)}$. This implies $\|v_N - v_N^*\|_{L^2((0,T)\times(0,1))} \le C\sqrt{\Delta t}$, and we have that in the limit, as $\Delta t \to 0$, $v = v^*$. □

Naturally, our goal is to prove that $\mathbf{G} = \nabla^\eta \mathbf{u}$. However, to achieve this goal we will need some stronger convergence properties of approximate solutions. Therefore, we postpone the proof until Sect. 2.6.7.

Strong Convergence of Approximate Sequences

Due to the nonlinearity of our FSI problem, to show that the limits obtained in the previous Lemma satisfy the weak form of problem (2.111)–(2.118), we will need to show that the approximate sequences converge strongly in the appropriate function spaces. The strong convergence results will be achieved by using the following compactness result by Simon [137]:

Theorem 2.1 ([137]). *Let X be a Banach space and $F \hookrightarrow L^q(0,T;X)$ with $1 \le q < \infty$. Then F is a relatively compact set in $L^q(0,T;X)$ if and only if*

(i) $\left\{\int_{t_1}^{t_2} f(t)dt : f \in F\right\}$ *is relatively compact in X, $0 < t_1 < t_2 < T$,*

(ii) $\|\tau_h f - f\|_{L^q(h,T;X)} \to 0$ *as h goes to zero, uniformly with respect to $f \in F$.*

This result was used in [119] to show compactness, but the proof was simpler because of the higher regularity of the lateral boundary of the fluid domain, namely, of the fluid–structure interface. In the present case we need to obtain some additional regularity for the fluid velocity $\mathbf{u}_N$ on Ω_F and its trace $\mathbf{v}_N$ on the lateral boundary, before we can use Theorem 2.1 to show strong convergence of approximate sequences. Notice, we only have that our fluid velocity on Ω_F is uniformly bounded in $L^2(\Omega_F)$, plus a condition that the transformed gradient $\nabla^{\tau_{\Delta t}\eta_N}\mathbf{u}_N$ is uniformly bounded in L^2. Since η is not Lipschitz, we cannot get that the gradient $\nabla \mathbf{u}_N$ is uniformly bounded in L^2 on Ω_F. This lower regularity of η_N causes additional problems in obtaining regularity of $\mathbf{u}_N$ on Ω_F, namely it will imply lower regularity of $\mathbf{u}_N$ in the sense that $\mathbf{u} \in H^s(\Omega_F)$, for $0 < s < 1/2$, and not $H^1(\Omega_F)$. Luckily, according to the trace theorem in [118], this will still allow us to make sense of the trace of $\mathbf{u}_N$ on Γ. More precisely, we prove the following Lemma.

Lemma 2.3. *The following statements hold:*

1. $(\mathbf{u}_N)_{N\in\mathbb{N}}$ *is uniformly bounded in* $L^2(0,T;H^s(\Omega_F))$, $0 < s < 1/2$;
2. $(v_N)_{N\in\mathbb{N}}$ *is uniformly bounded in* $L^2(0,T;H^{s/2}(0,1))$, $0 < s < 1/2$.

Proof. We start by mapping the fluid velocity $\mathbf{u}_N$ defined on Ω_F, back to the physical fluid domain with the lateral boundary $\tau_{\Delta t}\eta_N(t,z) = \eta_N(t-\Delta t, z)$. We denote by $\mathbf{u}^N(t,.)$ the fluid velocity on the physical domain $\Omega_{\tau_{\Delta t}\eta_N}$:

$$\mathbf{u}^N(t,.) = \mathbf{u}_N(t,.) \circ A^{-1}_{\tau_{\Delta t}\eta_N}(t), \; N \in \mathbb{N}.$$

As before, we use sub-script N to denote fluid velocity defined on the physical space. From (2.109) we see that

$$\nabla \mathbf{u}^N = \nabla^{\tau_{\Delta t}\eta_N} \mathbf{u}_N.$$

Proposition 2.6, statement 3, implies that the sequence $(\nabla \mathbf{u}^N)_{N\in\mathbb{N}}$ is uniformly bounded in L^2, and so we have that $\|\mathbf{u}^N\|_{L^2(0,T;H^1(\Omega_{\tau_{\Delta t}\eta}))}$ is uniformly bounded.

Now, from the fact that the fluid velocities $\mathbf{u}^N$ defined on the physical domains are uniformly bounded in H^1, we would like to obtain a similar result for the velocities $\mathbf{u}_N$ defined on the reference domain Ω_F. For this purpose, we recall that the functions η_N, $N \in \mathbb{N}$ that are involved in the ALE mappings $A_{\tau_{\Delta t}\eta_N}(t)$, $N \in \mathbb{N}$, are uniformly bounded in $H^1(0,1)$. This is, unfortunately, not sufficient to obtain uniform boundedness of the gradients $(\nabla u_N)_{N\in\mathbb{N}}$ in $L^2(\Omega_F)$. However, from the Sobolev embedding $H^1(0,1) \hookrightarrow C^{0,1/2}(0,1)$ we have that the sequence $(\eta_N)_{N\in\mathbb{N}}$ is uniformly bounded in $L^\infty(0,T;C^{0,1/2}(0,1))$. This will help us obtain uniform boundedness of $(\mathbf{u}_N)_{n\in\mathbb{N}}$ in a slightly lower-regularity space, namely in the space $L^2(0,T;H^s(\Omega_F))$, $0 < s < 1/2$. To see this, we first notice that $\mathbf{u}_N$ on Ω_F can be expressed in terms of function $\mathbf{u}^N$ defined on $\Omega_{\tau_{\Delta t}\eta_N}$ as

$$\mathbf{u}_N(t,\tilde{z},\tilde{r}) = \mathbf{u}^N(t,\tilde{z},(1+\tau_{\Delta t}\eta_N)(t,\tilde{z}))\tilde{r}), \; (\tilde{z},\tilde{r}) \in \Omega_F. \tag{2.157}$$

Therefore, $\mathbf{u}_N$ can be written as an H^1-function $\mathbf{u}^N$ composed with a $C^{0,1/2}$-function η_N, in the way described in (2.157). The following Lemma, proved in [118], implies that $\mathbf{u}_N$ belongs to a space with asymmetric regularity (more regular in $\tilde{r}$ than in $\tilde{z}$) in the sense that $\mathbf{u}_N \in L^2(0,1;H^s(0,1)), 0 < s < 1/2$, and $\partial_{\tilde{r}}\mathbf{u}_N \in L^2(0,1;L^2(0,1))$. We use notation from Lions and Magenes [112], p. 10, to denote the corresponding function space by

$$W(0,1;s) = \{f : f \in L^2(0,1;H^s(0,1))), \; \partial_{\tilde{r}} f \in L^2(0,1;L^2(0,1))\}.$$

More precisely, Lemma 3.3 from [118] states the following:

Lemma 2.4 ([118]). *Let* $\eta \in C^{0,\alpha}$, $0 < \alpha < 1$, *and let* $u \in H^1(\Omega_\eta)$. *Define*

$$\tilde{u}(\tilde{r}, \tilde{z}) = u(\tilde{z}, (1 + \eta(\tilde{z}))\tilde{r}), \quad (\tilde{z}, \tilde{r}) \in \Omega_F. \tag{2.158}$$

Then $\tilde{u} \in W(0, 1; s)$ *for* $0 < s < \alpha$.

Thus, Lemma 2.4 implies that $\mathbf{u}_N(t, .) \in W(0, 1; s)$ for $0 < s < 1/2$. Now, using the fact $W(0, 1; s) \hookrightarrow H^s(\Omega_F)$ we get

$$\|\mathbf{u}_N(t, .)\|^2_{H^s(\Omega_F)} \leq C\|\mathbf{u}^N(t, .)\|^2_{H^1(\Omega_{\eta(t-\Delta t)})}, \ a.a.\ t \in (0, T),\ 0 < s < 1/2.$$

By integrating the above inequality w.r.t. t we get the first statement of Lemma 2.3.

To prove the second statement of Lemma 2.3 we use Theorem 3.1 of [118], which states that the notion of trace for the functions of the form (2.157) for which $\mathbf{u}^N \in H^1$ and $\eta_N \in C^{0,1/2}$, can be defined in the sense of $H^{s/2}$, $0 < s < 1/2$. For completeness, we state Theorem 3.2 of [118] here.

Theorem 2.2 ([118]). *Let* $\alpha < 1$ *and let* η *be such that*

$$\eta \in C^{0,\alpha}(0, 1),\ \eta(z) \geq \eta_{\min} > -1,\ z \in [0, 1],\ \eta(0) = \eta(1) = 1.$$

Then, the trace operator

$$\gamma_\eta : C^1(\overline{\Omega_\eta}) \to C(\Gamma)$$

that associates with each function $u \in C^1(\overline{\Omega_\eta})$ *its "Lagrangian trace"* $u(\tilde{z}, 1 + \eta(\tilde{z})) \in C(\Gamma)$, *defined via* (2.158) *for* $\tilde{r} = 1$,

$$\gamma_\eta : u \mapsto u(\tilde{z}, 1 + \eta(\tilde{z})),$$

can be extended by continuity to a linear operator from $H^1(\Omega_\eta)$ *to* $H^s(\Gamma)$ *for* $0 \leq s < \alpha/2$.

By recalling that $v_N = (\mathbf{u}_N)_{|\Gamma}$, this proves the second statement of Lemma 2.3. □

Notice that the difficulty associated with bounding the gradient of $\mathbf{u}_N$ is somewhat artificial, since the gradient of the fluid velocity $\mathbf{u}^N$ defined on the physical domain is, in fact, uniformly bounded (by Proposition 2.6). Namely, the difficulty is imposed by the fact that we decided to work with the problem defined on a fixed domain Ω_F, and not on the family of moving domains. This decision, however, simplifies other parts of the main existence proof. The "expense" that we had to pay for this decision is embedded in the proof of Lemma 2.3.

We are now ready to use Theorem 2.1 to prove compactness of the sequences $(v_N)_{N\in\mathbb{N}}$ and $(\mathbf{u}_N)_{N\in\mathbb{N}}$.

Theorem 2.3. *Sequences* $(v_N)_{N\in\mathbb{N}}$ *and* $(\mathbf{u}_N)_{N\in\mathbb{N}}$ *are relatively compact in* $L^2(0,T;L^2(0,1))$ *and* $L^2(0,T;L^2(\Omega_F))$*, respectively.*

Proof. We use Theorem 2.1 with $q = 2$, and $X = L^2$. We verify that both assumptions (i) and (ii) hold.

Assumption (i): To show that the sequences $(v_N)_{N\in\mathbb{N}}$ and $(\mathbf{u}_N)_{N\in\mathbb{N}}$ are relatively compact in $L^2(0,1)$ and $L^2(\Omega_F)$, respectively, we use Lemma 2.3 and the compactness of the embeddings $H^s(\Omega_F) \hookrightarrow L^2(\Omega_F)$ and $H^{s/2}(0,1) \hookrightarrow L^2(0,1)$, respectively, for $0 < s < 1/2$. Namely, from Lemma 2.3 we know that sequences $(\mathbf{u}_N)_{N\in\mathbb{N}}$ and $(v_N)_{N\in\mathbb{N}}$ are uniformly bounded in $L^2(0,T;H^s(\Omega_F))$ and $L^2(0,T;H^{s/2}(0,1))$, respectively, for $0 < s < 1/2$. The compactness of the embeddings $H^s(\Omega_F) \hookrightarrow L^2(\Omega_F)$ and $H^{s/2}(0,1) \hookrightarrow L^2(0,1)$ verify Assumption (i) of Theorem 2.1.

Assumption (ii): We prove that the "integral equicontinuity," stated in assumption (ii) of Theorem 2.1, holds for the sequence $(v_N)_{N\in\mathbb{N}}$. Analogous reasoning can be used for $(\mathbf{u}_N)_{N\in\mathbb{N}}$. Thus, we want to show that for each $\varepsilon > 0$, there exists a $\delta > 0$ such that

$$\|\tau_h v_N - v_N\|^2_{L^2(\omega;L^2(0,1))} < \varepsilon, \quad \forall |h| < \delta, \text{ independently of } N \in \mathbb{N}, \tag{2.159}$$

where ω is an arbitrary compact subset of Ω. Indeed, we will show that for each $\varepsilon > 0$, the following choice of δ:

$$\delta := \min\{\text{dist}(\omega, \partial\Omega)/2, \varepsilon/(2C)\}$$

provides the desired estimate, where C is the constant from Lemma 2.1 (independent of N).

Let h be an arbitrary real number whose absolute value is less than δ. We want to show that (2.159) holds for all $\Delta t = T/N$. This will be shown in two steps. First, we will show that (2.159) holds for the case when $\Delta t \geq h$ (Case 1), and then for the case when $\Delta t < h$ (Case 2).

A short remark is in order: For a given $\delta > 0$, we will have $\Delta t < \delta$ for infinitely many N, and both cases will apply. For a finite number of functions (v_N), we will, however, have that $\Delta t \geq \delta$. For those functions (2.159) needs to be proved for all Δt such that $|h| < \delta \leq \Delta t$, which falls into Case 1 below. Thus, Cases 1 and 2 cover all the possibilities.

Case 1: $\Delta t \geq h$. We calculate the shift by h to obtain (see Fig. 2.11, left):

$$\|\tau_h v_N - v_N\|^2_{L^2(\omega;L^2(0,1))} \leq \sum_{j=1}^{N-1} \int_{j\Delta t - h}^{j\Delta t} \|v_N^j - v_N^{j+1}\|^2_{L^2(0,1)}$$

$$= h \sum_{j=1}^{N-1} \|v_N^j - v_N^{j+1}\|^2_{L^2(0,1)} \leq hC < \varepsilon/2 < \varepsilon.$$

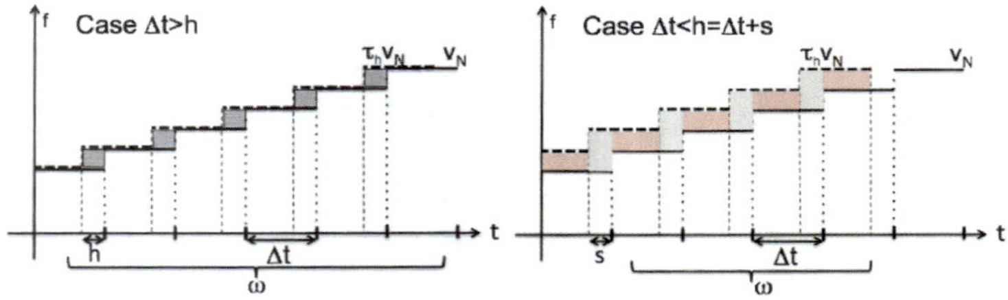

Fig. 2.11 *Left panel*—Case 1: $\Delta t \geq h$. The graph of v_N is shown in *solid line*, while the graph of the shifted function $\tau_h v_N$ is shown in *dashed line*. The *shaded area* denotes the nonzero contributions to the norm $\|\tau_h v_N - v_N\|^2_{L^2}$. *Right panel*—Case2: $\Delta t < h = \Delta t + s, 0 < s < \Delta t$. The graph of v_N is shown in *solid line*, while the graph of the shifted function $\tau_h v_N$ is shown in the *dashed line*. The *shaded areas* denote nonzero contributions to the norm $\|\tau_h v_N - v_N\|^2_{L^2}$. The *two colors* represent the contributions to the first and second integral in (2.160) separately

The last inequality follows from $|h| < \delta \leq \varepsilon/(2C)$.
Case 2: $\Delta t < h$. In this case we can write $h = l\Delta t + s$ for some ł $\in \mathbb{N}$, $0 < s \leq \Delta t$. Similarly, as in the first case, we get (see Fig. 2.11, right):

$$\|\tau_h v_N - v_N\|^2_{L^2(\omega;L^2(0,1))} = \sum_{j=1}^{N-l-1} \left(\int_{j\Delta t}^{(j+1)\Delta t - s} \|v_N^j - v_N^{j+l}\|^2_{L^2(0,1)} + \int_{(j+1)\Delta t - s}^{(j+1)\Delta t} \|v_N^j - v_N^{j+l+1}\|^2_{L^2(0,1)} \right). \tag{2.160}$$

Now we use the triangle inequality to bound each term under the two integrals from above by $\sum_{i=1}^{l+1} \|v_N^{j+i-1} - v_N^{j+i}\|^2_{L^2(0,1)}$. After combining the two terms together one obtains

$$\|\tau_h v_N - v_N\|^2_{L^2(\omega;L^2(0,1))} \leq \Delta t \sum_{j=1}^{N-l-1} \sum_{i=1}^{l+1} \|v_N^{j+i-1} - v_N^{j+i}\|^2_{L^2(0,1)}. \tag{2.161}$$

Lemma 2.1 now implies that the right-hand side of (2.161) is bounded by $\Delta t(l+1)C$. Now, since $h = l\Delta t + s$ we see that $\Delta t \leq h/l$, and so the right-hand side of (2.161) is bounded by $\frac{l+1}{l}hC$. Since $|h| < \delta$ and from the form of δ we get

$$\|\tau_h v_N - v_N\|^2_{L^2(\omega;L^2(0,1))} \leq \Delta t(l+1)C \leq \frac{l+1}{l}hC \leq \frac{l+1}{l}\frac{\varepsilon}{2} < \varepsilon.$$

Thus, if we set $\omega = [\delta/2, T - \delta/2]$ we have shown:

$$\|\tau_{\delta/2} v_N - v_N\|^2_{L^2(\delta/2,T-\delta/2;L^2(0,1))} < \varepsilon, \quad N \in \mathbb{N}.$$

To show that condition (ii) from Theorem 2.1 holds it remains to estimate $\|\tau_{\delta/2}v_N - v_N\|^2_{L^2(T-\delta/2,T;L^2(0,1))}$. From the first inequality in Lemma 2.1 (boundedness of $v_N^{n+\frac{i}{2}}, i = 1,2$ in $L^2(0,1)$) we have

$$\int_{T-\delta/2}^{T} \|\tau_{\delta/2}v_N - v_N\|^2_{L^2(0,1)} \leq \frac{\delta}{2}2C < \varepsilon, \quad N \in \mathbb{N}.$$

Thus, we have verified all the assumptions of Theorem 2.1, and so the compactness result for $(v_N)_{N\in\mathbb{N}}$ follows from Theorem 2.1. Similar arguments imply compactness of $(\mathbf{u}_N)_{N\in\mathbb{N}}$. □

To show compactness of $(\eta_N)_{N\in\mathbb{N}}$ we use the approach similar to that in [119], except that, due to the weaker regularity properties of η_N, we will have to use different embedding results (Hilbert interpolation inequalities). In the end, compactness of the sequence of lateral boundary approximation will follow due to the Arzelà–Ascoli Theorem.

As in [119], we start by introducing a slightly different set of approximate functions of $\mathbf{u}$, v, η and $\boldsymbol{V}$. Namely, for each fixed Δt (or $N \in \mathbb{N}$), define $\tilde{\mathbf{u}}_N$, $\tilde{\eta}_N$, $\tilde{v}_N$ and $\tilde{\boldsymbol{V}}_N$ to be continuous, *linear* on each sub-interval $[(n-1)\Delta t, n\Delta t]$, and such that for $n = 0,\ldots,N$:

$$\begin{aligned} &\tilde{\mathbf{u}}_N(n\Delta t,.) = \mathbf{u}_N(n\Delta t,.),\ \tilde{v}_N(n\Delta t,.) = v_N(n\Delta t,.),\\ &\tilde{\eta}_N(n\Delta t,.) = \eta_N(n\Delta t,.),\ \tilde{\boldsymbol{V}}_N(n\Delta t,.) = \boldsymbol{V}_N(n\Delta t,.), \end{aligned} \tag{2.162}$$

See Fig. 2.12. A straightforward calculation gives the following inequalities (see [141, p. 328])

$$\begin{aligned} \|v_N - \tilde{v}_N\|^2_{L^2(0,T;L^2(0,1))} &\leq \frac{\Delta t}{3}\sum_{n=1}^{N}\|v^{n+1} - v^n\|^2_{L^2(0,1)},\\ \|\mathbf{u}_N - \tilde{\mathbf{u}}_N\|^2_{L^2(0,T;L^2(\Omega_F))} &\leq \frac{\Delta t}{3}\sum_{n=1}^{N}\|\mathbf{u}^{n+1} - \mathbf{u}^n\|^2_{L^2(\Omega_F)},\\ \|\eta_N - \tilde{\eta}_N\|^2_{L^2(0,T;L^2(0,1))} &\leq \frac{\Delta t}{3}\sum_{n=1}^{N}\|\eta^{n+1} - \eta^n\|^2_{L^2(0,1)},\\ \|\boldsymbol{V}_N - \tilde{\boldsymbol{V}}_N\|^2_{L^2(0,T;L^2(\Omega_S))} &\leq \frac{\Delta t}{3}\sum_{n=1}^{N}\|\boldsymbol{V}^{n+1} - \boldsymbol{V}^n\|^2_{L^2(\Omega_S)}, \end{aligned} \tag{2.163}$$

Now from

$$\partial_t\tilde{\eta}_N(t) = \frac{\eta^{n+1} - \eta^n}{\Delta t} = \frac{\eta^{n+1/2} - \eta^n}{\Delta t} = v^{n+\frac{1}{2}},\ t \in (n\Delta t, (n+1)\Delta t),$$

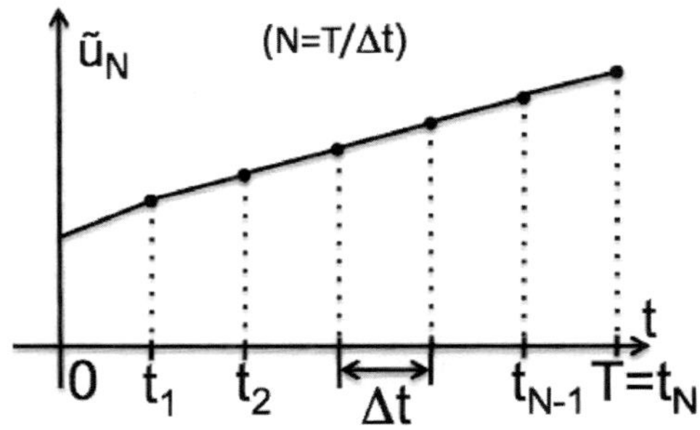

Fig. 2.12 A sketch of $\tilde{\mathbf{u}}_N$

since v_N^* was defined in (2.148) as a piece-wise constant function defined via $v_N^*(t,\cdot) = v^{n+\frac{1}{2}}$, for $t \in (n\Delta t, (n+1)\Delta t]$, we see that

$$\partial_t \tilde{\eta}_N = v_N^* \ a.e. \text{ on } (0,T). \tag{2.164}$$

Lemma 2.1 (the boundedness of $E_N^{n+\frac{i}{2}}$) then implies

$$(\tilde{\eta}_N)_{N\in\mathbb{N}} \text{ is bounded in } L^\infty(0,T;H_0^1(0,1)) \cap W^{1,\infty}(0,T;L^2(0,1)).$$

We now use the following result on continuous embeddings:

$$L^\infty(0,T;H_0^1(0,1)) \cap W^{1,\infty}(0,T;L^2(0,1)) \hookrightarrow C^{0,1-\alpha}([0,T];H^\alpha(0,1)), \tag{2.165}$$

for $0 < \alpha < 1$. This result follows from the standard Hilbert interpolation inequalities, see [112]. A slightly different result (assuming higher regularity) was also used in [79] to deal with a set of mollifying functions approximating a solution to a moving-boundary problem between a viscous fluid and an elastic plate. From (2.165) we see that $(\tilde{\eta}_N)_{N\in\mathbb{N}}$ is also bounded (uniformly in N) in $C^{0,1-\alpha}([0,T];H^\alpha(0,1))$. Now, from the continuous embedding of $H^\alpha(0,1)$ into $H^{\alpha-\epsilon}$, and by applying the Arzelà–Ascoli Theorem, we conclude that sequence $(\tilde{\eta}_N)_{N\in\mathbb{N}}$ has a convergent subsequence, which we will again denote by $(\tilde{\eta}_N)_{N\in\mathbb{N}}$, such that

$$\tilde{\eta}_N \to \tilde{\eta} \text{ in } C([0,T];H^s(0,1)),\ 0 < s < 1.$$

Since (2.163) implies that $(\tilde{\eta}_N)_{N\in\mathbb{N}}$ and $(\eta_N)_{N\in\mathbb{N}}$ have the same limit, we have $\eta = \tilde{\eta} \in C([0,T];H^s(0,1))$, where η is the weak* limit of $(\eta_N)_{N\in\mathbb{N}}$, discussed in (2.155). Thus, we have

$$\tilde{\eta}_N \to \eta \text{ in } C([0,T];H^s(0,1)),\ 0 < s < 1.$$

We can now prove the following lemma:

Lemma 2.5. $\eta_N \to \eta$ *in* $L^\infty(0,T;H^s(0,1))$, $0 < s < 1$.

Proof. The proof is similar to the proof of Lemma 3 in [119]. The result follows from the continuity in time of η, and from the fact that $\tilde{\eta}_N \to \eta$ in $C([0,T];H^s(0,1))$, for $0 < s < 1$, applied to the inequality

$$\begin{aligned}
\|\eta_N(t) - \eta(t)\|_{H^s(0,1)} &= \|\eta_N(t) - \eta(n\Delta t) + \eta(n\Delta t) - \eta(t)\|_{H^s(0,1)} \\
&= \|\eta_N(n\Delta t) - \eta(n\Delta t) + \eta(n\Delta t) - \eta(t)\|_{H^s(0,1)} \\
&\leq \|\eta_N(n\Delta t) - \eta(n\Delta t)\| + \|\eta(n\Delta t) - \eta(t)\|_{H^s(0,1)} \\
&= \|\tilde{\eta}_N(n\Delta t) - \eta(n\Delta t)\|_{H^s(0,1)} + \|\eta(n\Delta t) - \eta(t)\|_{H^s(0,1)}.
\end{aligned}$$

□

The strong convergence results obtained in Theorem 2.3 and Lemma 2.5 can be summarized as follows: there exist subsequences $(\mathbf{u}_N)_{N\in\mathbb{N}}$, $(v_N)_{N\in\mathbb{N}}$ and $(\eta_N)_{N\in\mathbb{N}}$ such that

$$\begin{aligned}
\mathbf{u}_N &\to \mathbf{u} \text{ in } L^2(0,T;L^2(\Omega_F)), \\
v_N &\to v \text{ in } L^2(0,T;L^2(0,1)), \\
\tau_{\Delta t}\mathbf{u}_N &\to u \text{ in } L^2(0,T;L^2(\Omega_F)), \\
\tau_{\Delta t}v_N &\to v \text{ in } L^2(0,T;L^2(0,1)), \\
\eta_N &\to \eta \text{ in } L^\infty(0,T;H^s(0,1)),\ 0 \leq s < 1.
\end{aligned} \tag{2.166}$$

Because of the uniqueness of derivatives, we also have $v = \partial_t \eta$ in the sense of distributions. The statements about the convergence of $(\tau_{\Delta t}\mathbf{u}_N)_{N\in\mathbb{N}}$ and $(\tau_{\Delta t}v_N)_{N\in\mathbb{N}}$ follow directly from

$$\|\tau_{\Delta t}\mathbf{u}_N - \mathbf{u}_N\|^2_{L^2((0,T)\times\Omega_F)} + \|\tau_{\Delta t}v_N - v_N\|^2_{L^2((0,T)\times(0,1))} \leq C\Delta t, \tag{2.167}$$

which is obtained after multiplying the third equality of Lemma 2.1 by Δt.

Furthermore, one can also show that subsequences $(\tilde{v}_N)_N$, $(\tilde{\mathbf{u}}_N)_N$ and $(\tilde{V}_N)_N$ also converge to v, $\mathbf{u}$ and V, respectively. More precisely,

$$\begin{aligned}
\tilde{\mathbf{u}}_N &\to \mathbf{u} \text{ in } L^2(0,T;L^2(\Omega_F)), \\
\tilde{v}_N &\to v \text{ in } L^2(0,T;L^2(0,1)), \\
\tilde{V}_N &\rightharpoonup V \text{ weakly}* \text{ in } L^\infty(0,T;L^2(\Omega_S))
\end{aligned} \tag{2.168}$$

This statement follows directly from the inequalities (2.163) and Lemma 2.1, which provides uniform boundedness of the sums on the right-hand sides of the inequalities.

We conclude this section by showing one last convergence result that will be used in the next section to prove that the limiting functions satisfy weak formulation of the FSI problem. Namely, we want to show that

$$\begin{aligned}
\eta_N &\to \eta \text{ in } L^\infty(0,T;C[0,1]), \\
\tau_{\Delta t}\eta_N &\to \eta \text{ in } L^\infty(0,T;C[0,1]).
\end{aligned} \tag{2.169}$$

The first statement is a direct consequence of Lemma 2.5 in which we proved that $\eta_N \to \eta$ in $L^\infty(0,T;H^s(0,1))$, $0<s<1$. For $s>\frac{1}{2}$ this implies

$$\eta_N \to \eta \text{ in } L^\infty(0,T;C[0,1]). \tag{2.170}$$

To show convergence of the shifted displacements $\tau_{\Delta t}\eta_N$ to the same limiting function η, we recall that

$$\tilde{\eta}_N \to \eta \text{ in } C([0,T];H^s[0,L]),\ 0<s<1,$$

and that $(\tilde{\eta}_N)_{N\in\mathbb{N}}$ is uniformly bounded in $C^{0,1-\alpha}([0,T];H^\alpha(0,1))$, $0<\alpha<1$. Uniform boundedness of $(\tilde{\eta}_N)_{N\in\mathbb{N}}$ in $C^{0,1-\alpha}([0,T];H^\alpha(0,1))$ implies that there exists a constant $C>0$, independent of N, such that

$$\|\tilde{\eta}_N((n-1)\Delta t)-\tilde{\eta}_N(n\Delta t)\|_{H^\alpha(0,1)} \le C|\Delta t|^{1-\alpha}.$$

This means that for each $\varepsilon>0$, there exists an $N_1>0$ such that

$$\|\tilde{\eta}_N((n-1)\Delta t)-\tilde{\eta}_N(n\Delta t)\|_{H^\alpha(0,1)} \le \frac{\varepsilon}{2}, \text{ for all } N\ge N_1.$$

Here, N_1 is chosen by recalling that $\Delta t=T/N$, and so the right-hand side implies that we want an N_1 such that

$$C\left(\frac{T}{N}\right)^{1-\alpha} < \frac{\varepsilon}{2} \text{ for all } N\ge N_1.$$

Now, convergence $\tilde{\eta}_N \to \eta$ in $C([0,T];H^s[0,1])$, $0<s<1$, implies that for each $\varepsilon>0$, there exists an $N_2>0$ such that

$$\|\tilde{\eta}_N(n\Delta t)-\eta(t)\|_{H^s(0,1)} < \frac{\varepsilon}{2}, \text{ for all } N\ge N_2.$$

We will use this to show that for each $\varepsilon>0$ there exists an $N^*\ge\max\{N_1,N_2\}$, such that

$$\|\tau_{\Delta t}\tilde{\eta}_N(t)-\eta(t)\|_{H^s(0,1)} < \varepsilon, \text{ for all } N\ge N^*.$$

Indeed, let $t\in(0,T)$. Then there exists an n such that $t\in((n-1)\Delta t,n\Delta t]$. We calculate

$$\begin{aligned}
\|\tau_{\Delta t}\tilde{\eta}_N(t)-\eta(t)\|_{H^s(0,1)} &= \|\tau_{\Delta t}\tilde{\eta}_N(t)-\tilde{\eta}_N(n\Delta t)+\tilde{\eta}_N(n\Delta t)-\eta(t)\|_{H^s(0,1)}\\
&= \|\tilde{\eta}_N((n-1)\Delta t)-\tilde{\eta}_N(n\Delta t)+\tilde{\eta}_N(n\Delta t)-\eta(t)\|_{H^s(0,1)}\\
&\le \|\tilde{\eta}_N((n-1)\Delta t)-\tilde{\eta}_N(n\Delta t)\|_{H^s(0,1)}+\|\tilde{\eta}_N(n\Delta t)-\eta(t)\|_{H^s(0,1)}.
\end{aligned}$$

The first term is less than $\varepsilon/2$ by the uniform boundedness of $(\tilde{\eta}_N)_{N\in\mathbb{N}}$ in $C^{0,1-\alpha}([0,T];H^{\alpha}(0,1))$, while the second term is less than $\varepsilon/2$ by the convergence of $\tilde{\eta}_N$ to η in $C([0,T];H^s[0,1])$, $0<s<1$.

Now, since $\tau_{\Delta t}\tilde{\eta}_N = \widetilde{(\tau_{\Delta t}\eta_N)}$, we can use the same argument as in Lemma 2.5 to show that sequences $\widetilde{(\tau_{\Delta t}\eta_N)}$ and $\tau_{\Delta t}\eta_N$ both converge to the same limit η in $L^\infty(0,T;H^s(0,1))$, for $0<s<1$.

2.6.7 The Limiting Problem and Weak Solution

Next we want to show that the limiting functions satisfy the weak form (2.139) of the full fluid–structure iteration problem. In this vein, one of the things that needs to be considered is what happens in the limit as $N\to\infty$, i.e., as $\Delta t\to 0$, of the weak form of the fluid sub-problem (2.146). Before we pass to the limit we must observe that, unfortunately, the velocity test functions in (2.146) depend on N! More precisely, they depend on η_N^n because of the requirement that the transformed divergence-free condition $\nabla^{\eta_N^n}\cdot\mathbf{q}=0$ must be satisfied. This is a consequence of the fact that we mapped our fluid sub-problem onto a fixed domain Ω_F. Therefore, we need to take special care when constructing suitable velocity test functions and passing to the limit in (2.146).

2.6.8 Construction of the Appropriate Test Functions

We begin by recalling that test functions $(\mathbf{q},\psi,\boldsymbol{\psi})$ for the limiting problem are defined by the space $\mathcal{Q}$, given in (2.133), which depends on η. Similarly, the test spaces for the approximate problems depend on N through the dependence on η_N.

To deal with the dependence of test functions on N, we follow the same ideas as those presented in [32, 119]. We restrict ourselves to a dense subset $\mathcal{X}$ of all test functions in $\mathcal{Q}$ that is independent of η_N even for the approximate problems. We construct the set $\mathcal{X}$ to consist of the test functions $(\mathbf{q},\psi,\boldsymbol{\psi})\in\mathcal{X}=\mathcal{X}_F\times\mathcal{X}_W\times\mathcal{X}_S$, such that the velocity components $\mathbf{q}\in\mathcal{X}_F$ are smooth, independent of N, and $\nabla\cdot\mathbf{q}=0$. Such functions can be constructed as an algebraic sum of the functions $\mathbf{q}_0$ that have compact support in $\Omega_\eta\cup\Gamma_{in}\cup\Gamma_{out}\cup\Gamma_b$, plus a function $\mathbf{q}_1$, which captures the behavior of the solution at the boundary Γ_η. More precisely, let $\Omega_{\min}$ and $\Omega_{\max}$ denote the fluid domains associated with the radii $R_{\min}$ and $R_{\max}$, respectively.

1. *Definition of test functions* $(\mathbf{q}_0,0,\mathbf{0})$ *on* $(0,T)\times\Omega_{\max}\times\Omega_S$: Consider all smooth functions $\mathbf{q}$ with compact support in $\Omega_\eta\cup\Gamma_{in}\cup\Gamma_{out}\cup\Gamma_b$, and such that $\nabla\cdot\mathbf{q}=0$. Then we can extend $\mathbf{q}$ by 0 to a divergence-free vector field on $(0,T)\times\Omega_{\max}$. This defines $\mathbf{q}_0$.

   Notice that since η_N converge uniformly to η, there exists an $N_q>0$ such that $\mathrm{supp}(\mathbf{q}_0)\subset\Omega_{\tau_{\Delta t}\eta_N}$, $\forall N\geq N_q$. Therefore, $\mathbf{q}_0$ is well defined on infinitely many approximate domains $\Omega_{\tau_{\Delta t}\eta_N}$.

2. *Definition of test functions* $(\mathbf{q}_1, \psi, \boldsymbol{\psi})$ *on* $(0,T) \times \Omega_{\max} \times \Omega_S$: Consider $\psi \in C_c^1([0,T); H_0^2(0,1))$. Define

$$\mathbf{q}_1 := \begin{cases} \left.\begin{array}{l} \text{A constant extension in the vertical} \\ \text{direction of } \psi \mathbf{e}_r \text{ on } \Gamma_\eta : \mathbf{q}_1 := (0, \psi(z))^T; \\ \text{Notice } \mathrm{div}\mathbf{q}_1 = 0. \end{array}\right\} \text{ on } \Omega_{\max} \setminus \Omega_{\min}, \\ \\ \left.\begin{array}{l} \text{A divergence-free extension to } \Omega_{\min} \\ \text{(see, e.g., [70, p. 127]).} \end{array}\right\} \text{ on } \Omega_{\min}. \end{cases}$$

From the construction it is clear that $\mathbf{q}_1$ is also defined on $\Omega_{\tau_{\Delta t}\eta_N}$ for each N, and so it can be mapped onto the reference domain Ω by the transformation $A_{\tau_{\Delta t}\eta_N}$. We take $\boldsymbol{\psi} \in H^1(\Omega_S)$ such that $\boldsymbol{\psi}(t,z,1) = \psi(t,z)$.

For any test function $(\mathbf{q}, \psi, \boldsymbol{\psi}) \in \mathcal{Q}$ it is easy to see that the velocity component $\mathbf{q}$ can then be written as $\mathbf{q} = \mathbf{q} - \mathbf{q}_1 + \mathbf{q}_1$, where $\mathbf{q} - \mathbf{q}_1$ can be approximated by divergence-free functions $\mathbf{q}_0$ that have compact support in $\Omega_\eta \cup \Gamma_{\text{in}} \cup \Gamma_{\text{out}} \cup \Gamma_b$. Therefore, one can easily see that functions $(\mathbf{q}, \psi) = (\mathbf{q}_0 + \mathbf{q}_1, \psi)$ in $\mathcal{X}$ satisfy the following properties:

- $\mathcal{X}$ is dense in the space $\mathcal{Q}$ of all test functions defined on the physical, moving domain Ω_η, defined by (2.133); furthermore, $\nabla \cdot \mathbf{q} = 0, \forall \mathbf{q} \in \mathcal{X}_F$.
- For each $\mathbf{q} \in \mathcal{X}_F$, define

$$\tilde{\mathbf{q}} = \mathbf{q} \circ A_\eta.$$

  The set $\{(\tilde{\mathbf{q}}, \psi, \boldsymbol{\psi}) | \tilde{\mathbf{q}} = \mathbf{q} \circ A_\eta, \mathbf{q} \in \mathcal{X}_F,\ \psi \in \mathcal{X}_S,\ \boldsymbol{\psi} \in \mathcal{X}_S\}$ is dense in the space $\mathcal{Q}_\eta$ of all test functions defined on the fixed, reference domain Ω_F, defined by (2.138).
- For each $\mathbf{q} \in \mathcal{X}_F$, define

$$\mathbf{q}_N := \mathbf{q} \circ A_{\tau_{\Delta t}\eta_N}.$$

  Functions $\mathbf{q}_N$ are defined on the fixed domain Ω_F, and they satisfy $\nabla^{\tau_{\Delta t}\eta_N} \cdot \mathbf{q_N} = 0$.

Functions $\mathbf{q}_N$ will serve as test functions for approximate problems associated with the sequence of domains $\Omega_{\tau_{\Delta t}\eta_N}$, while functions $\tilde{\mathbf{q}}$ will serve as test functions associated with the domain Ω_η. Both sets of test functions are defined on Ω_F.

Lemma 2.6. *For every* $(\mathbf{q}, \psi, \boldsymbol{\psi}) \in \mathcal{X}$ *we have* $\mathbf{q}_N \to \tilde{\mathbf{q}}$ *uniformly in* $L^\infty(0,T; C(\Omega_F))$.

Proof. By the Mean-Value Theorem we get:

$$\begin{aligned} |\mathbf{q}_N(t,z,r) - \tilde{\mathbf{q}}(t,z,r)| &= |\mathbf{q}(t,z,(1+\tau_{\Delta t}\eta_N)r) - \mathbf{q}(t,z,(1+\eta)r)| \\ &= |\partial_r \mathbf{q}(t,z,\zeta)r|\, |\eta(t,z) - \eta_N(t-\Delta t,z)|. \end{aligned}$$

The uniform convergence of $\mathbf{q}_N$ follows from the uniform convergence of η_N, since $\mathbf{q}$ are smooth. □

We are now ready to identify the weak limit $\mathbf{G}$ from Lemma 2.2.

Proposition 2.7. $\mathbf{G} = \nabla^\eta \mathbf{u}$, *where* $\mathbf{G}$, $\mathbf{u}$ *and* η *are the weak and weak* limits given by Lemma 2.2.*

Proof. As in Lemma 2.3, it will be helpful to map the approximate fluid velocities and the limiting fluid velocity onto the physical domains. For this purpose, we introduce the following functions

$$\mathbf{u}^N(t,.) = \mathbf{u}_N(t,.) \circ A^{-1}_{\tau_{\Delta t}\eta_N}(t), \qquad \tilde{\mathbf{u}}(t,.) = \mathbf{u}(t,.) \circ A^{-1}_{\eta}(t),$$

$$\chi^N \mathbf{f}(t,\mathbf{x}) = \begin{cases} \mathbf{f}, & \mathbf{x} \in \Omega_{\tau_{\Delta t}\eta_N}(t) \\ 0, & \mathbf{x} \notin \Omega_{\tau_{\Delta t}\eta_N}(t) \end{cases}, \quad \chi\mathbf{f}(t,\mathbf{x}) = \begin{cases} \mathbf{f}, & \mathbf{x} \in \Omega_{\eta}(t) \\ 0, & \mathbf{x} \notin \Omega_{\eta}(t) \end{cases},$$

where A is the ALE mapping defined by (2.106), η is the weak* limit $\eta_N \rightharpoonup \eta$ in $L^\infty(0,T;H^1_0(0,1))$ satisfying the uniform convergence property (2.169), and $\mathbf{f}$ is an arbitrary function defined on the physical domain. Notice, again, that superscript N is used to denote a function defined on the physical domain, while subscript N is used denote a function defined on the fixed domain Ω_F.

The proof consists of three main steps: (1) we will first show that $\chi^N \mathbf{u}^N \to \chi\tilde{\mathbf{u}}$ strongly in $L^2((0,T)\times\Omega_{\max})$, then, by using step (1), we will show (2) $\chi^N \nabla \mathbf{u}^N \to \chi\nabla\tilde{\mathbf{u}}$ weakly in $L^2((0,T)\times\Omega_{\max})$, and, finally by using (2) we will show (3) $\int_0^T \int_{\Omega_F} \mathbf{G} : \tilde{\mathbf{q}} = \int_0^T \int_{\Omega_F} \nabla^\eta \mathbf{u} : \tilde{\mathbf{q}}$ for every test function $\tilde{\mathbf{q}} = \mathbf{q} \circ A_\eta$.

STEP 1. We will show that $\|\chi^N \mathbf{u}^N - \chi\tilde{\mathbf{u}}\|_{L^2((0,T)\times\Omega_{\max})} \to 0$. To achieve this goal, we introduce the following auxiliary functions

$$\tilde{\mathbf{u}}^N(t,.) = \mathbf{u}_N(t,.) \circ A^{-1}_{\eta}(t),$$

which will be used in the following estimate

$$\begin{aligned} &\|\chi^N \mathbf{u}^N - \chi\tilde{\mathbf{u}}\|_{L^2((0,T)\times\Omega_{\max})} \\ &\quad \le \|\chi^N \mathbf{u}^N - \chi\tilde{\mathbf{u}}^N\|_{L^2((0,T)\times\Omega_{\max})} + \|\chi\tilde{\mathbf{u}}^N - \chi\tilde{\mathbf{u}}\|_{L^2((0,T)\times\Omega_{\max})}. \end{aligned}$$

The second term on the right-hand side converges to zero because of the strong convergence of $\mathbf{u}_N$ to $\mathbf{u}$ on the reference domain Ω_F, namely,

$$\|\chi\tilde{\mathbf{u}}^N - \chi\tilde{\mathbf{u}}\|^2_{L^2(\Omega_{\max})} = \int_{\Omega_F} (1+\eta)|\mathbf{u}_N - \mathbf{u}|^2 \to 0.$$

To show that the first term on the right-hand side converges to zero, first notice that

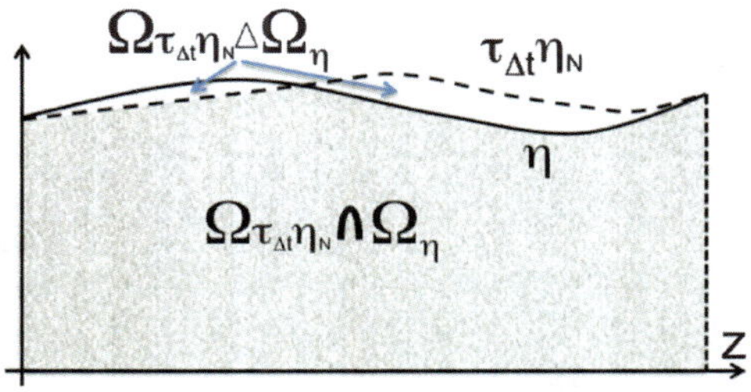

Fig. 2.13 A sketch of the fluid domains in STEP 1

$$\int_0^T \int_{\Omega_{\max}} |\chi \tilde{\mathbf{u}}^N - \chi^N \mathbf{u}^N|^2$$
$$= \Big(\int_0^T \int_{\Omega_\eta(t)\Delta\Omega_{\tau_{\Delta t}\eta_N}(t)} + \int_0^T \int_{\Omega_\eta(t)\cap\Omega_{\tau_{\Delta t}\eta_N}(t)}\Big)|\chi \tilde{\mathbf{u}}^N - \chi^N \mathbf{u}^N|^2.$$

Here $A\Delta B := (A \cup B) \setminus (A \cap B)$. See Fig. 2.13. Because of the uniform convergence (2.169) we can make the measure $|\Omega_\eta(t)\Delta\Omega_{\tau_{\Delta t}\eta_N}(t)|$ arbitrary small. Furthermore, by Propositions 2.5 and 2.6 we have that the sequence $(\chi \tilde{\mathbf{u}}^N - \chi^N \mathbf{u}^N)_{N\in\mathbb{N}}$ is uniformly bounded in $L^2((0,T)\times\Omega_{\max})$. Therefore, for every $\varepsilon > 0$, there exists an $N_0 \in \mathbb{N}$ such that for every $N \geq N_0$ we have

$$\int_0^T \int_{\Omega_\eta(t)\Delta\Omega_{\tau_{\Delta t}\eta_N}(t)} |\chi \tilde{\mathbf{u}}^N - \chi^N \mathbf{u}^N|^2 < \frac{\varepsilon}{2}. \tag{2.171}$$

To estimate the second term, we need to measure the relative difference between the function $\mathbf{u}_N$ composed with $A_\eta^{-1}(t)$, denoted by $\tilde{\mathbf{u}}^N$, and the same function $\mathbf{u}_N$ composed with $A^{-1}_{\tau_{\Delta t}\eta_N(t)}$, denoted by $\mathbf{u}^N$. We will map them both on the same domain and work with one function $\mathbf{u}_N$, while the convergence of the L^2-integral will be obtained by estimating the difference in the ALE mappings. More precisely, we introduce the set $\omega = A_\eta^{-1}(\Omega_\eta(t) \cap \Omega_{\tau_{\Delta t}\eta_N}(t)) \subset \Omega_F$. Now, we use the properties of the ALE mapping A_η and the definitions of $\tilde{\mathbf{u}}^N$, $\mathbf{u}^N$ to get

$$\int_0^T \int_{\Omega_\eta(t)\cap\Omega_{\tau_{\Delta t}\eta_N}(t)} |\chi \tilde{\mathbf{u}}^N - \chi^N \mathbf{u}^N|^2 = \int_0^T \int_\omega \frac{1}{1+\eta}|\mathbf{u}_N - \mathbf{u}_N \circ A^{-1}_{\tau_{\Delta t}\eta_N(t)} \circ A_{\eta(t)}|^2$$
$$= \int_0^T \int_\omega \frac{1}{1+\eta(t,z)}|\mathbf{u}_N(t,z,r) - \mathbf{u}_N(t,z,\frac{1+\eta(t,z)}{1+\tau_{\Delta t}\eta_N(t,z)}r)|^2$$
$$= \int_0^T \int_\omega \left|\partial_r \mathbf{u}_N(t,z,\zeta)r\left(1 - \frac{1+\eta(t,z)}{1+\tau_{\Delta t}\eta_N(t,z)}\right)\right|^2$$

Now because of the uniform convergence (2.169) of the sequence $(\tau_{\Delta t}\eta_N)_{N\in\mathbb{N}}$, and the uniform boundedness of $(\|\partial_r \mathbf{u}_N\|_{L^2(\Omega_F)})_{N\in\mathbb{N}}$, which is consequence of Proposition 2.6, we can take $N_1 \geq N_0$ such that

$$\int_0^T \int_{\Omega_\eta(t)\cap\Omega_{\tau_{\Delta t}\eta_N}(t)} |\chi\tilde{\mathbf{u}}^N - \chi^N \mathbf{u}^N|^2 < \frac{\varepsilon}{2},\ N \geq N_1.$$

This inequality, together with (2.171) proofs that $\chi^N \mathbf{u}^N \to \chi\tilde{\mathbf{u}}$ strongly in $L^2((0,T)\times\Omega_{\max})$.

STEP 2. We will now show that $\chi^N \nabla\mathbf{u}^N \rightharpoonup \chi\nabla\tilde{\mathbf{u}}$ weakly in $L^2((0,T)\times\Omega_{\max})$. First notice that from

$$\nabla\mathbf{u}^N = \nabla^{\tau_{\Delta t}\eta_N}\mathbf{u}_N$$

and from uniform boundedness of $(\nabla^{\tau_{\Delta t}\eta_N}\mathbf{u}_N)_{N\in\mathbb{N}}$ in $L^2((0,T)\times\Omega_F)$, established in Proposition 2.6, we get that the sequence $(\chi^N\nabla\mathbf{u}^N)_{N\in\mathbb{N}}$ converges weakly in $L^2((0,T)\times\Omega_{\max})$. Let us denote the weak limit of $(\chi^N\nabla\mathbf{u}^N)_{N\in\mathbb{N}}$ by $\tilde{\mathbf{G}}$. Therefore,

$$\int_0^T\int_{\Omega_{\max}} \tilde{\mathbf{G}}\cdot\boldsymbol{\phi} = \lim_{N\to\infty}\int_0^T\int_{\Omega_{\max}} \chi^N\nabla\mathbf{u}^N\cdot\boldsymbol{\phi},\quad \boldsymbol{\phi}\in C_c^\infty((0,T)\times\Omega_{\max}).$$

We want to show that $\tilde{\mathbf{G}} = \chi\nabla\tilde{\mathbf{u}}$.

For this purpose, we first consider the set $(\Omega_{\max}\setminus\Omega_\eta(t))$ and show that $\tilde{\mathbf{G}} = 0$ there, and then the set $\Omega_\eta(t)$ and show that $\tilde{\mathbf{G}} = \nabla\tilde{\mathbf{u}}$ there.

Let $\boldsymbol{\phi}$ be a test function such that $\text{supp}\boldsymbol{\phi} \subset (0,T)\times\Big(\Omega_{\max}\setminus\Omega_\eta(t)\Big)$. Using the uniform convergence of the sequence $\tau_{\Delta t}\eta_N$, obtained in (2.169), there exists an $N_{\boldsymbol{\phi}}$ such that $\chi^N(\mathbf{x}) = 0$, $N \geq N_{\boldsymbol{\phi}}$, $\mathbf{x}\in\text{supp}\boldsymbol{\phi}$. Therefore, we have

$$\int_0^T\int_{\Omega_{\max}} \tilde{\mathbf{G}}\cdot\boldsymbol{\phi} = \lim_{N\to\infty}\int_0^T\int_{\Omega_{\max}} \chi^N\nabla\mathbf{u}^N\cdot\boldsymbol{\phi} = 0.$$

Thus, $\tilde{\mathbf{G}} = 0$ on $(0,T)\times\Big(\Omega_{\max}\setminus\Omega_\eta(t)\Big)$.

Now, let us take a test function $\boldsymbol{\psi}$ such that $\text{supp}\boldsymbol{\psi}\subset(0,T)\times\Omega_\eta(t)$. Again using the same argument as before, as well as the uniform convergence of the sequence $\tau_{\Delta t}\eta_N$, obtained in (2.169), we conclude that there exists an $N_{\boldsymbol{\psi}}$ such that $\chi^N(\mathbf{x}) = 1$, $N\geq N_{\boldsymbol{\psi}}$, $\mathbf{x}\in\text{supp}\boldsymbol{\psi}$. Therefore, we have

$$\int_0^T\int_{\Omega_{\max}} \tilde{\mathbf{G}}\cdot\boldsymbol{\psi} = \lim_{N\to\infty}\int_0^T\int_{\Omega_{\max}} \chi^N\nabla\mathbf{u}^N\cdot\boldsymbol{\psi} = \lim_{N\to\infty}\int_0^T\int_{\Omega_\eta(t)} \nabla\mathbf{u}^N\cdot\boldsymbol{\psi}.$$

From the strong convergence $\chi^N\mathbf{u}^N \to \chi\tilde{\mathbf{u}}$ obtained in STEP 1, we have that on the set $\text{supp}\boldsymbol{\psi}$, $\mathbf{u}^N \to \tilde{\mathbf{u}}$ in the sense of distributions, and so, on the same set $\text{supp}\boldsymbol{\psi}$,

$\nabla \mathbf{u}^N \to \nabla \tilde{\mathbf{u}}$ in the sense of distributions. Therefore we have

$$\int_0^T \int_{\Omega_{\max}} \tilde{\mathbf{G}} \cdot \boldsymbol{\psi} = \lim_{N \to \infty} \int_0^T \int_{\Omega_\eta(t)} \nabla \mathbf{u}^N \cdot \boldsymbol{\psi} = \int_0^T \int_{\Omega_\eta(t)} \nabla \tilde{\mathbf{u}} \cdot \boldsymbol{\psi}.$$

Since this conclusion holds for all the test functions $\boldsymbol{\psi}$ supported in $(0, T) \times \Omega_{\eta(t)}$, from the uniqueness of the limit, we conclude $\tilde{\mathbf{G}} = \nabla \tilde{\mathbf{u}}$ in $(0, T) \times \Omega_{\eta(t)}$.

Therefore, we have shown that

$$\chi^N \nabla \mathbf{u}^N \rightharpoonup \chi \nabla \tilde{\mathbf{u}} \text{ weakly in } L^2((0, T) \times \Omega_{\max}).$$

STEP 3. We want to show that $\int_0^T \int_{\Omega_F} \mathbf{G} : \tilde{\mathbf{q}} = \int_0^T \int_{\Omega_F} \nabla^\eta \mathbf{u} : \tilde{\mathbf{q}}$ for every test function $\tilde{\mathbf{q}} = \mathbf{q} \circ A_\eta$, $\mathbf{q} \in \mathcal{X}_F$. This will follow from STEP 2, the uniform boundedness and convergence of the gradients $\nabla^{\tau_{\Delta t} \eta_N} \tilde{\mathbf{u}}_N$ provided by Lemma 2.2, and from the strong convergence of the test functions $\mathbf{q}_N \to \tilde{\mathbf{q}}$ provided by Lemma 2.6. More precisely, we have that for every $\tilde{\mathbf{q}} = \mathbf{q} \circ A_\eta$, $\mathbf{q} \in \mathcal{X}_F$

$$\begin{aligned}
\int_0^T \int_{\Omega_F} \mathbf{G} : \tilde{\mathbf{q}} &= \lim_{N \to \infty} \int_0^T \int_{\Omega_F} \nabla^{\tau_{\Delta t} \eta_N} \mathbf{u}_N : \mathbf{q}_N \\
&= \lim_{N \to \infty} \int_0^T \int_{\Omega_{\max}} \frac{1}{1 + \tau_{\Delta t} \eta_N} \chi^N \nabla \mathbf{u}^N : \mathbf{q} \\
&= \int_0^T \int_{\Omega_\eta} \frac{1}{1 + \eta} \nabla \tilde{\mathbf{u}} : \mathbf{q} = \int_0^T \int_{\Omega_F} \nabla^\eta \mathbf{u} : \tilde{\mathbf{q}}.
\end{aligned}$$

Here, we have used from (2.110) that $\nabla \mathbf{u}^N = \nabla^{\tau_{\Delta t} \eta_N} \mathbf{u}_N$, and $\nabla \tilde{\mathbf{u}} = \nabla^\eta \mathbf{u}$. This completes proof. □

Corollary 2.1. *For every* $(\mathbf{q}, \psi, \boldsymbol{\psi}) \in \mathcal{X}$ *we have*

$$\nabla^{\tau_{\Delta t} \eta_N} \mathbf{q}_N \to \nabla^\eta \tilde{\mathbf{q}}, \text{ in } L^2((0, T) \times \Omega_F).$$

Proof. Since $\tau_{\Delta t} \eta_N \mathbf{q}_N$ and $\tilde{\mathbf{q}}$ are the test functions for the velocity fields, the same arguments as in Proposition 2.7 provide weak convergence of $(\nabla^{\tau_{\Delta t} \eta_N} \mathbf{q}_N)_{N \in \mathbb{N}}$. To prove strong convergence it is sufficient to prove the convergence of norms $\|\nabla^{\tau_{\Delta t} \eta_N} \mathbf{q}_N\|_{L^2(\Omega_F)} \to \|\nabla^\eta \tilde{\mathbf{q}}\|_{L^2(\Omega_F)}$. This can be done, by using the uniform convergence of $(\tau_{\Delta t} \eta_N)_{N \in \mathbb{N}}$, in the following way:

$$\begin{aligned}
\|\nabla^{\tau_{\Delta t} \eta_N} \mathbf{q}_N\|^2_{L^2(\Omega_F)} &= \int_0^T \int_{\Omega_{\max}} \chi^N \frac{1}{1 + \tau_{\Delta t} \eta_N} |\nabla \mathbf{q}|^2 \to \int_0^T \int_{\Omega_{\max}} \chi \frac{1}{1 + \eta} |\nabla \mathbf{q}|^2 \\
&= \int_0^T \int_{\Omega_F} |\nabla^\eta \tilde{\mathbf{q}}|^2 = \|\nabla^\eta \tilde{\mathbf{q}}\|^2_{L^2(\Omega_F)}.
\end{aligned}$$

The notation used here is analogous to that used in the proof of Proposition 2.7. □

Before we can pass to the limit in the weak formulation of the approximate problems, there is one more useful observation that we need. Namely, notice that although $\mathbf{q}$ are smooth functions both in the spatial variables and in time, the functions $\mathbf{q}_N$ are discontinuous at $n\Delta t$ because $\tau_{\Delta t}\eta_N$ is a step function in time. As we shall see below, it will be useful to approximate each discontinuous function $\mathbf{q}_N$ in time by a piece-wise constant function, $\bar{\mathbf{q}}_N$, so that

$$\bar{\mathbf{q}}_N(t,.) = \mathbf{q}(n\Delta t-,.), \quad t \in [(n-1)\Delta t, n\Delta t),\ n = 1,\dots,N,$$

where $\mathbf{q}_N(n\Delta t-)$ is the limit from the left of $\mathbf{q}_N$ at $n\Delta t$, $n = 1,\dots,N$. By using Lemma 2.6, and by applying the same arguments in the proof of Lemma 2.5, we get

$$\bar{\mathbf{q}}_N \to \tilde{\mathbf{q}} \text{ uniformly on } [0,T]\times\Omega.$$

Passing to the Limit

To get to the weak formulation of the coupled problem, take the test functions $(\psi(t), \boldsymbol{\psi}(t)) \in \mathcal{X}_W \times \mathcal{X}_S$ as the test functions in the weak formulation of the structure sub-problem (2.6.5) and integrate the weak formulation (2.6.5) with respect to t from $n\Delta t$ to $(n+1)\Delta t$. Notice that the construction of the test functions is done in such a way that $(\psi(t), \boldsymbol{\psi}(t))$ do not depend on N, and are continuous. Then, consider the weak formulation (2.146) of the fluid sub-problem and take the test functions $(\mathbf{q}_N(t), \psi(t))$ (where $\mathbf{q}_N = \mathbf{q}\circ A_{\tau_{\Delta t}\eta_N}$, $\mathbf{q} \in \mathcal{X}_F$). Integrate the fluid sub-problem (2.146) with respect to t from $n\Delta t$ to $(n+1)\Delta t$. Add the two weak formulations together, and take the sum from $n = 0,\dots,N-1$ to get the time integrals over $(0,T)$ as follows:

$$\begin{aligned}
&\int_0^T\int_{\Omega_F}(1+\tau_{\Delta t}\,\eta_N)\Big(\partial_t\tilde{\mathbf{u}}_N\cdot\mathbf{q}_N + \frac{1}{2}(\tau_{\Delta t}\mathbf{u}_N - \mathbf{w}_N)\cdot\nabla^{\tau_{\Delta t}\eta_N}\mathbf{u}_N\cdot\mathbf{q}_N \\
&\quad -\frac{1}{2}(\tau_{\Delta t}\mathbf{u}_N - \mathbf{w}_N)\cdot\nabla^{\tau_{\Delta t}\eta_N}\mathbf{q}_N\cdot\mathbf{u}_N\Big) + \frac{1}{2}\int_0^T\int_{\Omega_F} v_N^*\mathbf{u}_N\cdot\mathbf{q}_N \\
&\quad +\int_0^T\int_{\Omega_F}(1+\tau_{\Delta t}\eta_N)2\mathbf{D}^{\tau_{\Delta t}\eta_N}(\mathbf{u_N}):\mathbf{D}^{\tau_{\Delta t}\eta_N}(\mathbf{q}_N) + \int_0^T\int_0^1\partial_t\tilde{v}_N\psi \qquad (2.172)\\
&\quad +\int_0^T\int_0^1\partial_z\eta_N\partial_z\psi + \int_0^T\int_{\Omega_S}\partial_t\tilde{\boldsymbol{V}}_N\cdot\boldsymbol{\psi} + \int_0^T\int_{\Omega_S}a_S(\boldsymbol{d}_N,\boldsymbol{\psi}) \\
&\quad = \int_0^T P_{\text{in}}^N dt\int_0^1 q_z(t,0,r)dr - \int_0^T P_{\text{out}}^N dt\int_0^1 q_z(t,L,r)dr,
\end{aligned}$$

with

$$\nabla^{\tau_{\Delta t}\eta}\cdot\mathbf{u}_N=0,\quad v_N=((u_r)_N)_{|\Gamma},\ \eta_N=(\boldsymbol{d}_N)_{|\Gamma},$$
$$\mathbf{u}_N(0,.)=\mathbf{u}_0,\ \eta(0,.)_N=\eta_0,\ v_N(0,.)=v_0. \tag{2.173}$$

Here $\tilde{\mathbf{u}}_N$, $\tilde{v}_N$, and $\tilde{\boldsymbol{V}}_N$ are the piecewise linear functions defined in (2.162), $\tau_{\Delta t}$ is the shift in time by Δt to the left, defined in (2.151), $\nabla^{\tau_{\Delta t}\eta_N}$ is the transformed gradient via the ALE mapping $A_{\tau_{\Delta t}\eta_N}$, defined in (2.110), and v_N^*, $\mathbf{u}_N$, v_N, η_N, $\boldsymbol{d}_N$, and $\boldsymbol{V}_N$ are defined in (2.148).

Using the convergence results obtained for the approximate solutions in Sect. 2.6.6, and the convergence results just obtained for the test functions $\mathbf{q}_N$, we can pass to the limit directly in all the terms except in the term that contains $\partial_t\tilde{\mathbf{u}}_N$. To deal with this term we notice that, since $\mathbf{q}_N$ are smooth on sub-intervals $(j\Delta t,(j+1)\Delta t)$, we can use integration by parts on these sub-intervals to obtain:

$$\begin{aligned}
&\int_0^T\int_{\Omega_F}(1+\tau_{\Delta t}\eta_N)\partial_t\tilde{\mathbf{u}}_N\cdot\mathbf{q}_N\\
&=\sum_{j=0}^{N-1}\int_{j\Delta t}^{(j+1)\Delta t}\int_{\Omega_F}(1+\eta_N^j)\partial_t\tilde{\mathbf{u}}_N\cdot\mathbf{q}_N\\
&=\sum_{j=0}^{N-1}\Big(-\int_{j\Delta t}^{(j+1)\Delta t}\int_{\Omega_F}(1+\tau_{\Delta t}\eta_N)\tilde{\mathbf{u}}_N\cdot\partial_t\mathbf{q}_N\\
&\quad+\int_{\Omega_F}(1+\eta^{j+1}-\eta^{j+1}+\eta^j)\mathbf{u}_N^{j+1}\cdot\mathbf{q}_N((j+1)\Delta t-)\\
&\quad-\int_{\Omega_F}(1+\eta^j)\mathbf{u}_N^j\cdot\mathbf{q}_N(j\Delta t+)\Big).
\end{aligned}\tag{2.174}$$

Here, we have denoted by $\mathbf{q}_N((j+1)\Delta t-)$ and $\mathbf{q}_N(j\Delta t+)$ the limits from the left and right, respectively, of $\mathbf{q}_N$ at the appropriate points.

The integral involving $\partial_t\mathbf{q}_N$ can be simplified by recalling that $\mathbf{q}_N=\mathbf{q}\circ A_{\eta_N}$, where η_N are constant on each sub-interval $(j\Delta t,(j+1)\Delta t)$. Thus, by the chain rule, we see that $\partial_t\mathbf{q}_N=\partial_t\mathbf{q}$ on $(j\Delta t,(j+1)\Delta t)$. After summing over all $j=0,\ldots,N-1$ we obtain

$$-\sum_{j=0}^{N-1}\int_{j\Delta t}^{(j+1)\Delta t}\int_{\Omega_F}(1+\tau_{\Delta t}\eta_N)\tilde{\mathbf{u}}_N\cdot\partial_t\mathbf{q}_N=-\int_0^T\int_{\Omega_F}(1+\tau_{\Delta t}\eta_N)\tilde{\mathbf{u}}_N\cdot\partial_t\mathbf{q}.$$

To deal with the last two terms in (2.174) we calculate

$$\sum_{j=0}^{N-1}\Big(\int_{\Omega_F}(1+\eta_N^{j+1}-\eta_N^{j+1}+\eta_N^j)\mathbf{u}_N^{j+1}\cdot\mathbf{q}_N((j+1)\Delta t-)$$
$$-\int_{\Omega_F}(1+\eta_N^j)\mathbf{u}_N^j\cdot\mathbf{q}_N(j\Delta t+)\Big)$$
$$=\sum_{j=0}^{N-1}\int_{\Omega_F}\Big((1+\eta_N^{j+1})\mathbf{u}_N^{j+1}\cdot\mathbf{q}_N((j+1)\Delta t-)$$
$$-(\eta_N^{j+1}-\eta_N^j)\mathbf{u}_N^{j+1}\cdot\mathbf{q}_N((j+1)\Delta t-)\Big)$$
$$-\int_{\Omega}(1+\eta_0)\mathbf{u}_0\cdot\mathbf{q}(0)-\sum_{j=1}^{N-1}\int_{\Omega_F}(1+\eta_N^j)\mathbf{u}_N^j\cdot\mathbf{q}_N(j\Delta t+)\Big)$$

Now, we can write $(\eta^{j+1}-\eta^j)$ as $v^{j+\frac{1}{2}}\Delta t$, and rewrite the summation indexes in the first term to obtain that the above expression is equal to

$$=\sum_{j=1}^{N}\int_{\Omega_F}(1+\eta_N^j)\mathbf{u}_N^j\cdot\mathbf{q}_N(j\Delta t-)-\int_0^T\int_{\Omega_F}v_N^*\mathbf{u}_N\cdot\bar{\mathbf{q}}_N-\int_{\Omega_F}(1+\eta_0)\mathbf{u}_0\cdot\mathbf{q}(0)$$
$$-\sum_{j=1}^{N-1}\int_{\Omega_F}(1+\eta_N^j)\mathbf{u}_N^j\cdot\mathbf{q}_N(j\Delta t+).$$

Since the test functions have compact support in $[0,T)$, the value of the first term at $j=N$ is zero, and so we can combine the two sums to obtain

$$=\sum_{j=1}^{N}\int_{\Omega_F}(1+\eta_N^j)\mathbf{u}_N^j\cdot(\mathbf{q}_N(j\Delta t-)-\mathbf{q}_N(j\Delta t+))-\int_{\Omega_F}(1+\eta_0)\mathbf{u}_0\cdot\mathbf{q}(0)$$
$$-\int_0^T\int_{\Omega_F}v_N^*\mathbf{u}_N\cdot\bar{\mathbf{q}}_N.$$

Now we know how to pass to the limit in all the terms except the first one. We continue to rewrite the first expression by using the Mean Value Theorem to obtain:

$$\mathbf{q}_N(j\Delta t-,z,r)-\mathbf{q}_N(j\Delta t+,z,r)$$
$$=\mathbf{q}(j\Delta t,z,(1+\eta_N^j)r)-\mathbf{q}(j\Delta t,z,(1+\eta_N^{j+1})r)=$$
$$=\partial_r\mathbf{q}(j\Delta t,z,\zeta)r(\eta_N^j-\eta_N^{j+1})=-\Delta t\,\partial_r\mathbf{q}(j\Delta t,z,\zeta)v_N^{j+\frac{1}{2}}r.$$

Therefore we have:

$$\sum_{j=1}^{N-1}\int_{\Omega_F}(1+\eta_N^j)\mathbf{u}_N^j\big(\mathbf{q}(j\Delta t-)-\mathbf{q}(j\Delta t+)\big)$$
$$=-\int_0^{T-\Delta t}\int_{\Omega_F}(1+\eta_N)\mathbf{u}_N r\tau_{-\Delta t}v_N^*\partial_r\bar{\mathbf{q}}.$$

We can now pass to the limit in this last term to obtain:

$$\int_0^{T-\Delta t}\int_{\Omega_F}(1+\eta_N)\mathbf{u}_N r\tau_{-\Delta t}v_N^*\partial_r\bar{\mathbf{q}}\to\int_0^{T}\int_{\Omega_F}(1+\eta)\mathbf{u}r\partial_t\eta\partial_r\mathbf{q}.$$

Therefore, by noticing that $\partial_t\tilde{\mathbf{q}}=\partial_t\mathbf{q}+r\partial_t\eta\partial_r\mathbf{q}$ we have finally obtained

$$\int_0^{T}\int_{\Omega_F}(1+\tau_{\Delta t}\eta_N)\partial_t\tilde{\mathbf{u}}_N\cdot\mathbf{q}_N\to-\int_0^{T}\int_{\Omega_F}(1+\eta)\mathbf{u}\cdot\partial_t\tilde{\mathbf{q}}-\int_0^{T}\int_{\Omega_F}\partial_t\eta\mathbf{u}\cdot\tilde{\mathbf{q}}$$
$$-\int_{\Omega_F}(1+\eta_0)\mathbf{u}_0\cdot\tilde{\mathbf{q}}(0),$$

where we recall that $\tilde{\mathbf{q}}=\mathbf{q}\circ A_\eta$.

Thus, we have shown that the limiting functions $\mathbf{u}$, η and $\boldsymbol{d}$ satisfy the weak form of problem (2.111)–(2.118) in the sense of Definition 2.2, for all test functions that belong to a dense subset of $\mathcal{Q}^\eta$. By density arguments, we have, therefore, shown the main result of this manuscript:

Theorem 2.4 (Main Theorem). *Suppose that the initial data $v_0\in L^2(0,1)$, $\mathbf{u}_0\in L^2(\Omega_{\eta_0})$, $\boldsymbol{V}_0\in L^2(\Omega_S)$, $\boldsymbol{d}_0\in H^1(\Omega_S)$, and $\eta_0\in H_0^1(0,1)$ are such that $1+\eta_0(z)>0$, $z\in[0,1]$ and compatibility conditions (2.102) are satisfied. Furthermore, let $P_{\rm in}$, $P_{\rm out}\in L^2_{loc}(0,\infty)$.*

Then, there exist a $T>0$ and a weak solution $(\mathbf{u},\eta,\boldsymbol{d})$ of problem (2.111)–(2.118) (or equivalently problem (2.91)–(2.102)) on $(0,T)$ in the sense of Definition 2.2 (or equivalently Definition 2.1), such that the following energy estimate is satisfied:

$$E(t)+\int_0^t D(\tau)d\tau\le E_0+C(\|P_{\rm in}\|^2_{L^2(0,t)}+\|P_{\rm out}\|^2_{L^2(0,t)}),\quad t\in[0,T],\ (2.175)$$

where C depends only on the coefficients in the problem, E_0 is the kinetic energy of initial data, and $E(t)$ and $D(t)$ are given by

$$E(t)=\frac{1}{2}\|\mathbf{u}\|^2_{L^2(\Omega_F)}+\frac{1}{2}\|\partial_t\eta\|^2_{L^2(0,1)}+\frac{1}{2}\|\boldsymbol{d}\|^2_{L^2(\Omega_S)}+\frac{1}{2}\big(\|\partial_z\eta\|^2_{L^2(0,1)}+a_S(\boldsymbol{d},\boldsymbol{d})\big),$$
$$D(t)=\|\mathbf{D}(\mathbf{u})\|^2_{L^2(\Omega_\eta(t)))}.$$

Furthermore, one of the following is true:

$$\text{either } T = \infty \ \textit{or} \ \lim_{t \to T} \min_{z \in [0,1]} (1 + \eta(z)) = 0. \tag{2.176}$$

Proof. It only remains to prove the last assertion, which states that our result is either global in time, or, in case the walls of the cylinder touch each other, our existence result holds until the time of touching. However, the proof of this argument follows the same reasoning as the proof of the Main Theorem in [119], and the proof of the main result in [32, pp. 397–398]. We avoid repeating those arguments here, and refer the reader to [32, 119]. □

2.7 Numerical Simulation

In this section we show how the Kinematically Coupled β-Scheme can be applied to FSI problems with multiple structural layers. We also present numerical arguments showing that the presence of a thin fluid–structure interface with mass regularizes solutions of the related FSI problems.

We first summarize the FSI problem that will be solved numerically, then present the numerical algorithm, and finally show the numerical results. The problem, the numerical method, and the results will be shown on an example in 2D.

2.7.1 Problem Definition

We consider the flow of an incompressible, viscous fluid in a two-dimensional channel of reference length L, and reference width $2R$, see Fig. 2.9. The channel is bounded by a two-layered deformable wall, which consists of a thin elastic layer with thickness h, and a thick elastic layer with thickness H. The thin structural layer serves as a fluid–structure interface with mass.

We are interested in simulating the normal stress-driven fluid flow through a deformable 2D channel with two-way coupling between the fluid and structure. Without loss of generality, we consider only the upper half of the fluid domain supplemented by a symmetry condition at the axis of symmetry. Thus, as before, the reference fluid and structure domains in our problem are given, respectively, by

$$\Omega_F := \{(z,r)|0 < z < L, 0 < r < R\},$$

$$\Omega_S := \{(z,r)|0 < z < L, R < r < R + H\}.$$

Here z and r denote the horizontal and vertical Cartesian coordinates, respectively (see Fig. 2.9).

The flow of an incompressible, viscous fluid is modeled by the Navier–Stokes equations:

$$\rho_F\left(\frac{\partial \mathbf{u}}{\partial t}+\mathbf{u}\cdot\nabla\mathbf{u}\right)=\nabla\cdot\boldsymbol{\sigma}(\mathbf{u},p) \qquad \text{in } \Omega_F(t)\times(0,T), \tag{2.177}$$

$$\nabla\cdot\mathbf{u}=0 \qquad \text{in } \Omega_F(t)\times(0,T), \tag{2.178}$$

where $\mathbf{u}=(u_z,u_r)$ is the fluid velocity, p is the fluid pressure, ρ_F is the fluid density, and $\boldsymbol{\sigma}$ is the fluid Cauchy stress tensor. For a Newtonian fluid the Cauchy stress tensor is given by $\boldsymbol{\sigma}(\mathbf{u},p)=-p\boldsymbol{I}+2\mu_F\boldsymbol{D}(\mathbf{u})$, where μ_F is the fluid viscosity and $\boldsymbol{D}(\mathbf{u})=(\nabla\mathbf{u}+(\nabla\mathbf{u})^\tau)/2$ is the rate-of-strain tensor.

Denote the inlet and outlet fluid boundaries by $\Gamma_{\text{in}}=\{0\}\times(0,R)$ and $\Gamma_{\text{out}}=\{L\}\times(0,R)$, respectively. At the inlet and outlet boundary we prescribe the normal stress:

$$\boldsymbol{\sigma}\boldsymbol{n}_{\text{in}}=-p_{\text{in}}(t)\boldsymbol{n}_{\text{in}} \qquad \text{on } \Gamma_{\text{in}}\times(0,T), \tag{2.179}$$

$$\boldsymbol{\sigma}\boldsymbol{n}_{\text{out}}=-p_{\text{out}}(t)\boldsymbol{n}_{\text{out}} \qquad \text{on } \Gamma_{\text{out}}\times(0,T), \tag{2.180}$$

where $\boldsymbol{n}_{\text{in}}$ and $\boldsymbol{n}_{\text{out}}$ are the outward normals to the inlet and outlet fluid boundaries, respectively. Even though not physiologically optimal, these boundary conditions are common in blood flow modeling [9, 125].

At the bottom fluid boundary $r=0$ we impose the symmetry conditions:

$$\frac{\partial u_z}{\partial r}(z,0,t)=0, \quad u_r(z,0,t)=0 \quad \text{on } (0,L)\times(0,T). \tag{2.181}$$

The lateral fluid boundary is bounded by a deformable, thin wall. We assume that the wall is linearly elastic, whose dynamics is modeled by the linearly elastic Koiter membrane model, specified in (2.25), Sect. 2.2.1:

$$\rho_K h\frac{\partial^2\eta_z}{\partial t^2}-C_2\frac{\partial\eta_r}{\partial z}-C_1\frac{\partial^2\eta_z}{\partial z^2}=f_z \qquad \text{on } \Gamma\times(0,T), \tag{2.182}$$

$$\rho_K h\frac{\partial^2\eta_r}{\partial t^2}+C_0\eta_r+C_2\frac{\partial\eta_z}{\partial z}=f_r \qquad \text{on } \Gamma\times(0,T), \tag{2.183}$$

where $\boldsymbol{\eta}(z,t)=(\eta_x(z,t),\eta_r(z,t))$ denotes the axial and radial displacement, $\boldsymbol{f}=(f_z,f_r)$ is the force surface density, ρ_K denotes the shell density and (see (2.25))

$$C_0=\frac{hE}{R^2(1-\sigma^2)},\quad C_1=\frac{hE}{1-\sigma^2},\quad C_2=\frac{hE\sigma}{R(1-\sigma^2)}.$$

The thick layer of the wall will be modeled by the equations of linear elasticity (2.59), with an added extra term $\gamma\mathbf{d}$ to account for circumferential strain whose effects are lost in the transition from 3D to 2D. This term corresponds to the non-differentiated term in the Koiter membrane equations (2.183) containing the coefficient C_0, which appears in these equations due to the cylindrical geometry of the domain. Adding the non-differentiated term $\gamma \mathbf{d}$ to the thick structure problem in 2D has been done by several authors, see [8,9,13,114]. If the structure is not fixed at the end points, this term helps keep the top and bottom portions of the structure domain together. The model reads:

$$\rho_S \frac{\partial^2 \mathbf{d}}{\partial t^2} + \gamma \mathbf{d} = \nabla \cdot \boldsymbol{S}(\mathbf{d}) \quad \text{in } \Omega_S \times (0, T), \tag{2.184}$$

with the first Piola–Kirschhoff stress tensor $\boldsymbol{S}$ given by

$$\boldsymbol{S}(\mathbf{d}) = 2\mu \boldsymbol{D}(\mathbf{d}) + \lambda(\nabla \cdot \mathbf{d})\boldsymbol{I},$$

where $\mathbf{d} = (d_z, d_r)$ is the structure displacement and ρ_S is the structure density.

As before, the structure is assumed to be fixed at the inlet and outlet boundaries:

$$\mathbf{d}(0, r, t) = \mathbf{d}(L, r, t) = 0 \quad \text{on } [R, R + H] \times (0, T), \tag{2.185}$$

and the external structure boundary $\Gamma_{\text{ext}} = \{R + H\} \times (0, L)$ is exposed to zero external ambient pressure, while the axial displacement remains fixed:

$$\boldsymbol{S}\boldsymbol{n}_{\text{ext}} \cdot \boldsymbol{n}_{\text{ext}} = 0 \qquad \text{on } \Gamma_{\text{ext}} \times (0, T), \tag{2.186}$$

$$d_z = 0 \qquad \text{on } \Gamma_{\text{ext}} \times (0, T), \tag{2.187}$$

where $\boldsymbol{n}_{\text{ext}}$ is the outward unit normal vector on Γ_{ext}.

Initially, the fluid and the structure are assumed to be at rest, with zero displacement from the reference configuration

$$\boldsymbol{v} = 0, \quad \boldsymbol{\eta} = 0, \quad \frac{\partial \boldsymbol{\eta}}{\partial t} = 0, \quad \mathbf{d} = 0, \quad \frac{\partial \mathbf{d}}{\partial t} = 0, \quad \text{at } t = 0. \tag{2.188}$$

The fluid and the multi-layered structure are coupled via the kinematic and dynamic boundary conditions (2.62), (2.63):

Continuity of the velocity:

$$\mathbf{u}(z + \eta_z(z, t), R + \eta_r(z, t), t) = \frac{\partial \boldsymbol{\eta}}{\partial t}(z, t) \quad \text{on } (0, L) \times (0, T), \tag{2.189}$$

Continuity of displacement:

$$\boldsymbol{\eta}(z, t) = \mathbf{d}(z, R, t) \quad \text{on } (0, L) \times (0, T). \tag{2.190}$$

Balance of forces:

$$\begin{pmatrix} \rho_K h \dfrac{\partial^2 \eta_z}{\partial t^2} - C_2 \dfrac{\partial \eta_r}{\partial z} - C_1 \dfrac{\partial^2 \eta_z}{\partial z^2} \\ \rho_K h \dfrac{\partial^2 \eta_r}{\partial t^2} + C_0 \eta_r + C_2 \dfrac{\partial \eta_z}{\partial z} \end{pmatrix} = \boldsymbol{S}\mathbf{e}_r|_\Gamma - J\ \boldsymbol{\sigma n}|_{\Gamma(t)} \quad \text{on } (0,L)\times(0,T), \tag{2.191}$$

where J is the Jacobian of the transformation from the Eulerian to Lagrangian framework, $\boldsymbol{n}$ is the outward unit normal to the deformed fluid domain, and $\mathbf{e}_r$ is the unit vector pointing in the vertical direction.

2.7.2 The Energy of the Coupled Problem

The coupled problem (2.177)–(2.191) with *dynamic inlet and outlet pressure data* satisfies the following energy equality:

$$\begin{aligned} &\frac{1}{2}\frac{d}{dt}\Big\{\rho_F \|\mathbf{u}\|^2_{\Omega_F(t)} + \rho_K h \|\partial_t \boldsymbol{\eta}\|^2_{L^2(\Gamma)} + \rho_S \|\partial_t \mathbf{d}\|^2_{L^2(\Omega_S)} + E^{\text{mem}}_{\text{el}}(\boldsymbol{\eta}) \\ &\quad + \gamma \|\mathbf{d}\|^2_{L^2(\Omega_S)} + 2\mu \|\mathbf{D}(\mathbf{d})\|^2_{L^2(\Omega_S)} + \lambda \|\nabla\cdot \mathbf{d}\|^2_{L^2(\Omega_S)}\Big\} + \mu_F \|\mathbf{D}(\mathbf{u})\|^2_{\Omega_F(t)} \\ &\quad = \int_0^R p_{\text{in}}(t) u_z|_{z=0} - \int_0^R p_{\text{out}}(t) u_z|_{z=L} \end{aligned}$$

where

$$\begin{aligned} E^{\text{mem}}_{\text{el}}(\boldsymbol{\eta}) &= \frac{h}{2}\int_0^L \mathcal{A}\mathbf{G}(\boldsymbol{\eta}) : \mathbf{G}(\boldsymbol{\eta}) \\ &= \frac{h}{2}\Bigg[\frac{4E}{1+\sigma}\left\|\frac{\eta_r}{R}\right\|^2_{L^2(0,L)} + \frac{4E}{1+\sigma}\left\|\frac{\partial \eta_z}{\partial z}\right\|^2_{L^2(0,L)} \\ &\quad + \frac{4E\sigma}{1-\sigma^2}\left\|\frac{\partial \eta_z}{\partial z} + \frac{\eta_r}{R}\right\|^2_{L^2(0,L)}\Bigg] \end{aligned}$$

Furthermore, the following energy estimate holds:

$$\begin{aligned} &\frac{1}{2}\frac{d}{dt}\Big\{\rho_F \|\mathbf{u}\|^2_{\Omega_F(t)} + \rho_K h \|\partial_t \boldsymbol{\eta}\|^2_{L^2(\Gamma)} + \rho_S \|\partial_t \mathbf{d}\|^2_{L^2(\Omega_S)} + E^{\text{mem}}_{\text{el}}(\boldsymbol{\eta}) \\ &\quad + \gamma \|\mathbf{d}\|^2_{L^2(\Omega_S)} + 2\mu \|\mathbf{D}(\mathbf{d})\|^2_{L^2(\Omega_S)} + \lambda \|\nabla\cdot \mathbf{d}\|^2_{L^2(\Omega_S)}\Big\} + \mu_F \|\mathbf{D}(\mathbf{u})\|^2_{\Omega_F(t)} \\ &\quad \le C(P_{\text{in}}(t), P_{\text{out}}(t)). \end{aligned}$$

2.7.3 The ALE Formulation

As mentioned in Sect. 2.3, to deal with the motion of the fluid domain we use the ALE approach. An ALE mapping $\mathcal{A}$ maps the reference domain Ω_F into the current domain $\Omega_F(t)$:

$$\mathcal{A} : \Omega_F \to \Omega_F(t) \subset \mathbb{R}^2, \quad \boldsymbol{x} = \mathcal{A}(\boldsymbol{x}) \in \Omega_F(t), \quad \text{for } \boldsymbol{x} \in \Omega_F.$$

We will use ALE mapping to deal with the deformation of the mesh, and to resolve the issues related to the approximation of the time-derivative $\partial \boldsymbol{v}/\partial t \approx (\boldsymbol{v}(t^{n+1}) - \boldsymbol{v}(t^n))/\Delta t$,, which due to the fact that $\Omega_F(t)$ depends on time, is not well defined. In particular, we will be using the ALE mapping which is defined as the harmonic extension of the boundary, determined by the current position of $\boldsymbol{\eta}$, to the entire fluid domain:

$$\begin{aligned} \Delta \mathcal{A} &= 0, \text{ in } \Omega_F, \\ \mathcal{A}|_\Gamma &= \boldsymbol{\eta}, \\ \mathcal{A}|_{\partial\Omega_F \setminus \Gamma} &= 0. \end{aligned}$$

To solve the Navier–Stokes equations numerically on a moving domain, we transform the time derivative of the fluid velocity using the chain rule

$$\left.\frac{\partial \mathbf{u}}{\partial t}\right|_{\Omega_F} = \frac{\partial \mathbf{u}}{\partial t} + \mathbf{w} \cdot \nabla \mathbf{u}, \tag{2.192}$$

where $\mathbf{w} = \dfrac{\partial \mathcal{A}}{\partial t}$ denotes the domain velocity, and consider the rest of the problem defined on the moving domain $\Omega_F(t)$. See Sect. 2.5.2. Therefore, with a slight abuse of notation, the Navier–Stokes problem (2.177)–(2.178) that will be solved numerically can be written in ALE formulation as follows: find $\mathbf{u} = (u_z, u_r)$ and p such that

$$\rho_F \left(\left.\frac{\partial \mathbf{u}}{\partial t}\right|_{\Omega_F} + (\mathbf{u} - \mathbf{w}) \cdot \nabla \mathbf{u} \right) = \nabla \cdot \boldsymbol{\sigma}(\boldsymbol{v}, p), \qquad \text{in } \Omega_F(t) \times (0, T), \tag{2.193}$$

$$\nabla \cdot \mathbf{u} = 0 \qquad \text{in } \Omega_F(t) \times (0, T), \tag{2.194}$$

satisfying the corresponding initial and boundary conditions.

The structure problems remain the same since the equations are defined on the reference, fixed domains Ω_S and Γ.

To perform the Lie splitting, described in Sect. 2.5.3, the coupled problem is rewritten as a first-order system in time. For this purpose, the trace of the fluid velocity on $\Gamma(t)$ will be denoted by $\mathbf{v} := \mathbf{u}|_{\Gamma(t)}$, while the trace of the thick structure velocity on Γ will be denoted by $\mathbf{V} = \partial \mathbf{d}/\partial t$. The kinematic coupling condition $\partial \boldsymbol{\eta}/\partial t = \mathbf{v}$ is then used to rewrite the system.

Notice again that $\mathbf{u}|_{\Gamma(t)}$ is defined on $\Gamma(t)$, namely, at $R+\eta(z,t)$. More precisely, $\mathbf{u}|_{\Gamma(t)} = \mathbf{u}(R+\eta(z,t),z,t)$. Therefore, $\mathbf{v}$, which is defined on Γ is equal to

$$\mathbf{v}(z,t) := \mathbf{u}(R+\eta(z,t),z,t).$$

The resulting problem is given by the following:

$$\rho_f \left(\left.\frac{\partial \mathbf{u}}{\partial t}\right|_{\Omega_F} + (\mathbf{u}-\mathbf{w}) \cdot \nabla \mathbf{u} \right) = \nabla \cdot \boldsymbol{\sigma} \qquad \text{in } \Omega_F(t) \times (0,T), \tag{2.195a}$$

$$\nabla \cdot \mathbf{u} = 0 \qquad \text{in } \Omega_F(t) \times (0,T), \tag{2.195b}$$

$$\rho_K h \frac{\partial v_z}{\partial t} - C_2 \frac{\partial \eta_r}{\partial z} - C_1 \frac{\partial^2 \eta_z}{\partial z^2} = f_z \qquad \text{on } \Gamma \times (0,T), \tag{2.195c}$$

$$\rho_K h \frac{\partial v_r}{\partial t} + C_0 \eta_r + C_2 \frac{\partial \eta_z}{\partial z} = f_r \qquad \text{on } \Gamma \times (0,T), \tag{2.195d}$$

$$\frac{\partial \boldsymbol{\eta}}{\partial t} = \mathbf{v} \qquad \text{on } \Gamma \times (0,T), \tag{2.195e}$$

$$\rho_S \frac{\partial \boldsymbol{V}}{\partial t} + \gamma \mathbf{d} = \nabla \cdot \boldsymbol{S}(\mathbf{d}) \qquad \text{in } \Omega_S \times (0,T), \tag{2.195f}$$

$$\frac{\partial \mathbf{d}}{\partial t} = \boldsymbol{V} \qquad \text{in } \Omega_S \times (0,T), \tag{2.195g}$$

with the coupling conditions at the fluid–structure interface

$$\mathbf{v} = \mathbf{u}|_{\Gamma(t)}, \qquad \boldsymbol{\eta} = \mathbf{d}|_{\Gamma}, \tag{2.196}$$

$$\rho_K h \frac{\partial v_z}{\partial t} - C_2 \frac{\partial \eta_r}{\partial z} - C_1 \frac{\partial^2 \eta_z}{\partial z^2} + J\, \boldsymbol{\sigma} \boldsymbol{n}|_{\Gamma(t)} \cdot \boldsymbol{e}_z + \boldsymbol{S}\mathbf{e}_z|_{\Gamma} \cdot \boldsymbol{e}_z = 0, \tag{2.197}$$

$$\rho_K h \frac{\partial v_r}{\partial t} + C_0 \eta_r + C_2 \frac{\partial \eta_z}{\partial z} + J\, \boldsymbol{\sigma} \boldsymbol{n}|_{\Gamma(t)} \cdot \boldsymbol{e}_r + \boldsymbol{S}\mathbf{e}_r|_{\Gamma} \cdot \boldsymbol{e}_r = 0. \tag{2.198}$$

Notice, again, that $\mathbf{v} = \mathbf{u}|_{\Gamma(t)}$ means $\mathbf{v}(z,t) = \mathbf{u}(R+\eta(z,t),z,t)$ on $(0,L)\times(0,T)$. This problem is supplemented with the boundary and initial conditions presented in Sect. 2.7.1.

Before we continue with the Lie splitting algorithm applied to the problem in ALE form (2.195)–(2.198), we introduce the notion of weak solutions for the problem studied in this section, namely, for problem (2.177)–(2.191). The corresponding function spaces on domains $\Omega_F(t)$ will be defined in terms of the functions defined on the fixed, reference domain Ω_F, where the association between the two is done via the ALE mapping, defined above.

2.7.4 Weak Formulation of FSI Problem (2.177)–(2.191)

For $t \in [0,T)$ introduce the following test function spaces: the fluid velocity space is defined by

$$\mathcal{V}_F(t) = \{\boldsymbol{\varphi} : \Omega_F(t) \to \mathbb{R}^2 |\ \boldsymbol{\varphi} = \hat{\boldsymbol{\varphi}} \circ (\mathcal{A})^{-1}, \hat{\boldsymbol{\varphi}} \in (H^1(\Omega_F))^2, \\ \varphi_r|_{r=0} = 0,\ \boldsymbol{\varphi}|_{z=0,L} = 0\},,$$

the fluid pressure:

$$Q(t) = \{q : \Omega_F(t) \to \mathbb{R} |\ q = \hat{q} \circ (\mathcal{A})^{-1}, \hat{q} \in L^2(\Omega_F)\},,$$

the test space for the thin structure problem:

$$\mathcal{V}_K = \{\boldsymbol{\zeta} : (0,L) \to \mathbb{R}^2 |\ \boldsymbol{\zeta} \in (H_0^1(\Gamma))^2\},$$

and the test space for the thick structure problem

$$\mathcal{V}_S = \{\boldsymbol{\psi} : \Omega_S \to \mathbb{R}^2 |\ \boldsymbol{\psi} \in (H^1(\Omega_S))^2, \boldsymbol{\psi}|_{z=0,L} = 0, \psi_z|_{\Gamma_{S_{\text{ext}}}} = 0\}.$$

The test space for the coupled FSI problem is given by:

$$\mathcal{Q}(t) = \{(\boldsymbol{\varphi}, \boldsymbol{\zeta}, \boldsymbol{\psi}) \in \mathcal{V}_F(t) \times \mathcal{V}_K \times \mathcal{V}_S \ |\ \boldsymbol{\varphi}|_{\Gamma(t)} = \boldsymbol{\zeta},\ \boldsymbol{\zeta} = \boldsymbol{\psi}|_{\Gamma}\}, \tag{2.199}$$

The variational formulation of the coupled FSI problem now reads: for $t \in (0,T)$, find $(\mathbf{u}, p, \boldsymbol{\eta}, \mathbf{d}) \in \mathcal{V}_F(t) \times Q(t) \times \mathcal{V}_K \times \mathcal{V}_S$ such that the kinematic coupling conditions (2.189) and (2.190) hold, and such that for all $(\boldsymbol{\varphi}, \boldsymbol{\zeta}, \boldsymbol{\psi}, q) \in \mathcal{Q}(t) \times Q(t)$ the following equations are satisfied:

$$\begin{aligned}
&\rho_F \int_{\Omega_F(t)} \frac{\partial \mathbf{u}}{\partial t} \cdot \boldsymbol{\varphi} + \int_{\Omega_F(t)} (\mathbf{u} \cdot \nabla)\mathbf{u} \cdot \boldsymbol{\varphi} + 2\mu_F \int_{\Omega_F(t)} \boldsymbol{D}(\mathbf{u}) : \boldsymbol{D}(\boldsymbol{\varphi}) \\
&- \int_{\Omega_F(t)} p \nabla \cdot \boldsymbol{\varphi} \int_{\Omega_F(t)} q \nabla \cdot \mathbf{u} + \rho_K h \int_0^L \frac{\partial^2 \eta_z}{\partial t^2} \zeta_z + \rho_K h \int_0^L \frac{\partial^2 \eta_r}{\partial t^2} \zeta_r \\
&- C_2 \int_0^L \frac{\partial \eta_r}{\partial z} \zeta_z + C_1 \int_0^L \frac{\partial \eta_z}{\partial z} \frac{\partial \zeta_z}{\partial z} + C_0 \int_0^L \eta_r \zeta_r + C_2 \int_0^L \frac{\partial \eta_z}{\partial z} \zeta_r \\
&+ \rho_S \int_{\Omega_S} \frac{\partial^2 \mathbf{d}}{\partial t^2} \cdot \boldsymbol{\psi} + 2\mu \int_{\Omega_S} \boldsymbol{D}(\mathbf{d}) : \boldsymbol{D}(\boldsymbol{\psi}) + \lambda \int_{\Omega_S} (\nabla \cdot \mathbf{d})(\nabla \cdot \boldsymbol{\psi}) \\
&+ \gamma \int_{\Omega_S} \mathbf{d} \cdot \boldsymbol{\psi} = \int_0^R p_{\text{in}}(t) \varphi_z|_{z=0} dr - \int_0^R p_{\text{out}}(t) \varphi_z|_{z=L} dr. \\
&\int_{\Omega_F} q \nabla \cdot \mathbf{u} = 0.
\end{aligned} \tag{2.200}$$

2.7.5 *Numerical Implementation of the Splitting Scheme*

The splitting is performed on the first-order system written in ALE form (2.195)–(2.198). System (2.195)–(2.198) is split into two sub-problems, the fluid and the structure sub-problem, as described in Sect. 2.5.4. We notice that in this splitting, we also separated the viscous part of the structure problem from the purely elastic part of the structure problem, so that in the final structure sub-problem, denoted by Problem A1 in Sect. 2.5.4, we only solve non-dissipative, hyperbolic part of the structure problem using appropriate solvers. Namely, it has been our experience that for the stability and accuracy of the splitting scheme, it is beneficial to separate the parabolic from the hyperbolic features of the coupled FSI problem, and apply non-dissipative solvers to the non-dissipative, hyperbolic sub-problems. While in the example studied in this section we do not have viscous dissipation in the structure problem, we, however, use the same logic to numerically solve the fluid sub-problem, which contains, in itself, dissipative and non-dissipative features. More precisely, we will split the fluid sub-problem into the pure advection sub-problem (non-dissipative), and the remaining, time-dependent Stokes problem capturing viscous dissipation. This will give rise to a splitting algorithm with three main steps:

A1. An elastodynamics sub-problem for the structure;
A2(a). A time-dependent Stokes problem for the fluid;
A2(b). A fluid and ALE advection problem.

To achieve higher accuracy, we implement the Kinematically Coupled β-Scheme, described in Sect. 2.5.4, in which the normal fluid stress is further split into two parts:

$$\sigma \boldsymbol{n} = \underbrace{\sigma \boldsymbol{n} + \beta p \boldsymbol{n}}_{(Part\ I)} \underbrace{-\beta p \boldsymbol{n}}_{(Part\ II)},$$

where $\beta \in [0, 1]$. It was shown in [21] that the accuracy of the scheme increases as the value of β increases from 0 to 1. Part I of the fluid stress will be taken into account in the fluid sub-problem, while Part II of the fluid stress will be used as loading to the structure in the structure sub-problem, and will appear as a Robin boundary condition for the thick structure equations. Details of the scheme are as follows:

Problem A1: The Elastodynamics Problem

This step involves solving the thick structure problem together with membrane elastodynamics. The membrane elastodynamics problem appears in this step as a Robin boundary condition on Γ for the thick structure problem defined on Ω_S, where we have used continuity of displacement (kinematic coupling condition) to write the problem this way. The Robin boundary condition also includes Part II of the normal fluid stress, which enters explicitly in the sense that the pressure is taken from the fluid sub-problem at the time step n (Problem A1(a)). The Jacobian of the transformation from the Eulerian to Lagrangian framework J is calculated based on the data obtained in the previous time step.

In this step we also compute the domain velocity $\mathbf{w}$ and use it in Problem A2(b) below to solve the fluid and ALE advection problem. The initial data for the structure velocity on Γ is taken to be the trace of the fluid velocity $\mathbf{v}$ calculated in the previous time step. Thus, the structure communicates with the fluid sub-problem through this initial data and through the pressure exerted by the fluid onto the fluid–structure interface. In turn, the updated structure velocity is then taken in Problem A2(a) as the initial data for the trace of the fluid velocity on $\Gamma(t)$.

In this step the fluid velocity $\mathbf{u}$ and fluid pressure p remain unchanged, and so

$$\mathbf{u}^{n+1/3} = \mathbf{u}^n,\ p^{n+1/3} = p^n.$$

The structure sub-problem reads: Find $\boldsymbol{\eta}$, $\mathbf{d}$, $\mathbf{v}$, and $\boldsymbol{V}$, such that for $t \in (t^n, t^{n+1})$

$$\begin{aligned}
&\rho_S \frac{\partial \boldsymbol{V}}{\partial t} + \gamma \mathbf{d} = \nabla \cdot \boldsymbol{S}(\mathbf{d}) && \text{in } \Omega_S \times (t^n, t^{n+1}),\\
&\frac{\partial \mathbf{d}}{\partial t} = \boldsymbol{V} && \text{in } \Omega_S \times (t^n, t^{n+1}),\\
&\rho_K h \frac{\partial v_z}{\partial t} - C_2 \frac{\partial \eta_r}{\partial z} - C_1 \frac{\partial^2 \eta_z}{\partial z^2} + \boldsymbol{S}\mathbf{e}_r \cdot \mathbf{e}_z = J^n \beta p^n \boldsymbol{n}|_{\Gamma(t)} \cdot \mathbf{e}_z && \text{on } \Gamma \times (t^n, t^{n+1}),\\
&\rho_K h \frac{\partial v_r}{\partial t} + C_0 \eta_r + C_2 \frac{\partial \eta_z}{\partial z} + \boldsymbol{S}\mathbf{e}_r \cdot \mathbf{e}_r = J^n \beta p^n \boldsymbol{n}|_{\Gamma(t)} \cdot \boldsymbol{e}_r && \text{on } \Gamma \times (t^n, t^{n+1}),\\
&\frac{\partial \boldsymbol{\eta}}{\partial t} = \mathbf{v} && \text{on } \Gamma \times (t^n, t^{n+1}),\\
&\boldsymbol{\eta} = \mathbf{d}|_\Gamma && \text{on } \Gamma \times (t^n, t^{n+1}),\\
&\boldsymbol{V}|_\Gamma = \mathbf{v} && \text{on } \Gamma \times (t^n, t^{n+1}),
\end{aligned}$$

with the following boundary conditions:

$$\mathbf{d}|_{z=0,L} = 0, \text{ and } d_z = 0, \quad \boldsymbol{n}^s_{\text{ext}} \cdot \boldsymbol{S}\boldsymbol{n}^s_{\text{ext}} = 0 \quad \text{on } \Gamma_{\text{ext}} \times (t^n, t^{n+1}).$$

The initial conditions are given by:

$$\mathbf{d}(t^n) = d^n, \boldsymbol{\eta}(t^n) = \boldsymbol{\eta}^n,\ \mathbf{v}(t^n) = \mathbf{v}^n,\ \boldsymbol{V}(t^n) = \boldsymbol{V}^n.$$

Then set

$$\mathbf{d}^{n+1/3} = \mathbf{d}(t^{n+1}), \boldsymbol{\eta}^{n+1/3} = \boldsymbol{\eta}(t^{n+1}), \mathbf{v}^{n+1/3} = \mathbf{v}(t^{n+1}),\ \boldsymbol{V}^{n+1/3} = \boldsymbol{V}(t^{n+1}).$$

After the new position of the structure has been calculated, and the new fluid domain updated, we calculate the ALE mapping $\mathcal{A}^{n+1}$ as the harmonic extension of the structure displacement $\boldsymbol{\eta}^{n+1}$ onto the whole domain Ω_F

$$\begin{aligned} \Delta \mathcal{A}^{n+1} &= 0 \quad \text{in } \Omega_F, \\ \mathcal{A}^{n+1}|_{\Gamma} &= \boldsymbol{\eta}^{n+1}, \\ \mathcal{A}^{n+1}|_{\partial\Omega_F\setminus\Gamma} &= 0. \end{aligned}$$

From here we calculate the domain velocity $\mathbf{w}^{n+1} = \dfrac{\partial \mathcal{A}^{n+1}}{\partial t}$, based on the updated location of the structure, and use it in the advection problem, Problem A2(b) below.

Remark 1. Note that in Problem A1, we can rewrite the membrane equations by using the kinematic coupling conditions in the following way:

$$\begin{aligned} &\rho_K h \frac{\partial V_z}{\partial t} - C_2 \frac{\partial d_r}{\partial z} - C_1 \frac{\partial^2 d_z}{\partial z^2} + \boldsymbol{S}\mathbf{e}_r \cdot \boldsymbol{e}_z = 0 && \text{on } \Gamma \times (t^n, t^{n+1}), \\ &\rho_K h \frac{\partial V_r}{\partial t} + C_0 d_r + C_2 \frac{\partial d_z}{\partial z} + \boldsymbol{S}\boldsymbol{e}_r \cdot \boldsymbol{e}_r = 0 && \text{on } \Gamma \times (t^n, t^{n+1}). \end{aligned}$$

In this way the membrane equations serve as Robin boundary conditions for the thick structure problem.

Problem A2(a): The Stokes Problem

This step involves solving a time-dependent Stokes problem on (t^n, t^{n+1}), with a Robin-type boundary condition involving the thin structure inertia and Part I of the fluid stress. This problem is solved on the fixed fluid domain $\Omega_F(t^n)$, determined by the structure position in the previous time step. Using the updated fluid domain calculated in Problem A1 is also an option. In the proof of stability of this scheme, using $\Omega_F(t^n)$ is more convenient for the proof. In this step the structure position and the velocity of the thick structure do not change, and so

$$\boldsymbol{\eta}^{n+2/3} = \boldsymbol{\eta}^{n+1/3}, \mathbf{d}^{n+2/3} = \mathbf{d}^{n+1/3}, \boldsymbol{V}^{n+2/3} = \boldsymbol{V}^{n+1/3}.$$

The problem reads as follows:

Find $\mathbf{u}$, p, and $\mathbf{v}$ such that for $t \in (t^n, t^{n+1})$, with p^n denoting the pressure obtained at the previous time step, the following holds:

$$\begin{aligned} &\rho_F \left.\frac{\partial \mathbf{u}}{\partial t}\right|_{\Omega_F} = \nabla \cdot \boldsymbol{\sigma}, \quad \nabla \cdot \mathbf{u} = 0 && \text{in } \Omega_F(t^n) \times (t^n, t^{n+1}), \\ &\rho_K h \frac{\partial (\mathbf{u}|_{\Gamma(t)})}{\partial t} + J\big(\boldsymbol{\sigma}\boldsymbol{n}|_{\Gamma(t)} + \beta p^n \boldsymbol{n}|_{\Gamma(t)}\big) = 0 && \text{on } \Gamma \times (t^n, t^{n+1}), \\ &\mathbf{v} = \mathbf{u}|_{\Gamma(t)} && \text{on } \Gamma \times (t^n, t^{n+1}), \end{aligned}$$

where $\mathbf{v} = \mathbf{u}|_{\Gamma(t)}$ means $\mathbf{v}(z,t) = \mathbf{u}(R + \eta(z,t), z, t)$ on Γ. This is supplemented with the following boundary conditions:

$$\frac{\partial u_z}{\partial r}(z,0,t) = \quad u_r(z,0,t) = 0 \quad \text{on } (0,L), \quad \mathbf{u}(0,R,t) = \mathbf{u}(L,R,t) = 0,$$

$$\boldsymbol{\sigma}\boldsymbol{n}_{\text{in}} = -p_{\text{in}}(t)\boldsymbol{n}_{\text{in}} \text{ on } \Gamma_{\text{in}}, \quad \boldsymbol{\sigma}\boldsymbol{n}_{\text{out}} = -p_{\text{out}}(t)\boldsymbol{n}_{\text{out}} \text{ on } \Gamma_{\text{out}},$$

and initial conditions: $\mathbf{u}(t^n) = \mathbf{u}^n,\ \mathbf{v}(t^n) = \mathbf{v}^{n+1/3}$. Then set

$$\mathbf{u}^{n+2/3} = \mathbf{u}(t^{n+1}),\ p^{n+2/3} = p(t^{n+1}),\ \mathbf{v}^{n+2/3} = \mathbf{v}(t^{n+1}).$$

Problem A2(b): The Advection Problem

Solve the fluid and ALE advection sub-problem defined on the fixed domain $\Omega(t^n)$, with the domain velocity $\mathbf{w}^{n+1}$ just calculated in Problem A1. The displacement of the structure, the velocity of the thick structure, the velocity of the thin structure, and the fluid pressure do not change in this step, so that

$$\eta^{n+1} = \eta^{n+2/3}, \mathbf{d}^{n+1} = \mathbf{d}^{n+2/3}, V^{n+1} = V^{n+2/3}, \mathbf{v}^{n+1} = \mathbf{v}^{n+2/3}, p^{n+1} = p^{n+2/3}.$$

The advection problem reads: Find $\mathbf{u}$ such that for $t \in (t^n, t^{n+1})$

$$\begin{aligned} &\left.\frac{\partial \mathbf{u}}{\partial t}\right|_{\Omega_F} + (\mathbf{u}^{n+2/3} - \mathbf{w}^{n+1}) \cdot \nabla \mathbf{u} = 0, && \text{in } \Omega_F(t^n) \times (t^n, t^{n+1}), \\ &\mathbf{u} = \mathbf{v}^{n+2/3}, && \text{on } \Gamma \times (t^n, t^{n+1}), \end{aligned}$$

with the inlet/outlet conditions:

$$\mathbf{u} = \mathbf{u}^{n+2/3} \quad \text{on} \quad \Gamma_{-}^{n+2/3} = \{\boldsymbol{x} \in \mathbb{R}^2 | \boldsymbol{x} \in \partial\Omega_F(t^n), (\mathbf{u}^{n+2/3} - \mathbf{w}^{n+1}) \cdot \boldsymbol{n} < 0\},$$

and initial conditions $\mathbf{u}(t^n) = \mathbf{u}^{n+2/3}$. Then set

$$\mathbf{u}^{n+1} = \mathbf{u}(t^{n+1}).$$

Set $n = n + 1$ and return to Problem A1.

2.7.6 Discretized Scheme in Weak Form

To discretize the problem in time, sub-divide the time interval $(0, T)$ into N sub-intervals of width Δt, and let $t^n = n\Delta t$, where $n \leq N$. The Backward Euler scheme is implemented to discretize the time-derivatives. For the space discretization, we use the finite element method approach. Thus, we define the finite element spaces $\mathcal{V}_F^h(t^n) \subset \mathcal{V}_F(t^n)$, $Q^h(t^n) \subset Q(t^n)$, $\mathcal{V}_K^h \subset \mathcal{V}_K$ and $\mathcal{V}_S^h \subset \mathcal{V}_S$, and introduce the following bilinear forms

$$a_F^n(\mathbf{u}, \boldsymbol{\varphi}) := 2\mu_F \int_{\Omega_F(t^n)} \boldsymbol{D}(\mathbf{u}) : \boldsymbol{D}(\boldsymbol{\varphi}),$$

$$b_F^n(p, \boldsymbol{\varphi}) := \int_{\Omega_F(t^n)} p \nabla \cdot \boldsymbol{\varphi},$$

$$a_K(\eta_r, \zeta_r) := C_0 \int_0^L \eta_r \zeta_r,$$

$$a_S(\mathbf{d}, \boldsymbol{\psi}) := 2\mu \int_{\Omega_S} \boldsymbol{D}(\mathbf{d}) : \boldsymbol{D}(\boldsymbol{\psi}) + \lambda \int_{\Omega_S} (\nabla \cdot \mathbf{d})(\nabla \cdot \boldsymbol{\psi}).$$

A weak formulation of the fully discrete loosely coupled algorithm applied to the simplified problem is given as follows:

Problem A1 (The Structure Problem). To discretize the structure problem in time we use the second-order Newmark scheme. The problem reads as follows: Find $(\mathbf{d}_h^{n+1/3}, \boldsymbol{V}_h^{n+1/3}) \in \mathcal{V}_S^h \times \mathcal{V}_S^h$ such that for all $(\boldsymbol{\psi}_h, \boldsymbol{\phi}_h) \in \mathcal{V}_S^h \times \mathcal{V}_S^h$

$$\rho_S \int_{\Omega_S} \frac{\boldsymbol{V}_h^{n+1/3} - \boldsymbol{V}_h^n}{\Delta t} \cdot \boldsymbol{\psi}_h + \gamma \int_{\Omega_S} \frac{\mathbf{d}_h^n + \mathbf{d}_h^{n+1/3}}{2} \cdot \boldsymbol{\psi}_h$$

$$+\rho_K h \int_\Gamma \frac{V_{r,h}^{n+1/3} - V_{r,h}^n}{\Delta t} \psi_{r,h} + a_K\left(\frac{d_{r,h}^n + d_{r,h}^{n+1/3}}{2}, \psi_{r,h}\right) + a_S\left(\frac{\mathbf{d}_h^n + \mathbf{d}_h^{n+1/3}}{2}, \boldsymbol{\psi}_h\right)$$

$$+\rho_S \int_{\Omega_S} \left(\frac{\boldsymbol{V}_h^n + \boldsymbol{V}_h^{n+1/3}}{2} - \frac{\mathbf{d}_h^{n+1/3} - \mathbf{d}_h^n}{\Delta t}\right) \cdot \boldsymbol{\phi}_h$$

$$+ \rho_K h \int_\Gamma \left(\frac{V_{r,h}^n + V_{r,h}^{n+1/3}}{2} - \frac{d_{r,h}^{n+1/3} - d_{r,h}^n}{\Delta t}\right) \cdot \phi_{r,h} = 0. \tag{2.201}$$

Note that in this step we take all the kinematic coupling conditions into account. More precisely:

1. Initially we set $V_{r,h}^n|_\Gamma = v_{r,h}^n = u_{r,h}^n|_{\Gamma(t^n)}$;
2. Once $\mathbf{d}_h^{n+1/3}$ and $\boldsymbol{V}_h^{n+1/3}$ are computed, $\eta_{r,h}^{n+1/3}, v_{r,h}^{n+1/3}$ and $u_{r,h}^{n+1/3}|_{\Gamma(t^n)}$ are recovered via

$$\eta_{r,h}^{n+1/3} = d_{r,h}^{n+1/3}|_\Gamma,\ v_{r,h}^{n+1/3} = u_{r,h}^{n+1/3}|_{\Gamma(t^n)} = V_{r,h}^{n+1/3}|_\Gamma.$$

In this step the fluid velocity does not change, and so

$$\mathbf{u}_h^{n+1/3} = \mathbf{u}_h^n.$$

In this step we also update the fluid domain velocity $\mathbf{w}$. As mentioned earlier, after the new position of the structure has been calculated, we calculate the ALE mapping $\mathcal{A}^{n+1} : \Omega_F \to \Omega_F(t^{n+1})$ as the harmonic extension of the structure displacement $\boldsymbol{\eta}_h^{n+1/3} = \boldsymbol{\eta}_h^{n+1}$ onto the whole fluid domain, and obtain the domain velocity $\mathbf{w}_h^{n+1}$ as the difference quotient between the new location of points associated with $\mathcal{A}^{n+1}$, minus the old location of points associated with $\mathcal{A}^n$, divided by Δt. This will be used it in the advection problem, i.e., Problem A2(b) below.

Problem A2(a) (The Time-Dependent Stokes Problem). Find $(\mathbf{u}_h^{n+2/3}, p_h^{n+2/3}) \in \mathcal{V}_F^h(t^n) \times Q^h(t^n)$ such that for all $(\boldsymbol{\varphi}_h, q_h) \in \mathcal{V}_F^h(t^n) \times Q^h(t^n)$

$$\begin{aligned}
&\rho_F \int_{\Omega_F(t^n)} \frac{\mathbf{u}_h^{n+2/3} - \mathbf{u}_h^{n+1/3}}{\Delta t} \cdot \boldsymbol{\varphi}_h + a_F^n(\mathbf{u}_h^{n+1}, \boldsymbol{\varphi}_h) - b_F^n(p_h^{n+2/3}, \boldsymbol{\varphi}_h) \\
&+\rho_K h \int_0^L \frac{u_{r,h}^{n+2/3}|_{\Gamma(t^n)} - u_{r,h}^{n+1/3}|_{\Gamma(t^n)}}{\Delta t} \varphi_{r,h} dx + b_F^n(q_h, \mathbf{u}_h^{n+2/3}) \\
&= \int_0^R p_{\text{in}}(t^{n+1}) \varphi_{z,h}|_{z=0} dr - \int_0^R p_{\text{out}}(t^{n+1}) \varphi_{z,h}|_{z=L} dr.
\end{aligned} \tag{2.202}$$

This step is computed on the fixed domain $\Omega^f(t^n)$. The only updated variables are the fluid velocity and pressure. For higher accuracy, this step can be computed on the updated domain $\Omega(t^{n+1})$.

Problem A2(b) (The Advection Problem). As mentioned earlier, it is convenient to write the fluid and ALE advection term in symmetric form, giving rise to the following weak formulation: Find $\mathbf{u}_h^{n+1} \in \mathcal{V}_F^h(t^n)$ such that for all $\boldsymbol{\varphi}_h \in \mathcal{V}_F^h(t^n)$

$$\begin{aligned}
&\rho_F \int_{\Omega_F(t^n)} \frac{\mathbf{u}_h^{n+1} - \mathbf{u}_h^{n+2/3}}{\Delta t} \cdot \boldsymbol{\varphi}_h + \frac{\rho_F}{2} \int_{\Omega_F(t^n)} (\nabla \cdot \boldsymbol{w}_h^{n+1}) \mathbf{u}_h^{n+1} \cdot \boldsymbol{\varphi}_h \\
&+ \frac{\rho_F}{2} \int_{\Omega_F(t^n)} \left(((\mathbf{u}_h^n - \boldsymbol{w}_h^{n+1}) \cdot \nabla) \mathbf{u}_h^{n+1} \cdot \boldsymbol{\varphi}_h - ((\mathbf{u}_h^n - \boldsymbol{w}_h^{n+1}) \cdot \nabla) \boldsymbol{\varphi}_h \cdot \mathbf{u}_h^{n+1} \right) = 0.
\end{aligned} \tag{2.203}$$

In this step all the other variables, except for the fluid velocity, remain unchanged so that

$$\mathbf{d}_h^{n+1} = \mathbf{d}_h^{n+2/3}, \boldsymbol{V}_h^{n+1} = \boldsymbol{V}_h^{n+2/3}, \eta_{r,h}^{n+1} = \eta_{r,h}^{n+2/3}, \mathbf{v}_{r,h}^{n+1} = \mathbf{v}_{r,h}^{n+2/3},$$

with

$$u_{r,h}^{n+1}|_{\Gamma(t^n)} = u_{r,h}^{n+2/3}|_{\Gamma(t^n)} = v_{r,h}^{n+2/3}.$$

It was shown in [24] that an energy estimate associated with unconditional stability of this scheme holds for the full nonlinear FSI problem. Therefore, we expect that this scheme is unconditionally stable for all the parameters in the problem.

2.7.7 Numerical Examples

We present two numerical examples. One is a simplified problem for which there exists an exact solution against which we can test our numerical scheme. The other one is a fully nonlinear FSI problem with a thin and thick structural layer. Since there are no numerical results in literature on FSI problems with multiple structural layers against which we could test our solution, in this second example we calculated solutions to a sequence of problems for which the thickness of the thin structure converges to zero, and showed that the limiting solution is the same as the solution of the FSI problem in which the structure consists of only one thick structural layer. This was proved using analytical methods in [24]. The solution of the limiting problem was then numerically tested against the solution of the FSI problem with only one thick structural layer, which was obtained using a different solver. We show below that the two solutions, obtained with two different solvers, are in good agreement.

Example 1

We consider a simplified FSI problem with multiple structural layers that satisfies the following simplifying assumptions:

1. The fluid problem is defined on the fixed, reference domain of width R, and length L (the coupling is linear).
2. The fluid problem is driven by the constant inlet and outlet pressure data p_{in} and $p_{\text{out}} = 0$ (the pressure drop is constant).
3. Only radial displacement of the thin and thick structure is assumed to be different from zero.

Assumption 3 implies that the thin structure membrane model takes the form:

$$\rho_K h \frac{\partial^2 \eta_r}{\partial t} + C_0 \eta_r = f_r,$$

while the thick structure problem simplifies as follows:

$$\rho_s \frac{\partial^2 d_r}{\partial t^2} = \mu \frac{\partial^2 d_r}{\partial x^2} + (\mu + \lambda) \frac{\partial^2 d_r}{\partial y^2}.$$

Finally, the coupling conditions between the fluid and the multi-layered structure are given by

$$\begin{aligned} f_r &= p + (\lambda + \mu) \frac{\partial d_r}{\partial y} && \text{on } \Gamma \times (0, T), \\ \frac{\partial \eta_r}{\partial t} &= u_r && \text{on } \Gamma \times (0, T), \\ \eta_r &= U_r && \text{on } \Gamma \times (0, T). \end{aligned}$$

The exact solution to this problem is given by the following. The fluid flow through the fixed cylinder with constant pressure drop is given by the Poiseuille velocity profile:

$$u_z^e(z, r) = u_z^e(r) = \frac{p_{\text{in}} - p_{\text{out}}}{2\mu_F L}(R^2 - r^2), \quad u_r^e = 0,$$

and the fluid pressure is linear within the channel:

$$p^e(z, r) = p^e(z) = \frac{p_{\text{out}} z + p_{\text{in}}(L - z)}{L}, \ z \in (0, L), \ r \in (0, R).$$

The radial displacements of the thin and thick structure are given by:

$$\eta_r^e(z) = \frac{p^e(z)}{C_0}, \quad d_r^e(z, r) = d_r^e(z) = \eta_r^e(z).$$

We solve this problem numerically using the parameters given in Table 2.5. The initial data was taken to be

$$\mathbf{u} = 0, p = p_{\text{out}}, \eta_r = 0, d_r = 0, \text{ at } t = 0,$$

while at the inlet and outlet boundaries we kept both structures fixed, with the inlet and outlet displacement data tailored so that the final solution does not exhibit a boundary layer:

$$\eta_r|_{z=0} = d_r|_{z=0} = \frac{p_{\text{in}}}{C_0}, \quad \eta_r|_{z=L} = d_r|_{z=L} = \frac{p_{\text{out}}}{C_0} = 0, \forall t > 0.$$

The numerical scheme with $\beta = 1$ was implemented, and the problem was solved until the steady state was achieved. With the time step $\Delta t = 10^{-5}$ it took 200 iterations to achieve the accuracy of less than 0.08%. Namely, the maximum

Table 2.5 Geometry, fluid and structure parameters used in Example 1

Parameters	Values	Parameters	Values
Radius R (cm)	0.5	Length L (cm)	6
In. press. p_{in} (dyne/cm^2)	250	Out. press. p_{out} (dyne/cm^2)	0
Fluid density ρ_f (g/cm^3)	1	Dyn. viscosity μ (g/cm s)	0.35
Thin wall:			
Density ρ_m (g/cm^3)	1.1	Thickness h (cm)	0.02
Lamé coeff. μ_m (dyne/cm^2)	1.07×10^6	Lamé coeff. λ_m (dyne/cm^2)	4.29×10^6
Thick wall:			
Density ρ_s(g/cm^3)	1.1	Thickness H (cm)	0.1
Lamé coeff. μ_s(dyne/cm^2)	1.07×10^6	Lamé coeff. λ_s (dyne/cm^2)	4.29×10^6
Spring coeff. γ (dyne/cm^4)	0		

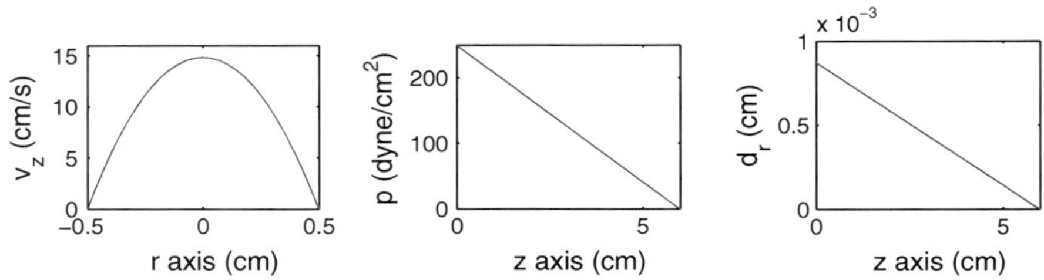

Fig. 2.14 Comparison between the computed solution (in *blue*) and the exact solution (in *red*). The two are superimposed. *Left*: Axial velocity. *Middle*: Fluid pressure. *Right*: Radial displacement

relative error between the computed and exact solution was less than 0.08% (namely, 0.000778).

Figure 2.14 shows a comparison between the computed (blue) and the exact solution (red) for axial velocity (left), fluid pressure (middle), and radial displacement (right), showing excellent agreement. The corresponding relative errors are given by the following:

$$\frac{||\mathbf{u}^e - \mathbf{u}||_{L^2(\Omega^f)}}{||\mathbf{u}^e||_{L^2(\Omega^f)}} = 7.78 \times 10^{-4}, \quad \frac{||p^e - p||_{L^2(\Omega^f)}}{||p^e||_{L^2(\Omega^f)}} = 1.17 \times 10^{-4},$$

$$\frac{||\eta_r^e - \eta_r||_{L^2(0,L)}}{||\eta_r^e||_{L^2(0,L)}} = 3.82 \times 10^{-5}, \quad \frac{||d_r^e - d_r||_{L^2(\Omega^s)}}{||d_r^e||_{L^2(\Omega^s)}} = 3.82 \times 10^{-5}.$$

We conclude that the scheme behaves well for this simplified FSI problem with multiple structural layers.

Example 2

In this example we solve the full, nonlinear FSI problem (2.177)–(2.191) with the structure consisting of two layers, using the data that correspond to a benchmark problem in FSI with a single thick structure. Moreover, we solve a sequence of

Table 2.6 Geometry, fluid and structure parameters that are used in Example 2

Parameters	Values	Parameters	Values
Radius R (cm)	0.5	Length L (cm)	6
Fluid density ρ_f (g/cm^3)	1	Dyn. viscosity μ (g/cm s)	0.035
Thin wall:			
Density ρ_m(g/cm^3)	1.1	Thickness h (cm)	0.02
Lamé coeff. μ_m (dyne/cm^2)	5.75×10^5	Lamé coeff. λ_m (dyne/cm^2)	1.7×10^6
Thick wall:			
Density ρ_s (g/cm^3)	1.1	Thickness H (cm)	0.1
Lamé coeff. μ_s (dyne/cm^2)	5.75×10^5	Lamé coeff. λ_s (dyne/cm^2)	1.7×10^6
Spring coeff. γ (dyne/cm^4)	4×10^6		

FSI problems (2.177)–(2.191) in which the thickness of the thin layer converges to zero. The limiting solution is then compared with the solution of the benchmark problem with a single, thick structure, obtained using a different solver. In the sequence of FSI problem with two structural layers, the combined thickness of the entire structure is set to be constant, and equal to the thickness of the thick structure from the benchmark problem. Furthermore, the elastic properties (i.e., the Young's modulus of elasticity and the Poisson ratio) of the thin and thick structure are all set to be equal to the elastic properties of the thick structure in the benchmark problem.

The elastodynamics of the thin structural layer is modeled using the linearly elastic Koiter membrane equations with both radial and longitudinal displacement (2.182), (2.183), while the elastodynamics of the thick structure is modeled using the equations of 2D linear elasticity (2.184). The same 2D linear elasticity model (2.184) is used to capture the elastodynamics of the thick structure in the FSI benchmark problem. In both cases the flow is driven by the time-dependent pressure data:

$$p_{\text{in}}(t) = \begin{cases} \frac{p_{\max}}{2}\left[1 - \cos\left(\frac{2\pi t}{t_{\max}}\right)\right] & \text{if } t \leq t_{\max} \\ 0 & \text{if } t > t_{\max} \end{cases}, \qquad p_{\text{out}}(t) = 0 \ \forall t \in (0, T),$$

where $p_{\max} = 1.333 \times 10^4$ (dyne/cm^2) and $t_{\max} = 0.003$ (s). The values of the parameters used in this example are given in Table 2.6. The same parameters were used to test partitioned FSI schemes in [9].

We assume that the combined thickness of the two-layered structure is fixed, and equal to $h + H = 0.12$ cm, which is set to be the same as the thickness of the single thick structure in the benchmark problem. Our kinematically coupled β scheme, described in Sect. 2.7.5, was used to solve the multi-layered FSI problem with $\beta = 1$, while the scheme presented in [22] was used to solve the single-layered FSI benchmark problem. The problem was solved over the time interval $[0, 0.012]$s, using the time step $\Delta t = 5 \times 10^{-5}$. Figure 2.15 shows the axial and radial displacement at time $t = 8$ ms obtained using the multi-layered model (left) and the single-layered model (right) for the arterial wall. Figure 2.16 shows the corresponding fluid velocity and pressure. One can notice significant smoothing of

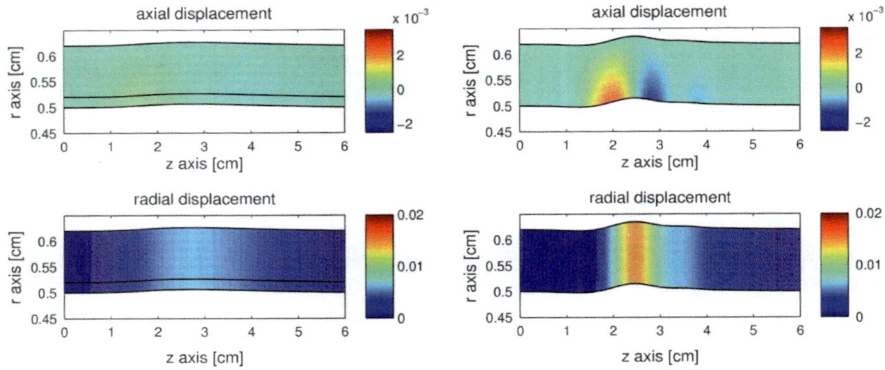

Fig. 2.15 Axial displacement (*top*) and radial displacement (*bottom*) at time $t = 8$ ms obtained using the model capturing two structural layers (*left*), and the model capturing FSI with a single thick structural layer [22] (*right*)

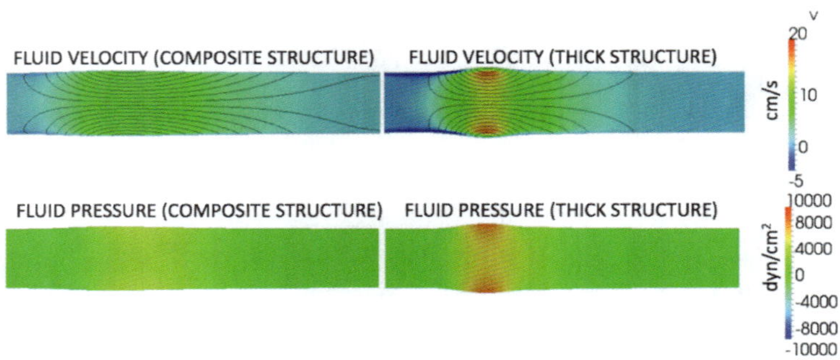

Fig. 2.16 Fluid velocity (*top*) and fluid pressure (*bottom*) at time $t = 8$ ms obtained using the model capturing two structural layers (*left*), and the model capturing FSI with a single thick structural layer [22] (*right*)

both the displacement as well as the fluid velocity and pressure in the composite, i.e., multi-layered structure case. Same data are used for both simulations. We further compared the results of the multi-layered model with the single-layered model as the thickness of the thin structure h goes to zero. As we decreased h, we increased H to maintain the constant combined thickness $h + H = 0.12$ cm. Figures 2.17, 2.18, and 2.19 show the flowrate, mean pressure, and displacement of the fluid–structure interface obtained using different values of h. The results obtained using the single-layered wall model correspond to the label $h = 0$. Indeed, we can see that as we decrease the thickness of the fluid–structure interface, the numerical results obtained using our multi-layered model approach the results obtained using the single-layered FSI model! Notice how for $h = 0.025$ cm the solutions obtained using the multi-layered model and the single thick structure model ($h = 0$ in Figs. 2.17, 2.18, and 2.19) are almost identical.

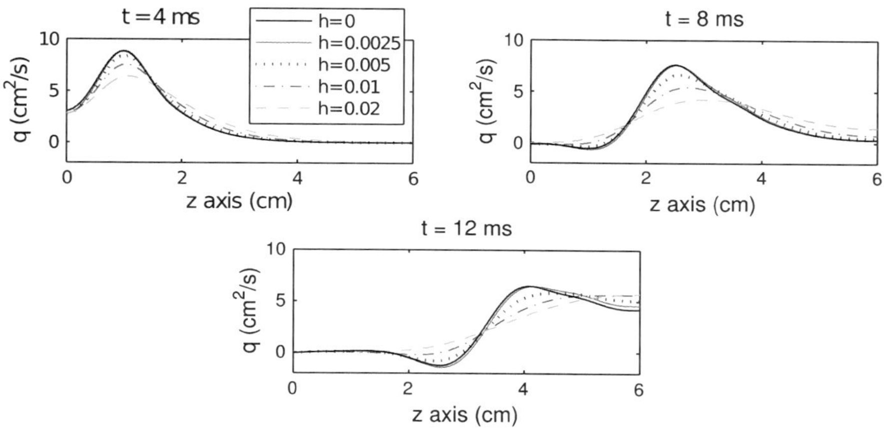

Fig. 2.17 Flowrate computed using two different models: the model in [22] containing a single thick structural layer ($h = 0$), and the model considered in this chapter, consisting of two layers. The thickness of the thin membrane layer was decreased from $h = 0.02$ to $h = 0.0025$ cm. The combined thickness of the two-layered structure was kept constant at $h + H = 0.12$ cm. Convergence of the solutions to the FSI solution containing a single, thick layered model ($h = 0$) can be observed

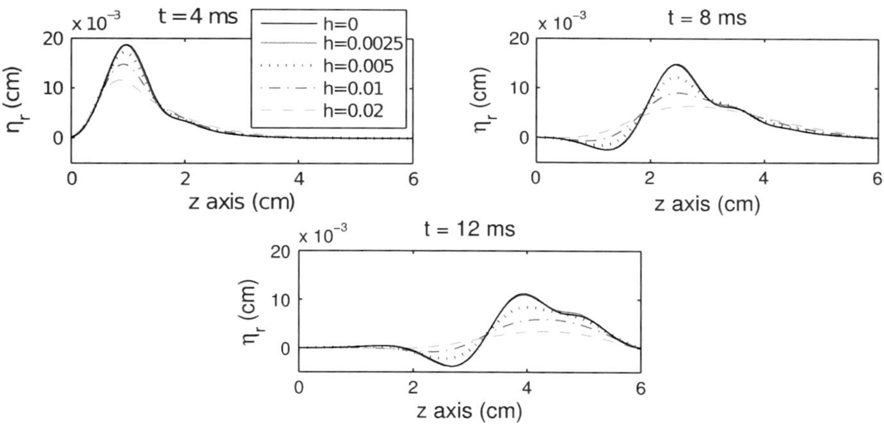

Fig. 2.18 Displacement of the fluid–structure interface obtained under the same conditions as those described in Fig. 2.17

Regularizing Effects by Thin Fluid–Structure Interface with Mass

We conclude this section with a remark on the regularizing effects of the thin fluid–structure interface with mass. Figures 2.17, 2.18, and 2.19 indicate that as we increase inertia of the thin fluid–structure interface with mass by increasing its thickness, the solution of the entire FSI problem is damped, or regularized. More precisely, if one looks at the FSI problem with a single thick structural layer, the fluid–

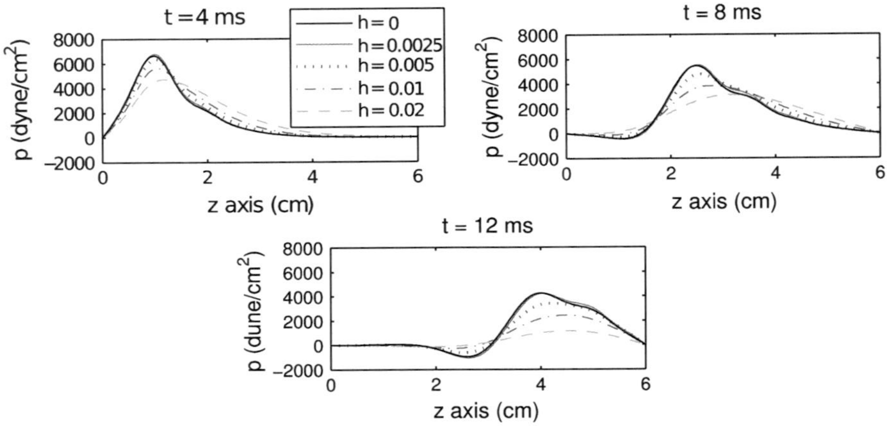

Fig. 2.19 Mean pressure obtained under the same conditions as those described in Fig. 2.17

structure interface is simply the massless trace of the thick structure that is in contact with the fluid. Mathematically, in that case the trace of the structure displacement is not well defined (assuming regularity of the data consistent with weak solutions), and using energy estimates it is not possible to even show that the fluid–structure interface is continuous. In the case when the fluid–structure interface has mass, we showed in Proposition 2.1 that not only is the fluid–structure interface continuous, but its evolution can be controlled by the energy norm of the time derivative of its displacement. We see effects of this in the solutions presented in Figs. 2.17, 2.18, and 2.19, and in Fig. 2.20 below. In Fig. 2.20 below we focus on the displacement and displacement velocity of the fluid–structure interface, which measures the effects of inertia. In the first row of Fig. 2.20 three snap-shots of the fluid–structure interface are shown as the inlet pressure wave travels down the tube. In the second row of Fig. 2.20 the same three snap-shots are shown, but for the fluid–structure interface velocity. The red solid line in these figures corresponds to the massless fluid–structure interface in the FSI problem with a single thick structural layer. The black dashed line corresponds to the fluid–structure interface with mass in the FSI problem with two structural layers. We see significant damping of the traveling wave in the case when the fluid–structure interface has mass. This indicates that inertia of the fluid–structure interface with mass regularizes solutions of FSI problems.

This is reminiscent of the results by Hansen and Zuazua [90] in which the presence of a point mass at the interface between two linearly elastic strings with solutions in asymmetric spaces (different regularity on each side) allowed the proof of well-posedness due to the regularizing effects by the point mass. In particular, in [90] two linearly elastic strings were considered, meeting at a point mass. The elastodynamics of each string was modeled by the linear wave equation. It was shown that as the wave with the displacement in $H^1(0, L)$ and velocity in $L^2(0, L)$ passes through the point mass, a reflected and a transmitted waves

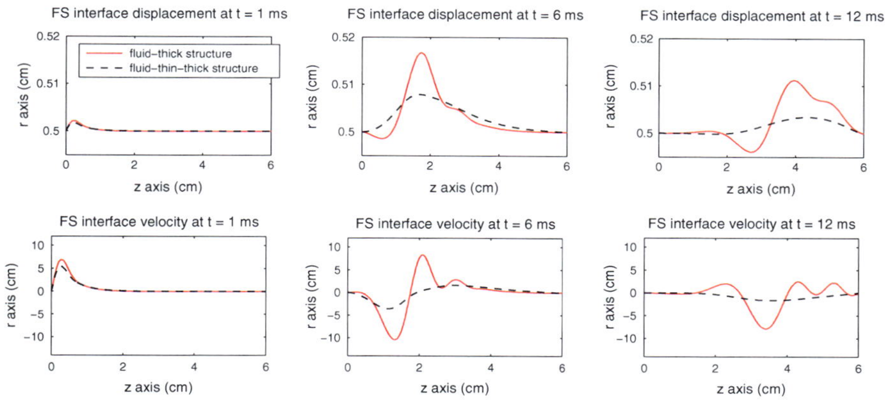

Fig. 2.20 Fluid–structure interface displacement (*top*) and velocity (*bottom*) obtained using a multi-layered wall model and the single-layered model from [22], shown at times $t = 1\,\text{ms}$, $t = 6\,\text{ms}$, and $t = 12\,\text{ms}$

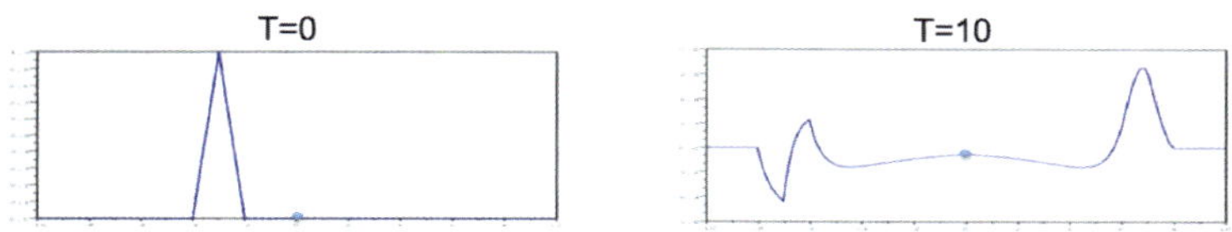

Fig. 2.21 Regularizing effects of point mass. These figures originally appeared as Fig. 1 and Fig. 2 (left) in [107] and are reprinted by permission of © American Mathematical Society 2014. The initial data (*left panel*) is smoothed out as the transmitted wave traveling to the right passes through the point mass (*right panel*)

form. The transmitted wave, which passes through the point mass, gets smoothed out to $H^2(0, L)$ regularity in displacement, and $H^1(0, L)$ regularity in velocity. A numerical simulation of this phenomenon was shown in [107]. Figure 2.21 shows one of the results from [107]. The panel on the left shows the initial displacement in $H^1(0, L)$ with zero initial velocity, located just left from the point mass. The panel on the right shows the solution at time $T = 10s$ at which the reflected and transmitted waves have formed, with the displacement of the reflected wave on the left of the point mass still in $H^1(0, L)$, but with the displacement of the transmitted wave, shown to the right of the point mass, belonging to $H^2(0, L)$. For a reader with further interest in the area of simplified coupled problems we mention [91, 123, 134, 145].

2.8 Conclusions

This chapter addresses an operator splitting approach to study multi-physics problems related to FSI with application to hemodynamics. The methodology is based on the Lie splitting scheme, also known as the Marchuk–Yanenko scheme. The splitting discussed in this chapter deals successfully with the added mass effect which is known to be responsible for instabilities in loosely coupled Dirichlet–Neumann schemes for FSI problems in which the density of the structure is comparable to that of the fluid, which is the case in blood flow applications. Particular attention was payed to a multi-physics FSI problem in which the structure is composed of multiple structural layers. Problems of this kind arise, for example, in modeling blood flow through human arteries which are composed of several layers, each with different mechanical characteristics and thickness. A benchmark problem was studied in which the structure consists of two layers: a thin layer which is in contact with the fluid, and a thick layer which sits on top of the thin layer. The thin layer serves as a fluid–structure interface with mass. Both analytical (existence of a weak solution) and numerical results were studied for the underlying benchmark problem. In particular, it was shown that the proposed scheme converges to a weak solution to the full nonlinear fluid-multi-layered structure interaction problem. Two academic examples were considered to test the performance of the numerical scheme.

The analytical and numerical methods presented here apply with slight modifications to a larger class of problems. They include, for example, a study of FSI with one structural layer (thin [20, 21, 85], or thick [22]), FSI with poroelastic structures [23], FSI between a mechanical device called stent, arterial wall and fluid [122], and FSI involving a non-Newtonian fluid [94–96].

This chapter provides the basic mathematical tools for further development of analytical and computational methods based on the Lie operator splitting approach, to study various multi-physics problems involving FSI.

Acknowledgements The research of the authors has been supported in part by the National Science Foundation under the following grants: NIGMS DMS-1263572, DMS-1318763, DMS-1311709, DMS-1262385, DMS-1109189, and by the Texas Higher Education Board under grant ARP-003652-0023-2009.

References

1. R.A. Adams, *Sobolev Spaces*. Pure and Applied Mathematics, vol. 65 (Academic [A subsidiary of Harcourt Brace Jovanovich, Publishers], New York/London, 1975)
2. Å.R. Ahlgren, M. Cinthio, S. Steen, H.W. Persson, T. Sjöberg, K. Lindström, Effects of adrenaline on longitudinal arterial wall movements and resulting intramural shear strain: a first report. Clin. Physiol. Funct. Imaging **29**, 353–359 (2009)
3. R.L. Armentano, J.G. Barra, J. Levenson, A. Simon, R.H. Pichel, Arterial wall mechanics in conscious dogs: assessment of viscous, inertial, and elastic moduli to characterize aortic wall behavior. Circ. Res. **76**, 468–478 (1995)
4. R.L. Armentano, J.L. Megnien, A. Simon, F. Bellenfant, J.G. Barra, J. Levenson, Effects of hypertension on viscoelasticity of carotid and femoral arteries in humans. Hypertension **26**, 48–54 (1995)
5. M. Astorino, F. Chouly, M.A. Fernández, An added-mass free semi-implicit coupling scheme for fluid-structure interaction. C. R. Math. **347**(1–2), 99–104 (2009)

6. M. Astorino, F. Chouly, M.A. Fernández Varela, Robin based semi-implicit coupling in fluid-structure interaction: stability analysis and numerics. SIAM J. Sci. Comput. **31**, 4041–4065 (2009)
7. F.P.T. Baaijens, A fictitious domain/mortar element method for fluid-structure interaction. Int. J. Numer. Methods Fluids **35**, 743–761 (2001)
8. S. Badia, A. Quaini, A. Quarteroni, Splitting methods based on algebraic factorization for fluid-structure interaction. SIAM J. Sci. Comput. **30**(4), 1778–1805 (2008)
9. S. Badia, F. Nobile, C. Vergara, Fluid-structure partitioned procedures based on Robin transmission conditions. J. Comput. Phys. **227**, 7027–7051 (2008)
10. S. Badia, F. Nobile, C. Vergara, Robin-robin preconditioned Krylov methods for fluid-structure interaction problems. Comput. Methods Appl. Mech. Eng. **198**(33–36), 2768–2784 (2009)
11. V. Barbu, Z. Grujić, I. Lasiecka, A. Tuffaha, Existence of the energy-level weak solutions for a nonlinear fluid-structure interaction model, in *Fluids and Waves*. Contemporary Mathematics, vol. 440 (American Mathematical Society, Providence, 2007), pp. 55–82
12. V. Barbu, Z. Grujić, I. Lasiecka, A. Tuffaha, Smoothness of weak solutions to a nonlinear fluid-structure interaction model. Indiana Univ. Math. J. **57**(3), 1173–1207 (2008)
13. A.T. Barker, X.C. Cai, Scalable parallel methods for monolithic coupling in fluid-structure interaction with application to blood flow modeling. J. Comput. Phys. **229**(3), 642–659 (2010)
14. R.D. Bauer, R. Busse, A. Shabert, Y. Summa, E. Wetterer, Separate determination of the pulsatile elastic and viscous forces developed in the arterial wall in vivo. Pflugers Arch. **380**, 221–226 (1979)
15. Y. Bazilevs, V.M. Calo, Y. Zhang, T.J.R. Hughes, Isogeometric fluid-structure interaction analysis with applications to arterial blood flow. Comput. Mech. **38**(4–5), 310–322 (2006)
16. Y. Bazilevs, V.M. Calo, T.J.R. Hughes, Y. Zhang, Isogeometric fluid-structure interaction: theory algorithms and computations. Comput. Mech. **43**, 3–37 (2008)
17. H. Beirão da Veiga, On the existence of strong solutions to a coupled fluid-structure evolution problem. J. Math. Fluid Mech. **6**(1), 21–52 (2004)
18. J. Bemelmans, G.P. Galdi, M. Kyed, Capillary surfaces and floating bodies. Ann. Mat. Pura ed Appl. 1–16 (2013)
19. M. Boulakia, Existence of weak solutions for the motion of an elastic structure in an incompressible viscous fluid. C. R. Math. Acad. Sci. Paris **336**(12), 985–990 (2003)
20. M. Bukač, S. Čanić, Longitudinal displacement in viscoelastic arteries: a novel fluid-structure interaction computational model, and experimental validation. J. Math. Biosci. Eng. **10**(2), 258–388 (2013)
21. M. Bukač, S. Čanić, R. Glowinski, J. Tambača, A. Quaini, Fluid-structure interaction in blood flow allowing non-zero longitudinal structure displacement. J. Comput. Phys. **235**, 515–541 (2013)
22. M. Bukač, S. Čanić, R. Glowinski, B. Muha, A. Quaini, An operator splitting scheme for fluid-structure interaction problems with thick structures. Int. J. Numer. Methods Fluids (Accepted 2013)
23. M. Bukač, P. Zunino, I. Yotov, Explicit partitioning strategies for interaction of the fluid with a multilayered poroelastic structure: an operator-splitting approach (2013, submitted)
24. M. Bukac, S. Canic, B. Muha, A partitioned scheme for fluid-composite structure interaction problems (submitted, 2013)
25. E. Burman, M.A. Fernández, Stabilization of explicit coupling in fluid-structure interaction involving fluid incompressibility. Comput. Methods Appl. Mech. Eng. **198**, 766–784 (2009)
26. S. Čanić, E.H. Kim, Mathematical analysis of the quasilinear effects in a hyperbolic model of blood flow through compliant axisymmetric vessels. Math. Methods Appl. Sci. **26**(14), 1161–1186 (2003)

27. S. Čanić, J. Tambača, G. Guidoboni, A. Mikelić, C.J. Hartley, D. Rosenstrauch, Modeling viscoelastic behavior of arterial walls and their interaction with pulsatile blood flow. SIAM J. Appl. Math. **67**(1), 164–193 (2006)
28. S. Čanić, C.J. Hartley, D. Rosenstrauch, J. Tambača, G. Guidoboni, A. Mikelić, Blood flow in compliant arteries: an effective viscoelastic reduced model, numerics and experimental validation. Ann. Biomed. Eng. **34**, 575–592 (2006)
29. S. Čanić, B. Muha, M. Bukač, Stability of the kinematically coupled β-scheme for fluid-structure interaction problems in hemodynamics (submitted), arXiv:1205.6887v1
30. P. Causin, J. Gerbeau, F. Nobile, Added-mass effect in the design of partitioned algorithms for fluid-structure problems. Comput. Methods Appl. Mech. Eng. **194**(42–44), 4506–4527 (2005)
31. M. Cervera, R. Codina, M. Galindo, On the computational efficiency and implementation of block-iterative algorithms for nonlinear coupled problems. Eng. Comput. **13**, 4–30 (1996)
32. A. Chambolle, B. Desjardins, M.J. Esteban, C. Grandmont, Existence of weak solutions for the unsteady interaction of a viscous fluid with an elastic plate. J. Math. Fluid Mech. **7**(3), 368–404 (2005)
33. C.H.A. Cheng, D. Coutand, S. Shkoller, Navier-Stokes equations interacting with a nonlinear elastic biofluid shell. SIAM J. Math. Anal. **39**(3), 742–800 (2007)
34. C.H.A. Cheng, S. Shkoller, The interaction of the 3D Navier-Stokes equations with a moving nonlinear Koiter elastic shell. SIAM J. Math. Anal. **42**(3), 1094–1155 (2010)
35. P.G. Ciarlet, A two-dimensional nonlinear shell model of Koiter type. C. R. Acad. Sci. Paris Ser. I Math. **331**, 405–410 (2000)
36. C.H. Ciarlet, D. Cautnad, An existence theorem for nonlinearly elastic "flexural" shells. J. Elast. **50**(3), 261–277 (1998)
37. P.G. Ciarlet, V. Lods, Asymptotic analysis of linearly elastic shells. III. Justification of Koiter's shell equations. Arch. Ration. Mech. Anal. **136**, 191–200 (1996)
38. C.R. Ciarlet, A. Roquefort, Justification of a two-dimensional shell model of Koiter type. C. R. Acad. Sci. Paris, Ser. I Math. **331**(5), 411–416 (2000)
39. M. Cinthio, A.R. Ahlgren, T. Jansson, A. Eriksson, H.W. Persson, K. Lindstrom, Evaluation of an ultrasonic echo-tracking method for measurements of arterial wall movements in two dimensions. IEEE Trans. Ultrason. Ferroelectr. Freq. Control **52**(8), 1300–1311 (2005)
40. M. Cinthio, A. Ahlgren, J. Bergkvist, T. Jansson, H.W. Persson, K. Lindstrom, Longitudinal movements and resulting shear strain of the arterial wall. Am. J. Physiol. Heart Circ. Physiol. **291**(1), H394–H402 (2006)
41. C. Conca, J. San Martín, M. Tucsnak, Motion of a rigid body in a viscous fluid. C. R. Acad. Sci. Paris Sér. I Math. **328**(6), 473–478 (1999)
42. G.H. Cottet, E. Maitre, T. Milcent, Eulerian formulation and level set models for incompressible fluid-structure interaction. Math. Model. Numer. Anal. **42**(3), 471–492 (2008)
43. D. Coutand, S. Shkoller, Motion of an elastic solid inside an incompressible viscous fluid. Arch. Ration. Mech. Anal. **176**(1), 25–102 (2005)
44. D. Coutand, S. Shkoller, The interaction between quasilinear elastodynamics and the Navier-Stokes equations. Arch. Ration. Mech. Anal. **179**(3), 303–352 (2006)
45. P. Cumsille, T. Takahashi, Wellposedness for the system modelling the motion of a rigid body of arbitrary form in an incompressible viscous fluid. Czechoslov. Math. J. **58**(133)(4), 961–992 (2008)
46. H. Demiray, Small but finite amplitude waves in a prestressed viscoelastic thin tube filled with an inviscid fluid. Int. J. Eng. Sci. **35**(4), 353–363 (1997)
47. S. Deparis, M.A. Fernández, L. Formaggia, Acceleration of a fixed point algorithm for a fluid-structure interaction using transpiration condition. Math. Model. Numer. Anal. **37**, 601–616 (2003)
48. S. Deparis, M. Discacciati, G. Fourestey, A. Quarteroni, Fluid-structure algorithms based on Steklov-Poincaré operators. Comput. Methods Appl. Mech. Eng. **195**, 5797–5812 (2006)
49. B. Desjardins, M.J. Esteban, Existence of weak solutions for the motion of rigid bodies in a viscous fluid. Arch. Ration. Mech. Anal. **146**(1), 59–71 (1999)

50. B. Desjardins, M.J. Esteban, C. Grandmont, P. Le Tallec, Weak solutions for a fluid-elastic structure interaction model. Rev. Mat. Complut. **14**(2), 523–538 (2001)
51. P. Destuynder, A classification of thin shell theories. Acta Appl. Math. **4**, 15–63 (1985)
52. J. Donea, Arbitrary Lagrangian-Eulerian finite element methods, in *Computational Methods for Transient Analysis* (North-Holland, Amsterdam, 1983)
53. Q. Du, M.D. Gunzburger, L.S. Hou, J. Lee, Analysis of a linear fluid-structure interaction problem. Discrete Contin. Dyn. Syst. **9**(3), 633–650 (2003)
54. Q. Du, M.D. Gunzburger, L.S. Hou, J. Lee, Analysis of linear fluid-structure interaction problem. Discrete Contin. Dyn. Syst. **9**, 633–650 (2003)
55. Q. Du, M.D. Gunzburger, L.S. Hou, J. Lee, Semidiscrete finite element approximations of a linear fluid-structure interaction problem. SIAM J. Numer. Anal **42**(1), 1–29 (2004)
56. H. Fang, Z. Wang, Z. Lin, M. Liu, Lattice Boltzmann method for simulating the viscous flow in large distensible blood vessels. Phys. Rev. E **65**, 051925 (2002)
57. L.J. Fauci, R. Dillon, Biofluidmechanics of reproduction. Ann. Rev. Fluid Mech. **38**, 371–394 (2006)
58. E. Feireisl, On the motion of rigid bodies in a viscous compressible fluid. Arch. Ration. Mech. Anal. **167**(4), 281–308 (2003)
59. Z.-G. Feng, E.E. Michaelides, The immersed boundary-lattice Boltzmann method for solving fluid-particles interaction problem. J. Comput. Phys. **195**, 602–628 (2004)
60. M.A. Fernández, Incremental displacement-correction schemes for the explicit coupling of a thin structure with an incompressible fluid. C. R. Math. Acad. Sci. Paris **349**(7–8), 473–477 (2011)
61. M.A. Fernández, Incremental displacement-correction schemes for incompressible fluid-structure interaction: stability and convergence analysis. Numer. Math. **123**(1), 21–65 (2013)
62. M.A. Fernández, M. Moubachir, A Newton method using exact Jacobians for solving fluid-structure coupling. Comput. Struct. **83**(2–3), 127–142 (2005)
63. M.A. Fernández, J. Mullaert, Displacement-velocity correction schemes for incompressible fluid-structure interaction. C. R. Math. Acad. Sci. Paris **349**(17–18), 1011–1015 (2011)
64. M.A. Fernández, J.F. Gerbeau, C. Grandmont, A projection algorithm for fluid-structure interaction problems with strong added-mass effect. C. R. Math. **342**(4), 279–284 (2006)
65. C. Figueroa, I. Vignon-Clementel, K.E. Jansen, T. Hughes, C. Taylor, A coupled momentum method for modeling blood flow in three-dimensional deformable arteries. Comput. Methods Appl. Mech. Eng. **195**, 5685–5706 (2006)
66. A.L. Fogelson, R.D. Guy, Platelet-wall interactions in continuum models of platelet thrombosis: formulation and numerical solution. Math. Med. Biol. **21**, 293–334 (2004)
67. L. Formaggia, J.F. Gerbeau, F. Nobile, A. Quarteroni, On the coupling of 3D and 1D Navier-Stokes equations for flow problems in compliant vessels. Comput. Methods Appl. Mech. Eng. **191**(6–7), 561–582 (2001)
68. Y.C. Fung, *Biomechanics: Circulation*, 2nd edn. (Springer, New York, 1984)
69. Y.C. Fung, *Biomechanics: Mechanical Properties of Living Tissues*, 2nd edn. (Springer, New York, 1993)
70. G.P. Galdi, *An Introduction to the Mathematical Theory of the Navier-Stokes Equations. Vol. I.* Linearized Steady Problems, Springer Tracts in Natural Philosophy, vol. 38 (Springer, New York, 1994)
71. G.P. Galdi, On the motion of a rigid body in a viscous liquid: a mathematical analysis with applications, in *Handbook of Mathematical Fluid Dynamics*, vol. I (North-Holland, Amsterdam, 2002), pp. 653–791
72. G.P. Galdi, Mathematical problems in classical and non-Newtonian fluid mechanics, in *Hemodynamical Flows*. Oberwolfach Seminar, vol. 37 (Birkhäuser, Basel, 2008), pp. 121–273
73. G.P. Galdi, M. Kyed, Steady flow of a Navier-Stokes liquid past an elastic body. Arch. Ration. Mech. Anal. **194**(3), 849–875 (2009)
74. G.P. Galdi, A.L. Silvestre, The steady motion of a Navier-Stokes liquid around a rigid body. Arch. Ration. Mech. Anal. **184**(3), 371–400 (2007)

75. G.P. Galdi, A.L. Silvestre, On the motion of a rigid body in a Navier-Stokes liquid under the action of a time-periodic force. Indiana Univ. Math. J. **58**(6), 2805–2842 (2009)
76. G.P. Galdi, A. Vaidya, M. Pokorný, D.D. Joseph, J. Feng, Orientation of symmetric bodies falling in a second-order liquid at nonzero Reynolds number. Math. Models Methods Appl. Sci. **12**(11), 1653–1690 (2002)
77. J.F. Gerbeau, M. Vidrascu, A quasi-Newton algorithm based on a reduced model for fluid-structure interactions problems in blood flows. Math. Model. Numer. Anal. **37**, 631–648 (2003)
78. R. Glowinski, Finite element methods for incompressible viscous flow, in *Handbook of Numerical Analysis*, vol. 9, ed. by P.G.Ciarlet, J.-L.Lions (North-Holland, Amsterdam, 2003)
79. C. Grandmont, Existence of weak solutions for the unsteady interaction of a viscous fluid with an elastic plate. SIAM J. Math. Anal. **40**(2), 716–737 (2008)
80. B.E. Griffith, On the volume conservation of the immersed boundary method. Commun. Comput. Phys. **12**, 401–432 (2012)
81. B.E. Griffith, Immersed boundary model of aortic heart valve dynamics with physiological driving and loading conditions. Int. J. Numer. Method Biomed. Eng. **28**, 317–345 (2012)
82. B.E. Griffith, X. Luo, Hybrid finite difference/finite element version of the immersed boundary method (submitted)
83. B.E. Griffith, R.D. Hornung, D.M. McQueen, C.S. Peskin, An adaptive, formally second order accurate version of the immersed boundary method. J. Comput. Phys. **223**, 10–49 (2007)
84. B.E. Griffith, X. Luo, D.M. McQueen, C.S. Peskin, Simulating the fluid dynamics of natural and prosthetic heart valves using the immersed boundary method. Int. J. Appl. Mech. **1**, 137–177 (2009)
85. G. Guidoboni, R. Glowinski, N. Cavallini, S. Čanić, Stable loosely-coupled-type algorithm for fluid-structure interaction in blood flow. J. Comput. Phys. **228**(18), 6916–6937 (2009)
86. G. Guidoboni, N. Cavallini, R. Glowinski, S. Čanić, S. Lapin, A kinematically coupled time-splitting scheme for fluid-structure interaction in blood flow. Appl. Math. Lett. **22**(5), 684–688 (2009)
87. G. Guidoboni, M. Guidorzi, M. Padula, Continuous dependence on initial data in fluid-structure motions. J. Math. Fluid Mech. **14**(1), 1–32 (2012)
88. J.D. Hamphrey, Mechanics of the arterial wall: review and directions. Crit. Rev. Biomed. Eng. **23**(1&2), 1–162 (1995)
89. P. Hansbo, Nitsche's method for interface problems in computational mechanics. GAMM-Mitt. **28**(2), 183–206 (2005)
90. S. Hansen, E. Zuazua, Exact controllability and stabilization of a vibrating string with an interior point mass. SIAM J. Control Optim. **33**(5), 1357–1391 (1995)
91. H. Koch, E. Zuazua, A hybrid system of PDE's arising in multi-structure interaction: coupling of wave equations in n and $n-1$ space dimensions, in *Recent Trends in Partial Differential Equations*. Contemporary Mathematics, vol. 409 (American Mathematical Society, Providence, 2006), pp. 55–77
92. M. Heil, An efficient solver for the fully coupled solution of large-displacement fluid-structure interaction problems. Comput. Methods Appl. Mech. Eng. **193**(1–2), 1–23 (2004)
93. T.J.R. Hughes, W.K Liu, T.K. Zimmermann, Lagrangian-Eulerian finite element formulation for incompressible viscous flows. Comput. Methods Appl. Mech. Eng. **29**(3), 329–349 (1981)
94. A. Hundertmark-Zaušková, M. Lukáčová-Medvidová, Š. Nečasová, On the existence of weak solution to the coupled fluid-structure interaction problem for non-newtonian shear-dependent fluid (2013, submitted)
95. A. Hundertmark-Zauskova, M. Lukacova-Medvidova, G. Rusnakova, Kinematic splitting algorithm for fluid-structure interaction in hemodynamics. Comput. Methods Appl. Mech. Eng. **265**, 83–106 (2013)
96. A. Hundertmark-Zauskova, M. Lukacova-Medvidova, G. Rusnakova, *Fluid-Structure Interaction for Shear-Dependent Non-Newtonian Fluids. Topics in Mathematical Modeling and Analysis.* Lecture Notes, vol. 7 (Necas Center for Mathematical Modeling, The Check Republic, 2012), pp. 109–158

97. W.T. Koiter, A consistent first approximation in the general theory of thin elastic shells. Part 1: foundations and linear theory. Technological University, Delft, 5 August 1959
98. W.T. Koiter, On the foundations of the linear theory of thin elastic shells. I, II. Nederl. Akad. Wetensch. Proc. Ser. B **73**, 169–182 (1970)
99. M. Krafczyk, M. Cerrolaza, M. Schulz, E. Rank, Analysis of 3D transient blood flow passing through an artificial aortic valve by Lattice-Boltzmann methods. J Biomech. **31**, 453–462 (1998)
100. M. Krafczyk, J. Tolke, E. Rank, M. Schulz, Two-dimensional simulation of fluid-structure interaction using Lattice-Boltzmann methods. Comput. Struct. **79**, 2031–2037 (2001)
101. I. Kukavica, A. Tuffaha, M. Ziane, Strong solutions for a fluid structure interaction system. Adv. Differ. Equ. **15**(3–4), 231–254 (2010)
102. I. Kukavica, A. Tuffaha, Solutions to a fluid-structure interaction free boundary problem. DCDS-A **32**(4), 1355–1389 (2012)
103. I. Kukavica, A. Tuffaha, Well-posedness for the compressible Navier-Stokes-Lamé system with a free interface. Nonlinearity **25**(11), 3111–3137 (2012)
104. P. Le Tallec, J. Mouro, Fluid structure interaction with large structural displacements. Comput. Methods Appl. Mech. Eng. **190**(24–25), 3039–3067 (2001)
105. D. Lengeler, Global weak solutions for an incompressible, generalized Newtonian fluid interacting with a linearly elastic Koiter shell (2012). arXiv:1212.3435
106. D. Lengeler, M. Ružička, Global weak solutions for an incompressible newtonian fluid interacting with a linearly elastic Koiter shell (2012). arXiv:1207.3696v1
107. V. Lescarret, E. Zuazua, Numerical approximation schemes for multi-dimensional wave equations in asymmetric spaces. Mathematics of Computation. (fall 2014) (in press)
108. J. Lequeurre, Existence of strong solutions to a fluid-structure system. SIAM J. Math. Anal. **43**(1), 389–410 (2011)
109. J. Lequeurre, Existence of strong solutions for a system coupling the Navier-Stokes equations and a damped wave equation. J. Math. Fluid Mech. **15**(2), 249–271 (2013)
110. A. Leuprecht, K. Perktold, M. Prosi, T. Berk, W. Trubel, H. Schima, Numerical study of hemodynamics and wall mechanics in distal end-to-side anastomoses of bypass grafts. J. Biomech. **35**(2), 225–236 (2002)
111. S. Lim, C.S. Peskin, Simulations of the whirling instability by the immersed boundary method. SIAM J. Sci. Comput. **25**, 2066–2083 (2004)
112. J.-L. Lions, E. Magenes, *Non-homogeneous Boundary Value Problems and Applications*, vol. I (Springer, New York, 1972) Translated from the French by P. Kenneth, Die Grundlehren der mathematischen Wissenschaften, Band 181
113. P. Luchini, M. Lupo, A. Pozzi, Unsteady Stokes flow in a distensible pipe. Z. Angew. Math. Mech. **71**, 367–378 (1991)
114. X. Ma, G.C. Lee, S.G. Lu, Numerical simulation for the propagation of nonlinear pulsatile waves in arteries. ASME J. Biomech. Eng. **114**, 490–496 (1992)
115. H. Matthies, J. Steindorf, Numerical efficiency of different partitioned methods for fluid-structure interaction. Z. Angew. Math. Mech. **2**, 557–558 (2000)
116. C. Michler, S.J. Hulshoff, E.H. van Brummelen, R. de Borst, A monolithic approach to fluid-structure interaction. Comput. Fluids **33**(5–6), 839–848 (2004)
117. L.A. Miller, C.S. Peskin, A computational fluid dynamics study of 'clap and fling' in the smallest insects. J. Exp. Biol. **208**, 195–212 (2005)
118. B. Muha, A note on the trace theorem for domains which are locally subgraph of a Hölder continuous function. Netw. Heterogeneous Media **9**(1), 191–196 (2014)
B. Muha, A note on the trace theorem for domains which are locally subgraph of a Hölder continuous function. Netw. Heterogeneous Media (2013, accepted)
119. B. Muha, S. Čanić, Existence of a weak solution to a nonlinear fluid–structure interaction problem modeling the flow of an incompressible, viscous fluid in a cylinder with deformable walls. Arch. Ration. Mech. Anal. **207**(3), 919–968 (2013)
120. B. Muha, S. Čanić, Existence of a solution to a fluid-multi-layered-structure interaction problem. J. Differ. Equ. (in print 2013). arXiv:1305.5310

121. B. Muha, S. Canic, A nonlinear, 3D fluid-structure interaction problem driven by the time-dependent dynamic pressure data: a constructive existence proof, Communications in Information and Systems (CIS) **13**(3), 357–397 (2013)
122. B. Muha, S. Čanić, A fluid-stent-artery interaction problem (in preparation)
123. B. Muha, A note on optimal regularity and regularizing effects of point mass coupling for a heat-wave system (submitted, 2014)
124. C.M. Murea, S. Sy, A fast method for solving fluid-structure interaction problems numerically. Int. J. Numer. Methods Fluids **60**(10), 1149–1172 (2009)
125. F. Nobile, Numerical approximation of fluid-structure interaction problems with application to haemodynamics. Ph.D. thesis, EPFL Switzerland, 2001
126. F. Nobile, C. Vergara, An effective fluid-structure interaction formulation for vascular dynamics by generalized Robin conditions. SIAM J. Sci. Comput. **30**(2), 731–763 (2008)
127. C. Peskin, Numerical analysis of blood flow in the heart. J. Comput. Phys. **25**, 220–252 (1977)
128. C. Peskin, D.M. McQueen, A three-dimensional computational method for blood flow in the heart I. Immersed elastic fibers in a viscous incompressible fluid. J. Comput. Phys. **81**(2), 372–405 (1989)
129. M. Persson, R. Ahlgren, T. Jansson, A. Eriksson, H.W. Persson, K. Lindstrom, A new non-invasive ultrasonic method for simultaneous measurements of longitudinal and radial arterial wall movements: first in vivo trial. Clin. Physiol. Funct. Imaging **23**(5), 247–251 (2003)
130. G. Pontrelli, A mathematical model of flow in a liquid-filled visco-elastic tube. Med. Biol. Eng. Comput. **40**(5), 550–556 (2002)
131. A. Quaini, Algorithms for fluid-structure interaction problems arising in hemodynamics. Ph.D. thesis, EPFL Switzerland, 2009
132. A. Quaini, A. Quarteroni, A semi-implicit approach for fluid-structure interaction based on an algebraic fractional step method. Math. Models Methods Appl. Sci. **17**(6), 957–985 (2007)
133. A. Quarteroni, M. Tuveri, A. Veneziani, Computational vascular fluid dynamics: problems, models and methods. Survey article. Comput. Vis. Sci. **2**, 163–197 (2000)
134. J. Rauch, X. Zhang, E. Zuazua, Polynomial decay for a hyperbolic-parabolic coupled system. J. Math. Pures Appl. (9) **84**(4), 407–470 (2005)
135. J.A. San Martín, V. Starovoitov, M. Tucsnak, Global weak solutions for the two-dimensional motion of several rigid bodies in an incompressible viscous fluid. Arch. Ration. Mech. Anal. **161**(2), 113–147 (2002)
136. D.U. Silverthorn, *Human Physiology: An Integrated Approach*, 4th edn. (Pearson Education, San Francisco, 2007)
137. J. Simon, Compact sets in the space $L^p(0,T;B)$. Ann. Math. Pura Appl. (4) **146**, 65–96 (1987)
138. S. Svedlund, L.M. Gan, Longitudinal wall motion of the common carotid artery can be assessed by velocity vector imaging. Clin. Physiol. Funct. Imaging **31**(1), 32–38 (2011)
139. J. Tambača, Notes on the derivation of the cylindrical Koiter shell (2004) Private Communication
140. R. Temam, Sur la résolution exacte et approchée d'un problème hyperbolique non linéaire de T. Carleman. Arch. Ration. Mech. Anal. **35**, 351–362 (1969)
141. R. Temam, *Navier-Stokes Equations. Theory and Numerical Analysis*. Studies in Mathematics and its Applications, vol. 2 (North-Holland, Amsterdam, 1977)
142. R. van Loon, P. Anderson, J. de Hart, F. Baaijens, A combined fictitious domain/adaptive meshing method for fluid-structure interaction in heart valves. Int. J. Numer. Methods Fluids **46**, 533–544 (2004)
143. I. Velčić, Nonlinear weakly curved rod by Γ-convergence. J. Elast. **108**(2), 125–150 (2012)
144. S.Z. Zhao, X.Y. Xu, M.W. Collins, The numerical analysis of fluid-solid interactions for blood flow in arterial structures Part 2: development of coupled fluid-solid algorithms. Proc. Inst. Mech. Eng. Part H **212**, 241–252 (1998)
145. X. Zhang, E. Zuazua, Long-time behavior of a coupled heat-wave system arising in fluid-structure interaction. Arch. Ration. Mech. Anal. **184**(1), 49–120 (2007)

S. Čanić (✉)
Department of Mathematics, University of Houston, Houston, TX, USA
e-mail: canic@math.uh.edu

B. Muha
Faculty of Science, Department of Mathematics, University of Zagreb, Zagreb, Croatia
e-mail: borism@math.hr

M. Bukač
Department of Mathematics, University of Pittsburgh, Pittsburgh, PA, USA
e-mail: martinab@pitt.edu

Chapter 3
Hyperbolic–Parabolic Coupling and the Occurrence of Resonance in Partially Dissipative Systems

Giovanni Paolo Galdi, Mahdi Mohebbi, Rana Zakerzadeh, and Paolo Zunino

Abstract It is well known that elastic solids, when subjected to a time-periodic load of frequency ω, may respond with a drastic increase of the magnitude of basic kinematic and dynamic quantities, such as displacement, velocity and energy, whenever ω is near to one of the "proper frequencies" of the solid. This phenomenon is briefly described as resonance. Objective of our analysis is to investigate whether the interaction of an elastic solid with a dissipative agent can affect and possibly prevent the occurrence of resonance. We shall study this problem in a broad class of dynamical systems that we call *partially dissipative*, and whose dynamics is governed by strongly continuous semigroups of contractions. For such systems we will provide sharp necessary and sufficient conditions for the occurrence of resonance. Afterward, we shall furnish a number of applications to physically relevant problems including thermo- and magneto-elasticity, as well as several liquid–structure interaction models.

Keywords Asymptotic behavior • Elastic solid • Linear magnetoelasticity • Linear thermoelasticity • Liquid–structure models • Mean ergodic theorem • Resonance • Stokes equations • Strongly continuous semigroup of contractions

MSC2010: 76D07, 35M99, 35Q61, 74B05, 74F05, 74F10, 74F15

3.1 Introduction

Resonance in elastic solids is a well-known phenomenon occurring in Nature that can be roughly described as follows: when the frequency of an applied time-periodic load approaches one (or a multiple of one) of the natural frequencies of oscillations of the solid, then the basic kinematic and dynamic parameters of the solid, such as displacement, velocity, and energy become increasingly large, resulting into damage and even rupture of the structure. In this respect, perhaps the most striking

T. Bodnár et al. (eds.), *Fluid-Structure Interaction and Biomedical Applications*, Advances in Mathematical Fluid Mechanics, DOI 10.1007/978-3-0348-0822-4_3,

representative is the wind-induced collapse of the Tacoma Narrows bridge in 1940; see [9] for a thorough analysis of this event.

Even though resonance is observed in both linear and nonlinear solids [46], the most common and simplest example showing the occurrence of such a phenomenon is that of materials modeled by the equations of elasticity in the small gradient approximation (linearized elasticity). For the sake of argument, *it is on these materials that we shall focus our present investigation.*

As is well known, the equations of linearized elasticity have the following characteristic properties: (1) they are of hyperbolic type, and (2) they conserve the total initial energy at each following instant of time. Basically, it is these two combined issues that are responsible for resonance. As a matter of fact, if we "parabolize" these equations by adding a *suitable* dissipative term (representative, for instance, of an appropriate damping factor due to the decay of the purely elastic properties of the material) then, at least mathematically, it is readily seen that the event of resonance is ruled out; see Sect. 3.2.

Our focus in this paper is different. We are not interested in how resonance can be removed by altering the constitutive properties of the material; rather, we would like to study whether and how the *interaction* of an elastic solid with a *dissipative* agent, for example heat loss or *viscous* liquid, can affect and even prevent the happening of resonance. As a matter of fact, we would like to *characterize* the conditions under which resonance may or may not take place in a physical system where a (linear) elastic material interacts with a phenomenon or another material whose dynamics is of *dissipative* nature. We will refer to systems of this type as *partially dissipative systems*.

It is quite remarkable that the question above, even though of great relevance in applied science, has received only very little attention from the mathematical community. In fact, we are only aware of the paper [16] where aspects of the problem are analyzed in a one-dimensional model of linear thermoelasticity.

The motivation for our study came originally from a quite relevant liquid–solid interaction problem, namely, that involving arterial blood flow. In this case, blood is pumped at pulsatile rate and, therefore, for certain frequencies it could be in principle able to produce harmful resonant effects on the arterial walls. However, the analysis we shall perform here will be of larger breadth. In fact, we shall characterize the occurrence of resonance in a vast class of partially dissipative systems that include models of liquid–solid interaction as a particular case.

To this end, we observe that a characteristic feature of the above systems is that the total energy $\mathscr{E}(t)$ is a non-increasing function of time $t \geq 0$. Therefore, denoting by $U = U(t)$, $t \geq 0$, the associated semigroup on the appropriate Banach space X, we must require

$$\|U(t)x\|^2 \equiv \mathscr{E}(t) \leq \mathscr{E}(0) \equiv \|x\|^2\,, \quad \text{for all } x \in X \text{ and all } t \geq 0\,,$$

where $\|\cdot\|$ denotes the norm of X. As a result, a partially dissipative system is naturally defined as one whose dynamics is *governed by a linear strongly continuous semigroup of contractions.*

Occurrence of resonance in partially dissipative systems is performed in Sect. 3.4. In order to summarize the main results in some details, we begin with the following simple observation, that resonance is ruled out if for any $T > 0$ and for any "force" f (in a suitable function class) of period T (shortly: T-periodic), we can always determine a corresponding motion of the dynamical system of the same period T. Therefore, we define a *resonant period* to be a positive number T for which there is at least one T-periodic f such that there is no corresponding T-periodic motion. With the help of Yosida's mean ergodic theorem we thus show that a resonant period T can exist *if and only if* if at least one of the following two conditions is *violated* (see Theorem 3.3)

$$\text{(a) } \mathsf{N}(I - U(T)) = \{0\}\,; \quad \text{(b) } \mathsf{R}(I - U(T)) \text{ is closed}\,, \tag{3.1}$$

where $\mathsf{N}(\cdot)$ and $\mathsf{R}(\cdot)$ denote null space (kernel) and range, respectively, and I is the identity. Of particular relevance in many concrete applications is the case when we know that (a) holds but not necessarily (b). In this situation, in fact, resonance can be "generically" excluded. More precisely, fix arbitrarily $T > 0$. We then show that there is a dense $\mathcal{Q} \subset L^1(0,T;X)$[1] such that, for any $f \in \mathcal{Q}$ we can find a unique corresponding T-periodic solution. Consequently, for any T-periodic force $f \in L^1(0,T;X)$, and any $\varepsilon > 0$, we may select a unique T-periodic solution corresponding to $f + f_\varepsilon$, where the magnitude of f_ε is less than ε. In fact, in a suitable subclass of partially dissipative systems we are able to provide a "concrete" representation of the dense set $\mathcal{Q}$, and show that it may be taken as the set of all T-periodic functions possessing a finite Fourier series; see Theorem 3.5. An important consequence of this result is that for such partially dissipative systems resonance cannot be produced by T-periodic forces having a *finite* number of modes.

As it turns out, the validity of properties (a) and (b) can be directly related to the asymptotic behavior (in "time" t, that is) of $U(t)$. From the physical viewpoint the latter means asymptotic behavior of the *total* energy $\mathcal{E} = \mathcal{E}(t)$ of the system in absence of external forces. In this respect, it should be emphasized that an important feature of partially dissipative systems is that, in absence of forces, $\mathcal{E}(t)$ is a decreasing function of time. Different scenarios may thus occur. (1) The dissipative mechanism is sufficiently "strong" as to force $\mathcal{E}(t) \to 0$ as $t \to \infty$, provided only $\mathcal{E}(0) < \infty$; namely, dissipation damps out all possible self-oscillations of the material. In terms of semigroup, this amounts to say that $U(t)$ is *strongly stable*. In turn, the latter ensures the validity of condition (a) in (3.1) and, therefore, by what we mentioned above, it "generically" excludes the event of resonance (Theorem 3.4). (2) $\mathcal{E}(t) \to 0$ as $t \to \infty$ whenever $\mathcal{E}(0) < \infty$, with an *explicit order* of decay rate. This means that $U(t)$ is *uniformly stable*, which guarantees that both conditions (a) and (b) in (3.1) are satisfied, so that for such partially dissipative systems, resonance is completely ruled out (Theorem 3.6). (3) Even though $\mathcal{E}(t)$ is decreasing, part of the energy (related to the conservative elastic component) is

[1]For notation, see the end of this introductory section.

stored and only part is dissipated. A typical example is furnished in Sect. 3.3.2 for three-dimensional linearized thermoelasticity with periodic boundary conditions. With respect to $U(t)$, this means that there exists an invariant subspace $X_0 \subseteq X$ such that $\|U(t)x\| = \|x\|$, for all $x \in X_0$ and all $t \geq 0$. In this case we show that resonance always occurs (Theorem 3.7).

The above general results are then applied to the study of resonance in several partially dissipative systems, including three-dimensional linear thermoelasticity (Sect. 3.5.1) and magnetoelasticity (Sect. 3.5.2). In both models, the interaction between solid and dissipative agent (the latter being heat or magnetic losses) takes place in the *bulk* of the material. Precisely, in the thermoelastic case, we show that resonance can be "generically" excluded in "most" bounded and sufficiently regular domains. Moreover, for such domains, resonance never occurs if the T-periodic forcing terms have a finite (but otherwise arbitrary) number of modes. However, resonance *does* occur for some familiar and simply shaped domains such as a ball; see Theorem 3.9 and Remark 3.3. Likewise, in linear magnetoelasticity resonance can be "generically" ruled out for *any* (sufficiently smooth) bounded domains, but only and only if the Lamé coefficients of the elastic solid satisfy appropriate conditions[2]; see Theorem 3.10. The above results may be concisely summarized by stating that resonance in linear thermoelasticity is due to the particular "geometry" of the reference configuration of the solid, while in magnetoelasticity is a consequence of its specific constitutive properties.[3]

Another significant application of the theory developed in Sect. 3.4 concerns the analysis of resonance for certain liquid–structure interaction models, and is worked out in Sect. 3.5.3. From the physical standpoint, the principal difference between these partially dissipative systems and those considered previously is that now the interaction takes place at the interface solid–liquid, through the continuity of stress and velocity fields, rather than in the bulk of the solid. From the mathematical point of view, this leads to a hyperbolic–parabolic coupled problem of completely different nature. As for the models we shall consider, besides being linearized versions of the original ones (namely, Navier equations for elasticity and Stokes equations for the liquid), they all show the specific feature of having the interface *fixed*. Though at first sight restrictive, such an assumption is certainly appropriate in a number of physically interesting situations, as well as useful from a strictly mathematical perspective.

In fact, from the physical point of view it is certainly valid within a good approximation whenever the volume occupied by the fluid is little affected by the deflection of the elastic structure. For example, as in a cylinder-like drum whose

[2]These conditions amount to say that the Poisson ratio must not be "too large," a condition that is verified in most common materials; see, e.g., [34].

[3]It is worth remarking that, as recently shown in [33], resonance is excluded in *nonlinear* magnetoelasticity without the above restrictions on the Lamé coefficients, in presence of a suitable *nonlinear* damping for the elastic material.

shell is completely filled up with a viscous liquid, and its height is much bigger than the diameter of the drumhead.

From the mathematical viewpoint, investigating a problem by these simplified models may have an important bearing on the resolution of the same problem for the original complete model. For instance, strong solutions to liquid–structure interaction problems are constructed around the solution to a set of "approximate equations" obtained by disregarding the nonlinearity and *fixing* the interface as the reference configuration for the structure; see, e.g., [7, 19] and the bibliography therein enclosed. The models we shall consider are exactly described by these "approximate equations."

There is a vast amount of literature dedicated to the mathematical analysis of such models, too long to be included here. We limit ourselves to refer the reader to [4, 12] and the bibliography therein.

The first liquid–structure problem we investigate is presented in Sect. 3.5.3 and consists of a viscous liquid (within the Stokes approximation) occupying a three-dimensional bounded domain, Ω_F, and surrounding a linearized elastic material occupying another bounded and simply connected domain Ω_S. Combining the result of Sect. 3.4 with those of [4] regarding the existence of a strongly continuous semigroup of contractions for the problem, we prove necessary and sufficient conditions for the occurrence of resonance. More precisely we show that resonance cannot "generically" occur if and only if Ω_S is such that a suitable overdetermined boundary-value problem in Ω_S has only the trivial solution.

As we already remarked, in this type of problems the dissipative interaction occurs at the interface liquid–structure, Γ. In mathematical terms this can be described as follows. Because of the adherence condition imposed on the liquid at Γ, the velocity fields of both liquid ($\boldsymbol{v}$) and structure ($\boldsymbol{u}_t$) must there coincide. Now, by the effect of viscosity ν, the coupled system liquid–solid dissipates its energy at a rate $\mathscr{D} := \nu \int_{\Omega_F} |\boldsymbol{D}(\boldsymbol{v})|^2$, where $\boldsymbol{D}$ is the stretching tensor. By the trace theorem and Korn's inequality, we can show, in particular, $\mathscr{D} \geq \nu \int_\Gamma |\boldsymbol{u}_t|^2$. As a result, the liquid "induces" into the structure some sort of dissipation, but only at the interface, which, however, need not diffuse inside the solid in such a way as to rule out the occurrence of resonance.

In view of the above observations, one may wonder whether this type of dissipation could prevent resonance at least in the case of "thin" structures, like plate or membranes. The answer to this question is positive, and the corresponding analysis is carried out in Sects. 3.5.4 and 3.5.5. The typical situation here [12, 37] is that the liquid is contained in a simply connected, smooth three-dimensional domain Ω with a part, Γ, of its boundary that is elastic. The displacement field on Γ is assumed to be directed in the normal direction. In the case Γ flat, in [12] it is shown, among other things, that the relevant system of equations defines a strongly continuous semigroup of contractions that is also *uniformly* stable. Consequently, we conclude the absence of resonance for this model (Theorem 3.12). Successively, we analyze the case when Γ may be "curved." In this situation, the methods of [12] are seemingly not applicable. Nevertheless, using some arguments developed in Sect. 3.4, we are able to conclude also in this case that resonance cannot occur Theorem 3.13.

One important consequence of the above results on liquid–solid interactions is that, at least for the class of models we have considered, *"thick" elastic materials may favor the occurrence of resonance.*

We conclude this introductory section with a further consideration. A puzzling issue that follows from our analysis regards those systems whose total energy decays to zero for sufficiently large times, but with an unspecified order of decay. In this situation, as we already mentioned, existence and uniqueness of T-periodic solutions (arbitrary fixed $T > 0$) is established for a dense set, $\mathscr{Q}$, of data or, in other words, resonance is "generically" excluded. The mathematical reason why we cannot take $\mathscr{Q} = X$ is because of lack of an a priori estimate of the solution in terms of the data. Thus, if really $\mathscr{Q} \neq X$, we should be able to find $f \in X - \mathscr{Q}$ and a sequence $\{f_k\} \subset \mathscr{Q}$ with $f_k \to f$, such that the "energy" $\|U(t)x_k\|^2$ of the corresponding (unique) solutions becomes increasingly large. We have performed this kind of investigation in Sect. 3.6 from a numerical point of view, for a model of solid–liquid interaction of the type analyzed in Sect. 3.5.3. The results obtained, however, do not show the expected trend of the energy, at least for the kind of periodic loads that we have chosen (squared profile). This leaves room for future investigation.

Notation. We end this introductory section by collecting the basic notations we shall use throughout the paper, referring the reader to, e.g., [18] for those (quite standard) not reported here. For A a domain of the real Euclidean space $\mathbb{R}^n$, $n \geq 1$, and s a real positive number, we set $H^s(A) := W^{s,2}(A)$, $H_0^s(A) := W_0^{s,2}(A)$, with corresponding norm $\|\cdot\|_{s,2,A}$, where the subscript "A" will be omitted if confusion does not arise. Here, as customary, $W^{s,2}(A)$, $W_0^{s,2}(A)$ denote usual Sobolev spaces, so that $H^0(A) \equiv H_0^0(A) = L^2(A)$, the classic Lebesgue space. Unless otherwise noticed, we shall use the same symbol X for a space of scalar-, vector-, and tensor-valued functions. The scalar product in $L^2(A)$ is denoted by $(\cdot,\cdot)_A$, where, again, the subscript "A" will be included whenever necessary. If X is a Banach space, with associate norm $\|\cdot\|_X$, and $T > 0$ we denote by $L^q(0,T;X)$, $q \geq 1$ [resp. $C([0,T];X)$] the class of functions $u : [0,T] \to X$ such that $\int_0^T \|u(t)\|_X dt < \infty$ [resp. $\max_{[0,T]} \|u(t)\|_X < \infty$].

3.2 Resonance in a Linearized Elastic Solid

Before addressing the question for a coupled hyperbolic–parabolic system, we deem it useful to recall some basic and well-known facts about the occurrence of resonance in linearly elastic solids. For the sake of simplicity, we shall assume that the material is isotropic and homogeneous. In such a case the governing equations are the Navier equations, with the subscript t denoting differentiation with respect to time:

$$\rho \boldsymbol{u}_{tt} - \mu \Delta \boldsymbol{u} - (\lambda + \mu)\nabla \mathrm{div}\, \boldsymbol{u} = \rho \boldsymbol{f} \quad \text{in } \Omega \times (0, \infty)\,. \tag{3.2}$$

Here $\boldsymbol{u} = \boldsymbol{u}(x,t)$ denotes the displacement field, $\Omega \subset \mathbb{R}^n$, $n = 2, 3$, the *bounded* reference configuration, $\rho > 0$ the constant density, $\mu > 0$ and λ the Lamé coefficients satisfying the condition

$$\lambda + 2\mu > 0\,, \tag{3.3}$$

and $\boldsymbol{f} = \boldsymbol{f}(x,t)$ the external load per unit mass. To (3.2) we append the boundary conditions, which we choose to be of displacement type:

$$\boldsymbol{u}(x,t) = \boldsymbol{0} \ \text{ at } \partial\Omega \times (0, \infty). \tag{3.4}$$

We now define the operator $\mathscr{L}$ as

$$\mathscr{L}(\boldsymbol{u}) = -\mu \Delta \boldsymbol{u} - (\lambda + \mu)\nabla \mathrm{div}\, \boldsymbol{u} \tag{3.5}$$

with domain $\mathsf{D}(\mathscr{L}) := H_0^1(\Omega) \cap H^2(\Omega) \subset L^2(\Omega)$ and range in $L^2(\Omega)$. By a direct calculation one shows that $\mathscr{L}$ is symmetric in $L^2(\Omega)$ and that, moreover, under the given assumption on λ and μ, there is a constant $\gamma > 0$ such that

$$(\mathscr{L}\boldsymbol{u}, \boldsymbol{u}) \geq \gamma\, \|\nabla \boldsymbol{u}\|_2^2\,, \quad \text{for all } \boldsymbol{u} \in D(\mathscr{L}).$$

The latter, along with Poincaré inequality implies that $\mathscr{L}$ is positive definite. Furthermore, from classical results on existence and regularity [13, Sect. 6.3], if Ω is of class C^2 for any $\boldsymbol{g} \in L^2(\Omega)$ there is one and only $\boldsymbol{u} \in D(\mathscr{L})$ such that

$$\|\boldsymbol{u}\|_{2,2} \leq c\, \|\boldsymbol{g}\|_2\,. \tag{3.6}$$

Hence $\mathscr{L}$ is surjective and, being also symmetric, is self-adjoint. Finally, from Rellich theorem and (3.6) it follows that $\mathscr{L}^{-1}$ is compact. In view of all the above properties, it follows that the spectrum of $\mathscr{L}$ is constituted only by a countable number of positive eigenvalues, $\{\mu_k\}$ each of finite multiplicity that can cluster only at infinity; see, e.g., [27, Theorem 6.26]. We will denote by $\{\boldsymbol{\psi}_k\}$ the corresponding eigenfunctions, which we may take as an orthonormal base in $L^2(\Omega)$.

Suppose now that the load $\boldsymbol{f}$ is a one-mode time-periodic function of period $T := 2\pi/\omega$ of the type

$$\boldsymbol{f}(x,t) = e^{\imath\omega t}\, \boldsymbol{g}(x)\,, \quad \boldsymbol{g} \in L^2(\Omega)\,.$$

We then look for a corresponding T-periodic solution to (3.2)–(3.4) of the form $\boldsymbol{u}(x,t) = e^{\iota\omega t}\,\boldsymbol{w}(x)$, where $\boldsymbol{w}$ satisfies

$$-\rho\,\omega^2 \boldsymbol{w} + \mathscr{L}(\boldsymbol{w}) = \rho\,\boldsymbol{g}\,. \tag{3.7}$$

Expanding $\boldsymbol{w}$ and $\boldsymbol{g}$ in the base $\{\boldsymbol{\psi}_k\}$ we deduce that $\boldsymbol{w}$ is a solution to (3.7) if and only if its Fourier coefficients, w_k, along $\{\boldsymbol{\psi}_k\}$ are related to those, g_k, of $\boldsymbol{g}$ by the formula

$$(\lambda_k - \rho\,\omega^2) w_k = \rho\, g_k\,, \quad k \in \mathbb{N}\,.$$

Thus, if $\rho\,\omega^2 \neq \lambda_k$, for all $k \in \mathbb{N}$, we infer

$$w_k = \frac{\rho g_k}{\lambda_k - \rho\omega^2}\,, \quad \text{for all } k \in \mathbb{N}\,,$$

in which case, the time-periodic solution exists and is given by

$$\boldsymbol{u}(x,t) = e^{\iota\omega t} \sum_{k=1}^{\infty} \frac{\rho g_k}{\lambda_k - \rho\,\omega^2} \boldsymbol{\psi}_k\,.$$

If, on the other hand, $\rho\,\omega^2 = \lambda_{\overline{k}}$ for some $\overline{k} \in \mathbb{N}$, the load $\boldsymbol{f}$ must satisfy the necessary condition $g_{\overline{k}} = 0$, and the solution is given by

$$\boldsymbol{u}(x,t) = e^{\iota\omega t} \sum_{k\in\mathbb{N}-\{\overline{k}\}} \frac{\rho g_k}{\lambda_k - \rho\,\omega^2} \boldsymbol{\psi}_k + C\, e^{\iota\omega t}\, \boldsymbol{\psi}_{\overline{k}}\,,$$

for arbitrary $C \in \mathbb{R}$. Notice that in the case $\rho\,\omega^2 \neq \lambda_k$ for all $k \in \mathbb{N}$, we deduce

$$\|\boldsymbol{u}(t)\|_2^2 = \rho^2 \sum_{k=1}^{\infty} \frac{|g_k|^2}{(\lambda_k - \rho\,\omega^2)^2}\,,$$

at each time t. As a result, the L^2-norm of the displacement becomes larger and larger the closer $\rho\,\omega^2$ gets to one of the eigenvalues. Analogous conclusions hold for the total (kinetic and strain) energy. These considerations describe in simple mathematical terms the phenomenon of resonance in a linear elastic material.[4]

It is now worth observing that resonance can be ruled out if the solid, instead of being perfectly elastic, possesses a suitable damping property. For example, in case of linear damping (proportional to the velocity, that is) (3.2) is replaced by the

[4]For the general case when the periodic load has an infinite number of modes, we refer to Remark 3.1.

following one

$$\rho \boldsymbol{u}_{tt} - \mu \Delta \boldsymbol{u} - (\lambda + \mu)\nabla \operatorname{div} \boldsymbol{u} + \alpha \boldsymbol{u}_t = \rho \boldsymbol{f} \quad \text{in } \Omega \times (0,\infty), \tag{3.8}$$

where $\alpha > 0$.[5] Therefore, if we repeat step by step the previous procedure, we obtain that the function $\boldsymbol{w}$ is, this time, a solution to the following equation

$$-\rho\,\omega^2 \boldsymbol{w} + \imath\,\omega\,\alpha\,\boldsymbol{w} + \mathscr{L}(\boldsymbol{w}) = \rho\,\boldsymbol{g}\,.$$

As a result, we prove

$$\boldsymbol{u}(x,t) = e^{\imath\omega t} \sum_{k=1}^{\infty} \frac{\rho g_k}{\lambda_k + \imath\,\alpha\,\omega - \rho\,\omega^2}\,\boldsymbol{\psi}_k\,,$$

implying

$$\|\boldsymbol{u}(t)\|_2^2 = \rho^2 \sum_{k=1}^{\infty} \frac{|g_k|^2}{[(\lambda_k - \rho\,\omega^2)^2 + \alpha^2\omega^2]^2}\,,$$

from which we deduce at once that resonance does not occur.

Another point that is worth remarking and that will be useful for future considerations regards the behavior in time of the total energy $\mathscr{E}$:

$$\mathscr{E}(t) := \tfrac{1}{2}\big(\|\boldsymbol{u}_t(t)\|_2^2 + \mu\|\nabla \boldsymbol{u}\|_2^2 + (\lambda + \mu)\|\operatorname{div} \boldsymbol{u}\|_2^2\big)\,,$$

in *absence* of external loads. Actually, in the purely elastic case described by (3.2), $\mathscr{E}$ is a constant function of time (conservation of energy):

$$\mathscr{E}(t) = \mathscr{E}(0)\,, \quad \text{for all } t \geq 0. \tag{3.9}$$

This property is well known and easy to show by dot-multiplying both sides of (3.2) with $\boldsymbol{f} \equiv \boldsymbol{0}$ by $\boldsymbol{u}$ and integrating by parts over Ω.

In the presence of linear damping, however, the total energy decays *exponentially fast*, a property that, as we shall see further on, can be intimately related to the lack of resonant effects. In order to sketch the proof of the above decay, we dot-multiply both sides of (3.8) with $\boldsymbol{f} \equiv \boldsymbol{0}$, by $\boldsymbol{u}_t$ and $\boldsymbol{u}$, respectively, and integrate by parts over Ω to obtain

$$\begin{aligned}
&\frac{d\mathscr{E}}{dt} = -\alpha\|\boldsymbol{u}_t\|_2^2\\
&\frac{d}{dt}\Big[\tfrac{\alpha}{2}\|\boldsymbol{u}\|_2^2 + (\boldsymbol{u}_t, \boldsymbol{u})\Big] = \|\boldsymbol{u}_t\|_2^2 - \mu\|\nabla \boldsymbol{u}\|_2^2 - (\lambda + \mu)\|\operatorname{div} \boldsymbol{u}\|_2^2\,.
\end{aligned}$$

---

[5]The calculations to follow show lack of resonance also in the case $\alpha < 0$. However, such an assumption is unacceptable from the physical viewpoint in that it would imply an increase of total energy in absence of external loads.

Multiplying the second of these equations by a parameter $\varepsilon > 0$ and summing the resulting relation to the first one we get

$$\frac{d\mathscr{V}}{dt} = (\varepsilon - \alpha)\|\boldsymbol{u}_t\|_2^2 - \varepsilon\,\mu\|\nabla\boldsymbol{u}\|_2^2 - \varepsilon\,(\lambda + \mu)\|\operatorname{div}\boldsymbol{u}\|_2^2 := \mathscr{F}\,, \tag{3.10}$$

with

$$\mathscr{V} := \mathscr{E} + \varepsilon\left[\tfrac{\alpha}{2}\|\boldsymbol{u}\|_2^2 + (\boldsymbol{u}_t, \boldsymbol{u})\right].$$

It is readily checked that by choosing ε sufficiently small, and using the assumptions on λ and μ along with the Poincaré inequality, both functionals $\mathscr{V}$ and $\mathscr{F}$ are equivalent to $\mathscr{E}$. As a consequence, from (3.10) and Gronwall's lemma we get

$$\mathscr{E}(t) \le M\,\mathscr{E}(0)\,\mathrm{e}^{-\delta t} \tag{3.11}$$

for suitable $M, \delta > 0$, which proves the desired property.

Remark 3.1. The result concerning the occurrence of resonance can be generalized to the case when the T-periodic load $\boldsymbol{f}$ has in general an infinite number of modes as follows. Let us expand $\boldsymbol{f}$ in a Fourier series in time

$$\boldsymbol{f}(x,t) = \sum_{|m|\in\mathbb{Z}} e^{\imath m \omega t}\boldsymbol{f}_m(x)\,,$$

and set $f_{mk} := (\boldsymbol{f}_m, \boldsymbol{\psi}_k)$. Then a T-periodic solution to (3.2) such that

$$\boldsymbol{u} \in W^{2,2}(0,T;L^2(\Omega)) \cap L^2 0,T;\mathsf{D}(\mathscr{L}))$$

exists if and only if the following conditions hold

(i) $f_{mk} \equiv 0$ for all $(m,k) \in \mathbb{Z}\times\mathbb{N}$ such that $\rho\, m^2\omega^2 = \lambda_k$;

(ii) $$\sum_{\{(m,n)\in\mathbb{Z}\times\mathbb{N};\rho m^2\omega^2\neq\lambda_k\}} \frac{m^4\omega^4|f_{mk}|^2}{(\rho\, m^2\omega^2 - \lambda_k)^2} < \infty\,.$$

3.3 An Interesting Case Study: Linear Thermoelasticity

In the previous section we have seen, among other things, how the introduction of damping in the material properties of the structure can eliminate the occurrence of resonance. Objective of this and the following sections is to investigate whether the *interaction* of the elastic structure with a dissipative phenomenon or material can produce the same outcome.

We will analyze this problem in the simple but instructive case when dissipation is due to thermal effects and nonlinear effects are negligible. Thus, within the linearized approximation for the elastic solid, and using a linear Fourier law for the heat flow, and homogeneous and isotropic coupling, the relevant equations governing the dynamics of the material reduce then to the classical equations of homogeneous linear thermoelasticity [28], [26, Chap. 1]:

$$\left.\begin{array}{l}\rho\,\boldsymbol{u}_{tt}-\mu\Delta\boldsymbol{u}-(\lambda+\mu)\nabla(\operatorname{div}\boldsymbol{u})=\sigma_1\nabla\theta+\rho\,\boldsymbol{f}\\ \theta_t-\kappa\Delta\theta=\sigma_2\operatorname{div}\boldsymbol{u}_t+Q\end{array}\right\}\ \text{in }\Omega\times(0,\infty)\,. \tag{3.12}$$

While keeping the notation introduced in the previous section, we further denote by θ the temperature field, $\kappa > 0$ the heat conduction coefficient, $Q = Q(x,t)$ a heat source, and σ_i, $i = 1,2$, coupling constants obeying the thermodynamical restriction $\sigma_1\sigma_2 > 0$. To (3.12) we append the following boundary conditions:

$$\boldsymbol{u}(x,t)=\boldsymbol{0}\,,\quad \theta(x,t)=0\ \text{ at }\partial\Omega\times(0,\infty). \tag{3.13}$$

As previously mentioned, the thermoelastic material is overall dissipative, and in fact, its total energy

$$\mathcal{E} := \tfrac{1}{2}\big(\rho\|\boldsymbol{u}_t(t)\|_2^2+\|\theta(t)\|_2^2+\mu\|\nabla\boldsymbol{u}\|_2^2+(\lambda+\mu)\|\operatorname{div}\boldsymbol{u}\|_2^2\big)$$

is a decreasing function of time. To see this, we formally dot-multiply both sides of $(3.12)_1$ by $\boldsymbol{u}$, multiply both sides of $(3.12)_2$ by θ, and then integrate by parts over Ω. By taking into account (3.13) and summing the resulting relations we thus deduce

$$\frac{d\mathcal{E}}{dt} = -\kappa\,\|\nabla\theta\|_2^2\,. \tag{3.14}$$

Since it can be shown [30] that if there is $t_0 \geq 0$ such that $\|\nabla\theta(t_0)\|_2 = 0$, then $\boldsymbol{u}$ and θ must be identically zero over the whole time interval $(0,\infty)$, from (3.14) we find, as a consequence, that the energy is a *strictly* decreasing function of time. However, this does not imply that each component of $\mathcal{E}$ is actually decreasing. Nevertheless, thanks to a classical result of Dafermos [15] one can proceed further and conclude that indeed $\mathcal{E}(t) \to 0$ as $t \to \infty$, for "most" (sufficiently smooth, bounded) domains of $\mathbb{R}^3$. More precisely, we have the following [15] (see also [24, 43]).

Proposition 3.1. *Let $\Omega \subset \mathbb{R}^3$ be a bounded domain of class C^2. If the problem*

$$\left.\begin{array}{l}-\sigma\,\boldsymbol{\psi}-\Delta\boldsymbol{\psi}=0\\ \operatorname{div}\boldsymbol{\psi}=0\end{array}\right\}\ \textit{in }\Omega\,,\qquad \boldsymbol{\psi}=\boldsymbol{0}\ \ \textit{on }\partial\Omega \tag{3.15}$$

has only the trivial solution $\boldsymbol{\psi} = 0$ for all $\sigma > 0$, then solutions to (3.12)–(3.13) *with $\boldsymbol{f} \equiv Q \equiv 0$ and $\mathcal{E}(0) < \infty$ satisfy $\mathcal{E}(t) \to 0$ as $t \to \infty$. Concerning problem*

(3.15), $\boldsymbol{\psi} \equiv \mathbf{0}$ *is the only solution for all* $\sigma > 0$, *whenever* Ω *is such that all eigenvalues of the Laplace operator are simple* [15], *which is indeed true for a residual set of (bounded) domains of class* C^2, *in the sense of the Baire category* [45].[6]

This classical result is of great physical relevance, in that it ensures that, for "most" (and sufficiently regular) bounded domains the dissipation due to thermal effects is able to damp out the free vibrations of the elastic body (within the above model, of course). Its proof can be obtained by LaSalle's invariance principle [23, 43], and is based on the following facts. Let

$$X := \left\{(\boldsymbol{u}, \boldsymbol{u}_t, \theta) \in [H^1(\Omega)]^3 \times [L^2(\Omega)]^3 \times L^2(\Omega)\right\},$$

be the "energy space." The set of cluster points (as $t \to \infty$) of all trajectories emanating from some $x \in X$ is contained in the subset, F, of X where the Liapounov functional $\mathcal{E}$ is constant. From (3.12) to (3.14) (with $\boldsymbol{f} \equiv Q \equiv 0$) we see that F is the space of solutions $(\boldsymbol{u}, \theta) \equiv (\boldsymbol{u}, 0)$ such that

$$\left.\begin{array}{l} \rho \boldsymbol{u}_{tt} - \mu \Delta \boldsymbol{u} - (\lambda + \mu)\nabla(\operatorname{div} \boldsymbol{u}) = \mathbf{0} \\ \operatorname{div} \boldsymbol{u}_t = 0 \end{array}\right\} \text{ in } \Omega \times (0, \infty) ,$$
$$\boldsymbol{u}_t = \mathbf{0} \text{ on } \partial\Omega .$$

Setting $\boldsymbol{w} := \boldsymbol{u}_t$, and expanding $\boldsymbol{w}$ in a time Fourier series, the above system may be equivalently reduced to (3.15). Therefore, if Ω satisfies the condition stated in Proposition 3.1, every trajectory converges to 0 in the X-norm (equivalent to $\mathcal{E}^{\frac{1}{2}}$).

The question that we would like to address next is whether such a dissipative mechanism is also able to rule out the occurrence of resonance.

In this regard, it must be emphasized that Proposition 3.1 does *not* guarantee, in general, any uniform rate of decay for $\mathcal{E}(t)$.[7] It should also be remarked that such a decay is not true if instead of Dirichlet boundary conditions, we use *periodic boundary conditions*.[8] We will analyze these questions in Sect. 3.3.2, and in a much broader context, in Sect. 3.4.

3.3.1 The One-Dimensional Case

In order to perform our study on resonance, we begin to consider the simple one-dimensional version of (3.12), where $\Omega = (0, \ell)$. By suitably rescaling the

[6] It is worth remarking that there are also "familiar" domains where (3.15) has an *infinite number* of linearly independent solutions. This happens, for example, when Ω is a ball [15, Remark 5.2], and, in fact, in the two-dimensional case, the circle is the only (sufficiently smooth) simply connected domain where (3.15) has an infinite number of linearly independent solutions [8].

[7] In this respect, see [31].

[8] See, however, also Remark 3.4.

appropriate variables, we then obtain, with the obvious meaning of the symbols, that the relevant equations take the following form

$$\begin{aligned} u_{tt} - u_{xx} + \alpha\, \theta_x &= f \\ \theta_t - \theta_{xx} + \beta\, u_{xt} &= Q \\ u(0,t) = u(\ell,t) = \theta(0,t) &= \theta(\ell,t) = 0 \;\; t > 0\,, \end{aligned} \tag{3.16}$$

where $\alpha\beta > 0$. We next expand u, θ, f, and Q in Fourier series in time with period $T = 2\pi/\omega$, namely

$$u(x,t) = \sum_{n\in\mathbb{Z}} u_n(x)\, e^{\imath n \omega t}\,, \text{ etc.}$$

From (3.16) we then infer that the corresponding coefficients u_n, θ_n, f_n, and Q_n must satisfy the following equations

$$\left.\begin{aligned} -n^2\omega^2 u_n - u_{n,xx} + \alpha\theta_{n,x} &= f_n(x) \\ \imath\, n\, \omega\, \theta_n - \theta_{n,xx} + \imath\, n\, \omega\, \beta u_{n,x} &= Q_n(x) \end{aligned}\right\} \quad 0 < x < \ell\,, \tag{3.17}$$

with boundary conditions

$$u_n(0) = u_n(\ell) = \theta_n(0) = \theta_n(\ell) = 0\,. \tag{3.18}$$

Our next result ensures the existence of solutions to (3.17)–(3.18) with corresponding estimates.

Proposition 3.2. *For any $f_n, Q_n \in L^2(\Omega)$, $n \in \mathbb{Z}$, there exists one and only one solution (u_n, θ_n) to (3.17)–(3.18) such that $(u_n, \theta_n) \in [H^2(\Omega) \cap H_0^1(\Omega)]^2$. Furthermore, there exists $C = C(n, \omega, \ell, \alpha, \beta) > 0$ such that*

$$\|u_n\|_{2,2} + \|\theta_n\|_{2,2} \le C\left(\|f_n\|_2 + \|Q_n\|_2\right). \tag{3.19}$$

Proof. Set $(D := d/dx)$

$$\mathscr{A} = \begin{pmatrix} -D^2 & 0 \\ 0 & -D^2 \end{pmatrix}, \quad \mathscr{K} = \begin{pmatrix} -n^2\omega^2 & \alpha\, D \\ \imath\, n\, \omega\beta\, D & \imath\, n\, \omega \end{pmatrix},$$

$$w_n = \begin{pmatrix} u_n \\ \theta_n \end{pmatrix}, \quad F_n = \begin{pmatrix} f_n \\ Q_n \end{pmatrix}.$$

The operators $\mathscr{A}$ and $\mathscr{K}$ are well defined from $\mathcal{X} := [H^2(\Omega) \cap H_0^1(\Omega)]^2$ in $L^2(\Omega)$, so that for each $n \in \mathbb{Z}$ our problem (3.17)–(3.18) can be written in the following abstract form

$$\mathscr{A}(w_n) + \mathscr{K}(w_n) = F_n \quad \text{in } L^2(\Omega)\,.$$

Clearly, $\mathscr{A}$ is a homeomorphism of $\mathcal{X}$ onto $L^2(\Omega)$, while, by Rellich theorem, $\mathscr{K}$ is compact. As a result, $\mathscr{A}+\mathscr{K}$ is Fredholm of index 0, and the desired existence result follows if we show that the only solution $w_n \in \mathcal{X}$ corresponding to $F_n \equiv 0$, is trivial. To this end, we multiply both sides of $(3.17)_1$, with $f_n \equiv 0$, by $\imath\,(\beta/\alpha)n\,\omega\overline{u_n}$ ("—"=c.c.) and integrate by parts over Ω to obtain, also with the help of (3.18)

$$-\,\imath\,|n|^3\omega^3\frac{\beta}{\alpha}\|u_n\|_2^2+\imath\,n\,\omega\frac{\beta}{\alpha}\|u_{n,x}\|_2^2+\imath\,n\omega\beta(\theta_{n,x},\overline{u_n})=0\,. \tag{3.20}$$

Likewise, taking the c.c. of $(3.17)_2$, with $Q_n \equiv 0$, and then multiplying both sides by θ_n and integrating over Ω allows us to deduce

$$\imath\,n\omega\|\theta_n\|_2^2+\|\theta_{n,x}\|_2^2-\imath\,n\omega\beta(\theta_{n,x},\overline{u_n})=0\,. \tag{3.21}$$

Summing side by side (3.20) and (3.21) then implies $\|\theta_{n,x}\|_2=0$, that is, in view of (3.18), $\theta_n \equiv 0$. As a consequence, from $(3.17)_2$ (with $Q_n \equiv 0$) and again (3.18) we show, since $\beta \neq 0$, that u_n must identically vanish, and the claimed existence result follows. The last statement in the proposition is a consequence of Banach closed range theorem. □

With the help of Proposition 3.2 we are now in a position to prove the following theorem which furnishes the non-occurrence of resonance for a sufficiently large class of data.

Theorem 3.1. *Let f, Q be time-periodic functions of arbitrarily fixed period $T>0$ with*

$$f \in W^{3,2}(0,T;L^2(\Omega))\,,\quad Q \in W^{2,2}(0,T;L^2(\Omega)).$$

Then, there exists one and only one time-periodic solution (u,θ) to (3.16) of period T such that

$$\begin{aligned}&u \in W^{2,2}(0,T;L^2(\Omega)) \cap W^{1,2}(0,T;H_0^1(\Omega)\cap L^2(0,T;H^2(\Omega))\\&\theta \in W^{2,2}(0,T;H_0^1(\Omega)) \cap W^{1,2}(0,T;H^2(\Omega))\end{aligned}$$

Proof. Our starting point is the system of equations (3.22) for *arbitrary* $n \in \mathbb{Z}$, whose corresponding solutions are provided in Proposition 3.2. The first objective is then to show estimates of the type (3.19) but with an *explicit* dependence of the involved constant on the number n. Clearly, in view of (3.19), it is enough to establish such estimates for all $|n| \geq 1$. For simplicity, throughout the proof we shall omit the subscript "n." Proceeding in the same way as we did to deduce (3.20) and (3.21) (this time with $f, Q \not\equiv 0$) we obtain

$$\begin{aligned}&-n^2\omega^2\,\|u\|_2^2+\|u_x\|_2^2+\alpha\,(\theta_x,\overline{u})=(f,\overline{u})\,,\\&\imath\,n\omega\|\theta\|_2^2+\|\theta_x\|_2^2-\imath\,n\omega\beta(\theta_x,\overline{u})=(\overline{Q},\theta)\,.\end{aligned} \tag{3.22}$$

By multiplying both sides of $(3.22)_1$ by $\imath\, n\, \omega\, \beta/\alpha$ and summing the resulting equation to $(3.22)_2$, we show, in particular,

$$\|\theta_x\|_2^2 = n\,\omega\,\frac{\beta}{\alpha}\Im\,(f,\overline{u}) + \Re\,(\overline{Q},\theta)\,,$$

which, in turn, by Cauchy–Schwarz and Poincaré inequalities, implies for all $\varepsilon > 0$

$$\|\theta\|_{1,2} \le c_0(|n|\|f\|_2 + \|Q\|_2) + \varepsilon\,\|u\|_2 \tag{3.23}$$

with $c_0 = c_0(\ell,\varepsilon,\omega) > 0$. We next multiply both sides of $(3.17)_2$ by $\overline{\theta}_{xx}$ and integrate over Ω to get

$$\|\theta_{xx}\|_2^2 = \imath\, n\,\omega\big[(\theta,\overline{\theta}_{xx}) + \beta(u_x,\overline{\theta}_{xx})\big] - (Q,\overline{\theta}_{xx})\,,$$

which, in turn, by Schwarz inequality, implies

$$\|\theta_{xx}\|_2 \le |n|\,\omega\,\big(\|\theta\|_2 + |\beta|\,\|u_x\|_2\big) + \|Q\|_2\,. \tag{3.24}$$

Since, by Poincaré's inequality,

$$|-\alpha(\theta_x,\overline{u}) + (f,\overline{u})| \le \gamma^2\alpha^2\|\theta_x\|_2^2 + \tfrac{1}{4}\,\|f\|_2^2 + \tfrac{1}{2}\|u_x\|_2^2\,,$$

with γ Poincaré constant, from $(3.22)_1$ and (3.24) we arrive at

$$\|\theta_{xx}\|_2 \le c_1\,\big[|n|\,\omega\,\|\theta\|_2 + \|\theta_x\|_2 + |n|\,\omega\,\|u\|_2 + \|f\|_2 + \|Q\|_2\big]\,, \tag{3.25}$$

with $c_1 = c_1(\alpha,\ell) > 0$. Let

$$U(x) := \int_0^x u(\xi)\,d\xi.$$

Multiplying both sides of $(3.17)_2$ by $\overline{U}$, integrating by parts, and using (3.18) we find

$$\imath\, n\,\omega(\theta,\overline{U}) + (\theta_x,u) - \theta_x(\ell)\int_0^\ell \overline{u}(\xi)\,d\xi - \imath\, n\,\omega\beta\,\|u\|_2^2 = (f,\overline{U})\,.$$

After a simple manipulation in this relation, we infer

$$|n|\,\omega\,|\beta|\,\|u\|_2 \le c_2\,\big(|n|\,\omega\,\|\theta\|_2 + \|\theta_x\|_2 + \|f\|_2 + |\theta_x(\ell)|\big)\,. \tag{3.26}$$

with $c_2 = c_2(\ell) > 0$. Starting from the identity

$$\theta_x^2(\ell) = \theta_x^2(x) + 2\int_x^\ell \theta_\xi(\xi)\theta_{\xi\xi}(\xi)\,d\xi\,,\ \ x \in (0,\ell)\,,$$

and using Schwarz inequality, one easily shows that

$$|\theta_x(\ell)| \le c_3\Big[(1+1/\eta)\|\theta_x\|_2 + \eta\,\|\theta_{xx}\|_2\Big]$$

where $c_3 = c_3(\ell) > 0$ and η is an arbitrary positive number. Replacing this inequality in (3.26) and employing (3.25) we derive the following

$$\big(|n|\,\omega\,|\beta| - c_4\,\eta\,|n|\,\omega\big)\|u\|_2 \le c_5\Big\{|n|\omega\|\theta\|_2 + (1+\frac{1}{\eta})\,\|\theta_x\|_2 + (1+\eta)\|f\|_2 + \eta\,\|Q\|_2\Big\}$$

where $c_4, c_5 > 0$ depend only on α, and ℓ. Choosing $\eta = |\beta|/2c_4$, from the previous inequality we show

$$|n|\,\omega\,\|u\|_2 \le c_6\big\{|n|\,\omega\,\|\theta\|_2 + \|\theta_x\|_2 + \|f\|_2 + \|Q\|_2\big\}, \tag{3.27}$$

with $c_6 = c_6(\alpha,\beta,\omega,\ell) > 0$. Recalling that $|n| \ge 1$, from (3.27) we deduce

$$\|u\|_2 \le c_7\,(\|\theta\|_{1,2} + \|f\|_2 + \|Q\|_2), \tag{3.28}$$

with $c_7 = c_7(\alpha,\beta,\omega,\ell) > 0$. Combining (3.23) with $\varepsilon = 1/2c_7$, and (3.28) we find

$$\|\theta\|_{1,2} \le c_8\,(|n|\,\|f\|_2 + \|Q\|_2)\,, \tag{3.29}$$

for another constant c_8 independent of n, which, once replaced in (3.27) allows us to conclude

$$|n|\,\|u\|_2 \le c_9\big[(1+|n|^2)\|f\|_2 + (1+|n|)\|Q\|_2\big], \tag{3.30}$$

with $c_9 > 0$ independent of n. From (3.30) and (3.25) we infer

$$\|\theta_{xx}\|_2 \le c_{10}\big[(1+|n|^2)\|f\|_2 + (1+|n|)\|Q\|_2\big], \tag{3.31}$$

while from $(3.17)_1$, and (3.29), (3.30) it follows that

$$\|u_{xx}\|_2 \le c_{11}\big[(1+|n|^3)(\|f\|_2 + (1+n^2)\|Q\|_2\big], \tag{3.32}$$

where, again, the constants involved do not depend on n. Finally, we recall the elementary inequality

$$\|u_x\|_2 \le \|u\|_2^{\frac{1}{2}}\,\|u_{xx}\|_2^{\frac{1}{2}}\,. \tag{3.33}$$

The desired result is then a consequence of (3.29)–(3.33) and Plancherel theorem. □

Remark 3.2. For future reference, it is important to observe that solutions to the initial-value problem associated with (3.16) with $f \equiv Q \equiv 0$ and data

$$(u(0), u_t(0), \theta(0)) \in H^1(\Omega) \times L^2(\Omega) \times L^2(\Omega) :\equiv X$$

decay *exponentially fast* to zero, as time goes to infinity, in the norm of X, as shown independently in [24] and [35].

3.3.2 The Higher Dimensional Case

For three-dimensional motions the picture is less clear.[9] Actually, as will be shown in a later section, one is only able to prove that resonance does not occur for system (3.12)–(3.13) whenever the external load and heat source possess a finite (but otherwise arbitrary) number of modes or, equivalently, belong to a dense linear space of the data space.

However, if instead of Dirichlet boundary conditions (3.13) we use (space) periodic boundary conditions, we may provide a somewhat more complete analysis.[10] To this end, we let

$$\mathcal{C} := \prod_{1=1}^{3} \left(-\frac{\ell_i}{2}, \frac{\ell_i}{2}\right), \quad \boldsymbol{k}^* := \left(\frac{k_1}{\ell_1}, \frac{k_2}{\ell_2}, \frac{k_3}{\ell_3}\right).$$

We are interested in time-periodic solutions to (3.12) that are also space-periodic of period $2\pi/\ell_i$ in the x_i-direction, $i = 1, 2, 3$. We thus introduce the following classes of functions, for $m \in \mathbb{N} \cup \{0\}$,

$$\begin{aligned}
&H^m_{\#}(\mathcal{C}) := \Big\{u = \sum_{\boldsymbol{k}\in\mathbb{Z}^3} u_{\boldsymbol{k}}\, e^{\imath\, \boldsymbol{k}^* \cdot \boldsymbol{x}} : u_{\boldsymbol{k}} = -\overline{u}_{-\boldsymbol{k}}\,,\ \sum_{\boldsymbol{k}\in\mathbb{Z}^3} (1 + |\boldsymbol{k}^*|)^{2m} |u_{\boldsymbol{k}}|^2 < \infty\Big\},\\
&\dot{H}^m_{\#}(\mathcal{C}) := \big\{u \in H^m_{\#}(\mathcal{C}) : u_0 = 0\big\}\,,\\
&H^0_{\#}(\mathcal{C}) \equiv H_{\#}(\mathcal{C})\,,\ \ \dot{H}^0_{\#}(\mathcal{C}) \equiv \dot{H}_{\#}(\mathcal{C}).
\end{aligned}$$

We also recall that, in the space-periodic framework, the well-known Helmholtz–Weyl decomposition of a generic vector function $\boldsymbol{u} \in H_{\#}(\mathcal{C})$ follows immediately from the vector identity

$$|\boldsymbol{k}|^2 \boldsymbol{u} = \boldsymbol{k}^* \times (\boldsymbol{u} \times \boldsymbol{k}^*) + (\boldsymbol{u} \cdot \boldsymbol{k}^*)\, \boldsymbol{k}^*\,,$$

[9] Considerations and results reported in this section remain valid also in dimension 2.

[10] In this regard, see also Remark 3.3.

and, therefore, we have

$$\boldsymbol{u} = \nabla\varphi + \operatorname{curl}\boldsymbol{v}$$
$$\varphi := \sum_{\boldsymbol{k}\in\mathbb{Z}^3} \frac{\boldsymbol{u}_{\boldsymbol{k}}\cdot\boldsymbol{k}^*}{|\boldsymbol{k}^*|^2}\, e^{\imath\,\boldsymbol{k}^*\cdot\boldsymbol{x}}\,, \quad \boldsymbol{v} := \sum_{\boldsymbol{k}\in\mathbb{Z}^3} \frac{\boldsymbol{u}_{\boldsymbol{k}}\times\boldsymbol{k}^*}{|\boldsymbol{k}^*|^2}\, e^{\imath\,\boldsymbol{k}^*\cdot\boldsymbol{x}}\,. \tag{3.34}$$

Theorem 3.2. *Assume $\boldsymbol{f}$ and Q are time-periodic functions of period T with $\boldsymbol{f}, Q \in L^2(0,T;H_\#(\mathcal{C}))$, and let*

$$\boldsymbol{f} = \nabla\Psi + \operatorname{curl}\boldsymbol{F} \tag{3.35}$$

be the Helmholtz–Weyl decomposition of $\boldsymbol{f}$. Then resonance may occur for solutions to system (3.12) satisfying space-periodic boundary conditions if and only if $\operatorname{curl}\boldsymbol{F} \not\equiv \boldsymbol{0}$. *On the other hand, if* $\operatorname{curl}\boldsymbol{F} \equiv \boldsymbol{0}$ *and, in addition,*

$$\Psi \in L^2(0,T;H^1_\#(\mathcal{C})) \cap W^{3,2}(0,T;H_\#(\mathcal{C}))\,, \quad Q \in W^{2,2}(0,T;\dot{H}_\#(\mathcal{C}))\,,$$

the system (3.12) has at least one periodic solution, $(\boldsymbol{u},\theta)$ of period T, arbitrary $T > 0$, with

$$\begin{aligned}
&\boldsymbol{u} \in W^{2,2}(0,T;H^1_\#(\mathcal{C})) \cap W^{1,2}(0,T;H^1_\#(\mathcal{C})) \cap L^2(0,T;H^2_\#(\mathcal{C}))\\
&\theta \in W^{2,2}(0,T;H^1_\#(\mathcal{C})) \cap W^{1,2}(0,T;H^2_\#(\mathcal{C}))\,.
\end{aligned}$$

This solution is also unique in its own class if and only if $\rho\,\omega^2 \neq \mu\,|\boldsymbol{k}^*|^2/n^2$, for all $(\boldsymbol{k},n) \in \mathbb{Z}^3\times\mathbb{Z}$.

Proof. By the Helmholtz–Weyl decomposition, we have

$$\boldsymbol{u} = \nabla\varphi + \operatorname{curl}\boldsymbol{v}\,,$$

which once replaced into (3.12), with the help of (3.34), delivers

$$\begin{aligned}
&\nabla\left[\rho\varphi_{tt} - (\lambda+2\mu)\Delta\varphi - \sigma_1\theta - \rho\Psi\right] + \operatorname{curl}\left[\rho\boldsymbol{v}_{tt} - \mu\Delta\boldsymbol{v} - \rho\boldsymbol{F}\right] = \boldsymbol{0}\\
&\theta_t - \kappa\Delta\theta = \sigma_2\Delta\varphi_t + Q\,.
\end{aligned} \tag{3.36}$$

By the uniqueness of the decomposition, (3.36) is equivalent to the following set of three equations

$$\begin{aligned}
&\rho\boldsymbol{v}_{tt} - \mu\Delta\boldsymbol{v} = \rho\boldsymbol{F}\\
&\rho\varphi_{tt} - (\lambda+2\mu)\Delta\varphi = \sigma_1\theta + \rho\Psi\\
&\theta_t - \kappa\Delta\theta = \sigma_2\Delta\varphi_t + Q\,.
\end{aligned} \tag{3.37}$$

Further on we shall show that, under the given assumptions on Ψ and Q, $(3.37)_{2,3}$ has one and only one T-periodic solution (φ,θ) in the appropriate class. As a consequence, from the considerations developed in Sect. 3.2 (by formally setting $\lambda + \mu = 0$) resonance may occur if and only if $\boldsymbol{F} \not\equiv \mathbf{0}$, which, from the general formula (3.34), is equivalent to curl $\boldsymbol{F} \not\equiv \mathbf{0}$. If, on the other hand, $\boldsymbol{F} \equiv \mathbf{0}$, then a T-periodic solution $\boldsymbol{v} \in L^2(0,T;H^2_\#(\mathcal{C})) \cap W^{2,2}(0,T;H_\#(\mathcal{C}))$ to $(3.37)_1$ will be identically zero if and only if $\rho\,\omega^2 \neq \mu\,|\boldsymbol{k}^*|^2/n^2$, for all $(\boldsymbol{k},n) \in \mathbb{Z}^3 \times \mathbb{Z}$. To show the theorem completely, it then remains to show the T-periodicity property of the solution (φ,θ) mentioned above. Applying Fourier expansion in time to both sides of $(3.37)_{2,3}$ we obtain

$$\begin{aligned} -\rho\, n^2\omega^2\varphi_n - (\lambda+2\mu)\Delta\varphi_n &= \sigma_1\theta_n + \rho\Psi_n \\ \imath\, n\,\omega\,\theta_n - \kappa\,\Delta\theta_n &= \imath\, n\,\omega\,\sigma_2\,\Delta\varphi_n + Q_n\,, \end{aligned} \tag{3.38}$$

where

$$\varphi(x,t) = \sum_{n\in\mathbb{Z}} \varphi_n(x)\, e^{\imath n\omega t}\,,\ \text{etc.}$$

We next apply to these equations Fourier expansion in space to get

$$\begin{aligned} \big[-\rho\, n^2\omega^2 + (\lambda+2\mu)|\boldsymbol{k}^*|^2\big]\,\varphi_{n\boldsymbol{k}} &= \sigma_1\theta_{n\boldsymbol{k}} + \rho\,\Psi_{n\boldsymbol{k}} \\ \big(\imath\, n\,\omega + \kappa\,|\boldsymbol{k}^*|^2\big)\theta_{n\boldsymbol{k}} &= -\imath\, n\,\omega\,|\boldsymbol{k}^*|^2\sigma_2\varphi_{n\boldsymbol{k}} + Q_{n\boldsymbol{k}}\,, \end{aligned} \tag{3.39}$$

where

$$\varphi_n = \sum_{\boldsymbol{k}\in\mathbb{Z}^3} \varphi_{n\boldsymbol{k}}\, e^{\imath\,\boldsymbol{k}^*\cdot\boldsymbol{x}}\,,\ \text{etc.}$$

Without loss, we may set $\Psi_{n,\mathbf{0}} = 0$ for all $n \in \mathbb{Z}$,[11] whereas, by assumption, $Q_{n,\mathbf{0}} = 0$, for all $n \in \mathbb{Z}$. Therefore, from (3.39), we infer that we may take

$$\varphi_{n,\mathbf{0}} = \theta_{n,\mathbf{0}} = 0\,, \quad \text{for all } n \in \mathbb{Z}$$

and

$$\begin{aligned} \varphi_{n\boldsymbol{k}} &= \frac{\rho(\imath\, n\,\omega + \kappa\,|\boldsymbol{k}^*|^2)\Psi_{n\boldsymbol{k}} + Q_{n\boldsymbol{k}}}{\mathscr{R}(n,\boldsymbol{k})} \\ \theta_{n\boldsymbol{k}} &= \frac{-\imath\, n\,\omega\,\sigma_2|\boldsymbol{k}|^2\varphi_{n\boldsymbol{k}} + Q_{n\boldsymbol{k}}}{\imath\, n\,\omega + \kappa|\boldsymbol{k}|^2} \end{aligned} \qquad (n,\boldsymbol{k}) \in \mathbb{Z}\times\mathbb{Z}^3 - \{(n,\mathbf{0})\}\,,$$

with

$$\begin{aligned} \mathscr{R}(n,\boldsymbol{k}) := \kappa\,|\boldsymbol{k}|^2\,(-\rho\, n^2\omega^2 + (\lambda+2\mu)|\boldsymbol{k}|^2) \\ + \imath\, n\,\omega\,(-\rho\, n^2\omega^2 + (\lambda+2\mu)|\boldsymbol{k}|^2 + \sigma_1\sigma_2|\boldsymbol{k}|^2)\,. \end{aligned}$$

[11]This means to modify Ψ by a function of time which, of course, does not affect the load $\boldsymbol{f}$.

Since $\sigma_1\sigma_2 > 0$ it follows $\mathscr{R}(n,\boldsymbol{k}) \neq 0$ for all $(n,\boldsymbol{k}) \in \mathbb{Z}\times\mathbb{Z}^3-\{(n,\boldsymbol{0})\}$, so that both Fourier coefficients φ_{nk} and θ_{nk} are well defined. Our next step is to obtain suitable bounds of these coefficients in terms of the data. To this end, we multiply both sides of $(3.39)_2$ by $\overline{\theta}_{nk}$, take the c.c. of both sides of $(3.39)_1$ and then multiply by φ_{nk}. If we sum the two resulting equations, we get

$$\kappa|\boldsymbol{k}^*|^2|\theta_{nk}|^2 = \rho\, n\, \omega\, |\boldsymbol{k}^*|^2\frac{\sigma_2}{\sigma_1}\Im\langle\Psi_{nk},\overline{\varphi}_{nk}\rangle + \Re\langle\bar{Q}_{nk},\theta_{nk}\rangle\,, \tag{3.40}$$

where $\langle\cdot,\cdot\rangle$ denotes the scalar product in $\mathbb{R}^3$. We next multiply $(3.39)_1$ by $|\boldsymbol{k}^*|^2$ and replace the term $|\boldsymbol{k}^*|^2\theta_{nk}$ from $(3.39)_2$ to show

$$\kappa\, n\, \omega\, |\boldsymbol{k}^*|^2\sigma_1\sigma_2|\varphi_{nk}|^2 = \kappa\, \sigma_1\, n\, \omega\, \Re\langle\theta_{nk},\overline{\varphi}_{nk}\rangle + \Im\langle Q_{nk},\overline{\varphi}_{nk}\rangle\,. \tag{3.41}$$

Recalling that $\sigma_1\sigma_2 > 0$, from (3.41) it immediately follows that

$$\kappa\, |n|\, \omega\, |\boldsymbol{k}^*|^2\sigma_1\sigma_2|\varphi_{nk}| \le \kappa\, |\sigma_1|\, |n|\, \omega\, |\theta_{nk}| + |Q_{nk}| \tag{3.42}$$

while from (3.40) we infer

$$\kappa|\boldsymbol{k}^*|^2|\theta_{nk}|^2 \le \rho\, |n|\, \omega\, |\boldsymbol{k}^*|^2\frac{\sigma_2}{\sigma_1}|\Psi_{nk}|\, |\varphi_{nk}| + |Q_{nk}|\, |\theta_{nk}|\,. \tag{3.43}$$

Employing (3.42) in (3.43) leads to the following relation

$$\kappa|\boldsymbol{k}^*|^2|\theta_{nk}|^2 \le \frac{\rho\, |n|\, \omega}{|\sigma_1|}\, |\Psi_{nk}|\, |\theta_{nk}| + \frac{\rho}{\kappa\sigma_1^2}|Q_{nk}|\, |\Psi_{nk}| + |Q_{nk}|\, |\theta_{nk}|\,,$$

which, in turn, by Cauchy inequality and recalling that $|n|, |\boldsymbol{k}| \ge 1$, implies

$$|\boldsymbol{k}^*|\, |\theta_{nk}| \le c_1\big[(1+|n|)|\Psi_{nk}| + |Q_{nk}|\big]\,, \tag{3.44}$$

where $c_1 > 0$ is independent of n and $\boldsymbol{k}$. Likewise, from (3.42) and (3.44) we deduce

$$|n|\, |\boldsymbol{k}^*|^2|\varphi_{nk}| \le c_2\big[(1+n^2)\, |\Psi_{nk}| + (1+|n|)|Q_{nk}|\,, \tag{3.45}$$

where also $c_2 > 0$ is independent of n and $\boldsymbol{k}$. Furthermore, combining $(3.39)_2$ with (3.44)–(3.45) we obtain

$$|\boldsymbol{k}^*|^2|\theta_{nk}| \le c_3\big[(1+n^2)|\Psi_{nk}| + (1+|n|)|Q_{nk}|\big]\,, \tag{3.46}$$

with $c_3 > 0$ independent of n and $\boldsymbol{k}$. Finally, from $(3.39)_1$, and again (3.44)–(3.45) it is not hard to show that

$$|\boldsymbol{k}^*|^3|\varphi_{nk}| \le c_4\big[(1+|n|^3+|\boldsymbol{k}^*|)|\Psi_{nk}| + (1+n^2)|Q_{nk}|\big]\,, \tag{3.47}$$

with $c_4 > 0$ independent of n and $\boldsymbol{k}$. The theorem then follows from (3.44) to (3.47) with the help of Plancherel theorem. □

Remark 3.3. A significant consequence of the argument adopted in the proof of the previous theorem is that, under the stated periodic boundary conditions, even though—in absence of external loads and heat sources—the total energy of the elastic body is strictly decreasing, the energy of the "divergence-free" part of the displacement field *remains constant in time*. This can be immediately seen from (3.37) with $\boldsymbol{F} \equiv \boldsymbol{0}$, $\Psi \equiv 0$ which after some simple manipulation furnishes

$$\begin{aligned}
\tfrac{1}{2}\frac{d\mathcal{E}_1}{dt} &\equiv \tfrac{1}{2}\frac{d}{dt}\left[\rho\,\|\operatorname{curl} \boldsymbol{v}_t(t)\|_2 + \mu\,\|\nabla(\operatorname{curl} \boldsymbol{v}(t))\|_2\right] = 0\,,\\
\tfrac{1}{2}\frac{d\mathcal{E}_2}{dt} &\equiv \tfrac{1}{2}\frac{d}{dt}\left[\rho\,\|\nabla\varphi_t(t)\|_2^2 + (\lambda + 2\mu)\|\Delta\varphi(t)\|_2^2 + \|\theta(t)\|_2^2\right] = -\|\nabla\theta(t)\|_2^2\,.
\end{aligned}$$

Notice also that, since one can prove that $\mathcal{E}_2(t) \to 0$ as $t \to \infty$, the energy remains stored *only* in the rotational part of the displacement field. As we shall see in details in the next section, "stored energy" is one of the main reasons (but not the only one) for the occurrence of resonance.

Remark 3.4. A result qualitatively analogous to that of Theorem 3.2 could be proved also in the case of a bounded and sufficiently smooth domain Ω, provided we use appropriate boundary conditions on the displacement field $\boldsymbol{u}$, as proposed in [32]; see also [42]. The latter require that on $\partial\Omega$ the tangential component of $\boldsymbol{u}$ as well as $\operatorname{div} \boldsymbol{u}$ identically vanish.

3.4 An Abstract Approach

From what we have seen in the previous section by analyzing the linear thermoelastic model, we may state that, roughly speaking, resonance in a mechanical system does not occur if and only if whatever the period of the applied driving mechanism can be, the system allows for a corresponding time-periodic motion of the same period. As a consequence, in order to furnish a general approach to the problem, it seems appropriate to investigate existence of time-periodic solutions for a sufficiently broad class of linear evolution equations, that include the relevant hyperbolic–parabolic systems as a special case. This will be the object of the present section, where we shall analyze this question with the help of the mean ergodic theorem and some of its main consequences, within a class of problems whose dynamics is governed by a strongly continuous semigroup of contractions.

Let X be a *reflexive* Banach space with associated norm $\|\cdot\|$, and let X^* denote its dual with norm $\|\cdot\|_*$. For a given operator $A : X \mapsto X$ we indicate by $\mathsf{D}(A)$ its domain of definition and by $\mathsf{R}(A)$ its range. Moreover, we set $\mathsf{N}(A) := \{x \in X : Ax = 0\}$. The identity operator on X is indicated by I.

Let $f : \mathbb{R} \mapsto X$ be an assigned periodic function of period $T > 0$. Our objective is to provide necessary and sufficient conditions for the occurrence of resonance (and, as a consequence, for the existence of a time-periodic solution) to the following equation

$$\frac{dx}{dt} = Ax + f(t) \ \ t \in \mathbb{R}, x \in X. \tag{3.48}$$

In this respect, we shall assume that A is the infinitesimal generator of a strongly continuous semigroup of contractions. As is well known, the latter means that there exists a one-parameter family of bounded, linear operators of X into itself, $\{U(t)\}_{t\geq 0}$, such that

$$Ax = \lim_{t\to 0+} \frac{1}{t}\|(T(t) - I)x\|, \text{ for all } x \in \mathsf{D}(A),$$

and satisfying the following properties

(i) $U(0) = I$;
(ii) $U(t+s) = U(t)U(s)$, all $t, s \geq 0$;
(iii) $\|U(t)x - U(t_0)x\| \to 0$ as $t \to t_0$, for all $x \in X$, and all $t_0 \geq 0$,

that is, $U(t)$ is *strongly continuous* and, in addition,

(iv) $\|U(t)x\| \leq \|x\|$, for all $x \in X$ and $t \geq 0$,

that is, $U(t)$ is a contraction at all $t \geq 0$.

Remark 3.5. The equations of thermoelasticity define a strongly continuous semigroup of contractions in the (energy) space $X := H_0^1(\Omega)\times L^2(\Omega)\times L^2(\Omega)$. In fact, setting $A_1 \equiv (1/\rho)\mathscr{L}$, $A_2 \equiv -\kappa\Delta$ with $\mathsf{D}(A_2) \equiv H^2(\Omega) \cap H_0^1(\Omega)$, $B \equiv (\sigma_1/\rho)\nabla$ with $\mathsf{D}(B) \equiv H_0^1(\Omega)$, and $B^* \equiv -\sigma_2 \mathrm{div}$ with $\mathsf{D}(B^*) \equiv H^1(\Omega)$. we can put (3.12)–(3.13) in the form (3.48) where

$$x := \begin{pmatrix} \boldsymbol{u} \\ \boldsymbol{v} \\ \theta \end{pmatrix}, \quad A := \begin{pmatrix} 0 & I & 0 \\ -A_1 & 0 & B \\ 0 & -B^* & -A_2 \end{pmatrix}, \quad f := \begin{pmatrix} 0 \\ \boldsymbol{f}/\rho \\ Q \end{pmatrix},$$

with $\mathsf{D}(A) := H^2(\Omega) \cap H^1(\Omega) \cap H_0^1(\Omega)$. Then, it can be shown that A is the infinitesimal generator of a strongly continuous semigroup of contractions; see, e.g., [24, Theorem 1].

As customary, we define $x = x(t)$ to be a *strong solution* to (3.48) if $x \in C([0,\tau]; X), dx/dt \in L^1(0,\tau; X)$, $\tau > 0$, and $x(t) \in \mathsf{D}(A)$ and satisfies (3.48) for a.a. $t \in [0,\tau]$. Likewise, $x = x(t)$ is a *mild solution* to (3.48) if it is continuous with values in X and satisfies the integrated *"mild formulation"* of (3.48):

$$x(t) = U(t)x(0) + \int_0^t U(t-s)f(s)\,ds, \quad t \geq 0. \tag{3.49}$$

Definition 3.1. A real number $T > 0$ is a *resonant period for (3.48)* [respectively, *for* (3.49)] if there exists a T-periodic function $f \in L^1(0, T; X)$ of period T for which (3.48) [respectively, (3.49)] does not have any corresponding T-periodic strong [respectively, mild] solution.

As we shall see further on, the existence or non-existence of a resonant period is strictly related to the asymptotic behavior as $t \to \infty$ of the generic solution to the *homogeneous* equation, that is, (3.48) with $f \equiv 0$. In particular, resonant periods do occur for systems possessing solutions not approaching 0 for large t (Theorem 3.3). A remarkable example in this direction is constituted by *conservative* systems where energy is kept constant at all $t \geq 0$ by all solutions (Theorem 3.7). On the opposite hand, resonance certainly does not occur if solutions decay exponentially fast to 0 as $t \to \infty$ (Theorem 3.6). In the intermediate case, where all solutions decay to 0 at an indefinite rate,[12] we can show that resonance may occur if and only if f belong to the complement of a dense set in the data space (Theorem 3.4). Furthermore, if the operator A obeys appropriate further assumptions, that are certainly met in several physically relevant models like linear thermo- and magneto-elasticity, as well as poroelastic diffusion, we can specialize this result to show that resonance does not occur whenever f possesses a finite (but otherwise arbitrary) number of modes or, equivalently, belongs to a *suitable* dense linear space of the data space (Theorem 3.2).

We begin to show the following result.

Theorem 3.3. *Let $T > 0$. The following properties hold.*

(i) *If*

$$\mathsf{N}(I - U(T)) \neq \{0\}, \tag{3.50}$$

then T is a resonant period for (3.49). *Moreover, if 0 is an eigenvalue of A, then every $T > 0$ is a resonant period for* (3.49).

(ii) *If $\mathsf{N}(I - U(T)) = \{0\}$, but $\mathsf{R}(I - U(T))$ is not closed, then T is a resonant period for* (3.49). *In addition, there is a dense subset, $\mathcal{Q}$, of $L^1(0, T; X)$ such that for every $f \in \mathcal{Q}$ there exists a unique corresponding T-periodic solution $x \in C([0, T]; X)$ to Eq. (3.49).*

(iii) *If*

$$\mathsf{N}(I - U(T)) = \{0\} \ \textit{and}\ \mathsf{R}(I - U(T)) \ \textit{is closed}, \tag{3.51}$$

then for every T-periodic function $f \in L^1(0, T; X)$ there exists one and only one corresponding T-periodic solution $x \in C([0, T]; X)$ to Eq. (3.49).

---

[12] Recall that if there exists $t_0 > 0$ such that $\|x(t_0)\| \equiv \|U(t_0)x(0)\| < \|x(0)\|$ for all $x(0) \in X$, then, by the semigroup property of $U(t)$, all solutions must decay exponentially fast [5, Remark at p. 178].

Therefore, a necessary and sufficient condition for the existence of a resonant period for (3.49) is that at least one of the conditions in (3.51) is violated.

Proof. Set $U := U(T)$, and denote by U_n, $n \in \mathbb{N}$, the Cesaro averages of U:

$$U_n := \frac{1}{n} \sum_{m=1}^{n} U^m \,. \tag{3.52}$$

For a fixed $x \in X$, by the property (iv) of the semigroup $U(t)$ we find

$$\|U_n x\| \leq \frac{1}{n}(n\|x\|)\,,$$

so that, since X is reflexive, for each $x \in X$ there exists $\{n'\} \subseteq \{n\}$ and $x_0 \in X$ such that

$$\text{weak} - \lim_{n' \to \infty} U_{n'} x = x_0 \,.$$

Moreover, again by the property (iv) of $U(t)$ the family $\{U^m\}$ is equi-continuous, in the sense that

$$\sup_{m \geq 1} \|U^m x\| \leq \|x\|\,, \quad \text{for each } x \in X\,.$$

As a result, from the mean ergodic theorem (e.g., [49, pp. 213–214]) we deduce

$$x_0 \in \mathsf{N}(I-U)\,, \ \lim_{n \to \infty} U_n x = x_0\,,$$

as well as the following characterizations

$$\begin{aligned} \overline{\mathsf{R}(I-U)} &= \{x \in X : \lim_{n \to \infty} U_n x = 0\} \\ \mathsf{N}(I-U) &= \{x_0 \in X : \lim_{n \to \infty} U_n x = x_0\}\,. \end{aligned} \tag{3.53}$$

Furthermore, we have the direct sum decomposition of the space X

$$X = \mathsf{N}(I-U) \oplus \overline{\mathsf{R}(I-U)}\,. \tag{3.54}$$

Now, suppose (3.50) holds and, by contradiction, that for any T-periodic $f \in L^1(0,T;X)$ Eq. (3.49) admits a corresponding T-periodic solution. The latter means that there is $x \in X$ such that

$$(I-U)x = \int_0^T U(T-s)f(s)ds\,. \tag{3.55}$$

Denote by U^* the conjugate of U, and by $\langle \cdot, \cdot \rangle$ the duality pair $X \leftrightarrow X^*$. Since (e.g., [21, p. 308])

$$\mathsf{N}(I - U^*) = \{x^* \in X^* : \langle x^*, x \rangle = 0, \text{ for all } x \in \mathsf{R}(I - U)\}, \tag{3.56}$$

and since

$$\overline{\mathsf{R}(I - U)} \neq X \tag{3.57}$$

we infer that $\mathsf{N}(I - U^*)$ is not trivial. Thus, picking $x^* \in \mathsf{N}(I - U^*) - \{0\}$, from (3.55) it follows

$$\langle (I - U^*)x^*, x \rangle = \langle x^*, (I - U)x \rangle = \int_0^T \langle x^*, U(T - s) f(s) \rangle \, ds$$

that is

$$\int_0^T \langle x^*, U(T - s) f(s) \rangle \, ds = 0 \,. \tag{3.58}$$

In view of (3.56) and (3.57) we deduce that there exists at least one $\overline{x} \in X - \overline{\mathsf{R}(I - U)}$ such that

$$\langle x^*, \overline{x} \rangle \neq 0 \,. \tag{3.59}$$

We now choose the T-periodic function f such that

$$f(t) = U(t)\overline{x} \;\; t \in [0, T] \,.$$

Clearly, $f \in L^\infty(\mathbb{R}; X)$. Using the semigroup property of $U(t)$ we thus have

$$\int_0^T \langle x^*, U(T - s) f(s) \rangle \, ds = \int_0^T \langle x^*, U\overline{x} \rangle \, ds = T \, \langle U^* x^*, \overline{x} \rangle \,.$$

Recalling that $x^* \in \mathsf{N}(I - U^*)$ and using (3.59), from the previous relation we conclude

$$\int_0^T \langle x^*, U(T - s) f(s) \rangle \, ds = T \, \langle x^*, \overline{x} \rangle \neq 0 \,,$$

which contradicts (3.58). This proves the first statement of the theorem. Suppose now 0 is an eigenvalue of the generator A. From the spectral mapping theorem it then follows that 1 is an eigenvalue of $U(t)$ for all $t \geq 0$ [38, Theorem 2.4], namely, $\mathsf{N}(I - U(t)) \neq \{0\}$ for all $t \geq 0$, and the second statement follows from the first one. We next prove property (ii). If $\overline{\mathsf{R}(I-U)} \neq X$, there exists $b \in X$ such that the equation

$$(I-U)x = b \tag{3.60}$$

does not have a solution. Therefore, since the existence of a T-periodic solution is equivalent to the resolution of (3.55), the first stated property in part (ii) follows provided we show that for any $b \in X$ there is $f \in L^1(0,T;X)$ such that

$$b = \int_0^T U(T-s)f(s)ds\,. \tag{3.61}$$

Once such an f is found, we may extend it periodically of period T to the whole real line. [13] For $\sigma > 0$ we set $\tilde{U}(t) := \mathrm{e}^{-\sigma t}U(t)$, and observe that since $\|\tilde{U}(T)\| < 1$, $I - \tilde{U}(T)$ is continuously invertible. Equation (3.61) becomes

$$b = \int_0^T \tilde{U}(T-s)w(s)ds\,,$$

where $w(s) := \mathrm{e}^{\sigma(T-s)}f(s)$. Replacing $w(s)$ by $w(T-s)$ and keeping the same notation, the question then reduces to show that the bounded linear operator

$$M : L^1(0,T;X) \to X$$

defined by

$$Mw := \int_0^T \tilde{U}(s)w(s)ds$$

is surjective. It is clear that $\mathsf{R}(M)$ is dense in X. To see this we observe that $w(t) = (\tilde{U}(T)-I)^{-1}(A-\sigma)b$ satisfies $Mw = b$, which in turn implies $\mathsf{D}(A) \subset \mathsf{R}(M)$, and, therefore, the density property. We shall next prove that the range of M is closed. To this end, by the Banach closed range theorem, this is equivalent to show that $\mathsf{R}(M^*)$ is closed, where M^* is the conjugate of M. From the identity

$$\langle Mw, x^*\rangle = \int_0^T \langle \tilde{U}(s)w(s), x^*\rangle ds = \int_0^T \langle w(s), \tilde{U}^*(s)x^*\rangle ds$$

it follows that $M^* : X^* \to L^1(0,T;X)^* \equiv L^\infty(0,T;X^*)$ is given by $M^*x^*(s) = \tilde{U}^*(s)x^*$, $s \in [0,T]$. This implies

$$\|x^*\|_* \leq \sup_{0\leq s\leq T} \|\tilde{U}^*(s)x^*\|_* = \|M^*x^*\|_{L^\infty(0,T;X^*)}\,.$$

[13]The argument that follows is due to Professor Jan Prüss, to whom we are indebted.

Hence M^* is injective and has closed range, and the proof of the first statement in (ii) is completed. In order to show the second one we introduce the quotient space

$$\mathscr{L}^1(0,T;X) := L^1(0,T;X)/\mathsf{N}(M)$$

where

$$\mathsf{N}(M) = \left\{ w \in L^1(0,T;X) : \int_0^T \tilde{U}(t)w(t)dt \equiv \int_0^T U(T-t)w(t) = 0 \right\}, \quad (3.62)$$

endowed with the usual norm

$$\|W\| := \inf_{z \in \mathsf{N}(M)} \|w - z\|, \; w \in W.$$

Recalling that M is surjective, the operator

$$\mathscr{M} : W \in \mathscr{L}^1(0,T;X) \mapsto Mw \in X, \; w \in W,$$

is then continuously invertible and, therefore, $\mathscr{M}^{-1}$ maps dense sets of X into dense sets of $\mathscr{L}^1(0,T;X)$. As a result, setting $\mathscr{D} = \mathscr{M}^{-1}(\mathsf{R}(I-U))$ we deduce, by assumption, that $\mathscr{D}$ is dense in $\mathscr{L}^1(0,T;X)$. Define

$$\mathcal{D} := \{v \in X : v \in V \text{ for some } V \in \mathscr{D}\},$$

and notice that for any $b \in \mathsf{R}(I-U)$ there is $v \in \mathcal{D}$ such that $Mv = b$. We claim that the set

$$\mathcal{Q} := \mathcal{D} + \mathsf{N}(M)$$

is dense in $L^1(0,T;X)$. In fact, pick $u \in X$, and $\varepsilon > 0$. By what we have just shown, there exists $u_\varepsilon \in \mathcal{D}$ such that

$$\inf_{z \in \mathsf{N}(M)} \|u - u_\varepsilon - z\| < \varepsilon,$$

which, in turn, by the property of the infimum, implies that we can find $\overline{z} \in \mathsf{N}(M)$ such that

$$\|u - u_\varepsilon - \overline{z}\| < 2\varepsilon,$$

which proves the desired property. We next observe that for any $b \in \mathsf{R}(I-U)$ (that, by hypothesis, is dense in X) (3.60) has one and only one solution x. Now, from what we have proved and (3.62), we conclude that for such a b there exists $f = f_1 + f_2 \in \mathcal{D} + \mathsf{N}(M)$ such that

$$b = \int_0^T U(T-s)f_1(s)ds = \int_0^T U(T-s)f(s)ds,$$

namely, for any f in the dense set $\mathcal{Q}$ there exists $x \in X$ such that

$$(I - U)x = \int_0^T U(T - s) f(s) ds ,$$

and this completes the proof of the second statement in (ii).

We finally prove property (iii). We recall that, for a given T-periodic function f, the existence of a corresponding T-periodic solution to (3.49) is *equivalent* to find $x \in X$ satisfying (3.55). Now, under the assumption (3.51), by the Banach closed range theorem $(I - U)$ is continuously invertible. Therefore, from (3.55) we obtain

$$x = (I - U)^{-1} \int_0^T U(T - s) f(s)\, ds ,$$

which provides the desired solution. Finally, the corresponding uniqueness result is an immediate consequence of the first condition in (3.51). □

Remark 3.6. As immediate consequence of Theorem 3.3, we deduce the following result. Necessary and sufficient condition in order that for any T-periodic $f \in L^1(0, T; X)$ Eq. (3.49) admits a unique T-periodic mild solution is that both conditions in (3.51) hold. This extends the analogous result of [41, Theorem 1] to the case of more general f.

Remark 3.7. If f is such that

$$\int_0^T U(T - s) f(s)\, ds = 0 ,$$

which, of course, is a non-generic property of the data, then the assumption on the closedness of the range in (3.51) is not required (since always $0 \in \mathsf{R}(I - U)$) and the unique T-periodic solution is given by

$$x(t) = \int_0^t U(t - s) f(s)\, ds ,$$

see [36, Theorem 3.3].

Remark 3.8. The first part of Theorem 3.3 can be equivalently restated by saying that if the homogeneous equation (namely (3.49) with $f \equiv 0$) has a non-trivial T-periodic solution, then T is a resonant period for the non-homogeneous equation (3.49). This is in complete agreement with what we have shown in Sect. 3.3 in the particular case of thermoelasticity with periodic boundary conditions.

Our next objective is to give sufficient conditions on the semigroup $U(t)$ that ensure the validity of (3.51) for all $T > 0$. As a by-product, we will then obtain existence of T-periodic solutions for arbitrary period, which in turn will rule out the occurrence of resonance. The above conditions are, as somewhat expected, related to the asymptotic behavior of $U(t)$ for large t.

To this end, we recall that a semigroup $S(t)$ is called *strongly stable* if

$$\lim_{t\to\infty} \|S(t)x\| = 0\,, \quad \text{for all } x \in X\,, \tag{3.63}$$

whereas it is called *uniformly stable* if

$$\lim_{t\to\infty} \|S(t)\| = 0\,. \tag{3.64}$$

The following result is well known; see, e.g., [5, Remark at p. 178].

Lemma 3.1. *A strongly continuous semigroup $S(t)$ is uniformly stable if and only if there exists $M, \delta > 0$ such that*

$$\|S(t)x\| \le M\,\mathrm{e}^{-\delta t}\|x\|\,, \text{ for all } t \ge 0 \text{ and all } x \in X. \tag{3.65}$$

The next theorem shows that strong stability prevents resonance for dense set of data but does not rule it out completely.

Theorem 3.4. *Suppose that $U(t)$ is strongly stable, and let $T > 0$ be arbitrarily given. Then we can find a dense set $\mathcal{Q} \subseteq L^1(0,T;X)$ such that for any T-periodic $f \in \mathcal{Q}$ there exists a unique corresponding T-periodic solution $x \in C([0,T];X)$ to Eq. (3.49).*

Proof. By Theorem 3.3(ii) it is enough to show

$$\mathsf{N}(I - U(T)) = \{0\} \text{ for all } T > 0\,. \tag{3.66}$$

Assume (3.63) holds and fix $T > 0$. Then, for any $\varepsilon > 0$ there is $\overline{n} \in \mathbb{N}$ such that

$$\|U(nT)x\| < \varepsilon\,, \text{ for all } n \ge \overline{n}\,. \tag{3.67}$$

We now have, for all $n > \overline{n}$,

$$U_n(T)x = \frac{1}{n}\left[U(T)x + U(2T)x + \cdots + U(\overline{n}T)x + \sum_{k=\overline{n}+1}^{n} U(kT)x\right].$$

Thus, using the property (iv) of the semigroup and (3.67)

$$\|U_n(T)x\| \le \frac{1}{n}\,[\overline{n} + \varepsilon(n - \overline{n})]\,\|x\|\,,$$

from which, by the arbitrariness of ε, we conclude

$$\lim_{n\to\infty} \|U_n(T)x\| = 0\,.$$

Therefore (3.66) follows.

An immediate consequence of the above theorem is the following result ensuring "generic existence" and uniqueness of T-periodic solutions for *arbitrary* $T > 0$, provided only $U(t)$ is strongly stable.

Corollary 3.1. *Assume $U(t)$ strongly stable, and let $f \in L^1(0,T;X)$ be T-periodic of arbitrary period $T > 0$. Then, for every $\varepsilon > 0$ there is $f_\varepsilon \in L^1(0,T;X)$ with*

$$\|f_\varepsilon\|_{L^1(0,T;X)} < \varepsilon, \tag{3.68}$$

such that (3.49), *with f replaced by $f + f_\varepsilon$, has one and only one corresponding T-periodic solution.*

Proof. Given f and $\varepsilon > 0$, we may find $\overline{f}_\varepsilon \in \mathcal{Q}$ satisfying (3.68). By Theorem 3.4, we then show the existence of one and only one corresponding T-periodic solution. Thus, the claimed property follows by writing $\overline{f}_\varepsilon = f + (\overline{f}_\varepsilon - f)$. □

In view of the classical result of Arendt and Batty [1], a sufficient condition in order to satisfy the assumption on $U(t)$ of Theorem 3.4 and Corollary 3.1 is that the infinitesimal generator A has a compact resolvent and no eigenvalues lying on the imaginary axis. These conditions are certainly met in a broad class of hyperbolic–parabolic problems that are relevant in many significant applications where the dissipative coupling is "sufficiently strong." In such a class, we are also able to provide a concrete realization of a dense set of "forces" with the property stated in Theorem 3.4. In particular, our result allows us to exclude the occurrence of resonance when f has a finite (but otherwise arbitrary) number of modes. Nevertheless, if the coupling is not "strong" enough, then resonance is shown to occur.

To state all the above in precise mathematical terms, we set

$$L^2_\sharp(0,T;X) = \left\{ r : t \in \mathbb{R} \mapsto X; r \text{ is } T\text{-periodic, and} \int_0^T \|r(t)\|^2 dt < \infty \right\}$$

and let

$$\mathcal{Q}_\sharp(0,T;X) := \left\{ q \in L^2_\sharp(0,T;X) : q(t) = \sum_{n=-N}^{N} q_n \, e^{\imath \frac{2n\pi}{T} t}, \; q_n \in X, \, N \in \mathbb{N} \right\} .$$

We recall the following result on vector-valued Fourier series (e.g., [25, Theorem 2.2]).

Lemma 3.2. *Let X be a Hilbert space. Then any $r \in L^2_\sharp(0,T;X)$ admits the following Fourier expansion (with $\omega := 2\pi/T$)*

$$r(t) = \sum_{k \in \mathbb{Z}} r_k \, \mathrm{e}^{\imath k \omega t}$$

where the Fourier coefficients r_k are given by

$$r_k := \frac{1}{T}\int_0^T r(t)\,\mathrm{e}^{\imath\,k\,\omega\,t}\,dt\,.$$

In the above, the integral is meant in the sense of Bochner and the convergence of the series in the following sense

$$\lim_{N\to\infty}\int_0^T \left\| r(t) - \sum_{k=-N}^{k=N} r_k(t)\,\mathrm{e}^{\imath\,k\,\omega\,t}\right\|^2 dt = 0\,.$$

Thus, in particular, $\mathcal{Q}_\sharp(0,T;X)$ *is dense in* $L^2_\sharp(0,T;X)$.

We are now in a position to prove the following result.

Theorem 3.5. *Let* $\mathcal{H}_1$ *and* $\mathcal{H}_2$ *be Hilbert spaces, and consider the equations*

$$\ddot{u} + A_1 u + \lambda_1 B w = f \;\text{ in } \mathcal{H}_1\,,\quad \dot{w} + A_2 w + \lambda_2 B^* \dot{u} = g \;\text{ in } \mathcal{H}_2\,, \tag{3.69}$$

where $\lambda_i \in \mathbb{R} - \{0\}$ *are coupling parameters,* $A_i : \mathsf{D}(A_i) \subset \mathcal{H}_i \to \mathcal{H}_i$, $i = 1, 2$, *are self-adjoint, strictly accretive operators with compact inverse, while* $B : \mathsf{D}(B) \to \mathcal{H}_1$, *is a densely defined operator with adjoint* B^*, *and* $\mathsf{D}(A_2) \subset \mathsf{D}(B)$, $\mathsf{D}(A_1) \subset \mathsf{D}(B^*)$. *Suppose, further, that* B *is* A_2*-compact, while* B^* *is* A_1*-compact. The following properties hold.*

(a) *Assume*

$$-\,\mu\, v + A_1 v = 0\,,\;\; B^* v = 0 \;\Longrightarrow v = 0 \tag{3.70}$$

for any value of the positive parameter μ, *and let* T *be an arbitrary positive number. Then, there exists a dense set* $\mathcal{Q} \subseteq L^2_\sharp(0,T;\mathcal{H}_1\times\mathcal{H}_2)$ *such that for any* $(f,g)\in\mathcal{Q}$, *the system* (3.69) *has at least one* T*-periodic solution* $(u,w)\in C^\infty([0,T];\mathsf{D}(A_1)\times\mathsf{D}(A_2))$. *More specifically, for any* f *and* g *such that*

$$f(t) = \sum_{n=-N}^{N} f_n\, e^{\imath n\omega t}\,,\quad g(t) = \sum_{n=-N}^{N} g_n\, e^{\imath n\omega t} \tag{3.71}$$

where $N\in\mathbb{N}$, $\omega := 2\pi/T$, *and* $(f_n, g_n)\in\mathcal{H}_1\times\mathcal{H}_2$, *for all* $0\le|n|\le N$, *the system* (3.69) *has a corresponding solution* (u,w) *of period* T *given by*

$$u(t) = \sum_{n=-N}^{N} u_n\, e^{\imath n\omega t}\,,\quad w(t) = \sum_{n=-N}^{N} w_n\, e^{\imath n\omega t}\,,$$

where $(u_n, w_n) \in \mathsf{D}(A_1) \times \mathsf{D}(A_2)$. *Furthermore*

$$\|A_1(u_n)\|_{\mathcal{H}_1} + \|A_2(w_n)\|_{\mathcal{H}_2} \le C(n)\,(\|f_n\|_{\mathcal{H}_1} + \|g_n\|_{\mathcal{H}_2})\,, \tag{3.72}$$

for all $0 \le |n| \le N$.

Finally, if (u^1, w^1) *is another* T*-periodic solution to* (3.69) *corresponding to the same data with*

$$u^1 \in W^{2,2}(0,T;\mathcal{H}_1) \cap L^2(0,T;\mathsf{D}(A_1))\,,$$
$$w^1 \in W^{2,2}(0,T;\mathcal{H}_2) \cap L^2(0,T;\mathsf{D}(A_2))$$

then necessarily $(u, w) = (u^1, w^1)$.

(b) *Conversely, assume there is* $\mu_0 > 0$ *such that the problem*

$$-\mu_0 v + A_1 v = 0, \quad B^* v = 0 \tag{3.73}$$

has a non-zero solution. Then, there exists at least one resonant period for (3.69).

Proof. Before proceeding, we would like to recall that, under the stated assumptions, by using LaSalle's invariance principle one can show that all solutions (u, w) to (3.69) with $f \equiv g \equiv 0$ having initial data in $\mathcal{H}_1 \times \mathcal{H}_2$, decay to zero as $t \to \infty$ in the $\mathcal{H}_1 \times \mathcal{H}_2$-norm; see [24, Theorem 2]. We now pass to the proof. In view of Lemma 3.2 it is enough to consider only f and g of the form (3.71). For each $0 \le |n| \le N$, and each $(f_n, g_n) \in \mathcal{H}_1 \times \mathcal{H}_2$ we look for solutions to the equations

$$-n^2\omega^2\, u_n + A_1 u_n + \lambda_1 B w_n = f_n\,, \quad \imath\, n\, \omega\, w_n + A_2 w_n - \imath\, n\, \omega\, \lambda_2 B^* u_n = g_n\,. \tag{3.74}$$

We shall show that (3.74) has one and only one solution $(u_n, w_n) \in \mathsf{D}(A_1) \times \mathsf{D}(A_2)$. To this end, we set

$$\mathscr{A} = \begin{pmatrix} A_1 & 0 \\ 0 & A_2 \end{pmatrix}, \quad \mathscr{K} = \begin{pmatrix} -n^2\omega^2 & \lambda_1 B \\ -\imath\, n\, \omega\, \lambda_2 B^* & \imath\, n\, \omega \end{pmatrix},$$
$$U_n = \begin{pmatrix} u_n \\ w_n \end{pmatrix}, \quad F_n = \begin{pmatrix} f_n \\ g_n \end{pmatrix},$$

so that (3.74) can be rewritten as

$$\mathscr{A} U_n + \mathscr{K} U_n = F_n \quad \text{in } \mathcal{H}$$

where $\mathcal{H} := \mathcal{H}_1 \times \mathcal{H}_2$. The operator $\mathscr{A} + \mathscr{L}$ can be thought of as defined in the space X with range in $\mathcal{H}$, where

$$X := \left\{ (u_1, u_2) \in \mathsf{D}(A_1) \times \mathsf{D}(A_2) : \sum_{i=1}^{2} \|A_i u_i\|_{\mathcal{H}_i} < \infty \right\}.$$

In view of the stated assumption on A_i, $i = 1, 2$, X is a Hilbert space, endowed with the appropriate scalar product, and, in addition, $\mathscr{A}$ is a homeomorphism of X onto $\mathcal{H}$. Furthermore, again by assumption, it follows that $\mathscr{K}$ is compact. In fact, let $\{(\phi_k, \psi_k)\}$ be a bounded sequence in X. Since A_i has a compact inverse in $\mathcal{H}_i$, $i = 1, 2$, there is a subsequence $\{(\phi_{k'}, \psi_{k'})\}$ that converges strongly in $\mathcal{H}_1 \times \mathcal{H}_2$. Moreover, in view of the assumption on B and B^*, so does $\{(B(\phi_{k'}), B^*(\psi_{k'}))\}$, which proves compactness. Consequently, $\mathscr{A}+\mathscr{K}$ is Fredholm of index 0. Denoting by $\langle\langle\cdot\rangle\rangle_i$ and $\langle\cdot,\cdot\rangle_i$, norm and scalar product in $\mathcal{H}_i$, respectively, from (3.69) with $f_n = g_n = 0$ we infer (with $\lambda := \lambda_2/\lambda_1$)

$$\begin{aligned} &-\imath\, |n|^3\omega^3\lambda\, \langle\langle u_n\rangle\rangle_1 + \imath\, n\, \omega\, \lambda\, \langle A_1(u_n), u_n\rangle_1 + \imath\, n\, \omega\, \lambda_2\langle B(w_n), u_n\rangle_1 = 0 \\ &-\imath\, n\, \omega\langle\langle w_n\rangle\rangle_2 + \langle A_2(w_n), w_n\rangle_2 - \imath\, n\, \omega\, \lambda_2\langle B^*(u_n), w_n\rangle_2 = 0\,. \end{aligned} \tag{3.75}$$

Summing side by side the latter two equations we get $\langle A_2(w_n), w_n\rangle_2 = 0$, which implies $w_n = 0$. Using this information back in (3.69) with $f_n = g_n = 0$, we conclude

$$-\, n^2\omega^2\, u_n + A_1 u_n = 0\,, \quad B^* u_n = 0\,, \tag{3.76}$$

which, by (3.70), furnishes $u_n = 0$. Therefore, $\mathscr{A} + \mathscr{L}$ is a homeomorphism of X onto $\mathcal{H}$, and the proof of existence is completed. The proof of (3.72) is a consequence of Banach closed range theorem. Finally, for uniqueness, after expanding (u^1, w^1) in a time-Fourier series with values in $\mathcal{H}_1 \times \mathcal{H}_2$ (see Lemma 3.2), we deduce that the pair $(u_n - u^1_n, w_n - w^1_n)$ must satisfy (3.72) wth $f_n = g_n = 0$ for all $n \in \mathbb{Z}$. As a consequence, arguing as in a previous part of this proof, we deduce that $(u_n - u^1_n, w_n - w^1_n)$ obeys (3.76), from which, in view of the assumption (3.70), we conclude $(u_n - u^1_n, w_n - w^1_n) = (0, 0)$. Part (a) of the theorem is proved. To show part (b), we have to prove the existence of $T > 0$ and of T-periodic f and g such that problem (3.69) does not admit a T-periodic solution. To this end, we observe that by assumption, μ_0 is an eigenvalue of A_1. We then choose $g = 0$ and $f = \mathrm{e}^{\imath\, \omega_0\, t} f_0$, where $\omega_0 = \sqrt{\mu_0}$ and f_0 is a non-zero solution to (3.73). Now, let (u, w) be a corresponding $\frac{2\pi}{\omega_0}$-periodic solution to (3.69). We can then expand it in a Fourier series and show that the (only nonzero) mode (u_0, w_0) corresponding to ω_0 must satisfy, in particular,

$$-\mu_0\, u_0 + A_1 u_0 + \lambda_1 B w_0 = f_0\,.$$

Taking the $\mathcal{H}_1$-scalar product of both sides of this equation by f_0, and recalling that A_1 is self-adjoint we get

$$\langle u_0, -\mu_0 f_0 + A_1 f_0\rangle_1 + \lambda_1\langle w_0, B^* f_0\rangle_1 = \langle\langle f_0\rangle\rangle_1^2\,.$$

However, by the choice of f_0 and (3.73) the left-hand side of this equation is zero, which furnishes a contradiction and concludes the proof of the theorem. □

The mere strong stability of the semigroup does not appear to be enough to secure the closedness property of $\mathsf{R}(I - U)$ and, therefore, by Theorem 3.3, to exclude the occurrence of resonance. In this respect, there are several characterizations of this property, and we quote the following one due to Browder [11, Lemma 5]

Lemma 3.3. $\mathsf{R}(I - U)$ *is closed in* X *if and only if*

$$\sup_{n\geq 1}\Big\|\sum_{m=1}^{n} U^m x\Big\| < \infty \quad \textit{for all } x \in X\,. \tag{3.77}$$

Using this result, we can give sufficient conditions on the semigroup $U(t)$ in order to verify (3.77). In fact we have the following.

Proposition 3.3. *Suppose that* $U(t)$ *is uniformly stable. Then* $\mathsf{R}(I - U(T))$ *is closed for all* $T > 0$.

Proof. By Lemma 3.1 $U(t)$ satisfies (3.65). As a result, for each $x \in X$ we find

$$\Big\|\sum_{m=1}^{n} U^m x\Big\| \leq M\|x\|\sum_{m=0}^{n} \mathrm{e}^{-m\delta t} = M\frac{1-\mathrm{e}^{-(n+1)\delta t}}{1-\mathrm{e}^{-\delta t}}\|x\| \leq M\,\|x\|\,,$$

and the claim follows from Lemma 3.3. □

We are now in a position to give sufficient conditions for the absence of resonant periods for (3.49).

Theorem 3.6. *Suppose* $U(t)$ *is uniformly stable. Then, given any* $T > 0$ *and a* T*-periodic function* $f \in L^1(0,T;X)$*, there exists one and only one corresponding* T*-periodic function* $x \in C([0,T];X)$ *satisfying* (3.49). *Moreover, there is* $C > 0$ *such that*

$$\max_{[0,T]}\|x(t)\| \leq C\,\|f\|_{L^1(0,T;X)}\,.$$

Proof. From the assumption and Theorem 3.4 we find that $\mathsf{N}(I - U(T)) = \{0\}$ for all $T > 0$, and this, again by assumption and Theorem 3.3 and Proposition 3.3, ensures the existence of T-periodic solutions corresponding to (sufficiently regular) T-periodic f of arbitrary period $T > 0$. The last statement is a consequence of the global invertibility of the operator $I - U(T)$. □

Remark 3.9. As an immediate application of the above finding, and in view of (3.11) we deduce the absence of resonant periods for a linearized elastic material with linear damping. In view of Remark 3.2 the same conclusion holds for one-dimensional thermoelasticity, in agreement with Theorem 3.1.

On the opposite side, we shall now investigate the occurrence of resonance for systems of equations that are (at least) *partially conservative*, namely, where the "energy" $\|U(t)x\|^2$, or "part" of it, is conserved at all times. In this situation we are expecting the existence of at least one resonant period. Actually, as we know from Sects. 3.2 and 3.3, Remark 3.4, this is indeed the case, for example, in linearized elasticity, where, in fact, the whole energy is conserved, and in linear thermoelasticity with periodic boundary conditions, where only a part of the energy is conserved. The next result confirms the above expectation, and makes it precise from the rigorous mathematical viewpoint.

Theorem 3.7. *Let X be a Hilbert space, and let X_0 be a subspaces that is left invariant by $U(t)$. Suppose that $\|U(t)x\| = \|x\|$, for all $t \geq 0$ and all $x \in X_0$. Then, there exists at least one resonant period for* (3.48).

Proof. Without loss, we can take $X_0 = X$. Assume, by contradiction that for any T-periodic force f, (3.48) admits a corresponding T-periodic strong solution. In view of Lemma 3.2, we may then expand both f and x in Fourier series:

$$f(t) = \sum_{k\in\mathbb{Z}} f_k \, e^{\imath\, k\, \omega\, t}\,, \quad x(t) = \sum_{k\in\mathbb{Z}} x_k \, e^{\imath\, k\, \omega\, t}\,.$$

Thus, from (3.48) we deduce that the Fourier coefficients x_k, f_k satisfy the equations

$$k\,\omega\, x_k - L\, x_k = \frac{1}{\imath} f_k\,, \quad k \in \mathbb{Z}\,, \tag{3.78}$$

where $L := \imath\, A$.[14] Since $U(t)$ is a semigroup of isometries, by the Cooper–Phillips theorem [14, Theorem 2(a)], [40, Theorem 1.1.4], L is maximal symmetric. As a result, at least one of the deficiency indices of L has to be zero and, therefore, the spectrum of L must contain (at least) a non-empty subset, $\mathscr{O}$, of the real line [27, p. 271]. We then choose T in such a way that $\overline{k}\omega \in \mathscr{O}$, for some $\overline{k} \in \mathbb{Z}$, and conclude that (3.78) does not have a solution for $k = \overline{k}$, provided we pick $f_{\overline{k}}$ appropriately. In fact, if $\overline{k}\omega$ is an eigenvalue, we take $f_{\overline{k}}$ in the corresponding eigenspace, while if $\overline{k}\omega$ is in the continuous or residual spectrum of L, we take $f_{\overline{k}} \notin \mathsf{R}(\overline{k}\omega\, I - L)$. The proof of the theorem is completed. □

3.5 Some Applications

Objective of this section is to provide several applications of the theory developed in Sect. 3.4 to a number of problems involving hyperbolic–parabolic couplings, including thermo- and magneto-elasticity and certain basic models of liquid–structure interaction.

3.5.1 *Three-Dimensional Linear Thermoelasticity (Revisited)*

As we know from the results of Sect. 3.3, while in the one-dimensional case one eliminates the occurrence of resonance (at least for a class of sufficiently regular

---

[14] Notice that, of course, $x_k \in \mathsf{D}(A)$, because $x(t) \in \mathsf{D}(A)$ for all $t \geq 0$.

data) in three dimensions the validity of an analogous result was left open with Dirichlet boundary conditions, whereas, in fact, in the case of (spatially) periodic boundary condition it was shown that resonance does take place.

However, by Proposition 3.1, we know that the semigroup generated by the linear elasticity equations is strongly stable, provided Ω is such that (3.15) has only the zero solution (and we know that this happens for "most" C^2-domains). Therefore, in view of Theorem 3.4 and Corollary 3.1 we deduce the following.

Theorem 3.8. *Let $\Omega \subset \mathbb{R}^3$ be a bounded domain of class C^2, and assume that problem* (3.15) *has only the trivial solution $\boldsymbol{\psi} = 0$ for all $\sigma > 0$. Then, there exists a dense set $\mathcal{Q} \subset L^1(0,T;[L^2(\Omega)]^3 \times L^2(\Omega))$ such that for any T-periodic $(\boldsymbol{f}, Q) \in \mathcal{Q}$, problem (3.12)–(3.13) has one and only one mild T-periodic solution. Furthermore, for any T-periodic $(\boldsymbol{f}, Q) \in L^1(0,T;[L^2(\Omega)]^3 \times L^2(\Omega))$ and any $\varepsilon > 0$, there is $(\boldsymbol{f}_\varepsilon, Q_\varepsilon) \in L^1(0,T;[L^2(\Omega)]^3 \times L^2(\Omega))$ with*

$$\|(\boldsymbol{f}_\varepsilon, Q_\varepsilon)\|_{L^1(0,T;[L^2(\Omega)]^3 \times L^2(\Omega))} < \varepsilon$$

such that (3.12)–(3.13), with $(\boldsymbol{f}, Q)$ replaced by $(\boldsymbol{f} + \boldsymbol{f}_\varepsilon, Q + Q_\varepsilon)$, has one and only one mild T-periodic solution.

Owing to the special structure of problem (3.12)–(3.13) we are, in fact, able to apply Theorem 3.5(a), and obtain a more precise information about the absence of resonance, as a consequence of the following result.

Theorem 3.9. *Let $\Omega \subset \mathbb{R}^3$ be a bounded domain of class C^2, and assume that problem* (3.15) *has only the trivial solution $\boldsymbol{\psi} = 0$ for all $\sigma > 0$. Moreover, let $\boldsymbol{f}, Q$ be periodic in time with period $T > 0$ and such that*

$$\boldsymbol{f}(x,t) = \sum_{n=-N}^{N} \boldsymbol{f}_n(x) e^{\imath n\omega t}, \quad Q(x,t) = \sum_{n=-N}^{N} Q_n(x) e^{\imath n\omega t}$$

with $N \in \mathbb{N}$, $\omega := 2\pi/T$, and $\boldsymbol{f}_n, Q_n \in L^2(\Omega)$, for all $0 \le |n| \le N$. As we know from Lemma 3.2, *the linear space constituted by the above finite sums is dense in $L^2_\sharp(0,T;L^2(\Omega))$. Then (3.12)–*(3.13) *has a corresponding solution $(\boldsymbol{u}, \theta)$ of period T given by*

$$\boldsymbol{u}(x,t) = \sum_{n=-N}^{N} \boldsymbol{u}_n(x) e^{\imath n\omega t}, \quad \theta(x,t) = \sum_{n=-N}^{N} \theta_n(x) e^{\imath n\omega t}$$

where $(\boldsymbol{u}_n, \theta_n) \in [H^2(\Omega) \cap H^1_0(\Omega)]^2$. Moreover, there exists a constant $C = C(n,\omega,\Omega,\rho,\lambda,\mu,\kappa,\sigma_1,\sigma_2) > 0$ such that

$$\|\boldsymbol{u}_n\|_{2,2} + \|\theta_n\|_{2,2} \le C\left(\|\boldsymbol{f}_n\|_2 + \|Q_n\|_2\right), \quad \textit{for all } 0 \le |n| \le N. \tag{3.79}$$

Finally, if $(\boldsymbol{w}, \tau)$ *is another* T*-periodic solution to* (3.16) *corresponding to the same data with*

$$\boldsymbol{w} \in W^{2,2}(0,T;L^2(\Omega) \cap W^{1,2}(0,T;H^1(\Omega)) \cap L^2(0,T;H^2(\Omega)),$$
$$\tau \in W^{1,2}(0,T;L^2(\Omega)) \cap L^2(0,T;H^2(\Omega))$$

then necessarily $(\boldsymbol{u}, \theta) = (\boldsymbol{w}, \tau)$.

Proof. Equations (3.12)–(3.13) can be written in the form (3.69), for a suitable choice of spaces and operators involved. To this end, we choose $\mathcal{H}_1 := [L^2(\Omega)]^3$, $\mathcal{H}_2 := L^2(\Omega)$. Moreover, we set $A_1 \equiv (1/\rho)\mathscr{L}$, $A_2 \equiv -\kappa\Delta$ with $\mathsf{D}(A_2) \equiv H^2(\Omega) \cap H_0^1(\Omega)$, $B \equiv \nabla$ with $\mathsf{D}(B) \equiv H_0^1(\Omega)$, $B^* \equiv -\mathrm{div}$ with $\mathsf{D}(B^*) \equiv H^1(\Omega)$, $\lambda_1 \equiv \sigma_1/\rho$, and $\lambda_2 \equiv -\sigma_2$. In view of the properties of $\mathscr{L}$ recalled in Sect. 3.2, and with the help of classical results on regularity for the Poisson equation along with Rellich theorem, it is readily established that the operators thus defined satisfy all the assumptions stated in Theorem 3.5. Furthermore, the requirement (3.70) is equivalent to the conditions that problem (3.15) has only the trivial solution for all $\sigma > 0$. The proof is then completed. □

Remark 3.10. If the domain Ω is such that problem (3.15) has a nonzero solution then, in view of Theorem 3.5(b), resonance *does* occur. As we previously noticed,[15] this is indeed the case for "few" but significant three-dimensional domains such as a ball.

Remark 3.11. It is interesting to observe that a result analogous to that of Theorem 3.9 continues to hold for linear, homogeneous poroelastic materials. We recall that the equations governing the dynamics of such materials are given by the classical Biot system [47]

$$\left.\begin{array}{l} \rho\boldsymbol{u}_{tt} - \mu\Delta\boldsymbol{u} - (\lambda+\mu)\nabla(\mathrm{div}\,\boldsymbol{u}) = \alpha_1\nabla p + \rho\boldsymbol{f} \\ p_t - c_0\Delta p = \alpha_2\,\mathrm{div}\,\boldsymbol{u}_t + h \end{array}\right\} \quad \text{in } \Omega\times(0,\infty)\ ,$$

where p is pressure, $c_0 > 0$ is the ratio of hydraulic conductivity to specific heat and α_i, $i = 1, 2$, are coupling constants. By direct inspection we see that the above equations formally coincide with those of linear thermoelasticity (3.12). As a consequence, the result proved in Theorem 3.9, as well as all results proved in Sect. 3.3 for (3.12) hold also for homogeneous, linear poroelastic materials.

3.5.2 *Linear Magnetoelasticity*

Another remarkable application of Theorem 3.4, Corollary 3.1, and Theorem 3.5 regards the occurrence of resonance in conducting non-ferromagnetic elastic bodies

[15]See Footnote 6.

subject to the action of a *constant* magnetic field $\boldsymbol{H}_0$, and whose motion is governed by the equations of magnetoelasticity. If we neglect the displacement current, assume that the material is homogeneous and isotropic and, finally, disregard nonlinear couplings, the above equations can be written as follows [6, Chap. 9]

$$\left.\begin{aligned}&\rho\boldsymbol{u}_{tt}-\mu\Delta\boldsymbol{u}-(\lambda+\mu)\nabla(\operatorname{div}\boldsymbol{u})=\tfrac{1}{\mu_0}\operatorname{curl}\boldsymbol{h}\times\boldsymbol{H}_0+\rho\boldsymbol{f}\\&\boldsymbol{h}_t-\tfrac{1}{\sigma\mu_0}\operatorname{curl}\operatorname{curl}\boldsymbol{h}=\operatorname{curl}(\boldsymbol{u}_t\times\boldsymbol{H}_0)\\&\operatorname{div}\boldsymbol{h}=0\end{aligned}\right\}\ \text{in } \Omega\times(0,\infty)\ , \tag{3.80}$$

where $\boldsymbol{h}$ is the magnetic field, $\sigma>0$ is the conductivity of the material, and $\mu_0>0$ the magnetic permeability of the vacuum. We shall consider the above equation with the following boundary conditions

$$\boldsymbol{u}(x,t)=\boldsymbol{0}\,,\ \ \boldsymbol{n}\cdot\boldsymbol{h}(x,t)=0\,,\ \ \boldsymbol{n}\times\operatorname{curl}\boldsymbol{h}(x,t)=\boldsymbol{0}\,,\ \ (x,t)\in\partial\Omega\times(0,\infty)\,, \tag{3.81}$$

where $\boldsymbol{n}$ is the unit outer normal to $\partial\Omega$.

We shall assume that Ω is of class C^2 and, for simplicity, also *simply connected.*

As in the thermoelastic case, also in the case at hand the relevant Eqs. (3.80)–(3.81) are partially dissipative. In fact, if we set $\boldsymbol{f}\equiv\boldsymbol{0}$, the total energy

$$\mathfrak{E}:=\tfrac{1}{2}\big(\rho\|\boldsymbol{u}_t(t)\|_2^2+\|\boldsymbol{h}(t)\|_2^2+\mu\|\nabla\boldsymbol{u}\|_2^2+(\lambda+\mu)\|\operatorname{div}\boldsymbol{u}\|_2^2\big)\,,$$

is shown to be a decreasing function of time as a consequence of the following equation

$$\frac{d\mathfrak{E}}{dt}=-\frac{1}{\sigma\mu_0}\|\operatorname{curl}\boldsymbol{h}\|_2^2\,.$$

The latter is easily established by dot-multiplying $(3.80)_1$ by $\boldsymbol{u}$, $(3.80)_2$ by $\boldsymbol{h}$ and integrating the resulting equations by parts over Ω. Moreover, in [39] it is shown that (3.80)–(3.81) generates a strongly continuous semigroup of contractions, $\mathfrak{U}(t)$, in the (Hilbert) space, $\mathfrak{H}$, of functions having finite energy, namely

$$\mathfrak{H}:=\big\{(\boldsymbol{u},\boldsymbol{u}_t,\boldsymbol{h})\in[H_0^1(\Omega)]^3\times[L^2(\Omega)]^3\times H(\Omega)\big\}\,, \tag{3.82}$$

with

$$H(\Omega):=\big\{\boldsymbol{h}\in L^2(\Omega):\operatorname{div}\boldsymbol{h}=0\ \ \boldsymbol{h}\cdot\boldsymbol{n}|_{\partial\Omega}=\boldsymbol{0}\big\}\,, \tag{3.83}$$

(where the trace is meant in the sense of $H^{-1/2}(\partial\Omega)$). In [39] it is also shown that $\mathfrak{E}(t)\to 0$, provided only that $\mathfrak{E}(0)<\infty$ and λ,μ satisfy $\lambda+\mu\neq 0$ (in addition to (3.3)). This result provides a rigorous proof of the damping effect of the magnetic field over the free vibration of the elastic material (within the model adopted). In addition, in view of Theorem 3.4 and Corollary 3.1 it also ensures the existence of T-periodic solutions corresponding to "generic" T-periodic loads $\boldsymbol{f}$.

We shall next show that (3.80)–(3.81) can be put in the abstract form (3.69), for an appropriate choice of spaces and operators involved, and that the assumptions of Theorem 3.5 are satisfied. As a consequence, resonance cannot occur, whenever the periodic load has a finite (but, in principle, arbitrarily large) number of modes. To achieve our goal, we begin to choose $\mathcal{H}_2 := [L^2(\Omega)]^3$, and $\mathcal{H}_1 := H(\Omega)$. as in the previous subsection, we set $A_1 := \frac{1}{\rho}\mathscr{L}$, with $\mathscr{L}$ the linearized elasticity operator introduced in Sect. 3.1, and

$$A_2 := \frac{1}{\sigma\mu_0}\,\mathrm{curl}\,\mathrm{curl}$$

with

$$\mathsf{D}(A_2) = \{\boldsymbol{h} \in H^2(\Omega) \cap H(\Omega) : \ \mathrm{curl}\,\boldsymbol{h} \times \boldsymbol{n} = 0 \text{ at } \partial\Omega\}\,.$$

We have $\mathsf{R}(A_2) \subseteq H(\Omega)$. In fact, for an arbitrary $\varphi \in C^\infty(\overline{\Omega})$ and $\boldsymbol{h} \in \mathsf{D}(A_2)$ we show, by integration by parts,

$$\int_\Omega \mathrm{curl}\,\mathrm{curl}\,\boldsymbol{h} \cdot \nabla\varphi = \int_{\partial\Omega} \nabla\varphi \cdot \mathrm{curl}\,\boldsymbol{h} \times \boldsymbol{n} = 0\,,$$

which by [18, Lemma III.2.1] proves the assertion. Actually, A_2 is a homeomorphism of $\mathsf{D}(A_2)$ (endowed with the $H^2(\Omega)$-norm) onto H. This property is a particular case of [20, Theorem 3.2.3]. Moreover, A_2 is strictly accretive. In fact, for all $\boldsymbol{h} \in \mathsf{D}(A_2)$,

$$\int_\Omega \mathrm{curl}\,\mathrm{curl}\,\boldsymbol{h} \cdot \boldsymbol{h} = \int_{\partial\Omega} \boldsymbol{n} \times \mathrm{curl}\,\boldsymbol{h} \cdot \boldsymbol{h} + \int_\Omega |\mathrm{curl}\,\boldsymbol{h}|^2 = \int_\Omega |\mathrm{curl}\,\boldsymbol{h}|^2 \geq 0\,.$$

However, choosing the equality sign in the last step would imply, in view of the assumption on Ω, $\boldsymbol{h} = \nabla\phi$ for some $\phi \in H^1(\Omega)$, which, since $\boldsymbol{h} \in H(\Omega)$ furnishes $\phi = \text{const}$ and $\boldsymbol{h} \equiv \boldsymbol{0}$, thus proving the desired property. We next define

$$B\boldsymbol{h} := \mathrm{curl}\,\boldsymbol{h} \times \boldsymbol{H}_0$$

with

$$\mathsf{D}(B) = \{\boldsymbol{h} \in H(\Omega) : \mathrm{curl}\,\boldsymbol{h} \times \boldsymbol{H}_0 \in [L^2(\Omega)]^3\}$$

and (clearly) $\mathsf{R}(B) \subset [L^2(\Omega)]^3$. Finally, we set

$$B^*\boldsymbol{v} = \mathrm{curl}\,(\boldsymbol{v} \times \boldsymbol{H}_0)$$

with

$$\mathsf{D}(B^*) = \{\boldsymbol{v} \in [L^2(\Omega)]^3 : \mathrm{curl}\,(\boldsymbol{v} \times \boldsymbol{H}_0) \in H(\Omega)\}.$$

We have that $\mathsf{D}(A_1) \subset \mathsf{D}(B^*)$. To show this, is of course enough to show that $H_0^1(\Omega) \subset \mathsf{D}(B^*)$. In turn, to prove the latter, we observe that for any $\varphi \in C^\infty(\overline{\Omega})$ and $\boldsymbol{v} \in H_0^1(\Omega)$, we have by integration by parts,

$$\int_\Omega \operatorname{curl}(\boldsymbol{v} \times \boldsymbol{H}_0) \cdot \nabla\varphi = \int_{\partial\Omega} \boldsymbol{n} \times (\boldsymbol{v} \times \boldsymbol{H}_0) \cdot \nabla\varphi = 0\,,$$

and the property follows from [18, Lemma III.2.1]. By another simple integration by parts, we prove that B^* is the adjoint of B. Furthermore, using the abovementioned functional properties of A_i, $i = 1, 2$, along with Rellich theorem we infer that the compactness properties of B and B^* stated in Theorem 3.5 are secured.

Thus, in order to apply Theorem 3.5(a) to the magnetoelastic case, it remains to investigate under which circumstances (3.70) holds. The latter, in our case, reduces to show that the following problem

$$\left.\begin{array}{l} -\mu\,\Delta\boldsymbol{w} - (\lambda+\mu)\nabla\operatorname{div}\boldsymbol{w} = \gamma^2\boldsymbol{w} \\ \operatorname{curl}(\boldsymbol{w}\times\boldsymbol{H}_0) = \boldsymbol{0} \end{array}\right. \text{ in } \Omega \tag{3.84}$$

with the boundary condition

$$\boldsymbol{w} = \boldsymbol{0} \ \text{ at } \partial\Omega\,, \tag{3.85}$$

has only the solution $\boldsymbol{w} \equiv \boldsymbol{0}$, for all $\gamma \in \mathbb{R}$; see [39]. In [39] and [44] it was shown that, if (in addition to (3.3)) λ, μ are such that $\lambda + \mu \neq 0$, then (3.84)–(3.85) has only the trivial solution.

We thus conclude with the following.

Theorem 3.10. *Let $\Omega \subset \mathbb{R}^3$ be a bounded, simply connected domain of class C^2. The following properties hold.*

(a) *Suppose the Lamé coefficients satisfy ((3.3) and) $\lambda + \mu \neq 0$. Let $\boldsymbol{f}$ be periodic in time with period $T > 0$ and such that*

$$\boldsymbol{f}(x,t) = \sum_{n=-N}^{N} \boldsymbol{f}_n(x)e^{in\omega t}\,,$$

with $N \in \mathbb{N}$, $\omega := 2\pi/T$, and $\boldsymbol{f}_n \in L^2(\Omega)$, for all $0 \le |n| \le N$. As we know from Lemma 3.2, *the linear space constituted by the above finite sums is dense in $L^2_\sharp(0,T;L^2(\Omega))$. Then (3.80)–(3.81) has a corresponding solution $(\boldsymbol{u},\boldsymbol{h})$ of period T given by*

$$\boldsymbol{u}(x,t) = \sum_{n=-N}^{N} \boldsymbol{u}_n(x)e^{in\omega t}\,,\quad \boldsymbol{h}(x,t) = \sum_{n=-N}^{N} \boldsymbol{h}_n(x)e^{in\omega t}$$

where $(\boldsymbol{u}_n, \boldsymbol{h}_n) \in [H^2(\Omega) \cap H_0^1(\Omega)] \times [H^2(\Omega) \cap H(\Omega)]$. *Moreover, there exists a constant* $C = C(n, \omega, \Omega, \rho, \lambda, \mu, \mu_0, \sigma, \boldsymbol{H}_0) > 0$ *such that*

$$\|\boldsymbol{u}_n\|_{2,2} + \|\boldsymbol{h}_n\|_{2,2} \le C\, \|\boldsymbol{f}_n\|_2\,, \quad \textit{for all } 0 \le |n| \le N. \tag{3.86}$$

Finally, if $(\boldsymbol{w}, \boldsymbol{b})$ *is another* T*-periodic solution to* (3.80) *corresponding to the same data with*

$$\begin{aligned}
&\boldsymbol{w} \in W^{2,2}(0,T; L^2(\Omega) \cap W^{1,2}(0,T; H^1(\Omega)) \cap L^2(0,T; H^2(\Omega))\,,\\
&\boldsymbol{b} \in W^{1,2}(0,T; L^2(\Omega)) \cap L^2(0,T; H^2(\Omega))
\end{aligned}$$

then necessarily $(\boldsymbol{u}, \boldsymbol{h}) = (\boldsymbol{w}, \boldsymbol{b})$.

(b) *Suppose the Lamé coefficients satisfy ((3.3) and)* $\lambda + \mu = 0$. *Then, there exists at least one resonant period for (3.80)–(3.81).*

Proof. In view of what we already proved in this section and of Theorem 3.5(a), we have to show only part (b). To this end, without loss we may take $\boldsymbol{H}_0 = H_0\, \boldsymbol{e}_3$, so that, setting $\boldsymbol{u} = (u_3, \boldsymbol{u}')$, Eq. (3.80) with $\boldsymbol{f} \equiv 0$ becomes

$$\left.\begin{aligned}
&\rho\,(u_3)_{tt} - \mu\Delta u_3 = 0\\
&\rho\,\boldsymbol{u}'_{tt} - \mu\Delta\boldsymbol{u}' = \tfrac{1}{\mu_0}\,\text{curl}\,\boldsymbol{h} \times \boldsymbol{H}_0\\
&\boldsymbol{h}_t - \tfrac{1}{\sigma\mu_0}\text{curl}\,\text{curl}\,\boldsymbol{h} = \text{curl}\,(\boldsymbol{u}'_t \times \boldsymbol{H}_0)\\
&\text{div}\,\boldsymbol{h} = 0
\end{aligned}\right\}\ \text{in } \Omega \times (0,\infty)\,, \tag{3.87}$$

Recalling (3.82), from (3.87) it is clear that the space

$$\mathfrak{H}_0 := \big\{(\boldsymbol{u}, \boldsymbol{u}_t, \boldsymbol{h}) \in \mathfrak{H} : \boldsymbol{u} = (u_3, \boldsymbol{0}),\ \boldsymbol{u}_t = ((u_3)_t, \boldsymbol{0}),\ \boldsymbol{h} = 0\}$$

is a subspace of $\mathfrak{H}$ left invariant by the action of the semigroup $\mathfrak{U}(t)$. Moreover, from $(3.87)_1$ and the fact that u_3 vanishes on $\partial\Omega$ at all times, by a simple procedure we infer

$$\rho\|u_3(t)\|_2^2 + \mu\|\nabla u_3(t)\|_2^2 = \rho\|u_3(0)\|^2 + \mu\|\nabla u_3(0)\|_2^2\,, \quad \text{for all } t \ge 0\,,$$

and the desired property follows from Theorem 3.7. □

3.5.3 A Liquid–Structure Interaction Problem Showing Generic Absence of Resonance

In the present and following sections we will investigate the occurrence of resonance within a certain class of linear models of liquid–structure interactions. The motion of the liquid is assumed to be governed by the Stokes (linearized) equations, while that of the structure by the usual Navier (or else membrane) equations, so that we are still in presence of a parabolic–hyperbolic coupling. However, the chief difference between this case and those treated previously is that the coupling (interaction) now takes place at the *interface* liquid–solid, rather than in the *"bulk"* of the structure, and in a totally different manner.

The main characteristic of the models we will consider throughout, besides the latter being a linearized version of the original ones, consists in the fact that the interface liquid–solid is assumed to be *fixed*. Even though such an assumption is at first sight questionable from the physical viewpoint, it can be still considered suitable and, to an extent, useful for the following reason. Typically, strong solutions to liquid–structure interaction problems are constructed around the solution to a set of "approximate equations" obtained by disregarding the nonlinearity and *fixing* the interface as the reference configuration for the structure; see [19] and the reference therein. Thus, our models are exactly described by these "approximate equations" and, therefore, their investigation may lay the foundation for results pertaining to more physically relevant situations. Mathematically, there is a vast amount of literature dedicated to problems of this type, too long to be included here. We limit ourselves to refer to [4, 12] and the bibliography therein.

It is also important to remark that the fact that the liquid–solid interface is kept fixed, brings in an entire one-parameter family of (non-zero) *steady-state solutions in absence of any driving mechanism*, constituting a one-dimensional space of the underlying (Hilbert) space. Being the problems linear, this amounts to say that 0 is an eigenvalue of the infinitesimal generator of the appropriate associated semigroup.

In view of Theorem 3.3(i), this circumstance may have, in principle, a significant bearing on our investigation on the occurrence of resonance, in that any steady-state solution is left invariant by the relevant dynamical semigroup. It is therefore worth spending a few words about this issue. In every incompressible liquid/elastic solid interaction problem, in absence of external forces there is always, regardless of the model, a steady-state solution describing the *equilibrium configuration* of the coupled system characterized by having velocity field of the liquid identically zero (the liquid is at rest), and a constant distribution of pressure (hydrostatic pressure). The latter, in turn, determines a corresponding deformation of the solid that fixes the "shape" of the (constant) volume occupied by the liquid [22]. Clearly, different (constant) hydrostatic pressures may generate different shapes. In the models we shall consider, however, the "shape" of the region occupied by the liquid is *fixed* and, as a consequence, any constant hydrostatic pressure is admissible, thus generating a one-dimensional space of equilibrium solutions corresponding to the *same* configuration. Therefore, in order to avoid this "unphysical" situation, one can work in the orthogonal complement of such solutions, which is left invariant by the semigroup.

The first liquid/structure model we shall consider is described as follows [4, 17]. Let Ω, Ω_S be smooth bounded domains of $\mathbb{R}^3$ with $\overline{\Omega_S} \subset \Omega$. We shall assume that Ω_S is the region occupied by the elastic structure, while the viscous liquid moves within the region $\Omega_F := \Omega - \overline{\Omega_S}$. Moreover, we suppose that the motion of the liquid is slow enough as to apply the Stokes approximation. We thus have

$$\begin{aligned} &\rho \boldsymbol{u}_{tt} - \mu\Delta \boldsymbol{u} - (\lambda + \mu)\nabla \mathrm{div}\, \boldsymbol{u} = \rho \boldsymbol{f} \ \text{ in } \Omega_S \times (0,\infty)\,, \\ &\left.\begin{aligned} &\boldsymbol{v}_t - \mathrm{div}\, \boldsymbol{T}(\boldsymbol{v},p) = \boldsymbol{0} \\ &\mathrm{div}\, \boldsymbol{v} = 0 \end{aligned}\right\} \ \text{in } \Omega_F \times (0,\infty) \end{aligned} \tag{3.88}$$

where $\boldsymbol{v}$, p are velocity and pressure fields of the liquid,

$$\boldsymbol{T}(\boldsymbol{v},p) = -p\mathbf{1} + \nu\big(\nabla\boldsymbol{v} + (\nabla\boldsymbol{v})^\top\big) \tag{3.89}$$

is the Cauchy stress tensor, and $\nu > 0$ the shear viscosity coefficient. To (3.88) we have to append the conditions at the interface $\Gamma := \partial\Omega_S$, of continuity of stress vector and velocity:

$$\boldsymbol{\sigma}(\boldsymbol{u})\cdot\boldsymbol{n} = -\boldsymbol{T}(\boldsymbol{v},p)\cdot\boldsymbol{n}\,, \quad \boldsymbol{v} = \boldsymbol{u}_t \ \text{ on } \Gamma\,, \tag{3.90}$$

where

$$\boldsymbol{\sigma} := \mu(\nabla\boldsymbol{u} + (\nabla\boldsymbol{u})^\top) + \lambda\,(\mathrm{div}\,\boldsymbol{u})\mathbf{1}\,,$$

is the (linearized) Cauchy stress tensor, and $\boldsymbol{n}$ is the unit outer normal to Ω_S, along with the adherence condition for $\boldsymbol{v}$

$$\boldsymbol{v} = \boldsymbol{0} \ \text{ on } \partial\Omega_F - \Gamma\,. \tag{3.91}$$

In [4] it is shown, among other things, that the problem (3.88)–(3.91) defines a strongly continuous semigroup of contractions, $\mathcal{U}(t)$, on the space of "finite energy" $X := \big\{(\boldsymbol{u},\boldsymbol{u}_t,\boldsymbol{v}) \in [H^1(\Omega)]^3 \times [L^2(\Omega)]^3 \times H(\Omega)\big\}$ where, we recall, $H(\Omega)$ is defined in (3.83). Moreover, in [4, Theorem 4.2(ii)] it is also proved that the infinitesimal generator, $\mathcal{A}$, of the semigroup has 0 as an eigenvalue, with corresponding one-dimensional eigenspace. As we mentioned earlier, this is due to the circumstance that the pressure field associated with steady-state solutions to (3.88)–(3.95) is determined only up to a constant, due to the (restrictive) hypothesis that the interface is *fixed*. To see this, if in the above equations with $\boldsymbol{f} \equiv \boldsymbol{0}$ we assume $\boldsymbol{u}, \boldsymbol{v}$ and p independent of t, $(3.88)_1$ and $(3.88)_{2,3}$ *decouple*. In particular, from (3.88), $(3.90)_2$, and (3.91) we get $\boldsymbol{v} \equiv \boldsymbol{0}$, $p = \kappa$, arbitrary $\kappa \in \mathbb{R}$, and $\boldsymbol{u} = \boldsymbol{u}_0(x)$ satisfying the following pure traction problem

$$\begin{aligned} \mathrm{div}\,\boldsymbol{\sigma}(\boldsymbol{u}_0) &= \boldsymbol{0} \ \text{ in } \Omega_S\,, \\ \boldsymbol{\sigma}(\boldsymbol{u}_0)\cdot\boldsymbol{n} &= -\kappa\boldsymbol{n} \ \text{ on } \Gamma\,. \end{aligned}$$

A way of avoiding this "unrealistic" family of solutions is to restrict the study of the evolution to the space orthogonal to $\mathsf{N}(\mathcal{A})$. This is exactly what is done in [3], where it is shown that, in fact, the restriction $\mathcal{U}_0(t)$ of $\mathcal{U}(t)$ to the space $X_0 :=$

$X \ominus \mathsf{N}(\mathcal{A})$ leaves X_0 invariant. Like in the thermoelastic case discussed in Sect. 3.5.1, the asymptotic behavior of $\mathcal{U}_0(t)$ and, in particular, its strong stability depends on whether an overdetermined boundary-value problem admits only the zero solution. More specifically, let

$$\begin{aligned} -\mathrm{div}\, \boldsymbol{\sigma}(\boldsymbol{\psi}) &= \mu\, \boldsymbol{\psi} \ \text{ in } \Omega_S\,, \\ \boldsymbol{\sigma}(\boldsymbol{\psi}) \cdot \boldsymbol{n} &= \kappa\, \boldsymbol{n}\,, \quad \boldsymbol{\psi} = \boldsymbol{0} \ \text{ on } \Gamma\,, \end{aligned} \tag{3.92}$$

where $\mu > 0$ and $\kappa \in \mathbb{R}$. Then, if Ω is such that for *all* $\mu > 0$ problem (3.92) has only the trivial solution $\boldsymbol{\psi} = \boldsymbol{0}$ and $\kappa = 0$, then $\mathcal{U}_0(t)$ is strongly stable [2, 4]. As shown in [3, Appendix] for a similar model problem, there is a large class of domains where the model-analogue of (3.92) has only the zero solution. However, in [3] it is also shown domains, like a ball, where the model-problem is an infinite number of nontrivial solutions.

These considerations lead to the following.

Theorem 3.11. *Let Ω, Ω_S be sufficiently regular domains, and $T > 0$. The following properties hold.*

(a) *Assume that for any given $\mu > 0$ problem (3.92) has only the solution $\boldsymbol{\psi} = \boldsymbol{0}$, $\kappa = 0$. Then, there exists a dense set $\mathcal{Q} \subset L^1(0, T; X_0)$ such that for any T-periodic $\boldsymbol{f} \in \mathcal{Q}$, problem (3.88)–(3.91) has one and only one mild T-periodic solution. Furthermore, for any T-periodic $\boldsymbol{f} \in L^1(0, T; X_0)$ and any $\varepsilon > 0$, there is $\boldsymbol{f}_\varepsilon \in L^1(0, T; X_0)$ with*

$$\|\boldsymbol{f}_\varepsilon\|_{L^1(0,T;X_0)} < \varepsilon$$

such that (3.88)–(3.91), with $\boldsymbol{f}$ replaced by $\boldsymbol{f} + \boldsymbol{f}_\varepsilon$, has one and only one mild T-periodic solution.

(b) *Conversely, assume that there is $\mu_0 > 0$ such that (3.92) has a solution $\boldsymbol{\psi} \not\equiv \boldsymbol{0}$, $\kappa \in \mathbb{R}$. Then, there exists at least one resonant period for (3.88)–(3.91).*

Proof. Part (a) is an immediate consequence of Theorem 3.4 and Corollary 3.1 since, under the given assumption, $\mathcal{U}_0(t)$ is strongly stable. To show part (b) we follow the same lines of the proof of Theorem 3.5(b). Let $(\boldsymbol{\psi}_0, \kappa_0)$ be the solution to (3.92) with $\boldsymbol{\psi}_0 \not\equiv \boldsymbol{0}$ and $\mu = \mu_0 > 0$. Choose $\boldsymbol{f} = \boldsymbol{\psi}_0 e^{\imath \omega_0 t}$. where $\omega_0^2 = \sqrt{\mu_0}/\rho$. If a corresponding $\frac{2\pi}{\omega_0}$-periodic solution $(\boldsymbol{v}, p, \boldsymbol{u})$ to (3.88)–(3.91) exists, then after expanding it in a Fourier series, we find that the (only nonzero) mode $(\boldsymbol{v}_0, p_0, \boldsymbol{u}_0)$ corresponding to the frequency ω_0 must satisfy, in particular, the following equations

$$\left.\begin{aligned} -\mu_0 \boldsymbol{u}_0 - \mathrm{div}\, \boldsymbol{\sigma}(\boldsymbol{u}_0) &= \boldsymbol{\psi}_0 \\ \mathrm{div}\, \boldsymbol{v}_0 &= 0 \end{aligned}\right\} \ \text{in } \Omega_S$$
$$\imath\, \omega_0 \boldsymbol{u}_0 = \boldsymbol{v}_0 \ \text{ on } \Gamma\,. \tag{3.93}$$

Notice that from $(3.93)_{2,3}$ it follows that

$$\int_\Gamma \boldsymbol{u}_0 \cdot \boldsymbol{n} = 0\,. \tag{3.94}$$

We next dot-multiply both sides of $(3.93)_1$ by $\boldsymbol{\psi}_0$ and integrate by parts over Ω_S. Recalling that $(\boldsymbol{\psi}_0, \kappa_0)$ satisfies (3.92), and using $(3.93)_3$ we thus deduce

$$\kappa_0 \int_\Gamma \boldsymbol{u}_0 \cdot \boldsymbol{n} = \|\boldsymbol{\psi}_0\|_2^2\,.$$

However, by (3.94) the left-hand side of this equation vanishes and we reach a contradiction that completes the proof of the theorem. □

3.5.4 Interaction of a Viscous Liquid with a Thin Structure: The Flat Case

The fundamental reason why, in the model described in the previous section, one cannot be sure that resonance does not occur, even when the domain is such that (3.92) has only the trivial solution, has to be ascribed to the fact that the dissipation due to the viscosity of the liquid, is not able to "propagate" from the liquid/solid interface into the bulk of the elastic solid. To explain this issue in more mathematical terms, we observe that setting

$$E(t) := \tfrac{1}{2}\big(\|\boldsymbol{u}_t(t)\|_2^2 + \mu\|\nabla \boldsymbol{u}(t)\|_2^2 + (\lambda+\mu)\|\operatorname{div} \boldsymbol{u}(t)\|_2^2 + \|\boldsymbol{v}(t)\|_2^2\big)\,,$$

from (3.88) to (3.91) with $\boldsymbol{f} \equiv \boldsymbol{0}$ we can readily show that

$$\frac{dE}{dt} = -2\nu \int_{\Omega_F} |\boldsymbol{D}(\boldsymbol{v})|^2\,, \tag{3.95}$$

where $\boldsymbol{D}(\boldsymbol{v})$ is the symmetric part of $\nabla \boldsymbol{v}$. Now, from (3.91) and the second Korn inequality (e.g. [48, Theorem 2]) it follows that

$$\|\boldsymbol{D}(\boldsymbol{v})\|_2 \geq C\,\|\boldsymbol{v}\|_{1,2}\,. \tag{3.96}$$

As a result, by the trace theorem and the continuity condition $(3.90)_2$, we deduce

$$\|\boldsymbol{D}(\boldsymbol{v})\|_2 \geq C\,\|\boldsymbol{u}_t\|_{1/2,2,\Gamma}\,,$$

which, once used into the energy equality (3.95) produces

$$\frac{dE}{dt} \leq -\nu\, K\big(\|\boldsymbol{D}(\boldsymbol{v})\|_{2,\Omega_F}^2 + \|\boldsymbol{u}_t\|_{1/2,2,\Gamma}^2\big)\,,$$

for a suitable choice of the positive constant K. Therefore, the liquid adds a further damping term into the energy equation, which, however, being restricted to the interface Γ, is not able to spread the dissipation also in the interior of the structure, in a way to ensure the *uniform decay* of the energy.

However, in the case of a "thin" structure that can be modeled as a two-dimensional manifold,[16] like a plate or a membrane, the interface liquid/solid *coincides* with the elastic structure, and one may expect that the viscosity effects of the liquid are "strong" enough as to prevent the occurrence of resonance.

In the present and following sections we shall show that this is indeed the case for two prototypical models mostly used in blood flow, and investigated in [12] and [37], respectively. This circumstance is indicative of the fact that *the "thickness" of the elastic wall may act in favor of the event of resonance*. In that regard, we refer the reader to Sect. 3.6, where numerical tests are presented that confirm the theoretical prediction.

In this section we shall consider the case when the "thin" structure is flat (a smooth portion of a plane). This model, introduced in [12], can be roughly regarded as a "drum completely filled with a viscous liquid," and will be specified next. Consider a sufficiently regular domain $\Omega \subset \mathbb{R}^3$ with a connected boundary constituted by two open components, Γ_1 and Γ such that $\Gamma_1 \cap \Gamma = \emptyset$. Moreover, we assume that Γ (the elastic structure) is flat, namely,

$$\Gamma \subset \{x = (x_1, x_2, 0) : \; x' := (x_1, x_2) \in \mathbb{R}^2\},$$

with a smooth boundary $\partial\Gamma$, while Γ_1 is a surface lying in the half-space $\{x_3 \le 0\}$. The domain Ω is completely filled with a viscous liquid that moves in the vanishing Reynolds number approximation, so that its motion is governed by the Stokes equations

$$\left.\begin{array}{l} \boldsymbol{v}_t - \operatorname{div} \boldsymbol{T}(\boldsymbol{v}, p) = \mathbf{0} \\ \operatorname{div} \boldsymbol{v} = 0 \end{array}\right\} \quad \text{in } \Omega \times (0, \infty). \tag{3.97}$$

As for the motion of the "plate" Γ, we assume that it can only undergo transversal displacements $u = u(x, t)$, therefore directed along the x_3 axis. In such a case, the governing equations become (e.g., [29])

$$u_{tt} + \Delta^2 u = -\boldsymbol{e}_3 \cdot \boldsymbol{T}(\boldsymbol{v}, p) \cdot \boldsymbol{e}_3 + f\,, \quad \text{in } \Gamma \times (0, \infty). \tag{3.98}$$

To (3.97)–(3.98) we have to append boundary conditions. As for the liquid, it adheres at the "rigid" as well as the elastic walls:

$$\boldsymbol{v}(x,t) = \mathbf{0}\,, \;\; (x,t) \in \Gamma_1 \times (0,\infty)\,; \;\; \boldsymbol{v}(x',t) = u_t(x',t)\boldsymbol{e}_3\,, \;\; (x',t) \in \Gamma \times (0,\infty)\,, \tag{3.99}$$

---

[16] In the case of three-dimensional flow, or else as a "string," in the two-dimensional case.

whereas the "plate" is clamped at its boundary:

$$u(x',t) = \boldsymbol{n} \cdot \nabla' u(x',t) = 0 \quad (x',t) \in \partial\Gamma \times (0,\infty)\,, \tag{3.100}$$

where ∇' operates only on the x'-variable. In view of $(3.97)_2$ and $(3.99)_2$, the plate equation (3.98) can be simplified. In fact, from (3.89) and $(3.99)_2$ deduce

$$\boldsymbol{e}_3 \cdot \boldsymbol{T}(\boldsymbol{v},p) \cdot \boldsymbol{e}_3 = 2\nu \frac{\partial v_3}{\partial x_3} - p \quad \text{on } \Gamma\,,$$

so that, again by $(3.99)_2$, and $(3.97)_2$ we conclude

$$\boldsymbol{e}_3 \cdot \boldsymbol{T}(\boldsymbol{v},p) \cdot \boldsymbol{e}_3 = -p \quad \text{on } \Gamma\,,$$

and (3.98) becomes

$$u_{tt} + \Delta^2 u = p + f\,, \quad \text{in } \Gamma \times (0,\infty). \tag{3.101}$$

We next observe that also this model, as the one discussed in the previous section, allows for a continuum of steady-state solutions (even) in the case $f \equiv 0$. In fact, if we take $\boldsymbol{v}$, p and u independent of t, say, $\boldsymbol{v} = \boldsymbol{v}_0(x)$, $p = p_0(x)$, $u = u_0(x')$ we deduce from (3.97) to (3.99) (with $f \equiv 0$) that $\boldsymbol{v}_0$ satisfies (3.97) with homogeneous boundary conditions on the whole $\partial\Omega$. Consequently, $\boldsymbol{v}_0 \equiv \boldsymbol{0}$, $p_0 = \kappa$, $\kappa \in \mathbb{R}$, and by (3.101),

$$\Delta^2 u_0 = \kappa\,, \quad \text{in } \Gamma. \tag{3.102}$$

The above observation suggests that the "interesting" dynamics should be restricted to the space orthogonal to the one-dimensional space characterized by $\boldsymbol{v}_0 = \boldsymbol{0}$ and $\Delta^2 u_0 = \text{const}$. To this end, following [12], we introduce the "energy space"[17]

$$\mathcal{H} := \{(\boldsymbol{v}, u.u_t) \in \tilde{H}(\Omega) \times H_0^2(\Gamma) \times \hat{L}^2(\Gamma)\}$$

where

$$\begin{aligned}\tilde{H}(\Omega) &:= \{\boldsymbol{v} \in L^2(\Omega) : \operatorname{div} \boldsymbol{v} = 0\,,\ \boldsymbol{v} \cdot \boldsymbol{n} = 0 \text{ on } \Gamma_1\}\,,\\ \hat{L}^2(\Gamma) &:= \{u \in L^2(\Gamma) :\ (u,1)_\Gamma = 0\}\,,\end{aligned}$$

along with

$$\hat{\mathcal{H}} := \{(\boldsymbol{v}, u.u_t) \in \tilde{H}(\Omega) \times \hat{H}_0^2(\Gamma) \times \hat{L}^2(\Gamma)\},$$

---

[17] Notice that the requirement $u_t \in \hat{L}^2(\Gamma)$ follows from $(3.97)_2$ and $(3.99)_2$.

where

$$\hat{H}_0^s(\Gamma) := H_0^s(\Gamma) \cap \hat{L}^2(\Gamma), \quad s > 0.$$

It is readily seen that $\hat{\mathcal{H}}$ is orthogonal to the one-dimensional space[18]

$$\mathcal{N} := \{(\mathbf{0}, u, 0) \in \mathcal{H} : \; \Delta^2 u = \kappa, \; \kappa \in \mathbb{R}\},$$

and that

$$\mathcal{H} = \hat{\mathcal{H}} \oplus \mathcal{N}. \tag{3.103}$$

Moreover, from [37, Theorem 2.2] it follows, in particular, that (3.97), (3.99)–(3.101) generates a strongly continuous semigroup of contractions in $\mathcal{H}$, $\mathcal{U}(t)$, whose restriction, $\hat{\mathcal{U}}(t)$, to $\hat{\mathcal{H}}$ in view (3.103), must leave $\hat{\mathcal{H}}$ invariant. Finally, in [12, Theorem 3.3] it is shown that $\hat{\mathcal{U}}(t)$ is uniformly stable. Therefore, from Theorem 3.6 we deduce the next theorem which, in particular, rules out the event of resonance for this particular model.

Theorem 3.12. *Let T an arbitrary positive number. For any T-periodic $f \in L^1(0, T; \hat{\mathcal{H}})$, problem (3.97), (3.99)–(3.101) has one and only one mild T-periodic solution.*

3.5.5 Interaction of a Viscous Liquid with a Thin Structure: The Curved Case

The model considered here is similar to that presented in the previous section, the only (main!) difference being that now the elastic part, Γ, of the boundary is no longer assumed flat. Therefore, the relevant equations for the liquid are still given by (3.97), while that of the structure (3.98) are replaced by the following one [37]

$$u_{tt} + Bu = -\boldsymbol{n} \cdot \boldsymbol{T}(\boldsymbol{v}, p) \cdot \boldsymbol{n} + f, \quad \text{in } \Gamma \times (0, \infty), \tag{3.104}$$

where $\boldsymbol{n}$ is the unit outer normal at Γ, while B is a fourth-order differential operator such that the associated bilinear form

$$b(u, w) := \int_\Gamma (Bu)w \tag{3.105}$$

[18] $H_0^2(\Gamma)$ is endowed with scalar product $(\Delta u_1, \Delta u_2)$.

is symmetric, and, in addition, coercive in $H_0^2(\Gamma)$, namely, there is $\alpha > 0$ such that

$$b(u,u) \geq \alpha \|u\|_{2,2,\Gamma}\,, \quad \text{for all } u \in H_0^2(\Gamma)\,. \tag{3.106}$$

Likewise, the boundary conditions (3.99) and (3.100) become

$$\boldsymbol{v}(x,t) = \boldsymbol{0}\,,\ \ (x,t) \in \Gamma_1 \times (0,\infty)\,;\ \ \boldsymbol{v}(x',t) = u_t(x',t)\boldsymbol{n}\,,\ \ (x',t) \in \Gamma \times (0,\infty)\,, \tag{3.107}$$

and

$$u(t) \in \hat{H}_0^2(\Gamma)\,,\quad \text{for all } t \in (0,\infty)\,, \tag{3.108}$$

respectively.

Our objective is to show that, also for this model, the occurrence of resonance is ruled out, thanks to the "thinness" of the structure. More precisely, we have the following result.

Theorem 3.13. *Let $T > 0$ and $f \in W^{1,2}(0,T;L^2(\Gamma))$ be T-periodic. Then, there exists one and only one corresponding T-periodic solution $(\boldsymbol{v}, p, u)$ to (3.104)–(3.108) such that*

$$\begin{aligned} &\boldsymbol{v} \in L^2(0,T;H^2(\Omega)) \cap W^{1,2}(0,T;H^1(\Omega))\,,\quad p \in L^2(0,T;H^1(\Omega)/\mathbb{R}) \\ &u \in L^2(0,T;\hat{H}_0^2(\Gamma)) \cap W^{1,2}(0,T;H^{1/2}(\Gamma)) \cap W^{2,2}(0,T;L^2(\Gamma))\,. \end{aligned}$$

Proof. The method used in [12] for the "flat" case to show the exponential decay of the semigroup—and that leads to the conclusion in Theorem 3.12—is seemingly not directly applicable when the structure is "curved." As a consequence, we shall use a different and more direct approach based on time-Fourier series development and some arguments employed in previous sections. To this end, we formally write

$$\boldsymbol{v}(x,t) = \sum_{k\in\mathbb{Z}} \boldsymbol{v}_k(x)\mathrm{e}^{\imath k\omega t}\,,\quad u(x,t) = \sum_{k\in\mathbb{Z}} u_k(x)\mathrm{e}^{\imath k\omega t}\,,\quad f(x,t) = \sum_{k\in\mathbb{Z}} f_k(x)\mathrm{e}^{\imath k\omega t}$$

with $\omega := 2\pi/T$, and replace these expressions back in (3.97), (3.104), (3.107)–(3.108) to obtain, for all $k \in \mathbb{Z}$,[19]

$$\begin{gathered} \left.\begin{aligned} &\imath\, k\,\omega\, \boldsymbol{v}_k - \nu\Delta\boldsymbol{v}_k + \nabla p_k = \boldsymbol{0} \\ &\operatorname{div} \boldsymbol{v}_k = 0 \end{aligned}\right\} \text{ in } \Omega \\ \boldsymbol{v}_k = \boldsymbol{0} \text{ on } \Gamma_1\,,\quad \boldsymbol{v}_k = \imath\, k\,\omega\, u_k\, \boldsymbol{n} \text{ on } \Gamma \\ -k^2\omega^2 u_k + Bu_k = -\boldsymbol{n}\cdot\boldsymbol{T}(\boldsymbol{v}_k, p_k)\cdot\boldsymbol{n} + f_k \text{ in } \Gamma\,, \\ u_k \in \hat{H}_0^2(\Gamma)\,. \end{gathered} \tag{3.109}$$

---

[19] Here and in the rest of the proof, all Banach spaces are meant over the field $\mathbb{C}$ of complex numbers.

Our next goal is to prove that, for each $k \in \mathbb{Z}$, problem (3.109) has one and only one solution (in appropriate spaces). To this end, let

$$\mathcal{V}(\Omega) := \{\boldsymbol{v} \in H^2(\Omega) : \operatorname{div} \boldsymbol{v} = 0 \text{ in } \Omega\,,\ \boldsymbol{v} = \boldsymbol{0} \text{ on } \Gamma_1\,, \boldsymbol{v} \cdot \boldsymbol{\tau} = 0 \text{ on } \Gamma\}\,,$$

where $\boldsymbol{\tau} = \boldsymbol{\tau}(y)$ is an arbitrary unit vector belonging to the tangent space to Γ at y, and set

$$\begin{aligned}\mathcal{H} &:= \{(\boldsymbol{v}, p, u) \in \mathcal{V}(\Omega) \times H^1(\Omega)/\mathbb{R} \times \hat{H}^2_0(\Gamma) : \ \boldsymbol{v} \cdot \boldsymbol{n} = \imath\, \beta\, u \text{ at } \Gamma\}\\ \mathcal{H}_0 &:= [L^2(\Omega)]^3 \times L^2(\Gamma)\,,\end{aligned}$$

where $\beta \in \mathbb{R}$ is given. We next define the operator $\mathscr{A}$:

$$\mathscr{A} : \begin{pmatrix} \boldsymbol{v} \\ p \\ u \end{pmatrix} \in \mathcal{H} \mapsto \begin{pmatrix} -\nu \Delta \boldsymbol{v} + \nabla p \\ Bu + \boldsymbol{n} \cdot \boldsymbol{T}(\boldsymbol{v}, p) \cdot \boldsymbol{n} \end{pmatrix} \in \mathcal{H}_0\,,$$

and begin to show that $\mathscr{A}$ is a homeomorphism. The latter amounts to say that for any given $(\boldsymbol{G}, f) \in \mathcal{H}_0$ there exists a unique triple $(\boldsymbol{v}, p, u) \in [H^2(\Omega)]^3 \times H^1(\Omega)/\mathbb{R} \times \hat{H}^2_0(\Gamma)$ such that

$$\left.\begin{aligned} -\nu \Delta \boldsymbol{v} + \nabla p &= \boldsymbol{G} \\ \operatorname{div} \boldsymbol{v} &= 0 \end{aligned}\right\} \text{ in } \Omega$$
$$\boldsymbol{v} = \boldsymbol{0} \ \text{ on } \Gamma_1\,, \quad \boldsymbol{v} = \imath\, \beta\, u\, \boldsymbol{n} \ \text{ on } \Gamma \tag{3.110}$$
$$Bu = -\boldsymbol{n} \cdot \boldsymbol{T}(\boldsymbol{v}, p) \cdot \boldsymbol{n} + f \ \text{ in } \Gamma\,.$$

Uniqueness is easily demonstrated. In fact, by dot-multiplying $(3.110)_1$ with $\boldsymbol{G} \equiv \boldsymbol{0}$ by $\overline{\boldsymbol{v}}$,[20] integrating by parts over Ω and then using $(3.110)_{2,3,4}$ and $(3.110)_5$ with $f \equiv 0$, we deduce

$$\nu \|\boldsymbol{D}(\boldsymbol{v})\|_2^2 + \imath\, \beta\, b(u, \overline{u}) = 0\,, \tag{3.111}$$

which, in turn, by Korn inequality (3.96), (3.106), and $(3.110)_3$ furnishes $\boldsymbol{v} \equiv \boldsymbol{0}$, $u \equiv 0$, $\|p\|_{H^1(\Omega)/\mathbb{R}} = 0$. Existence can be established by means of Leray–Schauder principle [50, Theorem 6.A]. In fact, consider the map

$$\mathcal{T} : w \in \hat{H}^{3/2}_0(\Gamma) \mapsto (\boldsymbol{v}, p) \mapsto u \in \hat{H}^{3/2}_0(\Gamma)\,,$$

where

[20] Recall that "—" = c.c.

$$\left.\begin{array}{r}-\nu\Delta\boldsymbol{v}+\nabla p=\boldsymbol{G}\\ \operatorname{div}\boldsymbol{v}=0\end{array}\right\}\ \text{in }\Omega$$
$$\boldsymbol{v}=\boldsymbol{0}\ \text{ on }\Gamma_1\,,\quad \boldsymbol{v}=\imath\,\beta\,w\,\boldsymbol{n}\ \text{ on }\Gamma \tag{3.112}$$
$$Bu=-\boldsymbol{n}\cdot\boldsymbol{T}(\boldsymbol{v},p)\cdot\boldsymbol{n}+f\ \text{ in }\Gamma\,.$$

Existence then follows from the above principle if we show that $\mathcal{T}$ is compact, and that there exists a constant, κ, depending only on the data such that all possible solutions $(\boldsymbol{v},p,u)\in[H^2(\Omega)]^3\times H^1(\Omega)/\mathbb{R}\times\hat{H}_0^2(\Gamma)$ to the problem

$$\left.\begin{array}{r}-\nu\Delta\boldsymbol{v}+\nabla p=\lambda\,\boldsymbol{G}\\ \operatorname{div}\boldsymbol{v}=0\end{array}\right\}\ \text{in }\Omega$$
$$\boldsymbol{v}=\boldsymbol{0}\ \text{ on }\Gamma_1\,,\quad \boldsymbol{v}=\imath\,\beta\,\lambda\,u\,\boldsymbol{n}\ \text{ on }\Gamma \tag{3.113}$$
$$Bu=-\boldsymbol{n}\cdot\boldsymbol{T}(\boldsymbol{v},p)\cdot\boldsymbol{n}+f\ \text{ in }\Gamma\,.$$

with $\lambda\in(0,1)$, obey the a priori estimate

$$\|(\boldsymbol{v},p,u)\|_{[H^2(\Omega)]^3\times H^1(\Omega)/\mathbb{R}\times\hat{H}_0^2(\Gamma)}\leq\kappa\,. \tag{3.114}$$

The compactness property of $\mathcal{T}$ is shown as follows. By well-known results on the Stokes problem [18, Theorem IV.6.1], and the assumption on w and $\boldsymbol{G}$, we prove the existence of a unique solution $(\boldsymbol{v},p)\in[H^2(\Omega)]^3\times H^1(\Omega)/\mathbb{R}$ to $(3.112)_{1,2,3}$. By the trace theorem, we then infer (at least) $\boldsymbol{T}(\boldsymbol{v},p)|_\Gamma\in L^2(\Gamma)$, so that, from $(3.112)_4$ we deduce that u satisfies $Bu=F\in L^2(\Gamma)$. This equation, in turn, in view of (3.107) and the other properties of b, by the Lax–Milgram theorem has one and only one solution $u\equiv\mathcal{T}(w)\in\hat{H}_0^2(\Gamma)\subset\subset\hat{H}_0^{3/2}(\Gamma)$, and the compactness of $\mathcal{T}$ is achieved. In order to prove procedure leading to (3.111) we can show

$$\nu\|\boldsymbol{D}(\boldsymbol{v})\|_2^2+\imath\,\beta\,b(u,\overline{u})=\lambda\left[(\boldsymbol{G},\overline{\boldsymbol{v}})+\imath\,\beta\,(f,\overline{u})\right],$$

from which, recalling (3.107), Korn's and Poincaré's inequalities, and that $\lambda<1$, we deduce

$$\begin{aligned}\|\boldsymbol{v}\|_{1,2}^2&\leq c_1\big[|(\boldsymbol{G},\overline{\boldsymbol{v}})|+|(f,\overline{u})|\big]\\ \|u\|_{2,2,\Gamma}^2&\leq c_2\big[|(\boldsymbol{G},\overline{\boldsymbol{v}})|+|(f,\overline{u})|\big],\end{aligned}$$

for suitable positive c_1,c_2 independent of λ. From these inequalities it easily follows that

$$\|\boldsymbol{v}\|_{1,2}+\|u\|_{2,2,\Gamma}\leq c_3\big(\|\boldsymbol{G}\|_2+\|f\|_2\big)\,, \tag{3.115}$$

for some $c_3>0$. Again applying [18, Theorem IV.6.1] to $(3.112)_{1,2,3}$, we show that $(\boldsymbol{v},p)\in[H^2(\Omega)]^3\times H^1(\Omega)/\mathbb{R}$ along with the estimate:

$$\|\boldsymbol{v}\|_{2,2}+\|p\|_{H^1(\Omega)/\mathbb{R}}\leq c_4\left(\|\boldsymbol{G}\|_2+\|u\|_{2/3,2,\Gamma}\right)$$

where $c_5 > 0$. Then, (3.114) is a consequence of the latter and of (3.115). This concludes the proof that the operator $\mathscr{A}$ is a homeomorphism. We next observe that setting

$$\boldsymbol{V}_k := \begin{pmatrix} \boldsymbol{v}_k \\ p_k \\ u_k \end{pmatrix}, \quad \boldsymbol{f} := \begin{pmatrix} \boldsymbol{0} \\ f \end{pmatrix}$$

problem (3.109) can be written in the following abstract form

$$\mathscr{A}\boldsymbol{V}_k + \mathscr{K}\boldsymbol{V}_k = \boldsymbol{f}\,, \text{ in } \mathcal{H}_0\,, \;\; k \in \mathbb{Z}\,, \tag{3.116}$$

where $\beta := k\omega$,[21] and

$$\mathscr{K} : \boldsymbol{V}_k \in \mathcal{H} \mapsto \begin{pmatrix} \imath\, k\, \omega\, \boldsymbol{v}_k \\ -k^2\omega^2 u_k \end{pmatrix} \in \mathcal{H}_0\,.$$

Clearly, by Rellich's theorem, $\mathscr{K}$ is compact, so that $\mathscr{M} := \mathscr{A} + \mathscr{K}$ is Fredholm of index 0. As a result, the surjectivity of $\mathscr{M}$ and, therefore, the existence of solutions to (3.109) for all $k \in \mathbb{Z}$ and any given $f_k \in L^2(\Gamma)$, is secured if and only if we can show $\mathsf{N}(\mathscr{M}) = \{\boldsymbol{0}\}$. This property is equivalent to show that $(\boldsymbol{v} = \boldsymbol{0}, \nabla p = \boldsymbol{0}, u = 0)$ is the only solution in $[H^2(\Omega)]^3 \times H^1(\Omega)/\mathbb{R} \times \hat{H}^2_0(\Gamma)$ to the following problem

$$\left.\begin{array}{c} \imath\, k\, \omega\, \boldsymbol{v} - \nu\Delta\boldsymbol{v} + \nabla p = \boldsymbol{0} \\ \operatorname{div} \boldsymbol{v} = 0 \end{array}\right\} \text{ in } \Omega$$
$$\boldsymbol{v} = \boldsymbol{0} \text{ on } \Gamma_1\,, \quad \boldsymbol{v} = \imath\, k\, \omega\, u\, \boldsymbol{n} \text{ on } \Gamma \tag{3.117}$$
$$-k^2\omega^2 u + Bu + \boldsymbol{n}\cdot\boldsymbol{T}(\boldsymbol{v}, p)\cdot\boldsymbol{n} = 0 \text{ in } \Gamma\,,$$

for all $k \in \mathbb{Z}$. However, this is easily established. Actually, if $k = 0$, by classical results on the Stokes problem, from $(3.117)_{1,2,3}$ we immediately get $\boldsymbol{v} \equiv \nabla p \equiv \boldsymbol{0}$, which, in turn, once replaced in $(3.117)_5$ delivers $Bu = \kappa$ on Γ, $\kappa \in \mathbb{R}$. Multiplying both sides of this equation by u, integrating over Γ and recalling that $(u, 1)_\Gamma = 0$, by (3.106) we deduce $u \equiv 0$, which completes the proof if $k = 0$. If $k \neq 0$, we dot-multiply both sides of $(3.117)_1$ by $\boldsymbol{v}$, integrate by parts over Ω and then use the other equations in (3.117) to show

$$\imath\, k\, \omega\, \|\boldsymbol{v}\|_2^2 + \nu\|\boldsymbol{D}(\boldsymbol{v})\|_2^2 - \imath\, k^3\omega^3\|u\|_{2,\Gamma}^2 + \imath\, k\, \omega\, b(u, \overline{u}) = 0\,. \tag{3.118}$$

From this and (3.96) we at once deduce $\boldsymbol{v} \equiv \boldsymbol{0}$ that, by $(3.117)_{1,4}$ gives $\nabla p \equiv \boldsymbol{0}$, $u \equiv 0$. This finding completes the proof of the desired property, that is, for any $f_k \in L^2(\Gamma)$ and any $k \in \mathbb{N}$, problem (3.109) admits one and only one solution $(\boldsymbol{v}_k, p_k, u_k) \in [H^2(\Omega)]^3 \times H^1(\Gamma)/\mathbb{R} \times \hat{H}^2_0(\Gamma)$. Our next goal is to prove estimates of this solution in terms of the data with the involved constants independent of

[21] See the definition of the space $\mathcal{H}$.

k. In what follows, we denote by C_i, $i \in \mathbb{N} \cup \{0\}$, generic positive constants possessing this property. We commence to observe that, if $k = 0$, from well-known results on the homogeneous Stokes problem and the property of the bilinear form b, from (3.109) we readily show

$$\boldsymbol{v}_0 = \nabla p_0 = \boldsymbol{0}\,, \quad \|u_0\|_{2,2,\Gamma} \leq C_0 \,\|f_0\|_{2,\Gamma}\,. \tag{3.119}$$

If $k \neq 0$, proceeding exactly as in the proof of (3.118), from $(3.109)_{1-5}$ we conclude

$$\imath\, k\,\omega\,\|\boldsymbol{v}_k\|_2^2 + \nu\,\|\boldsymbol{D}(\boldsymbol{v}_k)\|_2^2 - \imath\, k^3\omega^3\,\|u_k\|_{2,\Gamma}^2 + \imath\, k\,\omega\, b(u_k,\overline{u_k}) = (f_k,\overline{u_k})_\Gamma\,, \tag{3.120}$$

which, by Korn's, Poincaré's, and Schwarz's inequalities implies

$$\|\boldsymbol{v}_k\|_{1,2}^2 \leq C_1\,\|f_k\|_{2,\Gamma}\,\|u_k\|_{2,\Gamma}\,.$$

Moreover, by $(3.109)_4$ and the trace theorem, we have

$$|k|\;\|u_k\|_{1/2,2,\Gamma} \leq C_2\,\|\boldsymbol{v}_k\|_{1,2} \tag{3.121}$$

that once combined with the previous inequality furnishes

$$|k|^2\;\|u_k\|_{1/2,2,\Gamma} + |k|\;\|\boldsymbol{v}_k\|_{1,2} \leq C_3\,\|f_k\|_{2,\Gamma}\,. \tag{3.122}$$

If we employ (3.122) into (3.120), with the help of (3.106), and recalling that $|k| \geq 1$ we show that

$$\|u_k\|_{2,2,\Gamma} \leq C_4\,\|f_k\|_{2,\Gamma}\,. \tag{3.123}$$

Finally, combining (3.123) with classical estimates for solutions to the Stokes system $(3.109)_{1-4}$, we conclude

$$\|\boldsymbol{v}_k\|_{2,2} + \|p_k\|_{H^1(\Omega)/\mathbb{R}} \leq C_5\,|k|\;\|f_k\|_{2,\Gamma}\,. \tag{3.124}$$

The existence result stated in the theorem is then a direct consequence of (3.119), (3.122)–(3.124) and of Plancharel theorem. Finally, uniqueness is discussed exactly as in Theorem 3.5, and its proof will therefore be omitted. □

3.6 Numerical Experiments

Objective of this section is to present some numerical experiments aimed at investigating the "generic" absence of resonance in liquid–structure models of the type considered in Sect. 3.5.3 (see Theorem 3.11(a)). The idea is that if the absence of resonance is only "generic," namely, for T-periodic forces only a dense set of the

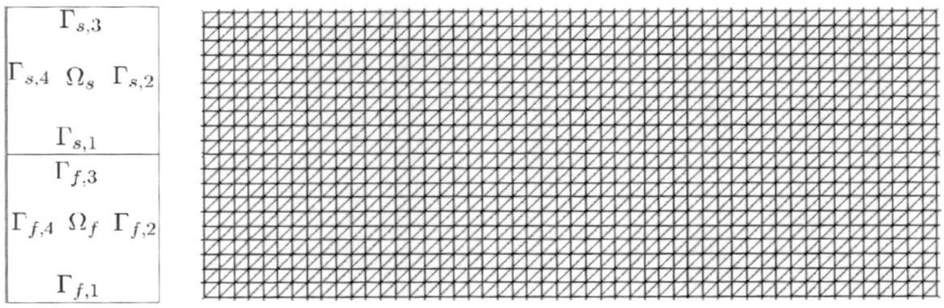

Fig. 3.1 An idealized sketch of the domains Ω_f, Ω_s and of the corresponding partition of their boundaries is reported on the *left*. The actual configuration of the domains as well as the computational grids used in the simulations is shown on the *right*

data space, we should be able to find (at least) one particular force f and a sequence $\{f_N\}$ with the following property: The total energy of the unique T-periodic solution corresponding to f_N should increase as $N \to \infty$. However, our finding does not seem to corroborate this view, at least for the type of forces that we have considered (squared profile). The only (expected) phenomenon observed is that the oscillation energy of the solid may increase as the viscosity of the liquid decreases.

For the sake of clarity, we recall below the precise problem formulation, including initial, boundary, and interface conditions. For practical reasons, we slightly modify the definition of the fluid Cauchy stress using the dynamic viscosity, namely $\boldsymbol{T}(\boldsymbol{v}, p) = \mu_f(\nabla\boldsymbol{v}+(\nabla\boldsymbol{v})^\top - p\mathbf{1}$. Then, the problem solved in the numerical simulation is the following,

$$\begin{array}{ll}
\rho_s\boldsymbol{u}_{tt} - \mu_s\Delta\boldsymbol{u} - (\lambda_s + \mu_s)\nabla\mathrm{div}\,\boldsymbol{u} = \rho_s\boldsymbol{f} & \text{in } \Omega_s \times (0,\infty) \\
\rho_f\boldsymbol{v}_t - \nabla\cdot\boldsymbol{T}(\boldsymbol{v},p) = \mathbf{0}, \quad \mathrm{div}\,\boldsymbol{v} = 0 & \text{in } \Omega_f \times (0,\infty) \\
\boldsymbol{T}(\boldsymbol{v},p)\cdot\boldsymbol{n} = \boldsymbol{\sigma}(\boldsymbol{u})\cdot\boldsymbol{n}, \quad \boldsymbol{v} = \boldsymbol{u}_t & \text{on } \Gamma \times (0,\infty) \\
\boldsymbol{u} = \mathbf{0} & \text{on } \Gamma_{s,2} \cup \Gamma_{s,4} \times (0,\infty) \\
\boldsymbol{\sigma}(\boldsymbol{u})\cdot\boldsymbol{n} = \mathbf{0} & \text{on } \Gamma_{s,3} \times (0,\infty) \\
\boldsymbol{T}(\boldsymbol{v},p)\cdot\boldsymbol{n} = \mathbf{0} & \text{on } \Gamma_{f,2} \cup \Gamma_{f,4} \times (0,\infty) \\
\boldsymbol{v}\cdot\boldsymbol{n} = 0, \quad \boldsymbol{t}\cdot\boldsymbol{T}(\boldsymbol{v},p)\cdot\boldsymbol{n} = 0 & \text{on } \Gamma_{f,1} \times (0,\infty) \\
\boldsymbol{u} = \boldsymbol{u}_t = \mathbf{0} & \text{in } \Omega_s \times \{0\} \\
\boldsymbol{v} = \mathbf{0} & \text{in } \Omega_f \times \{0\}
\end{array} \tag{3.125}$$

The fluid and solid domains are represented by two rectangular adjacent regions, of vertical and horizontal sides a, b, respectively. We refer to Fig. 3.1 for a sketch of the domains and a precise definition of the boundaries $\Gamma_{f,i}$, $\Gamma_{s,i}$, $i = 1,\dots,4$. We observe that the interface Γ coincides with $\Gamma_{s,1}$ and $\Gamma_{f,3}$. These domains are embedded in $\mathbb{R}^2$, endowed with a Cartesian coordinate system x, y with origin in the bottom left corner of Ω_f.

As discussed in Sect. 3.2, resonance effects emerge only when some precise vibration frequencies of the system are excited. For the particular case of the linear elasticity equation on a rectangular domain, the characteristic frequencies and the corresponding modes of vibration can be analytically determined by means of separation of variables. At the boundaries, the structure is clamped on the left and right sides and subject to zero traction forces on the upper and lower sides, in agreement with the boundary conditions of system (3.125). In order to facilitate the calculations, it is convenient to drop the term $(\lambda_s + \mu_s)\nabla \text{div}\boldsymbol{u}$ in the elasticity equation. This can be easily done by adopting a particular choice of the second Lamé coefficient λ_s, that is $\lambda_s = -\mu_s$. As a result, the operator $\rho_s \boldsymbol{u}_{tt} - \mu_s \Delta \boldsymbol{u}$ can be split and analyzed independently for the x and y components of the displacement and the characteristic modes of vibration of each component are,

$$u_{m,n}(x, y, t) = \psi_{m,n}(x, y)\phi_{m,n}(t),$$
$$\psi_{m,n}(x, y) = \sin\left(\tfrac{m\pi}{a}x\right)\cos\left(\tfrac{n\pi}{b}y\right), \quad \phi_{m,n}(t) = \sin\left(2\pi\frac{t}{T_{m,n}}\right).$$

The characteristic period of each vibration mode can be calculated as

$$T_{m,n} = \frac{2\pi}{\lambda_{m,n}}\sqrt{\frac{\rho_s}{\mu_s}}, \quad \lambda_{m,n}^2 = \left(\frac{m\pi}{a}\right)^2 + \left(\frac{n\pi}{b}\right)^2.$$

The fundamental vibration mode is obtained by setting $m = n = 1$ and the corresponding period can be quantified as follows, in terms of the physical parameters of the problem,

$$T = \frac{2ab}{\sqrt{a^2 + b^2}}\sqrt{\frac{\rho_s}{\mu_s}}.$$

According to these considerations, to excite the fundamental vibration mode of the structure, the right-hand side $\boldsymbol{f}$ of the elasticity equation in (3.125) is defined as follows:

$$\boldsymbol{f} = \{0, g_N(t)\sin(\pi x)\}, \quad g_N(t) = \sum_{n=1}^{N} \frac{2}{\pi n}\sin\left(\frac{\pi}{2}n\right)\cos\left(2\pi n\frac{t}{T}\right), \tag{3.126}$$

where $g_N(t)$ is a truncated Fourier series of functions whose frequency is equal or multiple of the fundamental one. To investigate the impact of Theorem 5.4 on the occurrence of resonance, we design $g_N(t)$ such that in the limit $N \to \infty$ it approximates a singular function. In the case of (3.126), the limit function (almost everywhere) is a T-periodic square wave $g(t) \in \{-\frac{1}{2}, \frac{1}{2}\}$, as shown in Fig. 3.2. For a finite number of modes $g_N(t)$ belongs to the dense set $\mathcal{D}$. As explained at the beginning of this section, according to the generic absence or resonance we posit that for $N \to \infty$ the forcing term $\boldsymbol{f}$ may no longer belong to this special set in which resonance is excluded.

For the numerical simulations we use the finite element method for the spatial approximation and a finite difference scheme to discretize the time derivatives. After defining an uniform, triangular grid on each subdomain, shown in Fig. 3.1, such

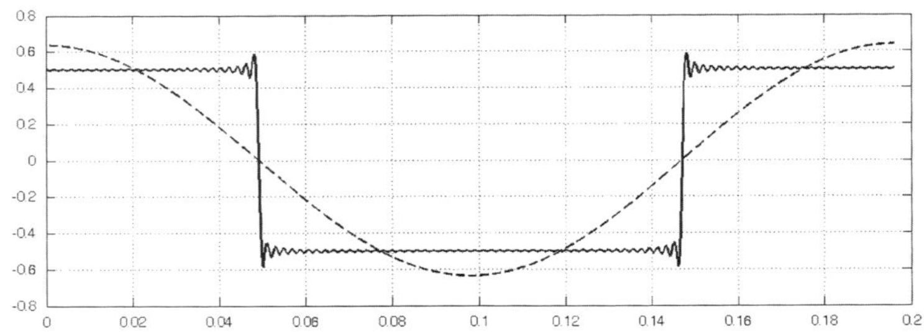

Fig. 3.2 Plots of the functions $g_1(t)$ and $g_{101}(t)$ used to excite the structure vibration modes (*dashed line* and *continuous line*, respectively)

that the conformity of grid edges is preserved at the interface, we discretize the fluid velocity field using piecewise quadratic finite elements, while the pressure is approximated by means of piecewise linear functions. These are the classical Taylor-Hood mixed finite elements. The structure displacement is discretized with the same finite elements used for the velocity field, in order to facilitate the approximation of the interface conditions and in particular the exchange of discrete functions between the two subdomains.

The main difficulty at the numerical level is to appropriately approximate the interaction between the fluid and the structure. This is a well-explored topic in the recent numerical literature. In particular, we apply here the algorithm proposed and analyzed in [10]. The distinctive feature of this algorithm consists in the weak enforcement of the interface conditions between the fluid and the structure. It is particularly interesting to our purpose because it allows to satisfy two important properties, usually incompatible for fluid-structure interaction methods: (1) unconditional stability with respect to the time and space approximation steps; (2) the ability to decouple the solution of the fluid and structure problems at each time step. For this reason, it is classified as a loosely coupled algorithm. Using this algorithm, we simulate the interaction of the fluid and the structure for a time interval equivalent to 40 periods of the fundamental vibration frequency of the structure. The time discretization step is appropriately chosen to guarantee an accurate approximation of each cycle.

The fundamental mechanism governing the possible occurrence of resonance in fluid-structure interaction is the ability of the elasticity equation to preserve the energy of the structure, competing with the dissipative nature of the adopted viscous flow model. For this reason, the numerical experiments are focused to investigate the behavior of the energy, when the viscosity of the fluid is varying. In particular, the energy of the fluid-structure system consists of the following terms:

$$E_f := \tfrac{1}{2}\rho_f \|\boldsymbol{v}\|^2_{L^2(\Omega_f)},$$
$$E_s' := \tfrac{1}{2}\rho_s \|\boldsymbol{u}_t\|^2_{L^2(\Omega_s)}, \quad E_s'' := \tfrac{1}{2}\left(\mu_s \|\nabla \boldsymbol{u}\|^2_{L^2(\Omega_s)} + \lambda_s \|\mathrm{div}\boldsymbol{u}\|^2_{L^2(\Omega_s)}\right),$$

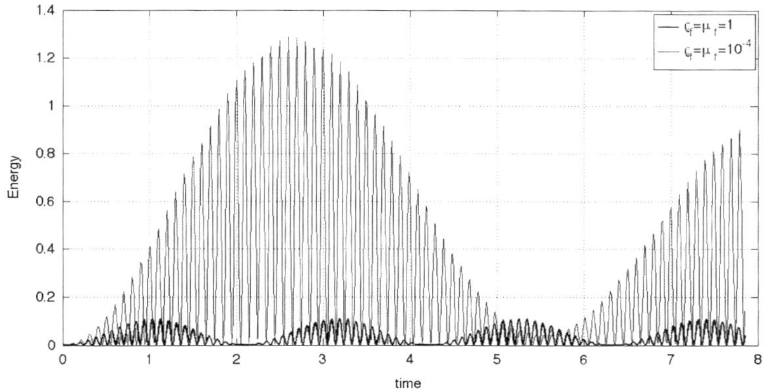

Fig. 3.3 Time evolution of the structure kinetic energy E_s' for different values of ρ_f, μ_f and for different functions $g_1(t)$, $g_{101}(t)$ defining the oscillating load

which represent the fluid kinetic energy, structure kinetic energy, and stored energy, respectively.

Figure 3.3 shows the temporal evolution of E_s' for different fluid properties, while the structure mechanical parameters remain unchanged. In particular we adopt $\mu_s = 100$, $\lambda_s = -\mu_s$, while the fluid density and viscosity decrease from $\rho_f = 1, \mu_f = 1$ to $\rho_f = 10^{-4}, \mu_f = 10^{-4}$. The force $\boldsymbol{f}$ is characterized by a single vibration mode, basically we choose $\boldsymbol{f} = \{0, g_1(t)\sin(\pi x)\}$, where $g_1(t)$ is proportional to $\cos\left(2\pi n \frac{t}{T}\right)$. The simulations confirm that the structure is subject to a periodic oscillatory motion under the action of the forcing term $\boldsymbol{f}$. However, the amplitude of the oscillations is not constant in time, but it features a periodic trend as well. Indeed, this is due to the competing role of the conservation and dissipation properties of structure and fluid, respectively. The energy of the structure tends to monotonically increase under the action of the force. In absence of dissipation, the structure displacement will eventually tend to an unbounded motion, that is the occurrence of resonance. The dissipation due to the viscous liquid acts against the latter, in such a way that the larger is the displacement of the structure, the more relevant becomes the dissipative effect. Figure 3.3 suggests that the system reaches a dynamic equilibrium between these opposing trends. The numerical investigation also confirms that the displacement of the structure significantly increases when the density and viscosity of the fluid decrease. Again, this confirms the fundamental role of viscous effects in the control of resonance. Finally, we study the case where the number of vibration modes contributing to $\boldsymbol{f}$ is increased up to $N = 101$. In practice, we switch from $g_1(t)$ to $g_{101}(t)$ in equation (3.126), as illustrated in Fig. 3.2. The perturbation due to the high frequency modes doesn't introduce any significant changes in the time evolution of E_s' reported in Fig. 3.3. Unfortunately, this result does not inform us on the impact of Theorem 5.4 on the occurrence of resonance, because it is open to two possible interpretations. On the one hand, it is possible that the sequence $g_N(t)$ is not an appropriate example to capture the effect of forcing terms that do not belong to the dense set $\mathscr{Q}$ defined in Theorem 3.11 The identification of functions in the complementary of this set and their appropriate approximation in the framework of a fluid-structure interaction simulator is a very challenging task, indeed. On the other hand, the current analysis may be still incomplete, and the occurrence of resonance for model (3.125) may be excluded for reasons that go beyond the theory established by Theorem 3.11.

Acknowledgements The work of G.P. Galdi and P. Zunino was partially supported by NSF DMS Grant-1311983.

References

1. W. Arendt, J.K. Batty, Tauberian theorems and stability of one-parameter semigroups. Trans. Am. Math. Soc. **306**, 837–852 (1988)
2. G. Avalos, The strong stability and instability of a fluid–structure semigroup. Appl. Math. Optim. **55**, 63–184 (2007)
3. G. Avalos, R. Triggiani, The coupled PDE system arising in fluid/structure interaction. I. Explicit semigroup generator and its spectral properties, in *Fluids and Waves. Contemporary Mathematics*, vol. 440 (American Mathematical Society, Providence, 2007), pp. 15–54
4. G. Avalos, R. Triggiani, Semigroup well-posedness in the energy space of a parabolic-hyperbolic coupled Stokes-Lamé PDE system of fluid-structure interaction. Discrete Continuous Dyn. Syst. Ser. S **2**, 417–447 (2009)
5. A.V. Balakrishnan, *Applied Functional Analysis*, 2nd edn. (Springer, New York, 1981)
6. D.I. Bardzokas, M.L. Filshtinsky, L.A. Filshtinsky, *Mathematical Methods in Electro-Magneto-Elasticity*. Lecture Notes in Applied and Computational Mechanics, vol. 32 (Springer, New York, 2007)
7. H. Beirão da Veiga, On the existence of strong solutions to a coupled fluid- structure evolution problem. J. Math. Fluid Mech. **6**, 21–52 (2004)
8. C.A. Berenstein, An inverse spectral theorem and its relation with the Pompeiu problem. J. Anal. Math. **37**, 128–144 (1980)
9. K. Billah, R. Scanlan, Resonance, tacoma narrows bridge failure, and undergraduate physics textbooks. Am. J. Phys. **59**, 118–124 (1991)
10. E. Burman, M.A. Fernández, Stabilization of explicit coupling in fluid-structure interaction involving fluid incompressibility. Comput. Methods Appl. Mech. Eng. **198**, 766–784 (2009)
11. F.E. Browder, On the iteration of transformations in noncompact minimal dynamical systems. Proc. Am. Math. Soc. **9**, 773–780 (1958)
12. I. Chueshov, I. Ryzhkova, A global attractor for a fluid-plate interaction model. Commun. Pure Appl. Anal. **12**, 1635–1656 (2013)
13. P.G. Ciarlet, *Mathematical Elasticity Vol. I. Three-Dimensional Elasticity*. Studies in Mathematics and its Applications, vol. 20 (North-Holland, Amsterdam, 1988)
14. L.J.B. Cooper, One-parameter semigroups of isometric operators in Hilbert space. Ann. Math. **48**, 827–842 (1947)
15. C. Dafermos, On the existence and the asymptotic stability of solutions to the equations of linear thermoelasticity. Arch. Ration. Mech. Anal. **29**, 241–271 (1968)
16. W.A. Day, Steady forced vibrations in coupled thermoelasticity. Arch. Ration. Mech. Anal. **93**, 323–334 (1986)
17. Q. Du, M.D. Gunzburger, L.S. Hou, J. Lee, Analysis of a linear fluid-structure interaction problem. Discrete Continuous Dyn. Syst. **9**, 633–650 (2003)

18. G.P. Galdi, *An Introduction to the Mathematical Theory of the Navier-Stokes Equations: Steady-State Problems* (Springer, New York, 2011)
19. G.P. Galdi, R. Rannacher (eds.), *Fundamental Trends in Fluid-Structure Interaction*. Contemporary Challenges in Mathematical Fluid Dynamics and its Applications, vol. 1 (World Scientific Publishing, Singapore, 2010)
20. V. Georgescu, Some boundary value problems for differential forms on compact Riemannian manifolds. Ann. Mat. Pura Appl. **122**, 159–198 (1979)
21. I. Gohberg, S. Goldberg, M.A. Kaashoek, *Basic Classes of Linear Operators* (Birkhäuser Verlag, Basel, 2003)
22. C. Grandmont, Existence for a three-dimensional steady state fluid-structure interaction problem. J. Math. Fluid Mech. **4**, 76–94 (2002)
23. J.K. Hale, Dynamical systems and stability. J. Math. Anal. Appl. **26**, 39–59 (1969)
24. D.B. Henry, A. Perissinitto, O. Lopes, On the essential spectrum of a semigroup of thermoelasticity. Nonlinear Anal. **21**, 65–75 (1993)
25. T.P. Hytönen, M.T Lacey, Pointwise convergence of vector-valued Fourier series. Math. Ann. **357**, 1329–1361 (2013)
26. J. Ignaczak, M. Ostoja-Starzewski, *Fundamentals of Linear Thermoelasticity with Finite Wave Speeds*. Oxford Mathematical Monographs (Oxford Univeristy Press, Oxford, 2009)
27. T. Kato, *Perturbation Theory for Linear Operators*. Classics in Mathematics (Springer, Berlin, 1995)
28. V.D. Kupradze, T.G. Gegelia, M.O. Basheleishvili, T.V. Burchuladze, Three-dimensional problems of the mathematical theory of elasticity and thermoelasticity, in *North-Holland Series in Applied Mathematics and Mechanics*, vol. 25, ed. by V.D. Kupradze (North-Holland, Amsterdam/New York, 1979)
29. J. Lagnese, J.L. Lions, *Modeling, Analysis and Control of Thin Plates* (Masson, Paris, 1988)
30. I. Lasiecka, M. Renardy, R. Triggiani, Backward uniqueness for thermoelastic plates with rotational forces. Semigroup Forum **62**, 217–242 (2001)
31. G. Lebeau, E. Zuazua, Decay rates for the three-dimensional linear system of thermoelasticity. Arch. Ration. Mech. Anal. **148**, 179–231 (1999)
32. R. Leis, Anfangs-Randwertaufgaben in der Theorie thermoelastischer Schwingungen. Z. Angew. Math. Mech. **64**, 12–17 (1984)
33. M. Mohebbi, J.C. Oliveira, Existence of time-periodic solutions for a magnetoelastic system in bounded domains. J. Elast. **113**, 113–133 (2013)
34. P.H. Mott, C.M. Roland, Limits to Poisson's ratio in isotropic materials -general result for arbitrary deformation. Phys. Scr. **87**, 055404 (2013)
35. J.E. Muñoz Rivera, Energy decay rates in linear thermoelasticity. Funkcialaj Ekvacioj **35**, 19–30 (1992)
36. T. Naito, J.-S. Shin, N.-V. Minh, Periodic solutions of linear differential equations. *Qualitative theory of functional equations and its application to mathematical science (Japanese) (Kyoto, 2000)*. Surikaisekikenkyusho Kokyuroku **1216**, 78–89 (2001)
37. A. Osses, J. Puel, Approximate controllability for a linear model of fluid structure interaction. ESIAM Control Optim. Calc. Var. **4**, 497–513 (1999)
38. A. Pazy, *Semigroups of Linear Operators and Applications to Partial Differential Equations* (Springer, New York/Berlin, 1983)
39. G. Perla-Menzala, E. Zuazua, Energy decay of magnetoelastic waves in a bounded conductive medium. Asymptotic Anal. **18**, 349–362 (1998)
40. R.S. Phillips, Dissipative operators and hyperbolic systems of partial differential equations. Trans. Am. Math. Soc. **90**, 193–254 (1959)
41. J. Prüss, On the spectrum of C_0-semigroups. Trans. Am. Math. Soc. **284**, 847–858 (1984)
42. R. Racke, Exponential decay for a class of initial boundary value problems in thermoelasticity. Comput. Appl. Math. **12**, 67–80 (1993)
43. M. Slemrod, E.F. Infante, An invariance principle for dynamical systems on Banach space: application to the general problem of thermoelastic stability, in *Instability of Continuous Systems, IUTAM Symposium*, Herrenalb, 1969 (Springer, Berlin, 1971), pp. 215–221

44. G. Sweers, E. Zuazua, On the nonexistence of some special eigenfunctions for the Dirichlet Laplacian and the Lamé system. J. Elast. **52**, 111–120 (1998/1999)
45. K. Uhlenbeck, Generic properties of eigenfunctions. Am. J. Math. **98**, 1059–1078 (1976)
46. T. von Kármán, The engineer grapples with nonlinear problems. Bull. Am. Math. Soc. **46**, 615–683 (1940)
47. H.F. Wang, *Theory of Linear Poroelasticity with Applications to Geomechanics and Hydrogeology* (Princeton University Press, Princeton, 2000)
48. L.-H. Wang, On Korn's inequality. J. Comput. Math. **21**, 321–324 (2003)
49. K. Yosida, *Functional Analysis*. Classics in Mathematics (Springer, Berlin, 1995)
50. E. Zeidler, *Nonlinear Functional Analysis and Applications*. Fixed-Point Theorems, vol.1 (Springer, New York, 1986)

G.P. Galdi (✉) • M. Mohebbi • R. Zakerzadeh • P. Zunino
Swanson School of Engineering, University of Pittsburgh, Pittsburgh, PA, USA
e-mail: galdi@pitt.edu; mam175@pitt.edu; raz25@pitt.edu; paz13@pitt.edu

Chapter 4
Topics in the Mathematical Theory of Interactions of Incompressible Viscous Fluid with Rigid Bodies

Matthieu Hillairet

Abstract In this paper, we review recent results devoted to the interactions between a collection of rigid bodies $(\mathcal{B}_i)_{i=1,\dots,n}$ and a surrounding viscous fluid $\mathcal{L}$, the whole system filling a container Ω. We assume that the motion of $\mathcal{L}$ (resp. the rigid bodies $\mathcal{B}_i$) is governed by the incompressible Navier Stokes equations (resp. Newton laws), and that velocities and stress tensors are continuous at the fluid/body interfaces. Our concern is the well-posedness of the associated Cauchy problem, with a specific eye towards the handling of contact between bodies or between one body and the container boundary.

Keywords Cauchy theory • Contact issue • Fluid–solid interactions

MSC2010: 35Q35, 35B44, 35Q74, 74F10, 76D03, 76D05

4.1 Introduction

Studying the motion of rigid bodies inside a viscous fluid is crucial to many natural and engineering problems such as sedimentation, filtration or slurry erosion, to mention a few. In biological flows also, a disperse phase containing rigid bodies appears in many contexts: in the modeling of rigid tracers [11] or sprays [17], in rheological studies on active suspensions [14, 36]. In all these cases, bodies/swimmers might exhibit a complex behavior because of their elastic properties or their elaborate retroaction on the fluid. They also can be numerous so that their collective behavior is efficiently described by an equation of Vlasov type (see [31, 32] and [63], for instance). Nevertheless, a toy-model to tackle such complex problems is to assume that the bodies/swimmers behave as a finite number of undeformable bodies submitted to Newton laws. As for the surrounding fluid, assumptions on its behavior might contain more-or-less complexity. In studies on micro swimmers, for instance, Reynolds numbers are so small that a stationary Stokes system is relevant. To keep

T. Bodnár et al. (eds.), *Fluid-Structure Interaction and Biomedical Applications*, Advances in Mathematical Fluid Mechanics, DOI 10.1007/978-3-0348-0822-4_4,

the unity of this paper, we consider herein that the fluid behavior is computed by integrating the incompressible Navier Stokes equations. We give some insights into the Stokes problem in Sect. 4.4.3. We refer the interested reader to [21,45] and the references therein for studies considering compressible or inviscid fluids.

4.2 Basic Equations

In this review, we consider three-dimensional as well as two-dimensional problems discriminating both cases with a dimension parameter $d \in \{2,3\}$. At first, we consider the time-evolution of the system under the action of gravity g on a time interval $(0,T)$, where $T \in (0,\infty]$.

4.2.1 Notations

As the bodies are rigid, their mechanical properties are characterized by their shapes $\mathcal{B}_i^0$ and their densities $\rho_i^0 : \mathcal{B}_i^0 \mapsto (0,\infty)$. Without further notifications, we assume that the shapes are bounded open subsets of $\mathbf{R}^d$ having smooth boundaries and that the densities are bounded and strictly positive, i.e.:

$$0 < c \leq \rho_i^0(x) \leq \frac{1}{c}, \quad \forall\, x \in \mathcal{B}_i^0, \quad \forall\, i = 1,\ldots,n, \tag{4.1}$$

When the bodies move, there exist two families of (sufficiently smooth) mappings $t \mapsto \mathbb{Q}_i(t) \in SO_d(\mathbf{R})$ and $t \mapsto G_i(t) \in \mathbf{R}^d$ such that the body domains at time t read:

$$\mathcal{B}_i(t) := \mathbb{Q}_i(t)\mathcal{B}_i^0 + G_i(t), \quad \forall\, t \in (0,T), \quad \forall\, i \in \{1,\ldots,n\}. \tag{4.2}$$

The corresponding isometries are denoted

$$\mathcal{M}_i^t : x \mapsto \mathbb{Q}_i(t)x + G_i(t), \quad \forall\, t \in [0,T). \tag{4.3}$$

To fix completely the description, we always assume in what follows that:

$$G_i(0) = 0, \quad \mathbb{Q}_i(0) = \mathbb{I}_d, \quad \forall\, i \in \{1,\ldots,n\}. \tag{4.4}$$

Introducing this convention, we identify the shapes of the rigid bodies with their initial position. The motions of the rigid bodies are associated with the eulerian velocities:

$$u_i(x,t) = \xi_i(t) + \omega_i(t) \times (x - G_i(t)), \quad \forall\, x \in \mathcal{B}_i(t), \tag{4.5}$$

where

$$\xi_i(t) = \dot{G}_i(t), \qquad \forall\, t \in (0,T), \tag{4.6}$$

$$\omega_i(t) \times x = \dot{\mathbb{Q}}_i(t)\mathbb{Q}_i(t)^\top x, \quad \forall\, x \in \mathbf{R}^d, \quad \forall\, t \in (0,T). \tag{4.7}$$

These identities stand for definitions of the translational (resp. angular) velocity ξ_i (resp. ω_i) of $\mathcal{B}_i$. In the two-dimensional case, there holds $\omega_i \in \mathbf{R}$ and

$$\omega_i \times x = \omega_i\, x^\perp, \quad \forall\, x \in \mathbf{R}^2,$$

with $\perp$ denoting the rotation with angle $\pi/2$. In the three-dimensional case $\omega_i \in \mathbf{R}^3$ and $\times$ stands for the usual vector product. To include this dimensional phenomenon, we shall write $\omega_i \in \mathbf{R}^{d^*}$ where $d^* = 3$ for $d = 3$ and $d^* = 1$ for $d = 2$. Finally, the body unknowns are $(\mathcal{B}_i(t), \xi_i, \omega_i)_{i=1,\dots,n}$. We note that, given $t \mapsto \mathcal{B}_i(t)$, we might compute (ξ_i, ω_i) through (4.2) and (4.6)–(4.7). Conversely, given our convention (4.4), the identities (4.6)–(4.7) represent differential equations which enable to compute G_i and $\mathbb{Q}_i$ in terms of ξ_i and ω_i. Eventually, one might restrict the set of body unknowns either to $(\xi_i, \omega_i)_{i=1,\dots,n}$ or to $(\mathcal{B}_i(t))_{i=1,\dots,n}$.

As for the fluid, we denote $u_f = (u_{f,1}, \dots, u_{f,d})$ the velocity-field and p_f the pressure of $\mathcal{L}$. These are the only fluid unknowns. They are defined over the (time–space) fluid domain

$$\mathcal{Q}_F = \bigcup_{t \in (0,T)} \{t\} \times \mathcal{F}(t),$$

where, for all $t \in (0,T)$, the $\mathcal{F}(t)$ stands for the complement in Ω of the body domains:

$$\mathcal{F}(t) = \Omega \setminus \overline{\mathcal{S}(t)}, \qquad \text{where} \quad \mathcal{S}(t) = \bigcup_{i=1}^{n} \mathcal{B}_i(t).$$

Corresponding to the time–space fluid domain, we denote $\mathcal{Q}_S$ the time–space body domain:

$$\mathcal{Q}_S = \bigcup_{t \in (0,T)} \{t\} \times \mathcal{S}(t).$$

For simplicity, we assume that the fluid has constant density $\rho_f = 1$. We denote ν its kinematic viscosity. We deal with newtonian fluids, so that the Cauchy stress tensor in the fluid $\mathbb{T}$ is given by Newton law:

$$\mathbb{T} = \mathbb{T}(u_f, p_f) := 2\nu D(u_f) - p_f \mathbb{I}_d, \tag{4.8}$$

where $D(u_f)$ stands for the symmetric part of ∇u_f.

4.2.2 Equations of Motion

The time-evolution of the full system is governed by the classical equations of fluid and solid mechanics. Interactions between the fluid and the bodies are fixed by assuming continuity of the velocity-field and of the stress tensor at the fluid/body interfaces.

First, the fluid unknowns (u_f, p_f) satisfy the incompressible Navier Stokes equations:

$$\left.\begin{aligned} \partial_t u_f + u_f \cdot \nabla u_f &= \nu \Delta u_f - \nabla p_f + g \\ \nabla \cdot u_f &= 0 \end{aligned}\right\} \text{ in } \mathcal{Q}_F. \tag{4.9}$$

The right-hand side of the momentum equation is also written $\nabla \cdot \mathbb{T}(u_f, p_f)$. As long as the bodies do not collide or do not collide with the container boundary, we have the following partition of the fluid-domain boundary:

$$\partial \mathcal{F}(t) = \partial \Omega \cup \bigcup_{i=1}^{n} \partial \mathcal{B}_i(t).$$

The Navier Stokes system is then completed with no-slip boundary conditions:

$$u_f = u_i, \qquad \text{on } \partial \mathcal{B}_i(t), \quad \forall\, i = 1, \dots, n, \tag{4.10}$$

$$u_f = 0, \qquad \text{on } \partial \Omega. \tag{4.11}$$

When contact occurs, these boundary conditions provide u_f with two values on some parts of $\partial \mathcal{F}(t)$. Depending on the contact dimension (whether the contact holds in a point, on a curve, or on a surface), this problem should be paid attention. We discuss this question in Sect. 4.4.2.

The Navier Stokes equations are associated with Newton laws for the rigid bodies. In these laws, we only consider the actions exerted by gravity and the viscous fluid. Hence, the Newton laws read:

$$\frac{\mathrm{d} m_i \xi_i}{\mathrm{d}t} = -\int_{\partial \mathcal{B}_i(t)} \mathbb{T} n \mathrm{d}\sigma + m_i g \tag{4.12}$$

$$\frac{\mathrm{d} \mathbb{J}_i \omega_i}{\mathrm{d}t} = -\int_{\partial \mathcal{B}_i(t)} (x - G_i(t)) \times \mathbb{T} n \mathrm{d}\sigma. \tag{4.13}$$

Here the symbol n stands for the normal to $\partial \mathcal{F}(t)$ directed outside the fluid domain. For all integers $i \in \{1, \dots, n\}$, we introduced the mass m_i of the body $\mathcal{B}_i$:

$$m_i := \int_{\mathcal{B}_i(t)} \rho_i^0([\mathcal{M}_i^t]^{-1}(x)) \mathrm{d}x, \tag{4.14}$$

and its inertia $\mathbb{J}_i$, which reads

$$\mathbb{J}_i := \int_{\mathcal{B}_i(t)} \rho_i^0([\mathcal{M}_i^t]^{-1}(x))(|x - G_i(t)|^2 \mathbb{I}_3 - (x - G_i(t)) \otimes (x - G_i(t)))\mathrm{d}x, \quad (4.15)$$

in the three-dimensional case, and

$$\mathbb{J}_i := \int_{\mathcal{B}_i(t)} \rho_i^0([\mathcal{M}_i^t]^{-1}(x))|x - G_i(t)|^2 \mathrm{d}x, \quad (4.16)$$

in the two-dimensional case. The mass m_i and the two-dimensional inertia $\mathbb{J}_i$ are constant scalars whereas the three-dimensional inertia is a 3×3 time-dependent positive-definite symmetric matrix. In the three-dimensional case, we recall that $\times$ stands for the classical vector product, whereas in the two-dimensional case we need to define a new operator:

$$a \times b = a^{\perp} \cdot b, \quad \forall\, (a, b) \in \mathbf{R}^2.$$

We underline that $\times$ is defined in two different ways in the two-dimensional case depending on whether the first operand is a vector or a scalar.

We denote (FRBI) (for fluid rigid-body interaction system) the full system (4.2)–(4.6)–(4.7)–(4.9)–(4.10)–(4.11)–(4.12)–(4.13). The unknowns of this system are $((\mathcal{B}_i(t), \xi_i, \omega_i)_{i=1,\dots,n}, u_f, p_f)$. It is completed with initial conditions:

$$\begin{array}{ll} \mathcal{B}_i(0) = \mathcal{B}_i^0, \quad u_f(0,\cdot) = u_f^0, \\ \xi_i(0) = \xi_i^0, \quad \omega_i(0) = \omega_i^0, \end{array} \quad \forall\, i = 1, \dots, n. \quad (4.17)$$

Multiplying formally (4.9) with u_f and combining with (4.12)–(4.13), we obtain that reasonable solutions to this system should satisfy:

$$\frac{1}{2}\frac{\mathrm{d}}{\mathrm{d}t}\left[\int_{\mathcal{F}(t)} |u_f|^2 + \sum_{i=1}^{n} \Big(m_i|\xi_i|^2 + \mathbb{J}_i\omega_i \cdot \omega_i + 2(m_i - |\mathcal{B}_i^0|)g \cdot G_i\Big)\right] + 2\nu \int_{\mathcal{F}(t)} |D(u_f)|^2 = 0. \quad (4.18)$$

This formal estimate states the decay of the total energy of the system. It is then natural to consider initial conditions with bounded kinetic energy:

$$\mathcal{E}_c^0 := \frac{1}{2}\left[\int_{\mathcal{F}^0} |u_f^0|^2 + \sum_{i=1}^{n} \Big(m_i|\xi_i^0|^2 + \mathbb{J}_i^0\omega_i^0 \cdot \omega_i^0\Big)\right],$$

which amounts to require that $u_f^0 \in L^2(\mathcal{F}^0)$. We shall restrict to this case throughout the paper.

4.2.3 Outline of the Paper

The Cauchy problem for (FRBI) has been the subject of many studies in the last 15 years. Several existence results have been obtained introducing little by little complexity in the geometrical configurations under consideration. As classical when dealing with Navier Stokes equations, two families of solutions are constructed: classical solutions for which (4.9) is satisfied almost everywhere, weak solutions for which a specific definition of solving (4.9)–(4.13) is introduced. In Sect. 4.3, we first describe the construction of weak solutions up to contact between bodies in the case of several bodies moving inside a bounded container. The second part of this section is devoted to the construction of classical solutions in the case of one rigid body inside a bounded container.

All the results of Sect. 4.3 concern solutions prior to contact between bodies or bodies and the container boundary. We analyze how contacts are handled by solutions to (FRBI) in the last section. Several questions are discussed. First, in the frame of classical solutions, we show that contact implies blow-up and discuss which norms of the solution blows up in case of contact. We envisage then the extension of classical solutions by weak solutions after contact. We show that such an extension is possible with the method described in Sect. 4.3 and reduces to completing (FRBI) with a sticky contact law. Finally, we discuss the possibility of contact occurrence in finite time for weak and strong solutions to (FRBI).

4.3 Existence and Uniqueness for the Initial Boundary-Value Problems

The first modern contributions to the study of the Cauchy problem associated with (FRBI) are the references [47, 52] which tackle the free-fall of one rigid body in an unlimited container ($\mathcal{L}$ fills the whole three-dimensional space). Following these seminal works, the (two-dimensional and three-dimensional) case of one rigid body moving inside a bounded container is considered in [5, 33–35, 43, 44]: in [33, 34], existence of classical solutions is obtained (under restrictive assumptions on the body densities), while the same problem is solved in a weak setting in [5, 35, 43, 44] (see also [4]). We point out that existence of solutions is obtained up to contact between the body and the container boundary in [5,34,35], while [43,44] are the first occurrences of existence results without this restriction. The case of several rigid bodies moving inside a bounded container is then tackled in [15, 16, 20, 51, 56, 57]. First, existence up to contact is proven in [15,16], for weak solutions, and in [56,57], for classical solutions. In these latter references, the method of [33, 34] is improved yielding existence of classical solutions without restriction on the magnitude of the body densities. Then, global existence of weak solutions regardless contact is also obtained for the "several rigid body" case [20, 51]. Uniqueness of weak solutions in the two-dimensional case is tackled in [30]. Finally, the configuration of one body moving in an unlimited container is provided with existence of classical solutions [8, 9, 24, 58]. Contrary to the assumptions herein, the densities of the rigid bodies are assumed to be constant in most of the references above.

4.3.1 Many Bodies in a Container: Weak Solutions

In the first part of this section, we detail the theory of weak solutions for (FRBI): we provide a definition that is adapted to the case of non-smooth bodies, we recall the main difficulties to be handled by a construction and we conclude by describing the answers given in the references above. The first part of this section relies on [4, 20, 43, 51].

Let $((\xi_i^0, \omega_i^0, \mathcal{B}_i^0)_{i=1,\dots,n}, u_f^0)$ be an initial condition with a smooth u_f^0 and consider that the collection $((\xi_i, \omega_i, \mathcal{B}_i)_{i=1,\dots,n}, u_f, p_f)$ represents an associated classical solution to (FRBI) on $(0,T)$ satisfying (4.18). To derive a weak formulation, we assume that rigid bodies remain far from contact i.e.:

$$\mathrm{dist}(\mathcal{B}_i^0, \mathcal{B}_j^0) > 0, \qquad \mathrm{dist}(\mathcal{B}_i^0, \partial\Omega) > 0, \qquad \forall\, i \neq j, \tag{4.19}$$

$$\mathrm{dist}(\mathcal{B}_i(t), \mathcal{B}_j(t)) > 0,\ \mathrm{dist}(\mathcal{B}_i(t), \partial\Omega) > 0,\ \forall\, i \neq j, \quad \forall\, t \in (0,T). \tag{4.20}$$

We also eliminate gravity for simplicity: $g = 0$.

We introduce the extended velocity-fields (see (4.5) for a definition of the body velocity-fields u_i):

$$u := \mathbf{1}_{\mathcal{F}(t)} u_f + \sum_{i=1}^{n} \mathbf{1}_{\mathcal{B}_i(t)} u_i, \qquad u^0 := \mathbf{1}_{\mathcal{F}^0} u_f^0 + \sum_{i=1}^{n} \mathbf{1}_{\mathcal{B}_i^0} u_i^0. \tag{4.21}$$

Because of the no-slip boundary conditions (4.10)–(4.11), we obtain divergence-free vector-fields which are defined on Ω and continuous through fluid–body interfaces. We first define Sobolev-like function spaces adapted to such velocity-fields. Namely, given an open domain $\mathcal{O}$, we introduce:

- the classical function spaces of incompressible hydrodynamics:

$$\mathcal{D}(\mathcal{O}) := \{u \in C_c^\infty(\mathcal{O}) \text{ s.t. } \nabla \cdot u = 0\}, \tag{4.22}$$

  and $H(\mathcal{O})$ (resp. $V(\mathcal{O})$) the closure of $\mathcal{D}(\mathcal{O})$ in $L^2(\mathcal{O})$ (resp. $H_0^1(\mathcal{O})$),
- the set of rigid velocity-fields:

$$\mathcal{R} := \left\{\xi + \omega \times x, \quad \forall\, x \in \mathbf{R}^d\ ; \quad (\xi, \omega) \in \mathbf{R}^d \times \mathbf{R}^{d^*}\right\}.$$

We define then:

Definition 4.1. Given $\mathcal{O}$ an open domain and an open set $\mathcal{S} \subset \mathcal{O}$, we denote:

$$K[\mathcal{S},\mathcal{O}] := \{\xi \in V(\mathcal{O}) \text{ s.t. } D(\xi) = 0 \text{ on } \mathcal{S}\},$$
$$K^0[\mathcal{S},\mathcal{O}] := \overline{K[\mathcal{S},\mathcal{O}]}^{L^2(\mathcal{O})}.$$

Given $T > 0$ and an open set $\mathcal{Q}_S \subset (0,T)\times\Omega$, let us denote

$$\mathcal{S}(t) := \{x \in \Omega \text{ s.t. } (t,x) \in \mathcal{Q}_S\}.$$

Then, we set:

$$L^p(0,T;K[\mathcal{Q}_S,\Omega])$$
$$:= \left\{\xi \in L^p(0,T;H^1_0(\Omega)) \text{ s.t. } \xi(t,\cdot) \in K[\mathcal{S}(t),\Omega] \text{ a.e. on } (0,T)\right\},$$
$$\mathcal{K}[\mathcal{Q}_S,\mathcal{O}] := \{\xi \in C^\infty_c([0,T);\mathcal{D}(\Omega)) \text{ s.t. } D(\xi) = 0 \text{ in a neighborhood of } \mathcal{Q}_S\}.$$

According to formal energy estimate (4.18), reasonable solutions and initial conditions should then satisfy:

$$u^0 \in K^0[\mathcal{S}^0,\Omega], \quad u \in L^\infty(0,T;L^2(\Omega)) \cap L^2(0,T;K[\mathcal{Q}_S,\Omega]).$$

Conversely, given a vector-field u having this latter regularity, one recovers the fluid and body velocity-fields thanks to the classical lemma (see [59, Lemma 1.1]):

Lemma 4.1. *Let Ω be an open subset of $\mathbf{R}^d$ and $\tilde{\xi} \in H^1_0(\Omega)$. Assume that there exists an open connected set $\tilde{\mathcal{S}} \subset \Omega$ such that $D(\tilde{\xi})$ vanishes on $\tilde{\mathcal{S}}$. Then, there exists $\xi \in \mathcal{R}$ such that $\tilde{\xi}_{|\tilde{\mathcal{S}}} = \xi_{|\tilde{\mathcal{S}}}$.*

Once the extended velocity-fields are introduced, it is classical to rephrase the body kinematics in terms of the indicator functions of their domains. Indeed, let us denote φ_i the indicator function of $\mathcal{B}_i(t)$. As $\mathcal{B}_i$ follows the characteristics associated with the extended (divergence-free) velocity-field u, this indicator function satisfies:

$$\left.\begin{aligned} \partial_t\varphi_i + \nabla\cdot(\varphi_i\, u) &= 0, && \text{on } (0,T)\times\Omega,\\ \varphi_i(0,\cdot) &= \mathbf{1}_{\mathcal{B}_i^0}, && \text{on } \Omega. \end{aligned}\right\} \tag{4.23}$$

For weak solutions, we also introduce the transported densities $\rho_i : (t,x) \mapsto \rho_i^0([\mathcal{M}_i^t]^{-1}(x))$, where the mappings $\mathcal{M}_i^t$ are the isometries associated with the body displacements (see (4.3)). They are solutions to:

$$\left.\begin{aligned} \partial_t\rho_i + \nabla\cdot(\rho_i\, u) &= 0, && \text{on } (0,T)\times\Omega,\\ \rho_i(0,\cdot) &= \rho_i^0\mathbf{1}_{\mathcal{B}_i^0}, && \text{on } \Omega. \end{aligned}\right\} \tag{4.24}$$

Hence, by combination, the extended density

$$\rho(t,x) = \sum_{i=1}^{n} \rho_i(t,x) + \mathbf{1}_{\mathcal{F}(t)}(x) = \sum_{i=1}^{n} \rho_i(t,x) + \Big(1 - \sum_{i=1}^{n} \varphi_i(t,x)\Big), \tag{4.25}$$

is also a solution to

$$\left.\begin{aligned} \partial_t \rho + \nabla\cdot(\rho\, u) &= 0, && \text{on } (0,T)\times\Omega, \\ \rho(0,\cdot) &= \rho^0, && \text{on } \Omega. \end{aligned}\right\} \tag{4.26}$$

where

$$\rho^0(x) = \sum_{i=1}^{n} \rho_i^0 \mathbf{1}_{\mathcal{B}_i^0}(x) + \mathbf{1}_{\mathcal{F}^0}(x), \quad \forall\, x \in \Omega. \tag{4.27}$$

As u vanishes on $\partial\Omega$, we might introduce its trivial extension to $\mathbf{R}^d$ and consider equations (4.24)–(4.26) as the restrictions to Ω of an equation in terms of the unknowns ρ and ρ_i on the whole $\mathbf{R}^d$. Existence, uniqueness, and stability of bounded solutions to such transport equations have been extensively studied. Given the a priori regularity of u, namely $u \in L^2(0,T;H^1_0(\Omega))$ with $\nabla\cdot u \in L^1(0,T;L^\infty(\Omega))$, the well-posedness and stability results of [18] shall apply in our case. For instance, these equations might be written in the following weak form:

$$\int_0^T \int_\Omega \varphi(\partial_t \eta + u\cdot\nabla\eta) = \int_\Omega \varphi^0 \eta(0,\cdot), \tag{4.28}$$

for all $\eta \in C_c^\infty([0,T)\times\Omega)$. Here φ denotes a general unknown standing for φ_i, ρ_i or ρ and φ^0 is the associated initial data.

Next, we compute a weak formulation for (4.9)–(4.13). The body domain $\mathcal{Q}_S$ being given as the disjoint union of the time–space domains occupied by the $\mathcal{B}_i$'s, we multiply the momentum equation in (4.9) by $w \in \mathcal{K}[\mathcal{Q}_S,\Omega]$. This yields:

$$\int_0^T \int_{\mathcal{F}(t)} (\partial_t u + u\cdot\nabla u)\cdot w = \int_0^T \int_{\mathcal{F}(t)} [\nabla\cdot\mathbb{T}]\cdot w.$$

As $\mathcal{F}(t)$ follows the characteristics associated with the velocity-field u, we integrate by parts the left-hand side in time:

$$\begin{aligned} \text{LHS} &= -\int_{\mathcal{F}^0} u^0\cdot w_{|t=0} - \int_0^T \int_{\mathcal{F}(t)} (\partial_t w + u\cdot\nabla w)\cdot u, \\ &= -\int_{\mathcal{F}^0} u^0\cdot w_{|t=0} - \int_0^T \int_{\mathcal{F}(t)} \partial_t w\cdot u - \int_0^T \int_\Omega u\otimes u : D(w). \end{aligned} \tag{4.29}$$

Integrating by parts with respect to space the right-hand side yields:

$$\begin{aligned}\text{RHS} &= \int_0^T \int_{\partial\mathcal{F}(t)} \mathbb{T}n \cdot w \mathrm{d}\sigma - 2\nu \int_0^T \int_{\mathcal{F}(t)} D(u) : D(w) \\ &= \sum_{i=1}^n \int_0^T \int_{\partial\mathcal{B}_i(t)} \mathbb{T}n \cdot w \mathrm{d}\sigma - 2\nu \int_0^T \int_\Omega D(u) : D(w), \end{aligned} \tag{4.30}$$

where we applied again that $D(w) = 0$ on $\mathcal{Q}_S$. According to Lemma 4.1, this last property yields also that there exist $(\ell_i, r_i)_{i=1,\dots,n} \in [\mathbf{R}^d \times \mathbf{R}^{d^*}]^n$ such that $w(t,x) = \ell_i(t) + r_i(t) \times (x - G_i(t))$ on $\mathcal{B}_i(t)$. Hence, introducing (4.12)–(4.13) yields:

$$\begin{aligned}\int_0^T \int_{\partial\mathcal{B}_i(t)} \mathbb{T}n \cdot w \mathrm{d}\sigma &= -\int_0^T \left(m_i \dot{\xi}_i \cdot \ell_i + \frac{\mathrm{d}}{\mathrm{d}t} [\mathbb{J}_i \omega_i] \cdot r_i \right) \\ &= m_i \xi_i^0 \cdot \ell_i(0) + \mathbb{J}_i^0 \omega_i^0 \cdot r_i(0) + \int_0^T \left(m_i \dot{\ell}_i \cdot \xi_i + \mathbb{J}_i \omega_i \cdot \dot{r}_i \right) \\ &= \int_{\mathcal{B}_i^0} \rho_i^0 u^0 \cdot w_{|_{t=0}} + \int_0^T \int_{\mathcal{B}_i(t)} \rho_i u \cdot \partial_t w. \end{aligned} \tag{4.31}$$

Combining (4.29)–(4.31) and introducing the extended density ρ, we obtain finally:

$$\int_0^T \int_\Omega (2\nu D(u) : D(w) - \rho u \cdot \partial_t w - \rho u \otimes u : D(w)) = \int_\Omega \rho^0 u^0 \cdot w_{|_{t=0}}. \tag{4.32}$$

All these computations motivate the following definition of weak solutions:

Definition 4.2. Let the following assumptions hold true:

- Ω is a bounded connected domain in $\mathbf{R}^d$,
- the initial shapes of rigid bodies $\mathcal{B}_i^0$ satisfy (4.19),
- the densities $\rho_i^0 \in L^\infty(\mathcal{B}_i^0)$ satisfy (4.1).

Let $u^0 \in H(\Omega)$ and $T \in (0, \infty)$. A weak solution to (FRBI) on $(0, T)$ with initial data u^0 is a collection $((\varphi_i, \rho_i)_{i=1,\dots,n}, u)$ satisfying

- For all $i = 1, \dots, n$, $\rho_i \in L^\infty((0,T) \times \Omega)$,
- For all $i = 1, \dots, n$, $\varphi_i \in L^\infty((0,T) \times \Omega) \cap C([0,T); L^1(\Omega))$, furthermore $\mathcal{B}_i(t) := \mathrm{Supp}(\varphi_i(t, \cdot))$ is isometric to $\mathcal{B}_i^0$ for all $t \in [0, T)$,
- $u \in L^\infty(0, T; L^2(\Omega)) \cap L^2(0, T; K[\mathcal{Q}_S, \Omega])$,
- For all $i = 1, \dots, n$ Eq. (4.28) is satisfied by $\varphi = \varphi_i$, for all $\eta \in C_c^\infty([0,T) \times \Omega))$, with $\varphi^0 = \varphi_i^0 := \mathbf{1}_{\mathcal{B}_i^0}$,
- For all $i = 1, \dots, n$ Eq. (4.28) is satisfied by $\varphi = \rho_i$, for all $\eta \in C_c^\infty([0,T) \times \Omega))$, with $\varphi^0 = \rho_i^0 := \rho_i^0 \mathbf{1}_{\mathcal{B}_i^0}$,
- Equation (4.32) is satisfied for all $w \in \mathcal{K}[\mathcal{Q}_S, \Omega]$ with ρ and ρ^0 defined by (4.25) and (4.27), respectively,

- For almost every $t \in (0, T)$ we have the energy estimate:

$$\frac{1}{2}\int_{\Omega} \rho(t,\cdot)|u(t,\cdot)|^2 + 2\nu \int_0^t \int_{\Omega} |D(u)|^2 \leq \frac{1}{2}\int_{\Omega} \rho^0|u^0|^2. \tag{4.33}$$

Definition 4.2 is an adaptation of [51] taking into account the improvements of [20, 21]. We replaced $\mathcal{B}_i(t)$ with its indicator function φ_i as unknown. We introduced the body densities ρ_i as unknowns because we consider rigid bodies having nonconstant densities here. We point out that the condition $u(t,\cdot) \in V(\Omega)$ implies that

$$2\int_{\Omega} |D(u)(t,\cdot)|^2 = \int_{\Omega} |\nabla u(t,\cdot)|^2, \quad \forall\, t \in (0, T).$$

So, we might prefer to rewrite the last item:

- For almost every $t \in (0, T)$ we have the energy estimate:

$$\frac{1}{2}\int_{\Omega} \rho(t,\cdot)|u(t,\cdot)|^2 + \nu \int_0^t \int_{\Omega} |\nabla u|^2 \leq \frac{1}{2}\int_{\Omega} \rho^0|u^0|^2, \tag{4.34}$$

We also remark that this definition requires neither that the shapes $\mathcal{B}_i^0$ of rigid bodies and the container Ω have smooth boundaries, nor that no contact occurs between solid boundaries. So, we apply this definition in these cases beyond its initial derivation. Finally, following [20], we extended the definition of weak solution to initial datum $u^0 \in H(\Omega)$. This is possible as the initial datum appears only through its norm in the energy estimate and through its multiplication by test-functions $w_{|_{t=0}} \in \mathcal{D}(\Omega)$ in the weak-form of (4.9)–(4.13). A drawback is that, if two initial velocity-fields share the same projection on $K^0[\mathcal{S}^0, \Omega]$, the associated weak solutions should be equal. All these extensions in the definition of weak solutions (considering non-smooth rigid boundaries and enlarging the set of initial data) have been added in order to facilitate the construction of global-in-time weak solutions without addressing the question of contact occurrence.

Compared to [51], we required the support of $\varphi_i(t,\cdot)$ to be isometric to $\mathcal{B}_i^0$. This enables to avoid splitting of the rigid bodies into sub-bodies. As is explained in [21, Sect. 3], in the definition above, the velocity-field u is constrained to be a rigid velocity-field on $\mathcal{B}_i(t)$ by "$D(u) = 0$ on $\mathcal{B}_i(t)$." However, this condition creates as many rigid velocity-fields as the number of connected components of $\mathcal{B}_i(t)$. We observe here a kind of nonlinearity: the fact that $\mathcal{B}_i(t)$ remains connected should arise from the fact that it is connected. However, the possibility of body-splitting has not been ruled out in full rigor, to our knowledge, requiring thus special care. There are other ways to avoid body-splitting. In [21], the author introduces a compatibility condition between $\mathcal{B}_i(t)$ and the velocity-field u. Briefly, this compatibility condition needs to introduce the isometries $\mathcal{M}_i^t$ and requires that $u_{|_{\mathcal{B}_i(t)}}$ is equal to the time-derivative of these isometries. When the body boundaries are smooth, there are other alternatives. In [26], the authors require $\mathcal{B}_i^0$ and $\mathcal{B}_i(t)$ to be connected. Applying Proposition 4.1, this implies that the rigid velocity on $\mathcal{B}_i(t)$ is uniquely defined, so that $\mathcal{B}_i$ moves as one block and cannot split. In [51], the condition that $\mathcal{B}_i(t)$ remains connected is obtained by construction. The authors apply that, for any radial function $\chi : \mathbf{R}^d \to [0,\infty)$ of unitary mass there holds $\xi * \chi = \xi$, for all $\xi \in \mathcal{R}$ (where $*$ stands for the classical convolution operator). Consequently, the authors choose to replace the unknown φ_i with the indicator function ψ_i of the δ-interior of $\mathcal{B}_i(t)$:

$$]\mathcal{B}_i(t)[_\delta := \{x \in \mathcal{B}_i(t) \text{ s.t. } B(x,\delta) \subset \mathcal{B}_i(t)\}. \tag{4.35}$$

This function is a weak solution to

$$\left.\begin{aligned} \partial_t \psi_i + \nabla \cdot (\psi_i [u]_\delta) &= 0, && \text{on } (0,T)\times\Omega, \\ \psi_i(0,\cdot) &= \mathbf{1}_{]\mathcal{B}_i^0[_\delta}, && \text{on } \Omega. \end{aligned}\right\} \tag{4.36}$$

where $[u]_\delta = u * \chi_\delta$ for a radial mollifier χ_δ such that $\mathrm{Supp}(\chi_\delta) \subset B(0,\delta)$. Then, φ_i is computed as the indicator of the δ-exterior of $\mathrm{Supp}(\psi_i)$:

$$[\mathrm{Supp}(\psi_i)]_\delta := \bigcup_{x \in \mathrm{Supp}(\psi_i)} B(x,\delta). \tag{4.37}$$

The parameter $\delta > 0$ is fixed sufficiently small in order that the operation δ-exterior is the converse operation of δ-interior on $\mathcal{B}_i^0$:

$$]\,[\mathcal{B}_i^0]_\delta[_\delta = [\,]\mathcal{B}_i^0[_\delta]_\delta = \mathcal{B}_i^0, \quad \forall\, i \in \{1,\dots,n\}. \tag{4.38}$$

Such a δ exists if $\mathcal{B}_i^0$ has a sufficiently smooth boundary. When u is a rigid velocity-field u_i on $\mathrm{Supp}(\varphi_i)$, there holds $[u]_\delta = u_i = u$ on $\mathrm{Supp}(\psi_i)$. Moreover as $[u]_\delta \in L^\infty(0,T; C^\infty_{loc}(\mathbf{R}^d))$, the flow $t \mapsto \mathcal{M}_i^t$ associated with $[u]_\delta$ is well defined and lipschitzian. In particular $\mathcal{B}_i(t) = \mathcal{M}_i^t(\mathcal{B}_i^0)$ is an open connected subset of Ω, for all $i \in \{1,\dots,n\}$ and the difficulty is overcome.

Finally, once it is obtained that the body domain $\mathcal{S}(t)$ is made of n rigid bodies, it is classical to show that if u is smooth in the remaining fluid domain and satisfies (4.32), then there exists a pressure p so that (4.9) holds true in a classical sense. As long as no contact occurs and the solid boundaries are smooth, it is also possible to lift any set of rigid velocities on the $\mathcal{B}_i(t)$'s into a test-function $w \in \mathcal{K}[\mathcal{Q}_S, \Omega]$ in order to prove that (4.12) and (4.13) are also satisfied with ξ_i and ω_i computed w.r.t. $u_{|\mathcal{B}_i(t)}$. If the body boundaries and container boundary are smooth, we might also apply trace arguments to recover (4.10)–(4.11).

We proceed with detailing the construction of weak solutions. As classical, the main points in such constructions are

(i) Definition of approximate problems;
(ii) Existence of solutions to the approximate problems;

(iii) Analysis of cluster points of solutions to approximate problems.

In the end of this section, we sketch methods to tackle the points (i) and (iii).

Construction of Weak Solutions I: Designing Approximate Problems

We recall that we do not consider contact for now and that we do not make any assumption on the regularity of body boundaries (having in mind the construction of global-in-time weak solutions without addressing the contact issue). In order to introduce an approximate problem that one can tackle either by a Fadeo–Galerkin method or by a semi-group approach, it is necessary to handle with care the following difficulties in Definition 4.2:

- the space of test-functions depends on the solution itself through the constraint that test-functions must be equal to a rigid velocity-field on body domains;
- the weak formulation of Navier Stokes equations contains a nonlinear term:

$$\int_0^T \int_\Omega \rho u \otimes u : D(w)\,;$$

- the shapes of the rigid bodies do not have smooth boundaries a priori.

The most intriguing part of the construction is to deal with the first nonlinearity. Classical methods enable to handle the nonlinear convective term. As for the regularity of the bodies and container boundaries, we argue by compactness introducing smoothened rigid bodies and container. We shall detail the regularization process in the compactness argument. Hence, we assume for now that the body shapes have smooth boundaries.

Concerning the convective term, a classical method is to linearize by replacing one u either with a regularized velocity-field (see [35, 39]) or with a previous guess of solution (in a fixed-point approach). We remark that this should be done with care. For instance, let us denote by ξ this other velocity-field and assume that the linearized problem is given by the following system:

$$\begin{cases} \partial_t \tilde{u}_f + \xi \cdot \nabla \tilde{u}_f = \nabla \cdot \mathbb{T}(\tilde{u}_f, \tilde{p}_f) & \text{in } \tilde{\mathcal{F}}(t), \\ \nabla \cdot \tilde{u}_f = 0, & \text{in } \tilde{\mathcal{F}}(t), \\ \tilde{u}_f = \tilde{\xi}_i + \tilde{\omega}_i \times (x - G_i), & \text{on } \partial \tilde{\mathcal{B}}_i(t), \\ \tilde{u}_f = 0, & \text{on } \partial\Omega, \end{cases}$$

where the body domains $\tilde{\mathcal{B}}_i(t)$ (and thus also the fluid domain $\tilde{\mathcal{F}}(t)$) are computed with respect to the body motion prescribed by $\tilde{u}_f$. The unknowns in this new system are $((\tilde{\mathcal{B}}_i(t), \tilde{\xi}_i, \tilde{\omega}_i)_{i=1,\dots,n}, \tilde{u}_f, \tilde{p}_f)$. Multiplying the linearized momentum equation by $\tilde{u}_f$ yields the following a priori estimate:

$$\frac{1}{2}\frac{\mathrm{d}}{\mathrm{d}t}\int_{\tilde{\mathcal{F}}(t)}|\tilde{u}_f|^2 + 2\nu\int_{\tilde{\mathcal{F}}(t)}|D(\tilde{u}_f)|^2$$

$$= \sum_{i=1}^{n}\int_{\partial\tilde{\mathcal{B}}_i(t)}\mathbb{T}(\tilde{u}_f,\tilde{p}_f)n\cdot u\mathrm{d}\sigma + \int_{\tilde{\mathcal{F}}(t)}(\tilde{u}_f-\xi)\cdot\nabla\frac{|\tilde{u}_f|^2}{2}$$

$$= \sum_{i=1}^{n}\int_{\partial\tilde{\mathcal{B}}_i(t)}\left[\mathbb{T}(\tilde{u}_f,\tilde{p}_f)n + ((\tilde{u}_f-\xi)\cdot n)\frac{\tilde{u}_f}{2}\right]\cdot\tilde{u}_f\mathrm{d}\sigma.$$

Hence, to close this a priori estimate (which entails that the linearized system is well-balanced) one has to transform Newton laws into:

$$\frac{\mathrm{d}m_i\tilde{\xi}_i}{\mathrm{d}t} = -\int_{\partial\tilde{\mathcal{B}}_i(t)}\left[\mathbb{T}(\tilde{u}_f,\tilde{p}_f)n + ((\tilde{u}_f-\xi)\cdot n)\frac{\tilde{u}_f}{2}\right]\mathrm{d}\sigma + m_i g$$

$$\frac{\mathrm{d}\mathbb{J}_i\tilde{\omega}_i}{\mathrm{d}t} = -\int_{\partial\tilde{\mathcal{B}}_i(t)}(x-G_i(t))\times\left[\mathbb{T}(\tilde{u}_f,\tilde{p}_f)n + ((\tilde{u}_f-\xi)\cdot n)\frac{\tilde{u}_f}{2}\right]\mathrm{d}\sigma.$$

The total kinetic energy of approximate solutions $((\tilde{\mathcal{B}}_i(t),\tilde{\xi}_i,\tilde{\omega}_i)_{i=1,\dots,n},\tilde{u}_f,\tilde{p}_f)$ is then decaying with time.

Several methods have been introduced to handle the dependence of the set of test-functions on the solution itself. One possibility is to remark that Definition 4.2 is similar to a classical definition of weak solutions to Navier Stokes equations on a cylindrical domain with supplementary conditions penalizing the presence of the body domains inside the container. In the case where the container contains only one rigid body, an alternative approach consists in fixing the body domain and make the container time-dependent, by applying a simple change of referential [5, 35]. Herein, we focus on the first approach.

Two ways to penalize the presence of the body domains have been introduced. The first one refers to the Lemma 4.1 or to the definition of $K[\mathcal{S}(t),\Omega]$ and imposes that the solution satisfies

$$\varphi D(u) = 0 \quad \text{with} \quad \varphi := \sum_{i=1}^{n}\mathbf{1}_{\mathcal{B}_i(t)}.$$

Taking into account that one needs to keep the body domains connected, this motivates the following definition of approximate solutions (adapted from [51]):

Definition 4.3. Let the following assumptions hold true:

- Ω is a container having a smooth boundary,
- the shapes of the bodies $\mathcal{B}_i^0$ are connected open subsets of Ω having smooth boundaries satisfying (4.19),
- the densities $\rho_i^0 \in L^\infty(\mathcal{B}_i^0)$ satisfy (4.1).

Let $\delta > 0$ such that (4.38) holds true, $T \in (0,\infty)$ and $\varepsilon > 0$. Given $u^0 \in H(\Omega)$, we call $\varepsilon - H^1$ penalized solution on $(0,T)$, with initial datum u^0, a collection $((\psi_i, \rho_i)_{i=1,\dots,n}, u)$ satisfying

PH1: for all $i \in \{1,\dots,n\}$, there holds $\psi_i \in L^\infty((0,T)\times\Omega)$, $\rho_i \in L^\infty((0,T)\times\Omega)$,

PH2: $u \in L^\infty(0,T;L^2(\Omega)) \cap L^2(0,T;V(\Omega))$,

PH3: for all $i \in \{1,\dots,n\}$ and all $\eta \in C_c^\infty([0,T)\times\Omega))$, there holds:

$$\int_0^T \int_\Omega \psi_i(\partial_t \eta + [u]_\delta \cdot \nabla \eta) = \int_\Omega \mathbf{1}_{]\mathcal{B}_i^0[_\delta} \eta_{|t=0},$$

where $[u]_\delta = u * \chi_\delta$ for a radial mollifier χ_δ such that $\mathrm{Supp}(\chi_\delta) \subset B(0,\delta)$;

PH4: for all $i \in \{1,\dots,n\}$ and all $\eta \in C_c^\infty([0,T)\times\Omega))$, there holds:

$$\int_0^T \int_\Omega \rho_i(\partial_t \eta + u \cdot \nabla \eta) = \int_\Omega \rho_i^0 \mathbf{1}_{\mathcal{B}_i^0} \eta_{|t=0},$$

PH5: for all divergence-free $w \in C_c^\infty([0,T)\times\Omega)$, there holds:

$$\int_0^T \int_\Omega \left(\left[2\nu + \frac{1}{\varepsilon}\varphi \right] D(u) : D(w) - \rho u \cdot \partial_t w - \rho u \otimes u : D(w) \right) = \int_\Omega \rho^0 u^0 \cdot w_{|t=0},$$

where

$$\varphi := \sum_{i=1}^n \mathbf{1}_{[\mathrm{Supp}(\psi_i)]_\delta}, \quad \rho := \sum_{i=1}^n \rho_i + \left(1 - \sum_{i=1}^n \varphi_i\right), \quad \rho^0 := \sum_{i=1}^n (\rho_i^0 - 1)\mathbf{1}_{\mathcal{B}_i^0} + 1,$$

PH6: for almost every $t \in (0,T)$ we have the energy estimate:

$$\frac{1}{2}\int_\Omega \rho(t,\cdot)|u(t,\cdot)|^2 + \int_0^t \int_\Omega 2\nu|\nabla u|^2 + \int_0^t \int_\Omega \frac{\varphi}{\varepsilon}|D(u)|^2 \le \frac{1}{2}\int_\Omega \rho^0 |u^0|^2.$$

For arbitrary $\varepsilon > 0$, existence of $\varepsilon - H^1$ approximate solutions reduces to a Navier Stokes problem for a nonhomogenous (nonconstant viscosity) fluid. We recall that the introduction of the new unknowns $(\psi_i)_{i=1,\dots,n}$ aims at ensuring the connectedness of the body domains $\mathcal{B}_i(t)$ and that δ is fixed w.r.t. the shapes $(\mathcal{B}_i^0)_{i=1,\dots,n}$. In this definition, letting ε go to 0 we obtain formally that

$$\int_0^T \int_{\mathcal{S}(t)} |D(u)|^2 = 0 \qquad \Longrightarrow \qquad D(u) = 0 \text{ on } \mathcal{Q}_S.$$

Simultaneously, for test-functions w which vanish on a neighborhood of $\mathcal{Q}_S$ we have:

$$\int_0^T \int_\Omega \frac{1}{\varepsilon}\varphi D(u) : D(w) = 0.$$

Consequently, we might expect to pass to the limit $\varepsilon \to 0$ in **PH5** for test-functions $w \in \mathcal{K}[\mathcal{Q}_S, \Omega]$ in order to obtain a weak solution in the sense of Definition 4.2. This H^1-penalization is applied in [16, 20, 51] and also in [43, 44] with supplementary regularizing terms.

The other approach consists in remarking that, for any open set $\mathcal{S} \subset \Omega$ made of n connected domains $\mathcal{B}_1, \dots, \mathcal{B}_n$, there holds:

$$v \in K[\mathcal{S}, \Omega] \iff v_{|\mathcal{B}_i} = P_{L^2(\Omega,\rho_i \mathrm{d}x)}[\mathcal{R}](v), \quad \forall\, i = 1, \dots, n.$$

where we introduced $L^2(\Omega, \rho_i \mathrm{d}x)$ the space of square-integrable functions w.r.t. the measure $\rho_i \mathrm{d}x$ and where $P_{L^2(\Omega,\rho_i \mathrm{d}x)}[\mathcal{R}](v)$ is the orthogonal projector from $L^2(\Omega, \rho_i \mathrm{d}x)$ onto $\{\xi_{|\mathcal{B}_i} ; \xi \in \mathcal{R}\}$. There exist explicit formulas for this projector: for all $\xi \in L^2(\Omega, \rho_i \mathrm{d}x)$ setting:

$$m_i := \int_\Omega \rho_i \mathrm{d}x, \qquad h_i := \frac{1}{m_i}\int_\Omega \rho_i x \mathrm{d}x, \tag{4.39}$$

$$\mathbb{J}_i := \int_\Omega \rho_i \left(|x - h_i|^2 - (x - h_i) \otimes (x - h_i)\right) \mathrm{d}x, \tag{4.40}$$

$$\xi_i := \frac{1}{m_i}\int_\Omega \rho_i \xi \mathrm{d}x \qquad \omega_i := \mathbb{J}_i^{-1}\int_\Omega \rho_i (x - h_i) \times \xi \mathrm{d}x, \tag{4.41}$$

there holds:

$$P_{L^2(\Omega,\rho_i \mathrm{d}x)}[\mathcal{R}](\xi) = \xi_i + \omega_i \times (x - h_i), \quad \forall\, x \in \mathcal{B}_i.$$

This leads to the following definition of approximate solutions that we adapt from [4]:

Definition 4.4. Let the following assumptions hold true:

- Ω is a container having a smooth boundary,
- the shapes of the bodies $\mathcal{B}_i^0$ are open connected subset of Ω having smooth boundaries satisfying (4.19),
- the densities $\rho_i^0 \in L^\infty(\mathcal{B}_i^0)$ satisfy (4.1).

Let $T \in (0, \infty)$ and $\varepsilon > 0$. Given $u^0 \in H(\Omega)$, we call $\varepsilon - L^2$ penalized solution on $(0, T)$ with initial datum u^0 a collection $((\varphi_i, \rho_i)_{i=1,\dots,n}, u)$ satisfying:

PL1: for all $i \in \{1, \dots, n\}$, $\rho_i \in L^\infty((0,T) \times \Omega)$ and $\varphi_i(t,x) \in L^\infty((0,T) \times \Omega) \cap C([0,T); L^1(\Omega))$ is such that $\mathcal{B}_i(t) := \mathrm{Supp}(\varphi_i)(t,\cdot)$ is a connected domain of Ω;

PL2: $u \in L^\infty(0,T; L^2(\Omega)) \cap L^2(0,T; V(\Omega))$;

PL3: for all $i \in \{1, \dots, n\}$ and all $\eta \in C_c^\infty([0,T) \times \Omega))$, there holds:

$$\int_0^T \int_\Omega \varphi_i (\partial_t \eta + u_i \cdot \nabla \eta) = \int_\Omega \mathbf{1}_{\mathcal{B}_i^0} \eta_{|t=0},$$

and

$$\int_0^T \int_\Omega \rho_i (\partial_t \eta + u_i \cdot \nabla \eta) = \int_\Omega \rho_i^0 \mathbf{1}_{\mathcal{B}_i^0} \eta_{|t=0},$$

where $u_i(t,\cdot) := P_{L^2(\Omega, \rho_i \mathrm{d}x)}[\mathcal{R}](u(t,\cdot))$ for all $t \in (0,T)$;

PL4': for all $\eta \in C_c^\infty([0,T) \times \Omega))$, there holds:

$$\int_0^T \int_\Omega \rho(\partial_t \eta + u \cdot \nabla \eta) = \int_\Omega \rho^0 \eta_{|t=0},$$

where

$$\rho^0 := \left[\sum_{i=1}^n (\rho_i^0 - 1) \mathbf{1}_{\mathcal{B}_i^0} + 1 \right];$$

PL5: for all divergence-free $w \in C_c^\infty([0,T) \times \Omega)$, there holds:

$$\int_0^T \int_\Omega \left(2\nu D(u) : D(w) - \rho u \cdot \partial_t w - \rho u \otimes u : D(w) + \sum_{i=1}^n \frac{\varphi_i}{\varepsilon} (u - u_i) \cdot w \right) = \int_\Omega \rho^0 u^0 \cdot w_{|t=0};$$

PL6: For almost every $t \in (0,T)$ we have the energy estimate:

$$\frac{1}{2} \int_\Omega \rho(t,\cdot) |u(t,\cdot)|^2 + \int_0^t \int_\Omega 2\nu |\nabla u|^2 + \sum_{i=1}^n \int_0^t \int_\Omega \frac{\varphi_i}{\varepsilon} |(u - u_i)|^2 \le \frac{1}{2} \int_\Omega \rho^0 |u^0|^2.$$

We emphasize that, in this definition, we compute the density ρ_i and the extended density ρ w.r.t. two different velocity-fields. When computing energy estimate, we need ρ to follow the characteristics associated with u, while computing ρ_i w.r.t. characteristics of the projection u_i guarantees that the domain $\mathcal{B}_i(t)$ remains isometric to $\mathcal{B}_i^0$. Given $\varepsilon > 0$, the construction of an $\varepsilon - L^2$ penalized solution is still a highly nonlinear problem. In **PL3** for instance, the computation of the velocity-field u_i depends itself on the value of the solution through formulas (4.39)–(4.41).

Existence and compactness of solutions to this nonlinear sub-problem are studied in details in [28, Sect. 3] in the case of constant densities. Extension of these results to general bounded densities is straightforward. Hence, a Galerkin method enables to construct an $\varepsilon - L^2$ penalized solution prior to contact between rigid bodies. We note that from **PL6**, all $\varepsilon - L^2$ penalized solutions for a same initial data have kinetic energy bounded by a same constant. This yields in particular that they all exist on a fixed time interval independent of ε.

To conclude this subsection, we mention that both penalization methods yield different numerical approximations. The properties of these different methods are analyzed in [1] in the case of fixed body domains for instance.

Construction of Weak Solutions II: Compactness Arguments

We proceed with the analysis of cluster points of a sequence of approximate solutions. The ingredients detailed in this section apply in two different cases. First, to construct weak solutions to (FRBI) when rigid bodies have smooth boundaries and the container is also smooth, one applies arguments similar to the one below in order to prove that a sequence of penalized solutions does converge to a weak solution to (FRBI). In this case, the shapes $\mathcal{B}_i^0$ and the container Ω do not depend on ε and $((\varphi_i^\varepsilon, \rho_i^\varepsilon)_{i=1,\dots,n}, u^\varepsilon)_{\varepsilon>0}$ stands for a sequence of ε-penalized solutions with initial data which do not depend on ε. Second, to construct weak solutions to (FRBI) when the shapes $\mathcal{B}_i^0$ and the container Ω do not have smooth boundaries, we also apply compactness arguments as the ones below. In this second case, we require that $\overline{\mathcal{B}_i^0}$ and $\mathbf{R}^d \setminus \Omega$ are connected. This yields that there exist sequences $(\mathcal{B}_i^{0,\varepsilon})_{\varepsilon>0}$ and $(\Omega^\varepsilon)_{\varepsilon>0}$ of open sets having smooth boundaries satisfying (4.19) such that (see (4.35) and (4.37) for notations):

- $\mathcal{B}_i^{\varepsilon,0}$ converges to $\mathcal{B}_i^0$ in the sense that, for all $\eta > 0$ there exists $\varepsilon_0 > 0$ such that, for all $\varepsilon < \varepsilon_0$, there holds:

$$\mathcal{B}_i^0 \subset \mathcal{B}_i^{\varepsilon,0} \subset [\mathcal{B}_i^0]_\eta;$$

- Ω^ε converges to Ω in the sense that, for all $\eta > 0$ there exists $\varepsilon_0 > 0$ such that, for all $\varepsilon < \varepsilon_0$, there holds:

$$]\Omega[_\eta \subset \Omega^\varepsilon \subset \Omega.$$

We also introduce sequences of densities $(\rho_i^{\varepsilon,0})_{\varepsilon>0}$ which are obtained from the ρ_i^0 by convolution and a sequence of approximate initial velocities $(u^{\varepsilon,0})_{\varepsilon>0}$. They satisfy:

- $\rho_i^{\varepsilon,0}$ is uniformly bounded and converges a.e. to $\rho_i^0 \mathbf{1}_{\mathcal{B}^0}$.
- $u^{\varepsilon,0} \in H[\Omega^\varepsilon] \subset H[\Omega]$ and $u^{\varepsilon,0}$ converges to u^0 in $H(\Omega)$.

In this second case also, $((\varphi_i^\varepsilon, \rho_i^\varepsilon)_{i=1,\dots,n}, u^\varepsilon)_{\varepsilon>0}$, is a sequence of weak solutions to (FRBI) associated with the container $(\Omega^\varepsilon)_{\varepsilon>0}$ and initial conditions given by $((\mathcal{B}_i^{\varepsilon,0}, \rho_i^{\varepsilon,0})_{i=1,\dots,n}, u^{0,\varepsilon})$. One then shows that a weak cluster point of this sequence of weak solutions to (FRBI) in the ε-geometry yields a weak solution to (FRBI) in the limit geometry.

For simplicity, we detail the computations in the second application only. We point out the arguments that need to be adapted when dealing with a sequence of penalized solutions. First, we extend all fields outside Ω^ε by 0. This yields sequences we still denote φ_i^ε, ρ_i^ε and u^ε. As initial velocity-fields are bounded in $H(\Omega)$ and body densities are also uniformly bounded, it yields that, up to extract a subsequence we do not relabel for simplicity, there holds:

$$\varphi_i^\varepsilon \to \varphi_i \text{ in } L^\infty((0,T)\times\Omega) - w*, \tag{4.42}$$

$$\rho_i^\varepsilon \to \rho_i \text{ in } L^\infty((0,T)\times\Omega) - w*, \tag{4.43}$$

$$u^\varepsilon \to u \text{ in } L^\infty(0,T;L^2(\Omega)) - w* \text{ and in } L^2(0,T;H_0^1(\Omega)) - w. \tag{4.44}$$

Our aim is to prove that the collection $((\varphi_i, \rho_i)_{i=1,\dots,n}, u)$ is a weak solution to (FRBI) for initial datum $((\mathcal{B}_i^0, \rho_i^0)_{i=1,\dots,n}, u^0)$ on $(0,T)\times\Omega$. The main ingredients of the proof are

- passage to the limit in the transport equation satisfied by $(\varphi_i^\varepsilon, \rho_i^\varepsilon)$, and construction of the isometries $\mathcal{M}_i^t$ associated with the body motions,
- passage to the limit in the nonlinear term for any test-function $w \in \mathcal{K}[\mathcal{Q}_S, \Omega]$:

$$\int_0^T \int_\Omega \rho^\varepsilon u^\varepsilon \otimes u^\varepsilon : D(w).$$

We consider these two steps separately. For the second step, we detail a method due to [43] well-adapted to this fluid/body problem. In this section, we consider solutions prior to contact. We recall that energy estimate (4.34) is sufficient to guarantee that the sequence $((\varphi_i^\varepsilon, \rho_i^\varepsilon)_{i=1,\dots,n}, u^\varepsilon)_{\varepsilon>0}$ remains far from contact uniformly on some time interval $(0,T)$ independent of $\varepsilon > 0$.

Step 1. Construction of Isometries. As we noticed already, the results of [18] apply to the sequence of divergence-free velocity-fields $(u^\varepsilon)_{\varepsilon>0}$. On the one hand, this sequence converges to u in $L^2(0,T;H_0^1(\Omega))-w$, on the other hand the sequence of initial data $(\varphi_i^{\varepsilon,0}, \rho_i^{\varepsilon,0})$ converge a.e. to $\mathbf{1}_{\mathcal{B}_i^0}$ and $\rho_i^0 \mathbf{1}_{\mathcal{B}_i^0}$, respectively. Consequently, there holds

$$\varphi_i^\varepsilon \to \varphi_i, \quad \text{in } C([0,T];L^q(\Omega)) \text{ for all finite } q, \tag{4.45}$$

$$\rho_i^\varepsilon \to \rho_i, \quad \text{in } C([0,T];L^q(\Omega)) \text{ for all finite } q, \tag{4.46}$$

and φ_i and ρ_i satisfy (4.28) with φ^0 given by their respective initial data, for all $\eta \in C_c^\infty([0,T)\times\Omega)$. This yields in particular that $\varphi_i(t,\cdot)$ is the indicator function of its support $\mathcal{B}_i(t,\cdot)$ and that $u \in L^2(0,T;K[\mathcal{Q}_S,\Omega])$. This yields also that $\mathcal{B}_i^\varepsilon(t) := \mathrm{Supp}(\varphi_i^\varepsilon(t,\cdot))$ converges to $\mathcal{B}_i(t)$ in the following sense:

$$\lim_{\varepsilon\to 0} \sup_{[0,T]} |\mathcal{B}_i^\varepsilon(t)\,\Delta\,\mathcal{B}_i(t)| = 0, \quad \forall\, t \in (0,T).$$

and that, when $\varepsilon \to 0$:

$$\rho^\varepsilon \to \rho := \sum_{i=1}^n \rho_i + \Big(1 - \sum_{i=1}^n \varphi_i\Big), \qquad \text{in } C([0,T];L^q(\Omega)), \text{ for all finite } q. \quad (4.47)$$

For all $\varepsilon > 0$, there exist affine isometries $\mathcal{M}_i^\varepsilon \equiv (\mathbb{Q}_i^\varepsilon, G_i^\varepsilon)_{i=1,\ldots,n}$ such that:

$$\mathcal{B}_i^\varepsilon(t) = \mathbb{Q}_i^\varepsilon(t)\mathcal{B}_i^0 + G_i^\varepsilon, \quad \forall\, i = \{1,\ldots,n\}.$$

As u^ε coincides with the eulerian velocity of $\mathcal{B}_i$ on $\mathcal{B}_i^\varepsilon(t)$, there holds:

$$\|\dot{G}_i^\varepsilon; L^\infty(0,T)\| + \|\dot{\mathbb{Q}}_i^\varepsilon[\mathbb{Q}_i^\varepsilon]^\top; L^\infty(0,T)\| \le C\,\|u; L^\infty(0,T;L^2(\Omega))\|,$$

where the constant C is independent of ε due to our uniform assumptions on initial data. Consequently, up to the extraction of a subsequence, we have convergence of the isometries to a mapping of isometries $\mathcal{M}_i^t \equiv (\mathbb{Q}_i, G_i)$ in $C([0,T])$ and in $W^{1,\infty}(0,T) - w*$. This entails that

$$\mathcal{B}_i(t) = \mathbb{Q}_i(t)\mathcal{B}_i^0 + G_i, \quad \forall\, i = \{1,\ldots,n\}.$$

In particular, the convergence of $\mathcal{B}_i^\varepsilon(t)$ to $\mathcal{B}_i(t)$ holds also in the following sense: for all $\eta > 0$, there exists $\varepsilon_0 > 0$ such that, given $\varepsilon < \varepsilon_0$, there holds:

$$\mathcal{B}_i^\varepsilon(t) \subset [\mathcal{B}_i(t)]_\eta \text{ and } \mathcal{B}_i(t) \subset [\mathcal{B}_i^\varepsilon(t)]_\eta, \quad \forall\, t \in [0,T]. \quad (4.48)$$

It is straightforward to adapt the above arguments to a sequence of $\varepsilon - L^2$ penalized solutions. In the case of $\varepsilon - H^1$ penalized solutions, there are no isometries at the level of the approximation. However, it is possible to introduce the flows $\mathcal{M}_\delta^{t,\varepsilon}$ associated with the velocity-fields $[u^\varepsilon]_\delta$. These latter velocity-fields enjoy better regularity and convergence properties. Eventually, one proves convergence of $\mathcal{M}_\delta^{t,\varepsilon}$ to some $\mathcal{M}_\delta^t$ in $C^{0,\alpha}((0,T);C^1(\Omega))$ (for arbitrary $\alpha < 1$). The penalization terms yielding that $D(u) = 0$ on all $\bigcup_{i=1}^n [\mathrm{Supp}(\psi_i)]_\delta$, all the restrictions $\mathcal{M}_i^t$ of $\mathcal{M}_\delta^t$ to $\mathcal{B}_i^0$ are isometric (see [51, Sect. 5.2]).

Next, any test function $w \in \mathcal{K}[\mathcal{Q}_S,\Omega]$ satisfies $\mathcal{O} := \bigcup_{t\in[0,T]} \mathrm{Supp}(w(t,\cdot)) \subset \overline{\mathcal{O}} \subset \Omega$. As a similar property holds for the domain where $D(w) = 0$, the convergence (4.48) of body domains yields that it is an admissible test-function for the ε-problem, given ε sufficiently small. We can then apply convergence results (4.44)–(4.47) to pass to the limit in all terms of the weak formulation except

$$\int_0^T \int_{\Omega^\varepsilon} \rho^\varepsilon u^\varepsilon \otimes u^\varepsilon : D(w) = \int_0^T \int_{\mathcal{O}} \rho^\varepsilon u^\varepsilon \otimes u^\varepsilon : D(w).$$

We need further properties of the limit $u^\varepsilon \to u$ to compute the asymptotics of this term. As $D(w) \in L^\infty((0,T) \times \mathcal{O})$ and $\rho^\varepsilon u^\varepsilon$ already converges weakly to ρu in $L^2((0,T) \times \mathcal{O})$ it would be sufficient here to obtain that u^ε converges strongly to u in $L^2((0,T) \times \Omega)$, for instance. The rest of this section is devoted to the proof of a variant of this statement.

Step 2. Asymptotics of the Nonlinear Terms. We extend herein to the several body case the method applied by Hoffmann and Starovoitov in [43].

Given an admissible test-function $w \in \mathcal{K}[\mathcal{Q}_S, \Omega]$ we first construct a cylindrical decomposition of the fluid-domain adapted to the test-function w. As $t \mapsto \mathcal{M}_i^t$ is continuous and the $t \mapsto \mathcal{M}_i^{t,\varepsilon}$ are uniformly bounded in $W^{1,\infty}((0,T) \times \Omega)$ and converge to $t \mapsto \mathcal{M}_i^t$, there exists a non-decreasing sequence $0 = t_0 < t_1 < \cdots < t_N = T$ and associated open sets $(\mathcal{O}_k)_{k=1,\dots,N}$ such that the $E_k := (t_k, t_{k+1}) \times \mathcal{O}_k$ satisfy:

- $\partial\mathcal{O}_k$ is the disjoint union of $\{\Sigma_k^i,\ i = 0,\dots,n\}$ for a collection of $(n+1)$ smooth Σ_k^i, each one surrounding a rigid boundary ($\partial\Omega$ or $\bigcup_{t \in (t_k, t_{k+1})} \partial\mathcal{B}_i(t)$ for $i = 1,\dots,n$);
- $\overline{\mathcal{O}_k} \subset \mathcal{F}^\varepsilon(t)$ for all $t \in (t_k, t_{k+1})$, for sufficiently small ε;
- $([0,T] \times \Omega) \setminus (\bigcup_{k=1}^N \overline{E_k}) \subset \{(t,x),\ \text{s.t. } D(w)(t,x) = 0\}$.

Consequently, E_k is a cylindrical domain for all $k \in \{1,\dots,N\}$ and w is a.e. a rigid velocity-field on any connected subset of the complement of $\bigcup_{k=1}^N E_k$ in $(0,T) \times \Omega$. In particular, there holds:

$$\int_0^T \int_{\Omega} \rho^\varepsilon u^\varepsilon \otimes u^\varepsilon : D(w) = \sum_{k=1}^N \int\int_{E_k} u^\varepsilon \otimes u^\varepsilon : D(w).$$

Our problem reduces then to computing the limit of the N integrals on the right-hand side of this last identity. We emphasize that N is fixed w.r.t. w independent of ε and that, in the integrals we want to compute now, we changed the space domain Ω into a $\mathcal{O}_k$ which has smooth boundaries. We might now apply all the classical results on hydrodynamic spaces $H(\mathcal{O}_k)$ and $V(\mathcal{O}_k)$: existence of traces, duality, (compact) embeddings... see [23, Chap. III].

Let $k \in \{1,\dots,N\}$. We note that, as w is equal to a rigid velocity-field on the smooth boundaries $(\Sigma_k^i)_{i=1,\dots,n}$ it is possible to adapt the construction in [49] (see Lemma 7.1 and more generally pp. 103–105) in order to obtain a divergence-free $W \in C^\infty([t_k, t_{k+1}] \times \overline{\mathcal{O}_k})$ such that $W|_{\Sigma_i^k} = w|_{\Sigma_i^k}$ for all $i = 1,\dots,n$ and

$$\left| \int\int_{E_k} u^\varepsilon \otimes u^\varepsilon : D(W) \right| \le \eta \| u^\varepsilon ; L^2((t_k, t_{k+1}); H^1(\Omega)) \|^2,$$

$$\left| \int \int_{E_k} u \otimes u : D(W) \right| \leq \eta \| u; L^2((t_k, t_{k+1}); H^1(\Omega)) \|^2,$$

for arbitrary small η. Hence, introducing $\tilde{w} = w - W$ we have

$$\begin{aligned} &\left| \int \int_{E_k} u^{\varepsilon} \otimes u^{\varepsilon} : D(w) - \int \int_{E_k} u \otimes u : D(w) \right| \\ &\qquad = \left| \int \int_{E_k} u^{\varepsilon} \otimes u^{\varepsilon} : D(\tilde{w}) - \int \int_{E_k} u \otimes u : D(\tilde{w}) \right| + C\eta. \end{aligned}$$

for arbitrary small η and a constant C uniform in ε. Our proof thus reduces to show that the first term on the right-hand side vanishes in the limit $\varepsilon \to 0$, for arbitrary $\tilde{w} \in C^{\infty}([t_k, t_{k+1}] \times \overline{\mathcal{O}_k})$ such that $\tilde{w}(t, \cdot) \in V(\mathcal{O}_k)$ for all $t \in (t_k, t_{k+1})$. We fix one $\tilde{w}$ in the remainder of this section and set:

$$I^{\varepsilon} := \int \int_{E_k} u^{\varepsilon} \otimes u^{\varepsilon} : D(\tilde{w}), \quad I := \int \int_{E_k} u \otimes u : D(\tilde{w}).$$

Next, we split velocity-fields u^{ε} and u in an appropriate way. We introduce $P[H(\mathcal{O}_k)]$ the orthogonal projector $L^2(\mathcal{O}_k) \to H(\mathcal{O}_k)$ and we set:

$$U^{\varepsilon} := P[H(\mathcal{O}_k)](u^{\varepsilon}), \quad U := P[H(\mathcal{O}_k)](u).$$

We have the following consequence of [23, Theorem 3.1.1]:

Proposition 4.1. *There exists $\zeta \in H^1(\mathcal{O}_k)$ (resp. $\zeta^{\varepsilon} \in H^1(\mathcal{O}_k)$) such that*

- *$u = U + \nabla\zeta$ (resp. $u^{\varepsilon} = U^{\varepsilon} + \nabla\zeta^{\varepsilon}$) on $\mathcal{O}_k$,*
- *$\Delta\zeta = 0$ (resp. $\Delta\zeta^{\varepsilon} = 0$) on $\mathcal{O}_k$.*

In particular, we might decompose $I := I_{uu} + I_{uz} + I_{zz}$ where:

$$\begin{aligned} I_{uu} &:= \int \int_{E_k} U \otimes U : D(\tilde{w}), \\ I_{uz} &:= \int \int_{E_k} (U \otimes \nabla z + \nabla z \otimes U) : D(\tilde{w}), \\ I_{zz} &:= \int \int_{E_k} \nabla z \otimes \nabla z : D(\tilde{w}), \end{aligned}$$

and similarly for I^{ε}.

The convergence of the last term is handled by the following algebraic trick:

Proposition 4.2. *Let $\mathcal{O}$ be an open bounded set having a $C^{1,1}$ boundary. Let w and ζ be two functions satisfying:*

- *$u \in H^1(\mathcal{O})$ is divergence-free and $w \cdot n = 0$ on $\partial\mathcal{O}$,*
- *$\zeta \in H^1(\mathcal{O})$ and $\Delta\zeta = 0$ on $\mathcal{O}$.*

Then, there holds:

$$\int_{\mathcal{O}} \nabla\zeta \otimes \nabla\zeta : D(w) = 0. \tag{4.49}$$

Proof. The proof is a combination of algebraic relations and integration by parts:

$$\begin{aligned}\int_{\mathcal{O}} \nabla\zeta \otimes \nabla\zeta : D(w) &= \int_{\mathcal{O}} \nabla\zeta \otimes \nabla\zeta : \nabla w \\ &= -\int_{\mathcal{O}} \nabla \cdot [\nabla\zeta \otimes \nabla\zeta] \cdot u \quad \text{as } w \cdot n = 0 \text{ on } \partial\mathcal{O} \\ &= -\frac{1}{2}\int_{\mathcal{O}} \nabla|\nabla\zeta|^2 \cdot w \qquad \text{as } \Delta\zeta = 0 \text{ on } \mathcal{O} \\ &= \frac{1}{2}\int_{\mathcal{O}} |\nabla\zeta|^2\ \nabla \cdot w = 0.\end{aligned}$$

□

This proposition entails $I^{\varepsilon}_{zz} = I_{zz} = 0$. Combining the weak convergence of u^{ε} with the projector $P[H(\mathcal{O}_k)]$ we obtain that ∇z^{ε} (resp. U^{ε}) converges weakly to ∇z (resp. U) in $L^2((t_k, t_{k+1}) \times \mathcal{O}_k)$. We obtain finally that I^{ε}_{uu} and I^{ε}_{uz} converge to I_{uu} and I_{uz}, respectively, by a classical application of the following strong-convergence result:

Proposition 4.3. *The sequence $(U^{\varepsilon})_{\varepsilon>0}$ is compact in $L^2((t_k, t_{k+1}) \times \mathcal{O}_k)$.*

Proof. We detail here the proof sketched by Hoffmann and Starovoitov in [43, p. 640].

First, we have by construction that U^{ε} is bounded in $L^2((t_k, t_{k+1}); L^2(\mathcal{O}_k))$. The convergence of body domains implies that $(t_k, t_{k+1}) \times \mathcal{O}_k \subset \mathcal{Q}^{\varepsilon}_F$ for sufficiently small ε, with obvious notations. We might then introduce the trivial extension of any $\xi \in C^{\infty}_c((t_k, t_{k+1}); V(\mathcal{O}_k))$ as test-function in (4.32). This yields:

$$\left|\int_{t_k}^{t_{k+1}} U^{\varepsilon} \cdot \partial_t \xi\right| = \left|\int_{t_k}^{t_{k+1}} \int_{\mathcal{O}_k} 2\nu D(u^{\varepsilon}) : D(\xi) - u^{\varepsilon} \otimes u^{\varepsilon} : D(\xi)\right|$$

$$\leq C_T \|\xi; L^2((t_k, t_{k+1}); V(\mathcal{O}_k))\| \|u^{\varepsilon}; L^2(0, T; H^1(\Omega))\| \big(1 + \|u^{\varepsilon}; L^{\infty}(0, T; L^2(\Omega))\|\big)$$

Hence $\partial_t U^{\varepsilon}$ is bounded in the dual space $L^2((t_k, t_{k+1}); V(\mathcal{O}_k)^*)$. As $H(\mathcal{O}_k) \subset V(\mathcal{O}_k)^*$ with a compact embedding, we apply the Aubin–Lions lemma [53, Corollary 4] and obtain that $(U^{\varepsilon})_{\varepsilon>0}$ is compact in $L^2((t_k, t_{k+1}); V(\mathcal{O}_k)^*)$.

Second, we prove by a contradiction argument that for all $\eta > 0$ there exists $K(\eta) > 0$ for which, for all $\xi \in H^1(\mathcal{O}_k)$, there holds:

$$\|P[H(\mathcal{O}_k)](\xi); L^2(\mathcal{O}_k)\| \leq \eta\|\xi; H^1(\mathcal{O}_k)\| + K(\eta)\|P[H(\mathcal{O}_k)](\xi); V(\mathcal{O}_k)^*\|. \tag{4.50}$$

We argue by contradiction. If this statement were false, there would exist $\eta > 0$ such that, for all $K \in \mathbf{N}$ there is $\xi_K \in H^1(\mathcal{O}_k)$ such that

$$\|P[H(\mathcal{O}_k)](\xi_K); L^2(\mathcal{O}_k)\| > \eta\|\xi_K; H^1(\mathcal{O}_k)\| + K\|P[H(\mathcal{O}_k)](\xi_K); V(\mathcal{O}_k)^*\|.$$

Without restriction, we assume that $\|\xi_K; H^1(\mathcal{O}_k)\| = 1$. On the one hand, the embedding $H^1(\mathcal{O}_k) \subset L^2(\mathcal{O}_k)$ being compact and $P[H(\mathcal{O}_k)]$ being continuous with respect to the L^2-topology, there would exist a subsequence, that we do not relabel, and $\xi \in L^2(\mathcal{O}_k)$ such that:

$$\lim_{K\to\infty} P[H(\mathcal{O}_k)](\xi_K) = P[H(\mathcal{O}_k)](\xi) \text{ in } L^2(\mathcal{O}_k).$$

In particular, the sequence $P[H(\mathcal{O}_k)](\xi_K)$ is bounded so that:

$$\eta + K\|P[H(\mathcal{O}_k)](\xi_K); V(\mathcal{O}_k)^*\| < \|P[H(\mathcal{O}_k)](\xi_K); L^2(\mathcal{O}_k)\| \leq C \tag{4.51}$$

Consequently, there holds:

$$\lim_{K\to\infty} \|P[H(\mathcal{O}_k)](\xi_K); V(\mathcal{O}_k)^*\| = 0, \text{ and } P[H(\mathcal{O}_k)](\xi) = 0.$$

Conversely, (4.51) also implies:

$$\|P[H(\mathcal{O}_k)](\xi_K); L^2(\mathcal{O}_k)\| > \eta, \quad \forall\, K \in \mathbf{N}.$$

We obtain a contradiction.

The expected result is then a corollary of [53, Theorem 1]. Indeed, as $(U^\varepsilon)_{\varepsilon>0}$ is compact in $L^2((t_k, t_{k+1}); V(\mathcal{O}_k)^*)$, this theorem entails that:

$$\lim_{h\to 0} \|U^\varepsilon(t+h,\cdot) - U^\varepsilon(t,\cdot); L^2((t_k, t_{k+1}-h); V(\mathcal{O}_k)^*)\| = 0, \qquad \text{uniformly in } \varepsilon$$

On the other hand, applying (4.50) to U^ε and recalling that thanks to (4.34), $(u^\varepsilon)_{\varepsilon>0}$ is already bounded in $L^2((t_k, t_{k+1}); H^1(\mathcal{O}_k))$, we obtain that there exists a constant $C > 0$ for which:

$$\begin{aligned}&\|U^\varepsilon(t+h,\cdot) - U^\varepsilon(t,\cdot); L^2((t_k, t_{k+1}-h) \times \mathcal{O}_k)\| \\ &\qquad \leq C\eta + K(\eta)\|U^\varepsilon(t+h,\cdot) - U^\varepsilon(t,\cdot); L^2((t_k, t_{k+1}-h); V(\mathcal{O}_k)^*)\|,\end{aligned}$$

for arbitrary $\eta > 0$. Consequently, we also have:

$$\lim_{h\to 0} \|U^{\varepsilon}(t+h,\cdot) - U^{\varepsilon}(t,\cdot); L^2((t_k, t_{k+1}-h)\times \mathcal{O}_k)\| = 0 \quad \text{uniformly in } \varepsilon.$$

Finally, for all $t_k < \tau_1 < \tau_2 < t_{k+1}$, there holds:

$$\int_{\tau_1}^{\tau_2} U^{\varepsilon}(t)\mathrm{d}t = P[H(\mathcal{O}_k)]\left(\int_{\tau_1}^{\tau_2} u^{\varepsilon}(t)\mathrm{d}t\right).$$

Again, because of the uniform bound on $(u^{\varepsilon})_{\varepsilon>0}$ in $L^2((t_k, t_{k+1}); H^1(\mathcal{O}_k))$, we have that $(\int_{\tau_1}^{\tau_2} u^{\varepsilon}(t)\mathrm{d}t)_{\varepsilon>0}$ is relatively compact in $L^2(\mathcal{O}_k)$. The mapping $P[H(\mathcal{O}_k)]$ being continuous, this entails that $(\int_{\tau_1}^{\tau_2} U^{\varepsilon}(t)\mathrm{d}t)_{\varepsilon>0}$ is relatively compact in $L^2(\mathcal{O}_k)$. This ends the proof. □

From the construction we presented here, we obtain the following theorem:

Theorem 4.1. *Let the following assumptions hold true:*

- *Ω is a container such that $\mathbf{R}^d \setminus \Omega$ is connected,*
- *the shapes of the bodies $\mathcal{B}_i^0$ are open subsets of Ω such that $\overline{\mathcal{B}_i}^0$ is connected and* (4.19) *holds true,*
- *the densities $\rho_i^0 \in L^{\infty}(\mathcal{B}_i^0)$ satisfy* (4.1).

Given $u^0 \in H(\Omega)$, there exists $T > 0$ and a weak solution $((\varphi_i, \rho_i)_{i=1,\dots,n}, u)$ to (FRBI) *on $(0,T)$. Furthermore, no contact occurs between rigid boundaries on $(0,T)$ in this solution.*

To conclude, we note that we have used here an algebraic trick (namely Proposition 4.2) which enables to rule out the question of the compactness of the velocity-fields $(u^{\varepsilon})_{\varepsilon>0}$ in $L^2((0,T)\times\Omega)$. In [51] or in [16], two methods are proposed to obtain such a compactness result (see also [20]). Briefly, in [20, 51] the proof consists in projecting locally in time the u^{ε} on a set of velocity-fields which coincides with a rigid velocity-field on a fixed neighborhood $\tilde{\mathcal{S}}$ of the $\mathcal{S}(t)$ (this neighborhood is in particular locally constant in time and independent of ε). Adapting the arguments of this section, compactness of the projected sequence is obtained in $L^2((0,T)\times\Omega)$. Finally, a trace result due to Fujita and Sauer [22] entails that the sequence of projections gets closer to u^{ε} ("uniformly" in ε) when $\tilde{\mathcal{S}}$ gets closer to $\mathcal{S}(t)$. This yields the expected compactness of $(u^{\varepsilon})_{\varepsilon>0}$.

4.3.2 Motion of One Body in a Bounded Domain: Classical Solutions

Up to now, we solved (FRBI) in a weak setting via the introduction of extended fields defined on the whole container. Another approach to tackle this system is to fix the fluid-domain by applying a change of unknown. This transforms (FRBI) into a quasilinear system completed with boundary conditions interacting with the nonlinearities. This method is applied in [8,9,24,33,34,56–58] to study the existence and uniqueness of classical solutions to (FRBI) in the two-dimensional and three-dimensional cases. Before going further, we mention that constructing classical solutions to (FRBI) is also possible by showing that weak solutions satisfy better regularity estimates [15]. We choose to detail the alternative approach here to give a broader picture of the methods that have been introduced in order to handle (FRBI).

Constructing classical solutions to (FRBI) requires the treatment of three major difficulties. First, in the very definition of classical solutions, it has to be noted that the fluid velocity-field is defined on a moving domain. Hence, one must find a new way for stating the usual regularity of classical solutions to Navier–Stokes equations in cylindrical domains:

$$u_f \in H^1(0,T;L^2(\mathcal{F})) \cap C([0,T];H^1(\mathcal{F})) \cap L^2(0,T;H^2(\mathcal{F})).$$

Second, a standard method to construct classical solutions to nonlinear PDEs is to apply a perturbation argument relying on a linearized problem having good properties. A genuine difficulty here is that, even if the geometry is fixed via a change of unknown, the fluid and body dynamics are still coupled at the linearized level. It is tempting to discouple both equations (as in [33, 34]). However, dealing with the full system then requires to make smallness assumptions on the coupling. A monolithic approach introduced in [56–58] is necessary to handle the general case. Finally, it is also standard that the perturbation argument entails existence and uniqueness of classical solutions locally-in-time. In the two-dimensional case, it is possible to extend classical solutions up to the first time of collision between rigid bodies or between rigid bodies and the container boundary. This result is based on the classical remark that the H^1-norm of solutions to Navier Stokes equations do not blow-up in finite time in the two-dimensional case. However, compared to the usual case of Navier Stokes equations in cylindrical domains, supplementary difficulties arise here as the velocity-field u_f satisfies nonhomogeneous boundary conditions at the moving fluid/body interfaces. In this paragraph, we detail the treatment of these three difficulties in the two-dimensional case of one rigid body moving inside a bounded container. This part of the section relies on [56].

Definition of Classical Solutions. Let us consider the case of one body $\mathcal{B}_1$ moving inside a bounded container Ω. A first way to measure the regularity of a classical solution $((\xi_1,\omega_1),u_f,p_f)$ is to introduce a mapping transforming a velocity-field and pressure defined on the moving domain $\mathcal{Q}_F$ into something defined on the cylindrical domain $Q_F := (0,T) \times \mathcal{F}^0$. One then measures the regularity of the images of u and p through this mapping. Of course, this mapping depends on the solution itself.

For instance, assuming, on the one hand, that $((\xi_1,\omega_1),u_f,p_f)$ is a classical solution on $(0,T)$ having sufficient time-regularity i.e.,

$$\xi_1 \in H^1(0,T), \quad \omega_1 \in H^1(0,T),$$

and, on the other hand, that the distance between $\mathcal{B}_1(t)$ and $\partial\Omega$ remains bounded from below by a strictly positive constant, it is possible to construct a vector-field $\Lambda \in H^1((0,T); C_c^\infty(\Omega))$ satisfying the following properties:

- $\mathrm{div}\Lambda = 0$ on $(0,T) \times \Omega$,
- $\Lambda(t,x) = \xi_1 + \omega_1(x - G_1)^\perp$ in a neighborhood of $\mathcal{B}_1(t)$, for all $t \in [0,T)$,
- $\mathrm{Supp}(\Lambda(t,\cdot)) \subset \Omega$ for all $t \in [0,T]$.

Classical results on differential systems entail that the associated flow $(t,x) \mapsto X(t,x)$ solution to

$$\begin{aligned} \partial_t X(t,y) &= \Lambda(t, X(t,y)), && \forall\,(t,y) \in (0,T) \times \Omega, \\ X(t,y) &= y, && \forall\, y \in \Omega, \end{aligned}$$

is then well defined globally and satisfies:

- $X \in C^1([0,T]; \mathrm{Diff}^\infty(\Omega))$ (where $\mathrm{Diff}^\infty(\Omega)$ stands for the set of smooth diffeomorphisms of Ω),
- $X(t, \mathcal{B}_1^0) = \mathcal{B}_1(t)$, for all $t \in (0,T)$,
- $X(t,\cdot)$ is an isometric mapping on $\mathcal{B}_1^0$ for all $t \in [0,T]$,
- $X(t,y) = y$ for all $(t,y) \in (0,T) \times \partial\Omega$.

Hence, we might define the associated change of unknown:

$$A(t,y) = a(t, X(t,y)), \quad \forall\,(t,y) \in Q_F. \tag{4.52}$$

We keep the convention that capital letters are associated with the change of unknown computed in this construction. Next, a classical solution is defined as follows:

Definition 4.5. Given $T > 0$, we call classical solution to (FRBI) on $(0,T)$ any collection $((\xi_1, \omega_1), u_f, p_f)$ satisfying

- $\xi_1 \in H^1(0,T)$ and $\omega_1 \in H^1(0,T)$;
- there exists $\delta > 0$ such that the associated body motion $t \mapsto \mathcal{B}_1(t)$ satisfies

$$\mathrm{dist}(\mathcal{B}_1(t), \partial\Omega) > \delta, \quad \forall\, t \in (0,T);$$

- $U_f \in H^1(0,T; L^2(\mathcal{F}^0)) \cap C([0,T]; H^1(\mathcal{F}^0)) \cap L^2(0,T; H^2(\mathcal{F}^0))$;
- $P_f \in L^2(0,T; H^1(\mathcal{F}^0))$;
- $((\xi_1, \omega_1), u_f, p_f)$ satisfies (FRBI) almost everywhere.

We note that, as long as no contact occurs, straightforward computations entail that the regularity statements in the above definition yield:

$$\partial_t u_f \in L^2(Q_F), \quad (u_f, \nabla u_f, \nabla^2 u_f) \in L^2(Q_F), \quad \nabla p_f \in L^2(Q_F). \tag{4.53}$$

Consequently, the last statement in this definition makes sense.

Definition 4.5 will be easy to handle as it fits the construction scheme we detail below. In particular, it will be obvious that the classical solutions we obtain satisfy the regularity statements of this definition. A drawback is that the definition requires the introduction of the change of variable X whose existence relies on the no-contact assumption. In particular, it seems more adapted to introduce an "eulerian" definition of classical solution in order to be able to consider contact in what follows. This is the aim of the following definition we adapt from [26, 39]:

Definition 4.6. Given $T > 0$, we call classical solution to (FRBI) on $(0, T)$ any collection $((\xi_1, \omega_1), u_f, p_f)$ such that:

- $\xi_1 \in H^1(0, T)$ and $\omega_1 \in H^1(0, T)$;
- there exists $\delta > 0$ such that the associated body motion $t \mapsto \mathcal{B}_1(t)$ satisfies

$$\mathrm{dist}(\mathcal{B}_1(t), \partial\Omega) > \delta, \quad \forall\, t \in (0, T);$$

- $u_f \in H^1(\mathcal{Q}_F)$ with $\nabla^2 u_f \in L^2(\mathcal{Q}_F)$;
- $p_f \in L^2_{loc}(\mathcal{Q}_F)$ satisfies $\nabla p_f \in L^2(\mathcal{Q}_F)$;
- $((\xi_1, \omega_1), u_f, p_f)$ satisfies (FRBI) almost everywhere.

We note that, in this second definition, we consider u_f and p_f as space/time functions defined on the (open) set $\mathcal{Q}_F$. In particular, the condition $u_f \in H^1(\mathcal{Q}_F)$ includes *time and space* derivatives. This has to be compared with the other regularity statements which do only involve space derivatives.

Fortunately, both definitions of classical solution are consistent. In order to avoid confusion, we provide the following proposition:

Proposition 4.4. *Definitions 4.5 and 4.6 are equivalent.*

Proof. Following the remark after Definition 4.5, and in particular (4.53), we have that any classical solution $((\xi_1, \omega_1), u_f, p_f)$ in the sense of Definition 4.5 is a classical solution in the sense of Definition 4.6.

Conversely, letting $((\xi_1, \omega_1), u_f, p_f)$ be a classical solution in the sense of Definition 4.6, similar computations to the ones entailing (4.53) yield that

$$U_f \in H^1(0, T; L^2(\mathcal{F}^0)) \cap L^2(0, T; H^2(\mathcal{F}^0)), \quad P_f \in L^2(0, T; H^1(\mathcal{F}^0)).$$

The only point requiring more care is the proof that $U_f \in C([0, T]; H^1(\mathcal{F}^0))$. To this end, we first note that

- on $\partial\mathcal{B}_1^0$, there holds

$$U_f(t, y) = \xi_1 + \omega_1 (y - G_1^0)^{\perp} \in H^1(0, T; C^{\infty}(\partial\mathcal{B}_1^0));$$

- on $\partial\Omega$, U_f vanishes.

Consequently, we might construct a velocity-field $U_{\mathrm{bdy}} \in H^1(0, T; H^2(\mathcal{F}^0))$ lifting these boundary conditions and such that $V := U_f - U_{\mathrm{bdy}}$ satisfies also

$$V \in H^1(0,T;L^2(\mathcal{F}^0)) \cap L^2(0,T;H^2(\mathcal{F}^0)).$$

We have then (up to a regularization argument we skip for conciseness):

$$\frac{\mathrm{d}}{\mathrm{d}t}\int_{\mathcal{F}^0} |\nabla V(t,y)|^2 \mathrm{d}y = -\int_{\mathcal{F}^0} \partial_t V \cdot \Delta V \in L^1(0,T).$$

Consequently, $\|V(t,\cdot); H^1(\mathcal{F}^0)\| \in C([0,T])$. Combining this relation with $\partial_t V \in L^2((0,T)\times\mathcal{F}^0)$ entails that $V \in C([0,T]; H^1_0(\mathcal{F}^0)) - w$ and then

$$V \in C([0,T]; H^1_0(\mathcal{F}^0)).$$

Adding the function U_{bdy} (for which we already have $U_{\mathrm{bdy}} \in C([0,T]; H^1(\mathcal{F}^0))$), we get finally that $U \in C([0,T]; H^1(\mathcal{F}^0))$. This ends the proof. □

Analysis of the Linearized Problem. As we mentioned above, to construct solutions to (FRBI) with the regularity of Definition 4.5 or Definition 4.6, one way is to go into a fixed geometry by introducing a change of unknown similar to (4.52). Actually, in order to keep the divergence-free condition, a better-adapted change of unknown reads:

$$\tilde{U}_f(t,y) = [\nabla X]^{-1}(t, X(t,y))\, u_f(t, X(t,y)),$$
$$\tilde{P}_f(t,y) = p_f(t, X(t,y)).$$

System (FRBI) is then transformed into a quasilinear PDE in terms of the collection $((\xi_1,\omega_1),\tilde{U}_f,\tilde{P}_f)$ that we do not write for conciseness. Let us only mention that, when $((\xi_1,\omega_1),u_f,p_f)$ is small, the diffeomorphisms $\{X(t,\cdot), t \in (0,T)\}$ are close to the identity. Consequently, the linearized version of this quasilinear system reads:

$$\left.\begin{aligned} \partial_t U - \nu\Delta U + \Delta P &= F_{NS} \\ \operatorname{div} U &= 0 \end{aligned}\right\} \text{ in } Q_F := (0,T)\times\mathcal{F}^0, \tag{4.54}$$

with boundary conditions

$$U(t,y) = \xi + \omega(y - G_1^0)^\perp, \quad \text{on } \partial\mathcal{B}_1^0, \tag{4.55}$$

$$U(t,y) = 0, \quad \text{on } \partial\Omega, \tag{4.56}$$

and coupled with Newton laws:

$$m_1\dot{\xi} = -\int_{\partial\mathcal{B}_1^0} \mathbb{T}(U,P)n\mathrm{d}\sigma + F_{LM}, \tag{4.57}$$

$$\mathbb{J}_1\dot{\omega} = -\int_{\partial\mathcal{B}_1^0} (y - G_1^0)^\perp \cdot \mathbb{T}(U,P)n\mathrm{d}\sigma + F_{AM}. \tag{4.58}$$

In this system F_{NS}, F_{LM}, F_{AM} are, respectively, source terms in the Navier–Stokes equations, linear momentum equation, and angular momentum equation. The system is completed with initial data:

$$U(0,\cdot) = U^0, \quad \xi(0) = \xi^0, \quad \omega(0) = \omega^0. \tag{4.59}$$

In the works of Grandmont and Maday [33, 34], this system is treated uncoupling at first the Stokes system (4.54) with boundary conditions (4.55)–(4.56) and the Newton laws (4.57)–(4.58). Namely, initial conditions and data (F_{NS}, F_{LM}, F_{AM}) being fixed, the Stokes system (resp. the Newton laws) is solved for given boundary data $t \mapsto (\xi(t), \omega(t))$, (resp. for a given fluid flow $(t,x) \mapsto (U(t,x), P(t,x))$). A solution to the coupled linearized system is then obtained by constructing a fixed point of the mapping:

$$\begin{aligned}(\xi,\omega) &\longrightarrow (U,P) \text{ solution to (4.54)–(4.55)–(4.56)}\\ &\longrightarrow (\bar{\xi},\bar{\omega}) \text{ solution to (4.57)–(4.58).}\end{aligned} \tag{4.60}$$

The shortcoming of this method appears computing rough energy estimates. Indeed, let us consider the case where source term F_{NS}, F_{LM}, and F_{AM} vanish. Then, given sufficiently smooth boundary data $t \mapsto (\xi(t), \omega(t))$, we introduce a smooth divergence-free vector-field U_{bdy} which lifts boundary conditions (4.55)–(4.56). The difference $V = U - U_{\text{bdy}}$ where U is the solution to (4.54)–(4.55)–(4.56) with initial data U^0 satisfies:

$$\left.\begin{aligned}\partial_t V - \nu\Delta V + \Delta P &= -\partial_t U_{\text{bdy}} + \nu\Delta U_{\text{bdy}}\\ \operatorname{div} V &= 0\end{aligned}\right\} \text{ in } Q_F, \tag{4.61}$$

and vanishes on boundaries. Classical energy methods yield the estimates:

$$\begin{aligned}\sup_{t\in(0,T)} \|V(t,\cdot); L^2(\mathcal{F}^0)\| &+ \nu \int_0^T \|\nabla V(t,\cdot); L^2(\mathcal{F}^0)|^2\\ &\leq C_0 + C_F \int_0^T \|\partial_t U_{\text{bdy}} - \nu\Delta U_{\text{bdy}}; L^2(\mathcal{F}^0)\|^2,\\ &\leq C_0 + C_F \left(\|\xi; H^1(0,T)\|^2 + \|\omega; H^1(0,T)\|^2\right),\end{aligned}$$

and

$$\begin{aligned}\sup_{t\in(0,T)} \nu\|\nabla V(t,\cdot); L^2(\mathcal{F}^0)\| &+ \int_0^T \left(\|V; H^2(\mathcal{F}^0)\|^2 + \|\nabla P; L^2(\mathcal{F}^0)\|^2\right)\\ &\leq C_0 + C_F \left(\|\xi; H^1(0,T)\|^2 + \|\omega; H^1(0,T)\|^2\right),\end{aligned}$$

with a constant C_F which depends only on the geometrical properties of the fluid domain $\mathcal{F}^0$. Next, one computes solutions to (4.57)–(4.58) with data $(t,x) \mapsto (U,P)$. Applying classical trace inequalities entails:

$$\begin{aligned}\|\bar{\xi}; H^1(0,T)\|^2 + \|\bar{\omega}; H^1(0,T)\|^2 & \\ &\leq C_0 + \frac{C_F}{m_1^2+\mathbb{J}_1^2}\int_0^T \|U; H^2(\mathcal{F}^0)\|^2 + \|\nabla P; L^2(\mathcal{F}^0)\|^2, \\ &\leq C_0 + \frac{C_F}{m_1^2+\mathbb{J}_1^2}\left(\|\xi; H^1(0,T)\|^2 + \|\omega; H^1(0,T)\|^2\right).\end{aligned}$$

These formal considerations show that constructing a fixed-point to (4.60) requires to control a little time-regularity of ξ and ω and to assume smallness of the coupling (i.e., here that m_1 and $\mathbb{J}_1$ are sufficiently large). Indeed, we need the coefficient $C_F/(m_1^2+\mathbb{J}_1^2)$ to be small enough in order that the mapping (4.60) fixes some ball of $H^1(0,T)$.

This limitation is overcome in the works of T. Takahashi and co-authors by solving the full system (4.54)–(4.55)–(4.56)–(4.57)–(4.58) at once. In [58], this linear problem is endowed with the semi-group structure associated with the generator A defined as follows. Let us denote:

$$D(A) = \left\{ U \in K[\mathcal{B}_1^0, \Omega] \text{ such that } U_{|_{\mathcal{F}^0}} \in H^2(\mathcal{F}^0) \right\},$$

and set $A = P[K^0[\mathcal{B}_1^0,\Omega]]\mathcal{A}U$ where:

$$\mathcal{A}U := \begin{cases} \nu\Delta U, & \text{in } \mathcal{F}^0, \\ F + T(y-G_1^0)^{\perp}, & \text{in } \mathcal{B}_1^0, \end{cases}$$

with the conventions that:

$$F := -\frac{2\nu}{m}\int_{\partial\mathcal{B}_1^0} D(U)n\mathrm{d}\sigma, \quad T := -\frac{2\nu}{\mathbb{J}_1}\int_{\partial\mathcal{B}_1^0}(z-G_1^0)^{\perp}\cdot D(U)n\mathrm{d}\sigma,$$

and that $P[K^0[\mathcal{B}_1^0,\Omega]]$ stands for the orthogonal projector $L^2(\Omega,\rho\mathrm{d}y) \to K^0[\mathcal{B}_1^0;\Omega]$.

We remark that in this "classical solution framework," we actually reintroduce the extended unknowns that were used in the "theory of weak solutions" developed in the previous section. In particular, we introduced here the extended density

$$\rho = \rho_1 \mathbf{1}_{\mathcal{B}_1^0} + \mathbf{1}_{\mathcal{F}^0},$$

and used abusively symbol U to denote extended velocity-fields. We recall that the spaces $K[\mathcal{B}_1^0,\Omega]$ and $K^0[\mathcal{B}_1^0,\Omega]$ are introduced in Definition 4.1 and that the restriction of the $L^2(\Omega, \rho \mathrm{d}y)$ scalar product to $K^0[\mathcal{B}_1^0, \Omega]$ reads:

$$\begin{aligned}(U, V) &= \int_{\mathcal{F}^0} U \cdot V + \int_{\mathcal{B}_1^0} \rho_1 U \cdot V, \\ &= \int_{\mathcal{F}^0} U \cdot V + m_1 \xi_V \cdot \xi_U + \mathbb{J}_1 \omega_V \omega_U,\end{aligned}$$

with the notations

$$U_{|_{\mathcal{B}_1^0}}(x) = \xi_U + \omega_U (x - G_1^0)^\perp, \qquad \text{resp.} \quad V_{|_{\mathcal{B}_1^0}}(x) = \xi_V + \omega_V (x - G_1^0)^\perp.$$

In [58], the authors prove that the operator $(D(A), A)$ is self-adjoint and positive yielding a contraction semi-group $t \mapsto S(t)$ on $K^0[\mathcal{B}_1^0, \Omega]$. We note that, as classical in incompressible fluid problems, the formalism presented here gets rid of the pressure P. This pressure is reintroduced afterwards as follows. In the case F_{NS}, F_{LM}, and F_{AM} vanish, let $U(t, \cdot) := S(t)U^0$. For sufficiently smooth data U^0, we have $U(t, \cdot) \in D(A)$ for all $t > 0$, so that we set

$$U(t, x) = \begin{cases} \xi_1(t) + \omega_1(t)(y - G_1^0)^\perp, & \text{on } \mathcal{B}_1^0, \\ U_f(t, x), & \text{on } \mathcal{F}^0. \end{cases}$$

By construction, U_f is divergence-free and satisfies boundary conditions:

$$U_f(t, y) = \xi_1 + \omega_1 (y - G_1^0)^\perp, \quad \forall\, y \in \partial\mathcal{B}_1^0, \quad U_f(t, y) = 0 \quad \text{on } \partial\Omega.$$

Then, we have

$$\partial_t U = AU \qquad \text{on } (0, T). \tag{4.62}$$

Multiplying this equation by the trivial extension of any $W \in \mathcal{D}(\mathcal{F}^0)$ yields that:

$$\int_{\mathcal{F}^0} (\partial_t U_f - \nu \Delta U_f) \cdot W = 0.$$

Applying DeRham theory, we construct a pressure P_f such that $((\xi_1, \omega_1), U_f, P_f)$ satisfies (4.54). Multiplying then (4.62) with any $W \in C_c^\infty(\Omega) \cap K[\mathcal{B}_1^0, \Omega]$ yields:

$$\begin{aligned}& m_1 \xi_W \cdot \dot{\xi}_1 + \mathbb{J}_1 \omega_W \dot{\omega}_1 + \int_{\mathcal{F}^0} \partial_t U \cdot W \\ &= \int_{\mathcal{F}^0} \nu \Delta U \cdot W - \int_{\partial\mathcal{B}_1^0} 2\nu D(U) n \mathrm{d}\sigma \cdot \xi_W - \int_{\partial\mathcal{B}_1^0} 2\nu (y - G_1^0)^\perp \cdot D(U) n \mathrm{d}\sigma \omega_W\end{aligned}$$

Introducing (4.54) and integrating by parts, we obtain finally:

$$m_1\xi_W \cdot \dot{\xi}_1 + \mathbb{J}_1\omega_W\dot{\omega}_1 = -\int_{\partial\mathcal{B}_1^0} \mathbb{T}(U,P)n\mathrm{d}\sigma \cdot \xi_W$$

$$-\int_{\partial\mathcal{B}_1^0} (y - G_1^0)^\perp \cdot \mathbb{T}(U,P)n\mathrm{d}\sigma\, \omega_W.$$

As ξ_W and ω_W are arbitrary, we obtain that Newton laws (4.57)–(4.58) are also satisfied.

Combining this analysis of the linearized problem with the computation of nonlinear terms (convective terms and geometrical terms due to the change of unknown), one obtains local-in-time existence and uniqueness of classical solutions (see [56, Proposition 6.1]):

Theorem 4.2. *Let* $\varepsilon > 0$ *and* $C_1 > 0$. *Given* $\mathcal{B}_1^0 \subset \Omega$ *and* $(\xi_1^0, \omega_1^0, u_f^0) \in \mathbf{R}^2 \times \mathbf{R} \times H^1(\mathcal{F}^0)$ *such that:*

$$\mathrm{dist}(\mathcal{B}_1^0, \partial\Omega) > \varepsilon \quad \|u^0; H^1(\mathcal{F}^0)\| + |\xi_1^0| + |\omega_1^0| < C_1.$$

Assume initial data satisfy compatibility conditions:

$$\mathrm{div}\, u_f^0 = 0, \quad on\ \mathcal{F}^0, \tag{4.63}$$

$$u_f^0(x) = \xi_1^0 + \omega_1^0(x - G_1^0)^\perp, \quad on\ \partial\mathcal{B}_1^0, \tag{4.64}$$

$$u_f^0(x) = 0, \quad on\ \partial\Omega, \tag{4.65}$$

then there exists T_0 *depending only on* ε *and* C_1 *such that there exists a unique classical solution* $((\xi_1, \omega_1), (u_f, p_f))$ *to (FRBI) with initial condition* $((\xi_1^0, \omega_1^0), u_f^0)$ *on* $(0, T_0)$.

Computation of Regularity Estimates. It is well known that Theorem 4.2 implies the existence of a unique maximal solution for arbitrary initial data $((\xi_1^0, \omega_1^0), u_f^0)$ satisfying compatibility conditions (4.63)–(4.64)–(4.65). Given the restrictions on the time T_0 in this local-in-time existence result, we also have the following blow-up alternative. Let $(T_{\max}, ((\xi_1, \omega_1), u_f, p_f))$ be a maximal solution to (FRBI) then, there holds:

- either $T_{\max} = +\infty$,
- either $T_{\max} < \infty$ and we have:

$$\limsup_{t \to T_{\max}} \left\{ \|u_f(t,\cdot); H^1(\mathcal{F}(t))\| + |\xi_1(t)| + |\omega_1(t)| + \frac{1}{\mathrm{dist}(\mathcal{B}_1(t), \partial\Omega)} \right\} = 0.$$

In the two-dimensional case that we detail here, we might prove that blow-up only arises because of contact between $\mathcal{B}_1(t)$ and $\partial\Omega$. Thanks to the previous alternative, this amounts to computing an a priori estimate on $\|u_f(t,\cdot); H^1(\mathcal{F}(t))\| + |\xi_1(t)| + |\omega_1(t)|$ locally-in-time. To this end, we go back in the moving geometry. As the Navier Stokes equations (4.9) are satisfied in $L^2(\mathcal{Q}_F)$, it can be multiplied by u_f yielding that classical solutions satisfy the basic kinetic-energy estimate (4.18). Consequently, we have a uniform bound on $|\xi_1(t)|+|\omega_1(t)|$ and the problem reduces to compute a bound on $\|u_f(t,\cdot); H^1(\mathcal{F}(t))\|$.

In the remainder of this paragraph, we fix one maximal solution to (FRBI), $(T_{\max}, ((\xi_1, \omega_1), u_f, p_f))$. When there are no rigid bodies inside the container, regularity estimate for the velocity-field is obtained by multiplying the Navier Stokes equations with $\partial_t u_f$. When there are bodies in the container, one way to extend this estimate is to multiply the Navier Stokes equations with:

$$v := v_1 - v_2 \quad v_1 := \partial_t u_f + \Lambda \cdot \nabla u_f, \quad v_2 := u_f \cdot \nabla \Lambda,$$

where Λ is an extension of the velocity-field of $\mathcal{B}_1$ such as defined in the first paragraph of this section (see p. 283). We point out that the first part of v stands for a material time-derivative of u_f while the second part compensates the fact that v_1 is not divergence-free. Both properties appear in the computations below. Indeed, assume for simplicity that u is smooth in the fluid domain. Multiplying (4.9) with v, we get:

$$\int_{\mathcal{F}(t)} \partial_t u_f \cdot v + \int_{\mathcal{F}(t)} \mathrm{div}\mathbb{T}(u_f, p_f) \cdot v = -R_0, \tag{4.66}$$

where

$$R_0 = \int_{\mathcal{F}(t)} (u_f \cdot \nabla u_f) \cdot v.$$

On the left-hand side, a straightforward decomposition of v yields:

$$\int_{\mathcal{F}(t)} \partial_t u_f \cdot v = \int_{\mathcal{F}(t)} |\partial_t u_f|^2 + R_1 \quad \text{with } R_1 := \int_{\mathcal{F}(t)} \partial_t u_f \cdot (\Lambda \cdot \nabla u_f - u_f \cdot \nabla \Lambda).$$

Next, we integrate by parts the second term:

$$\int_{\mathcal{F}(t)} \mathrm{div}\mathbb{T}(u_f, p_f) \cdot v = \int_{\partial\mathcal{F}(t)} \mathbb{T}(u_f, p_f)n \cdot v \mathrm{d}\sigma - \int_{\mathcal{F}(t)} \mathbb{T}(u_f, p_f) : \nabla v.$$

The construction of v yields that:

- $v(t,x) = (\dot{\xi}_1 - \omega_1 \xi_1^{\perp}) + \dot{\omega}_1 (x - G_1(t))^{\perp}$ on $\partial\mathcal{B}_1(t)$,
- v vanishes on $\partial\Omega$,
- $\mathrm{div}\, v = 0$ on $\mathcal{Q}_F$,
- $D(v) = D(\partial_t v) + \Lambda \cdot \nabla D(v) + \nabla u \nabla \Lambda - D(u \cdot \nabla \Lambda)$ on $\mathcal{Q}_F$.

Consequently, we have:

$$\int_{\partial\mathcal{F}(t)} \mathbb{T}(u_f, p_f)n \cdot v \mathrm{d}\sigma = -m_1|\dot{\xi}_1|^2 - \mathbb{J}_1|\dot{\omega}_1|^2 + R_1$$

with $R_1 := m_1\omega_1\xi_1^{\perp} \cdot \dot{\xi}_1$. Similarly, we have:

$$\begin{aligned}\int_{\mathcal{F}(t)} \mathbb{T}(u_f, p_f) : \nabla v &= 2\nu \int_{\mathcal{F}(t)} D(u_f) : D(v) \\ &= \nu \frac{\mathrm{d}}{\mathrm{d}t} \int_{\mathcal{F}(t)} |D(u_f)|^2 + R_2\end{aligned}$$

with:

$$R_2 := 2\nu \int_{\mathcal{F}(t)} D(u_f) : (\nabla u_f \nabla\Lambda - D(u_f \cdot \nabla\Lambda)).$$

Finally, (4.66) reads:

$$\nu \frac{\mathrm{d}}{\mathrm{d}t} \int_{\mathcal{F}(t)} |D(u_f)|^2 + \int_{\mathcal{F}(t)} |\partial_t u_f|^2 + m_1|\dot{\xi}_1(t)|^2 + J_1|\dot{\omega}_1(t)|^2 = R, \qquad (4.67)$$

with $R = R_0 + R_1 - R_2$. This remainder term is controlled by the energy on the right-hand side reproducing the classical arguments applied in the case of Navier Stokes equations on cylindrical domains: interpolation estimates to control the remainder terms with the H^2-norm of u_f and ellipticity of the stationary Stokes problem to bound this H^2-norm by the L^2-norm of $\partial_t u_f$. Several new terms are involved because of the presence of Λ. However, as Λ is smooth and depends on the solution only through ξ_1 and ω_1 they do not represent critical terms. We refer the reader to [8, 9] for more details. Beyond the control of these remainder terms, we also point out that these references apply a careful density argument to justify the formal integration by parts above. Indeed, we assume here that $\partial_t u_f$ is sufficiently smooth in the fluid domain. However the only a priori information we have is $\partial_t u_f \in L^2(\mathcal{F}(t))$ with $\mathrm{div}\partial_t u_f = 0$. Consequently, only the normal trace of $\partial_t u_f$ on $\partial\mathcal{B}_1(t)$ is a priori well defined in $H^{-1/2}(\partial\mathcal{B}_1(t))$. Computing the trace of v on $\partial B_1(t)$ requires thus special care.

To conclude this section, we mention that the construction of classical solutions has been extended to the case of bodies having singular shapes (namely, $C^{1,\alpha}$ with $\alpha \in (0,1)$) [26]. In this latter case, H^2-regularity of the velocity-field is out of reach as elliptic regularity for the stationary Stokes problem no longer holds in this function-space for non-smooth fluid domains. It is replaced with a BMO-gradient regularity. We also note that the study of the semi-group $S(t)$, that is introduced in the second paragraph, has been extended to a non-Hilbertian framework in [25] and [19]. The motivation in the first of these papers is to extend the Cauchy theory to non-newtonian fluids while the second reference aims at computing the long-time behavior of solutions to (FRBI) in an unbounded container ($\Omega = \mathbf{R}^2$).

4.4 The Contact Issue

We proceed by analyzing the way classical and weak solutions to (FRBI) handle contacts between rigid bodies or between a body and the container boundary. Several questions are in order. First, in the framework of classical solutions that we detailed in Sect. 4.3.2, contacts between rigid boundaries imply blow-up of the classical solution because they are excluded *by definition*. However, with our definitions, one may continue *a priori* a classical solution after contact by a weak solution. It is then critical to determine what regularity of the solution is lost in the extension process. It is also necessary to prove that weak solutions do exist after contact and to discuss the interpretation of these solutions. Indeed, we derived the definition of weak solution starting from a classical solution for which no contact occurs and explained the computations backwards as long as no contact occurs. We extended then arbitrarily the definition of weak solution after contact as it made sense. However, the computations backwards are no longer valid after a contact and a novel approach is required. These questions arise only if contact occurs. The last (but not least) question is to determine if and when contact between rigid boundaries do actually hold in (weak or classical) solutions to (FRBI).

The section is organized as follows. We first deal with the blow-up criterion of classical solutions. We then tackle the behavior of weak solutions with contact. These two parts rely on the two notes by Starovoitov [54, 55]. In the last part, we discuss contact occurrence in solutions to (FRBI). We detail in this part the studies [26, 38–40]. Similar results are obtained by T.I. Hesla in his Ph.D. dissertation for other two-dimensional configurations [37].

4.4.1 *Blow-Up Criterion for Classical Solutions*

We first mention a heuristic argument which shows that a solution might not remain smooth through a contact. Indeed, assuming that the fluid does not slip on solid boundaries, we imposed that the boundaries of rigid bodies and of the container follow the characteristics associated with the velocity-field u_f. Contact between bodies may then be interpreted as a crossing of two fluid-characteristics. Hence, if contact occurs in $T_* > 0$, contradicting the Lipschitz criterion for uniqueness in differential systems entails:

$$\int_0^{T*} \|u_f(t,\cdot); C^{0,1}(\mathcal{F}(t))\| \mathrm{d}t = +\infty.$$

We aim here at generalizing this condition in the framework of Sobolev spaces, keeping in mind that all constants in Sobolev imbeddings depend on the geometry of the domain. Our main result reads as follows:

Theorem 4.3. *Let* $((\xi_i, \omega_i)_{i=1,\dots,n}, u_f, p_f)$ *be a classical solution to (FRBI) on* $(0, T_*)$ *and assume that contact occurs in* T_*. *Then,*

(i) there holds:

$$\int_0^{T_*} \|\nabla u_f(t,\cdot); L^{d+1}(\mathcal{F}(t))\| \mathrm{d}t = +\infty; \tag{4.68}$$

(ii) if $n = 1$ *and* $\mathcal{B}_1^0$ *is a ball, there also holds:*

$$\int_0^{T_*} \|\nabla^2 u_f(t,\cdot); L^2(\mathcal{F}(t))\| \mathrm{d}t = +\infty. \tag{4.69}$$

We note that both criteria do not prevent from continuing a solution s.t. $\|u_f; L^2(\mathcal{F}(t))\|$ and $\|\nabla u_f; L^2(\mathcal{F}(t))\|$ remain bounded with time. However, the second implication prevents from requiring that $\nabla^2 u_f \in L^2(\mathcal{Q}_F)$ after collision.

Proof of Item (i). Let us consider $((\xi_i, \omega_i)_{i=1,\dots,n}, u_f, p_f)$ a classical solution to (FRBI) on some time-interval $(0, T_*)$. We define:

$$\mathcal{B}_0(t) := \mathbf{R}^d \setminus \overline{\Omega}, \quad \forall\, t \in [0, T_*),$$

and

$$d[\{\mathcal{B}_i(t)\}_{i=0,\dots,n}] := \inf \left\{ \mathrm{dist}(\mathcal{B}_i(t), \mathcal{B}_j(t)), i \neq j \in \{0, \dots, n\}^2 \right\}, \quad \forall\, t \in [0, T_*).$$

By assumption, we have

$$\liminf_{t \to T_*} d[\{\mathcal{B}_i(t)\}_{i=0,\dots,n}] = 0.$$

The cornerstone of the proof is the following proposition adapted from [54, Theorem 3.1]:

Proposition 4.5. *Let* $B_1 := B(X_1, \delta)$ *and* $B_2 := B(X_2, \delta)$ *be two disjoint balls of* $\mathbf{R}^d$ *and denote*

$$e_{12} := \frac{(X_2 - X_1)}{|X_2 - X_1|}, \qquad d_{12} := \mathrm{dist}(B_1, B_2).$$

Let $\xi \in W^{1,p}(\mathbf{R}^d)$ *such that*

- $\nabla \cdot \xi = 0$ *on* $\mathbf{R}^d$,
- $\xi(x) := \xi_i + \omega_i \times (x - X_i), \quad \forall\, x \in B_i$.

Then, if d_{12} is small, there exists a constant C depending only on the radius δ and the dimension d, for which:

$$|(\xi_2 - \xi_1) \cdot e_{12}| \leq C |d_{12}|^{p} \, \|\nabla \xi \, ; \, L^p(\mathbf{R}^d)\|, \tag{4.70}$$

where:

$$p = \frac{1}{2}\left(3 - \frac{d+1}{p}\right).$$

Proof. For the proof, we introduce $e_1, \ldots, e_d$ an orthonormal basis of $\mathbf{R}^d$ and $(x_1, \ldots, x_d)$ the corresponding components of $x \in \mathbf{R}^d$. Without further restriction, we assume that

$$B_1 := B(-\delta e_d, \delta), \quad B_2 := B((\delta + d_{12})e_d, \delta).$$

This yields in particular that $e_{12} = e_d$. For legibility, we denote $d_{12} =: h$ in what follows. In a neighborhood of 0, the boundaries of B_1 and B_2 are, respectively, parametrized by:

$$x_d = \gamma_1(|(x_1, \ldots, x_{d-1})|), \qquad x_d = \gamma_2(|(x_1, \ldots, x_{d-1})|),$$

where, for $s \in [0, \delta)$:

$$\gamma_1(s) = -\delta + \sqrt{\delta^2 - s^2}, \qquad \gamma_2(s) = h + \delta - \sqrt{\delta^2 - s^2}.$$

Finally, for arbitrary $l \in (0, \delta)$, we denote Ω_l the symmetric set between B_1 and B_2 with horizontal width $2l$:

$$\Omega_l := \Big\{(x_1, \ldots, x_d) \text{ s.t. } |(x_1, \ldots, x_{d-1})| < l, \\ x_d \in \Big(\gamma_1(|x_1, \ldots, x_{d-1}|), \gamma_2(|(x_1, \ldots, x_{d-1})|)\Big)\Big\}.$$

Given $l \in (0, \delta)$, integrating that ξ is divergence-free on Ω_l yields:

$$\int_{\partial\Omega_l \cap (\partial B_1 \cup \partial B_2)} \xi \cdot n \, \mathrm{d}\sigma + \int_{\partial\Omega_l \setminus (\partial B_1 \cup \partial B_2)} \xi \cdot n \, \mathrm{d}\sigma = 0. \tag{4.71}$$

On the one hand, we replace ξ with its explicit values on $B_1 \cup B_2$. This entails that:

$$\int_{\partial\Omega_l \cap (\partial B_1 \cup \partial B_2)} \xi \cdot n \, \mathrm{d}\sigma = \kappa_d l^{d-1}[(\xi_2 - \xi_1) \cdot e_d], \tag{4.72}$$

where $\kappa_2 = 2$ and $\kappa_3 = \pi$. On the other hand, we rewrite:

$$\int_{\partial\Omega_l \setminus (\partial B_1 \cup \partial B_2)} \xi \cdot n \mathrm{d}\sigma = \int_{\gamma_1}^{\gamma_2} \int_{|(x_1,\ldots,x_{d-1})|=l} \xi \cdot n\, l^{d-2} \mathrm{d}\sigma' \, \mathrm{d}x_d .$$

Here "$d\sigma' = \delta_{x_1=l} + \delta_{x_1=-l}$" in the two-dimensional case and "$\mathrm{d}\sigma' = \mathrm{d}\theta$" corresponding to cylindrical coordinates (r, θ, z) associated with (x_1, x_2, x_3) in the three-dimensional case. We note that:

$$\int_{|(x_1,\ldots,x_{d-1})|=l, x_d=\gamma_1(l)} \xi \cdot n \mathrm{d}\sigma' = 0.$$

Consequently, we bound from above the lateral flux by combining a Hölder inequality with a Poincaré inequality:

$$\left| \int_{\partial\Omega_l \setminus (\partial B_1 \cup \partial B_2)} \xi \cdot n \mathrm{d}\sigma \right| \\ \leq (\gamma_2(l) - \gamma_1(l))^{2-\frac{1}{p}} l^{(d-2)(1-\frac{1}{p})} \left[\int_{|(x_1,\ldots,x_{d-1})|=l} |\nabla \xi|^p \mathrm{d}\sigma \right]^{\frac{1}{p}} . \quad (4.73)$$

Finally, introducing (4.72) and (4.73) into (4.71) yields:

$$|\kappa_d l^{d-1} [(\xi_2 - \xi_1) \cdot e_d]| \\ \leq (\gamma_2(l) - \gamma_1(l))^{2-\frac{1}{p}} l^{(d-2)(1-\frac{1}{p})} \left[\int_{|(x_1,\ldots,x_{d-1})|=l} |\nabla \xi|^p \mathrm{d}\sigma \right]^{\frac{1}{p}} .$$

We can now integrate this identity between $l = 0$ and $l = r \in (0, \delta)$. As $\gamma_2 - \gamma_1$ is non-decreasing, this entails that:

$$|[(\xi_2 - \xi_1) \cdot e_d]| \leq C(\gamma_2(r) - \gamma_1(r))^{2-\frac{1}{p}} r^{(d-1)(1-\frac{1}{p})-d} \|\nabla \xi; L^p(\mathbf{R}^d)\|,$$

with a constant C depending on d. For h sufficiently small, we might choose $r = \sqrt{h}$ to obtain:

$$|[(\xi_2 - \xi_1) \cdot e_d]| \leq \tilde{C} h^{\frac{1}{2}(3-(d+1)\frac{1}{p})} \|\nabla \xi; L^p(\mathbf{R}^d)\|,$$

where $\tilde{C}$ depends now on d and δ. □

We conclude the proof of Theorem 4.3, item (i), as follows. Assume that the distance $d[\{\mathcal{B}_i(t)\}_{i=0,\ldots,n}]$ goes to 0 for a sequence of time $t \to T_*$. Then, by a compactness argument, there exist X_1^0 and X_2^0 such that $B(X_1^0, \delta) \subset \mathcal{B}_i^0$ and

$B(X_2^0,\delta) \subset \mathcal{B}_j^0$ for which

$$\liminf_{t\to T^*} \mathrm{dist}(B(\mathbb{Q}_i(t)X_1^0 + G_i(t),\delta), B(\mathbb{Q}_j(t)X_2^0 + G_j(t),\delta)) = 0.$$

Let us now denote by u the extension of u_f on Ω as defined in (4.21) (in full rigor, we should say that u corresponds to the trivial extension to $\mathbf{R}^d$ of the field defined in (4.21)). For all $t \in (0, T_*)$ this extended velocity-field satisfies the assumption of Proposition 4.5 with

$$B_1 := B(\mathbb{Q}_i(t)X_1^0 + G_i(t),\delta), \quad B_2 := B(\mathbb{Q}_j(t)X_2^0 + G_j(t),\delta).$$

Hence, we have, with $p = d + 1$ and the notations of this proposition, for t close to T_* (say $t \in (T_1, T_*)$):

$$|(\xi_2(t) - \xi_1(t)) \cdot e_{12}| \le C d_{12}(t)\, \|\nabla u(t,\cdot)\,;\, L^{d+1}(\mathbf{R}^d)\|.$$

In our case, $(\xi_2(t) - \xi_1(t)) \cdot e_{12} = \dot{d}_{12}(t)$, so that this last inequality reads:

$$\left| \frac{d}{\mathrm{d}t}\Big[\ln d_{12}\Big](t) \right| \le C\, \|\nabla u(t,\cdot)\,;\, L^{d+1}(\mathbf{R}^d)\|.$$

Integrating between T_1 and t going to T_* and introducing that d_{12} vanishes for a sequence of $t \to T_*$, we obtain:

$$\int_0^{T_*} \|\nabla u(t,\cdot)\,;\, L^{d+1}(\mathbf{R}^d)\| \mathrm{d}t = +\infty.$$

It remains then to remark that $\nabla u_{|\mathcal{B}_i(t)}$ coincides with the angular velocity of $\mathcal{B}_i$ so that the decay of the kinetic energy of the whole system implies that

$$\|\nabla u(t,\cdot)\,;\, L^{d+1}(\mathbf{R}^d)\| = \|\nabla u_f(t,\cdot)\,;\, L^{d+1}(\mathcal{F}(t))\| + O(1).$$

Proof of Item (ii). The proof of this second item is similar to the first one. It relies on the following variant of Proposition 4.5:

Proposition 4.6. *Let Ω be a smooth domain and $B_1 := B(X_1,\delta) \subset \Omega$ such that* $\mathrm{dist}(B_1, \partial\Omega) > 0$. *Given $X \in \partial\Omega$, let us denote:*

$$e_X := \frac{(X_1 - X)}{|X_1 - X|}, \quad d_X := \mathrm{dist}(B_1, X).$$

For arbitrary $\varepsilon > 0$ and $\xi \in H_0^1(\Omega) \cap H^2(\Omega \setminus \overline{B}_1)$ such that

- $\nabla \cdot \xi = 0$ *on* Ω,
- $\xi(x) := \xi + \omega \times (x - X_1), \quad \forall\, x \in B_1,$

there exists a constant C depending only on the radius δ, the dimension d and the parameter ε, for which:

$$|\xi \cdot e_X| \leq C|d_X|^{\frac{9-d}{4}-\varepsilon} \, \|\nabla^2 \xi \, ; \, L^2(\Omega \setminus \overline{B}_1)\|. \tag{4.74}$$

Proof. We note that, as ξ vanishes on $\partial\Omega$, inequality (4.74) is a straightforward consequence to classical trace inequalities for large values of d_X. This inequality is thus only relevant for small d_X. We restrict to this case in what follows.

The proof is then similar to the proof of Proposition 4.5. We introduce similar notations. We keep $(x_1, \ldots, x_d)$ as system of coordinates such that $B_1 = B((d_X + \delta)e_d, \delta)$ and replace d_X with h. In the neighborhood of the origin we keep γ_1 and γ_2 for the respective parametrizations of ∂B_1 and $\partial\Omega$. The only modification here is that γ_2 depends a priori on all variables $(x_1, \ldots, x_{d-1})$ and that we no longer have an explicit value for γ_2. Furthermore, we do not assume here that ∂B_1 and $\partial\Omega$ share the same tangent space in the origin. Consequently, γ_1 admits the following expansion in $(x_1, \ldots, x_{d-1}) = 0$:

$$\gamma_2(x_1, \ldots, x_{d-1}) = L^\top \cdot (x_1, \ldots, x_{d-1}) + O(|(x_1, \ldots, x_{d-1})|^2).$$

for some $L \in \mathbf{R}^{d-1}$. We note that the orientation of our system of coordinates is fixed by B_1 so that L depends on B_1 (but the second order term does not and the expansion is valid when $(x_1, \ldots, x_{d-1})$ ranges a neighborhood $\mathcal{V}$ depending on X only). However, when B_1 is sufficiently close to X, we prove that all possible values of L satisfy a uniform bound. Indeed, as $\overline{B}_1 \subset \Omega$ we have that

$$\gamma_1(|(x_1, \ldots, x_{d-1})|) > \gamma_2(x_1, \ldots, x_{d-1}), \quad \forall \, (x_1, \ldots, x_{d-1}) \in \mathcal{V}.$$

For h sufficiently small, we might apply this inequality to all $(x_1, \ldots, x_{d-1}) = h^{\frac{1}{2}+\varepsilon}(y_1, \ldots, y_{d-1})$ with $|(y_1, \ldots, y_{d-1})|$ in $B(0,1)$. This yields that $|L| \leq Ch^{\frac{1}{2}-\varepsilon}$ for h sufficiently small w.r.t. ε.

We now introduce similar domains Ω_l as in the proof of Proposition 4.5 and integrate that ξ is divergence-free on these domains. Repeating the computations in the proof of Proposition 4.5 yields:

$$\int_{\partial\Omega_l \cap (\partial B_1 \cup \partial\Omega)} \xi \cdot n \, \mathrm{d}\sigma = \kappa_d l^{d-1} [\xi \cdot e_d], \tag{4.75}$$

and

$$\int_{\partial\Omega_l \setminus (\partial B_1 \cup \partial\Omega)} \xi \cdot n \mathrm{d}\sigma = \int_{\gamma_1}^{\gamma_2} \int_{|(x_1, \ldots, x_{d-1})| = l} \xi \cdot n \, l^{d-2} \mathrm{d}\sigma' \mathrm{d}x_d.$$

Here we note that ξ vanishes on $\partial\Omega$ and is equal to a rigid motion on ∂B_1. Symmetry arguments then yield that:

$$\int_{|(x_1,\dots,x_{d-1})|=l, x_d=\gamma_1(l)} \xi\cdot n\, l^{d-2}\mathrm{d}\sigma' = \int_{|(x_1,\dots,x_{d-1})|=l, x_d=\gamma_2(l)} \xi\cdot n\, l^{d-2}\mathrm{d}\sigma' = 0.$$

Consequently, we apply the Poincaré inequality two times to obtain:

$$\left|\int_{\partial\Omega_l\setminus(\partial B_1\cup\partial\Omega)} \xi\cdot n\mathrm{d}\sigma\right| \\ \leq (\gamma_2(l)-\gamma_1(l))^{\frac{5}{2}} l^{\frac{d-2}{2}} \left[\int_{|(x_1,\dots,x_{d-1})|=l} |\nabla^2\xi|^2\mathrm{d}\sigma\right]^{\frac{1}{2}}. \quad (4.76)$$

Combining (4.75) and (4.76) and integrating between $l=0$ and $l=r\in(0,\delta)$, we get:

$$|\xi\cdot e_d| \leq C\frac{(\gamma_2(r)-\gamma_1(r))^{\frac{5}{2}}}{r^{\frac{d+1}{2}}}.$$

We finally obtain the expected result by setting $r=h^{\frac{1}{2}-\tilde{\varepsilon}}$ for a sufficiently small $\tilde{\varepsilon}$. □

The end of the proof of Theorem 4.3, item (ii), follows the line of the previous item. Briefly, we first construct $X\in\partial\Omega$ such that contact holds in X when $t\to T_*$. Then we integrate the bound (4.74) for t between 0 and T_* noticing that the exponent on the right-hand side of (4.74) might be chosen greater than 1 for ε sufficiently small.

Comments. In [54], the author extends the content of Proposition 4.5 to more general geometries. For bodies having smooth shapes, this would enable to compute sharper estimates on the blow-up of the $W^{1,p}$ norm of the solution when contact occurs. For bodies having shapes with typical regularity $C^{1,\alpha}$, the inequality proven by V. Starovoitov yields that an equivalent blow-up criterion to (4.68) holds with $p=d+1$ replaced by $p_\alpha=(p+\alpha)/\alpha$. This criterion is applied in [26] to prove that classical solutions to (FRBI) blow up also when contact occurs if rigid bodies do not have smooth boundaries.

In the framework of weak solutions, inequality (4.70) furnishes other types of information. Indeed, reproducing the arguments in the end of the proof of item (i), we get that, if contact occurs in a weak solution at time T_*, then we can construct two balls B_1^0 and B_2^0 inside two distinct rigid bodies such that $h(t):=\mathrm{dist}(B_1(t),B_2(t))$ satisfies $\liminf h(t)\to 0$ when $t\to T_*$ and

$$\frac{|\dot{h}(t)|}{|h(t)|^{\frac{5-d}{4}}} \leq C\,\|\nabla u(t,\cdot);L^2(\Omega)\|.$$

We note here that $(5-d)/4$ is always smaller than 1 so that this last inequality does not rule out contact. Nevertheless, if contact occurs in T^* we might integrate the above inequality between t and T_* for arbitrary $t < T_*$, this yields that:

$$|h(t)|^{\frac{d-1}{4}} = o((T_* - t)^{\frac{1}{2}}), \text{ as } \int_0^{T_*} \|\nabla u(t,\cdot); L^2(\Omega)\|^2 \mathrm{d}t < \infty.$$

Adding that $d[\{\mathcal{B}_i(t)\}_{i=0,\dots,n}] \leq h(t)$, we finally obtain:

- $d[\{\mathcal{B}_i(t)\}_{i=0,\dots,n}] = o((T_* - t)^2)$ when $d = 2$, yielding that a contact is glueing in the sense that both relative velocity and acceleration of colliding bodies vanish at contact-time;
- $d[\{\mathcal{B}_i(t)\}_{i=0,\dots,n}] = o(T_* - t)$ when $d = 3$.

In the two-dimensional case, the content of this remark is due to [51].

The second criterion (4.69) appears in [40]. We generalize it here to the case of a container Ω having a smooth but arbitrary boundary. Prior to contact, a solution $((\xi_i, \omega_i)_{i=1,\dots,n}, u_f, p_f)$ with smooth initial datum satisfies:

$$\int_0^T \left[\|\nabla^2 u_f; L^2(\mathcal{F}(t))\|^2 + \|\nabla p_f; L^2(\mathcal{F}(t))\|^2\right] \mathrm{d}t < \infty.$$

The second criterion shows that this estimate does not hold any longer after a contact. We believe it is possible to generalize the computation to arbitrary geometries. The difficulties here are that, on the one hand, the second order derivatives of the extended velocity-field are not a priori L^2-functions on Ω, because of the discontinuities in the stress tensor on solid boundaries. So, we might not reduce our computations to the case of rigid disks as in the first criterion. On the other hand, we need the symmetries of the ball $\mathcal{B}_1$ in order that the lateral flux cancels on $\partial\mathcal{B}_1$ and $\partial\Omega$. To generalize the computations above to arbitrary configurations, it should be possible to define domains $(\Omega_l)_{l>0}$ adapted to the shape of the rigid bodies in order that this cancellation property is preserved.

4.4.2 *On Weak Solutions with Contact*

In Sect. 4.3.1, we presented a proof of existence of local-in-time weak solutions to (FRBI). Prior to studying the meaning of weak solutions in case of contact, we show in this section that it is possible to construct global-in-time weak solutions, whether contact occurs or not.

Global Existence of Weak Solutions. Following the arguments in [20], we obtain:

Theorem 4.4. *Let the following assumptions hold true:*

- *Ω is a container having a smooth boundary,*

- *the shapes of the bodies $\mathcal{B}_i^0$ are open connected subsets of Ω having smooth boundaries and such that* (4.19) *holds true,*
- *the densities $\rho_i^0 \in L^\infty(\mathcal{B}_i^0)$ satisfy* (4.1).

Given $u^0 \in H(\Omega)$ and $T > 0$, there exists a weak solution $((\varphi_i, \rho_i)_{i=1,\dots,n}, u)$ to (FRBI) *on* $(0, T)$.

The proof is a combination of [20, Lemma 2.2] and of the construction of weak solutions adapted to non-smooth body shapes that we presented in Sect. 4.3.1. We recall the main ingredients.

First Step: Extension of Weak Solutions up to Contact. In Theorem 4.1, existence of weak solutions is proven locally in time. In order to extend existence of weak solutions up to contact between rigid bodies, we remark that the time of existence is only limited by the L^2-norm of the initial data and the initial distance between rigid bodies. Existence of weak solutions up to contact then yields from a concatenation principle for weak solutions to (FRBI) that we adapt from [20, Lemma 2.2 and Sect. 4].

Let $((\varphi_i, \rho_i)_{i=1,\dots,n}, u)$ be a weak solution with initial data $((\mathcal{B}_i^0, \rho_i^0)_{i=1,\dots,n}, u^0)$ defined on $(0, T_0)$. Let us denote $\mathcal{M}_i^t$ the isometries such that

$$\mathcal{M}_i^t(\mathcal{B}_i^0) = \mathcal{B}_i(t), \quad \forall\, t \in (0, T_0).$$

As $u \in L^\infty(0, T; L^2(\Omega))$ and the eulerian velocity of $\mathcal{M}_i^t$ coincides with u, we have that $\mathcal{M}_i^t \in W^{1,\infty}(0, T) \subset C([0, T_0])$. Hence, we might define

$$\tilde{\mathcal{B}}_i^0 = \mathcal{M}_i^{T_0}(\mathcal{B}_i^0), \quad \forall\, i = \{1, \dots, n\}.$$

Assuming that we remain far from contact, these new shapes are open subsets of Ω having smooth boundaries and satisfying (4.19). Similarly, when $t \to T_0$ the densities $\rho_i(t, \cdot)$ converge almost everywhere (and thus in all L^p-spaces for finite p) to

$$\tilde{\rho}_i^0(x) = \rho_i^0([\mathcal{M}_i^{T_0}]^{-1} x)\mathbf{1}_{\tilde{\mathcal{B}}_i^0}.$$

Obviously, these new initial densities satisfy (4.1). Finally, for all $w \in \mathcal{D}(\Omega)$ s.t. $D(w) = 0$ in the neighborhood of $\mathcal{S}(T)$, we might introduce $w\chi$ as a test-function in the weak formulation (4.32), for arbitrary $\chi \in C_c^\infty(0, T_0)$, with support sufficiently close to T_0. This yields that

$$f(t) = \int_\Omega \rho(t, \cdot) u(t, x) \cdot w(t, x) \in H^1(0, T_0) \subset C([0, T_0]).$$

Because of energy estimate (4.34), there holds:

$$|f(t)| \leq \sqrt{\mathcal{E}_c(0)} \| \sqrt{\rho(t, \cdot)} w; L^2(\Omega) \|.$$

Hence, the limits of the functions f when $t \to T$ define a linear form

$$L : \{w \in \mathcal{D}(\Omega) \text{ s. t. } D(w) = 0 \text{ on } \mathcal{S}(T)\} \to \mathbf{R}$$

s.t.

$$|L(w)| \leq \sqrt{\mathcal{E}_c(0)} \|w; L^2(\Omega, \tilde{\rho}^0(x)\mathrm{d}x)\|.$$

where

$$\tilde{\rho}^0 := \sum_{i=1}^{n} \tilde{\rho}_i^0 + (1 - \sum_{i=1}^{n} \mathbf{1}_{\tilde{\mathcal{B}}_i^0}).$$

Applying the Hahn–Banach theorem in $(H(\Omega)$ endowed with the Hilbert structure associated with the norm on $L^2(\Omega, \tilde{\rho}^0 \mathrm{d}x)$, we finally construct $\tilde{u}^0 \in H(\Omega)$ such that:

$$L(w) = \int_{\Omega} \tilde{\rho}^0(x) \tilde{u}^0(x) \cdot w(x) \mathrm{d}x, \quad \forall\, w \in H(\Omega),$$

and such that:

$$\int_{\Omega} \tilde{\rho}^0(x) |\tilde{u}^0(x)|^2 \mathrm{d}x \leq \mathcal{E}_c(0).$$

The new initial data $((\tilde{\mathcal{B}}_i^0, \tilde{\rho}_i^0)_{i=1,\dots,n}, \tilde{u}^0)$ are suitable for applying Theorem 4.1. We obtain $\tilde{T}_0$ and a weak solution $((\tilde{\varphi}_i, \tilde{\rho}_i)_{i=1,\dots,n}, \tilde{u})$ without contact on $(0, \tilde{T}_0)$. Concatenating then the solutions $((\varphi_i, \rho_i)_{i=1,\dots,n}, u)$ and $((\tilde{\varphi}_i, \tilde{\rho}_i)_{i=1,\dots,n}, \tilde{u})$, classical computations yield a solution on $(0, T_0 + \tilde{T}_0)$.

Second Step: Extension of Weak Solutions Through a Contact. We now assume that a solution $((\varphi_i, \rho_i)_{i=1,\dots,n}, u)$ exist on $(0, T_0)$ and that contact occurs in T_0. Proceeding exactly as in the first step, we construct $((\tilde{\mathcal{B}}_i^0, \tilde{\rho}_i^0)_{i=1,\dots,n}, u^0)$. Following [20, Sect. 4], we then set $\tilde{\mathcal{B}}_0^0 = \mathbf{R}^d \setminus \overline{\Omega}$ and construct $(M_0, \dots, M_{n'})$ a partition of $\{0, \dots, n\}$ corresponding to the partition of the body domain $\overline{\tilde{\mathcal{S}}^0} \cup \overline{\tilde{\mathcal{B}}_0^0}$ into connected components:

$$\overline{\tilde{\mathcal{S}}^0} \cup \bigcup_{i=1}^{n} \overline{\tilde{\mathcal{B}}}_i^0 = \bigcup_{i=1}^{n'} \bigcup_{j \in M_i} \overline{\tilde{\mathcal{B}}}_j^0.$$

We then set:

$$\hat{\mathcal{B}}_i^0 = \bigcup_{j \in M_i} \tilde{\mathcal{B}}_j^0, \quad \hat{\rho}_i^0 = \sum_{j=1}^{n} \tilde{\rho}_j^0.$$

By construction, the new container $\Omega = \mathbf{R}^d \setminus \overline{\hat{\mathcal{B}}_0^0}$ and initial data $((\hat{\mathcal{B}}_i^0, \hat{\rho}_i^0)_{i=1,\dots,n'}, \hat{u})$ satisfy the assumptions of Theorem 4.1. This yields $\hat{T}_0$ and a weak solution to (FRBI) $((\hat{\varphi}_i, \hat{\rho}_i)_{i=1,\dots,n'}, \hat{u})$. Applying the associated isometries to the $\tilde{\mathcal{B}}_i^0$, we might then split

$$\hat{\mathcal{B}}_i(t) = \bigcup_{j\in M_i} \tilde{\mathcal{B}}_j(t), \quad \hat{\varphi}_i = \sum_{i=1}^{n} \tilde{\varphi}_i \quad \hat{\rho}_i = \sum_{j\in M_i} \tilde{\rho}_i .$$

where:

$$\tilde{\varphi}_j(t,\cdot) = \mathbf{1}_{\tilde{\mathcal{B}}_j(t)}, \quad \tilde{\rho}_j = \hat{\rho}_i \tilde{\varphi}_j \quad \forall\, i \in \{1,\dots,\tilde{n}\} \quad \forall\, j \in M_i .$$

Finally straightforward computations show that we obtain a weak solution on $(0, T_0 + \hat{T}_0)$ when concatenating $((\varphi_i, \rho_i)_{i=1,\dots,n}, u)$ and $((\tilde{\varphi}_i, \tilde{\rho}_i)_{i=1,\dots,n}, \hat{u})$. This ends the proof.

Properties of Weak Solutions with Contact. We extended the definition of weak solutions for (FRBI) after contact. However, even if the initial body shapes had smooth boundaries and the fluid velocity-field u_f of the weak solution were smooth, there would remain several discrepancies between the weak solution and what we expect a classical solution of (FRBI) to be.

The first difference comes from the set of test-functions that we have chosen. Indeed, we require that a test-function w coincides with a rigid velocity-field on *a neighborhood* of $\overline{\mathcal{S}(t)}$ for all t. This restriction does not enable to distinguish between rigid bodies which belong to the same connected component of $\overline{\mathcal{S}(t)}$, or equivalently, between rigid bodies in contact. For instance, let two indices i and j satisfy $\overline{\mathcal{B}_i(t)} \cap \overline{\mathcal{B}_j(t)} \neq \emptyset$ for $t \in (0,T)$. Then, we have that any test-function $w \in \mathcal{K}(\mathcal{Q}_S)$ satisfy:

$$w(t,x) = \xi + \omega \times x, \quad \forall\, x \in \mathcal{B}_i(t) \cup \mathcal{B}_j(t).$$

Consequently, setting $\omega = 0$, letting ξ take arbitrary values and reproducing formally the computations of Sect. 4.3.1, we do not get (4.12) for i and j separately but merely the equation:

$$\frac{\mathrm{d}}{\mathrm{dt}}\left[m_i \xi_i + m_j \xi_j\right](t) = -\int_{\partial\mathcal{B}_i(t)\cup\partial\mathcal{B}_j(t)} \mathbb{T} n \,\mathrm{d}\sigma.$$

Similarly, only one equation holds for a combination of ω_i and ω_j. Consequently, the system is algebraically underconstrained unless a compatibility condition between (ξ_i, ω_i) and (ξ_j, ω_j) is implied by the property that contact holds between $\mathcal{B}_i$ and $\mathcal{B}_j$.

This latter question is analyzed by Starovoitov in [54, Sect. 2]. In the case of bodies with smooth boundaries, he proves:

Proposition 4.7. *Let $\mathcal{B}_1$ and $\mathcal{B}_2$ be two disjoint open subsets having smooth boundaries and assume $X \in \overline{\mathcal{B}_1} \cap \overline{\mathcal{B}_2}$. Given $u \in H^1(\mathbf{R}^d)$ such that*

$$u = \xi_i + \omega_i \times (x - X), \quad \forall\, x \in \mathcal{B}_i, \quad \text{for } i = 1, 2,$$

there holds:

- *if $d = 2$: $\xi_1 = \xi_2$ and $\omega_1 = \omega_2$,*
- *if $d = 3$: $\xi_1 = \xi_2$ and the projections of ω_1 and ω_2 on the common tangent space to $\partial\mathcal{B}_1$ and $\partial\mathcal{B}_2$ in X coincide.*

The proof is a variant of the proof of Proposition 4.5. We refer the reader to [54, Theorems 2.1 and 2.2] for more details. In both dimensions, this result yields that, if contact occurs on a curve or on a surface, then the rigid bodies in contact have the same rigid velocity. In particular, in case of contact, the boundary conditions (4.10) might only prescribe different values in isolated points.

In the three-dimensional case, this proposition shows that we have indeed one more degree of freedom than the number of equations that are imposed by the weak formulation. The system is thus really underconstrained. We note also that, in the construction we presented above, if there is contact between two rigid bodies $\mathcal{B}_i$ and $\mathcal{B}_j$ at time $T > 0$, then

- the rigid bodies $\mathcal{B}_i$ and $\mathcal{B}_j$ remain stuck for ever after T,
- the rigid bodies $\mathcal{B}_i$ and $\mathcal{B}_j$ share the same velocity-field after T.

In particular, whereas it is possible that the rigid bodies roll one on the other right before contact occurs, we impose that they do not after contact. The construction above creates a discontinuity in the body motions that seems to be non-physical.

In the two-dimensional case, Proposition 4.7 yields that rigid bodies in contact share the same velocity-fields. Hence, the definition of weak solution is algebraically well-posed and the construction above corresponds to "sticky contacts." An open question is to determine whether weak solutions remain unique after contact taking into account this supplementary rebound law. We recall that uniqueness prior to contact is already proven in [30].

Without a supplementary rebound law, Starovoitov shows that uniqueness is lost [55]. We detail briefly his construction below. This construction holds in the two-dimensional space $\mathbf{R}^2$, endowed with the orthonormal basis (e_1, e_2), in the case $n = 1$, $\Omega = B(0, R)$ and $\mathcal{B}_1^0 = B(0, r)$. Let χ be a truncation function $[0, R] \to \mathbf{R}$ such that:

$$\chi(r) = 1 \quad \chi'(r) = 0 = \chi(R) = \chi'(R).$$

For arbitrary $b \in [0, (R - r))$, let us denote F_b a smooth diffeomorphism $B(0, R) \setminus B(0, r) \to B(0, R) \setminus B(be_1, r)$ and G_b its converse application (see [55, p. 4895] for explicit formulas). Let then introduce, for all $b \in [0, R-r)$:

$$\xi_b(x) = \begin{cases} \nabla^{\perp}(x_2\phi(|G_b(x)|)), & \forall\, x \in B(0,R) \setminus \overline{B(be_1, r)}, \\ e_1, & \forall\, x \in B(be_1, r). \end{cases}$$

The vector-field ξ satisfies the following straightforward properties:

$$\xi_b \in V(B(0,R)) \cap C^{\infty}(B(0,R) \setminus B(be_1, r)), \quad \forall\, b \in (0, R-r).$$

On the basis of explicit formulas for G, it satisfies also the supplementary ones:

Proposition 4.8. *There exists an absolute constant $C < \infty$, such that, for all $b \in [0, R-r)$:*

$$\|\xi_b; L^2(B(0,R))\| \le C$$
$$\|\nabla\xi_b; L^2(B(0,R))\| \le \frac{C}{(R-r-b)^{\frac{3}{4}}}$$

Then, one looks for a solution of the form:

$$u(t,x) = \dot{b}(t)\xi_{b(t)}(x), \quad \forall\, t \in (0, 2T), \quad \forall\, x \in B(0,R).$$

Given $T > 0$, and b given by:

$$\sigma(t) := \left(1 - \frac{t}{T}\right)^4, \quad b(t) = \sigma(t)(R-r), \quad \forall\, t \in (0, 2T),$$

it yields that, $(\mathbf{1}_{B(b(t)e_1, r)}, u)$ is a weak solution to (FRBI) with a source term

$$g \in L^2(0, T; [V(B(0,R))]^*).$$

We would also be able to construct a weak solution $(\bar{\varphi}, \bar{u})$ to (FRBI) with the initial condition $(\mathbf{1}_{B(0,r)}, u(0)))$ and the same source term $g \in L^2(0, T; [V(B(0,R))]^*)$ applying the method of the previous subsection. However both weak solutions might not coincide. Indeed, in our construction the body $\mathcal{B}_1$ remains stuck to the container boundary after contact whereas it splits from the boundary in the case of the solution constructed by V. Starovoitov. Hence, in the two-dimensional case, uniqueness of the weak solution including contact does not hold for arbitrary initial condition without a supplementary rebound law. We point out that, in this construction, existence of a contact is enforced by the introduction of a singular source term g.

4.4.3 Contact vs No-Contact

We discuss now on the possibility of contacts in weak and classical solutions to (FRBI). This question is tackled by Vázquez and Zuazua [61] on the following one-dimensional baby-model introduced in [60]:

$$\begin{cases} \partial_t u + \kappa u \partial_x u - \nu \partial_{xx} u = 0, & \text{in } \mathbf{R} \setminus \{h_i(t)\}_{i=1,\dots,n} \\ u(t, h_i(t)) = \dot{h}_i(t), & \text{for } i = 1, \dots, n, \\ [\nu \partial_x u](t, h_i(t)) = m_i \ddot{h}_i(t), & \text{for } i = 1, \dots, n. \end{cases} \tag{4.77}$$

In this system the h_i's stand for the position of the bodies (which are points in the one-dimensional case). For simplicity, we assume that they are numbered in increasing order: $h_i < h_{i+1}$. The two last equations mimic the no-slip boundary conditions and body dynamics, respectively. We used the convention that:

$$[u](t,h) = \lim_{x \to h, x > h} u(t,x) - \lim_{x \to h, x < h} u(t,x), \quad \forall\, t > 0, \quad \forall\, h \in \mathbf{R}.$$

In the first equation, of viscous Burgers type, we introduced the parameter $\nu > 0$ which stands for the viscosity of the fluid, and $\kappa \in \mathbf{R}$ a dimensionless parameter.

For this system, J.L. Vázquez and E. Zuazua prove among other results that, whatever the choice of the initial data:

$$u^0 \in L^2(\mathbf{R}), \quad (h_i^0, \dot{h}_i^0) \in \mathbf{R}^{2n}, \quad \text{s.t.} \inf_{i \in \{1,\dots,n-1\}} |h_{i+1}^0 - h_i^0| > 0,$$

the system (4.77) admits a unique global solution in which no contact between rigid bodies occurs in finite time:

$$\inf_{i \neq j} |h_{i+1}(t) - h_i(t)| > 0, \quad \forall\, t > 0.$$

In this one-dimensional case, the regularity estimate for classical solutions yields the bound:

$$\int_0^T \left[\sum_{i=1}^{n-1} \int_{h_i(t)}^{h_{i+1}(t)} |\partial_{xx} u(t,z)|^2 \mathrm{d}z + \sum_{i=1}^{n} m_i |\ddot{h}_i|^2 \right] \mathrm{d}t < C_0(T), \tag{4.78}$$

where C_0 depends on initial data and T only. Applying classical arguments to the no-slip boundary condition seen as a differential equation, this control on $\partial_{xx} u$ induces that h_{i+1} might not collide h_i in finite time. The result extends to the case of L^2 initial data due to the classical smoothening properties of the viscous Burgers equations. In the multi-dimensional case, the regularity estimate for solutions to (FRBI) relies on ellipticity of the Stokes problem which depends itself on the minimal distance between two rigid boundaries (see [15, Sect. 4.1]).

As a consequence, it is unlikely to obtain an equivalent bound to (4.78) prior to a detailed analysis of the contact issue.

First studies tackling the question on contact occurrence, in the multi-dimensional case, go back to the 1960s. In a series of papers, M.E. O'Neill and his collaborators consider a rigid sphere moving close to a plane wall and compute the forces exerted by a solution to the stationary Stokes system on the sphere [6, 7, 13, 50]. They apply methods previously developed by Brenner [3] to the case where the ratio distance between the ramp and the body *vs* radius of the body is small. They show that the drag, which the sphere undergoes, diverges rapidly when the ratio goes to 0 and prevents the sphere from touching the wall in finite time. This is called afterwards the *no-collision paradox*. Similar computations in the lubrication approximation are gathered in [10]. In the more recent papers [26, 38–40], the authors show that the *no-collision paradox* for the Stokes system extends to solutions to (FRBI) in many cases. In [38, 39], the two-dimensional and three-dimensional cases of a sphere, or a cylinder, moving close to a plane wall, are considered. As in the studies of M.E. O'Neill and his collaborators, it is proven that no-contact between the rigid body and the wall occurs in finite time. In [39], an example of a three-dimensional configuration in which contact occurs is exhibited. However, this construction is limited to a very peculiar configuration so that the *no-collision paradox* seems to hold generically.

All the results concerning the full system (FRBI) are obtained by applying a multiplier method. This multiplier is constructed thanks to a detailed analysis of solutions to the stationary Stokes problem and the associated drag. In this section, we first present an efficient method to compute the drag in the frame of the Stokes problem. We then recall the Lorentz formula associated with the Stokes problem and discuss its application in the extension of the no-collision paradox to solutions to (FRBI) yielding the results in [26, 38–40].

To conclude this introductory part on the contact issue, we note that all the mentioned results underline that the system (FRBI) is not relevant to describe the close-contact interactions between rigid bodies. In real life, one expects contact between rigid bodies to occur in much more general contexts than the one exhibited in [40], see [46] for experiments. In order to derive (FRBI) we assumed implicitly that the relative velocity of two bodies is slow w.r.t. their distance so that we might neglect several phenomena in the fluid and bodies behaviors. In particular, when the distance between rigid bodies becomes very small with the bodies having fast relative velocities:

- the fluid pressure diverges so that non-newtonian properties [2] and compressibility [48] might become critical in the fluid behavior;
- the fluid strain-tensor becomes large so that slip at the fluid/bodies interface [42] and also elasticity in the bodies equations [12] should be considered;
- the fluid-layer is thin so that asperities in the description of the bodies surfaces and container boundaries should be included [62].

In [26–29], the influence of various roughness models for fluid/bodies interactions (singular body boundaries and slip at the fluid/body interface) is discussed. With any of these models, the no-collision paradox is ruled out.

Analysis of the Stokes System. Let us first consider the same system as in [7]. Namely, we assume that $\Omega = \mathbf{R}^d_+$, that $n = 1$ and that $\mathcal{B}_1(t) = B(G(t), 1)$ has constant density $\rho_1 > 0$. Without restricting the generality, we set the radius of the moving body to be 1. The fluid/disk system evolves according to the following simplified system:

$$\nu \Delta u_f - \nabla p_f = 0, \quad \text{in } \mathcal{Q}_F, \tag{4.79}$$

$$\nabla \cdot u_f = 0, \quad \text{in } \mathcal{Q}_F, \tag{4.80}$$

$$u_f = \xi + \omega \times (x - G), \quad \text{on } \partial\mathcal{B}_1(t), \tag{4.81}$$

$$u_f = 0, \quad \text{on } \partial\Omega \text{ and at infinity}, \tag{4.82}$$

$$\dot{G} = \xi, \tag{4.83}$$

$$m_1 \dot{\xi} = -\int_{\partial\mathcal{B}_1(t)} \mathbb{T}(u_f, p_f) n \mathrm{d}\sigma + mg \tag{4.84}$$

$$J\dot{\omega} = -\int_{\partial\mathcal{B}_1(t)} (x - G(t)) \times \mathbb{T}(u_f, p_f) n \mathrm{d}\sigma. \tag{4.85}$$

This system is obtained from (FRBI) by deleting the convective terms in the Navier Stokes system (yielding a stationary Stokes system). We normalized the pressure in order that the gravity does not appear in the fluid equation. This yields the term mg corresponding to Archimedes' force, we denoted $m = m_1 - 4/3\rho_f\pi$ where m_1 is the mass of the rigid body and $\rho_f = 1$. We also applied that the body is a homogeneous sphere so that its inertia reduces to a scalar matrix $\mathbb{J}_1 = J\mathbb{I}_d$.

In (4.79)–(4.85), the unknowns are $((\mathcal{B}_1(t), \xi, \omega), u_f, p_f)$ a priori. However, we remark that the body domain is completely fixed by $G(t)$ so that we might reduce the unknown $\mathcal{B}_1(t)$ to $G(t)$. Furthermore, we note that (4.79)–(4.82) is a Stokes system with unknowns (u_f, p_f) and data $(G(t), \xi, \omega)$. As this system is well-posed (see [23, Chap. V]), (u_f, p_f) might be seen as a function of these data. Finally (4.79)–(4.85) reduces to an autonomous differential system:

$$\dot{G} = \xi \tag{4.86}$$

$$m\dot{\xi} = F(G, \xi, \omega) + m_1 g \tag{4.87}$$

$$J\dot{\omega} = T(G, \xi, \omega) \tag{4.88}$$

in the unknown (G, ξ, ω) and where $F(G, \xi, \omega)$ (resp. $T(G, \xi, \omega)$) is the force (resp. torque) exerted on $\mathcal{B}_1$ by the solution (u_f, p_f) to the Stokes system (4.79)–(4.82).

We complete (4.86)–(4.88) with initial conditions

$$G(0) = G^0, \quad \xi(0) = \xi^0, \quad \omega(0) = \omega^0, \tag{4.89}$$

where initial data satisfy the initial no-contact assumption:

$$(\xi^0, \omega^0) \in \mathbf{R}^d \times \mathbf{R}^{d*} \quad G^0 \in \{X = (X_1, \ldots, X_d) \in \mathbf{R}^d \text{ s.t. } X_d > 1\}. \tag{4.90}$$

Because of a symmetry arguments that we skip for conciseness, discussing the possibility of contact in solutions to (4.86)–(4.89) reduces to determine whether in solutions of the form $G(t) = (0, \ldots, h(t) + 1)$, $\xi(t) = (0, \ldots, \dot{h}(t))$, $\omega(t) = 0$, the distance function h might vanish in finite time or not. For such solutions (4.86)–(4.89) reduces to

$$m_1 \ddot{h} = -\dot{h}\, F(h) + mg \cdot e_d, \tag{4.91}$$

where, denoting $B_h := B((0, \ldots, 0, (1 + h)), 1)$ for arbitrary $h > 0$, we have:

$$F(h) = \int_{\partial B_h} \mathbb{T}(u_h, p_h) n \mathrm{d}\sigma \cdot e_d,$$

with (u_h, p_h) solution to

$$\nu \Delta u_h - \nabla p_h = 0, \quad \text{in } \mathbf{R}^d_+ \setminus \overline{B_h}, \tag{4.92}$$

$$\nabla \cdot u_h = 0, \quad \text{in } \mathbf{R}^d_+ \setminus \overline{B_h}, \tag{4.93}$$

$$u_h = e_d, \quad \text{on } \partial B_h, \tag{4.94}$$

$$u_h = 0, \quad \text{on } \partial \mathbf{R}^d_+ \text{ and at infinity.} \tag{4.95}$$

An analytical expression for the drag force $F(h)$ is provided in [7] and asymptotic expansions on the basis of lubrication approximations are also given in [10]. We present here an alternative approach for extracting the size of this drag when $h << 1$ (see [27, 41]):

Proposition 4.9. *Let $h > 0$. Then setting*

$$\mathcal{Y}_h := \left\{ u \in C^\infty_c(\mathbf{R}^d_+ \setminus \overline{B}_h) \text{ s.t. } \nabla \cdot u = 0 \text{ and } u_{|\partial B_h} = e_d \right\},$$

there holds:

$$F(h) = \inf \left\{ \nu \int_{\mathbf{R}^d_+ \setminus \overline{B}_h} |\nabla u|^2 \ ; \ u \in \mathcal{Y}_h \right\}. \tag{4.96}$$

Proof. Before going into the proof, we note that, up to extend any element of $\mathcal{Y}_h$ by e_d inside B_h and 0 outside $\mathbf{R}^d$, we might rewrite Korn equality on $\mathbf{R}^d$ as:

$$\int_{\mathbf{R}^d_+ \setminus \overline{B}_h} |\nabla u|^2 = 2 \int_{\mathbf{R}^d_+ \setminus \overline{B}_h} |D(u)|^2, \quad \forall\, u \in \mathcal{Y}_h.$$

We deal indifferently with both forms of this quantity in what follows.

We recall that from the theory for the Stokes problem in [23, Chap. V]:

- there exists a unique vector-field u_h in the closure of $\mathcal{Y}_h$ (for the norm $\|\nabla \cdot;\, L^2(\mathbf{R}^d_+ \setminus \overline{B}_h)\|$) such that, for all $w \in \mathcal{D}(\mathbf{R}^d_+ \setminus \overline{B}_h)$:

$$2\nu \int_{\mathbf{R}^d_+ \setminus \overline{B}_h} D(u_h) : D(w) = 0. \tag{4.97}$$

- there exists a unique pressure p_h (up to a constant) such that $(u_h, p_h) \in C^\infty(\mathbf{R}^d \setminus B_h)$ and (4.92)–(4.95) holds true pointwise.

We point out that the lack of uniqueness in the pressure has no influence on the computation of $F(h)$. Hence, we fix a pressure p_h in what follows. Because of the two items above, there holds:

$$\int_{\mathbf{R}^d_+ \setminus \overline{B}_h} |\nabla u_h|^2 < \infty,$$

and there exists a sequence of vector-fields u_n in $\mathcal{Y}_h$ converging to u_h for the topology associated with $\|\nabla \cdot;\, L^2(\mathbf{R}^d_+ \setminus \overline{B}_h)\|$. For this sequence, we have on the one hand:

$$\int_{\mathbf{R}^d_+ \setminus \overline{B}_h} |\nabla u_h|^2 = \lim_{n \to \infty} \int_{\mathbf{R}^d_+ \setminus \overline{B}_h} |\nabla u_n|^2$$

On the other hand, for arbitrary n, multiplying (4.92) by u_n and integrating by parts yields that:

$$\begin{aligned} 0 &= \int_{\mathbf{R}^d_+ \setminus \overline{B}_h} [\nabla \cdot \mathbb{T}(u_h, p_h)] \cdot u_n \\ &= \int_{\partial B_h} \mathbb{T}(u_h, p_h) \mathrm{d}\sigma \cdot e_d - 2\nu \int_{\overline{B}_h} D(u_h) : D(u_n). \end{aligned}$$

Hence, in the limit $n \to \infty$, we obtain:

$$F(h) = 2\nu \int_{\mathbf{R}^d_+ \setminus \overline{B}_h} |D(u_h)|^2 = \nu \int_{\mathbf{R}^d_+ \setminus \overline{B}_h} |\nabla u_h|^2.$$

Now, given $u \in \mathcal{Y}_h$, we denote $u = u_h + \xi$ with $\xi \in C^\infty(\mathbf{R}^d_+ \setminus B_h)$ and split the integral:

$$\nu \int_{\mathbf{R}^d_+ \setminus \overline{B}_h} |\nabla u|^2 = \nu \int_{\mathbf{R}^d_+ \setminus \overline{B}_h} |\nabla u|^2 + 4\nu \int_{\mathbf{R}^d_+ \setminus \overline{B}_h} D(u_h) : D(\xi) + \nu \int_{\mathbf{R}^d_+ \setminus \overline{B}_h} |\nabla \xi|^2.$$

By construction, we have $\xi = \lim_{n\to\infty} \xi_n$, with $\xi_n := u - u_n \in \mathcal{D}(\mathbf{R}^d_+ \setminus \overline{B}_h)$, for the topology associated with the norm $\|\nabla \cdot\,;\, L^2(\mathbf{R}^d_+ \setminus \overline{B}_h)\|$. Hence, passing to the limit in (4.97), we obtain:

$$2\nu \int_{\mathbf{R}^d_+ \setminus \overline{B}_h} D(u_h) : D(\xi) = 0.$$

Finally, there holds:

$$\begin{aligned} \nu \int_{\mathbf{R}^d_+ \setminus \overline{B}_h} |\nabla u_h|^2 &= \nu \int_{\mathbf{R}^d_+ \setminus \overline{B}_h} |\nabla u|^2 + \nu \int_{\mathbf{R}^d_+ \setminus \overline{B}_h} |\nabla \xi|^2 \\ &\geq \nu \int_{\mathbf{R}^d_+ \setminus \overline{B}_h} |\nabla u_h|^2. \end{aligned}$$

This ends the proof. □

Relying on this minimizing property, one might prove that $h \mapsto F(h)$ is locally lipschitz on $(0, \infty)$. This would justify rigorously that the Cauchy problem associated with (4.91) is locally well-posed. We do not go into the details and only write the subsequent existence result:

For arbitrary $h_0 > 0$ and $\dot{h}_0 \in \mathbf{R}$, there exists a unique maximal solution (T_*, h) to (4.91) *with initial data:*

$$h(0) = h_0, \quad \dot{h}(0) = \dot{h}^0.$$

Furthermore, we have the alternative:

- *either $T_* = \infty$,*
- *either $T_* < \infty$ and*

$$\limsup_{t \to T_*} \left[\frac{1}{h(t)} + h(t) + \dot{h}(t) \right] = +\infty.$$

We are more interested here in analyzing the blow-up of the maximal solutions:

Proposition 4.10. *Assume that the gravity brings the sphere to the wall, i.e., $mg \cdot e_d < 0$. Given $h_0 > 0$ and $\dot{h}_0 \in \mathbf{R}$ and (T_*, h) the associated maximal solution, there holds $T_* = \infty$. In particular, there exists a function $h_{min} : [0, \infty) \to (0, \infty)$*

depending only on initial data, such that:

$$h(t) \geq h_{min}(t), \quad \forall\, t \in [0, \infty).$$

Proof. Let (T_*, h) be a maximal solution. First, we multiply (4.86) with $\dot{h}$ and remark that F is always positive. This yields that the total energy of the system:

$$\mathcal{E}(t) := \frac{m_1}{2}|\dot{h}(t)|^2 - [mg \cdot e_d]h(t)$$

is decreasing. Assuming $mg \cdot e_d < 0$, we obtain that

$$|\dot{h}(t)| \leq \left(|\dot{h}^0|^2 - \frac{2[mg \cdot e_d]}{m_1} h^0 \right)^{\frac{1}{2}}, \quad h(t) \leq -\frac{m_1}{2[mg \cdot e_d]} |\dot{h}^0|^2 + h^0.$$

for all $t \in (0, T_*)$. Hence, blow-up of the maximal solution occurs if and only if $h(t)$ goes to 0. We integrate now once (4.91) between 0 and $t \in (0, T_*)$. This yields:

$$m_1(\dot{h}(t) - \dot{h}(0)) = -\int_{h(0)}^{h(t)} F(z)\mathrm{d}z + [mg \cdot e_d]t.$$

Hence:

$$\int_{h(0)}^{h(t)} F(h)\mathrm{d}h \geq -2m_1 \left(|\dot{h}^0|^2 - \frac{2[mg \cdot e_d]}{m_1} h^0 \right)^{\frac{1}{2}} + [mg \cdot e_d]t. \tag{4.98}$$

To prove contact never holds, we bound now from above the integral on the left-hand side by a function of $h(t)$ which diverges to $-\infty$ when $h(t) \to 0$. To this end, we note that, given $u \in \mathcal{Y}_h$ we might apply Proposition 4.5 with $B_1 := B_h$, $B_2 := B(-1, 1)$ and $p = 2$. Inequality (4.70) then yields that there exists an absolute constant C for which:

$$\frac{1}{h^{(3-\frac{d+1}{2})}} \leq C \int_{\mathbf{R}^d} |\nabla u|^2 = C \int_{\mathbf{R}^d_+ \setminus \overline{B}_h} |\nabla u|^2.$$

As for $d = 2, 3$ the exponent in the denominator of the left-hand side is greater than 1, this yields that there exists a constant C_0, depending only on the maximal distance reached by our solution, for which:

$$F(h(t)) \geq \frac{C_0}{h(t)}, \quad \forall\, t \in (0, T_*).$$

Introducing this information into (4.98), we obtain that there exists a constant C_0 depending only on initial data for which:

$$h(t) \geq C_0 \exp\Big([mg \cdot e_d]t\Big), \quad \forall\, t \in (0, T_*).$$

This yields the expected results. □

This proposition states in particular that no-contact between the sphere, or the cylinder, and the ramp occurs in finite time. In the three-dimensional case, it corresponds to the *no-collision paradox* that was pointed out in [7].

We note that, in the last proof, applying Proposition 4.5 furnishes a bound from below that may only enable to prove a no-contact result. In order to obtain contact, or simply to ensure that the bound from below which we obtain is asymptotically equivalent to $F(h)$, we compute a bound from above for $F(h)$. To this end, we apply again the characterization (4.96). This yields that any $u \in \mathcal{Y}_h$ satisfies:

$$\int_{\mathbf{R}^d_+ \setminus \overline{B}_h} |\nabla u|^2 \geq F(h),$$

so that extracting a sharp bound reduces to construct a good candidate u (of course, the solution to the Stokes problem (4.92)–(4.95) would be the best choice) We detail here the construction provided in [27, 41] in the two-dimensional case.

We recall notations from the proof of Proposition 4.5. We have $\partial\mathbf{R}^2_+ := \{(x_1, 0),\ x_1 \in \mathbf{R}\}$ and, close to the origin, $\partial B_h := \{(x_1, \gamma_h(x_1)),\ |x_1| < 1/2\}$ with

$$\gamma_h(s) = h + 1 - \sqrt{1 - s^2}, \quad \forall\, s \in (-1, 1). \tag{4.99}$$

We also denote:

$$\Omega_{1/2} := \{(x_1, x_2) \in \mathbf{R}^2_+ \text{ s.t. } |x_1| < 1/2 \quad 0 < x_2 < \gamma_h(x_1)\}.$$

We focus on the construction in $\Omega_{1/2}$. As B_h remains away from $\partial\mathbf{R}^2_+$ outside this domain, one may extend the constructed vector-field with something independent of h and yielding $O(1)$ terms when computing the H^1-norm to be minimized. In $\Omega_{1/2}$, any $u \in \mathcal{Y}_h$ reads $u = \nabla^\perp \psi$, where, normalizing ψ to vanish in the origin, boundary conditions satisfied by u imply that:

$$\psi(x_1, 0) = 0, \quad \psi(x_1, \gamma_h(x_1)) = x_1 + C. \tag{4.100}$$

Extracting the dominating part of the minimizing problem characterizing $F(h)$, we look for ψ solution to the approximate minimizing problem:

$$\psi := \text{argmin}\Big\{\int_{\Omega_{1/2}} |\partial_{22}\psi|^2,\ \psi \in C^\infty(\Omega_{1/2}),\ \psi \text{ satisfies (4.100)}\Big\}.$$

Explicit computations yield:

$$\psi(x_1, x_2) = x_1 \chi\left(\frac{x_2}{\gamma_h(x_1)}\right) \text{ with } \chi(t) = t^2(3 - 2t). \tag{4.101}$$

Setting $\tilde{u}_h$, an extension of $\nabla^\perp\psi$ we obtain that:

$$\int_{\mathbf{R}^2_+ \setminus \overline{B}_h} |\nabla \tilde{u}_h|^2 = \int_{\Omega_{1/2}} |\partial_{22}\psi|^2 + O(1) \leq \frac{C}{h^{\frac{3}{2}}}.$$

Consequently, the rate of divergence of $F(h)$ when $h \to 0$ is indeed $h^{-3/2}$ in the two-dimensional case. A comparable construction in the three-dimensional case yields that $F(h)$ diverges like $1/h$ when $h \to 0$. This yields that Proposition 4.5 gives the sharp rate of divergence of $F(h)$ when $h \to 0$ and it also confirms the asymptotic computations in [7, 10]. We give very rough information on the asymptotics of $F(h)$ herein. We note that one can extract much better information on the comparison of the asymptotics of $F(h)$ and the norm of the constructed ψ (see [27]).

The method for computing bounds on $F(h)$ can be transposed to many geometries. When considering spheres in the three-dimensional case, it enables to exhibit configurations for which contact occurs in finite time. Indeed, consider the case of a sphere falling down between two spheres (as is suggested in [37, p. 152] and computed in [40]). To fix ideas, we set

$$\Omega := \mathbf{R}^3 \setminus \overline{B((-2,0,0),1) \cup B((2,0,0,),1)},$$

with $\mathcal{B}_1(t) = B(G(t), 1)$ and look for a solution to (4.79)–(4.82). For symmetry reasons, we restrict to the case where $\mathcal{B}_1$ does not rotate and $G(t) = (0, 0, a(t))$. Then, (4.79)–(4.82) reduces to:

$$m_1 \ddot{a} = -\int_{\partial\mathcal{B}_1(t)} \mathbb{T}(u_f, p_f) n \mathrm{d}\sigma \cdot e_3 + mg \cdot e_3. \tag{4.102}$$

We still assume that the gravity brings the body toward $\partial\Omega$, i.e., $a(0) > 0$ and $mg \cdot e_3 =: f < 0$. Arguing with the variational approach of Proposition 4.9, one might show that, when a gets close to 0, a good approximation of the force is

$$-\int_{\partial\mathcal{B}_1(t)} \mathbb{T}(u_f, p_f) n \mathrm{d}\sigma = \dot{h}\tilde{F}(h) e_+ + \dot{h}\tilde{F}(h) e_-, \tag{4.103}$$

where $h := \text{dist}(\mathcal{B}_1(t), B((-2,0,0),1))$ $(= \text{dist}(\mathcal{B}_1(t), B((2,0,0),1))$ by symmetry), the drag $\tilde{F}(h)$ diverges like $1/h$ when $h \to 0$ and the vectors e_+, e_- are defined by:

$$e_+ = (0,0,a) - (2,0,0), \quad e_- = (0,0,a) - (-2,0,0).$$

In (4.103) the first contribution on the right-hand side is an approximation of the drag force exerted by the fluid on $\mathcal{B}_1$ due to the presence of the hole $B((-2,0,0),1)$ whereas the second term stands for the drag force due to the presence of the hole $B((2,0,0),1)$. Consequently, a good approximation of (4.102) reads:

$$m_1 \ddot{a} = -2\dot{h}\tilde{F}(h)(e_+ \cdot e_3)e_3 + f.$$

Integrating once, we obtain:

$$m_1[\dot{a}(t) - \dot{a}(0)] + \int_0^t 2\dot{h}\tilde{F}(h)(e_+ \cdot e_3)\mathrm{d}s = ft.$$

where, introducing that the total energy of the system decreases with time, we obtain that $|[\dot{a}(t) - \dot{a}(0)]| \leq C_0$ is bounded w.r.t. initial data only before contact. Finally, computing $e_+ \cdot e_3$ w.r.t. h, we obtain that:

$$0 \leq \phi(h) = \tilde{F}(h)(e_+ \cdot e_3) \leq \frac{C}{\sqrt{h}} \quad \text{for small } h.$$

This entails that before contact:

$$2\int_{h(0)}^{h(t)} \phi(h)\mathrm{d}h \leq C_0 + ft.$$

and we prove that a contact must occur by a contradiction argument. Indeed, $\phi \in L^1(0, h(0))$ so that the left-hand side of this last inequality is bounded from below by a fixed constant depending on initial data, whereas the right-hand side goes to $-\infty$ when $t \to \infty$. Hence, contact must occur in finite time. More details and rigorous estimates for remainder terms in this construction (for the full system (FRBI)) can be found in [40].

From the Stokes System to (FRBI). As shown in the previous construction, lack of contact in solutions to the Stokes problem comes from overestimating the drag force that rigid bodies undergo. This property is transferred to solutions to (FRBI) thanks to the Lorentz reciprocal theorem:

Proposition 4.11. *Let $\mathcal{O}$ be an open bounded set of $\mathbf{R}^d$ having a smooth boundary. Let $(u_i, p_i) \in H^2(\mathcal{O}) \times H^1(\mathcal{O})$ $(i = a, b)$ such that:*

$$\nabla \cdot u_a = \nabla \cdot u_b = 0, \quad \textit{on } \mathcal{O}.$$

Let $(f_i, u_i^*) \in L^2(\mathcal{O}) \times H^{3/2}(\mathcal{O})$ *be defined by:*

$$\begin{aligned} \nu \Delta u_i - \nabla p_i &= f_i, \quad \text{on } \mathcal{O}, \\ u_i &= u_i^*, \quad \text{on } \partial\mathcal{O}. \end{aligned}$$

Then, there holds:

$$\int_{\partial\mathcal{O}} \mathbb{T}(u_a, p_a)n \cdot u_b^* \mathrm{d}\sigma - \int_{\partial\mathcal{O}} \mathbb{T}(u_b, p_b)n \cdot u_a^* \mathrm{d}\sigma = \int_{\mathcal{O}} f_a \cdot u_b - f_b \cdot u_a.$$

This proposition is a consequence of the symmetry of the Stokes operator. Its proof is a standard combination of integration by parts. In the case of contact issues, it applies in the following way. Let us consider for simplicity the case of a disk falling over a wall as in the previous section. Let $((\xi_1, \omega_1), u, p)$ be a (classical) solution to (FRBI). For simplicity, let also drop the index 1 in what follows and assume that the fall of the disk is vertical so that the distance $h(t)$ between the wall and the body domain $\mathcal{B}(t)$ fixes completely the geometrical configuration at time $t > 0$. Given $h > 0$, we introduce a (sufficiently regular) pair (ξ_h, q_h), such that:

$$\nu \Delta \xi_h - \nabla q_h = f_h, \quad \text{in } \mathbf{R}^d_+ \setminus \overline{B_h}, \tag{4.104}$$

$$\nabla \cdot \xi_h = 0, \quad \text{in } \mathbf{R}^d_+ \setminus \overline{B_h}, \tag{4.105}$$

$$\xi_h = e_d, \quad \text{on } \partial B_h, \tag{4.106}$$

$$\xi_h = 0, \quad \text{on } \partial\mathbf{R}^d_+ \text{ and at infinity.} \tag{4.107}$$

Then, for any $t > 0$ prior to contact, we apply Lorentz reciprocal theorem with $(u_a, p_a) = (\xi_{h(t)}, q_{h(t)})$ and $(u_b, p_b) = (u(t), p(t))$. This yields:

$$\begin{aligned} \int_0^t \dot{h}(s)\tilde{F}(h(s))\mathrm{d}s = \int_0^t \int_{\mathcal{F}(t)} f_{h(s)}(x) \cdot u(s,x) - \left[m_1 \dot{h}(t) + \int_{\mathcal{F}(s)} u(s) \cdot \xi_{h(s)} \right]_0^t \\ + mg \cdot e_d t - \int_0^t \int_{\mathcal{F}(s)} u \cdot \left(\partial_t \xi_{h(s)} + u \cdot \nabla \xi_{h(s)} \right). \end{aligned} \tag{4.108}$$

where:

$$\tilde{F}(h) = \int_{\partial B_h} \mathbb{T}(\xi_h, q_h) n \mathrm{d}\sigma \cdot e_d.$$

For instance, setting $(\xi_h, q_h) := (u_h, p_h)$, the solution to the Stokes system (4.92)–(4.95) yields:

$$\int_0^t \dot{h}(s)F(h(s))\mathrm{d}s = -\left[m_1\dot{h}(t) + \int_{\mathcal{F}(s)} u(s)\cdot u_{h(s)}\right]_0^t$$
$$+ mg\cdot e_d t - \int_0^t \int_{\mathcal{F}(s)} u\cdot\left(\partial_t u_{h(s)} + u\cdot\nabla u_{h(s)}\right). \quad (4.109)$$

where F is computed by Proposition 4.9 and has been shown to diverge like $h^{-3/2}$ when $h \to 0$. Proving that the *no-collision paradox* extends to the nonlinear system (FRBI) then reduces to bound the remainder term on the right-hand side locally in time, having in mind that u satisfies (4.34). The main difficulty here is thus to extract fine properties of the solution to the Stokes problem in the limit $h \to 0$. Hesla applies this method in [37] to the case of one cylinder inside a cylindrical domain and proves no contact occurs between the cylinder and the domain boundary in finite time.

An alternative method, proposed in [38], is to set $\xi_h = \tilde{u}_h$ the approximate Stokes solution constructed in the previous section. One then note that, in the gap between the cylinder and the wall, there holds $\tilde{u}_h = \nabla^{\perp}\psi$ (see (4.101) for a definition of ψ) so that:

$$\nu\Delta\tilde{u}_h = \nu\begin{pmatrix} -\nu\partial_{112}\psi - \nu\partial_{222}\psi \\ \nu\partial_{111}\psi + \nu\partial_{221}\psi \end{pmatrix}$$

Setting

$$\tilde{p}_h(x_1,x_2) = -\nu\partial_1^{-1}\partial_{222}\psi(s,x_2)\mathrm{d}s - \nu\partial_{12}\psi$$

yields:

$$\nu\Delta\tilde{u}_h - \nabla\tilde{p}_h = \begin{pmatrix} 0 \\ \nu\partial_{111}\psi + 2\nu\partial_{221}\psi \end{pmatrix}.$$

We emphasize that a remarkable feature of ψ here is that it is polynomial of degree 3 in x_2 so that the $\partial_1^{-1}\partial_{222}\psi$ does not depend on x_2. The pressure is extended to the whole $\mathbf{R}_+^d \setminus B_h$ by truncation. This yields (see [26, Proposition 9] for instance):

Proposition 4.12. *There exists a constant $C < \infty$ s.t. for all $h < 1$ and $\xi \in K[B_h, \mathbf{R}_+^d]$ there holds:*

$$\left|\int_{\mathbf{R}_+^d\setminus\overline{B}_h} (\nu\Delta\tilde{u}_h - \nabla\tilde{p}_h)\cdot\xi\right| \le C\|\xi; H^1(\mathbf{R}_+^d)\|.$$

With these test-function $\tilde{u}_h$ and associated pressure $\tilde{p}_h$ we rewrite the right-hand side of (4.108) as

$$m_1\dot{h}(0) - m_1\dot{h}(t) + \int_{\mathcal{F}(0)} u^0 \cdot \tilde{u}_{h(0)} + mg \cdot e_d t + I_0 + I_1 + I_2 + I_3,$$

with:

$$I_0 = \int_{\mathcal{F}(t)} u(t) \cdot \tilde{u}_{h(t)},$$

$$I_1 = \int_0^t \int_{\mathcal{F}(t)} (\nu \Delta \tilde{u}_{h(s)} - \nabla \tilde{p}_{h(s)}) \cdot u(s, \cdot),$$

$$I_2 = \int_0^t \int_{\mathcal{F}(s)} u \cdot \partial_t \tilde{u}_{h(s)},$$

$$I_3 = \int_0^t \int_{\mathcal{F}(s)} u \cdot \nabla \tilde{u}_{h(s)} \cdot u.$$

Combining Proposition 4.12 with kinetic energy estimate, we obtain:

$$|I_1| \leq C \int_0^t \|\nabla u(s, \cdot); H^1(\mathcal{F}(s))\|, \leq C\sqrt{t\mathcal{E}_c(0)}.$$

Introducing Hölder inequalities together with Poincaré inequalities (taking into account that u vanishes on $x_2 = 0$ and that the geometry depends on x_1) the contribution of the gap to other remainder terms are bounded by introducing weighted norms of the test-function $\tilde{u}_h$ that one computes explicitly. This yields finally that

$$\left| \int_0^t \dot{h}(s) \tilde{F}(h(s)) \mathrm{d}s \right| \leq C(1 + \mathcal{E}_c(0) + \sqrt{t\mathcal{E}_c(0)})$$

Combining this inequality with (in the two-dimensional case)

$$\tilde{F}(h) \geq \frac{C}{h^{\frac{3}{2}}}$$

entails (see [38] or [26] for more details):

Proposition 4.13. *Let the shape $\mathcal{B}_1^0$ be $B(G_1^0, 1)$ with $G_1^0 \cdot e_2 > 0$, the initial density ρ_1^0 satisfy* (4.1), *and $\Omega := \mathbf{R}^2_+$. Given initial data $u^0 \in H(\Omega)$ and $T > 0$, any weak solution $((\rho, \varphi), u)$ to (FRBI) on $(0, T)$ satisfies:*

$$d[\{\mathcal{B}_1(t), \mathbf{R}^d \setminus \overline{\Omega}\}](t) > 0, \quad \forall\, t \in [0, T].$$

This result extends to the case of a sphere above a wall in [39]. Relying on a similar method, it is possible to study the influence of the body geometry [26] and of the boundary condition [29] for the full (FRBI) system (not only its

Stokes approximation). In particular, in these references, it is shown at the level of the nonlinear (FRBI) system that the *no-collision paradox* is ruled out either by considering asperities at the bodies surfaces or by including slip in the fluid/body interfaces.

References

1. P. Angot, C.-H. Bruneau, P. Fabrie, A penalization method to take into account obstacles in incompressible viscous flows. Numer. Math. **81**(4), 497–520 (1999)
2. G. Barnocky, R.H. Davis, The influence of pressure-dependent density and viscosity on the elastohydrodynamic collision and rebound of two spheres. J. Fluid Mech. **209**, 501–519 (1989)
3. H. Brenner, The slow motion of a sphere through a viscous fluid towards a plane surface. Chem. Eng. Sci. **16**, 242–251 (1961)
4. C. Bost, G.-H. Cottet, E. Maitre, Convergence analysis of a penalization method for the three-dimensional motion of a rigid body in an incompressible viscous fluid. SIAM J. Numer. Anal. **48**(4), 1313–1337 (2010)
5. C. Conca, J.A. San Martín, M. Tucsnak, Existence of solutions for the equations modelling the motion of a rigid body in a viscous fluid. Commun. Partial Differ. Equ. **25**(5–6), 1019–1042 (2000)
6. M.D.A. Cooley, M.E. O'Neill, On the slow rotation of a sphere about a diameter parallel to a nearby plane wall. J. Inst. Math. Appl. **4**, 163–173 (1968)
7. M.D.A. Cooley, M.E. O'Neill, On the slow motion generated in a viscous fluid by the approach of a sphere to a plane wall or stationary sphere. Mathematika **16**, 37–49 (1969)
8. P. Cumsille, T. Takahashi, Wellposedness for the system modelling the motion of a rigid body of arbitrary form in an incompressible viscous fluid. Bol. Soc. Esp. Mat. Apl. SeMA **41**, 117–126 (2007)
9. P. Cumsille, T. Takahashi, Wellposedness for the system modelling the motion of a rigid body of arbitrary form in an incompressible viscous fluid. Czechoslov. Math. J. **58**, 961–992 (2008)
10. R.G. Cox, The motion of suspended particles almost in contact. Int. J. Multiphase Flow **1**, 343–371 (1974)
11. M. Dashti, J.C. Robinson, The motion of a fluid-rigid disc system at the zero limit of the rigid disc radius. Arch. Ration. Mech. Anal. **200**(1), 285–312 (2011)
12. R.H. Davis, J.-M. Serayssol, E. Hinch, The elastohydrodynamic collision of two spheres. J. Fluid Mech. **163**, 479–497 (1986)
13. W.R. Dean, M.E. O'Neill, A slow motion of viscous liquid caused by the rotation of a solid sphere. Mathematika **10**, 13–24 (1963)
14. A. Decoene, S. Martin, B. Maury, Microscopic modelling of active bacterial suspensions. Math. Model. Nat. Phenom. **6**(5), 98–129 (2011)
15. B. Desjardins, M.J. Esteban, Existence of weak solutions for the motion of rigid bodies in a viscous fluid. Arch. Ration. Mech. Anal. **146**(1), 59–71 (1999)
16. B. Desjardins, M.J. Esteban, On weak solutions for fluid-rigid structure interaction: compressible and incompressible models. Commun. Partial Differ. Equ. **25**(7–8), 1399–1413 (2000)
17. L. Desvillettes, F. Golse, V. Ricci, The mean-field limit for solid particles in a Navier-Stokes flow. J. Stat. Phys. **131**(5), 941–967 (2008)
18. R.J. DiPerna, P.-L. Lions, Ordinary differential equations, transport theory and Sobolev spaces. Invent. Math. **98**(3), 511–547 (1989)
19. S. Ervedoza, M. Hillairet, C. Lacave, Long-time behavior for the two-dimensional motion of a dsk in a viscous fluid. Commun. Math. Phys. (2012, to appear). fr.arxiv.org/pdf/1301.4420
20. E. Feireisl, On the motion of rigid bodies in a viscous incompressible fluid. J. Evol. Equ. **3**(3), 419–441 (2003). Dedicated to Philippe Bénilan

21. E. Feireisl, On the motion of rigid bodies in a viscous compressible fluid. Arch. Ration. Mech. Anal. **167**(4), 281–308 (2003)
22. H. Fujita, N. Sauer, On existence of weak solutions of the Navier-Stokes equations in regions with moving boundaries. J. Fac. Sci. Univ. Tokyo Sect. I **17**, 403–420 (1970)
23. G.P. Galdi, *An Introduction to the Mathematical Theory of the Navier-Stokes Equations.* Steady-State Problems, Springer Monographs in Mathematics, 2nd edn. (Springer, New York, 2011)
24. G.P. Galdi, A.L. Silvestre, Strong solutions to the problem of motion of a rigid body in a Navier-Stokes liquid under the action of prescribed forces and torques, in *Nonlinear Problems in Mathematical Physics and Related Topics, I.* International Mathematical Series (N. Y.), vol. 1 (Kluwer/Plenum, New York, 2002), pp. 121–144
25. M. Geissert, K. Götze, M. Hieber, L^p-theory for strong solutions to fluid-rigid body interaction in Newtonian and generalized Newtonian fluids. Trans. Am. Math. Soc. **365**(3), 1393–1439 (2013)
26. D. Gérard-Varet, M. Hillairet, Regularity issues in the problem of fluid structure interaction. Arch. Ration. Mech. Anal. **195**(2), 375–407 (2010)
27. D. Gérard-Varet, M. Hillairet, Computation of the drag force on a sphere close to a wall: the roughness issue. ESAIM Math. Model. Numer. Anal. **46**(5), 1201–1224 (2012)
28. D. Gérard-Varet, M. Hillairet, Existence of weak solutions up to collision for viscous fluid-solid systems with slip. http://hal.archives-ouvertes.fr/hal-00713331 (2012)
29. D. Gérard-Varet, M. Hillairet, C. Wang, Influence of boundary conditions on the contact problem in a 3d incompressible flow. http://hal.archives-ouvertes.fr/hal-00795366 (2013)
30. O. Glass, F. Sueur, Uniqueness results for weak solutions of two-dimensional fluid-solid systems (March 2012). arXiv:1203.2894v1
31. T. Goudon, P.-E. Jabin, A. Vasseur, Hydrodynamic limit for the Vlasov-Navier-Stokes equations. I. Light particles regime. Indiana Univ. Math. J. **53**(6), 1495–1515 (2004)
32. T. Goudon, P.-E. Jabin, A. Vasseur, Hydrodynamic limit for the Vlasov-Navier-Stokes equations. II. Fine particles regime. Indiana Univ. Math. J. **53**(6), 1517–1536 (2004)
33. C. Grandmont, Y. Maday, Existence de solutions d'un problème de couplage fluide-structure bidimensionnel instationnarie. C. R. Acad. Sci. Paris Sér. I Math. **326**(4), 525–530 (1998)
34. C. Grandmont, Y. Maday, Existence for an unsteady fluid-structure interaction problem. M2AN Math. Model. Numer. Anal. **34**(3), 609–636 (2000)
35. M.D. Gunzburger, H.-C. Lee, G.A. Seregin, Global existence of weak solutions for viscous incompressible flows around a moving rigid body in three dimensions. J. Math. Fluid Mech. **2**(3), 219–266 (2000)
36. V. Gyrya, K. Lipnikov, I.S. Aranson, L. Berlyand, Effective shear viscosity and dynamics of suspensions of micro-swimmers from small to moderate concentrations. J. Math. Biol. **62**(5), 707–740 (2011)
37. T.I. Hesla, Collisions of smooth bodies in viscous fluids: a mathematical investigation. Ph.D. thesis, University of Minnesota, revised version, October 2005
38. M. Hillairet, Lack of collision between solid bodies in a 2D incompressible viscous flow. Commun. Partial Differ. Equ. **32**(7–9), 1345–1371 (2007)
39. M. Hillairet, T. Takahashi, Collisions in three-dimensional fluid structure interaction problems. SIAM J. Math. Anal. **40**(6), 2451–2477 (2009)
40. M. Hillairet, T. Takahashi, Blow up and grazing collision in viscous fluid solid interaction systems. Ann. Inst. H. Poincaré Anal. Non Linéaire **27**(1), 291–313 (2010)
41. M. Hillairet, A. Lozinski, M. Szopos, Simulation of particulate flow governed by lubrication forces and far-field hydrodynamic interactions. Discrete Continuous Dyn. Syst. Ser. B **11**, 935–956 (2011)
42. L. Hocking, The effect of slip on the motion of a sphere close to a wall and of two adjacent spheres. J. Eng. Mech. **7**, 207–221 (1973)
43. K.-H. Hoffmann, V. Starovoitov, On a motion of a solid body in a viscous fluid. Two-dimensional case. Adv. Math. Sci. Appl. **9**(2), 633–648 (1999)

44. K.-H. Hoffmann, V. Starovoitov, Zur Bewegung einer Kugel in einer zähen Flüssigkeit. Doc. Math. **5**, 15–21 (2000)
45. J.-G. Houot, J. San Martin, M. Tucsnak, Existence of solutions for the equations modeling the motion of rigid bodies in an ideal fluid. J. Funct. Anal. **259**(11), 2856–2885 (2010)
46. G. Joseph, Collisional dynamics of macroscopic particles in a viscous fluid. Ph.D. thesis, California Institute of Technology, Pasadena, May 2003
47. N.V. Judakov, The solvability of the problem of the motion of a rigid body in a viscous incompressible fluid. Dinamika Splošn. Sredy (Vyp. 18 Dinamika Zidkost. so Svobod. Granicami) **255**, 249–253 (1974)
48. H. Kytömaa, P. Schmid, On the collision of rigid spheres in a weakly compressible fluid. Phys. Fluids A **4**, 2683–2689 (1992)
49. J.-L. Lions, *Quelques méthodes de résolution des problèmes aux limites non linéaires* (Dunod, Paris, 1969)
50. M.E. O'Neill, K. Stewartson, On the slow motion of a sphere parallel to a nearby plane wall. J. Fluid Mech. **27**, 705–724 (1967)
51. J. San Martín, V. Starovoitov, M. Tucsnak, Global weak solutions for the two-dimensional motion of several rigid bodies in an incompressible viscous fluid. Arch. Ration. Mech. Anal. **161**(2), 113–147 (2002)
52. D. Serre, Chute libre d'un solide dans un fluide visqueux incompressible. Existence. Jpn. J. Appl. Math. **4**(1), 99–110 (1987)
53. J. Simon, Compact sets in the space $L^p(0, T; B)$. Ann. Mat. Pura Appl. (4) **146**, 65–96 (1987)
54. V. Starovoitov, Behavior of a rigid body in an incompressible viscous fluid near a boundary, in *Free Boundary Problems (Trento, 2002)*. International Series of Numerical Mathematics, vol. 147 (Birkhäuser, Basel, 2004), pp. 313–327
55. V. Starovoitov, Nonuniqueness of a solution to the problem on motion of a rigid body in a viscous incompressible fluid. J. Math. Sci. **130**(4), 4893–4898 (2005)
56. T. Takahashi, Analysis of strong solutions for the equations modeling the motion of a rigid-fluid system in a bounded domain. Adv. Differ. Equ. **8**(12), 1499–1532 (2003)
57. T. Takahashi, Existence of strong solutions for the problem of a rigid-fluid system. C. R. Math. Acad. Sci. Paris **336**(5), 453–458 (2003)
58. T. Takahashi, M. Tucsnak, Global strong solutions for the two-dimensional motion of an infinite cylinder in a viscous fluid. J. Math. Fluid Mech. **6**(1), 53–77 (2004)
59. R. Temam, *Problèmes mathématiques en plasticité*, volume 12 of *Méthodes Mathématiques de l'Informatique [Mathematical Methods of Information Science]* (Gauthier-Villars/Montrouge, Paris, 1983)
60. J. Vázquez, E. Zuazua, Large time behavior for a simplified 1D model of fluid-solid interaction. Commun. Partial Differ. Equ. **28**(9–10), 1705–1738 (2003)
61. J.L. Vázquez, E. Zuazua, Lack of collision in a simplified 1D model for fluid-solid interaction. Math. Models Methods Appl. Sci. **16**(5), 637–678 (2006)
62. O. Vinogradova, G. Yakubov, Surface roughness and hydrodynamic boundary conditions. Phys. Rev. E **73**, 479–487 (1986)
63. C. Yu, Global weak solutions to the incompressible Navier–Stokes–Vlasov equations. J. Math. Pures Appl. (9) **100**(2), 275–293 (2013)

M. Hillairet (✉)
Université Paris Dauphine, Paris, France
e-mail: hillairet@ceremade.dauphine.fr

Chapter 5
Numerical Simulation of Fluid–Structure Interaction Problems with Applications to Flow in Vocal Folds

Miloslav Feistauer, Petr Sváček, and Jaromír Horáček

Abstract Recently, the numerical solution of FSI problems has become important also in biomechanics, among others in voice modelling. The numerical analysis of this case is very complicated: Human voice is created by passage of air flow between vocal folds, where the constriction formed by the vocal folds induces acceleration of the flow and vocal fold oscillations, which generates the sound. The modelling of such a complex phenomenon encounters many difficulties as it is a result of coupling complex fluid dynamics and structural behavior. We focus on mathematical and numerical modelling of nonlinear coupled problems of fluid–structure interactions (FSI). The main attention is paid to the mathematical description of a relevant problem and to the description of the applied numerical methods. The mathematical description consists of the elasticity equations describing the motion of an elastic structure, and the air flow modelled by the Navier–Stokes equations. Both models are coupled via interface conditions.

The solution of dynamic elasticity equations is realized with the aid of conforming finite elements or the elastic structure motion is modelled by a simplified model of vibrating rigid body. Both compressible and incompressible fluid model is considered. The approximation of flow in moving domains is treated with the aid of the arbitrary Lagrangian–Eulerian method. The incompressible Navier–Stokes equations are approximated by the stabilized finite element method. The compressible Navier–Stokes equations are discretized by the discontinuous Galerkin finite element method. The time discretization based on a semi-implicit linearized scheme is described and the solution of the coupled problem of FSI is realized by a coupling algorithm.

Keywords Aeroelasticity • ALE method • Biomechanics of voice • Compressible flow • Coupling algorithm • Discontinuous Galerkin method • Dynamic elasticity problem • Fluid–structure interaction • Navier–Stokes equations • Stabilized finite element method • Time and space discretization • Two degrees of freedom model • Vocal folds

T. Bodnár et al. (eds.), *Fluid-Structure Interaction and Biomedical Applications*, Advances in Mathematical Fluid Mechanics, DOI 10.1007/978-3-0348-0822-4_5,

MSC2010: 74F10, 76Z05, 76M10, 74S05, 74H15

	Notation
t	time (s)
Ω_t	fluid current configuration at time instant t
Ω_0	fluid reference configuration
$x = (x_1, x_2)$	point of current configuration
ρ	fluid/air density
p	pressure
μ	dynamic fluid viscosity
$\boldsymbol{v} = (v_1, v_2)$	fluid velocity vector
$\boldsymbol{w} = (w_1, w_2, w_3, w_4)$	state vector
τ_{ij}^f	fluid stress tensor
Ω^s	structural reference configuration
$X = (X_1, X_2)$	point of reference configuration
$\boldsymbol{u} = (u_1, u_2)$	displacement vector
$z = (z_1, z_2)$	ALE domain velocity
$\overline{w}_1, \overline{w}_2$	vocal fold generalized coordinates
ρ^s	structural density
τ_{ij}^s	structural stress tensor
e_{ij}^s	structural strain tensor
λ^s, μ^s	Lamé coefficients of the structure

5.1 Introduction

The flow-induced vibrations of structures are usually associated with various technical applications, however recently fluid–structure interactions (FSI) problems become important also in biomechanics and especially here in biomechanics of human voice [93]. Voice is one of the basic characteristics of human beings, enabling their communication and playing an important role in the quality of human life. Understanding the basic principles of voice production is important for better interpretation of clinical findings, detection of laryngeal cancers or other pathologies and for treatment of laryngeal disorders, and in development of voice prostheses for laryngectomized patients, see, e.g., [91]. Despite all the efforts, all the details of the voice production mechanisms are not yet fully understood.

Voice production is a complicated and complex biomechanical process, which involves several basic factors: airflow coming from the lungs, vocal fold self-oscillations, and acoustic resonances of the cavities of the human vocal tract (see Fig. 5.1). Primary pressure fluctuations arise in the human larynx as a result of the airflow being modulated by the vibrating vocal folds (also called vocal cords). The narrow oscillating constriction between the two vocal folds is called glottis. The vocal fold tissue is composed of the muscle vocalis and ligament covered by epithelium. The vocal folds are fixed between the thyroid and cricoid cartilages enabling to change their longitudinal tension, which determines the fundamental

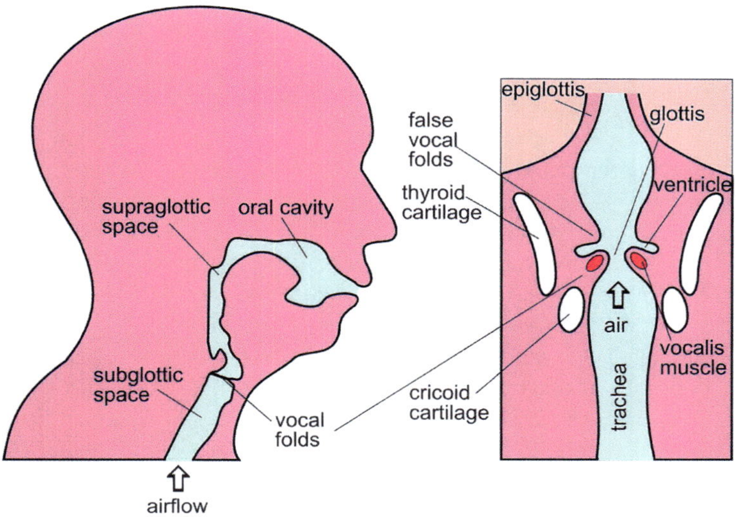

Fig. 5.1 Schema of the human vocal tract with a detail of the laryngeal part

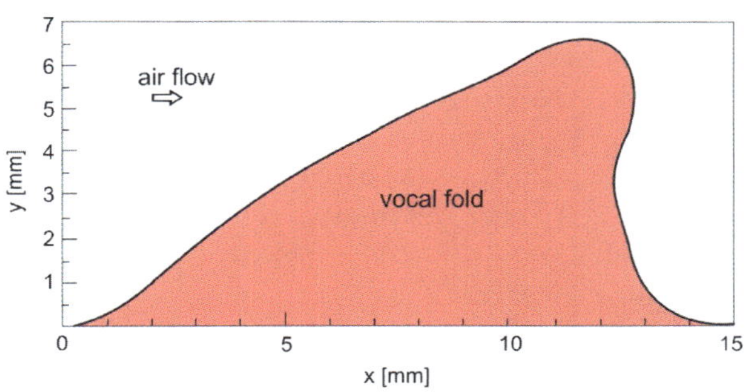

Fig. 5.2 Example of 2D shape of vocal fold measured in phonation position for female and fundamental frequency 304 Hz at a cross-section in the middle of the vocal fold length, see [71]

frequency of the human voice. The vocal folds can start to oscillate and close the channel periodically for certain airflow rate and a defined prephonatory position. For higher flow rate, the glottis is almost (or completely) closing during vibrations and the vocal folds collide generating the primary voice source sound. The modelling of such a complex phenomenon encounters many difficulties as it is a result of coupling complex fluid dynamics and structural behavior including contact and acoustic problems. Figure 5.2 shows an example of a real size and geometry of human vocal fold during phonation, i.e., when the vocal folds are pre-stressed in the longitudinal direction in an active phonation position. We should note that in addition to the fundamental phonation frequency the dimensions of the vocal folds depend on gender and human age, see [93].

The acoustic resonant supraglottal spaces, formed by the air cavities upstream the vocal folds, modify the sound and codetermine its quality [28, 80, 92]. The vocal folds, excited by the airflow, generate a primary laryngeal tone whose fundamental frequency corresponds to the vibration frequency of the vocal folds. In the airways above the vocal folds, i.e., in the vocal tract, the acoustic resonant phenomena modify the spectrum of the primary laryngeal tone, especially the higher harmonics. The acoustic resonances of the vocal tract create the so-called formants, which occur as peaks in the voice spectrum. The formants in human voice define vowels and cause differences in the voice timber. The formant frequencies are determined by the size and shape of the vocal tract cavities that can be varied, for instance, by changing the position of the tongue or by mouth opening.

Understanding the fluid–structure–acoustic interaction between the airflow, vibrating compliant vocal folds and sub- and supraglottal acoustic spaces relies on three physical domains: aerodynamics, acoustics, and geometrical and material properties of the vocal fold tissue. Considering the inaccessibility of the vocal folds, it is close to impossible to perform exact airflow or tissue measurements in vivo and it is rather complicated in vitro. Thus, most experimental data on the laryngeal airflow has been obtained using physical models [46].

The airflow coming from the trachea accelerates in the glottis. Near the narrowest cross-section, airflow separates from the surface of the vocal folds due to adverse pressure gradient and forms a jet. The glottal jet pulsates due to vocal fold oscillations. The jet inclination angle, jet core flow velocity, and the position of the flow separation point are highly sensitive to the geometry of the vocal folds (which is, in real subjects, never perfectly symmetric), to the position within the oscillation cycle and magnitude of the subglottal pressure. Further downstream of the glottis, the jet interacts with laryngeal walls and supraglottal large-scale vortical structures, which leads to a complex flow field.

There is a number of experimental papers using laser visualization techniques and Particle Image Velocimetry (PIV). These include results obtained on externally driven models [27, 49, 51, 95], and self-oscillating models [13, 44, 50, 58, 72, 90]. A comprehensive overview of the physical vocal fold models used in voice research is given in a recent paper of [46].

Due to the difficulties encountered in both in vivo and in vitro measurements, there have always been efforts to develop mathematical models. The computational modelling of voice production enables generating numerous model situations even if due to the lack of precise data in literature, most of the computational models have grossly simplified geometries and mechanical properties.

Among the computational models, the 2-mass model of the vocal folds of [45] is usually regarded as a cornerstone of vocal fold modelling. A number of subsequent works, e.g., [39, 40, 54, 61, 68, 76] used a similar concept: the vocal fold modelled by discrete masses connected by springs and dampers, coupled to 1D or 2D simplified airflow model (potential flow or using only Bernoulli equation). Simple mass-spring models coupled with a quasi-1D airflow proved very useful and its variants are widely used up to present [102, 103]. For example, the developed aeroelastic model [41] has applications in simulation of vowels phonation [5] and in estimation of the vocal folds loading by impact stress and inertial forces [42, 43]. The phonation onset was studied by using the potential flow model and three-mass lumped model for the vibrating vocal folds in [39, 40] and later in [103, 104]. The main advantage of these models is the fact that the equations may be solved either analytically, or using simple and very fast numerical methods for ordinary differential equations, making it possible to perform nearly real-time simulations on current computers.

Since the fluid models based on potential flow theory do not provide much information on the glottal airflow, increasing effort has been devoted to numerical solution of the 2D Navier–Stokes equations in computational domains approximating the glottal channel. We can note that for an inviscid incompressible flow model (e.g., using the Euler equations) the maximum flow velocity tends to infinity just before the glottis closes when the viscous forces are important and the use of such models is problematic.

The finite volume approximations of the Navier–Stokes equations on Cartesian grids were coupled with the two-mass [45] dynamic model of the vocal folds in [22]. The 2D finite element approximations of the Navier–Stokes equations were coupled in [81, 83, 84] with the vibrating vocal fold described by a mechanically equivalent two degrees of freedom system (see [39, 40]). There are many other similar studies focused on the unsteady flow field in the glottis and considering prescribed (forced) vocal fold oscillations (see, e.g., [2, 3, 9, 56, 66, 67, 69, 73, 105, 108]).

de Oliviera et al. [21] and Thomson et al. [90] can be considered as first authors that modelled the vocal fold fluid–structure interaction using the coupled finite element (FE) models of the vocal folds and airflow. Consecutive FE models with the airflow fully coupled with elastic tissue oscillations were published by Tao and Jiang [88], Luo et al. [55], Link et al. [52], Zheng et al. [106].

Only a few of the computational studies [21, 56, 73] solve the flow field in 3D. One of the most complex approaches to phonation modelling was recently published by Zheng et al. [107, 109], and Seo and Mittal [70], who use the immersed boundary method for incompressible low-Mach number flow coupled with a finite element solver for the viscoelastic tissue to calculate the 3D flow field and flow-induced vibrations of the vocal folds including glottal closure and contact forces. Using the aerodynamic–acoustic splitting technique, the acoustic field is then calculated by solving linearized perturbed compressible equations. An extensive overview of the computational vocal fold models can be found in a recent review paper [4].

The Reynolds numbers found in airflow past vocal folds range from 1,000 up to about 5,000–10,000 [27]. This implies that the subglottal flow may be laminar. However, Neubauer et al. [58] showed that near the glottal region, transition from laminar to turbulent flow occurs and that the supraglottal flow field, dominated by separated jet flow and recirculation, is rather turbulent. The Reynolds-Averaged Navier–Stokes (RANS) turbulence modelling was used in [67].

A current challenging question is a mathematical and physical description of the mechanism for transforming the airflow energy in the glottis into the acoustic energy representing the voice source in humans. The flow velocities in the human glottal region are lower than $100\,\mathrm{m\,s^{-1}}$ and thus the influence of fluid compressibility on the flow-induced instability of the vocal folds and their vibrations can be mostly neglected. However, the voice production is based on aero-acoustics phenomena associated with the fluid compressibility and a direct numerical simulation of the voice production using the compressible Navier–Stokes equations is problematic because the difference between the magnitudes of amplitudes of the aerodynamic pressure fluctuations, given by the channel closure at the glottis, are several orders higher than the acoustic pressure radiated from the mouth. This is why the hybrid methods, solving the flow and acoustic fields separately is option in computational aero-acoustics. Acoustic wave propagation in the vocal tract is usually modelled separately using linear acoustic perturbation theory, the wave equation for the potential flow [93] or the Lighthill approach on sound generated aerodynamically [52,74].

Suh and Frankel [79] solved the compressible Navier–Stokes equations in 3D by the FE method in order to study the flow–acoustic interaction in a rigid glottis. Tao et al. [89] used FLOTRAN coupled with a two-mass model programmed in ANSYS. Link et al. [52] solved the full fluid–solid–acoustic interaction in vocal folds using a FEM solver and Lighthills analogy. Numerical simulation of the 2D compressible flow field in the glottis region by finite volume method was recently published in [63, 65, 66]. The changes of the channel cross-section are prescribed and the channel is harmonically opening and nearly closing in the narrowest cross-section of the channel. Particular attention is paid to the analysis of the position of the flow separation point on the vibrating surface, and to the effect of nonsymmetric flow appearing in a symmetric channel due to the so-called Coanda effect.

Recently, the discontinuous Galerkin finite element method (DGFEM) for the space–time discretization of a nonstationary convection-diffusion initial-boundary value problem with nonlinear convection and linear diffusion applied separately in space and time using different space grids on different time levels was applied to the simulation of vibrations of vocal folds during phonation onset, see [17, 33, 37]. The time-dependence of the domain occupied by the fluid is treated by the ALE (Arbitrary Lagrangian–Eulerian) method, when the 2D compressible Navier–Stokes equations are formulated in the ALE form. The deformation of the elastic body is described by the dynamic elasticity equations. Both these systems are coupled by the transmission conditions. The DGFEM is used for the space discretization of the flow problem. The time discretization is realized by the backward difference formula. The structural problem is discretized by the conforming FE method and the Newmark method. The FSI is realized via weak or strong coupling algorithms.

The results of direct numerical simulation of vowels phonation using the originally developed full FSI model based on the FE solution of the 3D compressible Navier–Stokes equations by the CFD/ANSYS program code were recently published in [86, 87].

Acoustic wave propagation in the vocal tract is usually modelled separately using linear acoustic perturbation theory. The fundamentals of vocal tract acoustics are summarized in the monographs by Fant [28] and Titze [92]. Using magnetic resonance imaging (MRI), Story et al. [78] acquired 3D vocal tract shapes that correspond to particular vowels and consonants. A set of vowel area functions, based on MRI measurements, has been parameterized by Story and Titze [77]. Later Story [75] modelled vocal tract by the transfer matrix method using cylindrical elements.

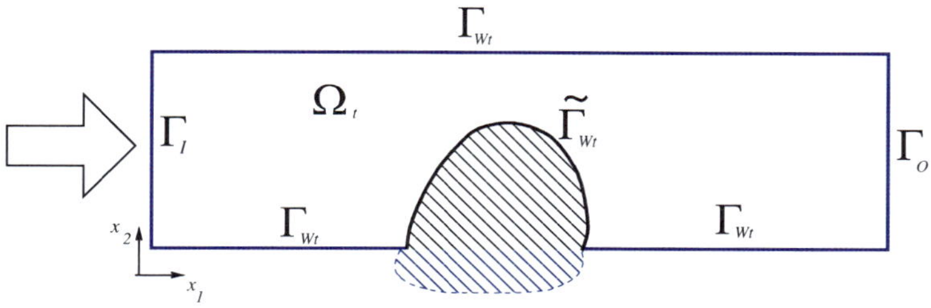

Fig. 5.3 Scheme of coupled fluid–structure problem

Adachi et al. [1] applied this method to tuning the vocal tract shape. Recently, a theory of interaction between the source of sound in phonation and the vocal tract filter was proposed by Titze [94]. The 3D FE models of the human vocal tract for vowels were developed by Švancara et al. [85] and Vampola et al. [96,97] and based on the MRI and computer tomography measurements during phonation.

Mathematical models for the human phonation process are valuable tools for providing insight into the basic mechanisms of phonation and in future could help with surgical planning, diagnostics, and voice rehabilitation. Our goal in this monograph is to present our original, recently developed numerical methods based on the finite element simulation of 2D incompressible and compressible laminar viscous flow described in the glottal region by the Navier–Stokes equations in interaction with a compliant tissue of the human vocal folds. The vocal folds are either modelled by a 2D elastic layered structure or as a vibrating rigid body. Some methods and results presented here were obtained in the works [17, 33, 37, 47, 48, 83, 84].

5.2 Incompressible Flow in Time-Dependent Domains

The mathematical description of the interaction of incompressible flow and vocal folds consists of equations of motion for the vocal folds coupled with the incompressible Navier–Stokes equations via interface conditions. First, for simplicity and clarity of aeroelastic principles, the vocal folds are modelled as rigid bodies with two degrees of freedom, elastically supported in the glottis, and then as linear 2D elastic continuum. The solution of the 2D dynamic elasticity equations for the vocal fold tissue is realized with the aid of conforming finite elements on a reference domain of Ω_t^s. The flow in a time-dependent domain Ω_t (see Fig. 5.3) is treated with the aid of ALE method. The incompressible Navier–Stokes equations are approximated by the stabilized finite element method. The time discretization based on a semi-implicit linearized scheme is described. The solution of the coupled problem of FSI is realized by coupling algorithms enforcing the interface conditions at a fluid–structure interface.

5.2.1 Incompressible Navier–Stokes Equations

Let us consider a two-dimensional computational domain Ω_t depending on time $t \in [0, T], T > 0$, with a Lipschitz-continuous boundary $\partial\Omega_t$. The flow of an incompressible viscous fluid in the domain Ω_t is described by the system of the incompressible Navier–Stokes equations (cf. [29])

$$\rho \frac{\partial v_i}{\partial t} + \rho(\boldsymbol{v} \cdot \nabla) v_i = \sum_{j=1}^{2} \frac{\partial \tau_{ij}^f}{\partial x_j}, \quad i = 1, 2, \tag{5.1}$$
$$\nabla \cdot \boldsymbol{v} = 0.$$

Here $\boldsymbol{v} = (v_1, v_2)$ is the fluid velocity vector, $\rho > 0$ is the constant fluid density, and τ_{ij}^f are the components of the fluid stress tensor given by

$$\tau_{ij}^f = -\frac{\partial p}{\partial x_i} \delta_{ij} + \mu \left(\frac{\partial v_i}{\partial x_j} + \frac{\partial v_j}{\partial x_i} \right), \tag{5.2}$$

where p is the pressure, $\mu > 0$ is the constant fluid viscosity and δ_{ij} denotes the Kronecker symbol.

Initial and Boundary Conditions. System (5.1) is equipped by the initial condition

$$\boldsymbol{v}(x, 0) = \boldsymbol{v}^0(x) \quad \text{for } x \in \Omega_0, \tag{5.3}$$

where $\boldsymbol{v}^0$ is a prescribed initial velocity, and boundary conditions. The boundary $\partial\Omega_t$ of the computational domain is assumed to be formed by mutually disjoint parts Γ_I—inlet, Γ_O—outlet, Γ_{Wt}—impermeable wall, whose part may move in dependence on time, and possibly Γ_S—symmetry axis $x_2 = \text{const}$. We consider the following boundary conditions:

$$\begin{aligned}
&\text{(a)} \quad \boldsymbol{v} = \boldsymbol{v}_I \ \text{ on } \Gamma_I, \quad \text{(b)} \quad \boldsymbol{v} = z_W \ \text{ on } \Gamma_{Wt}, \\
&\text{(c)(i)} \quad \mathrm{v}_2 = 0, \qquad \text{(ii)} \frac{\partial \mathrm{v}_1}{\partial \mathrm{x}_2} = 0 \text{ on } \Gamma_{\mathrm{S}}, \\
&\text{(d)} \quad -\sum_{j=1}^{2} \tau_{ij}^f n_j + \frac{1}{2} \rho (\boldsymbol{v} \cdot \boldsymbol{n})^- v_i = p_{\mathrm{ref}} n_i, \qquad i = 1, 2, \text{ on } \Gamma_O.
\end{aligned} \tag{5.4}$$

The following notation is used: $\boldsymbol{n}$—unit outer normal vector to $\partial\Omega_t$, $\boldsymbol{v}_I$—inlet velocity, p_{ref}—reference pressure value, z_W—velocity of the wall motion (of course, $z_W = 0$ on fixed parts of the wall). Further, α^- denotes the negative part of a real number α, i.e., $\alpha^- = \min(0, \alpha)$. (As for the condition (5.4), (d), see, e.g., [16, 38]).

On the inlet Γ_I condition (5.4), (a) can be replaced by the condition

$$-\sum_{j=1}^{2} \tau_{ij}^{f} n_j + \frac{1}{2}\rho(\boldsymbol{v}\cdot\boldsymbol{n})^{-} v_i = p_{\text{inlet}} n_i, \quad i = 1, 2. \tag{5.5}$$

This means that on the inlet we prescribe either the inlet flow velocity $\boldsymbol{v}_I$ or the inlet pressure $p_{\text{inlet}} = p_{\text{ref}} + \Delta p$, where Δp is the pressure drop between the inlet and outlet. The reference pressure p_{ref} can be chosen arbitrarily and, therefore, we set $p_{\text{ref}} = 0$ in what follows.

5.2.2 *Arbitrary Lagrangian–Eulerian Method*

In order to take into account the time dependence of the domain, we use the so-called *ALE* technique, proposed, e.g., in [59]. It is based on a regular one-to-one ALE mapping of the reference configuration Ω_0 onto the current configuration Ω_t:

$$\mathcal{A}_t : \overline{\Omega}_0 \longrightarrow \overline{\Omega}_t, \text{ i.e. } X \in \overline{\Omega}_0 \longmapsto x = x(X,t) = \mathcal{A}_t(X) \in \overline{\Omega}_t.$$

Here we use the notation X for points in $\overline{\Omega}_0$ and x for points in $\overline{\Omega}_t$.

Further, we define the domain velocity

$$\tilde{z}(X,t) = \frac{\partial}{\partial t}\mathcal{A}_t(X), \quad t \in (0,T),\ X \in \Omega_0, \tag{5.6}$$

$$z(x,t) = \tilde{z}(\mathcal{A}_t^{-1}(x),t), \quad t \in (0,T),\ x \in \Omega_t$$

and the ALE derivative of a function $f = f(x,t)$ defined for $x \in \Omega_t$ and $t \in (0,T)$:

$$\frac{D^A}{Dt} f(x,t) = \frac{\partial \tilde{f}}{\partial t}(X,t), \tag{5.7}$$

where

$$\tilde{f}(X,t) = f(\mathcal{A}_t(X),t),\ X \in \Omega_0,\ x = \mathcal{A}_t(X).$$

The following lemma formulates the relation between the partial time derivative and the ALE derivative.

Lemma 5.1. *Let the ALE mapping $\mathcal{A}_t(X) = \Phi(X,t)$ have continuous first order derivatives and let $\mathcal{A}_t = \Phi(\cdot,t) : \Omega_0 \to \Omega_t$ be a bijective mapping of Ω_0 onto Ω_t for any $t \in (0,T)$. Let a function $f = f(x,t)$, $x \in \Omega_t$, $t \in (0,T)$, be continuously* *differentiable. Then*

$$\frac{D^{\mathcal{A}} f}{Dt}(x,t) = \frac{\partial f}{\partial t}(x,t) + z(x,t) \cdot \nabla f(x,t), \tag{5.8}$$

where z is the domain velocity defined by (5.6).

Proof. Let us set $x = \Phi(X,t)$. Then by (5.7) and (5.6), we get

$$\begin{aligned} \frac{D^{\mathcal{A}} f}{Dt}(x,t) &= \frac{\partial \tilde{f}}{\partial t}(X,t) \\ &= \frac{d}{dt}\Big[f(\Phi(X,t),t) \Big] \\ &= \frac{\partial f}{\partial t}\big(\Phi(X,t),t\big) + \sum_{i=1}^{2} \frac{\partial f}{\partial x_i}\big(\Phi(X,t),t\big) \frac{\partial \Phi_i}{\partial t}(X,t) \\ &= \frac{\partial f}{\partial t}(x,t) + z(x,t) \cdot \nabla f\,(x,t)\,. \end{aligned}$$

□

Using the relation (5.8), we can rewrite the *Navier–Stokes system* (5.1) *in the ALE form*

$$\rho \frac{D^{\mathcal{A}} v_i}{Dt} + \rho((\boldsymbol{v} - z) \cdot \nabla) v_i = \sum_{j=1}^{2} \frac{\partial \tau_{ij}^f}{\partial x_j}, \quad i = 1,2, \quad \nabla \cdot \boldsymbol{v} = 0. \tag{5.9}$$

5.2.3 Numerical Approximation of the Incompressible Navier–Stokes Equations

This section will be concerned with the discretization of the flow problem (5.9), (5.3), (5.4) (where the condition (5.4), (a) can be replaced by (5.5)).

Time Discretization

First let us describe the time discretization of the problem. We consider a partition $0 = t_0 < t_1 < \cdots < T$, $t_k = k\Delta t$, with a constant time step $\Delta t > 0$, of the time interval $[0,T]$ and approximate the solution $\boldsymbol{v}(t_n)$, $p(t_n)$ (defined in Ω_{t_n}) at time t_n by $\boldsymbol{v}^n$, p^n. For the time discretization we use the second-order two-step scheme using the computed approximate solution $\boldsymbol{v}^{n-1}$, p^{n-1} in $\Omega_{t_{n-1}}$ and $\boldsymbol{v}^n$, p^n in Ω_{t_n} for the calculation of $\boldsymbol{v}^{n+1}$, p^{n+1} in the domain $\Omega := \Omega_{t_{n+1}}$.

We approximate the ALE velocity $z(t_{n+1})$ by z^{n+1}, where

$$z^{n+1}(x) = \frac{3\mathcal{A}_{t_{n+1}}(X) - 4\mathcal{A}_{t_n}(X) + \mathcal{A}_{t_{n-1}}(X)}{2\Delta t}, \tag{5.10}$$
$$x = \mathcal{A}_{t_{n+1}}(X) \in \Omega_{t_{n+1}}.$$

Further, it is necessary to approximate the ALE derivative of the velocity $\boldsymbol{v}$ at time t_{n+1}. To this end, for $i = n, n-1$ we set $\hat{\boldsymbol{v}}^i = \boldsymbol{v}^i \circ \mathcal{A}_{t_i} \circ \mathcal{A}_{t_{n+1}}^{-1}$, which is defined in the domain $\Omega_{t_{n+1}}$ (the symbol $\circ$ denotes the composite function). Then, by (5.7), for $x \in \Omega_{t_{n+1}}$ and $X = \mathcal{A}_{t_{n+1}}^{-1}(x)$, using the second-order backward difference formula, we can write

$$\begin{aligned} \frac{D^A \boldsymbol{v}}{Dt}(x, t_{n+1}) &= \frac{\partial \tilde{\boldsymbol{v}}(X, t_{n+1})}{\partial t} \\ &\approx \frac{3\tilde{\boldsymbol{v}}(X, t_{n+1}) - 4\tilde{\boldsymbol{v}}(X, t_n) + \tilde{\boldsymbol{v}}(X, t_{n-1})}{2\Delta t} \\ &\approx \frac{3\boldsymbol{v}^{n+1}(x) - 4\hat{\boldsymbol{v}}^n(x) + \hat{\boldsymbol{v}}^{n-1}(x)}{2\Delta t}. \end{aligned} \tag{5.11}$$

This leads to the problem of finding unknown functions $\boldsymbol{v} = \boldsymbol{v}^{n+1} : \Omega \to R^2$ and $p = p^{n+1} : \Omega \to R$ satisfying the equations

$$\begin{aligned} &\rho \frac{3\boldsymbol{v} - 4\hat{\boldsymbol{v}}^n + \hat{\boldsymbol{v}}^{n-1}}{2\Delta t} + \rho\left((\boldsymbol{v} - z^{n+1}) \cdot \nabla\right)\boldsymbol{v} - \nabla \cdot \left(\mu(\nabla \boldsymbol{v} + \nabla^T \boldsymbol{v})\right) + \nabla p = 0, \\ &\operatorname{div} \boldsymbol{v} = 0, \quad \text{in } \Omega, \end{aligned} \tag{5.12}$$

and the boundary conditions (5.4) (where condition (a) can be replaced by (5.5)—in what follows, this eventuality will not be emphasized).

Weak Formulation

The starting point for the finite element discretization of problem (5.12) with the boundary conditions (5.4) is the so-called weak formulation. To this end, we define the velocity spaces W, X and the pressure space Q:

$$W = (H^1(\Omega))^2, \quad X = \{\boldsymbol{v} \in W; \boldsymbol{v}|_{\Gamma_I \cup \Gamma_{W_t}} = 0,\ v_2|_{\Gamma_S} = 0\}, \tag{5.13}$$

$$Q = L^2(\Omega), \tag{5.14}$$

where $L^2(\Omega)$ is the Lebesgue space of square integrable functions over the domain Ω, and $H^1(\Omega)$ is the Sobolev space of square integrable functions together with their first-order derivatives.

Now, we multiply the first and second equation in (5.12) by any function $\boldsymbol{\varphi} \in X$ and $q \in Q$, respectively, sum them, integrate over Ω, transform the viscous term containing second-order derivatives of $\boldsymbol{v}$ and the term ∇p with the aid of Green's theorem and use the boundary conditions (5.4), (c)(ii), (d). We define the weak solution of the problem (5.12), (5.4) as a couple $U = (\boldsymbol{v}, p) \in W \times Q$ satisfying the conditions (5.4), (a), (b), (c)(i) and the identity

$$a(U, U, V) = f(V), \qquad \text{for all } V = (\boldsymbol{\varphi}, q) \in X \times Q. \tag{5.15}$$

We use the notation

$$(\alpha, \beta)_\omega = \int_\omega \alpha \cdot \beta \mathrm{d}x, \tag{5.16}$$

for the scalar product in $L^2(\omega)$ for a set ω. It generates the norm $\|\cdot\|_{L^2(\Omega)}$. The form $a(U^*, U, V)$ is defined by

$$\begin{aligned} a(U^*, U, V) = {} & \frac{3\rho}{2\Delta t} (\boldsymbol{v}, \boldsymbol{\varphi})_\Omega + \frac{\mu}{2} \left(\nabla \boldsymbol{v} + \nabla^T \boldsymbol{v}, \nabla \boldsymbol{\varphi} + \nabla^T \boldsymbol{\varphi}\right)_\Omega + \rho c(\boldsymbol{v}^*, \boldsymbol{v}, \boldsymbol{\varphi}) \\ & -\rho \left(\boldsymbol{z}^{n+1} \cdot \nabla \boldsymbol{v}, \boldsymbol{\varphi}\right)_\Omega - (p, \nabla \cdot \boldsymbol{\varphi})_\Omega + (\nabla \cdot \boldsymbol{v}, q)_\Omega, \qquad (5.17) \\ f(V) = {} & \frac{\rho}{2\Delta t} \left(4\hat{\boldsymbol{v}}^n - \hat{\boldsymbol{v}}^{n-1}, \boldsymbol{\varphi}\right)_\Omega - \int_{\Gamma_O} p_{\mathrm{ref}} \boldsymbol{\varphi} \cdot \boldsymbol{n} \, \mathrm{d}S, \\ U = {} & (\boldsymbol{v}, p), \; U^* = (\boldsymbol{v}^*, p) \in W, \; V = (\boldsymbol{\varphi}, q) \in X, \end{aligned}$$

where the convective form $c(\boldsymbol{v}^*, \boldsymbol{v}, \boldsymbol{\varphi})$ reads

$$c(\boldsymbol{v}^*, \boldsymbol{v}, \boldsymbol{\varphi}) = \frac{1}{2} \left((\boldsymbol{v}^* \cdot \nabla)\boldsymbol{v}, \boldsymbol{\varphi}\right)_\Omega - \frac{1}{2} \left((\boldsymbol{v}^* \cdot \nabla)\boldsymbol{\varphi}, \boldsymbol{v}\right)_\Omega + \int_{\Gamma_O} \frac{1}{2} (\boldsymbol{v}^* \cdot \boldsymbol{n})^+ \boldsymbol{v} \cdot \boldsymbol{\varphi} \, \mathrm{d}S. \tag{5.18}$$

The form $c(\boldsymbol{v}^*, \boldsymbol{v}, \boldsymbol{\varphi})$ is obtained from the convective term $(\boldsymbol{v}^* \cdot \nabla \boldsymbol{v}, \boldsymbol{\varphi})_\Omega$ by integration by parts, using the boundary conditions e (5.4) (d) at the outlet and the definition of the space X (i.e., $\boldsymbol{\varphi} = 0$ on $\Gamma_I \cup \Gamma_{W_t}$ and $\boldsymbol{\varphi} \cdot \boldsymbol{n} = 0$ on Γ_S). For $\alpha \in R$ we set $\alpha^+ = \max(0, \alpha)$.

Space Discretization and Stabilization

In order to apply the Galerkin finite element method (FEM) to the discretization of the problem (5.15), we approximate the spaces W, X, Q from the weak formulation by finite dimensional subspaces W_h, X_h, Q_h, $h \in (0, h_0)$, $h_0 > 0$,

$$X_h = \{\boldsymbol{v}_h \in W_h; \boldsymbol{v}_h|_{\Gamma_I \cap \Gamma_{W_t}} = 0, v_2|_{\Gamma_S} = 0\}. \tag{5.19}$$

Then we define the approximate solution as a couple $U_h = (\boldsymbol{v}_h, p_h) \in W_h \times Q_h$ such that $\boldsymbol{v}_h$ satisfies approximately conditions (5.4), (a), (b), (c)(i) and the identity

$$a(U_h, U_h, V_h) = f(V_h), \quad \text{for all } V_h = (\boldsymbol{\varphi}_h, q_h) \in X_h \times Q_h. \tag{5.20}$$

The couple (X_h, Q_h) of the finite element spaces has to satisfy the Babuška–Brezzi (BB) inf–sup condition (see, e.g., [35, 36] or [99]), i.e., we assume that there exists a constant $\beta > 0$ such that

$$\sup_{\boldsymbol{v} \in X_h} \frac{(p, \nabla \cdot \boldsymbol{v})}{\|\nabla \boldsymbol{v}\|_{L^2(\Omega)}} \geq \beta \|p\|_{L^2(\Omega)}, \qquad \forall p \in Q_h. \tag{5.21}$$

In practical computations we assume that the domain Ω is a polygonal approximation of the region occupied by the fluid at time t_{n+1}. The spaces W_h, X_h, Q_h are defined over a triangulation $\mathcal{T}_h$ of the domain Ω, formed by a finite number of closed triangles $K \in \mathcal{T}_h$ with the following properties:

(a) $\overline{\Omega} = \bigcup_{K \in \mathcal{T}_h} K$,
(b) the intersection of two different elements $K, K' \in \mathcal{T}_h$ is either empty or a common edge or a common vertex of these elements,
(c) the vertices lying on $\partial\Omega$ belong to $\partial\Omega_{t_{n+1}}$,
(d) the end points of Γ_I, Γ_O, and Γ_{Wt} are vertices of some elements $K \in \mathcal{T}_h$.

We shall denote by $\tilde{h}_K$ the length of the maximal side and we assume the index h is chosen as $h = \max_{K \in \mathcal{T}_h} \tilde{h}_K$. The spaces W_h, X_h, and Q_h are formed by piecewise polynomial functions. In our computations, the well-known Taylor–Hood P_2/P_1 conforming finite elements are used for the velocity/pressure approximation. This means that p_h is a linear function and $\boldsymbol{v}_h$ is a quadratic vector-valued function on each element $K \in \mathcal{T}_h$, i.e., the spaces W_h, X_h, and Q_h are defined by

$$\begin{aligned} H_h &= \{\varphi \in C(\overline{\Omega}); \varphi|_K \in P_2(K) \text{ for each } K \in \mathcal{T}_h\}, \\ W_h &= [H_h]^2, \qquad\qquad X_h = W_h \cap X, \\ Q_h &= \left\{\varphi \in C(\overline{\Omega});\ \varphi|_K \in P_1(K) \text{ for each } K \in \mathcal{T}_h\right\}, \end{aligned} \tag{5.22}$$

where $P_k(K)$ denotes the space of all polynomials on K of degree less or equal to k.

The standard Galerkin discretization (5.20) may produce approximate solutions suffering from spurious oscillations for high Reynolds numbers. In order to avoid this drawback, the *stabilization via streamline-diffusion/Petrov–Galerkin technique* is applied (see, e.g., [34, 53]). The stabilization terms are defined as

$$\mathcal{L}_h(U^*, U, V) = \sum_{K \in \mathcal{T}_h} \delta_K \left(\frac{3\rho}{2\Delta t} \boldsymbol{v} - \nabla \cdot \mu(\nabla \boldsymbol{v} + \nabla^T \boldsymbol{v}) + \rho \left((\boldsymbol{v}^* - \boldsymbol{z}^{n+1}) \cdot \nabla \right) \boldsymbol{v} \right.$$

$$+\nabla p, ((\boldsymbol{v}^* - z^{n+1}) \cdot \nabla)\boldsymbol{\varphi} + \nabla q \Big)_K,$$

$$\mathcal{F}_h(V) = \sum_{K \in \mathcal{T}_h} \delta_K \left(\frac{\rho}{2\Delta t}(4\hat{\boldsymbol{v}}^n - \hat{\boldsymbol{v}}^{n-1}), ((\boldsymbol{v}^* - z^{n+1}) \cdot \nabla)\boldsymbol{\varphi} + \nabla q \right)_K, \tag{5.23}$$

$$U = (\boldsymbol{v}, p), \quad V = (\boldsymbol{\varphi}, q), \quad U^* = (\boldsymbol{v}^*, p^*),$$

where $\delta_K \geq 0$ are suitable parameters. Moreover, the additional div–div stabilization form

$$\mathcal{P}_h(U, V) = \sum_{K \in \mathcal{T}_h} \tau_K (\nabla \cdot \boldsymbol{v}, \nabla \cdot \boldsymbol{\varphi})_K \tag{5.24}$$

is introduced with suitable parameters $\tau_K \geq 0$.

The *stabilized discrete problem* reads: Find $U_h = (\boldsymbol{v}_h, p_h) \in W_h \times Q_h$ such that $\boldsymbol{v}_h$ satisfies approximately conditions (5.4), (a), (b), (c)(i) and

$$a(U_h, U_h, V_h) + \mathcal{L}_h(U_h, U_h, V_h) + \mathcal{P}_h(U_h, V_h) = f(V_h) + \mathcal{F}_h(V_h)$$
$$\text{for all } V_h = (\boldsymbol{\varphi}_h, q_h) \in X_h \times Q_h. \tag{5.25}$$

Stabilization Parameters

The choice of the parameters δ_K and τ_K is carried out according to [53] and [82]. The parameter δ_K is defined on the basis of the local transport velocity $\boldsymbol{v}^* - z^{n+1}$ and local element size h_K of K measured in the direction of the vector $(\boldsymbol{v}^* - z^{n+1})(b_K)$, where b_K denotes the barycenter of K. In the case of the Taylor–Hood finite elements the following choice of parameters appears suitable:

$$\tau_K = \tau_*, \qquad \delta_K = \delta^* h_K^2, \tag{5.26}$$

where $\tau^* > 0$ and $\delta^* > 0$ are fixed constants.

The fully stabilized problem allows also the application of the equal-order P_1/P_1 finite elements, which do not satisfy the BB condition. In this case, we set $\nu = \mu/\rho$ and introduce the parameters

$$\tau_K = \nu \left(1 + Re^{\text{loc}} + \frac{h_K^2}{\nu\, \Delta t} \right), \qquad \delta_K = \frac{h_K^2}{\tau_K}, \tag{5.27}$$

where the local Reynolds number Re^{loc} is defined as

$$Re^{\text{loc}} = \frac{h_K \|\boldsymbol{v}\|_K}{2\nu}. \tag{5.28}$$

5.2.4 *Numerical Solution of the Nonlinear Discrete Problem*

Oseen Linearization Process

The nonlinear discrete problem (5.25) is solved on each time level t_{n+1} with the aid of the linearized Oseen iterative process

$$a(U_h^{(\ell)}, U_h^{(\ell+1)}, V_h) + \mathcal{L}_h(U_h^{(\ell)}, U_h^{(\ell+1)}, V_h) + \mathcal{P}_h(U_h^{(\ell+1)}, V_h) \quad (5.29)$$
$$= f(V_h) + \mathcal{F}_h(V_h) \qquad \text{for all } V_h \in X_h \times Q_h,$$

where we start from the initial approximation $U_h^{(0)} = (\hat{\boldsymbol{v}}^n, \hat{p}^n)$ or $U_h^{(0)} = (2\hat{\boldsymbol{v}}^n - \hat{\boldsymbol{v}}^{n-1}, 2\hat{p}^n - \hat{p}^{n-1})$. Numerical experiments show that it is usually enough to compute 5–8 Oseen iterations on each time level.

Solution of the Linear Algebraic System

The solution of the linear algebraic system equivalent to (5.29) can be realized by the direct solver UMFPACK [19], which works sufficiently fast for systems with up to 10^5 equations. For larger systems it is necessary to apply more robust and efficient iterative techniques, such as the domain decomposition approach and/or the multigrid method.

5.3 Structural Models

5.3.1 *Aeroelastic Model of Vocal Folds Vibration with Two Degrees of Freedom*

Original theoretical model for vibration onset of the vocal folds in the airflow coming from the human subglottal tract allows studying the influence of the physical properties of the vocal folds (e.g., geometrical shape, mass, and damping) on the natural frequencies, mode shapes of vibration, and the thresholds of instability [40,41].

The model of the vocal fold is designed as a simplified dynamic system with two degrees of freedom (rotation and translation) vibrating on an elastic foundation in the wall of a channel conveying air. The phonatory airflow is approximated by unsteady one-dimensional flow theory for inviscid incompressible fluid. A generally defined shape of the vocal fold surface is considered for expressing the unsteady aerodynamic forces in the glottis. The parameters of the mechanical part of the model, i.e., the mass, stiffness, and damping matrices are related to the geometry and material density of the vocal folds as well as to the fundamental natural frequency and damping known from the experiments. The coupled numerical solution yields

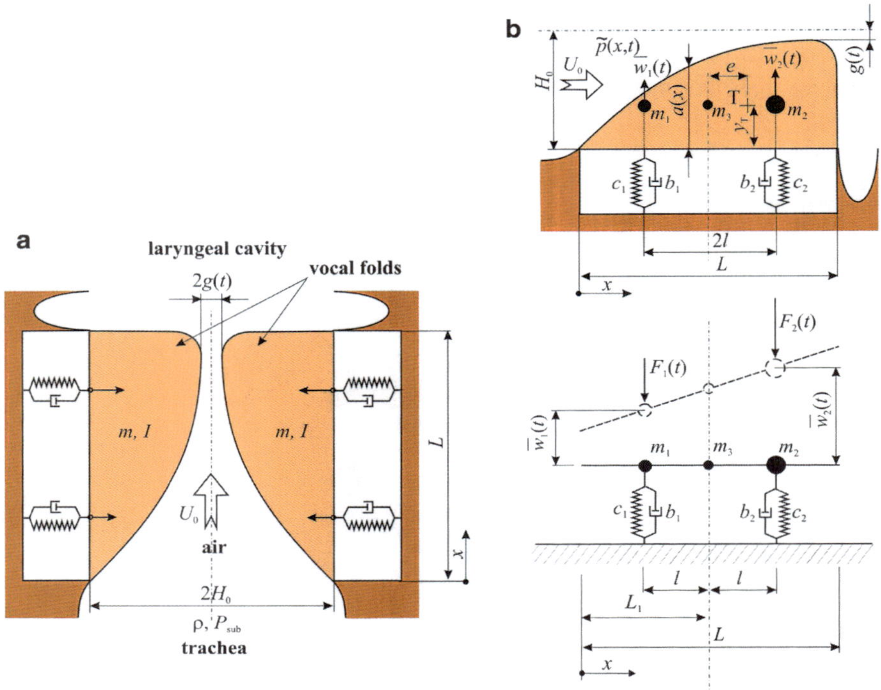

Fig. 5.4 (**a**) Simplified model of the glottis with vibrating vocal folds in the airflow coming from the trachea and (**b**) scheme of the equivalent mechanical system with two-degrees of freedom for one vocal fold

the instability thresholds of the aeroelastic system. The model is particularly suitable for studying the phonation threshold, i.e., the onset of vibration of the vocal folds.

The model of the glottis with the vocal folds is shown in Fig. 5.4a, the glottis forms a channel with planar symmetry conveying air. In this section, the first coordinate x_1 shall be denoted by x and the second coordinate x_2 by y for simplicity. The length L of the channel is measured parallel to both the plane of symmetry and the direction of airflow. The channel walls are created by two vocal-fold-shaped rigid bodies of mass m and moment of inertia I, which are vibrating symmetrically in the opposite phase with identical amplitudes on an elastic foundation modelled by two springs and dampers. The rigid bodies oscillate in the fluid of density ρ flowing from the trachea, where P_{sub} denotes the mean subglottal pressure. The mean airflow velocity at the inlet of the glottal channel ($x = 0$), where its cross-section equals $2H_0$, is denoted as U_0. The minimum cross-section of the channel, the so-called glottal width, is denoted as $2g$. The vibrating element has a smooth convergent glottal inlet and a short, not highly divergent outlet, where the flow separation is supposed. Symmetric oscillations of the vocal folds are assumed, allowing modelling only a half of the glottal region.

The simplified fluid-elastic system is schematically shown in Fig. 5.4b including a scheme of the equivalent mechanical part. The vocal fold can be approximated by a two-degrees-of-freedom rigid body element with a defined shape $a(x)$, where x is the axial coordinate. The element is supported by two discrete springs with stiffnesses c_1 and c_2 and its vibration is described by its rotation and translation. An equivalent three-mass system is used to formulate the equations of motion of the element, based on three conditions of identical total mass, static moment and moment of inertia of the rigid body. The vibrating rigid body with the center of gravity T at the location (x_T, y_T) is replaced by three masses m_1, m_2, and m_3 joined together by a rigid massless rod of the total length L.

The distance between the positions of the masses m_1 and m_2 is denoted as $2l$. The distance between the location of the mass m_3 from the upstream end of the rod is denoted by L_1. The displacements of the masses m_1 and m_2 are denoted as $\overline{w}_1(t)$ and $\overline{w}_2(t)$, where t is time. The length L should approximately correspond to the anatomical data, the lengths l and L_1 can, however, be varied for the purpose of tuning of the model.

Equations of Motion for the Equivalent Mechanical System of the Vocal-Fold-Shaped Vibrating Element

The three masses m_1, m_2, m_3 of the equivalent mechanical system shown in Fig. 5.4b can be calculated from the following equations: equivalent mass of the system:

$$m_1 + m_2 + m_3 = m, \tag{5.30}$$

equivalent static moment

$$-m_1 \frac{L}{2} + m_2 \frac{L}{2} = me, \tag{5.31}$$

equivalent moment of inertia

$$m_1 \left(\frac{L}{2}\right)^2 + m_2 \left(\frac{L}{2}\right)^2 = I + me^2, \tag{5.32}$$

which gives

$$m_{1,2} = \frac{1}{2l^2}\left(I + me^2 \mp mel\right), \quad m_3 = m\left(1 - \left(\frac{e}{l}\right)^2\right) - \frac{I}{l^2}. \tag{5.33}$$

The displacement of the rigid massless rod can be written as

$$\overline{w}(x,t) = (x - L_1)V_1(t) + V_2(t), \tag{5.34}$$

where the rotation and translation of the rigid body element was introduced as

$$V_1(t) = \frac{\overline{w}_2(t) - \overline{w}_1(t)}{2l}, \qquad V_2(t) = \frac{\overline{w}_2(t) + \overline{w}_1(t)}{2}. \tag{5.35}$$

The equivalent aerodynamic excitation forces $\overline{F}_1(t)$ and $\overline{F}_2(t)$ Fig. 5.4b are given by the integrals of perturbation aerodynamic pressure $\tilde{p}(x,t)$ along the vibrating body surface

$$\begin{aligned} \overline{F}_1(t) &= \frac{h^s}{2l}\int_0^L \tilde{p}(x,t)\big(l + L_1 - x - \big(a'(x) + V_1(t)\big) \\ &\quad \cdot \big(a(x) - (x - L_1)V_1(t) - y_T\big)\big)\,\mathrm{d}x, \\ \overline{F}_2(t) &= \frac{h^s}{2l}\int_0^L \tilde{p}(x,t)\big(l - L_1 + x - \big(a'(x) + V_1(t)\big) \\ &\quad \cdot \big(a(x) - (x - L_1)V_1(t) - y_T\big)\big)\,\mathrm{d}x, \end{aligned} \tag{5.36}$$

where h^s is the width of the channel measured perpendicular to the direction of airflow and parallel to the plane of symmetry, h^s is identical with the width of the rigid body. Then by calculating the integrals (5.36) the aerodynamic forces $\overline{F}_1(t)$ and $\overline{F}_2(t)$ can be expressed as functions of the displacements $V_1(t)$ and $V_2(t)$.

After expressing the potential and kinetic energies of the system in a similar way as in the article [40] and their substitution in the Lagrange equations (see, e.g., [15]), the equations of motion are obtained in the form:

After substitution in the Lagrange equations, the equations of motion are obtained in the form

$$\mathbb{M}\ddot{\mathbf{V}} + \mathbb{B}\dot{\mathbf{V}} + \mathbb{K}\mathbf{V} + \mathbf{F} = 0, \tag{5.37}$$

where the following displacement and excitation force vector were introduced

$$\mathbf{V} = \begin{pmatrix} V_1(t) \\ V_2(t) \end{pmatrix}, \quad \mathbf{F} = \begin{pmatrix} \overline{F}_1(t) \\ \overline{F}_2(t) \end{pmatrix}, \tag{5.38}$$

and where $\mathbb{M}$, $\mathbb{B}$, $\mathbb{K}$ are the structural mass, damping, and stiffness matrices:

$$\mathbb{M} = \begin{pmatrix} -lm_1 & m_1 + \frac{m_3}{2} \\ lm_2 & m_2 + \frac{m_3}{2} \end{pmatrix}, \qquad \mathbb{B} = \varepsilon_1\mathbb{M} + \varepsilon_2\mathbb{K}, \qquad \mathbb{K} = \begin{pmatrix} -c_1 l & c_1 \\ c_2 l & c_2 \end{pmatrix}. \tag{5.39}$$

The damping matrix B represents a proportional model of structural damping (see, e.g., [15]) and ε_1, ε_2 are constants adjusted according to the desired damping ratios for the two natural modes of vibration of the system. The structure of the matrices $\mathbb{M}$ and $\mathbb{K}$ reveals that a mass coupling caused by the mass m_3 is generally in the system even if $\mathbf{F} = 0$.

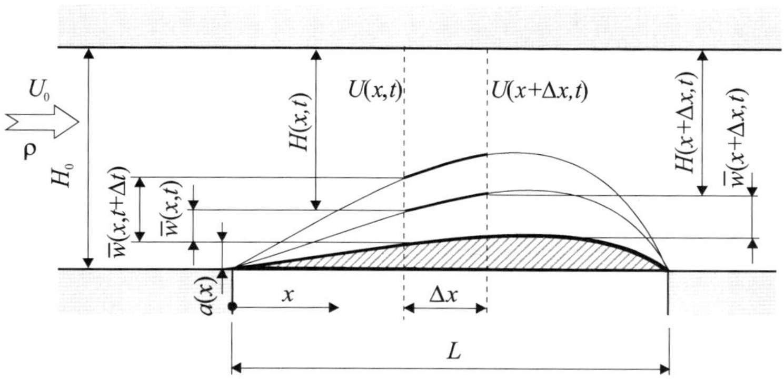

Fig. 5.5 Simplified model of a channel

Aerodynamic Forces

The one-dimensional (1D) model of the unsteady incompressible fluid flow in the rectangular channel of the height $H(x,t)$ and the width $h^s = \text{const.}$ is given by the following Euler (momentum) and the approximation of the continuity equations with a space increment Δx (see [60] and Fig. 5.5):

$$\rho \frac{\partial U}{\partial t}(x,t) + \rho U(x,t) \frac{\partial U}{\partial x}(x,t) + \frac{\partial P}{\partial x}(x,t) = 0, \tag{5.40}$$

$$\frac{\partial(\rho h^s H(x,t)\Delta x)}{\partial t} = \Big(\rho U(x,t) h^s H(x,t)\Big) - \Big(\rho U(x+\Delta x,t) h^s H(x+\Delta x,t)\Big), \tag{5.41}$$

where the flow velocity $U(x,t)$ and the pressure $P(x,t)$ can be considered to be composed of the steady and perturbation parts as follows:

$$U(x,t) = \overline{U}_o(x) + \tilde{u}(x,t), \qquad P(x,t) = P_0(x) + \tilde{p}(x,t). \tag{5.42}$$

According to the previous section and Fig. 5.5,

$$H(x,t) = H_0 - \overline{w}(x,t) - a(x). \tag{5.43}$$

Substitution of the expressions (5.42) into Eq. (5.40) yields

$$\frac{\partial \tilde{u}}{\partial t} + \frac{\partial(\tilde{u}\overline{U_0})}{\partial x} + \tilde{u}\frac{\partial \tilde{u}}{\partial x} = -\frac{1}{\rho}\frac{\partial \tilde{p}}{\partial x}, \tag{5.44}$$

where the Bernoulli equation

$$\frac{1}{2}\rho\overline{U}_0^2 + P_0(x) = \text{const.} \tag{5.45}$$

was used. Here, the attention is focused on the phonatory threshold states, for the purpose of which only small velocity perturbations ($\tilde{u}\,\frac{\partial\tilde{u}}{\partial x}$) can be considered. Introducing the velocity potential

$$\tilde{\Phi}(x,t) = \Phi(x,t) + \Xi(t) \tag{5.46}$$

where $\Xi(t)$ is a nonspecified (dummy) time function,

$$\tilde{u} = \frac{\partial\tilde{\Phi}}{\partial x} = \frac{\partial\Phi}{\partial x}, \tag{5.47}$$

the integration of Eq. (5.44) over x gives

$$\frac{\partial\Phi}{\partial t} + \frac{\partial\Xi}{\partial t} + \tilde{u}\overline{U_0} = -\frac{1}{\rho}\tilde{p} + \xi_{\Xi}(t). \tag{5.48}$$

When setting

$$\xi_{\Xi}(t) = \frac{\partial\Xi}{\partial t},$$

the perturbation pressure can be expressed as

$$\tilde{p} = -\rho\left(\frac{\partial\Phi}{\partial t} + \overline{U}_0(x)\frac{\partial\Phi}{\partial x}\right). \tag{5.49}$$

Substitution of the expressions (5.42) for $U(x,t)$ and (5.43) for $H(x,t)$ into (5.41) and the limit process for $\Delta x \to 0$ yield

$$\frac{\partial\overline{w}}{\partial t} = -\frac{\partial(\overline{w}\overline{U_0})}{\partial x} + H_0\frac{\partial\tilde{u}}{\partial x} - \frac{\partial(\tilde{u}\overline{w})}{\partial x} - \frac{\partial(a\tilde{u})}{\partial x}, \tag{5.50}$$

where the continuity equation

$$\overline{U}_0(x)\big(H_0 - a(x)\big) = \text{const.}, \tag{5.51}$$

for the steady flow was used. Integration of Eq. (5.50) over x gives

$$\tilde{u}(x,t) = \frac{1}{H_0 - a(x) - \overline{w}(x,t)} \tag{5.52}$$

$$\times\left(\frac{\partial}{\partial t}\left(\frac{x^2}{2}V_1(t)+\left(V_2(t)-\frac{L}{2}V_1(t)\right)x+C(t)\right)+\overline{w}(x,t)\overline{U}_0(x)\right).$$

For the phonatory threshold states only small displacements $\overline{w}(x,t) \ll H_0$ can be assumed, and substituting $\overline{w}$ from Eq. (5.34) into Eq. (5.52), the derivative (5.47) of the potential becomes

$$\frac{\partial\Phi}{\partial x}=\frac{1}{H_0-a(x)}\left(\frac{x^2}{2}\dot{V}_1(t)+\left(\dot{V}_2(t)-\frac{L}{2}\dot{V}_1(t)\right)x+\dot{C}(t)\right.\tag{5.53}$$
$$\left.+\left(\left(x-\frac{L}{2}\right)V_1(t)+V_2(t)\right)\overline{U}_0(x)\right),$$

where $C(t)$ is an unknown function of time. The integration gives the potential in the form

$$\Phi(x,t)=\dot{V}_1(t)I_1(x)+\dot{V}_2(t)I_2(x)+\dot{C}(t)I_3(x)+V_1(t)I_4(x)$$
$$+V_2(t)I_5(x)+\overline{C}(t),\tag{5.54}$$

where $\overline{C}(t)$ is another unknown time function and $I_i(x)$, $i=1,\ldots,5$, are the integrals given in [39, 40]. Using the boundary conditions for the flow at the inlet $(x=0)$ and outlet $(x=L)$:

$$\tilde{u}=\frac{\partial\Phi}{\partial x}=0|_{x=0},\qquad \text{and } \tilde{p}=0|_{x=L},\tag{5.55}$$

considering the displacement $\overline{w}(x,t) \ll H_0$ and using the same procedures as in the papers [39,40], the unknown time functions $C(t)$ and $\overline{C}(t)$ can be determined. Then the unsteady component of the pressure $\tilde{p}(x,t)$ can be obtained from Eq. (5.49) and the resulting aerodynamic forces $\overline{F}_1(t)$ and $\overline{F}_2(t)$ are computed from Eq. (5.36). We note that the mean flow velocity in the glottis for $x \in [0,L]$ can be expressed from the continuity equation as

$$\overline{U}(x)=\frac{U_0}{1-\frac{a(x)}{H_0}},\tag{5.56}$$

where U_0 and H_0 are the flow velocity and the height of the channel, respectively, at the inlet $(x=0)$ to the glottal region and the function $a(x)/H_0$ is given by the geometry of the vocal folds and larynx. The airflow velocity U_0 [m s^{-1}] is simply related to the so-called glottal volume velocity (glottal flux)

$$Q=2U_0H_0,\tag{5.57}$$

which belongs to the most important physiological characteristics of voiced sound production, and for the male adults it is normally in the range $Q \approx 0.08$–0.6 l/s.

Substituting the linearized aerodynamic forces $\overline{F}_1(t)$, $\overline{F}_2(t)$ in the equations of motion (5.37) and dividing this equation by $ml/2$ yields the following physically well-structured equations of motion of the coupled aeroelastic system in a linear approximation:

$$\overline{\mathbb{M}}\,\ddot{\overline{V}} + \overline{\mathbb{B}}\,\dot{\overline{V}} + \overline{\mathbb{K}}\,\overline{V} = \frac{\rho h^s L^3}{m H_0}\left(\hat{\mathbb{M}}\ddot{\overline{V}} + \frac{U_0}{L}\hat{\mathbb{B}}\dot{\overline{V}} + \frac{U_0^2}{L^2}\hat{\mathbb{K}}\overline{V}\right), \tag{5.58}$$

where

$$\overline{\mathbb{M}} = \begin{pmatrix} -\left(\frac{I}{ml^2} + \left(\frac{e}{l}\right)^2 - \frac{e}{l}\right) & 1 - \frac{e}{l} \\ \left(\frac{I}{ml^2} + \left(\frac{e}{l}\right)^2 - \frac{e}{l}\right) & 1 + \frac{e}{l} \end{pmatrix}, \overline{\mathbb{B}} = \varepsilon_1 \overline{\mathbb{M}} + \varepsilon_2 \overline{\boldsymbol{K}}, \mathbb{K} = \omega_0^2 \begin{pmatrix} -1 & 1 \\ \frac{c_2}{c_1} & \frac{c_2}{c_1} \end{pmatrix} \tag{5.59}$$

are the dimensionless mass, damping, and stiffness matrices for the rigid body vibrating *in vacuo*, $\omega_0^2 = 2c_1/m$. The expression

$$\overline{\boldsymbol{V}} = \begin{pmatrix} 1 & 0 \\ 0 & 1/l \end{pmatrix} \boldsymbol{V} \tag{5.60}$$

denotes the vector of dimensionless displacements. The matrices $\hat{\mathbb{M}}$, $\hat{\mathbb{B}}$, and $\hat{\mathbb{K}}$ are aerodynamic mass, aerodynamic damping, and aerodynamic stiffness matrices, respectively. Their elements are complicated functions of the channel geometry. Because of their complicated form, we do not introduce it. For details see [41].

The unsteady aerodynamic forces on the right-hand side of Eq. (5.58) are obviously proportional to the dimensionless added mass of fluid ($\frac{\rho h^s L^3}{m H_0}$) and they have a lucid physical meaning. The first term corresponds to the aerodynamic inertia forces, the second term to the aerodynamic damping forces ($\approx U_0$) related to the Coriolis forces and the third term to the aerodynamic stiffness forces ($\approx U_0^2$), which are related to the centrifugal forces. The Coriolis and centrifugal forces are increasing functions of the fluid flow velocity U_0 causing aeroelastic instability and self-oscillations.

Assuming $\overline{\boldsymbol{V}} = \overline{\boldsymbol{V}}_0 e^{st}$ for the dynamic response, the solution of Eq. (5.58) is given by the numerical computation of the eigenvalues $s = Re(s) + i Im(s)$ and eigenmodes $\overline{\boldsymbol{V}}_0 = (V_{01}, V_{02}/l)^T$ for the eigenvalue problem. In this way, it is possible to calculate the critical flow velocity $U_{0,\text{crit}}$ at which the real part of the eigenvalue changes the sign from a negative value $Re(s) < 0$ to a positive value $Re(s) > 0$. Here, the system either becomes unstable by divergence (when $Im(s) = 0$), or it becomes unstable by flutter simulating the start of phonation (when $Im(s) > 0$).

Vocal Fold Modelled as a Flexibly Supported Rigid Body Coupled with the Navier–Stokes Equations

The aerodynamic forces can also be computed from a numerical approximation of the Navier–Stokes equations (5.1). In order to solve the coupled problem, the motion equations (5.37)–(5.39) are rewritten with the aid of the two displacements $\overline{w}_1, \overline{w}_2$ (upward positive), see Fig. 5.4b. The vibrating part (vocal folds) of the channel walls (computational domain) is governed by the motion of Γ_{W_t} driven by the displacements $\overline{w}_1(t)$ and $\overline{w}_2(t)$ of the two masses m_1 and m_2, respectively (see Fig. 5.4b).

The displacements $\overline{w}_1(t)$ and $\overline{w}_2(t)$ are then described by the following equations:

$$\mathbb{M}\begin{pmatrix}\ddot{\overline{w}}_1\\ \ddot{\overline{w}}_2\end{pmatrix} + \mathbb{B}\begin{pmatrix}\dot{\overline{w}}_1\\ \dot{\overline{w}}_2\end{pmatrix} + \mathbb{K}\begin{pmatrix}\overline{w}_1\\ \overline{w}_2\end{pmatrix} = \begin{pmatrix}-\overline{F}_1\\ -\overline{F}_2\end{pmatrix}. \tag{5.61}$$

Here $\mathbb{M}$ and $\mathbb{K}$ are the mass and stiffness matrices, respectively, given by

$$\mathbb{M} = \begin{pmatrix} m_1 + \frac{m_3}{4} & \frac{m_3}{4} \\ \frac{m_3}{4} & m_2 + \frac{m_3}{4}\end{pmatrix}, \qquad \mathbb{K} = \begin{pmatrix} c_1 & 0 \\ 0 & c_2\end{pmatrix},$$

where m_1, m_2, m_3 are the equivalent masses, c_1, c_2 are the spring constants,

$$\mathbb{B} = \varepsilon_1 \mathbb{M} + \varepsilon_2 \mathbb{K}$$

is the matrix of the proportional structural damping and $\overline{F}_1, \overline{F}_2$ are the aerodynamic forces (downward positive), see Fig. 5.4b.

Time Discretization

Similarly as in Sect. 5.2.3, we consider the partition t_k of the time interval $[0, T]$ with the same constant time step $\Delta t > 0$ and use the approximations $\overline{w}_i^n \approx \overline{w}_i(t_n)$ and $\dot{\overline{w}}_i^n \approx \frac{d\overline{w}_i}{dt}(t_n)$. We transform the system (5.61) to the first order system, use the notation $\Lambda^n = (\overline{w}_1^n, \overline{w}_2^n, \dot{\overline{w}}_1^n, \dot{\overline{w}}_2^n)^T$, and discretize in time with the aid of BDF2 formula. Then we get the system

$$\begin{pmatrix}\mathbb{E} & \mathbb{O}\\ \mathbb{O} & \mathbb{M}\end{pmatrix} \frac{3\Lambda^{n+1} - 4\Lambda^n + \Lambda^{n-1}}{2\Delta t} = \begin{pmatrix}\mathbb{O} & \mathbb{E}\\ -\mathbb{K} & -\mathbb{B}\end{pmatrix} \Lambda^{n+1} + \begin{pmatrix}0\\ 0\\ -\overline{F}_1^{n+1}\\ -\overline{F}_2^{n+1}\end{pmatrix}, \tag{5.62}$$

where $\mathbb{E}$ is the unit 2×2 matrix and $\mathbb{O}$ is the zero matrix.

5.3.2 Dynamic Elasticity Problem

If the vocal folds are constructed as a compliant 2D structure and small displacements are supposed, we can formulate the governing equations of motion of the structure using linear elasticity theory. In what follows, $\Omega^s \subset R^2$ will denote a bounded open set representing the vocal folds as an elastic body. We denote by $\boldsymbol{u}(X,t)$ the displacement of the structure at a time instant $t \in (0,T)$ and a point $X = (X_1, X_2) \in \Omega^s$, and define the components of the small strain tensor

$$e_{ij}(\boldsymbol{u}) = \frac{1}{2}\left(\frac{\partial u_i}{\partial X_j} + \frac{\partial u_j}{\partial X_i}\right), \quad i,j = 1,2. \tag{5.63}$$

The deformation of the elastic body is modelled by the dynamic elasticity system of equations with the generalized Hooke law for isotropic material in the form

$$\tau^s_{ij} = \lambda^s \mathrm{div}\,\boldsymbol{u}\delta_{ij} + 2\mu^s e_{ij}, \quad i,j = 1,2, \tag{5.64}$$

where λ^s and μ^s are the so-called Lamé coefficients. In practice the Young modulus E^s of elasticity and the Poisson ratio σ^s,

$$E^s = \frac{\mu^s(3\lambda^s + 2\mu^s)}{\lambda^s + \mu^s}, \qquad \sigma^s = \frac{\lambda^s}{2(\lambda^s + \mu^s)}, \tag{5.65}$$

are often used.

The equations of motion of an elastic body have the form

$$\varrho^s \frac{\partial^2 u_i}{\partial t^2} + C\varrho^s \frac{\partial u_i}{\partial t} - \sum_{j=1}^{2} \frac{\partial \tau^s_{ij}}{\partial X_j} = f_i, \quad \text{in } \Omega^s \times (0,T), \quad i = 1,2. \tag{5.66}$$

Here $\mathbf{f} = (f_1, f_2)$ is the density of the volume force. In reality, mechanical systems dissipate mechanical energy. The expression $C\varrho^s \frac{\partial u_i}{\partial t}$, where $C \geq 0$, is a dissipative damping of the system. By ϱ^s we denote the density of the solid material. We complete the elasticity problem by initial and boundary conditions. The initial conditions read

$$\boldsymbol{u}(\cdot,0) = 0, \qquad \frac{\partial \boldsymbol{u}}{\partial t}(\cdot,0) = 0 \quad \text{in } \Omega^s. \tag{5.67}$$

By $\Gamma^s_W \subset \Gamma_{W0}$ we denote a moving interface between the vocal folds and the domain occupied by the fluid at the initial time $t = 0$. We assume that Γ^s_W and Γ^s_D are disjoint parts of $\partial\Omega^s$ and $\partial\Omega^s = \Gamma^s_W \cup \Gamma^s_D$. Let the aerodynamic surface force $\boldsymbol{T}^n = (T^n_1, T^n_2)$ be prescribed on the boundary Γ^s_W. This leads to the condition

$$\sum_{j=1}^{2} \tau^s_{ij} n_j = T^n_i \quad \text{on } \Gamma^s_W \times (0,T), \quad i = 1,2. \tag{5.68}$$

Here $\mathbf{n} = (n_1, n_2)$ denotes the outer normal to $\partial\Omega^s$.
On the boundary Γ_D^s we prescribe the zero displacement:

$$\boldsymbol{u} = 0 \quad \text{on } \Gamma_D^s \times (0, T). \tag{5.69}$$

In the dynamic elasticity problem we look for the displacement $\boldsymbol{u}$ satisfying Eq. (5.66), the initial conditions (5.67), and the boundary conditions (5.68) and (5.69).

5.3.3 Finite Element Space Discretization of the Elasticity Problem

In this section we shall be concerned with the discretization of the dynamic elasticity problem. We reformulate the problem in a weak sense and apply the Galerkin finite element method. The semidiscretized problem is equivalent to a second-order system of ordinary differential equations. For the time discretization we apply the Newmark scheme. In each time step we get a linear algebraic system with a symmetric positive definite matrix. The solution of this system can be realized by the preconditioned conjugate gradient method.

The space semidiscretization of the structural problem will be carried out by the conforming finite element method. By Ω_h^s we denote a polygonal approximation of the domain Ω^s. We construct a triangulation $\mathcal{T}_h^s$ of the domain Ω_h^s formed by a finite number of closed triangles with the following properties:

(a) $\overline{\Omega}_h^s = \bigcup_{K \in \mathcal{T}_h^s} K$,
(b) the intersection of two different elements $K, K' \in \mathcal{T}_h^s$ is either empty or a common edge or a common vertex of these elements,
(c) the vertices lying on $\partial\Omega_h^s$ belong to $\partial\Omega^s$,
(d) the set $\overline{\Gamma}_W^s \cap \overline{\Gamma}_D^s$ is formed by vertices of some elements $K \in \mathcal{T}_h^s$.

By Γ_{Wh}^s and Γ_{Dh}^s we denote the parts of $\partial\Omega_h^s$ approximating Γ_W^s and Γ_D^s. The approximate solution of the structural problem will be sought in the finite-dimensional space $\boldsymbol{X_h} = X_h \times X_h$, where

$$X_h = \left\{ v_h \in C(\overline{\Omega}_h^s);\ v_h|_K \in P_k(K),\ \forall K \in \mathcal{T}_h^s \right\}, \tag{5.70}$$

$k \geq 1$ is an integer and $P_k(K)$ denotes the set of all polynomials in x_1, x_2 of degree $\leq k$ on K. In $\boldsymbol{X}_h$ we define the subspace $\boldsymbol{V_h} = V_h \times V_h$, where

$$V_h = \left\{ y_h \in X_h;\ y_h|_{\overline{\Gamma}_{Dh}^s} = 0 \right\}. \tag{5.71}$$

The derivation of the space semidiscretization can be obtained in a standard way. Multiplying system (5.66) by any test function $y_{hi} \in V_h$, $i = 1, 2$, applying Green's theorem and using the boundary condition (5.68), we obtain the identity containing the forms defined for $\boldsymbol{u}_h = (u_{h1}, u_{h2})$, $\boldsymbol{y}_h = (y_{h1}, y_{h2}) \in \boldsymbol{X}_h$:

$$a_h(\boldsymbol{u}_h, \boldsymbol{y}_h) = \int_{\Omega_h^s} \lambda^s \operatorname{div} \boldsymbol{u}_h \operatorname{div} \boldsymbol{y}_h \, \mathrm{d}X + 2 \int_{\Omega_h^s} \mu^s \sum_{i,j=1}^{2} e_{ij}^s(\boldsymbol{u}_h) \, e_{ij}^s(\boldsymbol{y}_h) \, \mathrm{d}X, \quad (5.72)$$

and

$$(\boldsymbol{\varphi}, \boldsymbol{\psi})_{\Omega_h^s} = \int_{\Omega_h^s} \boldsymbol{\varphi} \cdot \boldsymbol{\psi} \, \mathrm{d}X, \quad (\boldsymbol{\varphi}, \boldsymbol{\psi})_{\Gamma_{Wh}} = \int_{\Gamma_{Wh}^s} \boldsymbol{\varphi} \cdot \boldsymbol{\psi} \, \mathrm{d}S. \quad (5.73)$$

We shall use the approximation $\boldsymbol{T}_h^n \approx \boldsymbol{T}^n$ and the notation $\boldsymbol{u}_h'(t) = \frac{\partial \boldsymbol{u}_h(t)}{\partial t}$ and $\boldsymbol{u}_h''(t) = \frac{\partial^2 \boldsymbol{u}_h(t)}{\partial t^2}$. Then we define the *approximate solution of the structural problem* as a function $t \in [0, T] \rightarrow \boldsymbol{u}_h(t) \in \boldsymbol{V}_h$ such that there exist the time derivatives $\boldsymbol{u}_h'(t), \boldsymbol{u}_h''(t)$ and the identity

$$(\varrho^s \boldsymbol{u}_h''(t), \boldsymbol{y}_h)_{\Omega_h^s} + (C\varrho^s \boldsymbol{u}_h'(t), \boldsymbol{y}_h)_{\Omega_h^s} + a_h(\boldsymbol{u}_h(t), \boldsymbol{y}_h) = (\boldsymbol{T}_h^n(t), \boldsymbol{y}_h)_{\Gamma_{Wh}},$$
$$\forall \boldsymbol{y}_h \in \boldsymbol{V}_h, \quad \forall t \in (0, T), \quad (5.74)$$

and the initial conditions

$$\boldsymbol{u}_h(X, 0) = 0, \quad \boldsymbol{u}_h'(X, 0) = 0, \quad X \in \Omega_h^s. \quad (5.75)$$

are satisfied.

The discrete problem (5.74), (5.75) is equivalent to the solution of a system of ordinary differential equations. Let functions $\varphi_1, \ldots, \varphi_m$ represent a basis of the space V_h. Then the system of $r = 2m$ of the vector functions $(\varphi_1, 0)$, ..., $(\varphi_m, 0)$, $(0, \varphi_1)$, ..., $(0, \varphi_m)$ form a basis of the space $\boldsymbol{V}_h$. Let us denote them by $\boldsymbol{\varphi}_1, \ldots, \boldsymbol{\varphi}_r$. Then the approximate solution $\boldsymbol{u}_h$ can be expressed in the form

$$\boldsymbol{u}_h(t) = \sum_{j=1}^{r} p_j(t) \boldsymbol{\varphi}_j, \quad t \in [0, T]. \quad (5.76)$$

Let us set $\boldsymbol{p}(t) = (p_1(t), \ldots, p_r(t))^T$. Using $\boldsymbol{\varphi}_j$, $j = 1, \ldots, r$, as test functions in (5.74), we get the following system of ordinary differential equations

$$\mathbb{M}\boldsymbol{p}'' = \boldsymbol{G} - \mathbb{K}\boldsymbol{p} - C\mathbb{M}\boldsymbol{p}', \quad (5.77)$$

where $\mathbb{M} = (m_{ij})_{i,j=1}^r$ is the mass matrix and $\mathbb{K} = (k_{ij})_{i,j=1}^r$ is the stiffness matrix with the elements $m_{ij} = (\rho^s \boldsymbol{\varphi}_i, \boldsymbol{\varphi}_j)_{\Omega_h^s}$ and $k_{ij} = a_h(\boldsymbol{\varphi}_i, \boldsymbol{\varphi}_j)_{\Omega_h^s}$, respectively. The aerodynamic force vector $\boldsymbol{G} = \boldsymbol{G}(t) = (G_1(t), \ldots, G_r(t))^T$ has the components

$G_i(t) = (\boldsymbol{T}_h^n(t), \boldsymbol{\varphi}_i)_{\Gamma_{Wh}}$, $i = 1, \ldots, r$. System (5.77) is equipped with the initial conditions

$$p_j(0) = 0, \quad p_j'(0) = 0, \quad j = 1, \ldots, r. \tag{5.78}$$

5.3.4 *Time Discretization of the Structural Problem*

The discrete initial value problem (5.77), (5.78) is solved by the Newmark method [18]. We consider the partition of the time interval $[0, T]$ formed by the time instants $t_k = k\Delta t$ with a time step Δt introduced in Sect. 5.2.3. Let us set $\boldsymbol{p}_0 = 0, z_0 = 0, \boldsymbol{G}_k = \boldsymbol{G}(t_k)$, and introduce the approximations $\boldsymbol{p}_k \approx \boldsymbol{p}(t_k)$ and $\boldsymbol{q}_k \approx \boldsymbol{p}'(t_k)$ for $k = 1, 2, \ldots$. The Newmark scheme can be written in the form

$$\begin{aligned} \boldsymbol{p}_{k+1} = \boldsymbol{p}_k + \Delta t \boldsymbol{q}_k + \Delta t^2 \Big(& \beta \left(\mathbb{M}^{-1} \boldsymbol{G}_{k+1} - \mathbb{M}^{-1} \mathbb{K} \boldsymbol{p}_{k+1} - C \boldsymbol{q}_{k+1} \right) \\ & + \left(\frac{1}{2} - \beta \right) \left(\mathbb{M}^{-1} \boldsymbol{G}_k - \mathbb{M}^{-1} \mathbb{K} \boldsymbol{p}_k - C \boldsymbol{q}_k \right) \Big), \end{aligned} \tag{5.79}$$

$$\begin{aligned} \boldsymbol{q}_{k+1} = \boldsymbol{q}_k + \Delta t \Big(& \gamma \left(\mathbb{M}^{-1} \boldsymbol{G}_{k+1} - \mathbb{M}^{-1} \mathbb{K} \boldsymbol{p}_{k+1} - C \boldsymbol{q}_{k+1} \right) \\ & + (1 - \gamma) \left(\mathbb{M}^{-1} \boldsymbol{G}_k - \mathbb{M}^{-1} \mathbb{K} \boldsymbol{p}_k - C \boldsymbol{q}_k \right) \Big), \end{aligned} \tag{5.80}$$

where $\beta, \gamma \in R$ are parameters. From Eq. (5.80) we get

$$\begin{aligned} \boldsymbol{q}_{k+1} = \frac{1}{1 + C\gamma\Delta t} \Big(& \boldsymbol{q}_k + \Delta t \Big(\gamma \left(\mathbb{M}^{-1} \boldsymbol{G}_{k+1} - \mathbb{M}^{-1} \mathbb{K} \boldsymbol{p}_{k+1} \right) \\ & + (1 - \gamma) \left(\mathbb{M}^{-1} \boldsymbol{G}_k - \mathbb{M}^{-1} \mathbb{K} \boldsymbol{p}_k - C \boldsymbol{q}_k \right) \Big) \Big). \end{aligned} \tag{5.81}$$

The substitution of (5.81) in (5.79) yields the relation

$$\begin{aligned} \boldsymbol{p}_{k+1} = \boldsymbol{p}_k + \Delta t \boldsymbol{q}_k + \beta \Delta t^2 \Big(& \mathbb{M}^{-1} \boldsymbol{G}_{k+1} - \mathbb{M}^{-1} \mathbb{K} \boldsymbol{p}_{k+1} - \frac{C}{1 + C\gamma\Delta t} \boldsymbol{q}_k \\ & - \frac{C\gamma\Delta t}{1 + C\gamma\Delta t} \left(\mathbb{M}^{-1} \boldsymbol{G}_{k+1} - \mathbb{M}^{-1} \mathbb{K} \boldsymbol{p}_{k+1} \right) \\ & - \frac{C\Delta t}{1 + C\gamma\Delta t} (1 - \gamma) \left(\mathbb{M}^{-1} \boldsymbol{G}_k - \mathbb{M}^{-1} \mathbb{K} \boldsymbol{p}_k - C \boldsymbol{q}_k \right) \Big) \\ & + \left(\frac{1}{2} - \beta \right) \Delta t^2 \left(\mathbb{M}^{-1} \boldsymbol{G}_k - \mathbb{M}^{-1} \mathbb{K} \boldsymbol{p}_k - C \boldsymbol{q}_k \right). \end{aligned} \tag{5.82}$$

Then we find that

$$\begin{aligned} \boldsymbol{p}_{k+1} &= \boldsymbol{p}_k + \Delta t \boldsymbol{q}_k - C\xi_k \boldsymbol{q}_k + \xi_k \left(\mathbb{M}^{-1}\boldsymbol{G}_{k+1} - \mathbb{M}^{-1}\mathbb{K}\boldsymbol{p}_{k+1}\right) \\ &+ \left(\left(\frac{1}{2} - \beta\right)\Delta t^2 - C\,(1-\gamma)\,\xi_k \Delta t\right)\left(\mathbb{M}^{-1}\boldsymbol{G}_k - \mathbb{M}^{-1}\mathbb{K}\boldsymbol{p}_k - C\boldsymbol{q}_k\right), \end{aligned} \tag{5.83}$$

which can be written in the form

$$\begin{aligned} \left(\mathbb{I} + \xi_k \mathbb{M}^{-1}\mathbb{K}\right)\boldsymbol{p}_{k+1} &= \boldsymbol{p}_k + (\Delta t - C\xi_k)\,\boldsymbol{q}_k + \xi_k \mathbb{M}^{-1}\boldsymbol{G}_{k+1} + \\ &+ \left(C\,(\gamma - 1)\,\xi_k \Delta t + \left(\frac{1}{2} - \beta\right)\Delta t^2\right)\left(\mathbb{M}^{-1}\boldsymbol{G}_k - \mathbb{M}^{-1}\mathbb{K}\boldsymbol{p}_k - C\boldsymbol{q}_k\right). \end{aligned} \tag{5.84}$$

where we set for the sake of simplicity

$$\xi_k = \beta \Delta t^2 \left(1 - \frac{C\gamma\Delta t}{1 + C\gamma\Delta t}\right) = \frac{\beta\Delta t^2}{1 + C\gamma\Delta t}. \tag{5.85}$$

If $\boldsymbol{p}_k$ and $\boldsymbol{q}_k$ are known, then $\boldsymbol{p}_{k+1}$ is obtained from system (5.84) and afterwards $\boldsymbol{q}_{k+1}$ is computed from (5.81).

In numerical examples presented in Sects. 5.7.2 and 5.10.2, the parameters $\beta = 1/4$ and $\gamma = 1/2$ were used. This choice yields the Newmark method of the second order.

5.4 Coupled FSI Problems

In the previous sections, the flow and structural problems were considered separately. In what follows, we shall be concerned with complete coupled FSI problems.

5.4.1 Coupled Problem of Incompressible Flow and Vocal Fold Rigid Body Model

The flow model (5.9), (5.4) is coupled with the equation of motion (5.61) by interface conditions. First, the boundary condition (5.4), (b) must be satisfied on $\tilde{\Gamma}_{Wt}$, where the domain velocity is given by Eq. (5.6). Further, the aerodynamic forces $\overline{F}_1, \overline{F}_2$ in (5.61) depend on the flow velocity $\boldsymbol{v}$ and the pressure p. The forces $\overline{F}_1$ and $\overline{F}_2$ are computed with the aid of the aerodynamic lift force L and the aerodynamic torsional moment M as

$$\overline{F}_1(t) = -\frac{L(t)}{2} - \frac{1}{2l}M(t), \qquad \overline{F}_2(t) = -\frac{L(t)}{2} + \frac{1}{2l}M(t), \tag{5.86}$$

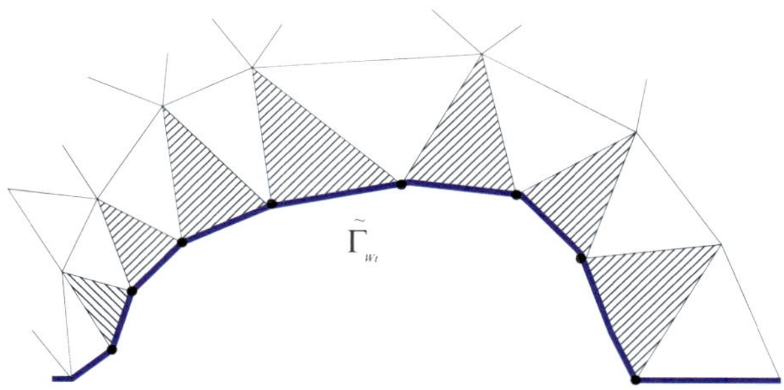

Fig. 5.6 Stress tensor extrapolation from elements K adjacent to the boundary $\tilde{\Gamma}_{Wt}$

where l denotes the distance of the masses m_1 and m_2 from the center mass m_3 (see Fig. 5.4b). The aerodynamic quantities are defined by

$$L(t) = h^s \int_{\Gamma_{Wt}} \sum_{j=1}^{2} \tau_{2j}^f n_j \, \mathrm{d}S, \qquad M(t) = -h^s \int_{\Gamma_{Wt}} \sum_{i,j=1}^{2} \tau_{ij}^f n_j r_i^{\mathrm{ort}} \, \mathrm{d}S, \quad (5.87)$$

where $r_1^{\mathrm{ort}} = -(x_2 - x_{C2})$, $r_2^{\mathrm{ort}} = x_1 - x_{C1}$, $(x_{C1}, x_{C2}) = (0, L/2)$ and the components τ_{ij}^f are defined by (5.2).

The evaluation of the lift L and the moment M at time $t = t_{n+1}$ from the approximate solution $U_h = (\boldsymbol{v}_h, p_h)$ can be carried out in two ways.

Stress Tensor Extrapolation

One possibility is to compute the components τ_{ij}^f of the stress tensor at time $t = t_{n+1}$ on the elements $K \in \mathcal{T}_h$ adjacent to the boundary $\tilde{\Gamma}_{Wt}$, extrapolate τ_{ij}^f to $\tilde{\Gamma}_{Wt}$, see Fig. 5.6, and then to compute L and M by the integration along $\tilde{\Gamma}_{Wt}$.

This approach can be improved by using the idea of superconvergence of gradient in the center of triangles, cf. [14,110]. It means that for any vertex $A \in \tilde{\Gamma}_{Wt}$ the value of the stress tensor τ_{ij}^f can be reconstructed using least squares approximation from the patches of neighboring elements. See Fig. 5.7.

Weak Formulation of Lift Force

In this paragraph $\tilde{\Gamma}_{Wt}$ represents again the interface between the structure and the domain Ω_t. The Navier–Stokes equations in the ALE form discretized with respect to time at instant $t := t_{n+1}$ can be expressed component-wise as

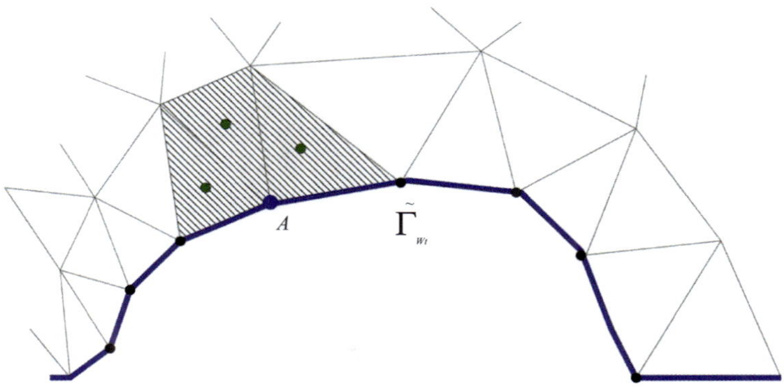

Fig. 5.7 Stress tensor extrapolation from the patches of elements adjacent to the point A of the boundary $\tilde{\Gamma}_{Wt}$

$$\rho \frac{3v_i - 4\hat{v}_i^n + \hat{v}_i^{n-1}}{2\tau} + \rho\left((\boldsymbol{v} - \boldsymbol{z}^{n+1}) \cdot \nabla\right) v_i = \sum_{j=1}^{2} \frac{\partial \tau_{ij}^f}{\partial x_j}, \tag{5.88}$$

$$\text{in } \Omega_t, \quad i = 1, 2.$$

Let us set

$$\tilde{\Omega}_{\Gamma_{Wt}} = \cup\{K \in \mathcal{T}_h;\ \ K \cap \tilde{\Gamma}_{Wt} \neq \emptyset\}. \tag{5.89}$$

This represents a one-layer strip around the vibrating body formed by finite elements. Multiplying Eq. (5.88) with $i = 2$ by a function $\varphi \in W_h$ such that

$$\begin{aligned} \varphi(x) &= 1 \quad \text{for } x \in \tilde{\Gamma}_{Wt}, \\ \varphi(x) &= 0 \quad \text{outside the set } \tilde{\Omega}_{\Gamma_{Wt}}, \end{aligned} \tag{5.90}$$

integrating over $\Omega_{\Gamma_{Wt}}$, applying Green's theorem to the terms with τ_{ij}^f and finally, writing the already known finite element approximations $\boldsymbol{v}_h$, $\boldsymbol{v}_h^n$ and $\boldsymbol{v}_h^{n-1}$ instead of the functions $\boldsymbol{v}$, $\boldsymbol{v}^n$ and $\boldsymbol{v}^{n-1}$, respectively, we arrive at the representation of the force L:

$$\frac{L}{h^s} = -\int_{\tilde{\Omega}_{\Gamma_{Wt}}} \left[\rho\left(\frac{3v_{h2} - 4\hat{v}_{h2}^n + \hat{v}_{h2}}{2\tau} + ((\boldsymbol{v}_h - \boldsymbol{w}^{n+1}) \cdot \nabla)v_{h2}\right)\varphi - \sum_{j=1}^{2} \tau_{2j}^f \frac{\partial \varphi}{\partial x_j}\right] \mathrm{d}x. \tag{5.91}$$

Let us remind that $\boldsymbol{v}_h = (v_{h1}, v_{h2})$, $\hat{\boldsymbol{v}}_h^n = (\hat{v}_{h1}^n, \hat{v}_{h2}^n)$, etc. Using the vector-valued function $\boldsymbol{\psi} = (\psi_1, \psi_2) = \varphi\,(r_1^{\text{ort}}, r_2^{\text{ort}})$, the aerodynamic moment M can be computed as

$$\frac{M}{h^s} = \int_{\tilde{\Omega}_{\Gamma_{Wt}}} \left[\rho \left(\frac{3\boldsymbol{v}_h - 4\hat{\boldsymbol{v}}_h^n + \hat{\boldsymbol{v}}_h}{2\tau} + \left((\boldsymbol{v}_h - \boldsymbol{w}^{n+1}) \cdot \nabla \right) \boldsymbol{v}_h \right) \cdot \boldsymbol{\psi} - \sum_{i,j=1}^{2} \tau_{ij}^f \frac{\partial \psi_i}{\partial x_j} \right] \mathrm{d}x. \tag{5.92}$$

5.4.2 *Coupled Problem of Incompressible Flow Problem and Elastic Structure*

In the same way as in Sect. 5.3.2, by $\boldsymbol{u} = \boldsymbol{u}(X, t)$, $X \in \Omega^s$, $t \in (0, T)$, we denote the displacement defining the deformation of the elastic body Ω^s, which represents the vocal folds. Then the common boundary $\tilde{\Gamma}_{Wt}$ between the fluid and the structure at time t is given by

$$\tilde{\Gamma}_{Wt} = \left\{ x \in R^2; x = X + \boldsymbol{u}(X, t), \Gamma_W^s \right\}. \tag{5.93}$$

It means that the domain Ω_t is determined by the displacement $\boldsymbol{u}$ of the part Γ_W^s at time t. If we know the domain Ω_t occupied by the fluid at time t, the flow problem can be solved and the aerodynamic surface force acting on the body on $\tilde{\Gamma}_{Wt}$ can be determined by extrapolation. Then the transformation of the surface force to the reference configuration, i. e., to the interface Γ_W^s is realized. In the case of the linear elasticity model, when only small deformations are considered, we get the transmission condition

$$\sum_{j=1}^{2} \tau_{ij}^s(X) n_j(X) = -\sum_{j=1}^{2} \tau_{ij}^f(x) n_j(X), \quad i, j = 1, 2, \quad x \in \tilde{\Gamma}_{Wt}, \ X \in \Gamma_W^s, \tag{5.94}$$

where τ_{ij}^f are the components of the fluid stress tensor given by (5.2) and the points x and X satisfy the relation

$$x = X + \boldsymbol{u}(X, t), \quad x \in \tilde{\Gamma}_{Wt}, \ X \in \Gamma_W^s. \tag{5.95}$$

By $\boldsymbol{n}(X) = (n_1(X), n_2(X))$ we denote the unit outer normal to the body Ω^s on Γ_W^s at the point X. Further, the fluid velocity is defined on the moving part of the boundary Γ_{Wt} by the transmission condition

$$\boldsymbol{v}(x, t) = z_W(x, t) := \frac{\partial \boldsymbol{u}(X, t)}{\partial t}, \tag{5.96}$$

where the points x and X satisfy (5.95). (Of course, $z_W(x, t) = 0$ on $\Gamma_{Wt} \setminus \tilde{\Gamma}_{Wt}$.)

5.5 Construction of the ALE Mapping

An important concept in the analysis of flow in the time-dependent domain Ω_t is the ALE mapping. Various strategies for its construction can be found, cf., e.g., [20] and [101]. Here we shall describe two possibilities, one based on analytical extension of the deformation given by two degrees of freedom, and the other one based on a linear elasticity analogy.

5.5.1 *Analytical Definition of the ALE Mapping for Two Degrees of Freedom*

First, let us define the ALE mapping for the case of the known geometry of the channel with vocal fold (see Fig. 5.8), which is displaced by the relation (5.34). Here we present the method, which can be used for the case of the flexibly supported structure with two degrees of freedom $\overline{w}_1$ and $\overline{w}_2$. In this case, the description of the ALE mapping differs in three subdomains of the undeformed computational domain Ω, subdomain Ω_G $(0 < X_1 < L)$, pre-glottal subdomain $\Omega_{\rm pre}$ $(X_1 < 0)$ and post-glottal subdomain $\Omega_{\rm post}$ $(X_1 > L)$, see Fig. 5.8. We define the vertical displacement U_2^G as

$$U_2^G(X,t) = \frac{1}{2l}\left[(L/2 + l - X_1)\overline{w}_1(t) + (X_1 - L/2 + l)\overline{w}_2(t)\right]. \tag{5.97}$$

The displacement U_2^G is then extended to the computational domain by

$$U_2(X,t) = U_2^G(X,t)\,\theta_G(X)\,\theta_{\rm pre}(X)\,\theta_{\rm post}(X)\,\theta_{\rm base}(X), \tag{5.98}$$

where the blending functions θ_G, $\theta_{\rm pre}$, $\theta_{\rm post}$, $\theta_{\rm base}$ are defined as follows:

$$\theta_G(X_1,X_2) = \begin{cases} \frac{X_2 - a(X_1)}{X_{2,\max} - a(X_1)}, & \text{for } X_2 \geq a(X_1),\ X_1 \in [0,L], \\ 1 & \text{otherwise,} \end{cases}$$

$$\theta_{\rm pre}(X_1,X_2) = \begin{cases} \max\{0, (X_1 + \Delta L)/\Delta L\}, & \text{for } X_1 < 0, \\ 1 & \text{otherwise,} \end{cases}$$

$$\theta_{\rm post}(X_1,X_2) = \begin{cases} \max\{0, (L + \Delta L - X_1)/\Delta L\}, & \text{for } X_1 > L, \\ 1 & \text{otherwise.} \end{cases}$$

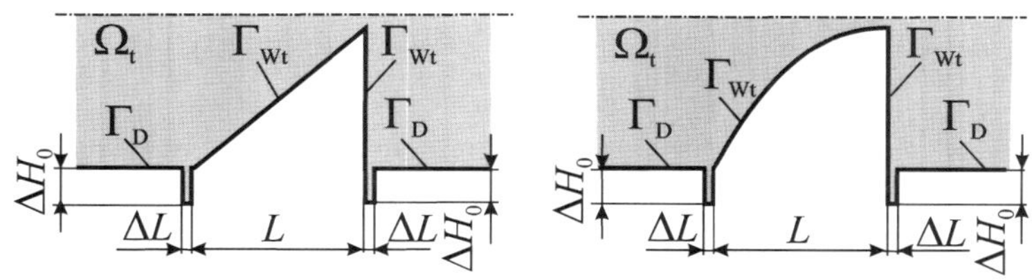

Fig. 5.8 The detail of buffer zone

The function θ_{post} measures the distance from the vocal fold region to the pre- or post-glottal part. Further, also the distance from the base of the vocal fold ($X_2 = 0$) can be used

$$\theta_{\text{base}}(X_1, X_2) = \begin{cases} \frac{\Delta H_0 + X_2}{\Delta H_0}, & \text{for } X_2 < 0, \\ 1 & \text{otherwise.} \end{cases}$$

Finally the ALE mapping can be defined for any $X = (X_1, X_2)$ by

$$\mathcal{A}_t(X_1, X_2) = [X_1, X_2 + U_2(X, t)]\,. \tag{5.99}$$

An extension of this approach to more complicated situation, as elastic structure deformation, is not possible. In the other approach, the ALE mapping can be constructed as an extension of the displacement of the vocal fold surface at time instant t given by $\mathcal{A}_t(X) = X + \boldsymbol{u}(X, t)$, where $X \in \tilde{\Gamma}^s_W$, $\boldsymbol{u} = (0, U_2^G(X, t))$ and $U_2^G(X, t)$ is given by (5.97).

5.5.2 *Artificial Elasticity Problem*

In case of more complicated flow domains, the ALE mapping $\mathcal{A}_t$ can be determined with the aid of an artificial stationary elasticity problem, where we seek $\boldsymbol{d} = (d_1, d_2)$ defined in Ω_0 as a solution of the elastic system

$$\sum_{j=1}^{2} \frac{\partial \tau^a_{ij}}{\partial X_j} = 0 \quad \text{in } \Omega_0, \quad i = 1, 2, \tag{5.100}$$

where τ^a_{ij} are the components of the artificial stress tensor

$$\tau^a_{ij} = \lambda^a \operatorname{div} \boldsymbol{d}\, \delta_{ij} + \mu^a e^a_{ij}(\boldsymbol{d}), \quad e^a_{ij}(\boldsymbol{d}) = \frac{1}{2}\left(\frac{\partial d_i}{\partial X_j} + \frac{\partial d_j}{\partial X_i}\right), \quad i, j = 1, 2. \tag{5.101}$$

The artificial Young modulus E^a and the artificial Poisson ratio σ^a can be derived from Lamé coefficients λ^a and μ^a in the same way as in (2.11). The problem is completed by the boundary conditions

$$\boldsymbol{d}|_{\Gamma_I \cup \Gamma_O} = 0, \quad \boldsymbol{d}|_{\Gamma_{W0} \setminus \Gamma_W^s} = 0, \quad \boldsymbol{d}(X,t) = \boldsymbol{u}(X,t), \quad X \in \Gamma_W^s, \tag{5.102}$$

where $\boldsymbol{u}$ is the solution of the elasticity problem treated in Sect. 5.3.2.

The solution of the problem (5.100)–(5.102) gives us the ALE mapping of $\overline{\Omega}_0$ onto $\overline{\Omega}_t$ in the form

$$\mathcal{A}_t(X) = X + \boldsymbol{d}(X,t), \quad X \in \overline{\Omega}_0, \tag{5.103}$$

for each time t.

System (5.100) is discretized by the conforming piecewise linear finite elements on the mesh $\mathcal{T}_{h0}$ used for computing the flow field in the beginning of the computational process in the polygonal approximation Ω_{h0} of the domain Ω_0. We introduce the finite element spaces

$$\mathcal{X}_h = \{\boldsymbol{d}_h = (d_{h1}, d_{h2}) \in C(\overline{\Omega}_{h0})^2; d_{hi}|_K \in P_1(K)\ \forall K \in \mathcal{T}_{h0},\ i = 1,2\}, \tag{5.104}$$
$$\mathcal{V}_h = \{\boldsymbol{\varphi}_h \in \mathcal{X}_h; \boldsymbol{\varphi}_h(Q) = 0 \text{ for all vertices } Q \in \partial\Omega_0\},$$

and the form

$$B_h(\boldsymbol{d}_h, \boldsymbol{\varphi}_h) = \int_{\Omega_{h0}} \lambda^a \operatorname{div} \boldsymbol{d}_h \operatorname{div} \boldsymbol{\varphi}_h \,\mathrm{d}X + 2\int_{\Omega_{h0}} \mu^a \sum_{i,j=1}^{2} e_{ij}^a(\boldsymbol{d}_h)\, e_{ij}^a(\boldsymbol{\varphi}_h)\,\mathrm{d}X. \tag{5.105}$$

The approximate solution of problem (5.100), (5.102) is defined as a function $\boldsymbol{d}_h \in \mathcal{X}_h$ satisfying approximately the Dirichlet boundary conditions (5.102) and the identity

$$B_h(\boldsymbol{d}_h, \boldsymbol{\varphi}_h) = 0 \quad \forall \boldsymbol{\varphi}_h \in \mathcal{V}_h. \tag{5.106}$$

Then the approximate ALE mapping $\mathcal{A}_{ht}$ is given by (5.103) with $\boldsymbol{d} := \boldsymbol{d}_h$ and $X \in \overline{\Omega}_{hs}$:

$$\mathcal{A}_{ht}(X) = X + \boldsymbol{d}_h(X,t), \quad X \in \overline{\Omega}_{h0}. \tag{5.107}$$

The use of linear finite elements is sufficient, because we only need to know the movement of the points of the mesh. The domain velocity is approximated by (5.10).

In our computations we choose the Lamé coefficients λ^a and μ^a as constants corresponding to the Young modulus and Poisson ratio $E^a = 10{,}000$ and $\sigma^a = 0.45$.

5.6 FSI Algorithms

For the solution of coupled FSI problems various strategies are applied. One possibility is to use a de-coupled scheme (i.e., scheme, where both structural and flow field are resolved separately, using only the information from the previous time steps). However, such de-coupled schemes have only limited applicability. Here we shall use more sophisticated methods based on weak (loose) or strong coupling algorithms. In what follows, we shall describe these algorithms applied to the simulation of vocal folds vibrations.

5.6.1 Algorithms for Interaction of Fluid Flow and Vibrating Rigid Body

We consider the following coupled problem at time instant t_{n+1}.

Problem 5.6.1 (Flow Interacting with Two Degrees of Freedom Vocal Fold). Find the approximate flow velocity $\boldsymbol{v}_h^{n+1}$ and pressure p_h^{n+1} satisfying (5.25) on an approximation of the domain $\Omega_{t_{n+1}}$, which is defined with the aid of the approximation $\mathcal{A}_{ht_{n+1}}$ of the ALE mapping given either by (5.99) or (5.106). The domain velocity z_h^{n+1} is given by (5.10). The ALE mapping $\mathcal{A}_{n+1,h}$ depends on $\overline{w}_1^{n+1}$ and $\overline{w}_2^{n+1}$, where $\overline{w}_1^{n+1}, \overline{w}_2^{n+1}, \dot{\overline{w}}_1^{n+1}, \dot{\overline{w}}_2^{n+1}$ satisfy (5.62), with the forces $\overline{F}_i^{n+1}$ given by (5.86) evaluated at t_{n+1}.

Weak Coupling

0. **Initialization.** Start from the approximations $\overline{w}_i^n, \dot{\overline{w}}_i^n, \boldsymbol{v}_h^n, p_h^n, \overline{F}_1^n, \overline{F}_2^n$ at time t_n (as well as on previous time levels).
1. **Extrapolation step.** Extrapolate aerodynamic forces

$$\overline{F}_i^{n+1} := \overline{F}_i(t_n) \approx \overline{F}_i(t_{n+1}), \quad i = 1, 2.$$

2. **Structural step.** Solve (5.62) for $\overline{w}_1^{n+1}$ and $\overline{w}_2^{n+1}$.
3. **Mesh step.** Find the approximation of the ALE mapping $\mathcal{A}_{ht_{n+1}} : \Omega_{h0} \mapsto \Omega_{ht_{n+1}}$ at t_{n+1} given either by (5.99) or (5.106), and approximate the domain velocity z_h^{n+1} by (5.10).
4. **Fluid step.** Solve (5.25) for the approximations of $\boldsymbol{v}_h^{n+1}$, p_h^{n+1} on $\Omega_{ht_{n+1}}$.
5. $n := n + 1$, go to 1.

This algorithm can be sensitive to the choice of the time step and the structural parameters. In order to make the computational process robust, the strong coupling is applied.

Strong Coupling

0. **Initialization.** Start from the approximations $\overline{w}_i^n$, $\dot{\overline{w}}_i^n$, $\boldsymbol{v}_h^n$, p_h^n, $\overline{F}_1^n$, $\overline{F}_2^n$ known at time t_n (as well as on previous time levels).
1. **Extrapolation step.** Extrapolate aerodynamic forces

$$\overline{F}_i^{n+1} := \overline{F}_i(t_n) \approx \overline{F}_i(t_{n+1}), \quad i = 1, 2.$$

2. **Structural step.** Solve (5.62) for $\overline{w}_1^{n+1}$ and $\overline{w}_2^{n+1}$.
3. **Mesh step.** Find the approximation of ALE mapping $\mathcal{A}_{ht_{n+1}} : \Omega_{h0} \mapsto \Omega_{ht_{n+1}}$ at t_{n+1} given either by (5.99) or (5.106), and approximate the domain velocity z_h^{n+1} by (5.10).
4. **Fluid step.** Solve (5.25) for the approximations $\boldsymbol{v}_h^{n+1}$, p_h^{n+1} on $\Omega_{ht_{n+1}}$.
5. **Aerodynamic forces.** Using the approximations $\boldsymbol{v}_h^{n+1}$, p_h^{n+1}, compute the aerodynamic forces $\overline{F}_1^{n+1}$, $\overline{F}_2^{n+1}$.
6. **Structural step.** Solve (5.62) for $\tilde{\overline{w}}_1^{n+1}$ and $\tilde{\overline{w}}_2^{n+1}$. Compute the difference $\varepsilon := \sum_{i=1}^2 |\tilde{\overline{w}}_i^{n+1} - \tilde{\overline{w}}_i^{n+1}|$.
7. If the difference ε is sufficiently small, go to 8. Else, set $\overline{w}_1^{n+1} := \tilde{\overline{w}}_1^{n+1}$, $\overline{w}_2^{n+1} := \tilde{\overline{w}}_2^{n+1}$ and go to 3.
8. $n := n + 1$, go to 1.

In our computations, the strongly coupled scheme was used in order to guarantee the stability of the computational process. Usually, if the time step is small enough, only few iterations need to be computed.

5.6.2 Algorithms for Interaction of Fluid and Elastic Structure

In the case of the interaction of fluid and elastic structure we can formulate the following continuous FSI problem: Our aim is to determine the domain Ω_t, $t \in (0, T]$, and functions $\boldsymbol{v} = \boldsymbol{v}(x, t)$, $p = p(x, t)$, $x \in \overline{\Omega}_t$, $t \in [0, T]$ and $\boldsymbol{u} = \boldsymbol{u}(X, t)$, $X \in \overline{\Omega}^s$, $t \in [0, T]$ satisfying Eqs. (5.9), (5.66), the initial conditions (5.3), (5.67), the boundary conditions (5.4), (5.69), and the transmission conditions (5.94), (5.96).

Here we consider the following discrete coupled problem at time t_{n+1}.

Problem 5.6.2 (Flow Interacting with Elastic Structure). Find the approximate flow velocity $\boldsymbol{v}_h^{n+1}$ and pressure p_h^{n+1} satisfying (5.25) on the approximation $\Omega_{ht_{n+1}}$ of the domain $\Omega_{t_{n+1}}$, which is defined by the approximation of the ALE mapping $\mathcal{A}_{ht_{n+1}}$ given by (5.106). Find $\boldsymbol{u}_h^{n+1}$ defined by (5.76) with $\boldsymbol{p}_{k+1}$ satisfying (5.84). The domain velocity z_h^{n+1} is given by (5.10).

Weak Coupling

1. Compute the approximate solution of the flow problem (5.9), (5.4) on the time level t_n.
2. Compute the corresponding fluid stress tensor τ_{ij}^f and the aerodynamic force acting on the structure and transform it to the interface Γ_{Wh}^s by (5.94).
3. Solve the discrete elasticity problem (5.74)–(5.75) at time t_n (by the Newmark method) and compute the deformation $\boldsymbol{u}_h^n$ at time t_n. On the basis of (5.93) set

$$\tilde{\Gamma}_{Wt_{n+1},h} = \{x = X + \boldsymbol{u}_h^n(X);\ X \in \Gamma_{W_h^s}\}, \tag{5.108}$$

   and determine the domain $\Omega_{ht_{n+1}}$.
4. Determine the ALE mapping $\mathcal{A}_{ht_{n+1}}$ by (5.107) and approximate the domain velocity z_h^{n+1} by (5.10).
5. Set $n := n + 1$, go to 1.

Strong Coupling

Prescribe N—maximal number of inner iterations and $\varepsilon > 0$—maximal error in the iterative process. Now proceed in the following way:

1. Assume that the approximate solution U_h^n of the flow problem and the deformation $\boldsymbol{u}_h^n$ of the structure are known on the time level t_n.
2. Set $\boldsymbol{u}_{h,n+1}^0 := \boldsymbol{u}_h^n,\ k := 1$ and apply the iterative process:

   (a) Compute the fluid stress tensor τ_{ij}^f and the aerodynamic force acting on the structure and transform it to the interface Γ_{Wh}^s.
   (b) Solve the elasticity problem, compute the approximation of the deformation $\boldsymbol{u}_{h,n+1}^k$, and construct the approximation $\Omega_{ht_{n+1}}^k$ of the flow domain at time t_{n+1}.
   (c) Determine the approximations of the ALE mapping $\mathcal{A}_{ht_{n+1}}^k$ and the domain velocity $z_{h,n+1}^k$.
   (d) Solve the flow problem in $\Omega_{ht_{n+1}}^k$ and obtain the approximate solution $U_{h,n+1}^k$.
   (e) If the variation $|\boldsymbol{u}_{h,n+1}^k - \boldsymbol{u}_{h,n+1}^{k-1}| \geq \varepsilon$ and $k < N$, go to (a) and $k := k + 1$. Else $\Omega_{ht_{n+1}} := \Omega_{ht_n}^k,\ U_h^{n+1} := U_{h,n+1}^k,\ \boldsymbol{u}_h^{n+1} := \boldsymbol{u}_{h,n}^k,\ n := n + 1$ and go to 2.

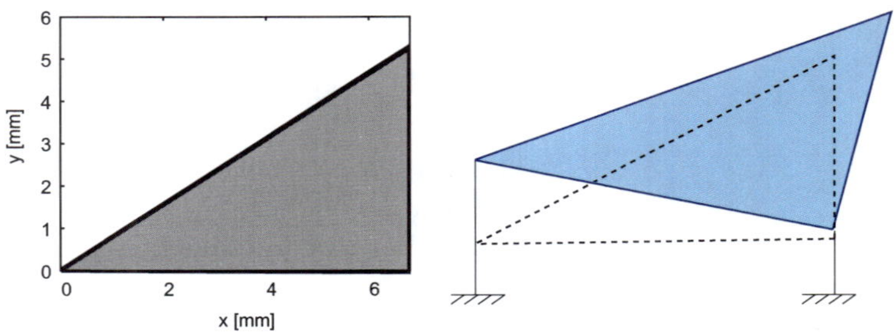

Fig. 5.9 The considered geometry of the vocal folds for the female model F (*left*) and example of the vocal fold in displaced position (*right*)

5.7 Numerical Examples

5.7.1 Interaction of Incompressible Flow with Rigid Body Models of Vocal Folds

For the numerical analysis the following parameters were used. The distance of the masses m_1 and m_2 from the center was $l = L/2$ (see Fig. 5.4b). The total mass m of the vocal folds, the inertia moment I, and the eccentricity e were computed using the vocal fold shape $a(x)$ and the material density $\rho^s = 1{,}020\,\mathrm{kg\,m^{-3}}$, length (depth of the channel) $h^s = 18\,\mathrm{mm}$ and thickness $L = 6.8\,\mathrm{mm}$, see [39] for details. The fluid density was $\rho^f = 1.2\,\mathrm{kg\,m^{-3}}$ and the dynamic viscosity $\mu = 1.896 \cdot 10^{-5}\,\mathrm{kg\,m^{-1}\,s^{-1}}$.

The geometry of vocal folds depends on the tension in the vocal folds and varies with the fundamental vibration frequency, loudness, and mode of phonation. For the purposes of numerical analysis in this paper the geometry of the vocal folds was chosen as linear function according to [40] as $a_f(x) = 0.77120x$ [m] (linear shape, approximation of the vocal fold for female—model F) or the vocal fold with intermediate bulging $a_m(x) = 1.858x - 159.861x^2$ [m] (parabolic shape, approximation of the vocal fold for male—model M), see Figs. 5.9 and 5.10, respectively. The channel half-height (at time $t = 0$) is chosen as $H_0 = \max_{x\in[0,L]} a(x) + g_0$, where g_0 is the initial half-gap, i.e., $g(0) = 2g_0$. The considered initial half-gap was chosen $g_0 = 0.25\,\mathrm{mm}$ and $g_0 = 0.3\,\mathrm{mm}$ for model F and model M, respectively. In all computations the subglottal length $L_0 = 1.5\,L$ and the supra-glottal length $L_2 = 5\,L$ were used. The computations were performed in the domains Ω_t shown in Fig. 5.11 with $H_0 = 5.54416\,\mathrm{mm}$ and $H_0 = 5.7\,\mathrm{mm}$ for model F and model M, respectively.

Aeroelastic Simulations for Model F with Inlet Pressure Condition

First, the problem of interaction of air flow with aeroelastic model F was considered with the inlet pressure boundary condition. The structural parameters and the

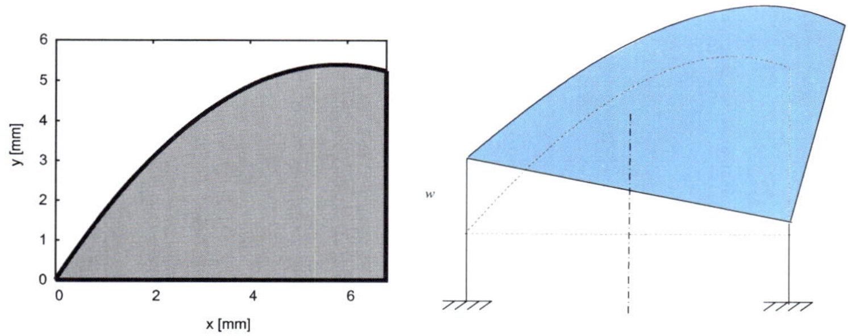

Fig. 5.10 The considered geometry of the vocal folds for the male model M (*left*) and example of the vocal fold in displaced position (*right*)

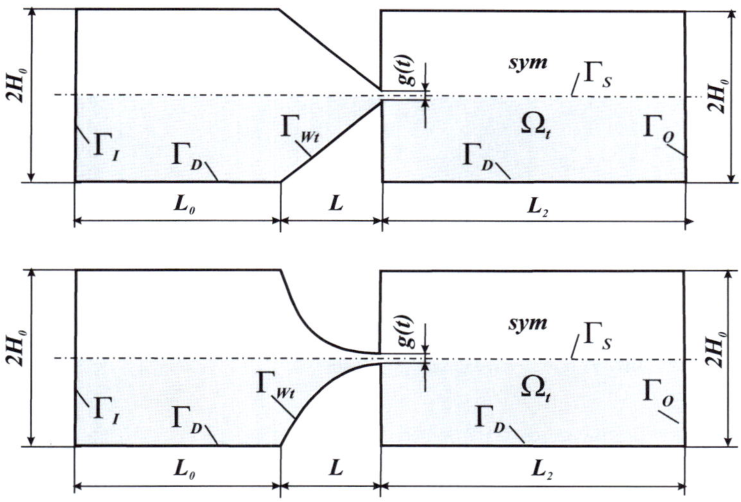

Fig. 5.11 The computational domain Ω_t for models F (*above*) and M (*below*) surrounding the vocal fold shape given by $a_f(x)$ and $a_m(x)$, respectively. The *lower shaded part* is used for computations assuming the symmetry boundary condition on the axis of symmetry

natural frequencies f_1, f_2 for the structure vibrating in vacuo are listed in Table 5.1. The results are shown in Figs. 5.12 and 5.13, where the aeroelastic response $\overline{w}_1(t)$, $\overline{w}_2(t)$ and the mean inlet velocity in dependence on time t are shown for several prescribed pressure differences $\Delta p \approx 100$–3,200 Pa (the inlet velocity V_0 oscillated in the interval $V_0 \approx 0.55$–$3.13\,\mathrm{m\,s^{-1}}$). In this range, the simplified method in [40] predicted the aeroelastic instability of flutter type for the inlet velocity $V_{\mathrm{crit}} = 0.87\,\mathrm{m\,s^{-1}}$, but in the mentioned paper the inlet/outlet velocity formulation was used.

Table 5.1 Structural parameters considered for the *model F* (female vocal fold)

Input data for model F			
Shape	$a_f(x)$	f_1 (Hz)	100
m (kg)	$3.274 \cdot 10^{-4}$	f_2 (Hz)	160
I (kg/m^2)	$1.341 \cdot 10^{-9}$	c_1 (N/m)	44.8
e (m)	$1.133 \cdot 10^{-3}$	c_2 (N/m)	84.6
ε_1 (s^{-1})	120.35	ε_2 (s)	$6.12 \cdot 10^{-5}$

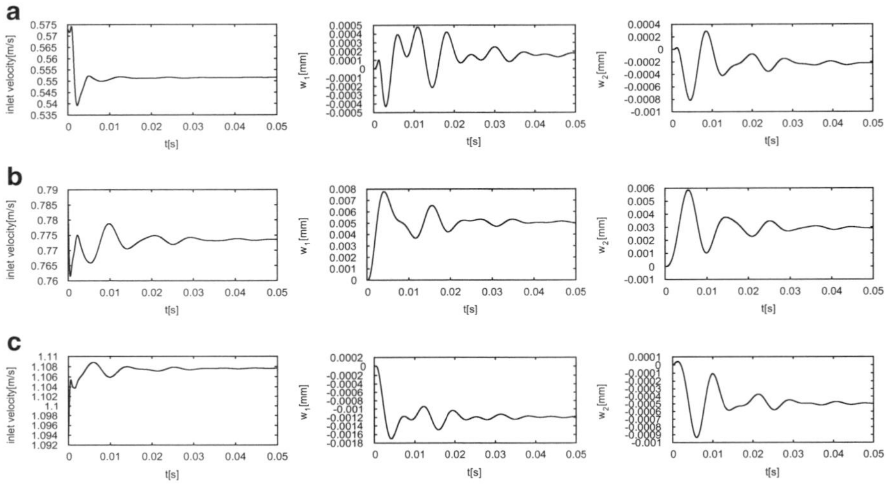

Fig. 5.12 The mean inlet velocity oscillations (*left column*) and the aeroelastic response of the system $\overline{w}_1(t)$, $\overline{w}_2(t)$ (*middle* and *right columns*) for *model F* and the prescribed inlet pressure boundary conditions. The inlet pressure was chosen (**a**) $\Delta p = 100$ Pa, (**b**) $\Delta p = 200$ Pa, (**c**) $\Delta p = 400$ Pa

The vibrations of the vocal fold in Figs. 5.12 and 5.13 dies out for all the values of the inlet pressure with no significant decrease or increase of the aerodynamic damping. Particularly, the aerodynamic damping is quite strong for all studied cases. The aeroelastic instability was never observed for the physically relevant values of the inlet pressure (values of Δp up to 5,000 Pa were tested). This behavior is probably caused by "additional damping" effects due to the prescribed inlet pressure boundary condition. Particularly, in Figs. 5.12 and 5.13 the inlet velocity oscillations are shown, where the inlet velocity is increasing with a wider opening of the glottal part $g(t)$ and similarly decreasing with a narrower enclosure of $g(t)$. The inlet velocity oscillations (as well as consequently the flow rate oscillations) influence the aerodynamic forces and are leading to damped vibrations of the structure.

Aeroelastic Simulations for Model F with Inlet Velocity Condition

The problem with the same input parameters used in previous section was also numerically analyzed with the inlet velocity boundary condition.

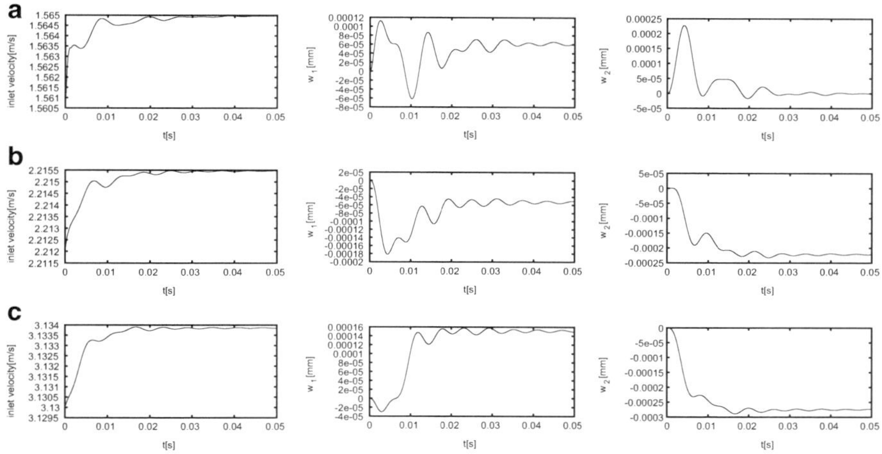

Fig. 5.13 The mean inlet velocity oscillations (*left column*) and the aeroelastic response of the system $\overline{w}_1(t)$, $\overline{w}_2(t)$ (*middle* and *right columns*) for *model F* and the prescribed inlet pressure boundary conditions. The inlet pressure was chosen (**a**) $\Delta p = 800\,\text{Pa}$, (**b**) $\Delta p = 1{,}600\,\text{Pa}$, (**c**) $\Delta p = 3{,}200\,\text{Pa}$

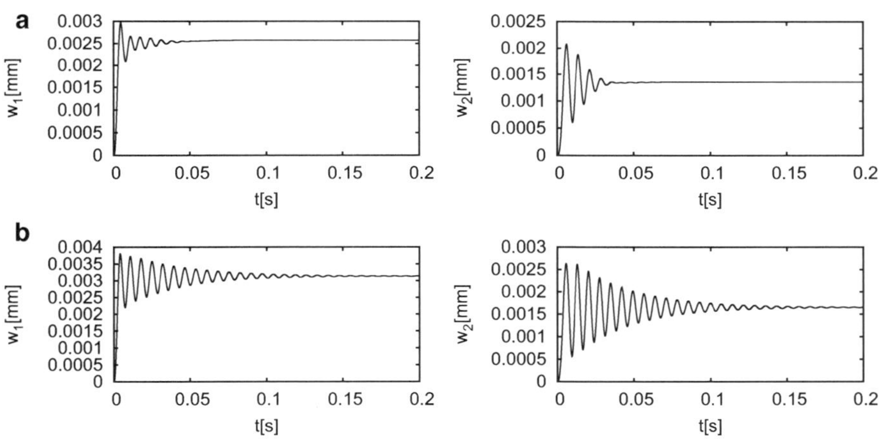

Fig. 5.14 The aeroelastic response of the aeroelastic system for *model F* with the prescribed inlet velocity; the graphs of $\overline{w}_1(t)$, $\overline{w}_2(t)$ in dependence on time t are shown for the different inlet velocities (**a**) $V_0 = 0.45\,\text{m}\,\text{s}^{-1}$ and (**b**) $V_0 = 0.55\,\text{m}\,\text{s}^{-1}$

The values of the inlet velocity V_0 were considered in the range 0.2–0.7 $\text{m}\,\text{s}^{-1}$ in order to detect the critical velocity for the flutter type of aeroelastic instability leading to self-sustained vibrations of the vocal fold. The aeroelastic responses are shown in Figs. 5.14 and 5.15 for the inlet flow velocities $V_0 = 0.45\,\text{m}\,\text{s}^{-1}$, $V_0 = 0.55\,\text{m}\,\text{s}^{-1}$, $V_0 = 0.6\,\text{m}\,\text{s}^{-1}$ and $V_0 = 0.65\,\text{m}\,\text{s}^{-1}$. For the velocities $0.45\,\text{m}\,\text{s}^{-1}$ and

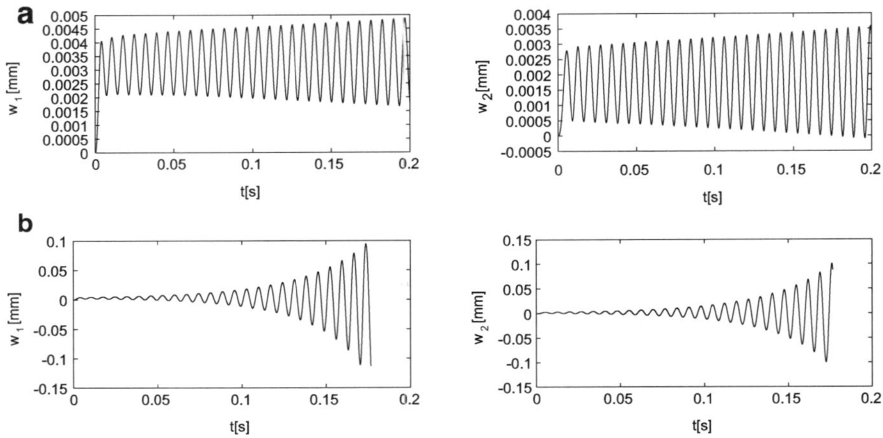

Fig. 5.15 The aeroelastic response of the aeroelastic system for *model F* with the prescribed inlet velocity; the graphs of $\overline{w}_1(t)$, $\overline{w}_2(t)$ in dependence on time t are shown for the different inlet velocities (**a**) $V_0 = 0.6\,\mathrm{m\,s^{-1}}$ and (**b**) $V_0 = 0.65\,\mathrm{m\,s^{-1}}$

Table 5.2 Structural parameters considered for the *model M* (male vocal fold)

Input data for model M			
Shape	$a_m(x)$	f_1 (Hz)	100
m (kg)	$4.812 \cdot 10^{-4}$	f_2 (Hz)	160
I (kg/m^2)	$2.351 \cdot 10^{-9}$	c_1 (N/m)	56
e (m)	$0.771 \cdot 10^{-3}$	c_2 (N/m)	174.3
ε_1 (s^{-1})	120.35	ε_2 (s)	$6.12 \cdot 10^{-5}$

$0.55\,\mathrm{m\,s^{-1}}$ the structural vibrations are damped in time and the aeroelastic system is stable. Nevertheless, the aerodynamic damping for the velocity $0.55\,\mathrm{m\,s^{-1}}$ is weaker compared to the lower inlet velocity. With the further increase of the inlet velocity to $V_0 = 0.6\,\mathrm{m\,s^{-1}}$ the self-oscillations can be observed in Fig. 5.15a. For the inlet velocity $V_0 = 0.65\,\mathrm{m\,s^{-1}}$ the vibrations of the vocal folds are growing very fast (see Fig. 5.15b). The simulation for $V_0 = 0.65\,\mathrm{m\,s^{-1}}$ is only shown in the time interval to 0.175 s, where the computations crashed due to the high distortion of the computational mesh.

Aeroelastic Simulations for Model M

Furthermore, the aeroelastic model of flow interaction with the vocal fold given by the parabolic shape $a_m(x)$ shown in Fig. 5.10 was analyzed. The structural parameters are listed in Table 5.2 (see also [39,40]). The aeroelastic response $w_1(t)$, $w_2(t)$ is shown in Figs. 5.16 and 5.17 and plotted over time in terms of displacements for the inlet flow velocities $V_0 = 1.0$–$1.2\,\mathrm{m\,s^{-1}}$. For the inlet velocities lower or

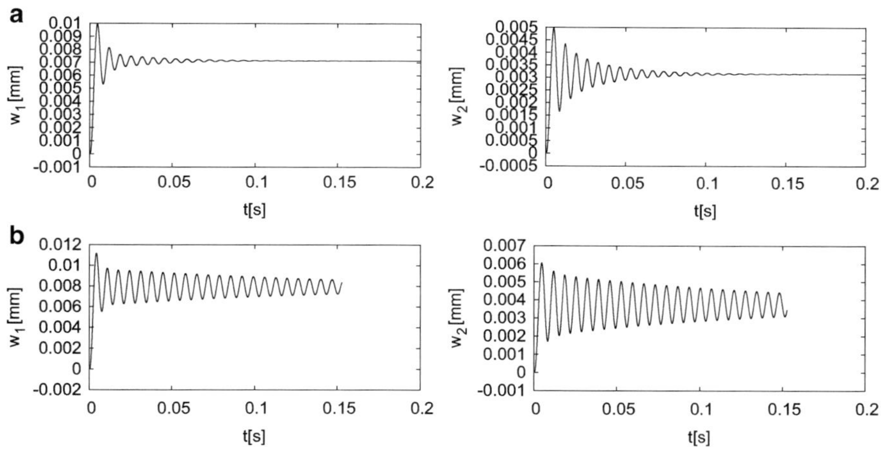

Fig. 5.16 The response of the aeroelastic system for *model M* with the prescribed inlet velocity boundary condition; the graphs of $\overline{w}_1(t)$, $\overline{w}_2(t)$ in dependence on time t are shown for the different inlet velocities (**a**) $V_0 = 1.0\,\mathrm{m\,s^{-1}}$, (**b**) $V_0 = 1.1\,\mathrm{m\,s^{-1}}$

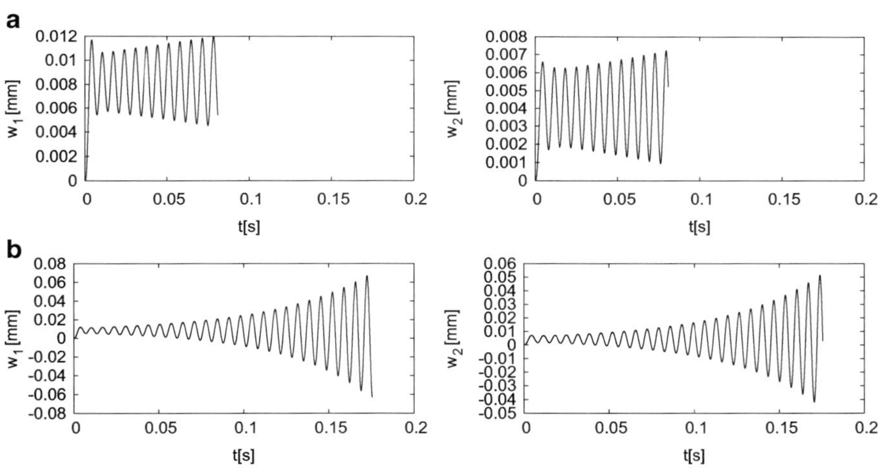

Fig. 5.17 The response of the aeroelastic system for *model M* with the prescribed inlet velocity boundary condition; the graphs of $\overline{w}_1(t)$, $\overline{w}_2(t)$ in dependence on time t are shown for the different inlet velocities (**a**) $V_0 = 1.15\,\mathrm{m\,s^{-1}}$ and (**b**) $V_0 = 1.2\,\mathrm{m\,s^{-1}}$

equal to $1.1\,\mathrm{m\,s^{-1}}$ the vocal fold oscillations die in time due to both structural and aerodynamic damping. For the flow velocity $V_0 = 1.15\,\mathrm{m\,s^{-1}}$ the self-oscillations of the vocal folds were obtained.

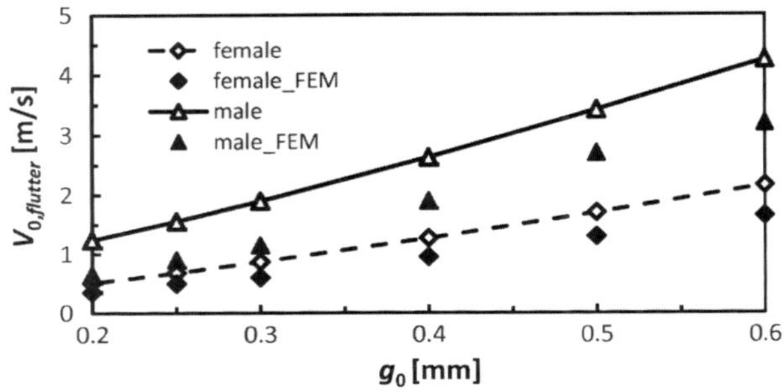

Fig. 5.18 Comparison of the flutter velocities computed by the FE method with the simplified flow theory [39–41]

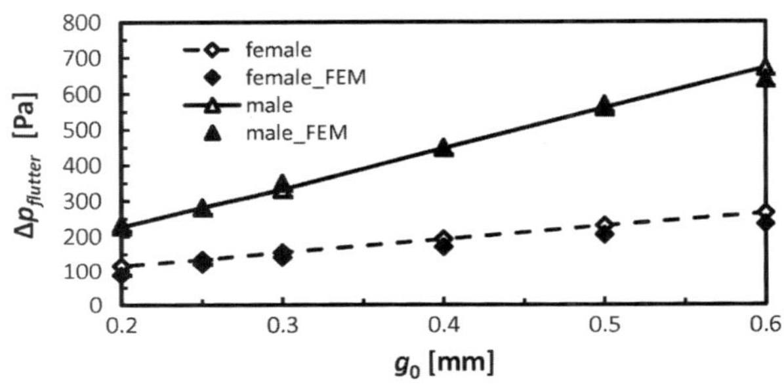

Fig. 5.19 Comparison of the subglottal pressures (phonation threshold pressures—PTP) computed by the FE method with the simplified flow theory [39–41]

Simulation examples of the flow velocity distribution in the glottis during the aeroelastic instability for $V_0 = 1.5\,\mathrm{m\,s^{-1}}$ are shown in Figs. 5.21, 5.22 at several time instants marked in the graph of $w_1(t)$ and $w_2(t)$. The maximal flow velocities in the channel are increasing when the glottal gap is becoming narrower, i.e., for high values of $w_2(t)$; the maximum flow velocity in the glottal gap is lower than $40\,\mathrm{m\,s^{-1}}$, which is in agreement with reality. Small changes in the position of the flow separation point on the vocal fold surface can be also detected in the flow field patterns in the glottal gap (see, e.g., the details in Fig. 5.22 at the time t_4 and t_5).

Comparison of the Results with Simplified Theory

The results obtained by the developed numerical method based on the FE solution of the 2D Navier–Stokes equations are compared with the results computed by the perturbation theory for 1D potential flow model [40] in Figs. 5.18, 5.19, and 5.20. The computed flutter airflow velocities $V_{0,\mathrm{flutter}}$, the pressure drop $\Delta p_{\mathrm{flutter}}$, i.e., the so-called phonation threshold pressures (PTP), and the flutter frequencies F_0, i.e.,

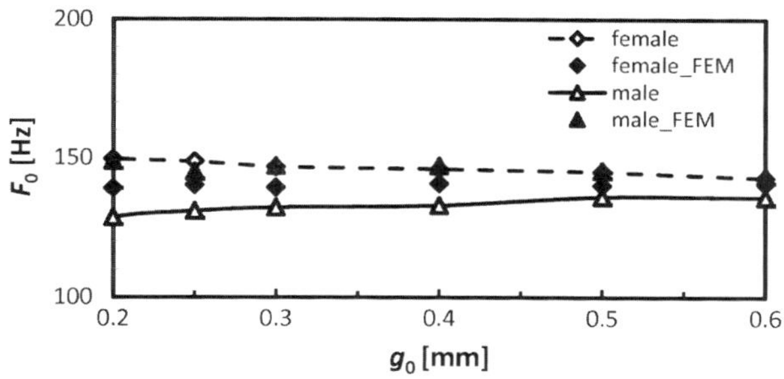

Fig. 5.20 Comparison of the flutter (phonation) frequencies computed by the FE method with the simplified flow theory [39–41]

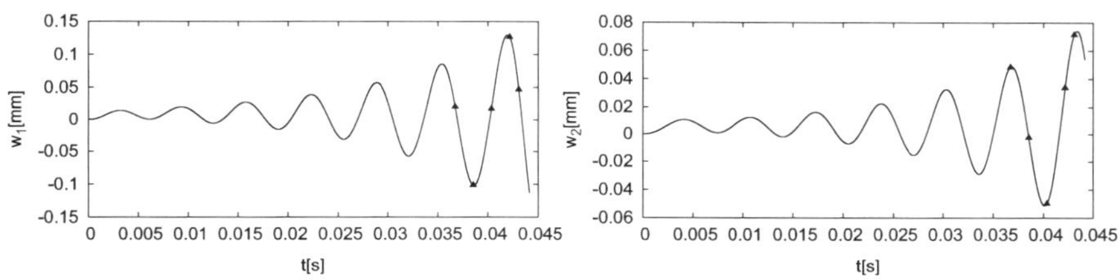

Fig. 5.21 The aeroelastic response $\overline{w}_1(t)$, $\overline{w}_2(t)$ for *model M* for the inlet velocity $1.5\,\mathrm{m\,s^{-1}}$

the so-called fundamental phonation frequencies, are shown in dependence on the prephonatory glottal half gap g_0 for the male and female models of the vocal folds. The computed results: $V_{0,\text{flutter}} \approx 0.5$–$4\,\mathrm{m\,s^{-1}}$ (corresponding to the flow rates $0.1 - 0.9$ l/s), $\Delta p_{\text{flutter}} \approx 100$–$700\,\mathrm{Pa}$ and $F_0 \approx 130$–$150\,\mathrm{Hz}$ are in physiologically relevant intervals for the phonation threshold in humans; the values for $V_{0,\text{flutter}}$ and $\Delta p_{\text{flutter}}$ for the female model are lower than for the male model and the opposite is valid for the phonation frequencies; the computed values $V_{0,\text{flutter}}$ and $\Delta p_{\text{flutter}}$ increase with the prephonatory glottal half-gap g_0 (see, e.g., [10, 41, 42]).

In general, the flutter velocities $V_{0,\text{flutter}}$ resulting from the FE simulations are lower than the flutter velocities computed according to the simplified theory presented in [40]. These differences (see Fig. 5.18) can be explained by the fluid viscosity considered in the FE simulations because the developed boundary layer on the surface of the vocal fold model results in narrowing of the glottal gap g_0. The second reason can be the position of the flow separation point which was in [40] artificially fixed at the vocal fold supraglottal edge, however, according to the FE computations the flow separation point was slightly moving (see Figs. 5.21, 5.22). These differences are smaller for the female model because the triangle shape satisfies better the conditions used in the simplified theory. The difference between the FE results and the results of the simplified theory [40] for the critical pressure drops $\Delta p_{\text{flutter}}$ and the flutter frequencies F_0 is much smaller (see Fig. 5.20).

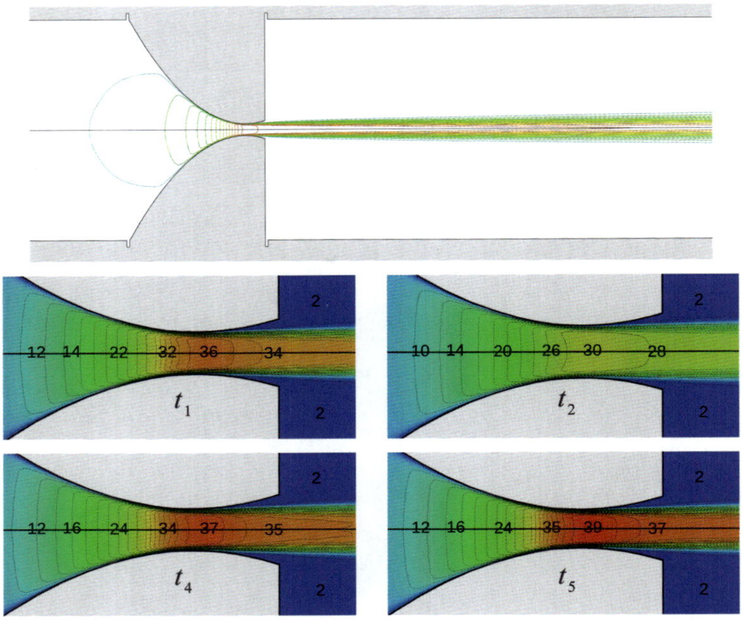

Fig. 5.22 The flow velocity isolines in the channel for *model M* for the inlet velocity $1.5\,\mathrm{m\,s^{-1}}$ (*top panel*) with details in the glottal gap (*bottom panel*) shown at time instants t_1, t_2, t_4, and t_5 marked in the graphs of $\overline{w}_1(t)$ and $\overline{w}_2(t)$ in Fig. 5.21

5.7.2 Interaction of Incompressible Flow with Elastic Model of Vocal Folds

Here we present numerical results obtained with the aid of the coupling of incompressible flow with the elasticity model described in Sect. 5.4.2. We consider model of human vocal folds and a simplified vocal tract, see also [47]. The vocal folds are isotropic bodies represented by the set Ω^s, which have different material characteristics in different subdomains (see Fig. 5.23 and Table 5.3). The material density $\rho^s = 1{,}040\,\mathrm{kg\,m^{-3}}$ for all subdomains. The domain Ω_t is occupied by air at time t. By $\tilde{\Gamma}_{Wt}$ we denote the movable common part of both domains. The other parts of the boundary are fixed. The method for the solution of the coupled air flow and elasticity problem use the same time step $\tau = 5 \cdot 10^{-5}\,\mathrm{s}$. The computational process started by the solution of the flow problem in the domain Ω_{t_α} at the initial time $t_\alpha = -10^{-4}\,\mathrm{s}$. We use the data $\mu = 1.755 \cdot 10^{-5}\,\mathrm{kg\,m^{-1}\,s^{-1}}$, $\varrho^f = 1.17\,\mathrm{kg\,m^{-3}}$, initial velocity $\boldsymbol{v}^0 = 0$. The boundary conditions on the walls of the channel are zero for impermeable fixed parts and on the elastic boundary are given by the domain velocity, (5.4) (b). On the inlet and outlet the boundary

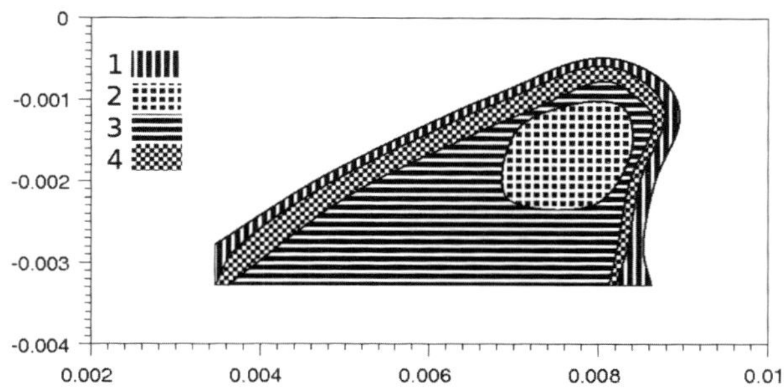

Fig. 5.23 The model of vocal fold

Table 5.3 Material characteristics

Subdomain	E^s	σ^s
Ω_1^s	100,000	0.4
Ω_2^s	12,000	0.4
Ω_3^s	8,000	0.4
Ω_4^s	1,000	0.495

conditions (5.4)(d) and (5.5) are used with the pressure drop $\Delta p = p_{\text{inlet}} = 600\,\text{Pa}$, $p_{\text{ref}} = 0\,\text{Pa}$.

At time $t = 0$ the structure was released and the solution of the complete FSI started. As for the domain Ω^s, we distinguish subdomains with different material characteristics.

The part of the boundary Γ_D^s is fixed, i.e., $\boldsymbol{u}^0 = 0$. On the other part of boundary Γ_W^s we prescribe the conditions for the movable boundary. We shall retrieve the surface forces $\boldsymbol{T}^n$ by solving the fluid flow problem. We neglect the outer forces and prescribe the initial displacement and the deformation velocity $\boldsymbol{u}^0 = 0, z^0 = 0$, in Ω^s.

Figures 5.24 and 5.25 show the computed velocity of fluid flow and the displacement of the computational domain at several time instants. Figure 5.26 shows the horizontal and vertical displacement of a sensor point.

In another example we consider the model of human vocal folds and the vocal tract as in [48]. In this case, the vocal folds have different material characteristics in three different subdomains. The material density $\varrho^s = 1{,}040\,\text{kg}\,\text{m}^{-3}$ is constant over all subdomains, but the values of the Young modulus E^s and the Poisson ratio σ^s are variable, see Table 5.4 and Fig. 5.27. We use the same time step $\tau = 5\cdot 10^{-5}\,\text{s}$ for the solution of the coupled flow and elasticity problem and the input data $\mu = 1.755 \cdot 10^{-5}\,\text{kg}\,\text{m}^{-1}\,\text{s}^{-1}$, $\varrho^f = 1.17\,\text{kg}\,\text{m}^{-3}$, structural damping coefficient $C = 0.1\,\text{s}^{-1}$, the initial velocity $\boldsymbol{v}^0 = 0\,\text{m}\,\text{s}^{-1}$. Similarly as in the previous example, on the inlet and outlet the boundary conditions (5.4)(d) and (5.5) are used with the pressure drop $\Delta p = p_{\text{inlet}} = 600\,\text{Pa}$, $p_{\text{ref}} = 0\,\text{Pa}$.

The computational process started by the solution of the flow problem in the domain Ω_{t_0} at the initial time $t_\alpha = -10^{-4}\,\text{s}$. At time $t = 0$ the structure was released and the solution of the complete FSI started.

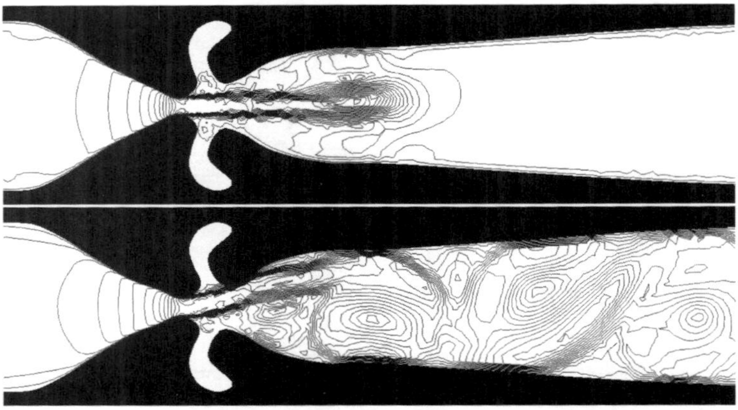

Fig. 5.24 Flow velocity isolines at time instants $t = 0.00125$ and 0.025 s

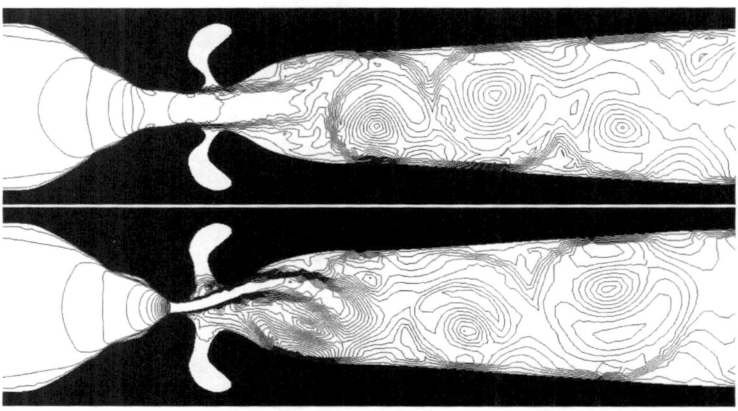

Fig. 5.25 Flow velocity isolines at time instants $t = 0.05275$ and 0.5375 s

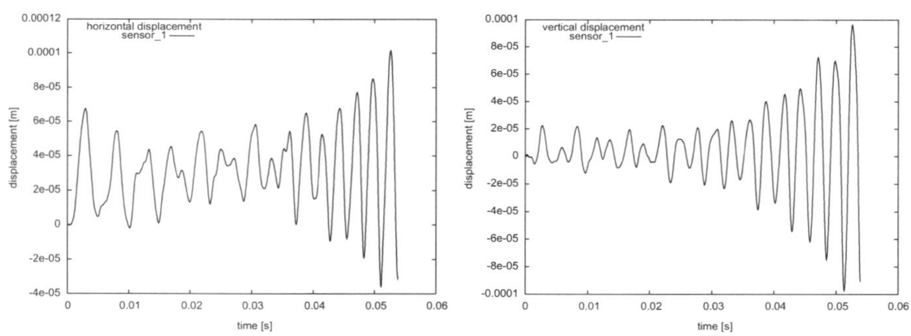

Fig. 5.26 Horizontal and vertical displacement of a point with the initial position [0.008, 0.001]

Table 5.4 Material characteristics for the second considered case

Subdomain	E	σ
Ω_1^s	25,000	0.49
Ω_2^s	65,000	0.4
Ω_3^s	8,000	0.49

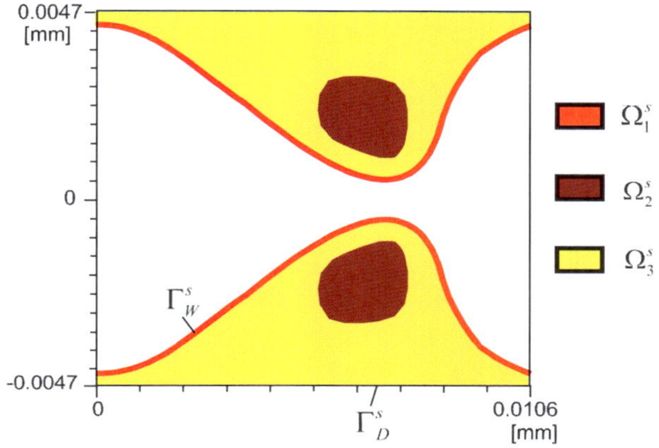

Fig. 5.27 Vocal folds model with three types of the tissue for epithelium (Ω_1^s), muscle (Ω_2^s), and ligament (Ω_3^s)

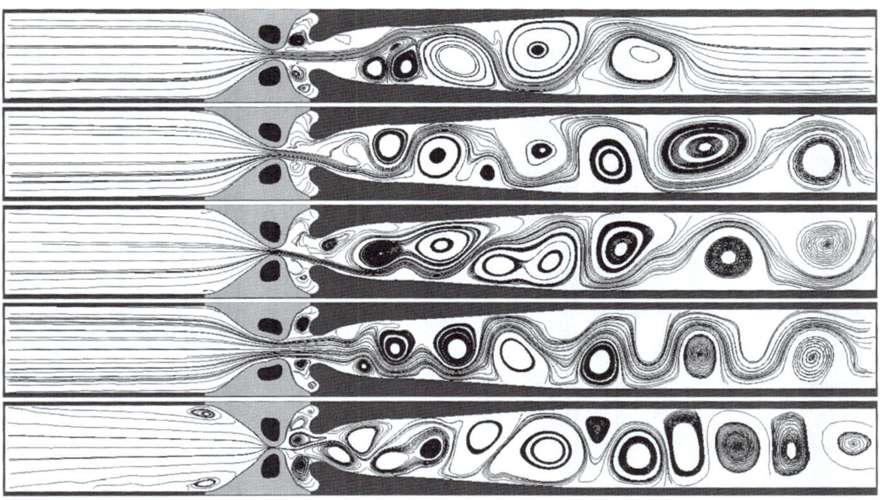

Fig. 5.28 Streamlines at time instants $t = 0.005$, 0.0425, 0.04615, 0.0597 and 0.0605 s

Figure 5.28 shows streamlines and the deformation of the computational domain at several time instants, when the vibrating vocal folds are nearly closing the channel creating a pulsating jet, and large eddies traveling behind the glottis to the outlet end of the channel, similarly as in the previous example.

5.8 Interaction of Compressible Flow with Elastic Structure

In the numerical solution of compressible flow it is necessary to overcome difficulties caused by nonlinear convection dominating over diffusion, which leads to sharp boundary layers and wakes for large Reynolds numbers, shock waves and contact discontinuities for high Mach numbers and instabilities caused by acoustic effects for low Mach numbers. There are various numerical techniques for the solution of a compressible flow—see, e.g., [31]. It appears that a suitable numerical method for the solution of a compressible flow, which overcomes successfully the abovementioned obstacles, is the DGFEM. It employs piecewise polynomial approximations without any requirement on the continuity on interfaces between neighboring elements. The theory of the DGFEM is treated in a number of works devoted to the solution of scalar equations. Let us mention, e.g., [7, 8] concerned with linear elliptic problems. The DGFEM applied to nonlinear parabolic problems is analyzed, for example, in [6, 25].

The DGFEM was used for the numerical simulation of the compressible Euler equations, for example, by Bassi and Rebay in [11], where the space DG discretization was combined with explicit Runge–Kutta time discretization. In [12] Baumann and Oden describe an *hp* version of the space DG discretization with explicit time stepping applied to compressible flow. Van der Vegt and van der Ven apply space–time discontinuous Galerkin method to the solution of the Euler equations in [98], where the discrete problem is solved with the aid of a multigrid accelerated pseudo-time-integration. The papers [24, 32], and [23] are concerned with a semi-implicit DG unconditionally stable technique for the solution of an inviscid and viscous compressible flow. In [30], this method was extended so that the resulting scheme is robust with respect to the magnitude of the Mach number.

In this section we describe the numerical technique for the solution of the compressible Navier–Stokes equations in time-dependent domains. The main ingredients of the method is the discontinuous Galerkin space semidiscretization of the Navier–Stokes equations written in the ALE form, semi-implicit time discretization, suitable treatment of boundary conditions so that they are transparent for acoustic waves at the inlet and outlet and the shock capturing avoiding Gibbs phenomenon at discontinuities. First, the numerical experiments were carried out for a compressible flow in a channel representing a model of a part of the vocal tract, with a prescribed motion of the channel walls. They prove the stability and robustness of the method and its applicability to complicated fluid–structure problems. Then the DGFEM solution of compressible flow is coupled with dynamic elasticity problem and applied to the flow-induced vocal folds vibrations.

5.8.1 Compressible Navier–Stokes Equations

We consider compressible flow in a bounded domain $\Omega_t \subset R^2$ depending on time $t \in [0, T]$. Let the boundary of Ω_t consist of three different parts: $\partial\Omega_t = \Gamma_I \cup \Gamma_O \cup \Gamma_{W_t}$, where Γ_I is the inlet, Γ_O is the outlet, and Γ_{Wt} denotes the impermeable walls that may move in dependence on time.

The system describing compressible flow consisting of the continuity equation, the Navier–Stokes equations, and the energy equation can be written in the form

$$\frac{\partial \boldsymbol{w}}{\partial t} + \sum_{s=1}^{2} \frac{\partial \mathbf{f}_s(\boldsymbol{w})}{\partial x_s} = \sum_{s=1}^{2} \frac{\partial \boldsymbol{R}_s(\boldsymbol{w}, \nabla \boldsymbol{w})}{\partial x_s}, \tag{5.109}$$

where

$$\begin{aligned}
\boldsymbol{w} &= (w_1, \dots, w_4)^T = (\rho, \rho v_1, \rho v_2, E)^T \in R^4, &(5.110)\\
\boldsymbol{w} &= \boldsymbol{w}(x,t),\ x \in \Omega_t,\ t \in (0,T),\\
\mathbf{f}_i(\boldsymbol{w}) &= (f_{i1}, \dots, f_{i4})^T = (\rho v_i, \rho v_1 v_i + \delta_{1i}\, p, \rho v_2 v_i + \delta_{2i}\, p, (E+p)v_i)^T,\\
\boldsymbol{R}_i(\boldsymbol{w}, \nabla \boldsymbol{w}) &= (R_{i1}, \dots, R_{i4})^T = \left(0, \tau_{i1}^V, \tau_{i2}^V, \tau_{i1}^V\, v_1 + \tau_{i2}^V\, v_2 k \partial\theta/\partial x_i\right)^T,\\
\tau_{ij}^V &= \lambda \operatorname{div}\boldsymbol{v}\, \delta_{ij} + 2\mu\, d_{ij}(\boldsymbol{v}),\ d_{ij}(\boldsymbol{v}) = \frac{1}{2}\left(\frac{\partial v_i}{\partial x_j} + \frac{\partial v_j}{\partial x_i}\right).
\end{aligned}$$

We use the following notation: ρ—density, p—pressure, E—total energy, $\boldsymbol{v} = (v_1, v_2)$—velocity, θ—absolute temperature, $\gamma > 1$—Poisson adiabatic constant, $c_v > 0$—specific heat at constant volume, $\mu > 0, \lambda = -2\mu/3$—viscosity coefficients, $k > 0$—heat conduction. The vector-valued function $\boldsymbol{w}$ is called the state vector, the functions $\mathbf{f}_i$ are the so-called inviscid fluxes, $\boldsymbol{R}_i$ represent the viscous terms, and τ_{ij}^V are the components of the viscous part of the fluid stress tensor.

The above system is completed by the thermodynamic relations

$$p = (\gamma - 1)(E - \rho|\boldsymbol{v}|^2/2), \quad \theta = \left(\frac{E}{\rho} - \frac{1}{2}|\boldsymbol{v}|^2\right)/c_v. \tag{5.111}$$

The complete system is equipped with the initial condition

$$\boldsymbol{w}(x,0) = \boldsymbol{w}^0(x), \quad x \in \Omega_0, \tag{5.112}$$

and the following boundary conditions:

$$\text{(a)}\ \rho|_{\Gamma_I} = \rho_I, \quad \text{(b)}\ \boldsymbol{v}|_{\Gamma_I} = \boldsymbol{v}_I = (v_{I1}, v_{I2})^{\mathrm{T}}, \tag{5.113}$$

$$\text{(c)} \quad \sum_{i,j=1}^{2} \tau_{ij}^{V} n_i v_j + k \frac{\partial \theta}{\partial n} = 0 \quad \text{on } \Gamma_I, \tag{5.114}$$

$$\text{(a)} \quad \boldsymbol{v}|_{\Gamma_{Wt}} = \boldsymbol{z}_W = \text{ velocity of the moving wall,} \tag{5.115}$$

$$\text{(b)} \quad \frac{\partial \theta}{\partial n}|_{\Gamma_{Wt}} = 0 \text{ on } \Gamma_{Wt},$$

$$\text{(a)} \quad \sum_{i=1}^{2} \tau_{ij}^{V} n_i = 0, \quad j = 1, 2, \quad \text{(b)} \frac{\partial \theta}{\partial n} = 0 \text{ on } \Gamma_O. \tag{5.116}$$

Here $\boldsymbol{w}^0$, ρ_I, $\boldsymbol{v}_I$ and $\boldsymbol{z}_W = (z_{W1}, z_{W2})$ are prescribed functions.

5.8.2 ALE Form of Compressible Navier–Stokes Equations

Similarly as in Sect. 5.2.2 we shall apply the ALE method to the formulation of flow in the domain Ω_t. By (5.8), we have

$$\frac{D^A f}{Dt} = \frac{\partial f}{\partial t} + \operatorname{div}(\boldsymbol{z} f) - f \operatorname{div} \boldsymbol{z}, \tag{5.117}$$

which allows us to write the compressible Navier–Stokes equations (5.109) in the ALE form

$$\frac{D^A \boldsymbol{w}}{Dt} + \sum_{s=1}^{2} \frac{\partial \boldsymbol{g}_s(\boldsymbol{w})}{\partial x_s} + \boldsymbol{w} \operatorname{div} \boldsymbol{z} = \sum_{s=1}^{2} \frac{\partial \boldsymbol{R}_s(\boldsymbol{w}, \nabla \boldsymbol{w})}{\partial x_s}, \tag{5.118}$$

where

$$\boldsymbol{g}_s(\boldsymbol{w}) := \mathbf{f}_s(\boldsymbol{w}) - z_s \boldsymbol{w}, \quad s = 1, 2,$$

are the ALE modified inviscid fluxes. We see that in the ALE formulation of the compressible Navier–Stokes equations the time derivative $\partial \boldsymbol{w} / \partial t$ is replaced by the ALE derivative $D^A \boldsymbol{w} / Dt$, the inviscid fluxes $\mathbf{f}_s$ are replaced by the ALE modified inviscid fluxes $\boldsymbol{g}_s$, and a new additional "reaction" term $\boldsymbol{w} \operatorname{div} \boldsymbol{z}$ appears.

5.9 Discretization of Viscous Compressible Flow

In what follows, the viscous compressible flow problem will be discretized in space and time.

5.9.1 Discontinuous Galerkin Space Semidiscretization

For the space semidiscretization we use the DFGEM. We construct a polygonal approximation Ω_{ht} of the domain Ω_t. By $\mathcal{T}_{ht}$ we denote a partition of the closure $\overline{\Omega}_{ht}$ of the domain Ω_{ht} into a finite number of closed triangles K with mutually disjoint interiors such that $\overline{\Omega}_{ht} = \bigcup_{K\in\mathcal{T}_{ht}} K$.

By $\mathcal{F}_{ht}$ we denote the system of all faces of all elements $K \in \mathcal{T}_{ht}$. Further, we introduce the set of all interior faces $\mathcal{F}_{ht}^{I} = \{\Gamma \in \mathcal{F}_{ht};\ \Gamma \subset \Omega\}$, the set of all boundary faces $\mathcal{F}_{ht}^{B} = \{\Gamma \in \mathcal{F}_{ht};\ \Gamma \subset \partial\Omega_{ht}\}$, and the set of all "Dirichlet" boundary faces $\mathcal{F}_{ht}^{D} = \left\{\Gamma \in \mathcal{F}_{ht}^{B};\ \text{a Dirichlet condition is prescribed on } \Gamma\right\}$. Each $\Gamma \in \mathcal{F}$ is associated with a unit normal vector $\boldsymbol{n}_\Gamma$ to Γ. For $\Gamma \in \mathcal{F}_{ht}^{B}$ the normal $\boldsymbol{n}_\Gamma$ has the same orientation as the outer normal to $\partial\Omega_{ht}$. We set $d(\Gamma) =$ length of $\Gamma \in \mathcal{F}_{ht}$.

For each $\Gamma \in \mathcal{F}_{ht}^{I}$ there exist two neighboring elements $K_\Gamma^{(L)}, K_\Gamma^{(R)} \in \mathcal{T}_h$ such that $\Gamma \subset \partial K_\Gamma^{(R)} \cap \partial K_\Gamma^{(L)}$. We use the convention that $K_\Gamma^{(R)}$ lies in the direction of $\boldsymbol{n}_\Gamma$ and $K_\Gamma^{(L)}$ lies in the direction opposite to $\boldsymbol{n}_\Gamma$. The elements $K_\Gamma^{(L)}$, $K_\Gamma^{(R)}$ are called neighbors. If $\Gamma \in \mathcal{F}_{ht}^{B}$, then the element adjacent to Γ will be denoted by $K_\Gamma^{(L)}$.

The approximate solution will be sought in the space of discontinuous piecewise polynomial functions

$$\mathbf{S}_{ht} = [S_{ht}]^4, \quad \text{with } S_{ht} = \{v; v|_K \in P_r(K)\ \forall\, K \in \mathcal{T}_{ht}\}, \tag{5.119}$$

where $r \geq 0$ is an integer and $P_r(K)$ denotes the space of all polynomials on K of degree $\leq r$. A function $\boldsymbol{\varphi} \in \mathbf{S}_{ht}$ is, in general, discontinuous on interfaces $\Gamma \in \mathcal{F}_{ht}^{I}$. By $\boldsymbol{\varphi}_\Gamma^{(L)}$ and $\boldsymbol{\varphi}_\Gamma^{(R)}$ we denote the values of $\boldsymbol{\varphi}$ on Γ considered from the interior and the exterior of $K_\Gamma^{(L)}$, respectively, and set $\langle\boldsymbol{\varphi}\rangle_\Gamma = (\boldsymbol{\varphi}_\Gamma^{(L)} + \boldsymbol{\varphi}_\Gamma^{(R)})/2$, $[\boldsymbol{\varphi}]_\Gamma = \boldsymbol{\varphi}_\Gamma^{(L)} - \boldsymbol{\varphi}_\Gamma^{(R)}$.

The discrete problem is derived in the following way: We

- multiply system (5.118) by a test function $\boldsymbol{\varphi}_h \in \mathbf{S}_{ht}$,
- integrate over $K \in \mathcal{T}_{ht}$,
- use Green's theorem,
- sum over all elements $K \in \mathcal{T}_{ht}$,
- introduce the concept of the numerical flux,
- introduce suitable terms vanishing for a regular exact solution.

In this way we get the following identity:

$$\sum_{K\in\mathcal{T}_{ht}} \int_K \frac{D^A\boldsymbol{w}}{Dt}\cdot\boldsymbol{\varphi}_h\,\mathrm{d}x + b_h(\boldsymbol{w},\boldsymbol{\varphi}_h) + a_h(\boldsymbol{w},\boldsymbol{\varphi}_h) + J_h(\boldsymbol{w},\boldsymbol{\varphi}_h) + d_h\boldsymbol{w},\boldsymbol{\varphi}) \tag{5.120}$$
$$= \ell_h(\boldsymbol{w},\boldsymbol{\varphi}_h).$$

Here

$$
\begin{aligned}
b_h(\boldsymbol{w}, \boldsymbol{\varphi}_h) = &- \sum_{K \in \mathcal{T}_{ht}} \int_K \sum_{s=1}^{2} \boldsymbol{g}_s(\boldsymbol{w}) \cdot \frac{\partial \boldsymbol{\varphi}_h}{\partial x_s} \, \mathrm{d}x \\
&+ \sum_{\Gamma \in \mathcal{F}_{ht}^I} \int_\Gamma \mathbf{H}_g(\boldsymbol{w}_\Gamma^{(L)}, \boldsymbol{w}_\Gamma^{(R)}, \boldsymbol{n}_\Gamma) \cdot [\boldsymbol{\varphi}_h]_\Gamma \, \mathrm{d}S \\
&+ \sum_{\Gamma \in \mathcal{F}_{ht}^B} \int_\Gamma \mathbf{H}_g(\boldsymbol{w}_\Gamma^{(L)}, \boldsymbol{w}_\Gamma^{(R)}, \boldsymbol{n}_\Gamma) \cdot \boldsymbol{\varphi}_{h\Gamma}^{(L)} \, \mathrm{d}S
\end{aligned}
\tag{5.121}
$$

is the convection form, defined with the aid of a numerical flux $\mathbf{H}_g$. We require that it is consistent with the fluxes $\boldsymbol{g}_s$: $\mathbf{H}_g(\boldsymbol{w}, \boldsymbol{w}, \boldsymbol{n}) = \sum_{s=1}^{2} \boldsymbol{g}_s(\boldsymbol{w}) n_s \quad (\boldsymbol{n} = (n_1, n_2), |\boldsymbol{n}| = 1)$, conservative: $\mathbf{H}_g(\boldsymbol{u}, \boldsymbol{w}, \boldsymbol{n}) = -\mathbf{H}_g(\boldsymbol{w}, \boldsymbol{u}, -\boldsymbol{n})$, and locally Lipschitz-continuous.

Further, we define the viscous form

$$
\begin{aligned}
a_h(\boldsymbol{w}, \boldsymbol{\varphi}_h) = &\sum_{K \in \mathcal{T}_{ht}} \int_K \sum_{s=1}^{2} \boldsymbol{R}_s(\boldsymbol{w}, \nabla \boldsymbol{w}) \cdot \frac{\partial \boldsymbol{\varphi}_h}{\partial x_s} \, \mathrm{d}x \\
&- \sum_{\Gamma \in \mathcal{F}_{ht}^I} \int_\Gamma \sum_{s=1}^{2} \langle \boldsymbol{R}_s(\boldsymbol{w}, \nabla \boldsymbol{w}) \rangle_\Gamma (\boldsymbol{n}_\Gamma)_s \cdot [\boldsymbol{\varphi}_h]_\Gamma \, \mathrm{d}S \\
&- \sum_{\Gamma \in \mathcal{F}_{ht}^D} \int_\Gamma \sum_{s=1}^{2} \boldsymbol{R}_s(\boldsymbol{w}, \nabla \boldsymbol{w}) (\boldsymbol{n}_\Gamma)_s \cdot \boldsymbol{\varphi}_{h\Gamma}^{(L)} \, \mathrm{d}S
\end{aligned}
\tag{5.122}
$$

(we use the incomplete discretization of viscous terms—the so-called IIPG version), the interior and boundary penalty terms and the right-hand side form, respectively, by

$$
J_h(\boldsymbol{w}, \boldsymbol{\varphi}_h) = \sum_{\Gamma \in \mathcal{F}_{ht}^I} \int_\Gamma \sigma [\boldsymbol{w}] \cdot [\boldsymbol{\varphi}_h]_\Gamma \, \mathrm{d}S + \sum_{\Gamma \in \mathcal{F}_{ht}^D} \int_\Gamma \sigma \boldsymbol{w} \cdot \boldsymbol{\varphi}_{h\Gamma}^{(L)} \, \mathrm{d}S, \tag{5.123}
$$

$$
\ell_h(\boldsymbol{w}, \boldsymbol{\varphi}_h) = \sum_{\Gamma \in \mathcal{F}_{ht}^D} \int_\Gamma \sum_{s=1}^{2} \sigma \boldsymbol{w}_B \cdot \boldsymbol{\varphi}_{h\Gamma}^{(L)} \, \mathrm{d}S. \tag{5.124}
$$

Here $\sigma|_\Gamma = C_W \mu / d(\Gamma)$ and $C_W > 0$ is a sufficiently large constant. The "reaction" form reads

$$
d_h(\boldsymbol{w}, \boldsymbol{\varphi}_h) = \sum_{K \in \mathcal{T}_{ht}} \int_K (\boldsymbol{w} \cdot \boldsymbol{\varphi}_h) \, \mathrm{div} z \, \mathrm{d}x. \tag{5.125}
$$

The boundary state $\boldsymbol{w}_B$ is defined on the basis of the Dirichlet boundary conditions and extrapolation:

$$\boldsymbol{w}_B = (\rho_I, \rho_I v_{I1}, \rho_I v_{I2}, c_v \rho_I \theta_\Gamma^{(L)} + \frac{1}{2}\rho_I |\boldsymbol{v}_I|^2) \quad \text{on } \Gamma_I, \tag{5.126}$$

$$\boldsymbol{w}_B = \boldsymbol{w}_\Gamma^{(L)} \quad \text{on } \Gamma_O,$$

$$\boldsymbol{w}_B = (\rho_\Gamma^{(L)}, \rho_\Gamma^{(L)} z_{W1}, \rho_\Gamma^{(L)} z_{W2}, c_v \rho_\Gamma^{(L)} \theta_\Gamma^{(L)} + \frac{1}{2}\rho_\Gamma^{(L)} |\boldsymbol{z}_W|^2) \quad \text{on } \Gamma_{Wt}.$$

The approximate solution is defined as $\boldsymbol{w}_h(t) \in \mathbf{S}_{ht}$ such that

$$\begin{aligned} &\sum_{K \in \mathcal{T}_{ht}} \int_K \frac{D^A \boldsymbol{w}_h(t)}{Dt} \cdot \boldsymbol{\varphi}_h \, \mathrm{d}x + b_h(\boldsymbol{w}_h(t), \boldsymbol{\varphi}_h) + a_h(\boldsymbol{w}_h(t), \boldsymbol{\varphi}_h) \\ &\quad + J_h(\boldsymbol{w}_h(t), \boldsymbol{\varphi}_h) + d_h(\boldsymbol{w}_h(t), \boldsymbol{\varphi}_h) = \ell_h(\boldsymbol{w}_h(t), \boldsymbol{\varphi}_h) \end{aligned} \tag{5.127}$$

holds for all $\boldsymbol{\varphi}_h \in \mathbf{S}_{ht}$, all $t \in (0, T)$, and $\boldsymbol{w}_h(0) = \boldsymbol{w}_h^0$ is an approximation of the initial state $\boldsymbol{w}^0$.

5.9.2 Time Discretization by the BDF Method

Let us construct a partition $0 = t_0 < t_1 < t_2 \cdots$ of the time interval $[0, T]$ with $t_k = k\Delta t$ and time step Δt. We use the approximations $\boldsymbol{w}_h(t_n) \approx \boldsymbol{w}_h^n \in \mathbf{S}_{ht_n}$, $\boldsymbol{z}(t_n) \approx \boldsymbol{z}^n$, $n = 0, 1, \ldots$ and introduce the function $\hat{\boldsymbol{w}}_h^k = \boldsymbol{w}_h^k \circ \mathcal{A}_{t_k} \circ \mathcal{A}_{t_{k+1}}^{-1}$, which is defined in the domain $\Omega_{ht_{k+1}}$. In order to approximate the ALE derivative at time t_{k+1}, we start from its definition (5.7) and then use the backward difference:

$$\begin{aligned} \frac{D^A \boldsymbol{w}_h}{Dt}(x, t_{k+1}) &= \frac{\partial \tilde{\boldsymbol{w}}_h}{\partial t}(X, t_{k+1}) \\ &\approx \frac{\tilde{\boldsymbol{w}}_h^{k+1}(X) - \tilde{\boldsymbol{w}}_h^k(X)}{\tau_k} = \frac{\boldsymbol{w}_h^{k+1}(x) - \hat{\boldsymbol{w}}_h^k(x)}{\tau_k}, \qquad x = \mathcal{A}_{t_{k+1}}(X) \in \Omega_{ht_{k+1}}. \end{aligned} \tag{5.128}$$

By the symbol $(\cdot, \cdot)$ we shall denote the scalar product in $L^2(\Omega_{ht_{k+1}})$. A possible full discretization reads

$$\text{(a)} \;\; \boldsymbol{w}_h^{k+1} \in \mathbf{S}_{ht_{k+1}}, \tag{5.129}$$

$$\text{(b)} \;\; \left(\frac{\boldsymbol{w}_h^{k+1} - \hat{\boldsymbol{w}}_h^k}{\tau_k}, \boldsymbol{\varphi}_h \right) + b_h(\boldsymbol{w}_h^{k+1}, \boldsymbol{\varphi}_h) + a_h(\boldsymbol{w}_h^{k+1}, \boldsymbol{\varphi}_h)$$

$$+J_h(\boldsymbol{w}_h^{k+1}, \boldsymbol{\varphi}_h) + d_h\left(\boldsymbol{w}_h^{k+1}, \boldsymbol{\varphi}_h\right) = \ell_h(\boldsymbol{w}_h^{k+1}, \boldsymbol{\varphi}_h)$$
$$\forall \boldsymbol{\varphi}_h \in \mathbf{S}_{ht_{k+1}}, \; k = 0, 1, \ldots .$$

However, this problem for $\boldsymbol{w}_h^{k+1}$ is equivalent to a strongly nonlinear algebraic system and its solution is rather difficult.

5.9.3 Semi-implicit BDF Scheme

Our goal is to develop a numerical scheme which would be accurate and robust, with good stability properties and efficiently solvable. Therefore, we proceed similarly to [24] and use a partial linearization of the forms b_h and a_h. This approach leads to a scheme that requires the solution of only one large sparse linear system on each time level.

The linearization of the first term of the form b_h is based on the relations

$$\boldsymbol{g}_s(\boldsymbol{w}_h^{k+1}) = (\mathbb{A}_s(\boldsymbol{w}_h^{k+1}) - z_s^{k+1}\mathbb{I})\boldsymbol{w}_h^{k+1} \approx (\mathbb{A}_s(\hat{\boldsymbol{w}}_h^{k}) - z_s^{k+1}\mathbb{I})\boldsymbol{w}_h^{k+1},$$

where $\mathbb{A}_s(\boldsymbol{w})$ is the Jacobi matrix of $\mathbf{f}_s(\boldsymbol{w})$, cf. [31]. The second term of b_h is linearized with the aid of the Vijayasundaram numerical flux (cf. [100]) defined in the following way. Taking into account the definition of $\boldsymbol{g}_s$, we have

$$\frac{D\boldsymbol{g}_s(\boldsymbol{w})}{D\boldsymbol{w}} = \frac{D\mathbf{f}_s(\boldsymbol{w})}{D\boldsymbol{w}} - z_s\mathbb{I} = \mathbb{A}_s(\boldsymbol{w}) - z_s\mathbb{I}, \tag{5.130}$$

and we can write

$$\mathbb{P}_g(\boldsymbol{w}, \boldsymbol{n}) = \sum_{s=1}^{2} \frac{D\boldsymbol{g}_s(\boldsymbol{w})}{D\boldsymbol{w}} n_s = \sum_{s=1}^{2} (\mathbb{A}_s(\boldsymbol{w})n_s - z_s n_s \mathbb{I}) . \tag{5.131}$$

By [31], this matrix is diagonalizable. It means that there exists a nonsingular matrix $\mathbb{T} = \mathbb{T}(\boldsymbol{w}, \boldsymbol{n})$ such that

$$\mathbb{P}_g = \mathbb{T}I\!\Lambda\mathbb{T}^{-1}, \quad I\!\Lambda = \mathrm{diag}(\lambda_1, \ldots, \lambda_4), \tag{5.132}$$

where $\lambda_i = \lambda_i(\boldsymbol{w}, \boldsymbol{n})$ are the eigenvalues of the matrix $\mathbb{P}_g$. Now we define the "positive" and "negative" parts of the matrix $\mathbb{P}_g$ by

$$\mathbb{P}_g^{\pm} = \mathbb{T}I\!\Lambda^{\pm}\mathbb{T}^{-1}, \quad I\!\Lambda^{\pm} = \mathrm{diag}(\lambda_1^{\pm}, \ldots, \lambda_4^{\pm}), \tag{5.133}$$

where $\lambda^+ = \max(\lambda, 0)$, $\lambda^- = \min(\lambda, 0)$. Using the above concepts, we introduce the modified Vijayasundaram numerical flux (cf. [100] or [31]) as

$$\boldsymbol{H}_g(\boldsymbol{w}_L, \boldsymbol{w}_R, \boldsymbol{n}) = \mathbb{P}_g^+\left(\frac{\boldsymbol{w}_L + \boldsymbol{w}_R}{2}, \boldsymbol{n}\right)\boldsymbol{w}_L + \mathbb{P}_g^-\left(\frac{\boldsymbol{w}_L + \boldsymbol{w}_R}{2}, \boldsymbol{n}\right)\boldsymbol{w}_R. \tag{5.134}$$

Using the above definition of the numerical flux, we introduce the approximation

$$\mathbf{H}_g(\boldsymbol{w}_{h\Gamma}^{k+1(L)}, \boldsymbol{w}_{h\Gamma}^{k+1(R)}, \boldsymbol{n}_\Gamma) \approx \mathbb{P}_g^+(\langle\hat{\boldsymbol{w}}_h^k\rangle_\Gamma, \boldsymbol{n}_\Gamma)\boldsymbol{w}_{h\Gamma}^{k+1(L)} + \mathbb{P}_g^-(\langle\hat{\boldsymbol{w}}_h^k\rangle_\Gamma, \boldsymbol{n}_\Gamma)\hat{\boldsymbol{w}}_{h\Gamma}^{k(R)}.$$

In this way we get the form

$$\begin{aligned} &\hat{b}_h(\hat{\boldsymbol{w}}_h^k, \boldsymbol{w}_h^{k+1}, \boldsymbol{\varphi}_h) \qquad (5.135)\\ &= -\sum_{K\in\mathcal{T}_{ht_{k+1}}} \int_K \sum_{s=1}^2 \left(\left(\mathbb{A}_s(\hat{\boldsymbol{w}}^k(x)) - z_s^{k+1}(x)\mathbb{I}\right)\boldsymbol{w}^{k+1}(x)\right)\cdot\frac{\partial\boldsymbol{\varphi}_h(x)}{\partial x_s}\,\mathrm{d}x,\\ &+ \sum_{\Gamma\in\mathcal{F}^I_{ht_{k+1}}} \int_\Gamma \left(\mathbb{P}_g^+(\langle\hat{\boldsymbol{w}}_h^k\rangle, \boldsymbol{n}_\Gamma)\boldsymbol{w}_h^{k+1(L)} + \mathbb{P}_g^-(\langle\hat{\boldsymbol{w}}_h^k\rangle, \boldsymbol{n}_\Gamma)\boldsymbol{w}_h^{k+1(R)}\right)\cdot[\boldsymbol{\varphi}_h]\,\mathrm{d}S\\ &+ \sum_{\Gamma\in\mathcal{F}^B_{ht_{k+1}}} \int_\Gamma \left(\mathbb{P}_g^+(\langle\hat{\boldsymbol{w}}_h^k\rangle, \boldsymbol{n}_\Gamma)\boldsymbol{w}_h^{k+1(L)} + \mathbb{P}_g^-(\langle\hat{\boldsymbol{w}}_h^k\rangle, \boldsymbol{n}_\Gamma)\hat{\boldsymbol{w}}_h^{k(R)}\right)\cdot\boldsymbol{\varphi}_h\,\mathrm{d}S. \end{aligned}$$

The linearization of the form a_h is based on the fact that $\boldsymbol{R}_s(\boldsymbol{w}_h, \nabla\boldsymbol{w}_h)$ is linear in $\nabla\boldsymbol{w}$ and nonlinear in $\boldsymbol{w}$. We get the linearized viscous form

$$\begin{aligned} \hat{a}_h(\hat{\boldsymbol{w}}_h^k, \boldsymbol{w}_h^{k+1}, \boldsymbol{\varphi}_h) &= \sum_{K\in\mathcal{T}_{ht_{k+1}}} \int_K \sum_{s=1}^2 \boldsymbol{R}_s(\hat{\boldsymbol{w}}_h^k, \nabla\boldsymbol{w}_h^{k+1})\cdot\frac{\partial\boldsymbol{\varphi}_h}{\partial x_s}\,\mathrm{d}x \qquad (5.136)\\ &- \sum_{\Gamma\in\mathcal{F}^I_{ht_{k+1}}} \int_\Gamma \sum_{s=1}^2 \langle\boldsymbol{R}_s(\hat{\boldsymbol{w}}_h^k, \nabla\boldsymbol{w}^{k+1})\rangle(\boldsymbol{n}_\Gamma)_s\cdot[\boldsymbol{\varphi}_h]\,\mathrm{d}S\\ &- \sum_{\Gamma\in\mathcal{F}^D_{ht_{k+1}}} \int_\Gamma \sum_{s=1}^2 \boldsymbol{R}_s(\hat{\boldsymbol{w}}_h^k, \nabla\boldsymbol{w}_h^{k+1})(\boldsymbol{n}_\Gamma)_s\cdot\boldsymbol{\varphi}_h\,\mathrm{d}S. \end{aligned}$$

5.9.4 Realization of the Boundary Conditions

If $\Gamma \in \mathcal{F}^B_{ht}$, it is necessary to specify the boundary state $\hat{\boldsymbol{w}}_{h\Gamma}^{k(R)}$ appearing in the definition of the inviscid form $\hat{b}_h$. For simplicity we shall use the notation $\boldsymbol{w}^{(R)}$ for values of the function $\hat{\boldsymbol{w}}_{h\Gamma}^{k(R)}$ which should be determined at individual integration points on the face Γ. Similarly, $\boldsymbol{w}^{(L)}$ will denote the values of $\hat{\boldsymbol{w}}_{h\Gamma}^{k(L)}$ at the corresponding points.

On the inlet and outlet, which are assumed fixed, we proceed in the same way as in [30], Sect. 4. Using the rotational invariance, we transform the Euler equations

$$\frac{\partial \boldsymbol{w}}{\partial t} + \sum_{s=1}^{2} \frac{\partial \mathbf{f}_s(\boldsymbol{w})}{\partial x_s} = 0$$

to the coordinates $\tilde{x}_1$, parallel with the normal direction $\boldsymbol{n} = (n_1, n_2) = \boldsymbol{n}_\Gamma$ to the boundary, and $\tilde{x}_2$, tangential to the boundary, neglect the derivative with respect to $\tilde{x}_2$ and linearize the system around the state $\boldsymbol{q}^{(L)} = \mathbb{Q}(\boldsymbol{n})\boldsymbol{w}^{(L)}$, where

$$\mathbb{Q}(\boldsymbol{n}) = \begin{pmatrix} 1, & 0, & 0, & 0 \\ 0, & n_1, & n_2, & 0 \\ 0, & -n_2, & n_1, & 0 \\ 0, & 0, & 0, & 1 \end{pmatrix} \tag{5.137}$$

is the rotational matrix. Then we obtain the linear system

$$\frac{\partial \boldsymbol{q}}{\partial t} + \mathbb{A}_1(\boldsymbol{q}^{(L)}) \frac{\partial \boldsymbol{q}}{\partial \tilde{x}_1} = 0 \tag{5.138}$$

for the transformed vector-valued function $\boldsymbol{q} = \mathbb{Q}(\boldsymbol{n})\boldsymbol{w}$, considered in the set $(-\infty, 0) \times (0, \infty)$ and equipped with the initial and boundary conditions

$$\boldsymbol{q}(\tilde{x}_1, 0) = \boldsymbol{q}^{(L)}, \ \tilde{x}_1 < 0, \qquad \text{and} \qquad \boldsymbol{q}(0, t) = \boldsymbol{q}^{(R)}, \ t > 0. \tag{5.139}$$

The goal is to choose $\boldsymbol{q}^{(R)}$ in such a way that this initial-boundary value problem is well posed, i.e., has a unique solution. The method of characteristics leads to the following process:

Let us put $\boldsymbol{q}^* = \mathbb{Q}(\boldsymbol{n})\boldsymbol{w}^*$, where $\boldsymbol{w}^*$ is a given boundary state at the inlet or outlet. We calculate the eigenvectors $\boldsymbol{r}_s$ corresponding to the eigenvalues λ_s, $s = 1, \dots, 4$, of the matrix $\mathbb{A}_1(\boldsymbol{q}^{(L)})$, arrange them as columns in the matrix $\mathbb{T}$ and calculate $\mathbb{T}^{-1}$. Now we set

$$\boldsymbol{\alpha} = \mathbb{T}^{-1}\boldsymbol{q}^{(L)}, \quad \boldsymbol{\beta} = \mathbb{T}^{-1}\boldsymbol{q}^* \tag{5.140}$$

and define the state $\boldsymbol{q}^{(R)}$ by the relations

$$\boldsymbol{q}^{(R)} := \sum_{s=1}^{4} \gamma_s \boldsymbol{r}_s, \quad \gamma_s = \begin{cases} \alpha_s, & \lambda_s \geq 0, \\ \beta_s, & \lambda_s < 0. \end{cases} \tag{5.141}$$

Finally, the sought boundary state $\boldsymbol{w}^{(R)}$ is defined as

$$\boldsymbol{w}^{(R)} = \mathbb{Q}^{-1}(\boldsymbol{n})\boldsymbol{q}^{(R)}. \tag{5.142}$$

On the impermeable moving wall we prescribe the normal component of the velocity

$$\boldsymbol{v} \cdot \boldsymbol{n} = z_W \cdot \boldsymbol{n}, \tag{5.143}$$

where $\boldsymbol{n}$ is the unit outer normal to Γ_{W_t} and z_W is the wall velocity. This means that two eigenvalues of $\mathbb{P}_g(\boldsymbol{w}, \boldsymbol{n})$ vanish, one is positive and one is negative. Then, in analogy to [31, Sect. 3.3.6], we should prescribe one quantity, namely $\boldsymbol{v} \cdot \boldsymbol{n}$, and extrapolate three quantities—tangential velocity, density, and pressure.

However, here we define the numerical flux on Γ_{W_t} as the physical flux through the boundary with the assumption (5.143) taken into account. Thus, on Γ_{W_t} we write

$$\sum_{s=1}^{2} \boldsymbol{g}_s(\boldsymbol{w}) n_s = (\boldsymbol{v} \cdot \boldsymbol{n} - z_W \cdot \boldsymbol{n})\boldsymbol{w} + p\,(0, n_1, n_2, \boldsymbol{v} \cdot \boldsymbol{n})^T \tag{5.144}$$

$$= p\,(0, n_1, n_2, z_W \cdot \boldsymbol{n})^T =: \boldsymbol{H}_g.$$

5.9.5 Stabilization by the Local Artificial Viscosity

In high-speed inviscid gas flow with large Mach numbers, discontinuities—called shock waves or contact discontinuities—appear. In viscous high-speed flow these discontinuities may be smeared due to viscosity and heat conduction. In both cases, near shock waves and contact discontinuities, the so-called Gibbs phenomenon, manifested by nonphysical spurious overshoots and undershoots, usually occurs in the numerical solution. The same phenomenon appears in the boundary layer in the case of a large Reynolds number. In order to avoid this undesirable phenomenon, it is necessary to apply a suitable stabilization procedure. Here we use the approach proposed in [30] using a local artificial viscosity. It is based on the discontinuity indicator

$$g^k(K) = \int_{\partial K} [\hat{\rho}_h^k]^2 \, \mathrm{d}S \big/ (h_K |K|^{3/4}), \quad K \in \mathcal{T}_{ht_{k+1}}, \tag{5.145}$$

introduced in [26]. By $[\hat{\rho}_h^h]$ we denote the jump of the function $\hat{\rho}_h^k$ on the boundary ∂K and $|K|$ denotes the area of the element K. Then we define the discrete discontinuity indicator

$$G^k(K) = 0 \quad \text{if } g^k(K) < 1, \quad G^k(K) = 1 \quad \text{if } g^k(K) \geq 1, \quad K \in \mathcal{T}_{ht_{k+1}}, \tag{5.146}$$

and the artificial viscosity forms

$$\hat{\beta}_h(\hat{\boldsymbol{w}}_h^k, \boldsymbol{w}_h^{k+1}, \boldsymbol{\varphi}_h) = \nu_1 \sum_{K \in \mathcal{T}_{ht_{k+1}}} h_K G^k(K) \int_K \nabla \boldsymbol{w}_h^{k+1} \cdot \nabla \boldsymbol{\varphi}_h \, \mathrm{d}x \tag{5.147}$$

and

$$\begin{aligned} &\hat{J}_h(\hat{\boldsymbol{w}}_h^k, \boldsymbol{w}_h^{k+1}, \boldsymbol{\varphi}_h) \\ &= \nu_2 \sum_{\Gamma \in \mathcal{F}_{ht}^I} \frac{1}{2}\big(G^k(K_\Gamma^{(L)}) + G^k(K_\Gamma^{(R)}\big) \int_\Gamma [\boldsymbol{w}_h^{k+1}] \cdot [\boldsymbol{\varphi}_h] \, \mathrm{d}S \end{aligned} \tag{5.148}$$

with parameters $\nu_1, \ \nu_2 = O(1)$.

It is important that the indicator $G^k(K)$ vanishes in regions where the solution is regular and the artificial viscosity is applied only locally in the vicinity of discontinuities or steep gradients. Therefore, the numerical solution does not lose the accuracy in regions, where the exact solution is regular and, moreover, it does not contain any nonphysical entropy production in these regions.

The complete resulting scheme has the following form:

$$\begin{aligned} &\text{(a) } \boldsymbol{w}_h^{k+1} \in \mathbf{S}_{ht_{k+1}}, \\ &\text{(b) } \left(\frac{\boldsymbol{w}_h^{k+1} - \hat{\boldsymbol{w}}_h^k}{\tau_k}, \boldsymbol{\varphi}_h \right) \\ &\qquad + \hat{b}_h(\hat{\boldsymbol{w}}_h^k, \boldsymbol{w}_h^{k+1}, \boldsymbol{\varphi}_h) + \hat{a}_h(\hat{\boldsymbol{w}}_h^k, \boldsymbol{w}_h^{k+1}, \boldsymbol{\varphi}_h) \\ &\qquad + J_h(\boldsymbol{w}_h^{k+1}, \boldsymbol{\varphi}_h) + d_h\left(\boldsymbol{w}_h^{k+1}, \boldsymbol{\varphi}_h\right) \\ &\qquad + \hat{\beta}_h(\hat{\boldsymbol{w}}_h^k, \boldsymbol{w}_h^{k+1}, \boldsymbol{\varphi}_h) + \hat{J}_h(\hat{\boldsymbol{w}}_h^k, \boldsymbol{w}_h^{k+1}, \boldsymbol{\varphi}_h) = \ell(\boldsymbol{w}_B^k, \boldsymbol{\varphi}) \\ &\qquad \forall \boldsymbol{\varphi}_h \in \mathbf{S}_{ht_{k+1}}, \ k = 0, 1, \ldots . \end{aligned} \tag{5.149}$$

The above scheme is of the first-order accuracy in time. The linearized second-order scheme can be written in the form

$$\begin{aligned} &\Big(\frac{3\boldsymbol{w}_h^{k+1} - 4\hat{\boldsymbol{w}}_h^k + \hat{\boldsymbol{w}}^{k-1}}{2\Delta t}, \boldsymbol{\varphi}_h \Big) + \hat{b}_h(\overline{\boldsymbol{w}}_h^{k+1}, \boldsymbol{w}_h^{k+1}, \boldsymbol{\varphi}_h) + \hat{a}_h(\overline{\boldsymbol{w}}_h^{k+1}, \boldsymbol{w}_h^{k+1}, \boldsymbol{\varphi}_h) \\ &+ J_h(\boldsymbol{w}_h^{k+1}, \boldsymbol{\varphi}_h) + d_h(\boldsymbol{w}_h^{k+1}, \boldsymbol{\varphi}_h) + \hat{\beta}_h(\hat{\boldsymbol{w}}_h^k, \boldsymbol{w}_h^{k+1}, \boldsymbol{\varphi}_h) \\ &+ \hat{J}_h(\hat{\boldsymbol{w}}_h^k, \boldsymbol{w}_h^{k+1}, \boldsymbol{\varphi}_h) = \ell(\overline{\boldsymbol{w}}_B^{k+1}, \boldsymbol{\varphi}_h), \quad \forall \boldsymbol{\varphi}_h \in \mathbf{S}_{ht_{k+1}}, \end{aligned} \tag{5.150}$$

where

$$\overline{\boldsymbol{w}}_h^{k+1} = 2\hat{\boldsymbol{w}}_h^k - \hat{\boldsymbol{w}}_h^{k-1}, \quad \overline{\boldsymbol{w}}_B^{k+1} = 2\hat{\boldsymbol{w}}_B^k - \hat{\boldsymbol{w}}_B^{k-1}. \tag{5.151}$$

In practical computations, integrals appearing in the definitions of the forms $\hat{a}_h, \hat{b}_h, \ldots$ are evaluated with the aid of quadrature formulae.

The linear algebraic system equivalent to (5.149) or (5.150) is solved by a direct solver UMFPACK [19] or by the GMRES method with a block diagonal preconditioning.

5.9.6 Coupling Procedure

Similarly as in the case of incompressible flow, in the solution of the complete coupled FSI problem we apply the coupling algorithms.

Weak Coupling

1. Compute the approximate solution $\boldsymbol{w}_h^n$ of the compressible flow problem on the time level t_n (from (5.149) or (5.150)).
2. Compute the corresponding fluid stress tensor τ_{ij}^f and the aerodynamic force acting on the structure and transform it to the interface Γ_{Wh}^s by (5.94).
3. Solve the elasticity problem (5.74)–(5.75) and compute the deformation $\boldsymbol{u}_h^n$ at time t_n. On the basis of (5.93) set

$$\tilde{\Gamma}_{W t_{n+1},h} = \{x = X + \boldsymbol{u}_h^n(X);\ X \in \Gamma_{Wh}^s\}, \tag{5.152}$$

   and determine the domain $\Omega_{t_{n+1},h}$.
4. Determine the ALE mapping $\mathcal{A}_{ht_{n+1}}$ by (5.106) and approximate the domain velocity z_h^{n+1} by (5.10).
5. Set $n := n + 1$, go to 1.

Strong Coupling

Prescribe N-maximal number of inner iterations and $\varepsilon > 0$-maximal error in the iterative process. Now proceed in the following way:

1. Assume that the approximate solution $\boldsymbol{w}_h^n$ of the compressible flow problem and the deformation $\boldsymbol{u}_h^n$ of the structure are known on the time level t_n.
2. Set $\boldsymbol{u}_{h,n+1}^0 := \boldsymbol{u}_h^n,\ k := 1$ and apply the iterative process:

   (a) Compute the fluid stress tensor τ_{ij}^f and the aerodynamic force acting on the structure and transform it to the interface Γ_{Wh}^s.

   (b) Solve the elasticity problem, compute the approximation of the deformation $\boldsymbol{u}_{h,n+1}^k$, and construct the approximation $\Omega_{ht_{n+1}}^k$ of the flow domain at time t_{n+1}.

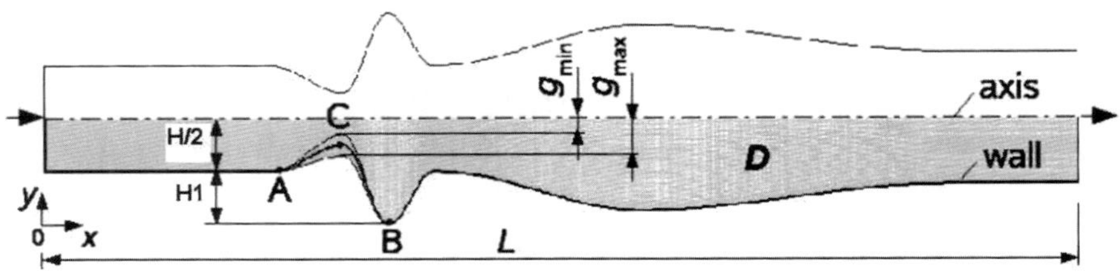

Fig. 5.29 Computational domain (cf. [64])

(c) Determine the approximations of the ALE mapping $\mathcal{A}^k_{ht_{n+1}}$ and the domain velocity $z^k_{h,n+1}$.
(d) Solve the flow problem in $\Omega^k_{ht_{n+1}}$ and obtain the approximate solution $\boldsymbol{w}^k_{h,n+1}$.
(e) If the variation $|\boldsymbol{u}^k_{h,n+1} - \boldsymbol{u}^{k-1}_{h,n+1}| \geq \varepsilon$ and $k < N$, go to (a) and $k := k+1$. Else $\Omega_{ht_{n+1}} := \Omega^k_{ht_n}$, $\boldsymbol{w}^{n+1}_h := \boldsymbol{w}^k_{h,n+1}$, $\boldsymbol{u}^{n+1}_h := \boldsymbol{u}^k_{h,n}$, $n := n+1$ and go to 2.

5.10 Numerical Results Obtained by the Discontinuous Galerkin Method

5.10.1 *Flow in a Channel with Prescribed Motion of Walls*

Here we present results of numerical experiments carried out for the flow in a channel with geometry inspired by the shape of the human glottis and a part of supraglottal spaces as shown in Fig. 5.29. The walls are moving in order to mimic the vibrations of vocal folds during the voice production. The lower channel wall between the points A and B and the upper wall symmetric with respect to the axis of the channel are vibrating up and down periodically with frequency 100 Hz. This movement is interpolated into the domain resulting in the ALE mapping $\mathcal{A}_t$. For the same geometry and similar data the computation was also carried out in [64] with the use of the finite volume method and assuming the symmetry of the flow field.

The width of the channel at the inlet (left part of the boundary) is $H = 0.016$ m and its length is $L = 0.16$ m. The width of the narrowest part of the channel (at the point C) oscillates between 0.0004 and 0.0028 m. We consider the following input parameters and boundary conditions: magnitude of the inlet velocity $v_{\text{in}} = 4\,\text{m}\,\text{s}^{-1}$, the viscosity $\mu = 15 \cdot 10^{-6}\,\text{kg}\,\text{m}^{-1}\,\text{s}^{-1}$, the inlet density $\rho_{\text{in}} = 1.225\,\text{kg}\,\text{m}^{-3}$, the outlet pressure $p_{\text{out}} = 97{,}611$ Pa, the Reynolds number $Re = \rho_{\text{in}} v_{\text{in}} H/\mu = 5{,}227$, heat conduction coefficient $k = 2.428 \cdot 10^{-2}\,\text{kg}\,\text{m}\,\text{s}^{-2}\,\text{K}^{-1}$, the specific

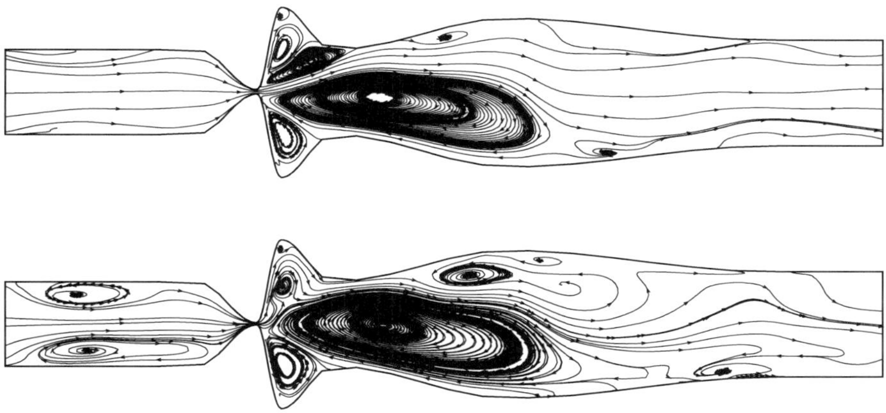

Fig. 5.30 Streamlines at time instants $t = 29, 31$ ms

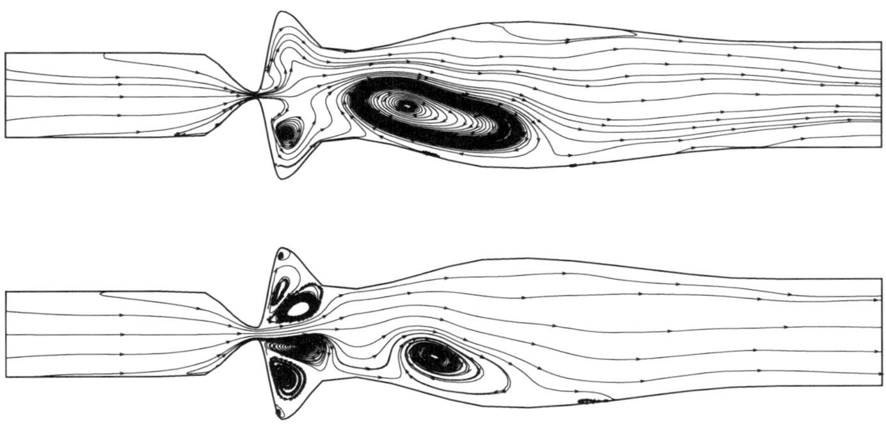

Fig. 5.31 Streamlines at time instants $t = 33, 34$ ms

heat $c_v = 721.428\,\mathrm{m^2\,s^{-2}\,K^{-1}}$, the Poisson adiabatic constant $\gamma = 1.4$. The inlet Mach number is $M_{\mathrm{in}} = 0.012$. In the numerical tests, piecewise quadratic elements ($r = 2$) are used.

Figures 5.30, 5.31, and 5.32 show the computed streamlines at different time instants $t = 29, 31, 33, 34, 36, 37, 39$ms during the fourth period of the motion. In the solution we can observe a flapping jet in the glottis and large vortex formations convected through the domain downstream. The flow field is not periodic and not axisymmetric, although the computational domain is axisymmetric and the motion of the channel walls is periodic and symmetric as well.

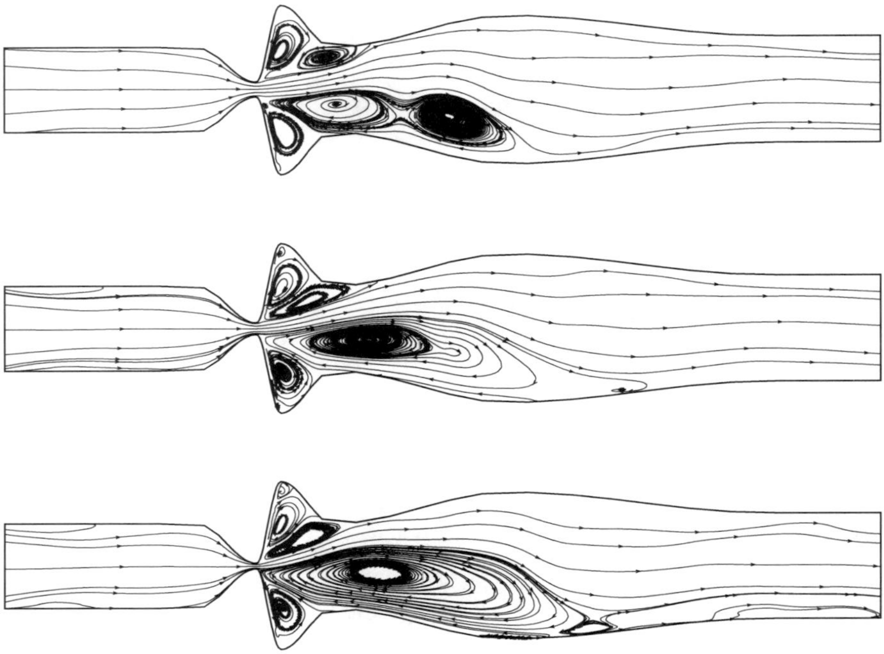

Fig. 5.32 Streamlines at time instants $t = 36, 37, 39$ ms

5.10.2 *Interaction of Compressible Flow with a Model of Elastic Vocal Folds*

We consider the model of flow through a channel with two bumps which represent time-dependent boundaries between the flow and a simplified model of vocal folds (see Figs. 5.33, 5.34). The numerical experiments were carried out for the following data: magnitude of the inlet velocity $v_{\text{in}} = 4\,\text{m}\,\text{s}^{-1}$, the viscosity $\mu = 15 \cdot 10^{-6}\,\text{kg}\,\text{m}^{-1}\,\text{s}^{-1}$, the inlet fluid density $\rho_{\text{in}} = 1.225\,\text{kg}\,\text{m}^{-3}$, the outlet pressure $p_{\text{out}} = 97{,}611\,\text{Pa}$, the Reynolds number $Re = \rho_{\text{in}} v_{\text{in}} H/\mu = 5{,}227$, heat conduction coefficient $k = 2.428 \cdot 10^{-2}\,\text{kg}\,\text{m}\,\text{s}^{-2}\,\text{K}^{-1}$, the specific heat $c_v = 721.428\,\text{m}^2\,\text{s}^{-2}\,\text{K}^{-1}$, the Poisson adiabatic constant $\gamma = 1.4$. The inlet Mach number is $M_{\text{in}} = 0.012$. The parameter of the computational accuracy of the GMRES solver was 10^{-10}. The Young modulus and the Poisson ratio of the structure have the values $E^s = 25{,}000\,\text{Pa}$ and $\sigma^s = 0.4$, respectively, the structural damping coefficient is equal to the constant $C = 100\,\text{s}^{-1}$ and the material density $\rho^s = 1{,}040\,\text{kg}\,\text{m}^{-3}$. The artificial Young modulus $E^a = 10{,}000$ and the Poisson ratio $\sigma^a = 0.45$. The used time step was $8 \cdot 10^{-6}$ s.

We present here the flow-induced deformations of the vocal folds model. The character of the vocal folds vibrations can be indicated in Figs. 5.35 and 5.36,

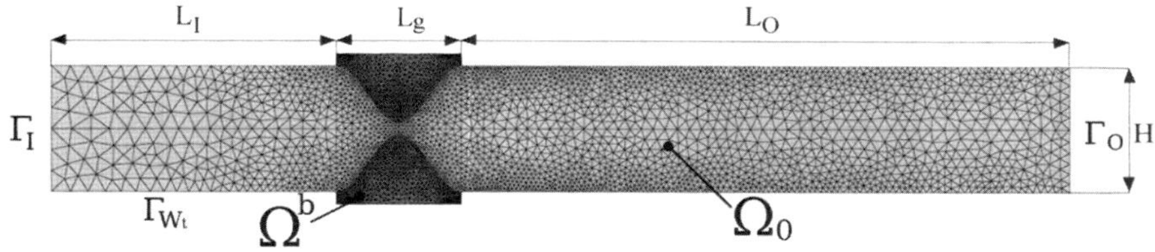

Fig. 5.33 Computational domain at time $t = 0$ with a finite element mesh and the description of its size: $L_I = 50\,\mathrm{mm}$, $L_g = 15.4\,\mathrm{mm}$, $L_O = 94.6\,\mathrm{mm}$, $H = 16\,\mathrm{mm}$. The width of the channel in the narrowest part is 1.6 mm

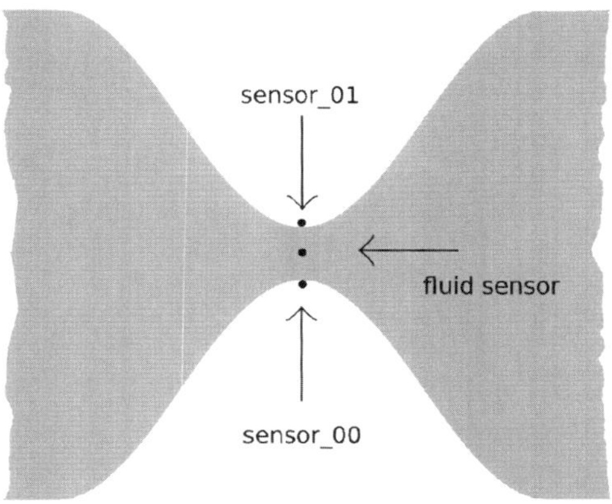

Fig. 5.34 Allocation of the sensors

which show the displacements d_x and d_y of the sensor points on the vocal folds surface (marked in Fig. 5.34) in the horizontal and vertical directions, respectively. Moreover, the fluid pressure fluctuations in the middle of the gap as well as the Fourier analysis of the signals are shown here. The vocal folds vibrations are not fully symmetric due to the "Coanda effect" (a flapping jet—see [57]) and are composed of the fundamental horizontal mode of vibration with the corresponding frequency 113 Hz and by the higher vertical mode with the frequency 439 Hz. The increase of vertical vibrations due to the aeroelastic instability of the system results in a fast decrease of the glottal gap. At about $t = 0.2\,\mathrm{s}$, when the gap is nearly closed, the fluid mesh deformation in this region is too high and the numerical simulation stopped. The dominant peak at 439 Hz in the spectrum of the pressure signal corresponds well to the vertical oscillations of the glottal gap, while the influence of the lower frequency 113 Hz associated with the horizontal vocal folds motion is in the pressure fluctuations negligible. The modelled flow-induced instability of the vocal folds is called phonation onset followed in reality by a complete closing of the glottis and consequently by the vocal folds collisions producing a source acoustic signal for voicing.

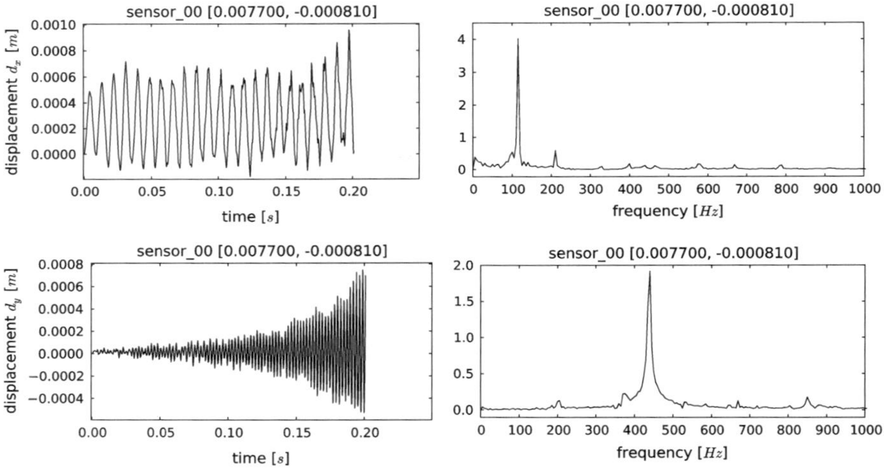

Fig. 5.35 Vibrations of sensor point 00 on the vocal folds in horizontal and vertical directions, and their Fourier analysis

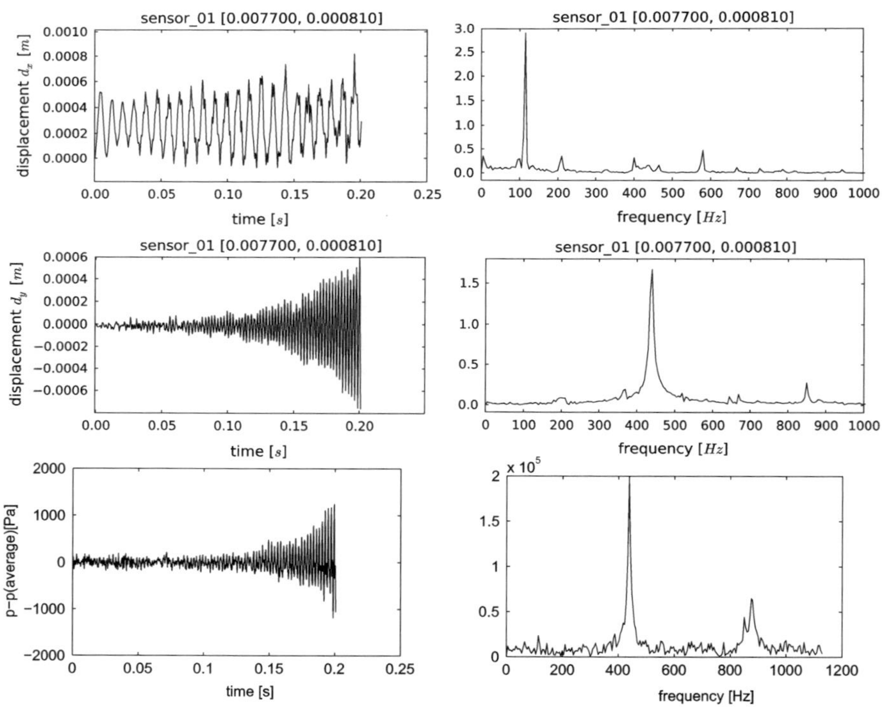

Fig. 5.36 Vibrations of sensor point 01 on the vocal folds in horizontal and vertical directions, and the fluid pressure fluctuations in the middle of the gap, and their Fourier analysis

5.11 Conclusion

5.11.1 Discussion of the Results

The coupled FSI problem of air flow through a vibrating glottal region was numerically analyzed and the comparison to the relevant results obtained by the theoretical analysis in [39] was presented. An attention was paid to the investigation of the flutter boundary for which self-sustained vibrations of vocal folds occur: the physical meaning of such instability is the so-called phonation onset which is an important voice production characteristic in humans.

In the case of the incompressible flow model, the aeroelastic response was studied in dependence on the type of the inlet boundary condition used. The inlet pressure and inlet velocity formulations were used, and the numerical results for the case with the same geometrical shape and the same structural parameters were compared. For the inlet pressure boundary condition the vibrations of the vocal fold were strongly damped for all values of the inlet pressure in the range corresponding to physiological flow rates and the self-oscillations were not observed in the considered range of the flow rates. In the same range of the flow rates with the prescribed inlet velocity the self-sustained oscillations were obtained. The results show that the presence of self-oscillations of vocal folds is influenced by the prescribed inlet boundary condition. This is extremely important as the inlet and outlet are the "artificial boundaries," where the values of aerodynamic quantities are only known approximately.

As a next part, we have been concerned with the numerical simulation of the interaction of fluid flow with an elastic body. The fluid flow was described by the incompressible or compressible Navier–Stokes equations formulated in the ALE form and the deformation of the elastic body was modelled by the dynamic elasticity system with the generalized Hooke law. The coupling of both systems was realized via suitable transmission conditions. The numerical methods for solving the problem have been developed and they were applied to the simulation of air flow in human vocal folds. Results of such computations can also be used as an acoustic source term in modelling of voicing, cf. [73], particularly at the phonation onset.

An efficient numerical method for the solution of the compressible Navier–Stokes equations in time-dependent domains is based on several important ingredients:

- the formulation of the Navier–Stokes system with the use of the ALE method,
- the application of the discontinuous Galerkin method for the space discretization,
- special treatment of boundary conditions,
- semi-implicit linearized time discretization,
- suitable limiting of the order of accuracy in the vicinity of discontinuities or steep gradients.

Numerical tests proved that the developed method is practically unconditionally stable. This means that the length of the time step is limited only by the requirement of the accuracy in time resolution. Moreover, it is robust with respect to the magnitude of the Mach number.

5.11.2 Open Problems and Topics for Future Work

Future work in computer modelling of voice production could be concentrated mainly on the following issues:

- Numerical simulation of the complete closure of the channel during the vocal folds collision including the remeshing of the flow region inside the closing glottis.
- Further analysis of various treatments of boundary conditions at the inlet and outlet of the channel modelling the vocal tract.
- Consideration of a real geometry of the vocal tract for specific vowels and consonants.
- Simulation of the full fluid–structure–acoustic interaction in a complete model consisting of the subglottal spaces, the vibrating vocal folds and the acoustic spaces of the supraglottal part of human vocal tract—numerical simulation of phonation and the human voice production in general.
- Consideration of nonlinear elasticity for large deformations of the vocal fold tissue.
- Modelling of contacts during vocal folds collisions and resulting stresses in the soft tissue of the vocal folds important for estimation of fatigue loading of the vocal folds during voicing.
- 3D modelling of problems in biomechanics of voice.

Acknowledgements This research was supported by grants No. 13-00522S and No. P101/11/0207 of the Czech Science Foundation. M. Feistauer acknowledges his membership in the Nečas Center for Mathematical Modeling (http://ncmm.karlin.mff.cuni.cz).

References

1. S. Adachi, H. Takemoto, T. Kitamura, P. Mokhtari, K. Honda, Vocal tract length perturbation and its application to male-female vocal tract shape conversion. J. Acoust. Soc. Am. **121**, 3874–3885 (2007)
2. F. Alipour, R.C. Scherer, Flow separation in a computational oscillating vocal fold model. J. Acoust. Soc. Am. **116**(3), 1710–1719 (2004)
3. F. Alipour, C. Fan, R.C. Scherer, A numerical simulation of laryngeal flow in a forced-oscillation glottal model. Comput. Speech Lang. **10**, 1637–1644 (1996)
4. F. Alipour, C. Brucker, D. Cook, A. Gommel, M. Kaltenbacher, W. Mattheus, Mathematical models and numerical schemes for the simulation of human phonation. Curr. Bioinform. **6**(3), 323–343 (2011)

5. P. Alku, J. Horáček, M. Airas, F. Griffond-Boitier, A.M. Laukkanen, Performance of glottal inverse filtering as tested by aeroelastic modelling of phonation and FE modelling of vocal tract. Acta Acust. United Acust. **92**(5), 717–724 (2006)
6. D.N. Arnold, An interior penalty finite element method with discontinuous elements. SIAM J. Numer. Anal. **19**, 742–760 (1982)
7. D.N. Arnold, F. Brezzi, B. Cockburn, D. Marini, Discontinuos Galerkin methods for elliptic problems, in *Discontinuous Galerkin Methods. Theory, Computation and Applications*, ed. by B. Cockburn et al. Lecture Notes in Computational Science and Engineering, vol. 11 (Springer, Berlin, 2000), pp. 89–101
8. D.N. Arnold, F. Brezzi, B. Cockburn, D. Marini, Unified analysis of discontinuous Galerkin methods for elliptic problems. SIAM J. Numer. Anal. **39**, 1749–1779 (2001)
9. Y. Bae, Y.J. Moon, Computation of phonation aeroacoustics by an INS/PCE splitting method. Comput. Fluids **37**(10), 1332–1343 (2008)
10. R.J. Baken, R.F. Orlikoff, *Clinical Measurement of Speech and Voice*, 2nd edn. (Singular Publishing Group, San Diego, 2000)
11. F. Bassi, S. Rebay, High-order accurate discontinuous finite element solution of the 2D Euler equations. J. Comput. Phys. **138**, 251–285 (1997)
12. C.E. Baumann, J.T. Oden, A discontinuous *hp* finite element method for the Euler and Navier-Stokes equations. Int. J. Numer. Methods Fluids **31**, 79–95 (1999)
13. S. Becker, S. Kniesburges, S. Mller, A. Delgado, G. Link, M. Kaltenbacher, Flow-structure-acoustic interaction in a human voice model. J. Acoust. Soc. Am. **125**(3), 1351–1361 (2009)
14. J. Brandts, M. Křížek, Gradient superconvergence on uniform simplicial partitions of polytopes. IMA J. Numer. Anal. **23**(3), 489–505 (2003)
15. R. Brepta, L. Půst, F. Turek, *Mechanical Vibrations* (Sobotáles, Praha, 1994, in Czech)
16. Ch.-H. Bruneau, P. Fabrie, Effective downstream boundary conditions for incompressible Navier–Stokes equations. Int. J. Numer. Methods Fluids **19**(8), 693–705 (1994)
17. J. Česenek, M. Feistauer, J. Horáček, V. Kučera, J. Prokopová, Simulation of compressible viscous flow in time-dependent domains. Appl. Math. Comput. **13**, 7139–7150 (2013)
18. A. Curnier, *Computational Methods in Solid Mechanics* (Kluwer Academic Publishing Group, Dodrecht, 1994)
19. T.A. Davis, I.S. Duff, A combined unifrontal/multifrontal method for unsymmetric sparse matrices. ACM Trans. Math. Softw. **25**, 1–19 (1999)
20. A. de Boer, M.S. van der Schoot, H. Bijl, Mesh deformation based on radial basis function interpolation. Comput. Struct. **85**, 784–795 (2007)
21. R.M. de Oliviera, J. Pereira, M. Grellet, A. Alwan, A contribution to simulating a three-dimensional larynx model using the finite element method. J. Acoust. Soc. Am. **114**(5), 2893–2905 (2003)
22. M.P. de Vries, H.K. Shutte, A.E.P. Veldman, G.J. Verkerke, Glottal flow through a two-mass model: comparison of Navier–Stokes solutions with simplified models. J. Acoust. Soc. Am. **111**(4), 1847–1853 (2002)
23. V. Dolejší, Semi-implicit interior penalty discontinuous Galerkin methods for viscous compressible flows. Commun. Comput. Phys. **4**, 231–274 (2008)
24. V. Dolejší, M. Feistauer, A semi-implicit discontinuous Galerkin finite element method for the numerical solution of inviscid compressible flow. J. Comput. Phys. **198**, 727–746 (2004)
25. V. Dolejší, M. Feistauer, Error estimates of the discontinuous Galerkin method for nonlinear nonstationary convection-diffusion problems. Numer. Funct. Anal. Optim. **26**, 349–383 (2005)
26. V. Dolejší, M. Feistauer, C. Schwab, On some aspects of the discontinuous Galerkin finite element method for conservation laws. Math. Comput. Simul. **61**, 333–346 (2003)
27. B. Erath, M. Plesniak, An investigation of asymmetric flow features in a scaled-up driven model of the human vocal folds. Exp. Fluids **49**(1), 131–146 (2010)
28. G. Fant, *Acoustic Theory of Speech Production*, 2nd edn. (Mouton, S'Gravenage, 1960)
29. M. Feistauer, *Mathematical Methods in Fluid Dynamics* (Longman Scientific & Technical, Harlow, 1993)

30. M. Feistauer, V. Kučera, On a robust discontinuous Galerkin technique for the solution of compressible flow. J. Comput. Phys. **224**, 208–221 (2007)
31. M. Feistauer, J. Felcman, J. Straškraba, *Mathematical and Computational Methods for Compressible Flow* (Clarendon Press, Oxford, 2003)
32. M. Feistauer, V. Dolejší, V. Kučera, On the discontinuous Galerkin method for the simulation of compressible flow with wide range of mach numbers. Comput. Vis. Sci. **10**, 17–27 (2007)
33. M. Feistauer, J. Hasnedlová-Prokopová, J. Horáček, A. Kosík, V. Kučera, DGFEM for dynamical systems describing interaction of compressible fluid and structures. J. Comput. Appl. Math. **254**, 17–30 (2013)
34. T. Gelhard, G. Lube, M.A. Olshanskii, J.-H. Starcke, Stabilized finite element schemes with LBB-stable elements for incompressible flows. J. Comput. Appl. Math. **177**, 243–267 (2005)
35. V. Girault, P.-A. Raviart, *Finite Element Methods for Navier-Stokes Equations* (Springer, Berlin, 1986)
36. P.M. Gresho, R.L. Sani, *Incompressible Flow and the Finite Element Method* (Wiley, Chichester, 1998)
37. J. Hasnedlová-Prokopová, M. Feistauer, J. Horáček, A. Kosík, V. Kučera, Numerical simulation of fluid-structure of compressible flow and elastic structure. Computing **95**, 573–585 (2013)
38. J.G. Heywood, R. Rannacher, S. Turek, Artificial boundaries and flux and pressure conditions for the incompressible Navier-Stokes equations. Int. J. Numer. Math. Fluids **22**, 325–352 (1992)
39. J. Horáček, J.G. Švec, Aeroelastic model of vocal-fold-shaped vibrating element for studying the phonation threshold. J. Fluids Struct. **16**(7), 931–955 (2002)
40. J. Horáček, J.G. Švec, Instability boundaries of a vocal fold modelled as a flexibly rigid body vibrating in a channel conveying fluid. Am. Soc. Mech. Eng. Appl. Mech. Div. AMD **253**(2), 1043–1054 (2002)
41. J. Horáček, P. Šidlof, J.G. Švec, Numerical simulation of self-oscillations of human vocal folds with Hertz model of impact forces. J. Fluids Struct. **20**(6), 853–869 (2005)
42. J. Horáček, A.M. Laukkanen, P. Šidlof, Estimation of impact stress using an aeroelastic model of voice production. Logoped. Phoniatr. Vocol. **37**, 185–192 (2007)
43. J. Horáček, A.M. Laukkanen, P. Šidlof, J.G. Švec, Comparison of acceleration and impact stress as possible loading factors in phonation: a computer modeling study. Folia Phoniatr. Logop. **61**(3), 137–145 (2009)
44. J. Horáček, V. Uruba, V. Radolf, J. Veselý, V. Bula, Airflow visualization in a model of human glottis near the self-oscillating vocal folds model. Appl. Comput. Mech. **5**, 21–28 (2011)
45. K. Ishizaka, J.L. Flanagan, Synthesis of voiced sounds from a two-mass model of the vocal coords. Bell Syst. Tech. J. **51**, 1233–1268 (1972)
46. S. Kniesburges, S.L. Thomson, A. Barney, M. Triep, P. Šidlof, J. Horáček, In vitro experimental investigation of voice production. Curr. Bioinform. **6**(3), 305–322 (2011)
47. A. Kosík, M. Feistauer, J. Horáček, P. Sváček, Numerical simulation of interaction of an elastic body and fluid flow, in *Proceedings of the Interaction and Feedbacks 2010* (Institute of Thermomechanics, Prague, 2010), pp. 49–56
48. A. Kosík, M. Feistauer, J. Horáček, P. Sváček, Numerical simulation of interaction of human vocal folds and fluid flow, in *Vibration Problems ICOVP 2011*, ed. by J. Náprstek, J. Horáček, M. Okrouhlík (Springer, Berlin, 2011), pp. 765–771
49. M. Krane, M. Barry, T. Wei, Unsteady behavior of flow in a scaled-up vocal folds model. J. Acoust. Soc. Am. **122**(6), 3659–3670 (2007)
50. F. Krebs, F. Silva, D. Sciamarella, G. Artana, A three-dimensional study of the glottal jet. Exp. Fluids **52**, 1–15 (2002)
51. B.R. Kucinschi, R.C. Scherer, K.J. DeWitt, T.T.M. Ng, An experimental analysis of the pressures and flows within a driven mechanical model of phonation. J. Acoust. Soc. Am. **119**(5), 3011–3021 (2006)
52. G. Link, M. Kaltenbacher, M. Breuer, M. Döllinger, A 2D finite element scheme for fluid-solid-acoustic interactions and its application to human phonation. Comput. Methods Appl. Mech. Eng. **198**, 3321–3334 (2009)

53. G. Lube, Stabilized Galerkin finite element methods for convection dominated and incompressible flow problems. Numer. Anal. Math. Model. **29**, 85–104 (1994)
54. J. Lucero, Dynamics of the two-mass model of the vocal folds: equilibria, bifurcations, and oscillation region. J. Acoust. Soc. Am. **94**(6), 3104–3111 (1993)
55. H. Luo, R. Mittal, X. Zheng, S.A. Bielamowicz, R.J. Walsh, J.K. Hahn, An immersed boundary method for flow-structure interaction in biological systems with application to phonation. J. Comput. Phys. **227**(22), 9303–9332 (2008)
56. W. Mattheus, C. Brücker, Asymmetric glottal jet deflection: differences of two and three-dimensional models. J. Acoust. Soc. Am. **130**(6), EL3739 (2011)
57. R. Mittal, B.D. Erath, M.W. Plesniak, Fluid dynamics of human phonation and speech. Ann. Rev. Fluid Mech. **45**, 437–467 (2013)
58. J. Neubauer, Z. Zhang, R. Miraghaie, D. Berry, Coherent structures of the near field flow in a self-oscillating physical model of the vocal folds. J. Acoust. Soc. Am. **121**(2), 1102–1118 (2007)
59. T. Nomura, T.J.R. Hughes, An arbitrary Lagrangian-Eulerian finite element method for interaction of fluid and a rigid body. Comput. Methods Appl. Mech. Eng. **95**, 115–138 (1992)
60. M.P. Norton, *Fundamentals of Noise and Vibration Analysis for Engineers* (Cambridge University Press, Cambridge, 1989)
61. X. Pelorson, A. Hirschberg, R. van Hassel, A. Wijnands, Y. Auregan, Theoretical and experimental study of quasisteady-flow separation within the glottis during phonation: application to a modified two-mass model. J. Acoust. Soc. Am. **96**(6), 3416–3431 (1994)
62. P. Pořízková, K. Kozel, J. Horáček, Numerical simulation of unsteady compressible flow in convergent channel: pressure spectral analysis. J. Appl. Math. **2012**, 9 pp. Article ID 545120 (2012)
63. P. Pořízková, K. Kozel, J. Horáček, Flows in convergent channel: comparison of numerical results of different mathematical models. Computing **95**, 573–585 (2013)
64. P. Punčochářová, J. Fürst, K. Kozel, J. Horáček, Numerical solution of compressible flow with low Mach number through oscillating glottis, in *Proceedings of the 9th International Conference on Flow-Induced Vibration (FIV 2008)* (Institute of Thermomechanics AS CR, Prague, 2008), pp. 135–140
65. P. Punčochářová-Pořízková, J.Fürst, J. Horáček, K. Kozel, Numerical solutions of unsteady flows with low inlet Mach numbers. Math. Comput. Simul. **80**(8), 1795–1805 (2010)
66. P. Punčochářová-Pořízková, K. Kozel, J. Horáček, Simulation of unsteady compressible flow in a channel with vibrating walls influence of the frequency. Comput. Fluids **46**(1), 404–410 (2011)
67. C. Renotte, V. Bouffioux, F. Wilquem, Numerical 3D analysis of oscillatory flow in the time-varying laryngeal channel. J. Biomech. **33**(12), 1637–1644 (2000)
68. D. Sciamarella, C. dAlessandro, On the acoustic sensitivity of a symmetrical two-mass model of the vocal folds to the variation of control parameters. Acta Acust. United Acust. **90**, 746–761 (2004)
69. D. Sciamarella, P.L. Qur, Solving for unsteady airflow in a glottal model with immersed moving boundaries. Eur. J. Mech. B/Fluids **27**, 42–53 (2008)
70. J.H. Seo, R. Mittal, A high-order immersed boundary method for acoustic wave scattering and low-Mach number flow-induced sound in complex geometries. J. Comput. Phys. **230**(4), 1000–1019 (2011)
71. P. Šidlof, J.G. Švec, J. Horáček, J. Veselý, I. Klepáček, R. Havlík, Geometry of human vocal folds and glottal channel for mathematical and biomechanical modeling of voice production. J. Biomech. **41**, 985–995 (2008)
72. P. Šidlof, O. Doaré, O. Cadot, A. Chaigne, Measurement of flow separation in a human vocal folds model. Exp. Fluids **51**(1), 123–136 (2011)
73. P. Šidlof, J. Horáček, V. Řidký, Parallel CFD simulation of flow in a 3D model of vibrating human vocal folds. Comput. Fluids **80**, 290–300 (2013)
74. P. Šidlof, S. Zörner, A. Hüppe, Numerical simulation of flow-induced sound in human voice production, Procedia Eng. **61**(0), 333–340 (2013)

75. B.H. Story, Technique for 'tuning' vocal tract area functions based on acoustic sensitivity functions. J. Acoust. Soc. Am. **119**, 715–718 (2006)
76. B. Story, I. Titze, Voice simulation with a body-cover model of the vocal folds. J. Acoust. Soc. Am. **97**(2), 1249–1260 (1995)
77. B.H. Story, I.R. Titze, Parameterization of vocal tract area functions by empirical orthogonal modes. J. Phon. **26**, 223–260 (1998)
78. B.H. Story, I.R. Titze, E.A. Hoffman, Vocal tract area functions from magnetic resonance imaging. J. Acoust. Soc. Am. **100**, 537–554 (1996)
79. J. Suh, S. Frankel, Numerical simulation of turbulence transition and sound radiation for flow through a rigid glottal model. J. Acoust. Soc. Am. **121**(6), 3728–3739 (2007)
80. J. Sundberg, *The Science of the Singing Voice* (Northern Illinois University Press, DeKalb, 1987)
81. P. Sváček, Numerical approximation of flow induced vibrations of channel walls. Comput. Fluids **46**(1), 448–454 (2011)
82. P. Sváček, M. Feistauer, Application of a stabilized FEM to problems of aeroelasticity, in *Numerical Mathematics and Advanced Application* (Springer, Berlin, 2004), pp. 796–805
83. P. Sváček, J. Horáček, Numerical approximation of flow induced vibration of vocal folds, in *BAIL 2010 - Boundary and Interior Layers, Computational and Asymptotic Methods* (Springer, Berlin, 2011), pp. 227–234
84. P. Sváček, J. Horáček, Numerical simulation of glottal flow in interaction with self oscillating vocal folds: comparison of finite element approximation with a simplified model. Commun. Comput. Phys. **12**(3), 789–806 (2012)
85. P. Švancara, J. Horáček, Numerical modelling of effect of tonsillectomy on production of Czech vowels. Acta Acust. United Acust. **92**(5), 681–688 (2006)
86. P. Švancara, J. Horáček, V. Hruza, FE modelling of the fluid-structure-acoustic interaction for the vocal folds self-oscillation, in *Vibration Problems ICOVP 2011*, ed. by J. Náprstek, J. Horáček, M. Okrouhlík (Springer, Berlin, 2011), pp. 801–807
87. P. Švancara, J. Horáček, J.G. Švec, Simulation of the self-oscillations of the vocal folds and of the resulting acoustic phenomena in the vocal tract, in *Advances in Mechanisms Design*, vol. 8, ed. by J. Beran, M. Bílek, M. Hejnová, P. Žabka (Springer, Dordrecht, 2012), pp. 357–363
88. C. Tao, J. Jiang, Mechanical stress during phonation in a self-oscillating finite element vocal fold model. J. Biomech. **40**(10), 2191–2198 (2007)
89. C. Tao, Y. Zhang, D. Hottinger, J. Jiang, Asymmetric airflow and vibration induced by the Coanda effect in a symmetric model of the vocal folds. J. Acoust. Soc. Am. **122**(4), 2270–2278 (2007)
90. S. Thomson, L. Mongeau, S. Frankel, Aerodynamic transfer of energy to the vocal folds. J. Acoust. Soc. Am. **118**(3 Pt 1), 1689–1700 (2005)
91. S. Thomson, J. Tack, G. Verkerke, A numerical study of the flow-induced vibration characteristics of a voice-producing element for laryngectomized patients. J. Biomech. **40**, 3598–3606 (2007)
92. I.R. Titze, *Principles of Voice Production* (Prentice-Hall, Englewood Cliffs, 1994)
93. I.R. Titze, *The Myoelastic Aerodynamic Theory of Phonation* (National Center for Voice and Speech, Denver, 2006)
94. I.R. Titze, Nonlinear source-filter coupling in phonation: theory. J. Acoust. Soc. Am. **123**, 2733–2749 (2008)
95. M. Triep, C. Brücker, W. Schröder, High-speed PIV measurements of the flow downstream of a dynamic mechanical model of the human vocal folds. Exp. Fluids **39**, 232–245 (2005)
96. T. Vampola, J. Horáček, J. Vokřál, FE modeling of human vocal tract acoustics. Part II: influence of velopharyngeal insufficiency on phonation of vowels. Acta Acust. United Acust. **94**, 448–460 (2008)
97. T. Vampola, J. Horáček, J. Švec, FE modeling of human vocal tract acoustics. Part I: production of Czech vowels. Acta Acust. United Acust. **94**, 433–447 (2008)
98. J.J.W. van der Vegt, H. van der Ven, Space-time discontinuous Galerkin finite element method with dynamic grid motion for inviscid compressible flow. J. Comput. Phys. **182**, 546–585 (2002)

99. R. Verfürth, Error estimates for mixed finite element approximation of the Stokes equations. R.A.I.R.O. Anal. Numer. Anal. **18**, 175–182 (1984)
100. G. Vijayasundaram, Transonic flow simulation using upstream centered scheme of Godunov type in finite elements. J. Comput. Phys. **63**, 416–433 (1986)
101. Z. Yang, D.J. Mavriplis, Unstructured dynamic meshes with higher-order time integration schemes for the unsteady Navier-Stokes equations, in *43rd AIAA Aerospace Sciences Meeting*, Reno (January 2005), 13 pp. (AIAA Paper 2005-1222)
102. A. Yang, J. Lohscheller, D.A. Berry, S. Becker, U. Eysholdt, D. Voigt, Biomechanical modeling of the three-dimensional aspects of human vocal fold dynamics. J. Acoust. Soc. Am. **127**(2), 1014–1031 (2010)
103. Z. Zhang, Influence of flow separation location on phonation onset. J. Acoust. Soc. Am. **124**(3), 1689–1694 (2008)
104. Z. Zhang, J. Neubauer, D. Berry, Physical mechanisms of phonation onset: a linear stability analysis of an aeroelastic continuum model of phonation. J. Acoust. Soc. Am. **122**, 2279–2295 (2007)
105. W. Zhao, C. Zhang, S. Frankel, L. Mongeau, Computational aeroacoustics of phonation, part I: computational methods and sound generation mechanisms. J. Acoust. Soc. Am. **112**, 2134–2146 (2002)
106. X. Zheng, S. Bielamowicz, H. Luo, R. Mittal, A computational study of the effect of false vocal folds on glottal flow and vocal fold vibration during phonation. Ann. Biomed. Eng. **37**(3), 625–642 (2009)
107. X. Zheng, Q. Xue, R. Mittal, S. Bielamowicz, A coupled sharp-interface immersed boundary-finite-element method for flow-structure interaction with application to human phonation. J. Biomech. Eng. **132**(11), 111003 (2010)
108. X. Zheng, R. Mittal, S. Bielamowicz, A computational study of asymmetric glottal jet deflection during phonation. J. Acoust. Soc. Am. **129**(4), 2133–2143 (2011)
109. X. Zheng, R. Mittal, Q. Xue, S. Bielamowicz, Direct-numerical simulation of the glottal jet and vocal-fold dynamics in a three-dimensional laryngeal model. J. Acoust. Soc. Am. **130**(1), 404–415 (2011)
110. O.C. Zienkiewicz, J.Z. Zhu, The superconvergent patch recovery and a posteriori error estimates. Part 1: the recovery technique. Int. J. Numer. Methods Eng. **33**, 1331–1364 (1992)

M. Feistauer (✉)
Faculty of Mathematics and Physics, Department of Numerical Mathematics, Charles University in Prague, Prague, Czech Republic
e-mail: feist@karlin.mff.cuni.cz

P. Sváček
Faculty of Mechanical Engineering, Department of Technical Mathematics, Czech Technical University in Prague, Prague, Czech Republic
e-mail: Petr.Svacek@fs.cvut.cz

J. Horáček
Institute of Thermomechanics, Academy of Sciences of the Czech Republic, Prague, Czech Republic
e-mail: jaromirh@it.cas.cz

Chapter 6
Data Assimilation in Cardiovascular Fluid–Structure Interaction Problems: An Introduction

Luca Bertagna, Marta D'Elia, Mauro Perego, and Alessandro Veneziani

Abstract Numerical methods for incompressible fluid dynamics have recently received a strong impulse from the applications to the cardiovascular system. In particular, fluid–structure interaction methods have been extensively investigated in view of an accurate and possibly fast simulation of blood flow in arteries and veins. This has been strongly motivated by the progressive interest in using numerical tools not only for understanding the general physiology and pathology of the vascular system. The opportunity offered by medical images properly preprocessed and elaborated to simulate blood flow in real patients highlighted the potential impact of scientific computing on the clinical practice. Therefore, *in silico* experiments are currently extensively used in bioengineering for completing (and sometimes driving) more traditional in vivo and in vitro investigations. Parallel to the development of numerical models, the need for quantitative analysis for diagnostic purposes has strongly stimulated the design of new methods and instruments for measurements and imaging. Thanks to these developments, a huge amount of data is nowadays available. *Data Assimilation* is the accurate merging of measures (including images) and numerical simulations for a mathematically sound integration of different sources of information. The outcome of this process includes both the patient-specific measures and the general principles underlying the development of mathematical models. In this way, simulations are adapted to the availability of individual data and are therefore supposed to be more reliable; measures are correspondingly filtered by the mathematical models assumed to describe the underlying phenomena, resulting in a (hopefully) significant reduction of the noise.

This chapter provides an introduction to methods for data assimilation, mostly developed in fields like meteorology, applied to computational hemodynamics. We focus mainly on two of them: methods based on stochastic arguments (Kalman filtering) and variational methods. We also address some examples that have been approached with different techniques, in particular the estimation of vascular compliance from displacement measures.

T. Bodnár et al. (eds.), *Fluid-Structure Interaction and Biomedical Applications*, Advances in Mathematical Fluid Mechanics, DOI 10.1007/978-3-0348-0822-4_6,

Keywords Biological flows • Data assimilation • Fluid–structure interaction • Kalman filter • Variational parameter identification

MSC2010: 49N45, 35Q35, 74F10, 76Z05, 62M20, 49M05, 49M15, 49M29, 60G25, 60G35, 90C46, 93E10, 93E11, 93E12

6.1 Preliminaries

Numerical methods for incompressible fluid dynamics have recently received a strong impulse from the applications to the cardiovascular system (see, e.g., [25, 30]). In particular, fluid–structure interaction (FSI) methods have been extensively investigated in view of an accurate and possibly fast simulation of blood flow in arteries and veins (see, e.g., the recent works by Y. Maday and by J.F. Gerbeau and M. Fernandez, Chaps. 8 and 9 of [25], respectively, or the chapter of the present book by C. Grandmont, M. Lukáčová, and Š. Nečasová). Beyond the intrinsic mathematical interest, the development of reliable tools for the numerical simulations of cardiovascular problems—and FSI in particular—has an impact on bioengineering and medical research. Thanks to the opportunity offered by improvements in imaging and measurement devices and the subsequent elaboration (see Chap. 4 of [25] authored by L. Antiga, D. Steinman and J. Peiró), nowadays scientific computing is not only a tool for investigating the physiopathology of the cardiovascular system at a general level, but also a way for analyzing in detail the single patient. Mathematical models, properly numerically approximated complete the patient-specific information provided by traditional (yet progressively improved) diagnostic tools. The complete patient-specific picture provided by numerical models may (and most likely will) have a diagnostic and prognostic impact and, more in general, provide a decision-making support in clinical practice. However, this fascinating perspective raises some important challenges. The general problem of quantifying and reducing uncertainty in mathematical models and to certify the quality of numerical simulations—common to any computer aided activity—is even more important when supporting the clinical practice, for its individual and social impact. This is related to the problem of *validating* numerical models. "Validation is the process of determining the extent to which the computer implementation corresponds to the real world. If solution verification has already been demonstrated, then validation asks whether the mathematical model is effective in simulating those aspects of the real world system under study" (from [22]).

Parallel to the development of numerical models, the need for quantitative analysis for diagnostic purposes has strongly stimulated the design of new methods and instruments for measurements and imaging. Thanks to these developments, a huge amount of data is nowadays available to bioengineers and medical doctors. Also, the reliability of these data and their significance in clinical practice needs to undergo a strict analysis and assessment, since they are typically affected by noise and errors.

Data Assimilation (DA) is a process for integrating the knowledge provided by numerical models and measurements with the purpose of improving the reliability of quantitative analysis. This approach has been developed since the mid of the twentieth century having as preferential application the weather forecasting. The rationale is that the predictions provided by numerical models, that we may call a *background knowledge*, being partially based on universal physical and constitutive laws, are affected by uncertainties in real-world problems. These are the consequence of simplifying assumptions as well as of an incomplete knowledge of parameters usually needed by the constitutive laws forming a mathematical model. For instance, referring to biomedical applications, blood viscosity (that in a Newton constitutive law is supposed to be constant) or compliance of an artery (that in a Hookean material is supposed to be represented by a parameter, the Young modulus) is available as estimated on samples, but when dealing with a specific patient they are in general not known, being impossible or inconvenient to measure. The integration with available measures, that we may call a *foreground knowledge*, since they are specific of the case, is expected to be beneficial to the quantitative analysis, reducing the uncertainty in the mathematical models. On the other hand, background models improve the knowledge extracted from the data, providing a way for filtering noise. In particular, this is important for at least three purposes,

1. *estimate* the state of a dynamical system (e.g., the velocity, the pressure) or its derivated quantities for which noisy data and mathematical models are available,
2. *predict* the state of a dynamical system for which data are available in the past,
3. *identify* one or more parameters involved in the mathematical model, adjusting their values on the basis of available data.

In the global picture—that we have represented in Fig. 6.1—DA reduces possibilities of failure in estimating, predicting, and identifying by merging background and foreground in a unique quantitative analysis. The necessity of this process in the traditional development of numerical models in cardiovascular mathematics becomes progressively more urgent with the increment of available data and, more importantly, of patients that may benefit from quantitative analysis.

In this chapter we want to provide an introduction to some topics brought in by DA in Cardiovascular Mathematics, with a particular emphasis to FSI problems. It is important to stress that, as such, this introduction cannot be complete. First, there are several ways for approaching DA and it is basically impossible to provide an exhaustive global picture of the possible methodologies. We refer to [8] as a more general introduction. Second, DA in cardiovascular modeling is a relatively recent topic and many questions and challenges are still open, so it is hard to draw conclusive statements about the adequacy of a methodology for a specific problem. Our perspective is to provide some examples that have been recently considered in the literature and a self-contained introduction to the methods used there. In particular, we have selected examples tackled with different approaches, providing different perspectives for solving the same problem. This is intended to give not only the idea of the complexity of the problems but also of the variety of approaches, the

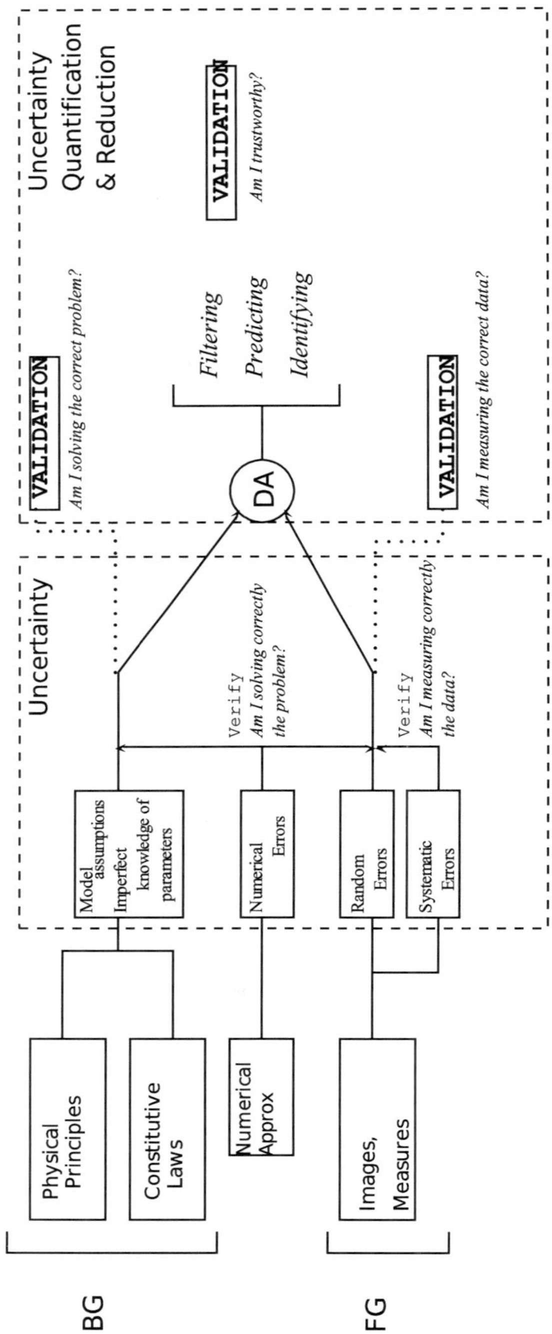

Fig. 6.1 The general framework of Data Assimilation as a process for improving the reliability of quantitative analyses (BG = background, FG = foreground)

differences and the complementary nature of the methods. In particular, we will consider two classes of methods,

1. *stochastic approaches*, when some probabilistic knowledge of the uncertainty affecting the model and the noise affecting the measures are available; in particular we refer to methods related to Kalman filtering and its extension to nonlinear problems; these methods will be addressed in Sect. 6.2;
2. *deterministic approaches*, when no clue on the statistical features of uncertainty is available; in particular, we will see variational methods based on the minimization of the mismatch between the data and numerical results, constrained by the background model; these methods are introduced in Sect. 6.3.

The above distinction is not strict. In fact, available statistical information can be included in variational models.

The FSI problem and more in general the problems involved in cardiovascular mathematics—usually represented by a system of partial differential equations—are complex and per se challenging. In the framework of DA, the issue of computational costs becomes even more important, as DA typically involves the solution of *inverse problems*. In practice, these problems can be solved by iterative approaches, where the solution of the FSI system (or more in general of the "forward" problem) needs to be performed at each iteration. It is promptly realized that this requires specific techniques to make the computational costs affordable. We address this issue in Sect. 6.3, in particular referring to methods for reducing the costs of each iteration by representing the solution on a "smart" low-dimensional basis functions set that strongly reduces the number of degrees of freedom required by a traditional numerical method (like finite element or spectral methods).

Detailed examples are provided in Sect. 6.4. In particular, we consider the assimilation of velocity measures with the numerical simulation of the Navier–Stokes equations for improving the estimate of blood velocity on an artery. We address two different deterministic approaches and how they can be reinterpreted or improved by a stochastic Bayesian perspective. Finally, we present in detail the problem of estimating vascular compliance by solving an inverse FSI problem. Again, we present both a stochastic approach based on Kalman filtering and a deterministic constrained minimization approach. In the latter case, we present a technique for reducing the computational costs by representing the solution on a low-dimensional basis obtained with a Proper Orthogonal Decomposition (POD) approach.

As we have pointed out, the methodological picture in the field of DA is pretty articulated, encompassing statistical as well as numerical issues for inverse problems [17, 71]. Here, we mention some references for the reader interested in having more details on the topics covered only partially in this introduction. The importance of uncertainty quantification in any field of scientific computing has been recently underlined in [22]. An excellent introduction to statistical methods for computational inverse problems is given in the books [13, 28, 47, 68]. A recent collection of contributions in the numerical solution of inverse problems and computational costs reduction is [7].

A classical book on methods for solving constrained minimization for flow problems is [38]. Fundamental contributions can be found in [33, 34], recently collected in [35]. Parameter estimation for partial differential equations has been extensively covered in [3] (see also the recent [2]).

6.1.1 An Introductory Example

To illustrate some basic concepts in DA, we refer to an oversimplified example. Let us assume to have a pipe where an incompressible fluid flows. We also suppose that N measures of velocity are available in N points P_i $(i = 1, \ldots, N)$. Our goal is to compute the shear stress at the wall of the pipe, which is defined as[1]

$$\mu\left(\nabla\mathbf{u} + \nabla\mathbf{u}^T\right)\mathbf{n} - \mu\left(\mathbf{n}\left(\nabla\mathbf{u} + \nabla\mathbf{u}^T\right)\mathbf{n}\right)\mathbf{n} \tag{6.1}$$

where μ is the blood viscosity, $\mathbf{u}$ is the velocity, $\mathbf{n}$ the unit vector normal to the surface. In particular, if we are interested in the estimate of the stress at the times for which velocity measures are available, this is an *estimate* or *filtering problem*.

There are different approaches for this.

1. *Data fitting approach*: a functional form for the velocity is decided (for instance, polynomial) and fitted with the data. Successively the WSS is obtained by applying (6.1) with the fitted velocity. The quality of the computation depends on the number and location of the measures, and the amount of noise. Fitting can be performed with either interpolation or least square approximation depending on how trustworthy the measures are considered. In this approach, we do not assume any background knowledge of fluid mechanics.
2. *Model based computation*: we may assume that blood flow is an incompressible Newtonian fluid. Under these assumptions, for a cylindrical pipe described by the coordinates z, r, θ, we can derive the Poiseuille solution (outlined in color in Fig. 6.2),

$$u_z = \frac{G_P}{4\mu}(R^2 - r^2), \quad u_r = u_\theta = 0, \quad p = G_P z + C \tag{6.2}$$

where G_P is the pressure gradient, R is the radius of the pipe, and C is an arbitrary constant. Should G_P and μ be available, we compute the WSS; such parameters are needed to get the final estimate of the stress, while measures are not needed. The quality of the estimate depends on the reliability of the model assumption. In general, the analytical solution may be replaced by a numerical

[1]We remind that the wall shear stress (WSS) is a quantity of great relevance in biomedical applications for its correlation with pathologies such as atherosclerosis—see, e.g., [14].

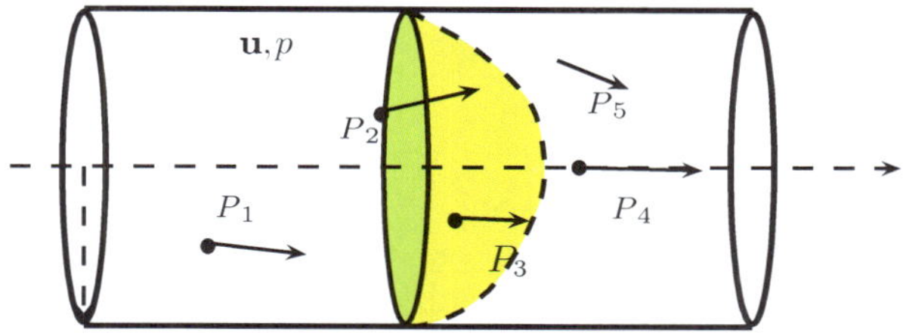

Fig. 6.2 Pipe where an incompressible fluid flows and velocity measures are available in the points P_i: how is it possible to reliably compute the wall shear stress at the wall?

solution. In such case, the accuracy of the estimated stress will also depend on the numerical approximation.

3. *Data assimilation procedure*: suppose that the assumptions behind the Poiseuille solution are acceptable but our knowledge is incomplete, for instance the viscosity μ and the pressure gradient G_P are unknown; in this case, we may take advantage of the velocity measures to fill the gap and eventually to compute the WSS by formulating the following problem. Find μ and G_P such that $\mathbf{u}$ minimizes the mismatch

$$\mathcal{J} = \sum_{i=1}^{N} \left(\mathbf{u}_m(P_i) - \mathbf{u}_p(P_i)\right)^2$$

where $\mathbf{u}_m(P_i)$ is the measured velocity and $\mathbf{u}_p$ is the Poiseuille solution (6.2). In this way, we are fitting the physical parameters μ and G_P so that the background model is matching the foreground knowledge. Once μ and G_P are computed, the WSS (both as an estimate or as a prediction) is quantified. Contextually, the noise affecting the data is filtered by our background knowledge of fluid mechanics in the physically driven least squares procedure. Notice that when quantifying the viscosity we are solving an *identification problem*.

In the more realistic case that the Poiseuille solution cannot be applied, we replace $\mathbf{u}_p$ with the (numerical) solution of the Navier–Stokes equations. In this case, the minimization procedure requiring to find the stationary points of $\mathcal{J}$ regarded as a function of μ and G_P is clearly nontrivial (as we will see in the next sections).

This simple example (that will be developed in Sect. 6.4) shows the relevance of DA in biomedical applications, especially related to the clinical practice. As a matter of fact, patient-specific knowledge of parameters that form a mathematical/numerical model is always incomplete. As for the boundary conditions, this has led to extensive investigation of the so-called defective boundary data problems (see, for instance, [24, 26]). Concerning parameter identification, we mention *elastography* as a method for detecting the rigidity of soft tissues by solving inverse elasticity problems [4, 5]. In this case, parameter identification is not only functional to the computation of a specific variable of interest, but it is by itself an important procedure for diagnostic purposes (e.g., breast cancer).

In the previous example, the DA procedure is performed without any real assumptions on the quantity we want to estimate and on the noise affecting the measures. However, in many cases some a priori knowledge is available and may be used to drive the assimilation process and eventually reduce the uncertainty affecting the final results. For instance, we speculate that fluid viscosity is a Gaussian variable whose average and variance are known. Similarly, we may assume that the noise features a probabilistic density function whose moments are available. Availability of trustworthy probabilistic information on quantities of interest and noise may be a discriminant for the choice of the DA methods. In Sect. 6.2 we address probabilistic approaches, while Sect. 6.3 is devoted to deterministic methods. As already pointed out, the two classes of approaches are somehow contiguous. As we will see in Sect. 6.4, solution obtained with one approach can be reinterpreted in the other framework, when the reliability of a priori *knowledge* tends to 0.

6.2 Probabilistic Approach

In this section, we consider the estimation/prediction/identification of quantities when we assume stochastic a priori information to be available. We may take therefore advantage of this knowledge to integrate models and measures. The latter are in turn considered to be the realization of a stochastic process, whose features are known.

The ingredients of the problem (see Fig. 6.3) are: (1) a variable $\mathbf{w}$—for generality we assume it is an n-dimensional random vector, whose probability density function (see below) is known—and (2) a set of observations $\mathbf{z}$—we assume to have p observations organized in an $n \times p$ matrix Z, regarded as p realizations drawn from a known probability density function. Our goal is to find an estimate $\hat{\mathbf{w}}$ of $\mathbf{w}$ based on both the a priori and the a posteriori knowledge we have. To introduce fundamental concepts, we start considering $\mathbf{w}$ as an instantaneous (or time independent) variable. Assume, for instance, that the distribution of the variable of interest is a Gaussian *probability density function* (p.d.f.) then the solution of the estimate problem is given by the average and the variance (and generally the statistical moments) of this distribution[2]. We will see several ways for obtaining these quantities, namely the *minimum variance* (MV), the *maximum a posteriori probability* (MAP), the *maximum likelihood* (ML).

Then, we consider the case when the variables of interest are part of a dynamical system. As a matter of fact, in the applications we are interested in there is a dynamics and we have a mathematical model describing how a variable of interest, that we call $\mathbf{u}$ (the *state* of the system), evolves according to a sequential parameter that will be in general the time. This may be either the fluid or the fluid coupled with the arterial wall, etc. In general, this is the knowledge given by mathematical

[2]Precise definitions of average and variance of a Gaussian variable will be given later on.

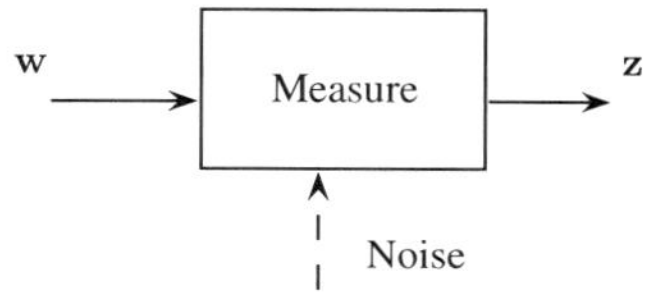

Fig. 6.3 In/Out system: **z** is the quantity measured, **w** is the quantity to be estimated

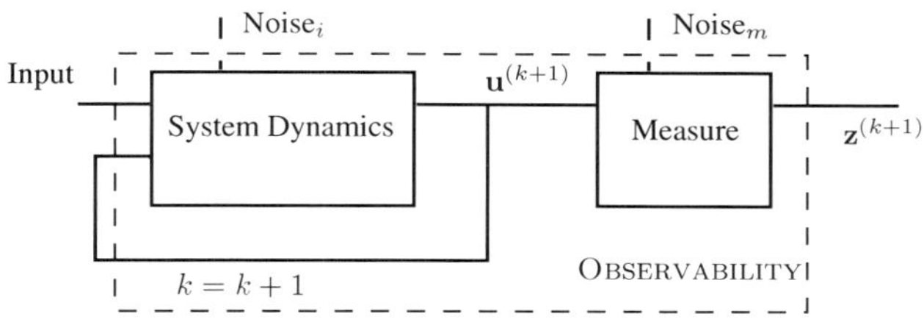

Fig. 6.4 Possible approaches for the estimate with a system dynamics. Here we rely upon the observability of the system and the knowledge of the statistical features of the stochastic process

modeling. In most of the cases, this is a system of partial differential equations and a numerical discretization procedure is necessary for its quantitative solution. The numerical model (i.e., the discretized mathematical model) reads

$$\mathbf{u}^{(k+1)} = \mathbf{u}^{(k+1)}(\text{Past, Input, Noise}).$$

Assume that we measure $\mathbf{z}^{(k)}$ and we want to estimate $\mathbf{u}^{(k)}$ (*filtering* problem) or $\mathbf{u}^{(k+p)}$ (p-step prediction problem). If the variable $\mathbf{u}$ that we want to estimate is a parameter of the original model (as it was the viscosity on the previous example), this is an *identification* problem. In the simplest case, assume that the system dynamics is linear and that noise affects both the input of the system and the measures. The system dynamics is regarded as the mechanism converting the p.d.f. of the input to the p.d.f. of the output. From the latter, we want to infer the state variable. If we apply the criterion of finding the estimate by minimizing the variance of the estimate, we will get a sequential estimate/prediction procedure called *Kalman filter*. Although the method strongly relies on the linearity of the system, that is not changing the nature of the p.d.f. , we will see that the method can be properly extended to non-linear cases. The effectiveness of the procedure relies upon the "observability" of the system, i.e., on how the output information is actually representative of the state of the system (Fig. 6.4).

As pointed out, our goal here is to give a short introduction to probabilistic estimation theory. For a more complete and extensive presentation, we refer to [68], Chaps. 4–6.

6.2.1 Basic Notation and Concepts

We summarize some fundamental concepts and notation that are useful in the remainder of this section. For a complete and rigorous introduction and explanation of these concepts, we refer, e.g., to [55].

Random Variables

For a random variable w, i.e., a variable whose value depends on a random experiment ω, we introduce the *distribution function*

$$F_W(w) \equiv P(\omega : \mathrm{w}(\omega) \leq w)$$

where the notation on the right-hand side represents the probability that the realization of w associated with ω is $\leq w$. Elementary properties of probability imply that $\lim_{w \to -\infty} F_W = 0$ and $\lim_{w \to +\infty} F_W = 1$ and that the function is non-decreasing. The corresponding p.d.f. is defined as

$$p_W(w) \equiv \frac{dF_W}{dw}.$$

For the properties of distribution, $p_W(w) \geq 0$ and $\int_{\mathbb{R}} p_W dw = 1$.

The Gaussian p.d.f. for instance reads

$$g_W(w) = \frac{1}{\sqrt{2\pi}\lambda} \exp\left(-\frac{(w-\mu)^2}{2\lambda^2}\right). \tag{6.3}$$

A p.d.f. can be characterized by its *moments*. In particular, we define the *expectation* operator $\mathcal{E}(\cdot)$ as

$$\mathcal{E}(\mathrm{w}) \equiv \int_{\mathbb{R}} w\, p_W(w) dw,$$

that associates the random variable with a number called *mean*. Similarly, we may consider the moments and the central moments of order m defined, respectively, as

$$\mathcal{E}(\mathrm{w}^m) \equiv \int_{\mathbb{R}} w^m p_W(w) dw, \qquad \mathcal{E}(\mathrm{w}^m) \equiv \int_{\mathbb{R}} (w - \mathcal{E}(\mathrm{w}))^m p_W(w) dw.$$

The central moment of order 2 is called *variance*. For the Gaussian p.d.f. g_W, the mean is μ and variance is λ^2.

Jointly Distributed Random Variables

We may consider the case of multiple random variables depending on ω according to a joint distribution

$$F_{W_1 W_2 \ldots W_n} \equiv P(\omega : \mathrm{w}_1 \leq w_1, \ldots, \mathrm{w}_n \leq w_n).$$

In this case, the joint p.d.f. reads

$$p_{W_1 W_2 \ldots W_n} \equiv \frac{\partial^n}{\partial x_1 \partial x_2 \ldots \partial x_n} F_{W_1 W_2 \ldots W_n}.$$

First order moments read

$$\mathcal{E}\left(w_j\right) = \int_{\mathbb{R}^n} w_j \, p_{W_1 W_2 \ldots W_n} d w_1 d w_2 \ldots d w_n.$$

Second order central moments form the symmetric *covariance* matrix

$$\begin{aligned} \lambda_{jk} &\equiv [\mathcal{E}\left((w_j - \mathcal{E}\left(w_j\right))(w_k - \mathcal{E}\left(w_k\right))\right)] \\ &= \int_{\mathbb{R}^n} (w_j - \mathcal{E}\left(w_j\right))(w_k - \mathcal{E}\left(w_k\right)) p_{W_1 W_2 \ldots W_n} d w_1 d w_2 \ldots d w_n \end{aligned}$$

where clearly $\lambda_{jj} = \lambda_j^2$, the variance of w_j. The *correlation coefficient* between w_j and w_k is defined as

$$\rho_{jk} \equiv \frac{\lambda_{jk}}{\lambda_j \lambda_k}. \tag{6.4}$$

For instance, two jointly distributed Gaussian variables have the distribution

$$p_{W_1 W_2}(w_1, w_2) = \frac{1}{2\pi \sqrt{|\Lambda|}} \exp\left(-\frac{1}{2} \boldsymbol{\delta}^T \Lambda^{-1} \boldsymbol{\delta}\right),$$

where $\boldsymbol{\delta} = \begin{bmatrix} w_1 - \mu_1 \\ w_2 - \mu_2 \end{bmatrix}$, $\Lambda = \begin{bmatrix} \lambda_1^2 & \lambda_{12} \\ \lambda_{12} & \lambda_2^2 \end{bmatrix}$ is the covariance matrix, $|\Lambda|$ stands for its determinant, and μ_1 and μ_2 are the means of the two variables.

In the sequel, this distribution is denoted by $\mathcal{G}(\boldsymbol{\mu}, \Lambda)$. In particular a distribution with $\boldsymbol{\mu} = \mathbf{0}$ and Λ diagonal (that means that the components of the vector are not correlated) is considered as a model for random disturbances or *white noise*.[3]

[3]The choice of Gaussian distribution for white noise is reasonable, but arbitrary. We could have considered other distributions for zero-mean, uncorrelated components.

The *marginal density function* of one of the random variables w_j in the pool may be obtained by the joint one after integration over the range of the other variables,

$$p_{W_j} = \int\limits_{\mathbb{R}^{n-1}} p_{W_1 W_2 \dots W_n} dw_1 \dots dw_{j-1} dw_{j+1} \dots dw_n$$

Conditional Probability

The *conditional* p.d.f. of the random vector $\mathbf{w}$ given the occurrence of the random vector $\mathbf{y}$ is defined as

$$p_{W|Y}(\mathbf{w}|\mathbf{y}) \equiv \frac{p_{W,Y}(\mathbf{w},\mathbf{y})}{p_Y(\mathbf{y})}.$$

Similarly, we have the definition

$$p_{Y|W}(\mathbf{y}|\mathbf{w}) \equiv \frac{p_{W,Y}(\mathbf{w},\mathbf{y})}{p_W(\mathbf{w})},$$

from which we obtain the *Bayes law*

$$p_{W|Y}(\mathbf{w}|\mathbf{y}) = \frac{p_{Y|W}(\mathbf{y}|\mathbf{w})\, p_W(\mathbf{w})}{p_Y(\mathbf{y})}. \tag{6.5}$$

The conditional expectation is defined consequently as

$$\mathcal{E}(\mathbf{w}|\mathbf{y}) = \int\limits_{\mathbb{R}^n} \mathbf{w}\, p_{W|Y}(\mathbf{w}|\mathbf{y}) d\mathbf{w}.$$

From the previous relations, it follows that

$$\mathcal{E}(\mathbf{w}) = \int\limits_{\mathbb{R}^n} \mathbf{w}\, p_W(\mathbf{w}) d\mathbf{w} = \int\limits_{\mathbb{R}^n}\int\limits_{\mathbb{R}^n} \mathbf{w}\, p_{W,Y}(\mathbf{w},\mathbf{y}) d\mathbf{y} d\mathbf{w}$$
$$\int\limits_{\mathbb{R}^n} \left(\int\limits_{\mathbb{R}^n} \mathbf{w}\, p_{W|Y} d\mathbf{w} \right) p_Y d\mathbf{y} = \mathcal{E}\left(\mathcal{E}(\mathbf{w}|\mathbf{y})\right).$$

6.2.2 Minimum Variance and Other In-Out Estimators

Let us consider a first example of estimator $\hat{\mathbf{w}}$ of the random vector $\mathbf{w}$ upon data $\mathbf{z}$ in the "steady" case—Fig. 6.3. Let $\mathbf{e} \equiv \hat{\mathbf{w}} - \mathbf{w}$ be the estimate error, and define

$$J(\mathbf{e}) = \mathbf{e}^T \mathrm{E}\mathbf{e}, \tag{6.6}$$

where E is a symmetric positive definite (s.p.d.) matrix. We assume J to be the measure of the estimate error or "risk." Here, $\hat{\mathbf{w}}$ depends on $\mathbf{z}$ and $\mathbf{w}$, and it is regarded as a stochastic process. With our definition of risk we may introduce the functional $\mathcal{J}(\hat{\mathbf{w}}) \equiv \mathcal{E}(J(\mathbf{e})) = \int_{\mathbb{R}^n} J(\mathbf{e})\, p_W\, d\mathbf{w}$ and in order to minimize the risk we refer to the estimator

$$\hat{\mathbf{w}} = \arg\min \mathcal{J}(\hat{\mathbf{w}}) \equiv \int_{\mathbb{R}^n}\int_{\mathbb{R}^n} J(\mathbf{e})\, p_{W,Z}(\mathbf{w},\mathbf{z})\, d\mathbf{z}\, d\mathbf{w}.$$

By exploiting the properties of p.d.f. recalled above, we rewrite the risk to minimize as

$$\mathcal{J}(\hat{\mathbf{w}}) = \int_{\mathbb{R}^n}\left(\int_{\mathbb{R}^n} \mathbf{e}^T \mathrm{E}\mathbf{e}\, p_{W|Z}\, d\mathbf{w}\right) p_Z\, d\mathbf{z} = \int_{\mathbb{R}^n} \mathcal{J}(\hat{\mathbf{w}}|\mathbf{z})\, p_Z\, d\mathbf{z},$$

for

$$\mathcal{J}(\hat{\mathbf{w}}|\mathbf{z}) \equiv \int_{\mathbb{R}^n} (\mathbf{w}-\hat{\mathbf{w}})^T \mathrm{E}(\mathbf{w}-\hat{\mathbf{w}})\, p_{W|Z}\, d\mathbf{w}.$$

Since the outer integral in the definition of the cost $\mathcal{J}(\hat{\mathbf{w}})$ does not involve $\hat{\mathbf{w}}$ and $p_Z \geq 0$, we minimize the risk by minimizing $\mathcal{J}(\hat{\mathbf{w}}|\mathbf{z})$.

Recall that for a generic vector $\mathbf{x}$ and a symmetric matrix A of proper size [60], we have $\dfrac{\partial \mathbf{x}^T \mathrm{A}\mathbf{x}}{\partial \mathbf{x}} = 2\mathrm{A}\mathbf{x}$. Then the minimization of $\mathcal{J}(\hat{\mathbf{w}}|\mathbf{z})$ leads to

$$0 = \frac{\partial \mathcal{J}(\hat{\mathbf{w}}|\mathbf{z})}{\partial \hat{\mathbf{w}}} = -2\mathrm{E}\int_{\mathbb{R}^n} (\mathbf{w}-\hat{\mathbf{w}})\, p_{W|Z}\, d\mathbf{w}.$$

Independently of E, we have the equation

$$\hat{\mathbf{w}}\int_{\mathbb{R}^n} p_{W|Z}\, d\mathbf{w} = \int_{\mathbb{R}^n} \mathbf{w}\, p_{W|Z}\, d\mathbf{w} = \mathcal{E}(\mathbf{w}|\mathbf{z}).$$

Since $\int_{\mathbb{R}^n} p_{W|Z}(\mathbf{w}|\mathbf{z})\, d\mathbf{w} = 1$, we have

$$\hat{\mathbf{w}} = \mathcal{E}(\mathbf{w}|\mathbf{z}). \tag{6.7}$$

This is called *minimum variance estimator*, hereafter denoted by $\hat{\mathbf{w}}_{MV}$. An important property of this estimator is that it is *unbiased*, i.e., $\mathcal{E}(\mathbf{e}) = \mathcal{E}(\hat{\mathbf{w}}_{MV} - \mathbf{w}) = 0$.

In fact, we have

$$\mathcal{E}(\hat{\mathbf{w}}_{MV}) = \int_{\mathbb{R}^n} \hat{\mathbf{w}}_{MV}\, p_Z d\mathbf{z} = \int_{\mathbb{R}^n}\int_{\mathbb{R}^n} \mathbf{w} p_{W|Z}\, d\mathbf{w}\, p_{\mathbf{z}} d\mathbf{z}$$

$$= \int_{\mathbb{R}^n}\int_{\mathbb{R}^n} \mathbf{w}\, p(\mathbf{w},\mathbf{z})\, d\mathbf{w} d\mathbf{z} = \mathcal{E}(\mathbf{w}).$$

It is also possible to verify that $\frac{\partial^2 \mathcal{J}(\hat{\mathbf{w}}_{MV})}{\partial \hat{\mathbf{w}}^2} = 2\text{E} > 0$, so $\hat{\mathbf{w}}_{MV}$ is indeed a minimum [68].

Maximum A Posteriori Estimator

Other estimators may be computed with a similar approach but referring to a different risk $\mathcal{J}(\hat{\mathbf{w}})$,

$$\mathcal{J}(\hat{\mathbf{w}}) = \mathcal{E}(J(\mathbf{e})) = \int_{\mathbb{R}^n} J(\mathbf{e}) p_W d\mathbf{w} = \int_{\mathbb{R}^n}\int_{\mathbb{R}^n} J(\mathbf{e}) p_{W,Z} d\mathbf{w} d\mathbf{z}$$

for different choices for $J(\cdot)$. For instance, another possible choice is the "uniform" (thresholding) cost:

$$J(\mathbf{e}) = \begin{cases} 0 & \text{for } \|\mathbf{e}\|_\infty < \varepsilon \\ \frac{1}{2\varepsilon} & \text{for } \|\mathbf{e}\|_\infty \geq \varepsilon \end{cases},$$

where $\|\cdot\|_\infty$ is the maximum norm. With this cost function, we obtain

$$\mathcal{J}(\hat{\mathbf{w}}) = \int_{\mathbb{R}^n}\int_{\mathbb{R}^n} J(\mathbf{e})\, p_{W|Z}\, d\mathbf{w} p_Z d\mathbf{z} = \int_{\mathbb{R}^n} \frac{1}{2\varepsilon} \left(\int_{\mathbb{R}^n \setminus \{\hat{\mathbf{w}}+[-\varepsilon,\varepsilon]^n\}} p_{W|Z} d\mathbf{w} \right) p_Z d\mathbf{z}.$$

By definition

$$\int_{\mathbb{R}^n \setminus \{\hat{\mathbf{w}}+[-\varepsilon,\varepsilon]^n\}} p_{W|Z} d\mathbf{w} = 1 - \int_{\hat{\mathbf{w}}+[-\varepsilon,\varepsilon]^n} p_{W|Z} d\mathbf{w}$$

so that

$$\mathcal{J}(\hat{\mathbf{w}}) = \frac{1}{2\varepsilon}\int_{\mathbb{R}^n} p_Z d\mathbf{z} - \frac{1}{2\varepsilon}\int_{\mathbb{R}^n} \int_{\hat{\mathbf{w}}+[-\varepsilon,\varepsilon]^n} p_{W|Z} d\mathbf{w} p_Z d\mathbf{z}.$$

The first term on the right-hand side is constant and does not affect the minimization process. Let us focus on the second term. The mean value theorem states that there exists a $\xi \in x + [-\varepsilon, \varepsilon]$ such that

$$\frac{1}{2\varepsilon}\int_{x-\varepsilon}^{x+\varepsilon} f(\chi)d\chi = \frac{1}{2\varepsilon}2\varepsilon f(\xi) = f(\xi).$$

For $\varepsilon \to 0$ we have $\frac{1}{2\varepsilon}\int_{x-\varepsilon}^{x+\varepsilon} f(\chi)d\chi = f(x)$. In particular, in our case, we obtain

$$\lim_{\varepsilon\to 0}\left(-\frac{1}{2\varepsilon}\int_{\hat{\mathbf{w}}+[-\varepsilon,\varepsilon]^n} p_{W|Z}d\mathbf{w}\right) = -p_{W|Z}(\hat{\mathbf{w}}|\mathbf{z}).$$

In other terms, the cost function to quantify the risk selected here leads to the maximization of the a posteriori p.d.f. $p_{W|Z}$. This estimator, such that $\frac{\partial p_{W|Z}}{\partial \mathbf{w}} = 0$, is denoted $\hat{\mathbf{v}}_{\mathrm{MAP}}$ and it is not necessarily unbiased.

Example. In this example, we assume that the scalar variables w and z are statistically related by having a joint Gaussian distribution $\mathcal{G}([0,0],\Lambda)$. The two variables features a Gaussian marginal p.d.f. with mean and variance $0, \lambda_w^2$ and $0, \lambda_z^2$, respectively.

As for the conditional p.d.f. we have

$$p_{W|Z} = \frac{p_{WZ}(w,z)}{p_Z} = \frac{\sqrt{\lambda_z^2}}{\sqrt{2\pi|\Lambda|}}\exp\left(-\frac{1}{2}[w \quad z]^T\Lambda^{-1}[w \quad z] + \frac{z^2}{2\lambda_z^2}\right).$$

Define $\lambda^2 \equiv \lambda_w^2 - \frac{\lambda_{wz}^2}{\lambda_z^2}$. Then, by direct inspection, one verifies that $(\mathbf{w}|\mathbf{z})$ is a random vector with Gaussian distribution $\mathcal{G}\left(\frac{\lambda_{wz}}{\lambda_z^2}z, \lambda^2\right)$.

Because for Gaussian distributions, the value where the maximum is achieved corresponds to the mean, we have that

$$\hat{w}_{MV} = \hat{w}_{\mathrm{MAP}} = \frac{\lambda_{wz}}{\lambda_z^2}z.$$

Using the definition of correlation coefficient given in (6.4), we have

$$\hat{w}_{MV} = \hat{w}_{\mathrm{MAP}} = \lambda_w^2\rho_{wz}z.$$

From here, we can check the consistency of our estimate with intuitive situations: if w is not correlated to z, the estimate is 0, which is the mean value of the marginal p.d.f. of w. In this case, the knowledge of z does not bring any advantage and the best estimate remains the a priori expected value of w.

Maximum Likelihood Estimate

Another reasonable approach is to compute the estimator $\hat{\mathbf{w}}$ as the value that maximizes the probability of measuring $\mathbf{z}$. In other terms, we get $\hat{\mathbf{w}}_{\mathrm{ML}} = \arg\max p_{Z|W}(\mathbf{z}|\mathbf{w})$ or

$$\hat{\mathbf{w}}_{\mathrm{ML}} : \frac{\partial p_{Z|W}}{\partial \mathbf{w}}|_{\hat{\mathbf{w}}_{\mathrm{ML}}} = 0.$$

The p.d.f. $p_{Z|W}$ is a measure of the *likelihood* that $\mathbf{z}$ is measured, so this estimate is called *maximum likelihood*.

It is interesting to establish a relation between this estimator and the previous ones. We do this for the case of Gaussian variables, even though the same conclusion holds in the general case.

We know that estimator $\hat{\mathbf{w}}_{\mathrm{MAP}}$ is such that $\frac{\partial p_{W|Z}}{\partial \mathbf{w}}|_{\hat{\mathbf{w}}_{\mathrm{MAP}}} = 0.$

Then, thanks to the Bayes Theorem, we have

$$p_{W|Z} = \frac{p_{WZ}}{p_Z} = \frac{p_{Z|W}\, p_W}{p_Z}.$$

Maximizing $p_{W|Z}$ is equivalent to the maximization of its logarithm, so we have

$$\frac{\partial p_{W|Z}}{\partial \mathbf{w}} = 0 \Rightarrow \frac{\partial \ln p_{W|Z}}{\partial \mathbf{w}} = \frac{\partial \ln \frac{p_{Z|W}\, p_W}{p_Z}}{\partial \mathbf{w}} = 0$$
$$\Rightarrow \frac{\partial \ln p_{Z|W}}{\partial \mathbf{w}} + \frac{\partial \ln p_W}{\partial \mathbf{w}} - \frac{\partial \ln p_Z}{\partial \mathbf{w}} = \frac{\partial \ln p_{Z|W}}{\partial \mathbf{w}} + \frac{\partial \ln p_W}{\partial \mathbf{w}} = 0$$

since p_Z is independent of $\mathbf{w}$.

Now, for a Gaussian variable, such that

$$p_W = \frac{1}{\sqrt{(2\pi)^n |\Lambda|}} \exp\left(-\frac{1}{2}(\mathbf{w} - \mathcal{E}(\mathbf{w}))^T \Lambda^{-1} (\mathbf{w} - \mathcal{E}(\mathbf{w})) \right),$$

we have

$$\frac{\partial \ln p_W}{\partial \mathbf{w}} = -\frac{1}{2}\frac{\partial (\mathbf{w} - \mathcal{E}(\mathbf{w}))^T \Lambda^{-1} (\mathbf{w} - \mathcal{E}(\mathbf{w}))}{\partial \mathbf{w}} = -\Lambda^{-1}(\mathbf{w} - \mathcal{E}(\mathbf{w})).$$

When the variance of a random variable is large, this means that our a priori knowledge is not trustworthy. The limit case of $\Lambda^{-1} \to \mathbf{0}$ corresponds to a total lack of a priori information on $\mathbf{w}$. Notice that in this condition

$$\frac{\partial \ln p_{W|Z}}{\partial \mathbf{w}} = \frac{\partial \ln p_{Z|W}}{\partial \mathbf{w}} + \frac{\partial \ln p_{W}}{\partial \mathbf{w}} \overset{(\Lambda^{-1}=\mathbf{0})}{=} \frac{\partial \ln p_{Z|W}}{\partial \mathbf{w}}.$$

We conclude therefore that $\hat{\mathbf{v}}_{\mathrm{ML}}$ can be considered the estimator in the "limit" case, when we do not have an a priori distribution for $\mathbf{w}$, i.e., when we assume that the variance of $\mathbf{w}$ is approaching ∞.

Example. Let us consider two scalar variables w and z with

$$z = Hw + \nu \tag{6.8}$$

where $w \sim \mathcal{G}(0, \lambda_w^2)$, and the noise $\nu \sim \mathcal{G}(0, r^2)$ is assumed to be uncorrelated with w. Then it is possible to verify that $z \sim \mathcal{G}(0, H^2\lambda_w^2 + r^2)$ and that w and z have a joint Gaussian distribution with $\lambda_{wz} = H\lambda_w^2$. In fact

$$\begin{aligned}
&\mathcal{E}(z) = H\mathcal{E}(w) + \mathcal{E}(\nu) = 0\\
&\lambda_z^2 = \mathcal{E}\left(z^2\right) = \mathcal{E}\left(H^2w^2 + 2Hw\nu + \nu^2\right) = H^2\lambda_w^2 + 0 + r^2\\
&\lambda_{wz} = \mathcal{E}(wz) = \mathcal{E}\left(Hw^2 + \nu w\right) = H\lambda_w^2.
\end{aligned}$$

Using the results of the previous example, in this case we can compute

$$\hat{\mathbf{w}}_{MV} = \hat{\mathbf{w}}_{\mathrm{MAP}} = \frac{H\lambda_w^2}{H^2\lambda_w^2 + r^2} z = \frac{z}{H}\frac{H^2\lambda_w^2}{H^2\lambda_w^2 + r^2} = \frac{z}{H}\left(1 - \frac{r^2}{\lambda_z^2}\right).$$

Using again the result of the previous example, we find that $p_{Z|W} = \dfrac{p_{WZ}}{p_Z}$ is a Gaussian distribution with mean $\dfrac{\lambda_{wz}}{\lambda_w^2} w$. The maximum of $p_{Z|W}$ is obtain in correspondence of its mean, i.e., for $z = \dfrac{\lambda_{wz}}{\lambda_w^2} w$ or, equivalently, for $w = \dfrac{\lambda_w^2}{\lambda_{wz}} z$. Therefore the ML estimator reads

$$\hat{\mathbf{w}}_{\mathrm{ML}} = \frac{\lambda_w^2}{\lambda_{wz}} z = \frac{z}{H}.$$

As expected, $\lim\limits_{\lambda_w \to \infty} \hat{\mathbf{w}}_{\mathrm{MAP}} = \hat{\mathbf{w}}_{\mathrm{ML}}$. The estimators coincide also when $r^2 = 0$, and return z/H. In fact, in this case the noise is 0, so the estimators give the deterministic relation (obtained by (6.8) for $\nu = 0$) $w = z/H$.

Example. Assume now that $\mathbf{w}$ and $\mathbf{z}$ are n-dimensional vectors, $\mathbf{w} \sim \mathcal{G}(\boldsymbol{\mu}, \Lambda)$ and

$$\mathbf{z} = \mathrm{H}\mathbf{w} + \boldsymbol{\nu}$$

where $\boldsymbol{\nu} \sim \mathcal{G}(0, \mathrm{R})$ is the noise independent of $\mathbf{w}$. H is called *observation matrix*.

It is possible to prove that $\mathbf{z} \sim \mathcal{G}(\boldsymbol{\mu}_{\mathbf{z}}, \Lambda_{\mathbf{z}})$ with

$$\boldsymbol{\mu}_{\mathbf{z}} = \mathcal{E}\left(\mathrm{H}\mathbf{w} + \boldsymbol{\nu}\right) = \mathrm{H}\boldsymbol{\mu}$$
$$\Lambda_{\mathbf{z}} = \mathcal{E}\left((\mathbf{z} - \boldsymbol{\mu}_{\mathbf{z}})^T(\mathbf{z} - \boldsymbol{\mu}_{\mathbf{z}})\right) = \mathrm{H}\Lambda\mathrm{H}^T + \mathrm{R}.$$

For the conditional probabilities, we find that

$$p_{\mathbf{w}|\mathbf{z}} = \frac{\sqrt{|\Lambda_{\mathbf{z}}|}}{\sqrt{(2\pi)^n|\Lambda||\mathrm{R}|}}\exp\left(-\frac{1}{2}\mathrm{J}\right)$$

where $\mathrm{J} = (\mathbf{w} - \hat{\mathbf{w}})^T\Lambda_{\mathbf{e}}^{-1}(\mathbf{w} - \hat{\mathbf{w}})$ and $\Lambda_{\mathbf{e}}^{-1} = \Lambda^{-1} + \mathrm{H}^T\mathrm{R}^{-1}\mathrm{H}$ and

$$\hat{\mathbf{w}}_{MV} \equiv \mathcal{E}\left(p_{\mathbf{w}|\mathbf{z}}\right) = \Lambda_{\mathbf{e}}\left(\mathrm{H}^T\mathrm{R}^{-1}\mathbf{z} + \Lambda^{-1}\boldsymbol{\mu}\right) = \hat{\mathbf{w}}_{\mathrm{MAP}}.$$

Moreover, we find that $p_{\mathbf{z}|\mathbf{w}}$ has average $\mathrm{H}\mathbf{w}$ and $\Lambda_{\mathbf{z}|\mathbf{w}} = \mathrm{R}$. If we maximize the likelihood, we obtain

$$\hat{\mathbf{w}}_{\mathrm{ML}} = \mathrm{H}^{-1}\mathbf{z}.$$

Again, it is possible to verify that the MV/MAP estimator is unbiased and the ML estimator is obtained by the MAP, when $\Lambda^{-1} \to 0$.

Remark 6.1. Contrarily to what previous examples may suggest, the coincidence of MV and MAP is not true in general.

6.2.3 *The Kalman Filter for Linear Problems*

Kalman filter [48] is one of the most important algorithms of the twentieth century, with an exceptional number of applications, ranging from robotics to mathematical finance. It is concerned with the case of a dynamical system, when the variable to be estimated is supposed to be the solution of a time-dependent linear system. Since for the biomedical applications of interest here, dynamics is in general given by the time discretization of a PDE (as we will see later on), here we consider a time-discrete case, even though the time-continuous case can be investigated as well. The time index will be denoted by k, and we represent the dynamics of interest (indexed by k) of the system as

$$\mathbf{u}^{(k)} = \mathrm{A}_{k-1}\mathbf{u}^{(k-1)} + \mathbf{b}^{(k-1)} \tag{6.9}$$

where $\mathbf{b}^{(k-1)}$ is a *Gaussian white noise* in time representing the model error, i.e., $\mathbf{b}^{(k)} \sim \mathcal{G}(\mathbf{0}, \mathrm{Q}_k)$, and the errors are not correlated in time, i.e.,

$$\mathcal{E}\left(\mathbf{b}^{(k)}\mathbf{b}^{(l),T}\right) = \mathrm{Q}_k\delta_{kl}.$$

Here δ_{kl} is the Kronecker delta ($= 1$ if $k = l$, 0 elsewhere).

The measurement process is denoted by

$$\mathbf{z}^{(k)} = \mathrm{H}_k \mathbf{u}^{(k)} + \boldsymbol{\nu}^{(k)}, \tag{6.10}$$

where $\boldsymbol{\nu}^{(\cdot)}$ is a Gaussian white noise with variance matrix R_k and assumed uncorrelated with $\mathbf{b}^{(\cdot)}$.

In absence of observations of $\mathbf{z}^{(k)}$, a natural prediction is simply the one obtained by dropping the noise in the system. In other terms a first reasonable prediction would be

$$\mathbf{u}_p^{(k)} = \mathrm{A}_{k-1} \mathbf{u}_*^{(k-1)}. \tag{6.11}$$

For the moment being, we assume that $\mathbf{u}_*^{(k-1)}$ is the "true" state $\mathbf{u}^{(k-1)}$.

This is a deterministic forecast that we take as starting point of our probabilistic measure. The fundamental questions now are: *how can the measure* $\mathbf{z}^{(k)}$ *improve this estimate? How can we reduce the error between the state and its prediction?*

As an arbitrary but reasonable choice, we postulate a *correction step* which is a linear combination between the prediction $\mathbf{u}_p^{(k)}$ and the data at the same instant $\mathbf{z}^{(k)}$,

$$\mathbf{u}_c^{(k)} = \mathrm{L}_k \mathbf{u}_p^{(k)} + \mathrm{K}_k \mathbf{z}^{(k)}.$$

The first term on the right-hand side measures how much the deterministic model is trustworthy, the latter defines the gain due to the observation. The weighting matrices L_k, K_k need to be determined. Let us introduce the estimate errors

$$\mathbf{e}_p^{(k)} = \mathbf{u}_p^{(k)} - \mathbf{u}^{(k)}, \quad \mathbf{e}_c^{(k)} = \mathbf{u}_c^{(k)} - \mathbf{u}^{(k)}.$$

Notice that $\mathbf{e}_p^{(k)} = -\mathbf{b}^{(k-1)}$ by construction (for $\mathbf{u}_*^{(k-1)} = \mathbf{u}^{(k-1)}$).

We have then

$$\begin{aligned}\mathbf{e}_c^{(k)} &= \mathrm{L}_k \mathbf{u}_p^{(k)} + \mathrm{K}_k \mathbf{z}^{(k)} - \mathrm{L}_k \mathbf{u}^{(k)} + \mathrm{L}_k \mathbf{u}^{(k)} - \mathbf{u}^{(k)} \\ &= \mathrm{L}_k \mathbf{e}_p^{(k)} + \mathrm{K}_k \boldsymbol{\nu}^{(k)} + (\mathrm{L}_k + \mathrm{K}_k \mathrm{H}_k - \mathrm{I})\, \mathbf{u}^{(k)}.\end{aligned}$$

In order to have an unbiased correction, i.e., $\mathcal{E}\left(\mathbf{e}_c^{(k)}\right) = 0$, we write

$$\begin{aligned}\mathcal{E}\left(\mathbf{e}_c^{(k)}\right) &= \mathrm{L}_k \mathcal{E}\left(\mathbf{e}_p^{(k)}\right) + \mathrm{K}_k \mathcal{E}\left(\boldsymbol{\nu}^{(k)}\right) + (\mathrm{L}_k + \mathrm{K}_k \mathrm{H}_k - \mathrm{I})\, \mathcal{E}\left(\mathbf{u}^{(k)}\right) \\ &= -\mathrm{L}_k \mathcal{E}\left(\mathbf{b}^{(k-1)}\right) + \mathrm{K}_k \mathcal{E}\left(\boldsymbol{\nu}^{(k)}\right) + (\mathrm{L}_k + \mathrm{K}_k \mathrm{H}_k - \mathrm{I})\, \mathcal{E}\left(\mathbf{u}^{(k)}\right) = 0.\end{aligned} \tag{6.12}$$

Because we assume that the noise has null mean, the first two terms are zero. To have an unbiased estimate we need to force

$$\mathrm{L}_k + \mathrm{K}_k \mathrm{H}_k - \mathrm{I} = 0 \Rightarrow \mathrm{L}_k = \mathrm{I} - \mathrm{K}_k \mathrm{H}_k.$$

With this position, we have

$$\mathbf{u}_c^{(k)} = \mathbf{u}_p^{(k)} + \mathrm{K}_k(\mathbf{z}^{(k)} - \mathrm{H}_k\mathbf{u}_p^{(k)}).$$

This representation is extremely expressive:

1. the first term on the right-hand side $\mathbf{u}_p^{(k)}$ is the deterministic estimate purely based on the *model*, with no observations;
2. $\mathbf{z}^{(k)} - \mathrm{H}_k\mathbf{u}_p^{(k)}$ is the *innovation*, i.e., what is new in $\mathbf{z}^{(k)}$ and that is not predictable by $\mathbf{u}_p^{(k)}$;
3. K_k, to be determined, is called the *gain matrix*, since it weighs the improvement brought to the deterministic estimate by the measures.

The two estimation errors are then related by the following equation

$$\begin{aligned}\mathbf{e}_c^{(k)} &= \mathbf{u}_c^{(k)} - \mathbf{u}^{(k)} = \mathbf{u}_p^{(k)} - \mathbf{u}^{(k)} + \mathrm{K}_k(\mathrm{H}_k\mathbf{u}^{(k)} + \boldsymbol{\nu}^{(k)} - \mathrm{H}_k\mathbf{u}_p^{(k)}) \\ &= (\mathrm{I} - \mathrm{K}_k\mathrm{H}_k)\,\mathbf{e}_p^{(k)} + \mathrm{K}_k\boldsymbol{\nu}^{(k)}.\end{aligned} \tag{6.13}$$

However, in the sequential process we do not know the true state $\mathbf{u}^{(k-1)}$ in (6.11). We replace $\mathbf{u}_*^{(k-1)}$ with what we consider sequentially our best estimation of the state, which is $\mathbf{u}_c^{(k-1)}$. The recursive *prediction* step reads therefore

$$\mathbf{u}_p^{(k)} = \mathrm{A}_{k-1}\mathbf{u}_c^{(k-1)}. \tag{6.14}$$

From this equation, we derive another relation for the errors

$$\begin{aligned}\mathbf{e}_p^{(k)} = \mathbf{u}_p^{(k)} - \mathbf{u}^{(k)} &= \mathrm{A}_{k-1}\mathbf{u}_c^{(k-1)} - \mathbf{u}^{(k)} = \\ \mathrm{A}_{k-1}\left(\mathbf{u}_c^{(k-1)} - \mathbf{u}^{(k-1)}\right) - \mathbf{b}^{(k-1)} &= \mathrm{A}_{k-1}\mathbf{e}_c^{(k-1)} - \mathbf{b}^{(k-1)}\end{aligned} \tag{6.15}$$

giving an evolution equation for the deterministic forecast error.

Remark 6.2. In this analysis, we are considering one-step prediction estimates, where the deterministic estimate is obtained by the previous step (6.14). We may consider also q-step predictions, obtained by dropping the noise at each step,

$$\mathbf{u}_p^{(k)} = \prod_{j=1}^{q} \mathrm{A}_{k-j}\mathbf{u}_c^{(k-q)}.$$

For the sake of simplicity, here we develop the case $q = 1$ and refer the interested reader to [68].

Let us compute the variance matrix of $\mathbf{e}_p^{(k)}$ and $\mathbf{e}_c^{(k)}$, i.e.,

$$\Lambda_p^{(k)} \equiv \mathcal{E}\left(\mathbf{e}_p^{(k)}\mathbf{e}_p^{(k,T)}\right), \qquad \Lambda_c^{(k)} \equiv \mathcal{E}\left(\mathbf{e}_c^{(k)}\mathbf{e}_c^{(k,T)}\right). \tag{6.16}$$

From (6.15), we have

$$\begin{aligned}\mathbf{e}_p^{(k)}\mathbf{e}_p^{(k),T} &= \mathrm{A}_{k-1}\mathbf{e}_c^{(k-1)}\mathbf{e}_c^{(k-1),T}\mathrm{A}_{k-1}^T + \mathbf{b}^{(k-1)}\mathbf{b}^{(k-1),T} \\ &\quad + \mathrm{A}_{k-1}\mathbf{e}_c^{(k-1)}\mathbf{b}^{(k-1),T} + \mathbf{b}^{(k-1)}\mathbf{e}_c^{(k-1),T}\mathrm{A}_{k-1}^T,\end{aligned} \tag{6.17}$$

leading to

$$\Lambda_p^{(k)} = \mathrm{A}_{k-1}\Lambda_c^{(k-1)}\mathrm{A}_{k-1}^T + \mathrm{Q}_{k-1} \tag{6.18}$$

because $\mathbf{b}^{(k-1)}$ has no correlation with $\mathbf{e}_c^{(k-1)}$.

Similarly, from (6.13) we obtain

$$\Lambda_c^{(k)} = (\mathrm{I} - \mathrm{K}_k\mathrm{H}_k)\Lambda_p^{(k)}(\mathrm{I} - \mathrm{K}_k\mathrm{H}_k)^T + \mathrm{K}_k\mathrm{R}_k\mathrm{K}_k^T. \tag{6.19}$$

usually called *Joseph formula*.

The Kalman Gain Matrix

Finally we determine the gain matrix. We will follow an optimality criterion. According to the (weighted) minimal variance approach, we could minimize $\mathcal{E}\left(\mathbf{e}_c^{(k,T)}\mathrm{E}_k\mathbf{e}_c^{(k)}\right)$, where E_k is an s.p.d. weight matrix. Note that

$$\begin{aligned}&\mathbf{e}_c^{(k),T}\mathrm{E}_k\mathbf{e}_c^{(k)} = \mathrm{Tr}(\mathrm{E}_k\mathbf{e}_c^{(k)}\mathbf{e}_c^{(k),T}) \\ &\quad\Rightarrow \mathcal{E}\left(\mathbf{e}_c^{(k),T}\mathrm{E}_k\mathbf{e}_c^{(k)}\right) = \mathcal{E}\left(\mathrm{Tr}(\mathrm{E}_k\mathbf{e}_c^{(k)}\mathbf{e}_c^{(k),T})\right) = \mathrm{Tr}(\mathrm{E}_k\Lambda_c^{(k)}).\end{aligned}$$

Using the following properties of the trace:

$$\mathrm{Tr}(\mathrm{A}+\mathrm{B}) = \mathrm{Tr}(\mathrm{A}) + \mathrm{Tr}(\mathrm{B}), \qquad \mathrm{Tr}(\mathrm{AB}) = \mathrm{Tr}(\mathrm{A}^T\mathrm{B}^T),$$

we get that

$$\mathrm{Tr}(\mathrm{E}_k\Lambda_c^{(k)}) = \mathrm{Tr}(\mathrm{E}_k\Lambda_p^{(k)}) - 2\mathrm{Tr}(\mathrm{E}_k\Lambda_p^{(k)}\mathrm{H}_k^T\mathrm{K}_k^T) + \mathrm{Tr}(\mathrm{E}_k\mathrm{K}_k(\mathrm{H}_k\Lambda_p^{(k)}\mathrm{H}_k^T + \mathrm{R}_k)\mathrm{K}_k^T).$$

Moreover, we recall that [60] for a generic matrix A and symmetric matrices B and C we have

$$\frac{\partial\mathrm{Tr}(\mathrm{AX}^T)}{\partial\mathrm{X}} = \mathrm{A}, \qquad \frac{\partial\mathrm{Tr}(\mathrm{BXCX}^T)}{\partial\mathrm{X}} = 2\mathrm{BCX}.$$

From these relations we obtain that the gain matrix K_k that minimizes the variance is such that

$$\frac{\partial \mathcal{E}\left(\mathbf{e}_c^{(k,T)} E_k \mathbf{e}_c^{(k)}\right)}{\partial K_k} = -2E_k \Lambda_p^{(k)} H_k^T + 2E_k K_k (H_k \Lambda_p^{(k)} H_k^T + R_k) = 0$$
$$\Rightarrow K_k^* = \Lambda_p^{(k)} H_k^T \left(H_k \Lambda_p^{(k)} H_k^T + R_k\right)^{-1}.$$

From the Joseph formula we have

$$\Lambda_c^{(k)} = (I - K_k H_k)\Lambda_p^{(k)} - \Lambda_p^{(k)} H_k^T K_k^T + K_k (H_k \Lambda_p^{(k)} H_k^T + R_k) K_k^T.$$

By using K_k^* in this formula, the last two terms cancel out and we are left with

$$\Lambda_c^{(k)} = (I - K_k H_k)\, \Lambda_p^{(k)}.$$

The matrix K_k^* is the so-called *Kalman gain matrix*.

The entire estimate process may be summarized as follows.

1. PREDICTION
   (a) $\mathbf{u}_p^{(k)} = A_{k-1} \mathbf{u}_c^{(k-1)}$
   (b) $\Lambda_p^{(k)} = A_{k-1} \Lambda_c^{(k-1)} A_{k-1}^T + Q_{k-1}$.
2. CORRECTION
   Kalman gain: $K_k^* = \Lambda_p^{(k)} H_k^T \left(H_k \Lambda_p^{(k)} H_k^T + R_k\right)^{-1}$.
   (a) State estimate:
   $$\mathbf{u}_c^{(k)} = \mathbf{u}_p^{(k)} + K_k^* (\mathbf{z}^{(k)} - H_k \mathbf{u}_p^{(k)}). \tag{6.20}$$
   (b) Covariance estimate:
   $$\Lambda_c^{(k)} = \left(I - K_k^* H_k\right) \Lambda_p^{(k)}. \tag{6.21}$$

This yields the MV *Kalman filter* estimation for a time-discrete system. When we want to *estimate* the state of the system by merging the mathematical model and the available measure, we refer to $\mathbf{u}_c^{(k)}$. When we want to *predict* the evolution of the state, using all the information we have at time t^k, we refer to the prediction[4] step $\mathbf{u}_p^{(k+1)} = A_k \mathbf{u}_c^{(k)}$.

[4]The third problem addressed in the Introduction, the *identification* of the system will be considered later on.

Remark 6.3. In many cases, the dynamical system features a deterministic input $\mathbf{f}^{(k)}$, so that (6.9) modifies in

$$\mathbf{u}^{(k)} = \mathrm{A}_{k-1}\mathbf{u}^{(k-1)} + \mathrm{C}_{k-1}\mathbf{f}^{(k-1)} + \mathbf{b}^{(k-1)}. \tag{6.22}$$

This reflects in a change of the prediction step, that reads

$$\mathbf{u}_p^{(k)} = \mathrm{A}_{k-1}\mathbf{u}_c^{(k-1)} + \mathrm{C}_{k-1}\mathbf{f}^{(k-1)}.$$

All the other steps drawn above remain unchanged.

Properties of the Kalman Filter

Orthogonality of the Estimate/Prediction and the Estimate/Prediction Error

Following an induction argument, it is possible to prove that when we select the Kalman gain matrix to compute the estimate, then

$$\mathcal{E}\left(\mathbf{u}_c^{(k)}\mathbf{e}_c^{(k),T}\right) = 0. \tag{6.23}$$

With this relation, it is possible to prove a similar relation between the prediction and the prediction error

$$\mathcal{E}\left(\mathbf{u}_p^{(k+1)}\mathbf{e}_p^{(k+1),T}\right) = \mathcal{E}\left(A_k\mathbf{u}_c^{(k)}\left(A_k\mathbf{e}_c^{(k)}\right)^T\right) = 0. \tag{6.24}$$

These relations have an interesting geometrical interpretation that provide a justification to the "optimal" nature of Kalman estimate/prediction. The two equations (6.23) and (6.24) state that the estimate and the prediction are "orthogonal" to their respective errors. Here we mean "orthogonal" in the sense of random vectors (see [68] page 9). (see Fig. 6.5).

Innovation

As we pointed out, the *innovation*

$$\mathbf{z}^{(k)} - \mathrm{H}_k\mathbf{u}_p^{(k)} = \mathbf{z}^{(k)} - \mathrm{H}_k\mathrm{A}_k\mathbf{u}_c^{(k-1)}$$

plays an important role in understanding how the Kalman estimate works. When we do not have other a priori information, from A_k and $\mathbf{u}_c^{(k-1)}$ the best we can do is

- to predict the state at k as $\mathbf{u}_p^{(k)} = \mathrm{A}_k\mathbf{u}_c^{(k-1)}$;
- to guess accordingly an "expected measure" $\mathrm{H}_k\mathrm{A}_k\mathbf{u}_c^{(k-1)}$.

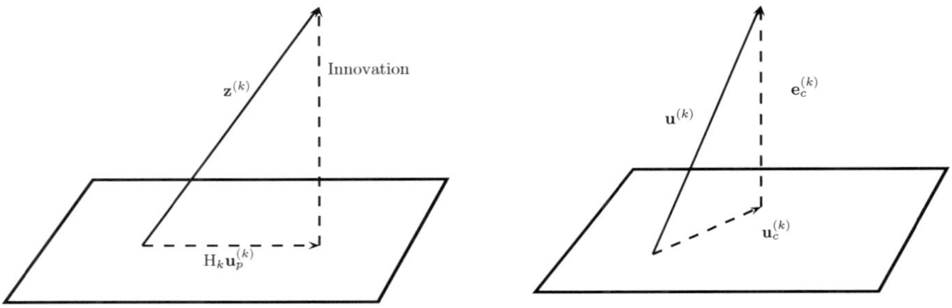

Fig. 6.5 The innovation is orthogonal to the past (*left*). The estimate error is orthogonal to the estimate itself (*right*)

This is the part of knowledge in the measure we could extract from the state at the previous time step, or *from the past*. We do expect that $\mathbf{z}^{(k)}$ is adding new information. The novel part of the information added by the measure is exactly the innovation $\mathbf{z}^{(k)} - \mathrm{H}_k \mathbf{u}_p^{(k)}$.

Notice that

$$\mathbf{z}^{(k)} - \mathrm{H}_k \mathbf{u}_p^{(k)} = \mathrm{H}_k \mathbf{u}^{(k)} + \boldsymbol{\nu}^{(k)} - \mathrm{H}_k \mathbf{u}_p^{(k)} = \boldsymbol{\nu}^{(k)} - \mathrm{H}_k \mathbf{e}_p^{(k)},$$

consequently

$$\mathcal{E}\left(\mathbf{z}^{(k)} - \mathrm{H}_k \mathbf{u}_p^{(k)}\right) = \mathcal{E}\left(\boldsymbol{\nu}^{(k)}\right) - \mathrm{H}_k \mathcal{E}\left(\mathbf{e}_p^{(k)}\right) = 0.$$

In addition, we compute the variance of the innovation

$$\begin{aligned}&\mathcal{E}\left((\mathbf{z}^{(k)} - \mathrm{H}_k \mathbf{u}_p^{(k)})(\mathbf{z}^{(k)} - \mathrm{H}_k \mathbf{u}_p^{(k)})^T\right)\\ &\quad = \mathcal{E}\left((\boldsymbol{\nu}^{(k)} - \mathrm{H}_k \mathbf{e}_p^{(k)})(\boldsymbol{\nu}^{(k)} - \mathrm{H}_k \mathbf{e}_p^{(k)})^T\right) = \mathrm{R}_k + \mathrm{H}_k \Lambda_p^{(k)} \mathrm{H}_k^T\end{aligned}$$

because the noise at k is not correlated to $\mathbf{e}_p^{(k)} = \mathrm{A}_k(\mathbf{u}_c^{(k)} - \mathbf{u}^{(k-1)}) - \mathbf{b}^{(k-1)}$.

It is also possible to prove [68] that for $j \geq 1$

$$\mathcal{E}\left((\mathbf{z}^{(k)} - \mathrm{H}_k \mathbf{u}_p^{(k)})(\mathbf{z}^{(k-j)} - \mathrm{H}_k \mathbf{u}_p^{(k-j)})^T\right) = 0.$$

This means that the innovation at time k has no correlation with the innovation at the previous time steps, so that we can conclude that the innovation is a *white process*.

Also in this case, we may give a geometrical interpretation to this equation, concluding that the splitting $\mathbf{z}^{(k)}$ = Predicted measure + Innovation is actually an orthogonal decomposition. Since the predicted measure $\mathrm{H}_k\mathbf{u}_p^{(k)}$ depends entirely on the past, we say that *innovation is orthogonal to the past* (see Fig. 6.5).

Variance Reduction

Let us establish a relation between the variance of $\mathbf{u}_p^{(k)}$ and of $\mathbf{u}_c^{(k)}$. We show that the Kalman correction in fact reduces the variance of the estimate. Let us introduce an auxiliary variable that we call *pseudo-observation*, i.e., an observation based on the prediction of the measure added by noise,

$$\mathbf{z}_{\mathrm{po}}^{(k)} = \mathrm{H}_k\mathbf{u}_p^{(k)} + \boldsymbol{\nu}^{(k)}.$$

It is possible to verify that

$$\Lambda_{\mathrm{po}}^{(k)} = \mathrm{H}_k\Lambda_p^{(k)}\mathrm{H}_k^T + \mathrm{R}_k, \quad \Lambda_{p,\mathrm{po}}^{(k)} := \mathcal{E}\left(\mathbf{u}_p^{(k)}\mathbf{z}_{\mathrm{po}}^{(k),T}\right) = \Lambda_p^{(k)}\mathrm{H}_k^T.$$

With this notation, we may rewrite the correction step of the Kalman filter as follows:

$$\mathrm{K}_k^* = \Lambda_{p,\mathrm{po}}^{(k)}(\Lambda_{\mathrm{po}}^{(k)})^{-1}, \quad \Lambda_c^{(k)} = \Lambda_p^{(k)} - \Lambda_{p,\mathrm{po}}^{(k)}(\Lambda_{\mathrm{po}}^{(k)})^{-1}\Lambda_{p,\mathrm{po}}^{(k),T}.$$

Since $\Lambda_{\mathrm{po}}^{(k)}$ is s.p.d., we have that $\Lambda_p^{(k)} - \Lambda_c^{(k)}$ is positive. This relation outlines the reduction of the variance induced by the correction step with respect to the variance of the prediction.

Recursive Formula for the Variance Equations, Riccati Equations

Let us eliminate $\Lambda_c^{(k)}$ from the equations of the Kalman filter, in particular we compute the variance $\Lambda_p^{(k+1)}$ as function of $\Lambda_p^{(k)}$. We get

$$\begin{aligned}\Lambda_p^{(k+1)} &= \mathrm{A}_k\Lambda_c^{(k)}\mathrm{A}_k^T + \mathrm{Q}_k \\ &= \mathrm{A}_k(\Lambda_p^{(k)} - \mathrm{K}_k\mathrm{H}_k\Lambda_p^{(k)})\mathrm{A}_k^T + \mathrm{Q}_k \\ &= \mathrm{A}_k\Lambda_p^{(k)}\mathrm{A}_k^T + \mathrm{Q}_k - \mathrm{A}_k\Lambda_p^{(k)}\mathrm{H}_k^T(\mathrm{H}_k\Lambda_p^{(k)}\mathrm{H}_k^T + \mathrm{R}_k)^{-1}\mathrm{H}_k\Lambda_p^{(k)}\mathrm{A}_k^T.\end{aligned} \tag{6.25}$$

In addition, by using the well-known Sherman–Morrison–Woodbury formula [36], we can write also the recursive variance matrix equation of the Kalman estimate

$$\Lambda_c^{(k)} = \left(\mathrm{I} - \mathrm{K}_k^*\mathrm{H}_k\right)\Lambda_p^{(k)} = \left((\Lambda_p^{(k)})^{-1} + \mathrm{H}_k^T\mathrm{R}_k^{-1}\mathrm{H}_k\right)^{-1}$$
$$= \left(\left(\mathrm{A}_{k-1}\Lambda_c^{(k-1)}\mathrm{A}_{k-1}^T + \mathrm{Q}_{k-1}\right)^{-1} + \mathrm{H}_k^T\mathrm{R}_k^{-1}\mathrm{H}_k\right)^{-1}. \tag{6.26}$$

Let us assume that the matrices A, H, R, Q do not depend on k. The latter equation in (6.25) reads then

$$\Lambda_p^{(k+1)} = \mathrm{A}\Lambda_p^{(k)}\mathrm{A}^T + \mathrm{Q} - \mathrm{A}\Lambda_p^{(k)}\mathrm{H}^T(\mathrm{H}\Lambda_p^{(k)}\mathrm{H}^T + \mathrm{R})^{-1}\mathrm{H}\Lambda_p^{(k)}\mathrm{A}^T.$$

This is called *Difference Riccati Equation* (DRE). A reasonable question related to this equation for time-independent dynamics refers to the existence of a stationary variance matrix. This is a variance matrix such that

$$\Lambda_p^{(k+1)} = \Lambda_p^{(k)} = \Lambda_p.$$

The latter can be clearly obtained as a fixed point of the DRE. This leads to solve the so-called *Algebraic Riccati Equation* (ARE)

$$\Lambda_p = \mathrm{A}\Lambda_p\mathrm{A}^T + \mathrm{Q} - \mathrm{A}\Lambda_p\mathrm{H}^T(\mathrm{H}\Lambda_p\mathrm{H}^T + \mathrm{R})^{-1}\mathrm{H}\Lambda_p\mathrm{A}^T.$$

This equation has been largely investigated by several authors [1, 46, 51], to determine under which conditions the solution Λ_{ARE} exists and it can be computed as the asymptotic limit of the corresponding DRE. In particular, let us assume that the dynamical system is asymptotically stable and converges to a steady solution. Clearly a good predictor is expected to follow the system dynamics, converging to the asymptotic estimate. Correspondingly, in this case we expect $\Lambda_p^{(k)}$ to converge to the asymptotic matrix Λ_{ARE}. Otherwise, our predictor would be unable to follow the system dynamics converging to the stationary solution. As a matter of fact, it is possible to prove that *if the system is stable, then the predictor is stable and its variance gets closer to the solution of the associated ARE* (see, e.g., [46] for a precise statement of the theorem).

In addition, we point out that this solution can be interpreted as an "approximate" Kalman filter, where the matrix $\Lambda_p^{(k)}$ is replaced by Λ_{ARE} to save the computational costs of computing $\Lambda_p^{(k)}$ at each step. This provides a stationary filter which is clearly sub-optimal, since the associated error is not orthogonal to the estimate. However, it may be computationally convenient.

Another possible use of Λ_{ARE} is to provide a bound to the variance of the "optimal case" when we apply the Kalman filter with no approximations.

Example. Let us consider the scalar case, with

$$u^{(k)} = u^{(k-1)} \quad (A = 1, b = 0)$$
$$z^{(k)} = u^{(k)} + v^{(k)} \quad (H = 1, v \sim \mathcal{G}(0,1)).$$

Assume also that the initial data $u^{(1)} \sim \mathcal{G}(\mu, 1)$. Set $u_p^{(1)} = \mu$. Then, the Kalman filter formulas read

$$u_p^{(k)} = u_c^{(k-1)}, \quad \lambda_p^{(k),2} = \lambda_c^{(k-1),2}$$

$$K_k = \frac{\lambda_p^{(k),2}}{\lambda_p^{(k),2} + 1}$$

$$u_c^{(k)} = u_p^{(k)} + \frac{\lambda_p^{(k),2}}{\lambda_p^{(k),2} + 1}(z^{(k)} - u_p^{(k)}) \quad = \frac{1}{\lambda_p^{(k),2} + 1} u_p^{(k)} + \frac{\lambda_p^{(k),2}}{\lambda_p^{(k),2} + 1} z^{(k)}$$

$$\lambda_c^{(k),2} = \frac{\lambda_p^{(k),2}}{\lambda_p^{(k),2} + 1} = \frac{\lambda_c^{(k-1),2}}{\lambda_c^{(k-1),2} + 1}.$$

We have therefore

$$u_p^{(1)} = \mu,\ \lambda_p^{(1)} = 1,\ K_1 = \frac{1}{2},\ u_c^{(1)} = \frac{\mu + z^{(1)}}{2} = u_p^{(2)}.$$

Notice that the prediction at $k = 2$ is just the sample average of the "past" and the new data. Similarly we obtain at a generic step k

$$u_p^{k+1} = u_c^k = \frac{\mu + \sum_{j=1}^k z^{(j)}}{k+1}.$$

Actually, we have the arithmetic average of the available data at t^k, that is somehow intuitively expected. Moreover, we have the recursive formula

$$\lambda_p^{(k+1),2} = \frac{\lambda_p^{(k),2}}{\lambda_p^{(k),2} + 1}, \quad \text{with } \lambda_p^{(1)} = 1.$$

By induction one can check that $\lambda_p^{(k),2} = \dfrac{1}{k}$. Consequently we have that

1. $\lim_{k \to \infty} \lambda_p^{(k)} = 0$, i.e., the prediction is asymptotically exact; similarly, the estimate is asymptotically exact;
2. the ARE $\lambda^2 = \lambda^2/(1 + \lambda^2)$ has only one solution, that is 0;
3. the Kalman filter is asymptotically stable, whereas the dynamic system is not asymptotically stable.

This example provides the case of an asymptotically stable estimator even when the dynamical system is not stable. As we have pointed out, the "reverse" situation (system is stable, estimator is unstable) is not possible: when the system is stable, the predictor is automatically stable.

An Alternative Look at the Kalman Filter

The Kalman filter can be obtained in different ways. Among the others, in particular here we mention a recent approach presented in [43], where the algorithm is the result of an application of the Newton root finding method with an appropriate initial guess. Beyond its intrinsic interest, this approach is actually oriented to extension to nonlinear systems in the form of an application of the Gauss–Newton method.

More precisely, assuming to have the exact initial state $\mathbf{u}^{(0)}$, let us consider the prediction-mismatch functional

$$\mathcal{J}_{k,p} = \frac{1}{2}\sum_{j=1}^{k} \|\mathbf{u}^{(j)} - \mathrm{A}_{j-1}\mathbf{u}^{(j-1)}\|^2_{\mathrm{Q}_j^{-1}} + \frac{1}{2}\sum_{j=1}^{k-1} \|\mathbf{z}^{(j)} - \mathrm{H}_j\mathbf{u}^{(i)}\|^2_{\mathrm{R}_j^{-1}}.$$

and the corresponding one for the estimate

$$\mathcal{J}_{k,c} = \mathcal{J}_{k,p} + \frac{1}{2}\|\mathbf{z}^{(k)} - \mathrm{H}_k\mathbf{u}^{(k)}\|^2_{\mathrm{R}_k^{-1}}.$$

The latter has the "natural" recursive formulation

$$\mathcal{J}_{k,c} = \mathcal{J}_{k-1,c} + \frac{1}{2}\|\mathbf{u}^{(k)} - \mathrm{A}_{k-1}\mathbf{u}^{(k-1)}\|^2_{\mathrm{Q}_k^{-1}} + \frac{1}{2}\|\mathbf{z}^{(k)} - \mathrm{H}_k\mathbf{u}^{(k)}\|^2_{\mathrm{R}_k^{-1}}.$$

We estimate $\mathbf{u}^{(k)}$ as the arg min of $\mathcal{J}_{k,c}$. When solving $\nabla\mathcal{J}_{k,c} = 0$, we apply the Newton method, that reads

$$\mathcal{H}\,(\mathbf{u}_{\text{new}} - \mathbf{u}_{\text{old}}) = -\nabla\mathcal{J}_{k,c}(\mathbf{u}_{\text{old}}) \tag{6.27}$$

where $\mathcal{H}$ is the Hessian matrix associated with $\mathcal{J}_{k,c}$. By selecting $\mathbf{u}_{\text{old}} = \mathbf{u}_p^{(k)} = \mathrm{A}_{k-1}\mathbf{u}_c^{(k-1)}$, it is possible to prove [43] that the Kalman estimate $\mathbf{u}_c^{(k)}$ is the solution $\mathbf{u}_{\text{new}}$ of (6.27). In the case of a linear system, $\mathbf{u}_{\text{new}} = \mathbf{u}_c^{(k)}$ minimizes $\mathcal{J}_{k,c}$. In fact, Newton method converges in one iteration when applied to linear equations.

Computational Issues Associated with the Kalman Filter

There are several issues associated with practical computation of the Kalman filter. Here we mention just a few.

From the numerical viewpoint the implementation of the filter following closely the formulas given above has a cost of $\mathcal{O}(n^3)$ operations at each time iteration, where n is the dimension of the matrix A. This cost is basically driven by the computation of the variance and gain matrices. For systems coming from the discretization of partial differential equations, n may be a fairly large number. However, the matrix is usually sparse and—as it is well known—this may reduce significantly the storage requirements and the number of operations. In addition, computational cost may still be an issue and specific methods for reducing the costs are mandatory. Among the others, we mention the replacement of the estimate covariance matrix with the asymptotic one (when the system dynamics is time independent) obtained by solving the ARE, as pointed out above.

Another computational issue is *numerical stability*. In particular, when computing the estimate variance matrix, Eq. (6.21) depends linearly on the computation error associated with the Kalman gain matrix K. In this respect, using Joseph formula is beneficial, since numerical errors are propagated quadratically. More in general, numerical errors may lead to computing non-positive covariance matrices. This problem can be faced by resorting to appropriate Cholesky or LDL^T factorizations of the covariance matrices that guarantee their numerical positiveness, leading to the so-called *square root form* of the filter.

6.2.4 Extension of the Kalman Filter to Nonlinear Problems

The most relevant limitation of the Kalman filter theory presented is that it relies upon linearity of the dynamical system, and that Gaussian densities remain Gaussian after linear transformations. However, in most of practical applications, the problem to solve is nonlinear. We show an important example hereafter (and many others later on, for the applications relevant to the contents of the present book).

We need therefore to find a way for extending the method to nonlinear cases, by properly approximating the procedures. We see methods based on both linearization and sampling.

Parameter Identification Via Kalman Filter

Consider a problem represented by a dynamical system with some parameters that we want to identify. We defined this as an *identification* problem. A possible approach to the problem is to add the parameter to the list of state variables and then to perform an estimation procedure. In general, this leads to a nonlinear dynamics. This approach is called *state augmentation technique*. We illustrate this in a case with a linear dynamics for the state variable $\mathbf{u}$

$$\begin{aligned} \mathbf{u}^{(k)} &= \mathrm{A}(\vartheta)\mathbf{u}^{(k-1)} + \mathbf{b}^{(k)} \\ \mathbf{z}^{(k)} &= \mathrm{H}(\vartheta)\mathbf{u}^{(k)} + \boldsymbol{\nu}^{(k)}. \end{aligned} \tag{6.28}$$

We assume for simplicity that ϑ is a time independent stochastic variable, so we have

$$\vartheta^{(k)} = \vartheta^{(k-1)} + \varepsilon^{(k)}$$

with $\varepsilon^{(k)} \sim \mathcal{G}(0, s_k)$, uncorrelated with other sources of noise. We augment the list of state variables of the parameter, so we have

$$\mathbf{v}^{(k)} = \begin{bmatrix} \mathbf{u}^{(k)} \\ \vartheta^{(k)} \end{bmatrix} \Rightarrow \mathbf{v}^{(k)} = \begin{bmatrix} \mathrm{A}(\vartheta^{(k-1)})\mathbf{u}^{(k-1)} \\ \vartheta^{(k-1)} \end{bmatrix} + \begin{bmatrix} \mathbf{b}^{(k)} \\ \varepsilon^{(k)} \end{bmatrix}, \tag{6.29}$$

with

$$\mathbf{z}^{(k)} = \mathrm{H}(\vartheta^{(k)})\mathbf{u}^{(k)} + \boldsymbol{\nu}^{(k)}.$$

In general, this is now a nonlinear augmented system so that the Kalman filter method presented above cannot be applied.

The Extended Kalman Filter

The Extended Kalman Filter (EKF) is the most immediate approach to extend the filter based on the *linearization* of both the system dynamics and of the observation process. Let us consider the nonlinear dynamic system

$$\begin{cases} \mathbf{u}^{(k)} = \mathcal{A}(\mathbf{u}^{(k-1)}) + \mathbf{b}^{(k)} \\ \mathbf{z}^{(k)} = \mathcal{H}(\mathbf{u}^{(k)}) + \boldsymbol{\nu}^{(k)}. \end{cases}$$

We still follow the minimal variance approach and introduce the *tangent operators*, i.e., the Jacobian matrices

$$\mathcal{A}'(\cdot) = \frac{\partial \mathrm{A}(\cdot)}{\partial \mathbf{u}}, \quad \mathcal{H}'(\cdot) = \frac{\partial \mathrm{H}(\cdot)}{\partial \mathbf{u}}.$$

After linearization, we get an extension of the Kalman filter. This reads

1. PREDICTION
   (a) $\mathbf{u}_p^{(k)} = \mathcal{A}(\mathbf{u}_c^{(k-1)})$,
   (b) $\Lambda_p^{(k)} = \mathcal{A}'(\mathbf{u}_c^{(k-1)})\Lambda_c^{(k-1)}(\mathcal{A}'(\mathbf{u}_c^{(k-1)}))^T + \mathrm{Q}_{k-1}$.
2. CORRECTION
   Kalman gain:
   $$\mathrm{K}_k = \Lambda_p^{(k)}\mathcal{H}'(\mathbf{u}_p^{(k)})^T \left(\mathcal{H}'(\mathbf{u}_p^{(k)})\Lambda_p^{(k)}(\mathcal{H}'(\mathbf{u}_p^{(k)}))^T + \mathrm{R}_k\right)^{-1}.$$
   (a) State estimate:
   $$\mathbf{u}_c^{(k)} = \mathbf{u}_p^{(k)} - \mathrm{K}_k \left(\mathbf{z}^{(k)} - \mathcal{H}(\mathbf{u}_p^{(k)})\right).$$
   (b) Covariance estimate:
   $$\Lambda_c^{(k)} = \Lambda_p^{(k)} - \mathrm{K}_k\, \mathcal{H}'(\mathbf{u}_p^{(k)})\Lambda_p^{(k)}.$$

As to be expected, most of the analysis holding for the linear case cannot be trivially extended to this case, since the covariance matrices associated with the errors depend on the linearization procedure. In particular, they depend on the set of observations so they are a random process. In addition, they are only an approximation of the error covariance and this leads to biased state estimates ($\mathcal{E}\left(\mathbf{e}_c^{(k)}\right) \neq 0$). Another drawback is the computational cost associated with the tangent operators, that for problems coming from the discretization of partial differential equations may be fairly expensive.

Nevertheless, we address the case of parameter estimation with the EKF.

EKF and Parameter Estimation

Let us apply EKF to (6.29), with

$$\mathcal{A}'(\mathbf{v}) = \begin{bmatrix} \mathrm{A}(\vartheta) & \dfrac{\partial \mathrm{A}}{\partial \vartheta}\mathbf{u} \\ O & \mathrm{I} \end{bmatrix}, \mathcal{H}'(\mathbf{v}) = \begin{bmatrix} \mathrm{H}(\vartheta) & \dfrac{\partial \mathrm{H}}{\partial \vartheta}\mathbf{u} \end{bmatrix}.$$

In many cases, H is independent of ϑ so that the last entry in $\mathcal{H}'(\mathbf{v})$ is zero. We assume that this is the case hereafter. In this form, the parameter estimation is performed following the EKF steps. The covariance matrix of the augmented status will be

$$\Lambda_{\mathrm{augm},\cdot} \equiv \begin{bmatrix} \Lambda_{u,\cdot} & \Lambda_{u\vartheta,\cdot} \\ \Lambda_{u\vartheta,\cdot} & \Lambda_{\vartheta,\cdot} \end{bmatrix}$$

where the dot can be either p in the prediction or c for the correction (or estimate). Then, we have the following steps.

1. PREDICTION

   (a) $\mathbf{u}_p^{(k)} = \mathcal{A}(\vartheta_c^{(k-1)})\mathbf{u}_c^{(k-1)}, \quad \vartheta_p^{(k)} = \vartheta_c^{(k-1)}$

   (b) $$\Lambda_{\mathrm{augm},p}^{(k)} = \begin{bmatrix} \mathrm{A}\Lambda_{u,c}^{(k-1)}\mathrm{A}^T + \mathrm{B} + \mathrm{B}^T + \dfrac{\partial \mathrm{A}}{\partial \vartheta}\mathbf{u}_c^{(k-1)}\Lambda_{\vartheta,c}^{(k-1)}\mathbf{u}_c^{(k-1),T}\dfrac{\partial \mathrm{A}^T}{\partial \vartheta} & \mathrm{C} \\ \mathrm{C}^T & \Lambda_{\vartheta,c}^{(k-1)} \end{bmatrix}$$

   where $\mathrm{B} = \mathrm{A}\Lambda_{u\vartheta,c}^{(k-1)}\mathbf{u}_c^{(k-1),T}\dfrac{\partial \mathrm{A}^T}{\partial \vartheta}$, $\mathrm{C} = \mathrm{A}\Lambda_{u\vartheta,c}^{(k-1)} + \dfrac{\partial \mathrm{A}}{\partial \vartheta}\mathbf{u}_c^{(k-1)}\Lambda_{\vartheta,c}^{(k-1)}$ and all the occurrences of A and its derivative are computed in $\vartheta_c^{(k-1)}$.

2. CORRECTION
   Kalman gain:

$$\mathrm{K}_k = \begin{bmatrix} \Lambda_{p,u}^{(k)}\mathrm{H}^T \\ \Lambda_{p,u\vartheta}^{(k)}\mathrm{H}^T \end{bmatrix}\left(\mathrm{H}\Lambda_{p,u}^{(k)}\mathrm{H}^T + \mathrm{R}_k\right)^{-1} = \begin{bmatrix} \mathrm{K}_{k,1} \\ \mathrm{K}_{k,2} \end{bmatrix}.$$

(a) State and Parameter:

$$\mathbf{u}_c^{(k)} = \mathbf{u}_p^{(k)} - \mathrm{K}_{k,1}\left(\mathbf{z}^{(k)} - \mathrm{H}(\vartheta_p^{(k)})\mathbf{u}^{(k)}\right),$$

$$\vartheta_c^{(k)} = \vartheta_p^{(k)} - \mathrm{K}_{k,2}\left(\mathbf{z}^{(k)} - \mathrm{H}(\vartheta_p^{(k)})\mathbf{u}^{(k)}\right).$$

(b) Covariance estimate:

$$\Lambda_{\mathrm{augm},c}^{(k)} = \Lambda_{\mathrm{augm},p}^{(k)} - \begin{bmatrix} \mathrm{K}_{k,1}\mathrm{H}(\vartheta_p^{(k)})\Lambda_{p,u}^{(k)} & \mathrm{K}_{k,1}\mathrm{H}(\vartheta_p^{(k)})\Lambda_{p,u\vartheta}^{(k)} \\ \mathrm{K}_{k,2}\mathrm{H}(\vartheta_p^{(k)})\Lambda_{p,u}^{(k)} & \mathrm{K}_{k,2}\mathrm{H}(\vartheta_p^{(k)})\Lambda_{p,u\vartheta}^{(k)} \end{bmatrix}.$$

It is worth noting that in this way we have a sort of adaptive filtering, since the improvement of the knowledge of the parameter affects the quality of the state estimate in a self-learning process.

As we have pointed out in the Introduction, there are several ways to perform parameter estimation (see, e.g., [2, 3]), this one is just an example. In Sect. 6.4.2 we present an example relevant to FSI. Since EKF suffers from the computation of the tangent operators, this can be avoided by resorting to a different extension of the Kalman Filter, that we introduce in the next section.

Remark 6.4. EKF can be regarded as the result of the application of one iteration of the Gauss–Newton method for the minimization of a suitable mismatch functional, as we have seen for the linear case. For more details, see [43]

6.2.5 The Unscented Kalman Filter

As pointed out above, errors associated with the linearization of EKF lead in general to sub-optimal performances. In the unscented Kalman filter (UKF) [45], the basic idea is to approximate the evolution of the nonlinear dynamic system not by linearization but by deterministic sampling, following the so-called *unscented transformation* (UT). The basic idea of UT is that "it is easier to approximate a Gaussian distribution than it is to approximate an arbitrary nonlinear function or transformation" [44]. For this reason, the nonlinear dynamics in UKF is statistically approximated by mean and covariance of samples suitably selected for the state variable to be estimated.

For instance, suppose to have a scalar Gaussian random variable $u^{(k)}$ with mean μ and variance λ^2. At the first step we determine two samples of $u^{(k)}$, as $s_{1,2} = \mu \pm \lambda$. If we need to approximate a nonlinear evolution $u^{(k+1)} = f(u^{(k)})$, we compute the samples $f_i \equiv f(s_i)$ and take

$$\begin{aligned} &\mathcal{E}\left(u^{(k+1)}\right) \approx w_1 f_1 + w_2 f_2 \equiv \overline{f}, \\ &\mathcal{E}\left(\left(f(u^{(k)}) - \mathcal{E}\left(f(u^{(k)})\right)\right)^2\right) \approx w_1(f_1 - \overline{f})^2 + w_2(f_2 - \overline{f})^2, \end{aligned}$$

where w_i are suitable weighting coefficients.

The selection of the sampling points (called σ-points) is clearly of paramount importance and can be done in different ways. In general [44, 45, 73], a canonical choice for a state variable $\mathbf{u}^{(k)}$ of size n, with Gaussian distribution with mean $\mathcal{E}\left(\mathbf{u}^{(k)}\right)$ and covariance matrix Λ reads

$$\begin{aligned} \mathbf{s}_0 &= \mathcal{E}\left(\mathbf{u}^{(k)}\right), \mathbf{s}_i = \mathcal{E}\left(\mathbf{u}^{(k)}\right) + \left(\sqrt{(n+\kappa)\Lambda}\right)_i, \\ \mathbf{s}_{i+n} &= \mathcal{E}\left(\mathbf{u}^{(k)}\right) - \left(\sqrt{(n+\kappa)\Lambda}\right)_i, \qquad\qquad i = 1, 2, \ldots, n, \\ w_0 &= \frac{\kappa}{\kappa+n}, \quad w_i = w_{i+n} = \frac{\kappa+n}{2}, \end{aligned}$$

where κ is a real scaling factor and $\left(\sqrt{(n+\kappa)\Lambda}\right)_i$ is the i-th row of the matrix $\sqrt{(n+\kappa)\Lambda}$. This can be computed by a Cholesky factorization of the s.p.d. matrix. Other criteria for sampling can be, however, pursued [6].

The UKF will eventually consist of a sampling step, followed by the "Kalman-like" prediction and correction steps.

1. SAMPLING: Let $\mathsf{C}(\cdot)$ denote the Cholesky decomposition of an s.p.d. matrix. We take

$$\begin{aligned} \mathrm{C}_{(k-1)} &= \sqrt{n+\kappa}\,\mathsf{C}(\Lambda_p^{(k-1)}) \\ \mathbf{u}_0^{(k-1)} &= \mathbf{u}_c^{(k-1)}, \\ \mathbf{u}_i^{(k-1)} &= \mathbf{u}_c^{(k-1)} + \mathrm{C}_{(k-1),i}, \quad i = 1, 2, \ldots, n, \\ \mathbf{u}_{i+n}^{(k-1)} &= \mathbf{u}_c^{(k-1)} + \mathrm{C}_{(k-1),i}, \quad i = 1, 2, \ldots, n. \end{aligned}$$

2. PREDICTION: Let w_i be the weight coefficients.

$$\begin{aligned} \mathbf{u}_{p,i}^{(k)} &= \mathrm{A}(\mathbf{u}_{p,i}^{(k-1)}) \quad \text{sample evolution} \\ \mathbf{u}_p^{(k)} &= \sum_i w_i \mathbf{u}_{p,i}^{(k)}, \quad \Lambda_p^{(k)} = \sum_i w_i \left(\mathbf{u}_{p,i}^{(k)} - \mathbf{u}_p^{(k)}\right)\left(\mathbf{u}_{p,i}^{(k)} - \mathbf{u}_p^{(k)}\right)^T \end{aligned}$$

3. CORRECTION:

$$\begin{aligned} \mathbf{z}_i^{(k)} &= \mathrm{H}(\mathbf{u}_{p,i}^{(k)}) \\ \Lambda_{\mathrm{po}}^{(k)} &= \sum_i w_i \left(\mathbf{z}_i^{(k)} - \mathrm{H}(\mathbf{u}_p^{(k)})\right)\left(\mathbf{z}_i^{(k)} - \mathrm{H}(\mathbf{u}_p^{(k)})\right)^T \\ \Lambda_{p,\mathrm{po}}^{(k)} &= \sum_i w_i \left(\mathbf{z}_i^{(k)} - \mathrm{H}(\mathbf{u}_p^{(k)})\right)\left(\mathbf{u}_{p,i}^{(k)} - \mathbf{u}_p^{(k)}\right)^T \\ \mathrm{K}_k &= \Lambda_{p,\mathrm{po}}^{(k)}\left(\Lambda_{\mathrm{po}}^{(k)}\right)^{-1} \\ \mathbf{u}_c^{(k)} &= \mathbf{u}_p^{(k)} + \mathrm{K}_k\left(\mathbf{z}^{(k)} - \mathrm{H}(\mathbf{u}_p^{(k)})\right) \\ \Lambda_c^k &= \Lambda_p^k - \Lambda_{p,\mathrm{po}}^{(k)}\left(\Lambda_{\mathrm{po}}^{(k)}\right)^{-1}\Lambda_{p,\mathrm{po}}^{(k),T}. \end{aligned}$$

Examples of this method can be found in [73]. *UKF dual estimation* is in particular the identification of parameters of the model simultaneous to the state estimation, similarly to what we have illustrated for the EKF. In this respect, we will see an example in Sect. 6.4.2. A smart implementation of the methods may be necessary for problems coming from the discretization of partial differential equations, using, for example, the so-called *Factorized UKF*. In particular, for parameter estimation, computational cost can be reduced by assuming that uncertainty affects only the parameter of interest and not the entire state. For more details, see [6, 54].

6.3 Deterministic Variational Assimilation Methods

In this section, we consider a different approach for data assimilation, based on a deterministic approach. We do not necessarily rely upon a priori statistical knowledge of the process and we formulate the problem as a minimization procedure, where the mathematical paradigm acts as a constraint. For instance, referring to Fig. 6.3, with an educated guess we can decide a functional form for $\mathbf{w}$ and then fit this form with the measures. In other words we find $\mathbf{w}$ belonging to some class of functions V such that

$$\mathrm{dist}(\mathbf{z}, \mathsf{Observation}(\mathbf{w})) \leq \mathrm{dist}(\mathbf{z}, \mathsf{Observation}(\mathbf{v}))$$

for all $\mathbf{v} \in V$, where $\mathsf{Observation}(\mathbf{v})$ is the application of the mathematical representation of the measure process to $\mathbf{v}$, to be compared with the "real" measure $\mathbf{z}$. If the measures are trustworthy, we could derive a model for capturing exactly the data pursuing an interpolation approach. In general measures are noisy and we resort to a Least Squares (LS) procedure, so that the model fits the data in a "weaker" sense. Notice that the definition of the distance is somehow arbitrary. For instance, it could include an a priori knowledge of the location of data more trustworthy than others by means of proper coefficients that give more relevance to these data.

In the general case of interest for our applications when we have a dynamical system evolving, we can recast the assimilation procedure as a *control problem*. In a very abstract setting, we may list the ingredients of this approach as follows (see Fig. 6.6).

A mathematical model describing the dynamics of interest for the variables or physical quantities describing the state of the system we are interested in. In our problems, this model or paradigm is given by a system of partial differential equations and, more precisely, a model describing FSI. In this case, the state variables are represented by the velocity, the pressure of the fluid, and the displacement of the structure.

A set of observations or measurements of the state variables or, more in general, of a function of the state.

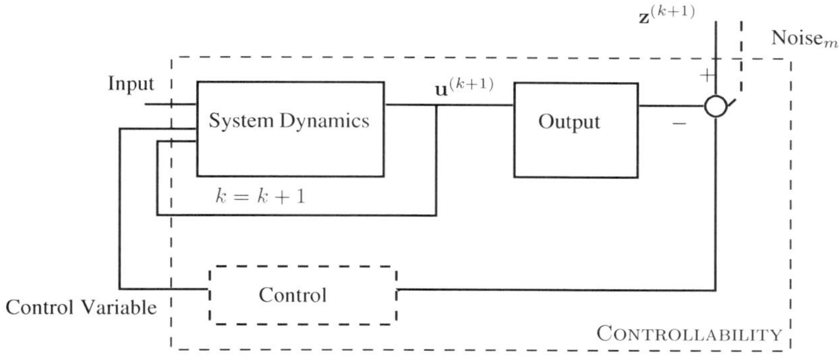

Fig. 6.6 Possible approaches for the estimate with a system dynamics: here we do not use stochastic knowledge and we refer to the concept of CONTROLLABILITY: the estimate is reformulated as a control process

A functional $\mathcal{J}$ to be minimized. In general, this is the discrepancy between the results obtained by the mathematical (numerical) model and the available data.

A control variable (CV), which is the variable that we tune to get the minimization done. Its choice strongly depends on the purpose of the assimilation. For instance, in identification (parameter estimation) problems, the parameter(s) to be identified will be the control variable(s) to drive the minimization.

Solution of constrained minimization problems with distributed models (partial differential equations) acting as constraint has been considered by several authors [3,35,66,69]. With no claim to be exhaustive, in the present section we provide some general solution methods with simple examples, that have been used in applications of interest for biomedical FSI problems. Since minimization procedures resort typically to iterative methods, the solution of the system of partial differential equation representing the model needs typically to be solved several times. This rapidly leads to high computational costs, in particular when working on unsteady problems, as the ones we are interested in. We need to address therefore the problem of reducing the computational costs.

The key concept in this case is *controllability*—which is the dual concept of observability advocated in the previous section—in other terms the effectiveness of the control strongly depends on the *sensitivity* of the functional to be minimized to the control variable. More the functional is sensitive to the control and most likely the minimization will be successful. This is somehow a change in the usual perspective of solving problems in engineering. As a matter of fact, *high sensitivity* comes from a lack of stability or robustness and the control action is intended to recover these properties. For instance, in the case of fluids we should expect a control to be more effective when the Reynolds number is high, because in this case, in general, the variable of interests are more sensible to perturbations and for this reasons they may be controlled.

The entire section is largely based on [38, Chaps. 2 and 5].

6.3.1 Least Squares Estimators

As we have done in the previous section, we start with some considerations on the "steady" case, when we perform an In-Out constrained minimization (Fig. 6.3).

Suppose that we have a sequence of measures of the same variable $\mathbf{w} \in \mathbb{R}^n$ affected by noise,

$$\mathbf{z}_i = \mathrm{H}_i \mathbf{w} + \text{noise}_i, \quad i = 1, 2, \ldots, m.$$

We do not postulate any a priori probabilistic knowledge of the noise. The problem of estimating $\mathbf{w}$ from these measures has a classical deterministic formulation given by the Least-Squares (LS) approach. More precisely, the problem is formulated as: find the optimal $\mathbf{w}$ such that

$$\mathbf{w} = \arg\min \mathcal{J},$$

where

$$\mathcal{J} = \frac{1}{2} \sum_{i=1}^{m} (\mathbf{z}_i - \mathrm{H}_i \mathbf{w})^T \Omega_m^{-1} (\mathbf{z}_i - \mathrm{H}_i \mathbf{w})$$

and Ω_m^{-1} is an $n \times n$ weight matrix, which is assumed to be s.p.d. Let

$$\hat{\mathrm{H}}_m = \begin{bmatrix} \mathrm{H}_1 \\ \mathrm{H}_2 \\ \cdots \\ \mathrm{H}_m \end{bmatrix} \in \mathbb{R}^{nm,n}, \quad \hat{\mathbf{z}} = \begin{bmatrix} \mathbf{z}_1 \\ \mathbf{z}_2 \\ \cdots \\ \mathbf{z}_m \end{bmatrix} \in \mathbb{R}^{nm}, \quad \hat{\Omega}_m = \begin{bmatrix} \Omega_m & & \mathrm{O} \\ & \ddots & \\ \mathrm{O} & & \Omega_m \end{bmatrix}$$

then,

$$\mathcal{J} = \frac{1}{2} (\hat{\mathbf{z}} - \hat{\mathrm{H}}_m \mathbf{w})^T \hat{\Omega}_m^{-1} (\hat{\mathbf{z}} - \hat{\mathrm{H}}_m \mathbf{w}).$$

Solving

$$\frac{\partial \mathcal{J}}{\partial \mathbf{w}} = 0$$

we find

$$\hat{\mathrm{H}}_m^T \hat{\Omega}_m^{-1} \left(\hat{\mathbf{z}} - \hat{\mathrm{H}}_m \mathbf{w}_{\mathrm{LS}} \right) = 0.$$

Thus,

$$\mathbf{w}_{\mathrm{LS}} = \left(\hat{\mathrm{H}}_m^T \hat{\Omega}_m^{-1} \hat{\mathrm{H}}_m \right)^{-1} \hat{\mathrm{H}}_m^T \hat{\Omega}_m^{-1} \hat{\mathbf{z}} = \Lambda_m \hat{\mathrm{H}}_m^T \hat{\Omega}_m^{-1} \hat{\mathbf{z}}.$$

where $\Lambda_m = (\hat{\mathrm{H}}_m^T \Omega_m^{-1} \hat{\mathrm{H}}_m)^{-1}$.

Remark 6.5. Let us consider a recursive formulation of this problem, obtained adjusting an available estimate (based on previous observations) when a new observation becomes available. Let us write recursively for $k > 1$

$$\hat{\mathrm{H}}_k = \begin{bmatrix} \hat{\mathrm{H}}_{k-1} \\ \mathrm{H}_k \end{bmatrix}, \qquad \hat{\mathbf{z}}_k = \begin{bmatrix} \hat{\mathbf{z}}_{k-1} \\ \mathbf{z}_k \end{bmatrix}.$$

We also write

$$\widehat{\Omega_k}^{-1} = \begin{bmatrix} \hat{\Omega}_{k-1}^{-1} & 0 \\ 0 & \Omega_k^{-1} \end{bmatrix},$$

where Ω_k^{-1} is an s.p.d. matrix. With this notation, we may write

$$\hat{\mathrm{H}}_k^T \hat{\Omega}_k^{-1} \hat{\mathrm{H}}_k = \hat{\mathrm{H}}_{k-1}^T \hat{\Omega}_{k-1}^{-1} \hat{\mathrm{H}}_{k-1} + \mathrm{H}_k^T \Omega_k^{-1} \mathrm{H}_k.$$

or, with the (suggestive) notation introduced above

$$\Lambda_k^{-1} = \Lambda_{k-1}^{-1} + \mathrm{H}_k^T \Omega_k^{-1} \mathrm{H}_k.$$

By the Sherman–Morrison–Woodbury formula we obtain

$$\Lambda_k = \left(\Lambda_{k-1}^{-1} + \mathrm{H}_k^T \Omega_k^{-1} \mathrm{H}_k\right)^{-1} = \Lambda_{k-1} - \Lambda_{k-1} \mathrm{H}_k^T \left(\Omega_k + \mathrm{H}_k \Lambda_{k-1} \mathrm{H}_k^T\right)^{-1} \mathrm{H}_k \Lambda_{k-1}.$$

Let us introduce the matrix

$$\mathrm{G}_k = \Lambda_{k-1} \mathrm{H}_k^T \left(\Omega_k + \mathrm{H}_k \Lambda_{k-1} \mathrm{H}_k^T\right)^{-1},$$

so we have

$$\Lambda_k = (\mathrm{I} - \mathrm{G}_k \mathrm{H}_k) \Lambda_{k-1}.$$

From here, we can obtain the recursive formula (see [68, Sect. 4.3])

$$\hat{\mathbf{w}}_k^{\mathrm{LS}} = \hat{\mathbf{w}}_{k-1}^{\mathrm{LS}} + \mathrm{G}_k \left(\mathbf{z}_k - \mathrm{H}_k \hat{\mathbf{w}}_{k-1}^{\mathrm{LS}}\right),$$

that has a formal analogy with Kalman filter formulas, even though in this case the dynamics is not related to the state (no dynamics occurs on $\mathbf{w}$) but just to the addition of new measures. In this respect, $\mathbf{z}_k - \mathrm{H}_k \hat{\mathbf{w}}_{k-1}^{\mathrm{LS}}$ represents the net content of new information brought by the new measure.

When we assume that the state evolves, the mathematical equation describing the dynamics may be used as a constraint to the minimization process. Algebraic aspects of constrained LS problems, with both equality and inequality constraints, have been addressed in [36, Chap. 12]. For instance, we may consider the problem:

find $\mathbf{x}$ such that

$$\mathbf{x} = \arg\min \|\mathrm{A}\mathbf{x} - \mathbf{b}\|_2, \quad \mathrm{B}\mathbf{x} = \mathbf{d}$$

where A is an $m \times n$ matrix, B is $p \times n$, $\mathbf{b} \in \mathbb{R}^m$, $\mathbf{d} \in \mathbb{R}^p$. We assume that the matrices are both full-rank. The problem can be solved by an application of the Generalized Singular Value Decomposition (GSVD),

$$\mathrm{U}^T\mathrm{AX} = \mathrm{diag}(\alpha_1, \alpha_2, \ldots, \alpha_n) = \mathrm{D}_A, \quad \mathrm{V}^T\mathrm{BX} = \mathrm{diag}(\beta_1, \beta_2, \ldots, \beta_p) = \mathrm{D}_B,$$

with U and V orthogonal matrices and $\mathbf{x}_i$ are the columns of X for $i = 1, 2, \ldots, n$. The solution to this problem then reads [36]

$$\mathbf{x} = \sum_{i=1}^{p} \frac{\mathbf{v}_i^T\mathbf{d}}{\beta_i}\mathbf{x}_i + \sum_{i=p+1}^{n} \frac{\mathbf{u}_i^T\mathbf{b}}{\alpha_i}\mathbf{x}_i.$$

We can consider the associated unconstrained LS problem

$$\mathbf{x} = \arg\min \| \begin{bmatrix} \mathrm{A} \\ \lambda\mathrm{B} \end{bmatrix} \mathbf{x} - \begin{bmatrix} \mathbf{b} \\ \lambda\mathbf{d} \end{bmatrix} \|_2$$

This can be solved with an ordinary LS procedure. Using the GSVD, it is possible to find the solution

$$\mathbf{x}(\lambda) = \sum_{i=1}^{p} \frac{\alpha_i\mathbf{u}_i^T\mathbf{b} + \lambda^2\beta_i\mathbf{v}_i^T\mathbf{d}}{\alpha_i^2 + \lambda^2\beta_i^2}\mathbf{x}_i + \sum_{i=p+1}^{n} \frac{\mathbf{u}_i^T\mathbf{b}}{\alpha_i}\mathbf{x}_i,$$

from which it is promptly realized that the solution to the constrained minimization problem $\mathbf{x} = \lim_{\lambda\to\infty}\mathbf{x}(\lambda)$. In the next section, we see a similar approach for solving unconstrained minimization when the constraint is represented by a partial differential equation.

6.3.2 Constrained Minimization Problems with PDEs: A Simple Working Example

To be concrete, we illustrate techniques of constrained minimization with partial differential equations on the following problem. Let $\Omega \subset \mathbb{R}^n$ ($n = 2, 3$) and $\partial\Omega$ be denoted by Γ. We assume that $u(\mathbf{x})$ is the state variable that obeys the following equation

$$-\Delta u + \mathbf{b}\cdot\nabla u + \gamma u + u^3 = \sum_{i=1}^{K}\alpha_i f_i, \quad \text{in } \Omega \tag{6.30}$$

γ and α_i ($i = 1, \dots, K$) are real coefficients. $\boldsymbol{\alpha} \in \mathbb{R}^K$ is the vector with entries α_i. $\mathbf{b}$ is a divergence-free vector function. In our applications, $\mathbf{b}$ may represent the blood flow, and u the concentration of some solute in the blood. We assume to have a *reference* or desired function $d(\mathbf{x})$ we would like to be approximated by $u(\mathbf{x})$. We assume moreover

$$u = 0 \quad \text{on } \partial\Gamma. \tag{6.31}$$

We assume that γ and α_i are unknown parameters. They need to be computed so to drive the state variable to the reference behavior. We formulate therefore the problem[5]: find $\gamma, \boldsymbol{\alpha}$ to minimize

$$\mathcal{J}(u) = \frac{1}{2}\int_\Omega (u(\mathbf{x}) - d(\mathbf{x}))^2 d\mathbf{x},$$

where $u(\mathbf{x})$ solves (6.30), (6.31).

Quite often $\mathcal{J}$ is added with a term depending explicitly on the parameters to be estimated in the form

$$\mathcal{J}_R(u, \boldsymbol{\alpha}, \gamma) = \mathcal{J}(u) + \frac{\sigma_1}{2}\|\boldsymbol{\alpha}\|^2 + \frac{\sigma_2}{2}\|\gamma - \gamma_{\text{ref}}\|^2, \tag{6.32}$$

where σ_1 and σ_2 are constants and $\|\cdot\|$ denotes a generic (convenient) norm; in the remainder of the section we assume $\|\cdot\| = \|\cdot\|_2$. This modification may have both practical and theoretical reasons.

1. First practical motivation: When the control variable corresponds to a physical control, like the coefficients $\boldsymbol{\alpha}$, it implies a practical cost (intended in a broad sense as the energy required to apply it). For this reason, the "size" of the control cannot be too large. The correction of $\mathcal{J}$ with $\frac{\sigma_1}{2}\|\boldsymbol{\alpha}\|^2$ is a "penalization" that includes the cost of the control.[6]
2. Second practical motivation: In some cases, in particular when identifying a parameter, a "nominal" reference guess is available, based, for instance, on averaging available measures or samples. This is denoted here by γ_{ref} and the real value is supposed to be "not too far" from this value. This leads to the term $\frac{\sigma_2}{2}\|\gamma - \gamma_{\text{ref}}\|^2$ that penalizes the difference respect to the nominal value.
3. Mathematical motivation: If we hypothetically consider only the terms $\mathcal{J}_R(u, d) = \frac{\sigma_1}{2}\|\boldsymbol{\alpha}\|^2 + \frac{\sigma_2}{2}\|\gamma - \gamma_{\text{ref}}\|^2$, the function to be minimized has excellent mathematical properties. It is actually quadratic and the minimization leads clearly to the solution $(\boldsymbol{\alpha}, \gamma) = (\mathbf{0}, \gamma_{\text{ref}})$. We infer therefore that the term $\mathcal{J}_R$

[5]A similar problem has been investigated as a simplified model of superconductivity in [69].

[6]This could be done also with unilateral constraints $\|\boldsymbol{\alpha}\| \leq$ max-cost-allowed.

has a regularizing effect on the minimization properties, balancing the bad (or not so good) properties of the original constrained minimization. As a matter of fact, in general the original problem may be ill-posed, featuring multiple local minima or none. The term $\mathcal{J}_R$ with a proper selection of the weights σ_1 and σ_2 allows us in general to have a well-posed problem. For this reason, when solving this kind of *inverse problems*, this term is often called *regularization* (*Tikhonov regularization* in the form in (6.32)). Other forms of regularization may be considered in practice, but they will not be addressed here (see [23, 39]).

The appropriate selection of weights $\sigma_{1,2}$ is not trivial. It is actually a trade-off between the minimization of the mismatch (that requires these weights to be small) and the regularization of the problem (that in general is improved for large positive values). Different strategies are possible. A general approach is to identify values such that the impact of additional terms on the non-regularized functional is bounded by the numerical errors, so to reduce the effects on mismatch minimization within the range acceptable after approximations, while improving the conditioning properties of the problem. See, e.g., [67, 72].

Gâteaux and Fréchet Derivatives

For the solution of a PDE constrained optimization problem, we need to be able to differentiate operators acting between functional spaces. In particular, let $\mathcal{F} : X \to Y$, X and Y being appropriate functional spaces and let $u, v \in X$. The derivative of $\mathcal{F}$, in the direction, v can be computed as

$$D\,\mathcal{F}(u;\, v) := \lim_{\varepsilon \to 0} \frac{\mathcal{F}(u + \varepsilon v) - \mathcal{F}(u)}{\varepsilon}.$$

Such derivative is called *Gâteaux* derivative. As an example, take $\mathcal{G}(u) = (u - f)^2$, then

$$D\,\mathcal{G}(u; v) = \lim_{\varepsilon \to 0} \frac{(u + \varepsilon v - f)^2 - (u - f)^2}{\varepsilon} = \lim_{\varepsilon \to 0} \frac{2\varepsilon(u - f)v + \varepsilon^2 v^2}{\varepsilon} = 2(u - f)v.$$

It is often possible to write the Gâteaux derivative of $\mathcal{F}$ in any direction v, as the application of a bounded linear operator $\left.\frac{D\mathcal{F}}{Du}\right|_u$ to v. Such operator is called *Fréchet* derivative. In the following we assume that the Fréchet derivative exists and we write

$$D\,\mathcal{F}(u;\, v) = \left.\frac{D\mathcal{F}}{Du}\right|_u (v).$$

In our example, $\left.\frac{D\mathcal{F}}{Du}\right|_u = 2(u - f)$. It is possible to show that the Gâteaux derivative of $\mathcal{G}(\boldsymbol{u}) = \boldsymbol{u}^T A\,\boldsymbol{u}$, in the direction $\boldsymbol{v}$, where $\boldsymbol{u}$ and $\boldsymbol{v}$ are vector functions and A is a constant square matrix of compatible dimensions, reads

$$D\,\mathcal{G}(\boldsymbol{u},\, \boldsymbol{v}) = \boldsymbol{u}^T\, A\, \boldsymbol{v} + \boldsymbol{v}^T\, A\, \boldsymbol{u} = \boldsymbol{u}^T(A + A^T)\boldsymbol{v},$$

while its Fréchet derivative reads

$$\left.\frac{D\mathcal{G}}{D\boldsymbol{u}}\right|_{\boldsymbol{u}} = \boldsymbol{u}^T(A + A^T),$$

with the understanding, in this case, that the application of the operator $\left.\frac{D\mathcal{G}}{D\boldsymbol{u}}\right|_{\boldsymbol{u}}$ to $\boldsymbol{v}$ is the usual matrix vector product of the one-row matrix $\left.\frac{D\mathcal{G}}{D\boldsymbol{u}}\right|_{\boldsymbol{u}}$ and the vector $\boldsymbol{v}$. As another example, consider $\mathcal{G}(u) = -\Delta u$, then $D\mathcal{G}(u,\, v) = -\Delta v$ and $\left.\frac{D\,\mathcal{G}}{Du}\right|_{u} = -\Delta$. In general, the derivative of a linear operator (the Laplacian operator in this case) is the linear operator itself.

The usual chain rule holds for the differentiation of composite functions

$$D\,(\mathcal{G} \circ \mathcal{F})(u,\, v) = D\,\mathcal{G}\,(\mathcal{F}(u),\; D\mathcal{F}(u, v))\,,$$

or

$$\left.\frac{D\,(\mathcal{G} \circ \mathcal{F})}{Du}\right|_{u} = \left.\frac{D\,(\mathcal{G} \circ F)}{D\mathcal{F}}\right|_{\mathcal{F}(u)} \left(\left.\frac{D\mathcal{F}}{Du}\right|_{u}\right).$$

Gradient-Based Optimization Approaches

A common and effective approach to deal with optimization constrained by partial differential equations is to include directly the constraint in the functional to be minimized. In this way, the minimization procedure is recast in an unconstrained case and the solution is obtained with classical arguments. In particular, the first order necessary conditions are obtained by setting to 0 the gradient of the functional.

In our case, this means that the solution u is computed as a function of the control variables $\boldsymbol{\alpha}$ and γ and the total derivative of $\mathcal{J}_R$, regarded as function of these variables, is set to 0. This procedure admits an iterative implementation. Let us denote the state problem (6.30), (6.31) with the abstract notation $\mathcal{F}(u, \boldsymbol{\alpha}, \gamma) = 0$.

Assume that an initial guess $\boldsymbol{\alpha}^{(0)}$ and $\gamma^{(0)}$ is given. Typically, we take $\gamma^{(0)} = \gamma_{\text{ref}}$. Then, we perform the following steps for $j = 0, 1, 2, \ldots$:

- find the state variable $u^{(j)}$ solution to $\mathcal{F}(u, \boldsymbol{\alpha}^{(j)}, \gamma^{(j)}) = 0$;
- compute $D\mathcal{J}_R(u^{(j)}, \boldsymbol{\alpha}^{(j)}, \gamma^{(j)})/D\boldsymbol{\alpha}\big|_{\boldsymbol{\alpha}^{(j)}}$ and $D\mathcal{J}_R u^{(j)}, \boldsymbol{\alpha}^{(j)}, \gamma^{(j)}/D\gamma\big|_{\gamma^{(j)}}$;
- **if** $\|D\mathcal{J}_R(u^{(j)}, \boldsymbol{\alpha}^{(j)}, \gamma^{(j)})/D[\boldsymbol{\alpha}, \gamma]\|$ is sufficiently small, solution is reached;

else compute a new guess $\boldsymbol{\alpha}^{(j+1)}$, $\gamma^{(j+1)}$, for instance by setting

$$\begin{aligned}\alpha_i^{(j+1)} &= \alpha_i^{(j)} - \omega_i^{(j)} \frac{D\mathcal{J}_R(u^{(j)}, \boldsymbol{\alpha}^{(j)}, \gamma^{(j)})}{D\alpha_i}, \qquad i = 1, 2, \ldots, K \\ \gamma^{(j+1)} &= \gamma^{(j)} - \omega_{K+1}^{(j)} \frac{D\mathcal{J}_R(u^{(j)}, \boldsymbol{\alpha}^{(j)}, \gamma^{(j)})}{D\gamma},\end{aligned} \tag{6.33}$$

where $\omega_i^{(j)}$, $i = 1, 2, \ldots, K+1$ are numerical coefficients that drive the convergence of the procedure.

This approach, based on (6.33), belongs to the family of *steepest descent methods* and the parameters ω_i define the step performed in updating the solution along the line identified by the gradient. These coefficients, in general, may be dynamically determined at each iteration. Other iterative methods may be considered for the sake of effectiveness. Among others, a method that usually outperforms the steepest descent approach is the *Broyden–Fletcher–Goldfarb–Shanno* (BFGS) method (see, e.g., [58]); another common choice is the Gauss–Newton method. The latter finds the roots of $D\mathcal{J}_R/D[\boldsymbol{\alpha}, \gamma] = 0$ using the Newton method, that means that, at each iteration j, the minimization of the paraboloid tangent to $\mathcal{J}_R$ in $\boldsymbol{\alpha}^{(j)}, \gamma^{(j)}$ is performed. The method is potentially second order, but it has the drawback that the Hessian of the functional $\mathcal{J}_R$ is needed.

The most troublesome step in the previous algorithm is the computation of the gradients $D\mathcal{J}_R(u^{(j)}, \boldsymbol{\alpha}^{(j)}, \gamma^{(j)})/D[\boldsymbol{\alpha}, \gamma]$. Let us address two possible methods.

Gradient Computation Through Sensitivities

A possible way for computing the gradients relies upon the chain rule

$$\left.\frac{D\mathcal{J}_R}{D\alpha_i}\right|_{\alpha^{(j)}} = \left.\frac{\partial \mathcal{J}_R}{\partial u}\right|_{u^{(j)}} \left(\left.\frac{\partial u}{\partial \alpha_i}\right|_{\alpha_i^{(j)}} \right) + \left.\frac{\partial \mathcal{J}_R}{\partial \alpha_i}\right|_{\alpha_i^{(j)}}, \qquad i = 1, 2, \ldots, K+1$$

where for easiness of notation we set $\alpha_{K+1} = \gamma$. We call *sensitivities* the derivatives

$$\phi_i \equiv \frac{\partial u}{\partial \alpha_i}, \qquad \forall i = 1, 2, \ldots, K+1$$

as they quantify the sensitivity of the solution to each control variable. From

$$\mathcal{F}(u^{(j)}, \boldsymbol{\alpha}^{(j)}) = 0 \Rightarrow \left.\frac{D\mathcal{F}}{D\alpha_i}\right|^{(j)} = \left.\frac{\partial \mathcal{F}}{\partial u}\right|_{u^{(j)}} \left(\phi_i^{(j)}\right) + \left.\frac{\partial \mathcal{F}}{\partial \alpha_i}\right|^{(j)} = 0,$$

we have

$$\left.\frac{\partial \mathcal{F}}{\partial u}\right|_{u^{(j)}} \left(\phi_i^{(j)}\right) = -\left.\frac{\partial \mathcal{F}}{\partial \alpha_i}\right|^{(j)}. \tag{6.34}$$

Sensitivities can be retrieved by solving this set of equations for $i = 1, 2, \ldots, K+1$.
In particular, for our working example we have

$$\left.\frac{D\mathcal{J}_R}{D\alpha_i}\right|_{\alpha_i^{(j)}} = \left.\frac{\partial \mathcal{J}_R}{\partial u}\right|_{u^{(j)}} \left(\phi_i^{(j)}\right) + \frac{\partial \mathcal{J}_R}{\partial \alpha_i}^{(j)} = \int_\Omega (u-d)\phi_i + \sigma_1 \alpha_i, \quad i = 1, 2, \ldots, K$$

$$\frac{D\mathcal{J}_R}{D\gamma}^{(j)} = \left.\frac{\partial \mathcal{J}_R}{\partial u}\right|_{u^{(j)}} \left(\phi_{K+1}^{(j)}\right) + \left.\frac{\partial \mathcal{J}_R}{\partial \alpha_i}\right|_{\alpha_{K+1}^{(j)}} = \int_\Omega (u-d)\,(\phi_{K+1}) + \sigma_2(\gamma - \gamma_{\text{ref}}).$$

Notice that from the state equations (6.30), (6.31), we have for $i = 1, 2, \ldots, K+1$

$$\left.\frac{\partial \mathcal{F}}{\partial u}\right|_{u^{(j)}} (\phi_i) = -\Delta\phi_i + \mathbf{b}\cdot\nabla\phi_i + \gamma\phi_i + 3(u^{(j)})^2\phi_i,$$

and

$$\left.\frac{\partial \mathcal{F}}{\partial \alpha_i}\right|^{(j)} = -f_i, \quad i = 1, 2, \ldots, K$$
$$\left.\frac{\partial \mathcal{F}}{\partial \gamma}\right|^{(j)} = u^{(j)}.$$

Then, the *sensitivities equations* read

$$\begin{cases} -\Delta\phi_i + \mathbf{b}\cdot\nabla\phi_i + \gamma\phi_i + 3(u^{(j)})^2\phi_i = f_i & \text{in } \Omega \\ -\Delta\phi_{K+1} + \mathbf{b}\cdot\nabla\phi_{K+1} + \gamma\phi_{K+1} + 3(u^{(j)})^2\phi_{K+1} = -u^{(j)} & \text{in } \Omega \\ \phi_i = 0, \quad i = 1, 2, \ldots, K+1 & \text{on } \Gamma. \end{cases} \tag{6.35}$$

Notice that these equations are linear in the sensitivities. Finally, we have

$$\left.\frac{D\mathcal{J}_R}{D\alpha_i}\right|^{(j)} = \int_\Omega (u^{(j)} - d)\phi_i + \sigma_1\alpha_i^{(j)}, \qquad \left.\frac{D\mathcal{J}_R}{D\gamma}\right|^{(j)} = \int_\Omega (u^{(j)} - d)\phi_{K+1} + \sigma_2\gamma^{(j)}.$$

Gradients of the functional with respect to the control variables following this approach requires therefore the solution of the $K+1$ sensitivity equations.

Gradient Computation Through Adjoint Equations

In the following, we omit the iteration index j for simplicity. In the previous section we computed the operator $\left.\frac{\partial \mathcal{F}}{\partial u}\right|_u$ applied to the sensitivities ϕ_i. Let us consider the *adjoint* of this operator, which is the operator $\left(\frac{\partial \mathcal{F}}{\partial u}\right)^* \Big|_u$ such that

$$< \left(\frac{\partial \mathcal{F}}{\partial u}\right)^* \Big|_u (\rho)\,,\, v > = < \rho\,,\, \frac{\partial \mathcal{F}}{\partial u}\Big|_u (v) >, \tag{6.36}$$

for any v belonging to an appropriate functional space. Here $< \cdot, \cdot >$ indicates a duality pairing. In particular, in a finite dimensional setting, $< \cdot, \cdot >$ typically denotes the usual Euclidean dot product, while in the continuous setting, it denotes one of the integrals

$$\begin{cases} < u,\, v > \equiv \int\limits_\Omega u\, v, & \text{for scalar functions,} \\ < \mathbf{u},\, \mathbf{v} > \equiv \int\limits_\Omega \mathbf{u}\cdot\mathbf{v}, & \text{for vector functions,} \\ < U,\, V > \equiv \int\limits_\Omega U : V & \text{for tensor functions.} \end{cases}$$

In our example, we have

$$< \rho, \frac{\partial \mathcal{F}}{\partial u}\Big|_u (v) >= \int\limits_\Omega \rho\left(-\Delta v + \mathbf{b}\cdot\nabla v + \gamma v + 3u^2 v\right).$$

Integrating by parts, and choosing ρ to vanish on Γ, we get[7]

$$< \rho, \frac{\partial \mathcal{F}}{\partial u}\Big|_u (v) >= \int\limits_\Omega \left(-\Delta\rho - \mathbf{b}\cdot\nabla\rho + \gamma\rho + 3u^2\rho\right) v.$$

Therefore, the adjoint operator reads

$$\left(\frac{\partial \mathcal{F}}{\partial u}\right)^* \Big|_u = -\Delta\rho - \mathbf{b}\cdot\nabla\rho + \gamma\rho + 3u^2\rho.$$

We consider the following adjoint problem, whose solution, as we will see later, is crucial to find the derivatives of $\mathcal{J}$ with respect to the parameters.

$$< \left(\frac{\partial \mathcal{F}}{\partial u}\right)^* \Big|_u (\rho)\,,\, v >= \frac{\partial \mathcal{J}_R}{\partial u}\Big|_u (v), \tag{6.37}$$

[7] We remind that we assumed **b** to be divergence free.

for any v belonging to appropriate functional spaces. In our specific example, this problem reads

$$\int_\Omega \left(-\Delta\rho - \mathbf{b}\cdot\nabla\rho + \gamma\rho + 3u^2\rho\right) v = \int_\Omega (u-d)\, v,$$

for any v, with ρ vanishing on Γ. Since such equation must hold for any v, we get the strong form of the adjoint equation

$$\begin{cases} -\Delta\rho - \mathbf{b}\cdot\nabla\rho + \gamma\rho + 3u^2\rho = u-d & \text{in } \Omega \\ \rho = 0 & \text{on } \Gamma \end{cases}.$$

Notice that once ρ is computed by solving this equation, we may write for $i = 1, 2, \ldots, K+1$

$$\begin{aligned} \frac{D\mathcal{J}_R}{D\alpha_i} &= \frac{\partial \mathcal{J}_R}{\partial u}\left(\frac{\partial u}{\partial \alpha_i}\right) + \frac{\partial \mathcal{J}_R}{\partial \alpha_i} =< \left(\frac{\partial \mathcal{F}}{\partial u}\right)^*\Bigg|_u (\rho)\,, \frac{\partial u}{\partial \alpha_i} > + \frac{\partial \mathcal{J}_R}{\partial \alpha_i} \\ &=< \rho\,, \frac{\partial \mathcal{F}}{\partial u}\Bigg|_u \left(\frac{\partial u}{\partial \alpha_i}\right) > + \frac{\partial \mathcal{J}_R}{\partial \alpha_i} = - < \rho\,, \frac{\partial \mathcal{F}}{\partial \alpha_i} > + \frac{\partial \mathcal{J}_R}{\partial \alpha_i}, \end{aligned} \tag{6.38}$$

where we exploit (6.37), (6.36), and (6.34). In other words, all the gradients needed by the iterative procedure are promptly computed after ρ is calculated. In the example, this reads for $i = 1, 2, \ldots, K$

$$\begin{aligned} \frac{D\mathcal{J}_R}{D\alpha_i} &= \sigma_1\alpha_i + \int_\Omega (u-d)\frac{\partial u}{\partial \alpha_i} = \sigma_1\alpha_i + \int_\Omega (u-d)\phi_i \\ &= \sigma_1\alpha_i + \int_\Omega \left(-\Delta\rho - \mathbf{b}\cdot\nabla\rho + \gamma\rho + 3u^2\rho\right)\phi_i \\ &\underset{\text{(by parts)}}{=} \sigma_1\alpha_i + \int_\Omega \left(-\Delta\phi_i + \mathbf{b}\cdot\nabla\phi_i + \gamma\phi_i + 3u^2\phi_i\right)\rho \\ &= \sigma_1\alpha_i + \int_\Omega f_i\,\rho, \end{aligned}$$

and similarly we obtain $\displaystyle\frac{D\mathcal{J}_R}{D\gamma} = \sigma_2(\gamma - \gamma_{\text{ref}}) - \int_\Omega u\rho.$

According to this procedure, it is enough to solve a differential problem in the adjoint operator to compute all the gradients needed by the iterative procedure. This approach is therefore more efficient, when the computation of the adjoint operator is possible (and doable).

The Lagrange Multiplier Approach and the KKT Conditions

Let us consider a different approach for reformulating the constrained minimization problem into an unconstrained one. It is a classical argument in which a companion functional is introduced to include the constraints. We stick to our simple working example to introduce the idea, referring to the mentioned references for a more complete presentation. Let us consider the functional

$$\mathcal{L}(u, \boldsymbol{\alpha}, \gamma, \chi) = \mathcal{J}_R(u, \boldsymbol{\alpha}, \gamma) - < \chi, \mathcal{F}(u, \boldsymbol{\alpha}, \gamma) >,$$

where χ is the adjoint (or co-state) function, the so-called *Lagrange multiplier*. The idea behind this approach is that *solutions of the constrained minimization problem are stationary points of* $\mathcal{L}$. As such they solve the following system of equations, representing the *first order necessary conditions* of optimality

$$\begin{cases} \dfrac{\partial \mathcal{L}}{\partial \chi} = 0 \text{ State equations} \\ \dfrac{\partial \mathcal{L}}{\partial u} = 0 \text{ Adjoint/Co-state equations} \\ \dfrac{\partial \mathcal{L}}{\partial \boldsymbol{\alpha}} = 0 \text{ Optimality conditions} \\ \dfrac{\partial \mathcal{L}}{\partial \gamma} = 0 \text{ Optimality condition.} \end{cases} \tag{6.39}$$

Here, each variable is independent of the others since no constraint holds. In our specific example, we have[8]

$$\mathcal{L}(u, \boldsymbol{\alpha}, \gamma, \chi_1, \chi_2) = \mathcal{J}_R(u, \boldsymbol{\alpha}, \gamma) - \int_\Omega \chi_1 \left(-\Delta u + \mathbf{b} \cdot \nabla u + \gamma u + u^3 - \sum_{i=1}^K \alpha_i f_i \right) - \int_\Gamma \chi_2 u.$$

Here, we considered the integral formulation of (6.30), (6.31), where χ_1 and χ_2 are the functions enforcing the constraint given by the state equation. To obtain the stationary points, we need to perform the Gateaux differentiation

$$\left. \frac{\partial \mathcal{L}}{\partial \chi_1} \right|_{\chi_1} (\delta_{\chi_1}) = \lim_{\varepsilon \to 0} \frac{1}{\varepsilon} \left(\mathcal{L}(u, \boldsymbol{\alpha}, \gamma, \chi_1 + \varepsilon \delta_{\chi_1}, \chi_2) - \mathcal{L}(u, \boldsymbol{\alpha}, \gamma, \chi_1, \chi_2) \right) \tag{6.40}$$

[8]Here we used the Lagrange multiplier χ_2 to prescribe the Dirichlet homogeneous boundary condition. Often, such condition is prescribed without using Lagrange multipliers but requiring directly that u and χ_1 vanish on the boundary.

where δ_{χ_1} is an admissible variation. We find

$$\int_\Omega \delta_{\chi_1}\left(-\Delta u + \mathbf{b}\cdot\nabla u + \gamma u + u^3 - \sum_{i=1}^{K}\alpha_i f_i\right) = 0.$$

Since δ_{χ_1} is arbitrary, from this equation we promptly obtain the state problem (6.30). Similarly,

$$\left.\frac{\partial\mathcal{L}}{\partial\chi_2}\right|_{\chi_2}(\delta_{\chi_2}) = \lim_{\varepsilon\to 0}\frac{1}{\varepsilon}\left(\mathcal{L}(u,\boldsymbol{\alpha},\gamma,\chi_1,\chi_2+\varepsilon\delta_{\chi_2}) - \mathcal{L}(u,\boldsymbol{\alpha},\gamma,\chi_1,\chi_2)\right) = \int_\Gamma \delta_{\chi_2} u = 0 \tag{6.41}$$

leading to (6.31).

Let us write explicitly now the adjoint equation

$$\begin{aligned}\left.\frac{\partial\mathcal{L}}{\partial u}\right|_{u}(\delta_u) &= \lim_{\varepsilon\to 0}\frac{1}{\varepsilon}\left(\mathcal{L}(u+\varepsilon\delta_u,\boldsymbol{\alpha},\gamma,\chi_1,\chi_2) - \mathcal{L}(u,\boldsymbol{\alpha},\gamma,\chi_1,\chi_2)\right)\\ &= \int_\Omega (u-d)\,\delta_u - \int_\Omega \chi_1\left(-\Delta\delta_u + \mathbf{b}\cdot\nabla\delta_u + \gamma\delta_u + 3u^2\delta_u\right)\\ &\quad -\int_\Gamma \chi_2\delta_u = 0.\end{aligned} \tag{6.42}$$

Let us factor out the arbitrary variation δ_u. If we integrate by parts the second and first order terms, we get

$$\begin{aligned}&\int_\Omega \delta_u\left(u - d + \Delta\chi_1 + \mathbf{b}\cdot\nabla\chi_1 - \gamma\chi_1 - 3u^2\chi_1\right)\\ &\quad + \int_\Gamma \chi_1\nabla\delta_u\cdot\mathbf{n} - \int_\Gamma (\nabla\chi_1\cdot\mathbf{n} + \chi_1\mathbf{b}\cdot\mathbf{n} + \chi_2)\delta_u = 0.\end{aligned}$$

Because δ_u is arbitrary, this equation is equivalent to

$$\begin{aligned}&-\Delta\chi_1 - \mathbf{b}\cdot\nabla\chi_1 + \gamma\chi_1 + 3u^2\chi_1 = u - d && \text{in } \Omega\\ &\chi_1 = 0 && \text{on } \Gamma\\ &\chi_2 = -\mathbf{b}\cdot\nabla\chi_1\cdot\mathbf{n} - \chi_1\mathbf{n} && \text{on } \Gamma.\end{aligned}$$

Notice that χ_2 does not affect the solution of the problem, therefore in the following we drop the last equation because we are not interested in the particular value assumed by χ_2. Finally, we compute the derivative with respect to the control variables

$$\frac{\partial \mathcal{L}}{\partial \alpha_i} = \int_\Omega \chi_1 f_i + \sigma_1 \alpha_i, \quad \frac{\partial \mathcal{L}}{\partial \gamma} = \int_\Omega -u\chi_1 + \sigma_2(\gamma - \gamma_{\text{ref}}). \tag{6.43}$$

Summarizing, the optimality system to be solved reads

$$\begin{cases} -\Delta u + \mathbf{b}\cdot\nabla u + \gamma u + u^3 = \sum_{i=1}^{K} \alpha_i f_i \quad \text{in } \Omega \\ u = 0 \quad \text{on } \partial\Omega \end{cases} \qquad \text{State equations}$$

$$\begin{cases} -\Delta \chi_1 - \mathbf{b}\cdot\nabla\chi_1 + \gamma\chi_1 + 3u^2\chi_1 = u - d \text{ in } \Omega \\ \chi_1 = 0 \text{ on } \partial\Omega \end{cases} \qquad \text{Adjoint equations}$$

$$\begin{cases} \alpha_i = -\dfrac{1}{\sigma_1}\displaystyle\int_\Omega \chi_1 f_i \quad i = 1,\ldots,K \\ \gamma = \gamma_{\text{ref}} + \dfrac{1}{\sigma_2}\displaystyle\int_\Omega \chi_1 u \end{cases} \qquad \text{Optimality conditions} \tag{6.44}$$

This set of equations represents the so-called *Karush–Khun–Tucker (KKT) conditions* [69].

In principle, this system provides the solution to the optimization problem in a monolithic or "one-shot" fashion. In practice, the cases of interest when the system can be solved directly are rare—in particular for nonlinear state problems, and we need again to resort to iterative procedures.

Let a guess for $\boldsymbol{\alpha}$ and γ be given at the iteration j. Again, typically, we take $\gamma^{(0)} = \gamma_{\text{ref}}$. A reasonable iterative procedure reads as follows.

1. Solve the state equations to compute $u^{(j+1)}$;
2. Solve the adjoint problem to compute $\chi_1^{(j+1)}$ and $\chi_2^{(j+1)}$.
3. Update the control variables using the optimality conditions. In this example it is natural to choose

$$\alpha_i^{(j+1)} = -\frac{1}{\sigma_1}\int_\Omega \chi_1^{(j+1)} f_i \text{ and } \gamma^{(j+1)} = \gamma_{\text{ref}} + \frac{1}{\sigma_2}\int_\Omega \chi_1^{(j+1)} u^{(j+1)},$$

until a convergence criterion is satisfied.

This procedure corresponds in fact to a fixed-step steepest descent method for $\mathcal{J}_R$ regarded as a function of the control variables. In fact, notice that the Lagrange multiplier χ introduced here corresponds to ρ introduced in the previous section. With this perspective, Eq. (6.38) reads

$$\frac{D\mathcal{J}_R}{D\alpha} = 0$$

that is exactly what we want to obtain when we are looking for a minimum of $\mathcal{J}_R$. As a matter of fact, the iterative algorithm introduced in the previous section to minimize $\mathcal{J}_{\mathcal{R}}$ is an iterative algorithm to solve the KKT conditions.

Sequential Quadratic Programming Algorithm

In contrast with what is done in the unconstrained approach considered so far, constrained algorithms try to compute the solution to the minimization problem through the convergence of the state and parameter variables $(u^{(j)}, \alpha^{(j)})$ simultaneously. This approach can be very effective in presence of nonlinear constraints, as the constraints need not to be solved at each iteration. In this section we consider one of these methods, the sequential quadratic programming (SQP) method [12] which consists of iteratively approximating the original problem with a quadratic problem subject to linear constraints. Such quadratic problem is then solved using quadratic programming (QP) algorithms. Assume that the problem is already discretized, and let the vector $\mathbf{x}^{(j)}$ include both the state ($\mathbf{u}^{(j)}$) and the parameter ($\boldsymbol{\alpha}^{(j)}$) vectors

$$\mathbf{x}^{(j)} = \begin{bmatrix} \mathbf{u}^{(j)} \\ \boldsymbol{\alpha}^{(j)} \end{bmatrix}.$$

The Lagrangian functional of the problem $\mathcal{L}(\mathbf{x}, \boldsymbol{\chi}) = \mathcal{J}_R(\mathbf{x}) - < \boldsymbol{\chi}, \mathcal{F}(\mathbf{x}) >$ is approximated at iteration j with the paraboloid tangent to the Lagrangian in $\mathbf{x}^{(j)}$, i.e.,

$$\mathcal{L}\left(\mathbf{x}, \boldsymbol{\chi}^{(j)}\right) \approx \mathcal{L}(\mathbf{x}^{(j)}, \boldsymbol{\chi}^{(j)}) + L_{\mathbf{x}}^{(j),T} \boldsymbol{\delta}_x^{(j)} + \frac{1}{2} \boldsymbol{\delta}_x^{(j),T} \mathrm{H}^{(j)} \boldsymbol{\delta}_x^{(j)},$$

where $L_{\mathbf{x}}^{(j)} = \left.\dfrac{\partial \mathcal{L}}{\partial \mathbf{x}}\right|_{\mathbf{x}^{(j)}}$, $\boldsymbol{\delta}_x^{(j)} = \mathbf{x} - \mathbf{x}^{(j)}$, and $\mathrm{H}^{(j)} = \left.\dfrac{\partial^2 \mathcal{L}}{\partial \mathbf{x}^2}\right|_{\mathbf{x}^{(j)}}$ is the *Hessian* matrix. Such approximation of the Lagrangian is minimized w.r.t. $\boldsymbol{\delta}_x^{(j)}$, subject to the linearization of the constraint $\mathcal{F}(\mathbf{x}) = 0$

$$\mathcal{F}\left(\mathbf{x}^{(j)}\right) + \mathrm{F}_{\mathbf{x}}^{(j),T} \boldsymbol{\delta}_x^{(j)} = 0. \tag{6.45}$$

where the matrix $\mathrm{F}_{\mathbf{x}}^{(j)} = \left.\dfrac{\partial \mathcal{F}}{\partial \mathbf{x}}\right|_{\mathbf{x}^{(j)}}$. Exploiting the fact that $\mathrm{F}_{\mathbf{x}}^{(j),T} \boldsymbol{\delta}_x^{(j)}$ is constant w.r.t. $\boldsymbol{\delta}_x^{(j)}$ because of (6.45), one can reformulate the quadratic programming problem as

$$\begin{aligned} \boldsymbol{\delta}_x^{(j)} &= \operatorname{argmin} \quad J_{\mathbf{x}}^{(j),T} \boldsymbol{\delta}_x^{(j)} + \frac{1}{2} \boldsymbol{\delta}_x^{(j),T} \mathrm{H}^{(j)} \boldsymbol{\delta}_x^{(j)} \\ \text{s.t. } \mathrm{F}_{\mathbf{x}}^{(j),T} \boldsymbol{\delta}_x^{(j)} &= -\mathcal{F}\left(\mathbf{x}^{(j)}\right), \end{aligned} \tag{6.46}$$

where the column vector $J_{\mathbf{x}}^{(j)} = \left.\dfrac{\partial \mathcal{J}_R}{\partial \mathbf{x}}\right|_{\mathbf{x}^{(j)}}$. The value $\mathbf{x}^{(j+1)}$ is obtained as $\mathbf{x}^{(j+1)} = \mathbf{x}^{(j)} + \zeta \boldsymbol{\delta}_x^{(j)}$, where the step length $\zeta \in (0, 1]$ is chosen using a line search method. The Lagrangian multiplier $\chi^{(j+1)}$ can be computed as $\chi^{(j+1)} = \chi^{(j)} + \gamma(\chi^{\text{opt}} - \chi^{(j)})$, where χ^{opt} is the optimal Lagrangian multiplier associated with problem (6.46). In order to avoid the computational costs associated with the evaluation of the Hessian, the matrix H can be replaced by a suitable approximation. A common approach is to use instead the matrix generated by the BFGS method. In general the effectiveness of the SQP method relies on the method used to solve the QP problem. Inequality constraints (e.g., the constraint that some parameters must be nonnegative) can be easily handled using SQP approach. In addition, we point out that when the exact Hessian is used, $\gamma = 1$, and only equality constraints are considered, the method is equivalent to solve the KKT conditions with the Newton method.

Notice that from the formulation of the SQP problem that the solution at iteration j does not need to be feasible, i.e., to satisfy the constraints. This approach allows to save a lot of time because we do not have to enforce the feasibility of the solution at each iteration. However, this lack of feasibility might affect the robustness of the method.

Unsteady Problems

The procedure illustrated above can be extended to unsteady problems, that are of major interest in FSI applications. However, in this case, it is important to notice that the adjoint problem (in any of the formulations we encountered) is a *final*-boundary value problem. This means that it is *backward* in time. This feature introduces high computational costs either when we solve the problem via the KKT system or we follow a gradient-based procedure based on the adjoint problem. In fact, the state problem (which is forward in time) and the adjoint problem need to be solved all together in space–time. The computational costs for this approach are therefore in many cases not affordable, not to mention the storage cost of the solutions at each time step. For this reason, different workarounds have been considered. For instance [38], the solution may be stored only on a predefined set of instants T_k (subset of the time discretization steps) called *checkpoints* and the state required by the optimization for computing the adjoint solution is locally recomputed or approximated.

Following a different approach, time discretization may be performed before the optimization, leading to a sequence of pseudo-steady optimization problems at each time step. An example of this approach will be provided in the next section for estimating the compliance of an artery.

Interplay Between Numerical Discretization and Solution of the Control Problem

In the numerical solution of control problems there is an usual dilemma, concerning the order of the steps for the optimization and the numerical approximation. We may summarize this as "Discretize then Optimize" (DO) vs. "Optimize then Discretize" (OD). The two operations are in general non-commutative and the solutions obtained with the two approaches are in general different. It is difficult to draw general indications about the most appropriate approach, which is basically a trade-off between accuracy and computational costs. The issue is extensively discussed in [38]. There are clearly pros and cons in both the sequences. With DO we may say that

- we avoid inconsistencies induced by the numerical differentiation of the KKT conditions; in other terms, the numerical approximation of the KKT conditions introduces a discrepancy with the real optimization condition;
- we can even use automatic differentiation software;
- we can split an unsteady problem into a sequence of pseudo-steady optimization problems.

On the other hand with OD:

- we do not deal with the differentiation of numerical artificial terms (like stabilization of advection terms for high Reynolds numbers);
- managing moving boundary problems as in shape optimization is easier, since we do not need the derivative of the grid with respect to the optimization parameters.

In the examples that follow we stick to a DO approach. A parameter estimation procedure based on OD can be found in [74] for the estimate of cardiac conductivities.

6.3.3 Reducing the Costs Via Solution Reduction

As we have pointed out several times, the optimization methods presented above suffer from high computational costs for different reasons. The state equations and possibly the adjoint problem need to be solved at each iteration, not to mention the additional costs in terms of computations and storage for unsteady problems, that need to be truly tackled in 4D (space and time).

In order to reduce the computing time we need to reduce either the number of iterations or the cost of each iteration (or both). The number of iterations may be reduced by using effective optimization algorithms as the BFGS method for updating the current solution. The cost of each iteration can be reduced by treating the constraints in a "flexible" way. This means that the fulfillment of the constraints may be relaxed in particular in the first iterations when this does not prevent the convergence to the admissible solution. This can be done by accepting a solution to the state equations featuring relatively large residuals or by replacing the state equations themselves with a simplified model. These approaches are mostly problem-dependent, being based on the possible simplifications offered by the problem at hand. For instance in electrocardiology, the Bidomain equations that describe the dynamics of the extra and intra-cellular potentials may be replaced by the simplified Monodomain system (see, e.g., [27]). When solving FSI problems in hemodynamics a fully 3D coupling may be downscaled to a 3D Fluid/2D Structure problem [57], as we see in the next section.

Here we address another (somehow complementary) way for reducing the computational costs, which is based on reducing more specifically the number of degrees of freedom required to give an accurate representation of the solution. As a matter of fact, a function in a (separable) Hilbert space (for instance, L^2 or H^1) admits the representation

$$u = \sum_{i=1}^{\infty} U_i \psi_i,$$

for a proper selection of the basis functions ψ_i. In the Galerkin approach for approximating the solution, we generally find a basis function set to represent the approximate solution $u_N(x)$ as

$$u_N = \sum_{i=1}^{N} U_{N,i} \varphi_i.$$

The basis functions may be piecewise polynomials as in the finite element method, or global polynomials as in spectral methods. In general, those basis functions can be defined to be *general purpose*, in the sense that they do not specifically rely on the feature of the problem to be solved and can be applied to a vast class of problems. This versatility has the drawback that, in general, to achieve accurate solutions the number N of degrees of freedom is high. This clearly implies high computational costs as the associated linear(ized) systems are large.

A somehow opposite approach would be to give up pursuing a general basis, using an "educated" basis that incorporates specific information of the problem. For instance, in *modal analysis* the solution is represented on the basis given by the eigenfunctions of the problem. The basis is therefore problem-dependent, bringing intrinsically information on the problem to be solved. The gain is that it is generally possible to achieve a good accuracy when truncating to a low number of degrees of freedom. However, this is not for free, as the basis needs to be specifically computed. In particular, computation of eigenfunctions is in general not trivial and quite costly [15].

Following the same idea of constructing an informed basis, we may consider *snapshot-based* approaches. In this case, the basis is the result of the elaboration of the solutions of the problem for particular configurations useful for the solution of the state problem in the optimization process. For instance, when the control variable is a parameter to be identified (as γ was in our working example), snapshots may be the solution of the state problem for a particular set of values of the parameter. The proper identification of this set is clearly crucial for the effectiveness of the entire procedure. This can be realized by considering that if the optimal value of the control variable falls within the range considered in the snapshots, the entire procedure configures as a sophisticated "interpolation," for which several convergence results are available. On the contrary, if the range of the snapshots computation is not well defined, we are actually performing an "extrapolation" and the convergence is not necessarily guaranteed. Again, the final goal is to keep the size N of the finite dimensional approximation of the solution as small as possible, thanks to the information contained in the basis.

From the computational standpoint, this snapshot-based approach relies on the *off-line/on-line* paradigm, namely

1. computation of the basis is "off-line," and it is intended to be an accurate (and therefore expensive) numerical approximation of the solution for different configurations that are considered to be relevant to the basis;
2. solution of the optimization problem, and in particular the computation of the coefficients $U_{N,i}$ along the iterations of the minimization procedure is "on-line," and contributes to the actual cost of the control procedure.

In this way, the computational costs are factorized, the major contribution being carried out in a step preliminary to the optimization. This paradigm clearly makes sense whenever the "off-line" part can be recycled for the solution of several optimization problems.[9]

Among the different snapshot-based strategies, we mention the *reduced basis method* and the *POD*. In the former, the snapshots are computed for values of the parameters that are evaluated to perform the best control of the error on the basis of rigorous error estimates (see [65, 70]). In particular, we mention [52] for an application of the reduced basis method to FSI problems. The latter is known also as Karhunen–Loève decomposition or *principal component analysis* and it is illustrated more in detail in the next paragraph.

POD Basis Selection

We start assuming that a set of size M of solutions is available, for instance, by computing snapshots for M different values of the parameter of interest after a uniform sampling of an appropriate range. We assume that M is still large for the purpose of reducing the computational costs and that a reduction of the size of the basis is required, by properly filtering redundancy in the snapshots set. Denote by $\boldsymbol{\rho}_j \in \mathbb{R}^N$ the M snapshots of the (approximate) solution, with $j = 1, \ldots, M$. Then, we perform the following steps.

1. *Sample average*: $\overline{\boldsymbol{\rho}} = \dfrac{1}{M}\sum_{j=1}^{M} \boldsymbol{\rho}_j$.
2. *Sample Covariance*: Compute $\mathrm{C} \in \mathbb{R}^{M\times M}$, whose elements are defined as $c_{ij} := \dfrac{1}{M}(\boldsymbol{\rho}_i - \overline{\boldsymbol{\rho}})^T(\boldsymbol{\rho}_j - \overline{\boldsymbol{\rho}})$. Matrix C is positive semidefinite and symmetric so the

[9] This can be problematic in a clinical context, where patient-specific geometries differ one from the other and the snapshot computation is not trivially recycled. Anatomical atlas mapping ideal to real geometries are required.

eigenvalues are all real and the eigenvectors $\{\mathbf{x}_j\}$ form an orthonormal basis in $\mathbb{R}^M$. We order the eigenvalues as

$$\lambda_1 \geq \lambda_2 \geq \cdots \geq \lambda_M \geq 0.$$

3. *Thresholding*: Select a tolerance $\tau \in [0, 1]$ and pick the minimum $\tilde{M}$ such that

$$\frac{\sum_{i=1}^{\tilde{M}} \lambda_i}{\sum_{i=1}^{M} \lambda_i} \geq \tau.$$

   Here τ is a threshold that identifies the "essential" information. Hopefully, this happens for $\tilde{M} \ll M$.
4. *New basis*. Let us select a new basis $\{\mathbf{y}_i\}$ consistent with the eigenvalues threshold. We take for $i = 1, \ldots, \tilde{M}$

$$\mathbf{y}_i = \sum_{j=1}^{\tilde{M}} (\mathbf{x}_i)_j (\boldsymbol{\rho}_j - \overline{\boldsymbol{\rho}}),$$

   where $(\mathbf{x}_i)_j$ is the j-th entry of the i-th eigenvector. Then, we normalize $\mathbf{y}_i^* = \dfrac{1}{\|\mathbf{y}_i\|}\mathbf{y}_i$.

This is by construction an orthonormal basis. In addition and more importantly, this basis fulfils an optimal property. As a matter of fact [15, 38], the space spanned by the POD basis is the best $\tilde{M}$-dimensional subspace approximation of the space spanned by the snapshots (in the 2-norm sense). A vector $\mathbf{x}$ in $\mathbb{R}^M$ can then be approximated in terms of the POD basis as

$$\mathbf{x} = \bar{\boldsymbol{\rho}} + \sum_{i=1}^{\tilde{M}} c_i \mathbf{y}_i^*$$

For particular problems, such as progressive waves, reduction of the size for the solution and eventually of the computational costs after this procedure may be not enough. In this case, other reduced solution techniques may be considered [31]. Nevertheless, an example of POD for the solution of an inverse FSI problem is illustrated in the next section.

Remark 6.6. Here we have presented a particular application of POD for reducing the dimension of the solution forward problem. However, POD can be used for reducing the dimensionality in different contexts. For instance, in [7, Chap. 7], POD is advocated also for reducing the dimensionality of the size of the parameter space, which is crucial when the parameter is a function represented by a large number of degrees of freedom.

Remark 6.7. Here we introduced the POD using the eigenvalue decomposition of the sample covariance matrix. Alternatively, one can perform the Singular Value Decomposition (SVD) of the snapshots matrix $X = [\boldsymbol{\rho}_1, \ldots, \boldsymbol{\rho}_M]$. This can be efficiently done by first performing a QR factorization of X and then by computing the SVD of the triangular factor. In other words, $X = QR = QU\Sigma V^T = \tilde{U}\Sigma V^T$. The POD basis is then made of the first $\tilde{M}$ left singular vectors (the columns of $\tilde{U}$), where $\tilde{M}$ is chosen with the same procedure as before, using the singular values of the snapshots matrix rather than the eigenvalues of the covariance matrix.

6.4 Some Applications of Data Assimilation in Hemodynamics Problems

In this section we consider some applications of DA and Parameter Estimation in computational hemodynamics.

First, we present the problem of reconstructing the blood flow in a vessel assimilating sparse noisy measures of the velocity with the numerical results obtained by solving the incompressible Navier–Stokes equations. Successively, we consider the problem of estimating the compliance of a vessel based on measures of the displacement retrieved from medical images. The solution to this problem leads to an *inverse fluid–structure interaction* (IFSI) problem. These are not the only examples of data assimilation in biomedical applications. We mention, for instance, the work in electrocardiology for the setup of patient-specific models in [19], and for estimating cardiac conductivities [32, 37, 74]. Other applications can be found, e.g., in [17, 29]. In particular, in [24, 26] DA methods are advocated for filling the gap between available boundary data and mathematical conditions required to solve the problem.

We have selected these examples because they offer the opportunity to see in action different methodologies based on the techniques illustrated in the previous sections.

6.4.1 Assimilation of Velocity Measures in Blood Flow Simulations

We consider the problem of merging velocity measures and the numerical simulation of blood flow. The DA problem can be addressed in several and diverse ways, as described in the previous sections. More precisely, we present two approaches introduced in two recent papers; in the first one [18] the problem is faced with a variational (control) method, where the control variable is the normal component of the stress at the inflow section of the vessel. In the second paper [40] the authors exploit a Least Squares Finite Element (LSFE) approximation treating internal

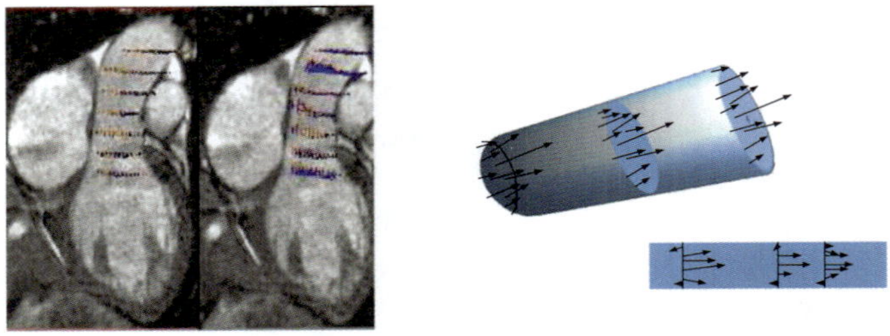

Fig. 6.7 On the *left*, view of blood measured velocities in an MRI of the ascending aorta. On the *right*, examples of a three-dimensional and two-dimensional domain for which data are collected on internal layers transversal to the flow

layers, where measures are available, as artificial boundaries. This approach can be reinterpreted as a MAP Bayesian estimate, as pointed out in [21].

We introduce the formal statement of the problem. Let us denote by Ω a domain in $\mathbb{R}^d$ ($d = 2, 3$; in real applications $d = 3$). We assume (see Fig. 6.7) that Ω features an inflow boundary Γ_{in}, an outflow boundary Γ_{out}, and the physical wall of the vessel Γ_{wall}. Γ_{in} and Γ_{out} can possibly consist of several sections. The variables of interest are the velocity $\mathbf{u}(\mathbf{x}) \in [H^1(\Omega)]^d$ and the pressure $p(\mathbf{x}) \in L_0^2(\Omega)$. Also, we assume to have some velocity measures as in, e.g., Fig. 6.7 or sparse in the domain.

Velocity and pressure are assumed to obey the incompressible Navier–Stokes equations (NSE)

$$
\begin{array}{ll}
\dfrac{\partial \mathbf{u}}{\partial t} - \mu\, \nabla \cdot (\nabla \mathbf{u} + \nabla \mathbf{u}^T) + (\mathbf{u} \cdot \nabla)\mathbf{u} + \nabla p = \mathbf{f} & \text{in } \Omega, \\
\nabla \cdot \mathbf{u} = 0 & \text{in } \Omega, \\
\mathbf{u} = \mathbf{0} & \text{on } \Gamma_{\text{wall}}, \\
-\mu\, (\nabla \mathbf{u} + \nabla \mathbf{u}^T) \cdot \mathbf{n} + p \cdot \mathbf{n} = \mathbf{h} & \text{on } \Gamma_{\text{in}}, \\
-\mu\, (\nabla \mathbf{u} + \nabla \mathbf{u}^T) \cdot \mathbf{n} + p \cdot \mathbf{n} = \mathbf{g} & \text{on } \Gamma_{\text{out}}.
\end{array}
\tag{6.47}
$$

A Newtonian rheology is supposed to hold, which is a common assumption in large and medium vessels [25], and μ is the constant kinematic viscosity. The choice of homogeneous Dirichlet boundary conditions on Γ_{wall} reflects the fact that we consider fixed geometries.

In this section we consider the steady case $\dfrac{\partial \mathbf{u}}{\partial t} = \mathbf{0}$.

Variational Approach

In [18] a variational DA procedure for the inclusion of velocity measures in the simulation of the NSE in hemodynamics is proposed. Sparse noisy velocity measures $d_1, \ldots, d_{N_s}$ are assumed to be available in the domain and possibly on the boundary at some sites[10] $\mathbf{x}_i^m \in \Omega$, $i = 1, \ldots, N_s$, that do not necessarily belong to a plane or a layer inside of Ω.

The assimilation technique in [18] is formulated as a control problem where the misfit between computed data (the NSE solution) and observed data is minimized. The equations of incompressible fluid dynamics are the constraint to the minimization procedure. The control variable is selected to be the inflow normal (or natural) stress $\mathbf{h}$; knowledge of this quantity is quite often not available in practice. The variational problem is formulated as

$$\begin{aligned} &\min_{\mathbf{h}} \mathcal{J}(\mathbf{u}, \mathbf{h}) = \|f(\mathbf{u}) - \mathbf{d}\|_{l^2} + \mathcal{R}(\mathbf{h}) \\ &\text{s.t. Steady version of (6.47).} \end{aligned} \tag{6.48}$$

Here, f is a filtering vector function that returns the value of the velocity field evaluated on the measurement sites; $\mathcal{R}$ is a regularization term added to prevent potential ill-posedness and ill-conditioning of the problem due to the location of data and the presence of noise.

For the numerical solution of the problem (6.48) we first consider the linearized NSE; then, we discuss the nonlinear case; in the linearized formulation the term $(\mathbf{u} \cdot \nabla)\mathbf{u}$ is substituted by $(\boldsymbol{\beta} \cdot \nabla)\mathbf{u}$, where $\boldsymbol{\beta}$ is a known advection field. We follow a DO approach (see Sect. 6.3.2), thus, after the discretization (via, e.g., the finite element method) of the functional and the linearized state equations we resort to the following algebraic optimization problem

$$\begin{aligned} &\min_{\mathbf{H}} J(\mathbf{V}, \mathbf{H}) = \frac{1}{2}\|\mathrm{D}\mathbf{V} - \mathbf{d}\|_2^2 + \frac{\alpha}{2}\|\mathrm{L}\mathbf{H}\|_2^2 \\ &\text{s.t. } \mathrm{S}\mathbf{V} = \mathrm{R}_{\text{in}}^T \mathrm{M}_{\text{in}}\mathbf{H} + \mathbf{F}. \end{aligned} \tag{6.49}$$

Here, $\mathbf{V} = [\mathbf{U}\ \mathbf{P}] \in \mathbb{R}^{N_u+N_p}$ is the vector of discretized velocity $\mathbf{U} \in \mathbb{R}^{N_u}$ and pressure $\mathbf{P} \in \mathbb{R}^{N_p}$; $\mathbf{H} \in \mathbb{R}^{N_{\text{in}}}$ is the discretization of the control variable $\mathbf{h}$; N_{in} is the number of degrees of freedom of the velocity on Γ_{in}; $\mathbf{d} = [d_1 \ldots d_{N_s}] \in \mathbb{R}^{d\,N_s}$ is the vector of the available measures. For $\alpha > 0$, $\frac{\alpha}{2}\|\mathrm{L}\mathbf{H}\|_2^2$ is a Tikhonov regularization term, see Sect. 6.3.2. The matrix S is defined as follows:

$$\mathrm{S} = \begin{bmatrix} \mathrm{C} + \mathrm{A} & \mathrm{B}^T \\ \mathrm{B} & \mathrm{O} \end{bmatrix}, \tag{6.50}$$

where $\mathrm{C}, \mathrm{A} \in \mathbb{R}^{N_u, N_u}$, and $\mathrm{B} \in \mathbb{R}^{N_p, N_u}$ are the discretization of the diffusion, advection, and divergence operators. D is the selection or observation matrix and it is defined as $\mathrm{D} = [\mathrm{Q} \quad \mathrm{O}]$, where $\mathrm{Q} \in \mathbb{R}^{dN_s, N_u}$ is such that $[\mathrm{Q}\mathbf{U}]_i$ is the numerical solution evaluated at the data sites. $\mathrm{R}_{\text{in}} \in \mathbb{R}^{N_{\text{in}}, N_u+N_p}$ is a restriction

[10]Notice that we use the word "sites" for the location of measurements, as opposed to the word "nodes" for points where velocities are computed. In general sites and nodes are different, but we do not exclude that the intersection of sites set and nodes set in non-empty.

matrix which selects the degrees of freedom of the velocity on Γ_{in}. $\mathrm{M}_{\text{in}} \in \mathbb{R}^{N_{\text{in}},N_{\text{in}}}$ is the discretization of the mass operator restricted to inlet boundary nodes.

For the solution of problem (6.49) we use the Lagrange multiplier approach, so we consider the Lagrange functional

$$L(\mathbf{V},\mathbf{H},\mathbf{X}) = \frac{1}{2}\|\mathrm{D}\mathbf{V}-\mathbf{d}\|_2^2 + \frac{\alpha}{2}\|\mathrm{L}\mathbf{H}\|_2^2 + \mathbf{X}^T(\mathrm{S}\mathbf{V} - \mathrm{R}_{\text{in}}^T\mathrm{M}_{\text{in}}\mathbf{H} - \mathbf{F}), \tag{6.51}$$

where $\mathbf{X} \in \mathbb{R}^{N_u+N_p}$ is the discrete Lagrange multiplier. The set of necessary conditions for optimality is given by the KKT system

$$\begin{cases} \dfrac{\partial L}{\partial \mathbf{V}} = \mathrm{D}^T(\mathrm{D}\mathbf{V}-\mathbf{d}) + \mathrm{S}^T\mathbf{X} = \mathbf{0} \\ \dfrac{\partial L}{\partial \mathbf{H}} = \alpha\mathrm{L}^T\mathrm{L}\,\mathbf{H} - \mathrm{M}_{\text{in}}^T\mathrm{R}_{\text{in}}\mathbf{X} = \mathbf{0} \\ \dfrac{\partial L}{\partial \mathbf{X}} = \mathrm{S}\mathbf{V} - \mathrm{R}_{\text{in}}^T\mathrm{M}_{\text{in}}\mathbf{H} - \mathbf{F} = \mathbf{0}. \end{cases} \tag{6.52}$$

By defining $\mathrm{Z} = \mathrm{D}\mathrm{S}^{-1}\mathrm{R}_{\text{in}}^T\mathrm{M}_{\text{in}}$ and $\mathrm{W} = \mathrm{Z}^T\mathrm{Z} + \alpha\,\mathrm{L}^T\mathrm{L}$ the reduced system, obtained by block elimination, reads $\mathrm{W}\mathbf{H} = \mathrm{Z}^T(\mathbf{d} - \mathrm{D}\mathrm{S}^{-1}\mathbf{F})$, where W is the so-called *reduced Hessian* matrix.

The following theorem states the necessary and sufficient conditions for the well-posedness of problem (6.49).

Theorem 6.1. *For* $\alpha = 0$, W *is non-singular, i.e.,* (6.49) *is well-posed,* $\Leftrightarrow$

$$\mathrm{Null}(\mathrm{D}) \cap \mathrm{Range}(\mathrm{S}^{-1}\mathrm{R}_{\text{in}}^T\mathrm{M}_{\text{in}}) = \{\mathbf{0}\}. \tag{6.53}$$

For the proof see [18]. Condition (6.53) is satisfied when "enough" data are available on the inflow section (number and location of the measures that guarantee the well-posedness depend on the discretization method used).

In order to consider the nonlinear advection term $(\mathbf{u}\cdot\nabla)\mathbf{u}$ and to solve the nonlinear PDE constrained optimization problem (6.48), we combine the DA procedure for the linearized case and classical fixed point schemes for the solution of the NSE. In particular, we refer to the Picard and Newton methods [62]. The assimilation problem is solved iteratively as follows.
Given a guess for the velocity field at iteration $k+1$, $\mathbf{U}_k$,

$$\textit{solve} \quad \begin{cases} \min\limits_{\mathbf{H}_k} \dfrac{1}{2}\|\mathrm{D}\mathbf{V}_{k+1} - \mathbf{d}\|_2^2 + \dfrac{\alpha}{2}\|\mathrm{L}\mathbf{H}_{k+1}\|_2^2 \\ \text{s.t.} \quad \mathrm{S}_k\mathbf{V}_{k+1} = \mathrm{R}_{\text{in}}^T\mathrm{M}_{\text{in}}\mathbf{H}_{k+1} + \mathbf{F}_k \end{cases} \tag{6.54}$$

until $\|\mathbf{V}_k - \mathbf{V}_{k+1}\| \leq \delta$, being δ a user defined tolerance. Here,

$$\mathrm{S}_k = \begin{bmatrix} \mathrm{C}+\mathrm{A}_k & \mathrm{B}^T \\ \mathrm{B} & \mathrm{O} \end{bmatrix}, \qquad \text{and} \qquad \mathbf{F}_k = \mathbf{F} + w\mathbf{Y}_k. \tag{6.55}$$

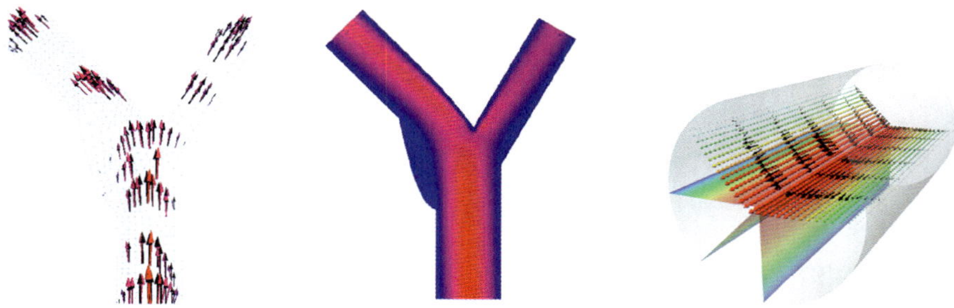

Fig. 6.8 On the *left*, the *colored* vector field consists in the available measures whereas the *black* one corresponds to the noise-free data. In the *center*, the magnitude of the assimilated vector field is reported. On the *right*, the *colored* vector field corresponds to the assimilated velocity, the *black* one to the noisy data and the *colored field* in the background corresponds to the magnitude of the velocity

A_k comes from the discretization of $(\overline{\mathbf{u}}_k \cdot \nabla)\mathbf{u}_{k+1} + w(\mathbf{u}_{k+1} \cdot \nabla)\overline{\mathbf{u}}_k$ ($w = 0$ for Picard method, 1 for Newton); $\mathbf{Y}_k$ is the discretization of $(\overline{\mathbf{u}}_k \cdot \nabla)\overline{\mathbf{u}}_k$. Here $\overline{\mathbf{u}}_k$ is defined as $\vartheta \mathbf{u}_{k-1} + (1-\vartheta)\mathbf{u}_k$, being $\vartheta \in [0, 1]$, w is a relaxation parameter, chosen empirically.

Numerical tests. In Fig. 6.8 we report the numerical results obtained on two geometries approximating blood vessels. In Fig. 6.8 (left) the computational grid is an approximation of a carotid artery; the colored vector field consists in the actual data used in the assimilation, these are generated adding Gaussian noise to a reference solution; to appreciate the presence of the noise the noise-free data are also reported in black. In the center, the magnitude of the assimilated velocity is displayed; a comparison with a reference solution (conducted in [18]) shows that the noise is filtered and that the assimilated solution is close to the reference one.

On the right, a three-dimensional cylindrical domain is reported, this case is treated with an axisymmetric formulation. On selected internal surfaces the assimilated field and its magnitude are reported; it is important to note that the noise affecting the components of the velocity parallel to the flow is significantly filtered.

Next, we consider the problem of estimating the WSS already described in the introductory example of Sect. 6.1. An accurate approximation of the WSS is fundamental in the investigation of cardiovascular pathologies since it is an index of the possibility of rupture of the vessel wall and formation of stenosis [25]. Approximations of the WSS retrieved from indirect measurements are in general unreliable because of the post-processing numerical errors and the noise affecting the measures. Including measurements in simulations is a way for improving the reliability of the computed solutions and, on the other hand, the introduction of the mathematical (numerical) model results in noise filtering. For the geometry of Fig. 6.8 (left), we compute the WSS on a selected internal wall. In order to quantify the accuracy of this solution we compare the assimilated WSS with the one associated with a reference solution, we introduce the index of accuracy $E_{\text{WSS}} = \|\text{WSS} - \text{WSS}_{\text{FE}}\|_2 / \|\text{WSS}_{\text{FE}}\|_2$ where WSS_{FE} is the value retrieved from

Table 6.1 Comparison of relative errors for the WSS computed with DA and forward solution

SNR	$E_{\text{WSS,DA}}$	$E_{\text{WSS,FW}}$
100	0.2536	0.2667
20	0.2591	0.3030
10	0.2738	0.3861
5	0.3149	0.6114

the reference solution. In correspondence of decreasing values of signal to noise ratio[11] (SNR), the WSS errors obtained with the assimilated velocity field, $E_{\text{WSS,DA}}$, compared with those obtained from a forward simulation on the same grid with the same noisy measures (used for DA) as boundary data on the inflow section are reported in Table 6.1. With high SNR the gain obtained with DA is not significant, however as we decrease SNR we can obtain up to the 50 % of gain with respect to the forward simulation.

Bayesian Approach

A *Variational Bayesian* approach to the assimilation problem is possible. This formulation features an overlap between statistical and variational techniques; both point estimators and confidence regions for the velocity are considered. Here, we recall the method for the computation of the MAP and ML estimators (see Sect. 6.2.2) and present some numerical results that illustrate how the knowledge of the nature of the measurement noise can significantly improve the quality of the estimation with respect to the deterministic estimates.

We assume to deal with discretized variables, all treated as random; in the remainder of this section the bold variables denote random vectors while the capital plain variables a specific realization. With an abuse of notation we introduce the random variable $\mathbf{H}$ which describes the normal stress of the fluid at the inflow section; $\mathbf{M}$ is the random variable that describes the measures and $\boldsymbol{\nu}$ the noise perturbing the measurements. We let p_H be the p.d.f. of $\mathbf{H}$, or its a priori distribution, and p_ν the one of $\boldsymbol{\nu}$; these distributions are assumed to be known. As described in Sect. 6.2.2 the purpose of the Bayesian procedure is to estimate the posterior distribution $p_{H|M}$ exploiting the Bayes formula (6.5) in the form

$$p_{H|M} = \frac{p_{M|H}\ p_H}{p_M}; \tag{6.56}$$

where p_M is the p.d.f. of the measures.

First, we assume that the relation between the random vectors $\mathbf{H}$ and $\mathbf{U}$, the random variable that describes the velocity, is linear (i.e., we consider the linearized NSE), then we treat the nonlinear case.

[11]We define the signal to noise ratio as the ratio between the maximum of the absolute value of the signal and the standard deviation of the noise

In the linearized formulation **H** and **M** are related by the following *additive noise* relation

$$\mathbf{U} + \boldsymbol{\nu} = \mathbf{M} \quad \Rightarrow \quad \mathrm{Z}\mathbf{H} + \boldsymbol{\nu} = \mathbf{M}. \tag{6.57}$$

Here, the *observation operator* between **H** and **M** (see the example in Sect. 6.2.2 for Gaussian vectors) is actually the inverse of a (discrete) differential operator; $\mathrm{Z} = \mathrm{D}\mathrm{S}^{-1}\mathrm{R}_{\mathrm{in}}\mathrm{M}_{\mathrm{in}}$ has been introduced after (6.52) and it describes the deterministic relation between the velocity and the normal stress. The random variable $\boldsymbol{\nu}$ accounts for the measurement noise. We make the assumption of mutual independence of **U** and $\boldsymbol{\nu}$. Since **H** and **U** are related by a linear relation this implies the independence of **H** and $\boldsymbol{\nu}$. As a consequence, the p.d.f. of $\boldsymbol{\nu}$ is independent of any realization of **H** and the likelihood function, $p_{M|H}$, can be expressed as

$$p_{M|H}(M) = p_{M|H}(\nu + ZH) = p_{\nu}(M - ZH). \tag{6.58}$$

Next, we consider the realization $M = \mathbf{d}$ (the vector of available velocity measures introduced previously), we have

$$p_{H|M}(H) \propto p_{M|H}(\mathbf{d})\, p_H(H) = p_{\nu}(\mathbf{d} - ZH)\, p_H(H). \tag{6.59}$$

Now we make the assumption that all variables are Gaussian and we define the a priori distribution and the noise distribution as follows:

$$\begin{aligned} p_H &= g_H \propto \exp\left\{-\frac{1}{2}(H - H_0)^T \Lambda_H^{-1}(H - H_0)\right\}, \\ p_{\nu} &= g_{\nu} \propto \exp\left\{-\frac{1}{2}(\nu - \nu_0)^T \Lambda_{\nu}^{-1}(\nu - \nu_0)\right\}; \end{aligned} \tag{6.60}$$

where H_0 and ν_0 are the expectation values and Λ_H and Λ_{ν} are the correlation matrices for **H** and $\boldsymbol{\nu}$, respectively. The analysis of Sect. 6.2.2 shows that the posterior distribution $p_{H|M}$ is a Gaussian distribution itself with covariance and mean given by

$$\begin{aligned} \Lambda_{H|M} &= (\Lambda_H^{-1} + \mathrm{Z}^T \Lambda_{\nu}^{-1}\mathrm{Z})^{-1}, \\ \mathcal{E}(\mathbf{H}) &= \Lambda_{H|M}^{-1}(\mathrm{Z}^T \Lambda_{\nu}^{-1}(\mathbf{d} - \nu_0) + \Lambda_H^{-1} H_0). \end{aligned} \tag{6.61}$$

We recall that the mean value of the posterior distribution is the value that maximizes $p_{H|M}$, and then, by definition, it is the MAP estimator of **H**, say $\hat{\mathbf{H}}_{\mathrm{MAP}}$. On the other hand, the value that maximizes the likelihood function, with respect to H, corresponds to the ML estimator for **H** and has the following expression

$$\hat{\mathbf{H}}_{\mathrm{ML}} = (\mathrm{Z}^T \Lambda_{\nu}^{-1}\mathrm{Z})^{-1}(\mathrm{Z}^T \Lambda_{\nu}^{-1}(\mathbf{d} - \nu_0)). \tag{6.62}$$

In treating the nonlinearity we consider an iterative approach similar to the deterministic one described in the previous section; in fact, also in this case, we rely on the Newton method for the NSE. The distribution $p_{H|M}$ for the nonlinear model is still Gaussian, the following algorithm is used to determine its mean and covariance.

Given a guess for the random vector $\mathbf{U}_k = Z_k \hat{\mathbf{H}}_{\mathrm{MAP},k}$ *at iteration* $k+1$,

$$\begin{array}{ll} (1)\ \textit{compute} & \Lambda_{H|M,k+1} = \Lambda_H^{-1} + Z_k^T \Lambda_\nu^{-1} Z_k \\ (2)\ \textit{solve} & \Lambda_{H|M,k+1}(\hat{\mathbf{H}}_{\mathrm{MAP},k+1}) = Z^T \Lambda_\nu^{-1}(\mathbf{d} - \nu_0) + \Lambda_H^{-1} h_0, \end{array} \tag{6.63}$$

until a convergence criterion is satisfied.

Here, for $Z_k = DS_k^{-1} R_{\mathrm{in}}^T M_{\mathrm{in}}$ we define

$$S_k = \begin{bmatrix} C + A_k & B^T \\ B & O \end{bmatrix}. \tag{6.64}$$

A_k is the discretization of the advection operator with advection field $\mathbf{U}_k$, the velocity vector associated with the normal stress $\mathcal{E}(H)_k$. Note that with this formulation $\mathbf{H}$ and $\mathbf{U}$, at each iteration, are related by a linear model and, for this reason, $\mathbf{U}$ can still be considered normally distributed.

Numerical tests. We assume to have an exact, analytic solution of the NSE and we compare the accuracy of the MAP and ML estimators vs. the "deterministic estimator" introduced in the previous section, i.e., the solution of the variational formulation. The index of accuracy is related to the velocity fields, $\hat{\mathbf{U}}$, retrieved from $\hat{\mathbf{H}}_{\mathrm{MAP}}$, $\hat{\mathbf{H}}_{\mathrm{ML}}$ and $\hat{\mathbf{H}}_{\mathrm{det}}$ (the deterministic estimate); it is defined as $E(\hat{\mathbf{U}}) = \frac{\|\hat{\mathbf{U}} - \mathbf{U}_{\mathrm{anl}}\|_2}{\|\mathbf{U}_{\mathrm{anl}}\|_2}$, where $\mathbf{U}_{\mathrm{anl}}$ is the discretized analytic solution. We also define an average error over a set of noise realizations $\{\boldsymbol{\nu}\}_{i=1}^n$, $\overline{E}(\hat{\mathbf{U}}) = \frac{1}{n}\sum_{i=1}^n E(\hat{\mathbf{U}}, i)$ where $E(\hat{\mathbf{U}}, i)$ is associated with the i-th realization of noise $\boldsymbol{\nu}_i$. In addition, we consider a measure of the gain, γ, in using statistical estimators as opposed to deterministic ones: $\gamma = 1 - \frac{\overline{E}(\hat{\mathbf{U}}_{\mathrm{stat}})}{\overline{E}(\hat{\mathbf{U}}_{\mathrm{det}})}$ where "stat" stands for either MAP or ML.

The details of the numerical tests are fully reported in [16].

In a square domain we consider data on Γ_{in} and internal data located on ten internal slices. In Table 6.2 we report results obtained in correspondence of SNR of 20 and 10. In the computation of $\hat{\mathbf{H}}_{\mathrm{MAP}}$ and $\hat{\mathbf{H}}_{\mathrm{det}}$ the regularization parameter $\alpha = 0.5$ is chosen empirically (left table in Table 6.2). In the computation of $\hat{\mathbf{H}}_{\mathrm{ML}}$ and $\hat{\mathbf{H}}_{\mathrm{det}}$ on the right table the regularization parameter α is set to 0. From the results we infer the following facts. (1) Compared to the deterministic estimator, the statistical estimators are always more accurate since they take into account additional information brought by statistical properties of the data. (2) The computational time required in solving the statistical formulations is, in average, 1.3 times bigger than the one required by the deterministic one. (3) The poor gain in correspondence of SNR $= 20$ means that statistical information associated with a low amount of noise is not significant enough to make a considerable difference with respect to deterministic estimates in terms of accuracy.

As a second example we consider the same problem setting of the previous section for the flow in a cylinder, see Fig. 6.8 (right); we consider measures on the

Table 6.2 Accuracy results for statistical and deterministic solutions for the NSE

SNR	$\overline{E}_{U,\mathrm{det}}$	$\overline{E}_{U,\mathrm{MAP}}$	γ
20	0.0822	0.07371	10 %
10	0.1394	0.1041	25 %

SNR	$\overline{E}_{U,\mathrm{det}}$	$\overline{E}_{U,\mathrm{ML}}$	γ
20	0.0855	0.0579	6 %
10	0.1675	0.1363	18 %

Table 6.3 Accuracy results for statistical and deterministic solutions for the axisymmetric case

SNR	$\overline{E}_{U,\mathrm{det}}$	$\overline{E}_{U,\mathrm{MAP}}$	γ
20	0.0396	0.0308	22 %
10	0.1423	0.0978	31 %

inflow boundary and internal data located on five internal slices. We only compute the MAP estimator (the problem for the computation of $\hat{\mathbf{H}}_{\mathrm{ML}}$ is ill-posed). In this experiment $\alpha = 1e{-}7$; results in Table 6.3 show that with the MAP estimator we have a significant gain in accuracy. Moreover, the computational time required by the statistical estimators is the same as for the deterministic one.

Weighted Least Squares Finite Element Method

Another approach to the assimilation of measured velocities has been proposed in [40]. This work is mainly inspired by the development of a new experimental technique, the particle imaging velocimetry [41], that can be used to determine two components of the blood velocity along a single plane within the ventricle of the heart. The proposed method relies therefore on the hypothesis that the measures are collected inside a three-dimensional region on a two-dimensional plane (as in Fig. 6.7); the latter is basically treated as a (artificial) boundary.

This variational technique exploits a weighted least squares finite element method (WLSFEM), based on the LSFEM [9–11]; the latter has been utilized in general for the solution of PDEs. It features great flexibility in the enforcement of various types of boundary conditions. However, the LSFE method has also been applied to inverse problems since the 1990s for the numerical solution of PDE constrained control problems; main contributors are Bochev and Gunzburger [9–11].

We consider the problem of solving the following generic boundary value problem

$$\begin{aligned} Lu &= f \qquad \text{in } \Omega \\ u &= g \qquad \text{on } \partial\Omega, \end{aligned} \tag{6.65}$$

where L is a first order linear differential operator and $J(u)$ is a cost functional defined as

$$J(u) = \|Lu - f\|^2_{L^2(\Omega)} + \|u - g\|^2_{H^{1/2}(\partial\Omega)}. \tag{6.66}$$

Then, the LSFE solution u is obtained as the minimal of $J(u)$.

Assume we have N_s measures $d_i(\mathbf{x})$ of the variable u on some layers, $\Gamma_1, \ldots, \Gamma_{N_s}$, internal to Ω. We want to perform DA for problem (6.65), i.e., we want to merge $\{d_i\}$ and the numerical solution of (6.65). The idea of the WLSFEM is to add penalization terms to the functional J. The internal layers are considered part of the boundary and the corresponding measures are treated as boundary data; these terms are then properly weighted according to the level of confidence of the measure. Thus, the cost functional is defined as

$$\hat{J}(u) = J(u) + w_1 \|u - d_1\|^2_{H^{1/2}(\Gamma_1)} + \cdots w_{N_s} \|u - d_{N_s}\|^2_{H^{1/2}(\Gamma_{N_s})}, \tag{6.67}$$

where $w_1, \ldots, w_{N_s}$ are the weights. Note that (1) with the introduction of these additional terms that penalize the difference between the observed data and the solution, the assimilation is weakly enforced; (2) DA introduces an additional extra cost to LSFE calculation.

When applying the WLSFE method to the NSE one has to keep in mind that it is designed for first order linear differential operators; thus, we must recast the fluid dynamic equations into a linearized first order differential system. To this end, we consider a non-primitive variable set: we introduce the variable $\omega = -\nabla \times \mathbf{u}$, the negative vorticity, and the variable

$$\mathbf{r} = \nabla p + \frac{\sqrt{\mathrm{Re}}}{2} \nabla |\mathbf{u}|^2,$$

commonly referred to as the "gradient of pressure," where Re is the Reynolds number. Then, we apply the WLSFE method to the equivalent problem in Ω (see [40] for details on how to derive the following system)

$$\begin{aligned}
&\nabla \times \mathbf{u} + \omega = 0 \\
&\nabla \cdot \mathbf{u} = 0 \\
&\frac{1}{\sqrt{\mathrm{Re}}} \nabla \times \omega - \mathbf{r} - \sqrt{\mathrm{Re}}(\mathbf{u} \times \omega) = 0 \\
&\nabla \cdot \omega = 0 \\
&\nabla \times \mathbf{r} = 0 \\
&\nabla \cdot \mathbf{r} - \sqrt{\mathrm{Re}}(\omega \cdot \omega) - \mathrm{Re}(\mathbf{u} \cdot \mathbf{r}) = 0.
\end{aligned} \tag{6.68}$$

The optimization problem, formulated as in (6.66) with $Lu = f$ given by Eq. (6.68), is then solved with standard techniques from the calculus of variations. Again, we stress the fact that, being L a linear operator, the cost of the WLSFE formulation is of the same order of the solution of the NSE. However, this approach, as opposed to primitive variables formulations might be less conducive than the straightforward inclusion of available measures.

Numerical tests. Consider a cylindrical geometry and assume the measures to be located on internal layers parallel to the flow; in Fig. 6.9 (left) the noisy data on the layer crossing the axis of symmetry of the cylinder are reported. It is also assumed

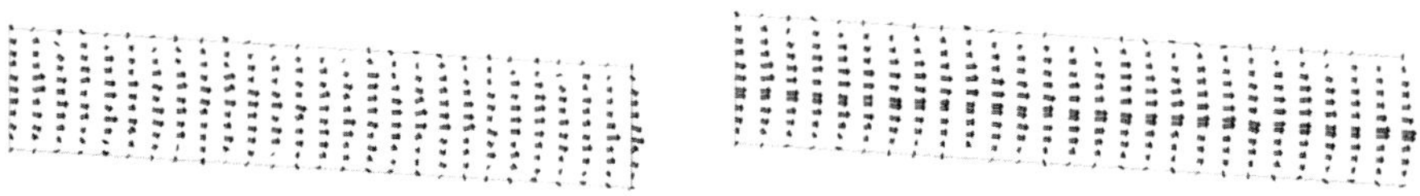

Fig. 6.9 On the *left* the synthetic measures generated adding Gaussian noise to an analytical solution are reported on a layer crossing the axis of symmetry. On the *right*, on a layer close to the one where the measures are collected, the assimilated velocity is reported. Adapted from [40], with permission of © Elsevier 2010

that the noise affects the boundary data (whereas in [18] they are considered exact), which is always the case in real applications. For the numerical solution with the WLSFE method the boundary and internal data are properly weighted according to the noise level (assumed known in this particular experiment); the assimilated solution, on an internal layer close to the measurement one, is reported in Fig. 6.9 (right). The filtering action of the DA on the noise is evident. Quantitative analysis, not reported here, reveals a good level of accuracy [40].

WLSFEM as a Bayesian Approach to DA

In [21] a reinterpretation of the WLSFEM in terms of Bayesian approach to DA is proposed; in fact, in [40] the method is not presented in an inverse problem framework. Here we show that the WLSFE solution can be interpreted as the *maximum a posteriori* (MAP) estimator in a variational Bayesian approach to DA, for a certain choice of a priori distribution and likelihood function. A statistical interpretation of the weights is also provided.

In describing the method we refer to the general boundary value problem (6.65). We recall that in a Bayesian approach to DA all variables are treated as random, the goal is to determine the p.d.f. of u conditioned on realizations of the measures $d_1, \ldots, d_{N_s}$ available on the internal layers $\Gamma_1, \ldots, \Gamma_{N_s}$. We assume that the measures are affected by the measurement noise $\nu_1, \ldots, \nu_{N_s}$ such that $d_i(\mathbf{x}) = u(\mathbf{x})|_{\Gamma_i} + \nu(\mathbf{x})_i$, for $i = 1, \ldots, N_s$. To apply the Bayes theorem we need to define an a priori distribution for u, p_u, based on our prior belief on u and a likelihood function for the measurement noise ν_i, $p_{\nu,i}$. In order to show the equivalence between the WLSFE deterministic solution, or WLSFE estimator, and the MAP estimator in the Bayesian setting we make the following choices.

We define a prior distribution which is large when u satisfies the governing equations (6.65) "well" and small otherwise; in this way the prior describes to what extent the equations are a good model for the observations. Formally

$$p_u(u) \propto \exp\{-J(u)\},$$

where J is defined as in (6.66).

Next, in defining the likelihood functions for ν_i, we assume that the measurement errors ν_i are independent and normally distributed with null mean and variance

$\frac{1}{2w_i}$, being w_i the weights introduced in the previous section. Applying the Bayes theorem we have

$$p_{u|d_1 \dots d_{N_s}} \propto \exp\left\{-J(u) - w_1 \|u - d_1\|^2_{H^{1/2}(\Gamma_1)} + \cdots w_{N_s} \|u - d_{N_s}\|^2_{H^{1/2}(\Gamma_{N_s})}\right\} \Rightarrow$$
$$p_{u|d_1 \dots d_{N_s}} \propto \exp\left\{-\hat{J}(u)\right\},$$

for $\hat{J}$ as in (6.67). The MAP estimator is then the value of u that maximizes the posterior distribution $p_{u|d_1 \dots d_{N_s}}$, thus

$$\hat{u}_{\text{MAP}} = \arg\max p_{u|d_1 \dots d_{N_s}} = \arg\min \hat{J}(u) = u_{\text{WLSFE}},$$

This leads to the conclusion that the WLSFE solution, u_{WLSFE}, is actually a Bayesian estimator for the variable u; thus, we have the following statistical interpretations

- the mathematical model encodes our prior belief on u;
- the data is a correcting likelihood;
- the weights reflect the variance of the measurement noise, i.e., are an index of the reliability of the measures.

This procedure, and the associated considerations, naturally apply to the first order form of the NSE so that the velocity estimated via WLSFEM is a Bayesian estimator. The latter differs from the one introduced in [16] in the choice of the prior distribution and likelihood function. The first approach is certainly more general as does not require the measures to be on a plane and more straightforward because formulated for the primitive variable, on the other hand, the second is computationally cheaper as it deals with the recast (linear) form of the NSE. As for the accuracy, an extensive comparison is still missing.

6.4.2 Estimation of the Arterial Compliance from Measurements of Displacement: An Inverse Fluid–Structure Interaction Problem

As a second example, we consider the estimation of the compliance of an artery. The problem consists of estimating the compliance of an artery wall, based on (noisy) data of the displacement of the wall, obtained using medical devices such as Magnetic Resonance Imaging (MRI) during a heart beat. We focus on two approaches that have been recently adopted in the literature. In [59] a variational approach is pursued: the compliance is used as control variable for minimizing the misfits between the results of a FSI problem and the displacement of the vessel (possibly retrieved from images). In [6] a Reduced Order UKF is advocated to solve the same problem. In the following we summarize these two approaches and present some results of these works. For details we refer the reader to the corresponding works.

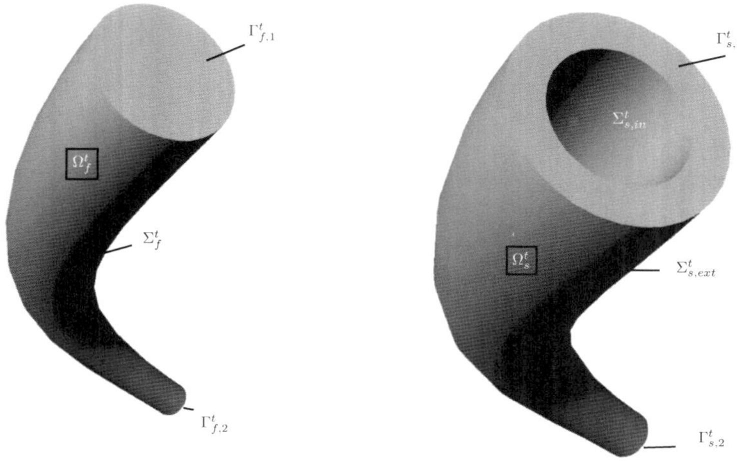

Fig. 6.10 Representation of the domain of the FSI problem: fluid domain on the *left*, structure domain on the *right*

Problem Formulation

We consider a domain of a vessel (structure domain) perfused by blood (fluid domain) as depicted in Fig. 6.10. We make the simplistic assumption that the vessel is (linearly) elastic, with the stress tensor $\boldsymbol{\sigma}_s$ depending on the vessel displacement $\boldsymbol{\eta}$ as

$$\boldsymbol{\sigma}_s(\boldsymbol{\eta}) \equiv \gamma_1(\nabla\boldsymbol{\eta} + (\nabla\boldsymbol{\eta})^T) + \gamma_2(\nabla\cdot\boldsymbol{\eta})\boldsymbol{I},$$

where

$$\gamma_1 := \frac{E}{2(1+\nu)}, \qquad \gamma_2 := \frac{E\nu}{(1+\nu)(1-2\nu)},$$

are the Lamé constants, $\boldsymbol{I}$ is the identity tensor, E is the Young's modulus, and ν is the Poisson's ratio. For the sake of notation, we factor the Young's modulus E out of the stress tensor, and so that we can write

$$\boldsymbol{\sigma}_s = E\,\tilde{\boldsymbol{\sigma}}_s, \qquad \tilde{\boldsymbol{\sigma}}_s := \frac{1}{2(1+\nu)}(\nabla\boldsymbol{\eta} + (\nabla\boldsymbol{\eta})^T) + \frac{\nu}{(1+\nu)(1-2\nu)}(\nabla\cdot\boldsymbol{\eta})\boldsymbol{I}.$$

The vessel deforms under the stress coming from the blood, and in turn, the elastic structure of the vessel affects the blood flow. This problem has been largely investigated in other chapters of this book (see also, e.g., [25]). For the sake of numerical approximation of the problem, the problem is formulated on a frame of reference moving with the physical wall of the artery and fixed on the artificial boundaries (inflow/outflow). This approach is known as the Arbitrary Lagrangian Eulerian (ALE) formulation, see, e.g., [20, 42]. We write the problem according to the ALE frame of reference. At time t the blood velocity $\mathbf{u}$ and pressure p live in the fluid domain ${\Omega^t}_f$, whereas the vessel displacement $\boldsymbol{\eta}$ lives in the structure vessel domain Ω^t_s. We denote the interface between the fluid and the solid domains with Σ^t (see Fig. 6.10). It is more convenient to model the structure displacement $\boldsymbol{\eta}$ in the reference configuration $\hat{\Omega}_s$; we denote a variable in the reference configuration with a $\hat{}$, e.g. $\hat{\boldsymbol{\eta}}$.

1. *Fluid–Structure problem*. Find fluid velocity $\mathbf{u}$, pressure p and structure displacement $\boldsymbol{\eta}$ such that

$$\begin{cases} \rho_f \dfrac{D^A \mathbf{u}}{Dt} + \rho_f((\mathbf{u}-\mathbf{w})\cdot\nabla)\mathbf{u} - \nabla\cdot\boldsymbol{\sigma}_f = \mathbf{f}_f & \text{in } {\Omega^t}_f \times (0,T), \\ \nabla\cdot\mathbf{u} = 0 & \text{in } {\Omega^t}_f \times (0,T), \\ \rho_s \dfrac{\partial^2 \hat{\boldsymbol{\eta}}}{\partial t^2} - \nabla\cdot(E\,\widehat{\boldsymbol{\sigma}}_s) = \hat{\mathbf{f}}_s & \text{in } \Omega_s \times (0,T), \\ \mathbf{u} = \dfrac{\partial \boldsymbol{\eta}}{\partial t} & \text{on } \Sigma^t \times (0,T), \\ \boldsymbol{\sigma}_s\,\mathbf{n} - \boldsymbol{\sigma}_f\,\mathbf{n} = \mathbf{0} & \text{on } \Sigma^t \times (0,T), \end{cases} \tag{6.69}$$

where $\boldsymbol{\sigma}_f(\mathbf{u}, p) = -p\boldsymbol{I} + \mu(\nabla\mathbf{u} + (\nabla\mathbf{u})^T)$, ρ_f and ρ_s are the fluid and structure density, μ is the constant blood viscosity, $\mathbf{f}_f$ and $\mathbf{f}_s$ are the forcing terms. Here, $\dfrac{D^A}{Dt}$ is the so-called ALE derivative and $\mathbf{w}$ is a lifting of the velocity at Σ^t in ${\Omega^t}_f$. Typically (but not necessarily) this lifting is obtained by solving a Poisson problem (harmonic lifting). At the inlet and outlet sections, proper boundary conditions have to be prescribed. In particular, it is important to use *absorbing* boundary conditions at the outlet to avoid unphysical solutions. The two matching conditions enforced at the interface are $(6.69)_4$ (*continuity of fluid and structure velocities*) and $(6.69)_5$ (*continuity of stresses*).

Before moving to the problem of the estimation of the parameters, to reduce computational costs a simplified set of equations for the FSI problem can be used. In fact, for large arteries, the wall thickness is in general significantly smaller than the dimension of the lumen, so that the arterial wall can be described as a 2D surface rather than a 3D structure. If we also assume that the displacement occurs only in the normal direction[12] it can be shown [57] that the structure equations reduce to

$$\rho_s h_s \frac{\partial^2 \eta}{\partial t^2} + E\beta\eta = f_s$$

where η refers to the normal displacement on the boundary of the vessel, β is a parameter embedding both geometrical and physical properties of the membrane, whose expression is given by $\beta = \frac{h_s}{1-\nu^2}(4k_m^2 - 2(1-\nu)k_g)$. Here k_m and k_g are,

[12]This assumption may be questionable for arteries close to the heart (like the aortic arch), however it is in general quite acceptable.

respectively, the mean and gaussian curvature of the membrane and h_s is the wall thickness. Discretizing the problem in time (using for instance backward Euler) and imposing the conservation of the normal stresses on the interface, the Fluid–Membrane Interaction (FMI) problem can be written as

$$\begin{cases} \frac{\rho_f}{\Delta t}(\mathbf{u}^n - \mathbf{u}^{n-1}) + \rho_f((\mathbf{u}^* - \mathbf{w}^*)\cdot\nabla)\mathbf{u}^n + \nabla\cdot\boldsymbol{\sigma}_f = \mathbf{f}_f & \text{in } \Omega^* \\ \nabla\cdot\mathbf{u}^n = 0 & \text{in } \Omega^* \\ \mathbf{u}^n \cdot \boldsymbol{\tau} = 0 & \text{on } \Sigma^* \\ \mathbf{n}\cdot\boldsymbol{\sigma}_f\mathbf{n} + \left(\frac{\rho_s h_s}{\Delta t} + E^n\beta\Delta t\right)\mathbf{u}^n\cdot\mathbf{n} = \left(\frac{\rho_s h_s}{\Delta t}\mathbf{u}^{n-1}\cdot\mathbf{n} - E^n\beta\eta^{n-1}\right) & \text{on } \Sigma^* \\ \eta^n = \eta^{n-1} + \Delta t\mathbf{u}^n\cdot\mathbf{n} & \text{on } \Sigma^* \end{cases} \tag{6.70}$$

where $\boldsymbol{\tau}$ is any unit versor in the tangent space to Σ^*. Notice how the effect of the structure is now expressed as a Robin-type boundary condition for the fluid equations. The superscript * denotes a suitable extrapolation of the quantity at the time t^n. Notice that if we use a semi-implicit scheme to deal with convective and geometric non-linearities, so that $\Omega^* = \Omega^{n-1}$ and the fluid and structure equations are then decoupled (within the time step). In particular, the equation for η^n can be promptly solved once the fluid equations have been solved. We will make use of this model in the following sections.

Parameter Estimation Problem

The displacement of a vessel can be retrieved from images properly segmented and registered in time. This means that at each available snapshots, the arterial wall is reconstructed as a triangulated surface; then, a map is properly computed to identify the image of each point at the subsequent snapshots (see, e.g., [53, 61]). The map is obtained by minimizing the mismatch between the image of its application to a snapshot and the successive one. In particular we denote by τ_k the instants when images are available and by $\Delta\tau$ the length of each time interval. Once the map is computed, the displacement is promptly available. In [61], for instance, the *Iterative Closest Point* criterion is used to quantify the mismatch and to compute the map. Our goal now is to estimate the compliance of the vessel such that the mismatch between the retrieved and the computed displacement is minimized. To this end, we introduce the following cost functional

$$\mathcal{J}_1 = \frac{1}{2}\sum_{k=1}^{N}\int_{\Sigma}(\boldsymbol{\eta}_{\text{meas}}(\boldsymbol{x},\tau_k) - \boldsymbol{\eta}(\mathbf{x},\tau_k))^2\,d\sigma. \tag{6.71}$$

where $\boldsymbol{\eta}(\mathbf{x},\tau_k)$ solves Eq. (6.69) at instants τ_k and $\boldsymbol{\eta}_{\text{meas}}$ is the (noisy) observed displacement. Here, we are assuming to have a continuous displacement field $\boldsymbol{\eta}_{\text{meas}}$ defined on Ω_s. In case we only have sparse measurements of the displacement, it is reasonable to use the following cost functional

$$\mathcal{J}_2 = \frac{1}{2} \sum_{k=1}^{N} \sum_{j=1}^{M} \|\Delta \eta_{j,k}\|^2_{\mathrm{R}_k^{-1}}, \tag{6.72}$$

where $\Delta \eta_{j,k} = \boldsymbol{\eta}_{\text{meas}}(\mathbf{x}_j, \tau_k) - \boldsymbol{\eta}(\mathbf{x}_j, \tau_k)$, R_k^{-1} is a weight s.p.d. matrix. Should probabilistic information on the displacement be available, R_k^{-1} is the covariance matrix of the noise of the displacement retrieval process.

As anticipated, we consider two approaches to solve this problem: a deterministic variational approach and a Kalman-based approach.

Remark 6.8. Typically, the time step Δt of the numerical scheme is smaller than the time sample $\Delta \tau$, requiring more observations than those available. A common practice is to recover the observation at needed time steps by interpolation. In the following we will use this approach.

Variational Approach

In order to minimize $\mathcal{J}_1$ we can use a gradient-based optimization approach as discussed in Sect. 6.3.2. However, as outlined there, the solution of an unsteady minimization problem, such as the FSI problem, would be very expensive because all the steps are coupled together, and it would also require the evaluation of shape derivatives since the geometry is evolving in time. To reduce the computational costs and the algorithm complexity, we exploit the fact that the parameter E does not change in time and solve the following suboptimal problem. First, we discretize the system in time. Then, at each time instant t^n we solve a steady suboptimal optimization problem, finding the value E^n which minimizes the functional

$$\mathcal{J}_3^n = \frac{1}{2} \int_{\Sigma} (\boldsymbol{\eta}_{\text{meas}}(\boldsymbol{x}, \tau_n) - \boldsymbol{\eta}(\mathbf{x}, \tau_n))^2 \, d\sigma, \tag{6.73}$$

constrained by the time-discrete FSI problem at time t^n. Finally, we compute E as the average of E^n: $E = \frac{1}{N} \sum_{n=1}^{N} E^n$.

Numerical Solution

For the sake of clarity, we focus on the simplified membrane model (6.70), already discretized in time. However, note that the optimization strategy described in the following has been applied to the original FSI problem (6.69) in [59]. When considering the membrane approximation, the cost functional J_3^n becomes

$$\mathcal{J}_m^n = \frac{1}{2} \int_{\Sigma} \left(\eta_{\text{meas}}^n - \eta^n\right)^2 d\sigma,$$

and the adjoint of problem (6.70) reads[13]

$$\begin{cases} \dfrac{\rho_f}{\Delta t}(\boldsymbol{\chi}^n - \mathbf{u}^{n-1}) - \rho_f((\mathbf{u}^* - \mathbf{w}^*)\cdot\nabla)\boldsymbol{\chi}^n + \nabla\cdot\boldsymbol{\sigma}_f(\boldsymbol{\chi}^n) = \Delta t(\eta^n_{\text{meas}} - \eta^n)\mathbf{n} & \text{in } \Omega^* \\ \nabla\cdot\boldsymbol{\chi}^n = 0 & \text{in } \Omega^* \\ \boldsymbol{\chi}^n \cdot \boldsymbol{\tau} = 0 & \text{on } \Sigma^* \\ \mathbf{n}\cdot\boldsymbol{\sigma}_f(\boldsymbol{\chi})\mathbf{n} + \left(\dfrac{\rho_s h_s}{\Delta t} + E^n\beta\Delta t\right)\boldsymbol{\chi}^n\cdot\mathbf{n} = 0 & \text{on } \Sigma^* \\ \eta^n = \eta^{n-1} + \Delta t\,\boldsymbol{\chi}^n\cdot\mathbf{n} & \text{on } \Sigma^*. \end{cases}$$

The gradient of the cost functional with respect to the parameter E^n is obtained using the adjoint variable $\boldsymbol{\chi}$ and relation (6.38), which, for the problem at hand reads

$$\left.\frac{D\mathcal{J}_m^n}{DE}\right|_{E^n} = -\int_{\Sigma^*} \beta\left(\Delta t\mathbf{u}^n\cdot\mathbf{n} + \eta^{n-1}\right)(\boldsymbol{\chi}^n\cdot\mathbf{n})d\sigma.$$

The optimization is performed using the BFGS method. In particular, at each time step, for a given initial guess of the parameter $E^{n,(0)}$, the BFGS method iteratively provides parameter guesses $E^{n,(j)}$, based on the values of the cost functional $\mathcal{J}_m^n\left(E^{n,(j-1)}\right)$ and its derivative $\left.\dfrac{D\mathcal{J}_m}{DE}\right|_{E^{n,(j-1)}}$. The iterative procedure stops when the norm of $\left.\dfrac{D\mathcal{J}_m}{DE}\right|_{E^{n,(j-1)}}$ is less than a given tolerance.

Remark 6.9. BFGS is a method devised for unconstrained optimization, while the problem at hand features the constraint $E > 0$. Unilateral constraints can be managed as indicated in [58]. Here, we include this, with a simple change of variable, by using as a control variable $\psi = \log(E)$, so that $E = \exp(\psi) > 0$ for every ψ.

Numerical Results on a Simplified Geometry Representing an Abdominal Aneurysm

In the numerical results presented in this section, we will use the simplified membrane model (6.70). The optimization strategy depicted above, however, can be equally applied to this simplified problem.

We consider a 2D axisymmetric geometry which represents an abdominal aneurysm (see Fig. 6.11, top-left). The radius of the vessel varies from 1 to 2.5 cm and the vessel length is 6 cm. We perform a synthetic simulation in

[13]See (6.37), and note that here the adjoint variable is denoted with χ as in this context ρ is used for the density.

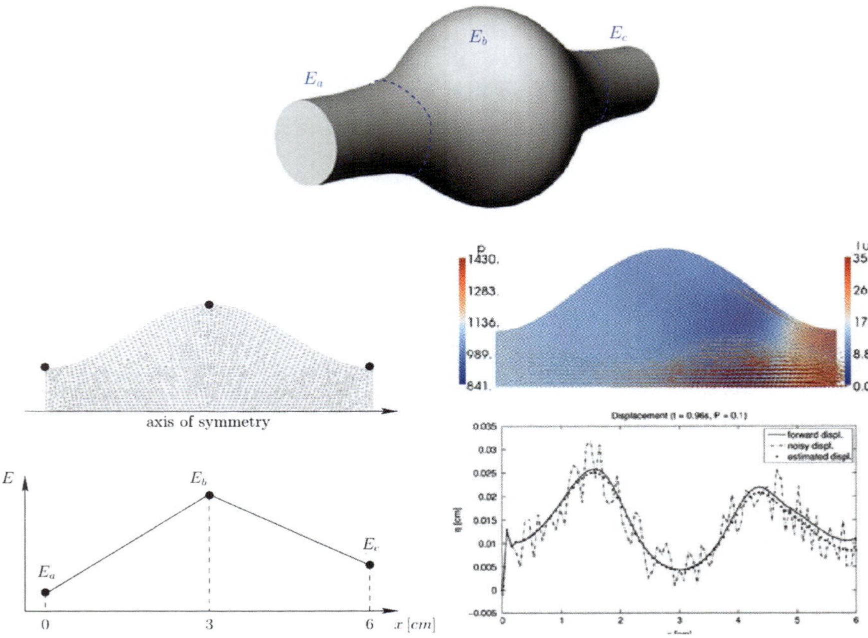

Fig. 6.11 Aneurysm simulation. *Top-left*: Mesh used for the simulation. *Bottom-left*: Piecewise linear approximation of the Young's modulus E in the forward simulation. *Top-right*: Velocity vectors and pressure at time $t = 0.96$ s. *Bottom-right*: Comparison between the displacement obtained with the forward simulation, the noisy data and the computed displacement, at time $t = 0.96$ s and for $P = 0.1$

which we prescribe the piecewise linear Young's modulus shown in Fig. 6.11 (bottom-left). For the forward simulation, we take $E_a = 4 \cdot 10^6$ dyne/cm^2, $E_b = 10^7$ dyne/cm^2, $E_c = 5 \cdot 10^6$ dyne/cm^2. We prescribe at the inlet a parabolic profile for the velocity, whose maximum $u_{\max}$ lies on the axis of symmetry and it is given by

$$u_{\max} = u^0_{\max} + A \max \left\{ \sin \left(\frac{2\pi t}{T} \right); 0 \right\},$$

where $u^0_{\max} = 5$ cm/s, $A = 55$ cm/s and $T = 0.6$ s. At the outlet we prescribe the absorbing boundary conditions proposed in [57]. We run the simulation for two heart beats, i.e., for $0 < t \leq 2T$. We add a uniform noise ν_P to the forward displacement η_{fwd} and we use the result as data for the control problem. In Fig. 6.11 (bottom-right) we report a comparison between the displacement obtained with the forward simulation, the noisy data and the computed displacement at time $t = 0.96$ s. The agreement is very good.

Table 6.4 Noisy case. Mean and standard deviation (to be multiplied by 10^6) of the ten estimates for E_a, E_b, E_c and number of state and adjoint iterations (bottom) for different values of the noise percentage P

SNR	E_a	E_b	E_c	iter.(state|adjoint)
34	4.047 ± 0.118 (1.2 %)	10.19 ± 0.295 (1.9 %)	5.194 ± 0.240 (3.9 %)	12.9|3.5
17	4.034 ± 0.281 (0.9 %)	10.40 ± 0.505 (4 %)	5.507 ± 0.584 (10 %)	14.8|3.8
8	4.200 ± 0.550 (5 %)	10.89 ± 0.850 (8.9 %)	–	16.0|4.2

The initial guess is $E_{a,0} = E_{b,0} = E_{c,0} = 2 \cdot 10^7$ dyne/cm^2

Table 6.5 Noisy case with regularization term

SNR	E_a	E_b	E_c	iter.(state|adjoint)
34	4.032 ± 0.119 (0.8 %)	10.15 ± 0.320 (1.5 %)	5.123 ± 0.129 (2.5 %)	13.1|3.7
17	4.222 ± 0.238 (5.5 %)	10.17 ± 0.510 (1.7 %)	5.349 ± 0.368 (7.0 %)	14.2|3.6
11	4.446 ± 0.426 (11 %)	10.57 ± 0.780 (5.7 %)	7.036 ± 3.90 (41 %)	15.5|4.1
8.3	4.386 ± 0.570 (9.6 %)	11.09 ± 1.519 (11 %)	7.802 ± 4.12 (56 %)	16.9|4.1

Mean and standard deviation (to be multiplied by 10^6) of the ten estimates for E_a, E_b, E_c and number of state and adjoint iterations (bottom) for different values of the noise percentage P. The initial guess is $E_{a,0} = E_{b,0} = E_{c,0} = 10^7$ dyne/cm^2

In Table 6.4, we report the average, over the ten realizations, of the estimated values of E_a, E_b, and E_c and the number of times the state and the adjoint problem have needed to be solved. Different noise percentage P are considered. The initial guess is $E_{a,0} = E_{b,0} = E_{c,0} = 2 \cdot 10^7$ dyne/cm^2. The estimated values for $P = 0.1$ and $P = 0.2$ are quite accurate. For $P = 0.3$ we do not find a converged value for E_c. To overcome this problem, we add a regularization term to the cost functional $\mathcal{J}_3^n$, penalizing values of E far from the initial guess. Table 6.5 shows that the regularization term is effective. The estimates for E_c are still the more sensible to the noise, but now the estimated values are acceptable. In the first time steps of the simulation, the displacements computed by the FSI solver are very small for $x > 3$ cm, hence the data is dominated by the noise in that region. This fact can be an explanation of the high sensibility to the noise of the estimated value for E_c.

Reduction of the Computational Costs Via POD

In this section we show an example of how the POD procedure explained in Sect. 6.3.3 can be used for reducing the computational costs of the problem of the estimation of the Young's modulus explained in the previous paragraph. We assume

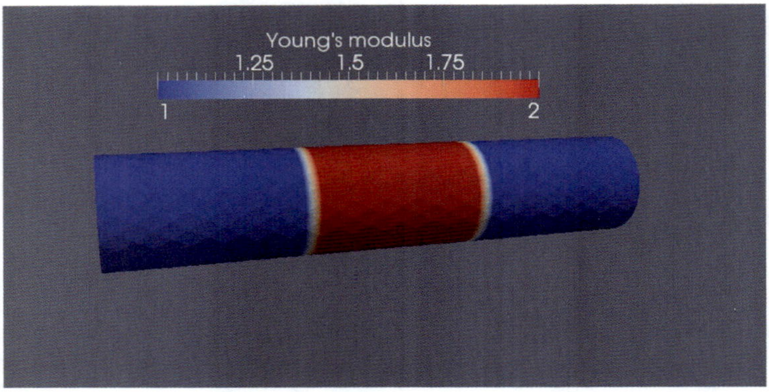

Fig. 6.12 Example of piecewise constant (along the axial direction) Young's modulus

that the structure is approximated as a 2D membrane, which allows us to use the simplified model (6.70). Furthermore, we divide the structure in a predetermined number k of regions along the axial direction of the vessel, and we consider the case where the Young's modulus is globally piecewise constant, with a constant value in each (predetermined) region (Fig. 6.12). This choice is driven by both practical and theoretical reasons. On the one hand, it can allow us to model the scenario where, due, for instance, to the presence of some pathology, the local properties of the tissue are altered. On the other hand, this choice guarantees the existence of a solution for the inverse problem, as shown in [59].

To show why a POD approach is reasonable, let us consider a flow in a cylinder, where the membrane has been divided into three regions in which the Young's modulus is constant, and let us consider for the inflow a sinusoidal pressure wave of the type

$$p(t) = 500 \sin(50\pi t).$$

We solve the forward problem for different values of each of the Young's moduli in the three regions and we compute the correlation matrix of the Finite Element snapshots for fluid velocity, denoted $\mathbf{u}_h$, and membrane displacement, denoted by η_h. The number of degrees of freedom is 9186 for the fluid velocity and 3540 for the membrane displacement.

The figures suggest that the unknowns can be well approximated by vectors belonging to spaces of dimensions much lower than the corresponding Finite Element ones. Therefore, it is natural of think to POD, as a possible strategy to reduce the computational costs.

At time t^n the fully discrete Inverse Fluid–Membrane Interaction (IFMI) problem reads: find E^n_{opt} such that

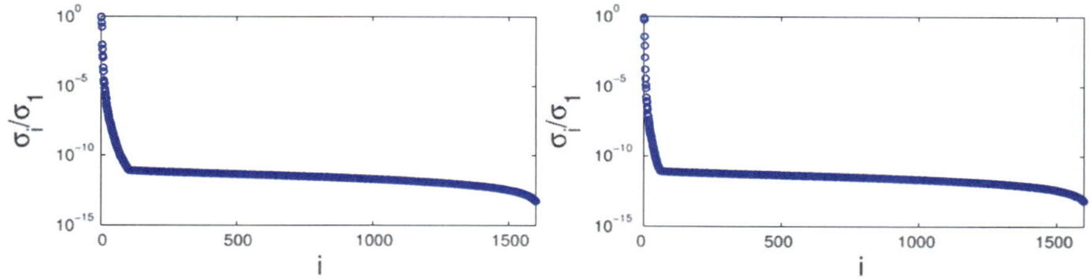

Fig. 6.13 Singular values of the velocity (*left*) and membrane displacement (*right*) snapshot matrix. Here we chose $E = (E_1, E_2, E_3)$, with $E_i \in \{1, 1.5, 2\} \cdot 10^6$ dyn/cm^2, and performed 60 time step

$$E^n_{\mathrm{opt},h} = \arg\min_{E \in \mathbb{R}^k_+} \mathcal{J}^n(E)$$
$$\text{s.t.} \quad \begin{bmatrix} \mathrm{C}(E) & \mathrm{B}^T & \mathrm{O} \\ \mathrm{B} & \mathrm{O} & \mathrm{O} \\ -\Delta t \mathrm{P} & \mathrm{O} & \mathrm{I} \end{bmatrix} \begin{bmatrix} \mathbf{u}^n_h \\ \delta p^n_h \\ \eta^n_h \end{bmatrix} = \begin{bmatrix} f^n_h(E_h) \\ 0 \\ \eta^{n-1}_h \end{bmatrix} \tag{6.74}$$

where P is a projection matrix, that extracts the normal component of the computed velocity on Σ. Here the dependence of the velocity matrix C and the velocity right-hand side f_h on E_h comes from the Robin boundary conditions in (6.70). We also point out that here we switched to an incremental formulation for the pressure, and we included the term $\mathrm{B}^T p^{n-1}_h$ in $f^n_h(E_h)$ for the sake of brevity.

To generate the POD basis, we solve the forward problem (6.70) for different values of the Young's modulus and we store the snapshots at each timestep. By this we mean that $\mathbf{u}^n_h$ and $\mathbf{u}^{n+1}_h$ are two different snapshots, even if they correspond to the same Young's modulus. However, we expect the solution to change smoothly in time, and therefore the singular values of the snapshots matrix should decay fast. This is confirmed by the numerical experiments in Fig. 6.13.

When building our reduced order model for the FMI problem, we want to exploit the divergence-free velocity snapshots. Assume we have stored the velocity vectors in the matrix W_u. When we project the momentum equation onto the range of W_u we obtain

$$\mathrm{W}^T_u \mathrm{C}\mathbf{u}^n + \mathrm{W}^T_u \mathrm{B}^T \delta p^n_h = \mathrm{W}^T_u f^n_h. \tag{6.75}$$

Should the geometry be constant in time, then the product $\mathrm{W}^T_u \mathrm{B}^T = (\mathrm{B}\mathrm{W}_u)^T$ would be identically zero, being the discrete space divergence-free, and the pressure increment term would disappear. When the geometry is moving, this is not true, since each snapshot is strictly speaking divergence free only in the geometry in which it was computed. However, for a small time step (and for small displacements) we do expect the increment δp^n to be small. For the sake of the computational costs, we drop the pressure correction term in the reduced problem. This can be regarded as an explicit treatment of the pressure in the time advancing scheme. Once the reduced momentum equation has been solved, the pressure can be recovered by solving the least square problem in the full Finite Element space, that is,

$$p_h^n = \min_{q_h \in Q_h} ||f_h^n - \mathrm{C}u_h^n - \mathrm{B}^T q||^2. \tag{6.76}$$

The solution to this problem exists and is unique, provided that the velocity and pressure FE spaces satisfy the *inf–sup* condition, which guarantees that B^T has full column rank. In order to have a representation of the pressure in the reduced space, one has to make sure that the reduced saddle point problem is non-singular. In literature this issue has been tackled by enriching the velocity reduced space [64].

We therefore construct the reduced basis only for the fluid velocity and membrane displacement fields. To this end, we solve the forward problem for a given set of Young moduli $E_1, \ldots, E_M$ and store the corresponding solutions (snapshots) $\mathbf{u}_{h,i}, \eta_{h,i}$. In order to deal with nonhomogeneous boundary conditions at the inflow/outflow sections, we modify the velocity snapshots in the following way

$$\hat{\mathbf{u}}_{h,i} = \mathbf{u}_{h,i} - \mathbf{u}_\ell \tag{6.77}$$

where $\mathbf{u}_\ell$ is the solution of a *steady* rigid-wall Stokes problem used as a *lifting function* for the nonhomogeneous boundary conditions. This choice allows us to preserve the divergence-free nature of the snapshots which are then collected (amended by the lifting) in the snapshots matrices X_u and X_η. We compute the SVD of these matrices and let W_α be the matrices containing the first k_α left singular vectors of X_α ($\alpha = u, \eta$), with k_α such that

$$\sum_{i=1}^{k_\alpha} \sigma_i \geq \tau \sum_{i=1}^{\mathcal{N}_\alpha} \sigma_i \tag{6.78}$$

where σ_i are the singular values of X_α, τ is the fraction of data variability that we want to capture (typically we take $\tau = 0.9, 0.95$ or 0.99) and $\mathcal{N}_\alpha$ is the dimension of the FE space. The columns of W_u and W_η form the reduced basis for the fluid velocity and membrane displacement spaces.

If we project the IFMI problem (6.74) onto the reduced space, we then obtain

$$\begin{aligned} E_h^n &= \arg\min_{E_h \in \mathbb{R}^k} \mathcal{J}_r^n(E_h) = \frac{1}{2}||\eta_{r,h}^n - d_r^n||_\Sigma^2 + \mathcal{R}(E_h) \\ &\text{s.t.} \begin{bmatrix} \mathrm{C}_r & \mathrm{O} \\ -\Delta t \mathrm{P}_r & \mathrm{I} \end{bmatrix} \begin{bmatrix} \mathbf{u}_r^n \\ \eta_r^n \end{bmatrix} = \begin{bmatrix} f_{r,h}^n \\ \eta_{r,h}^{n-1} \end{bmatrix} \end{aligned} \tag{6.79}$$

where $\mathrm{C}_r = \mathrm{W}_u^T \mathrm{C} \mathrm{W}_u$, $\mathrm{M}_r = \mathrm{W}_\eta^T \mathrm{M}_\Sigma \mathrm{W}_\eta$, $\mathrm{P}_r = \mathrm{W}_\eta^T \mathrm{P} \mathrm{W}_u$, $f_{r,h}^n = \mathrm{W}_u^T (f_h^n - \mathrm{B}^T p_h^*)$, $d_{r,h}^n = \mathrm{W}^T \eta_h \eta_{\text{meas}}^n$, and the dependence of C and f_h^n on E is understood for brevity.

The minimization problem is then solved as in the previous section, using the BFGS method. In particular, in order to evaluate the functional and its gradient, we solve the state and adjoint problems, respectively, which are given by

$$\begin{cases} \begin{bmatrix} C_r & O \\ -\Delta t P_r & I \end{bmatrix} \begin{bmatrix} \mathbf{u}^n_{r,h} \\ \eta^n_{r,h} \end{bmatrix} = \begin{bmatrix} f^n_{r,h} \\ \eta^{n-1}_{r,h} \end{bmatrix} & \text{(State)} \\ \begin{bmatrix} C_r^T & -\Delta t P_r^T \\ O & I \end{bmatrix} \begin{bmatrix} \lambda_u \\ \lambda_\eta \end{bmatrix} = \begin{bmatrix} 0 \\ -M_r(\eta^n_r - d^n_r) \end{bmatrix} & \text{(Adjoint)} \end{cases} \tag{6.80}$$

Numerical Results on an Idealized Aortic Arch

In this section we study the flow in a curved pipe resembling the shape of an idealized aortic arch. In particular, the geometry consists of a half torus joint with a cylinder. We chose the major and minor radii of the torus (i.e., the distance between the center of the torus and the centerline of the pipe and the radius of the pipe, respectively) to be $R = 1.5\,\text{cm}$ and $r = 0.5\,\text{cm}$, while the length of the cylindrical part is $L = 5\,\text{cm}$. At the inflow/outflow sections we prescribe the Neumann conditions

$$p\boldsymbol{n} - \nu\left(\nabla \boldsymbol{u} + \nabla \boldsymbol{u}^T\right) = g\boldsymbol{n}$$

with $g = 0$ at the outflow and $g(t) = 500\sin(100\pi t)$ at the inflow.

As in the previous section, we solve the forward problem for a given Young's modulus and we store the corresponding membrane displacement. This provides the synthetic measures to be used in the DA procedure. In order to not commit an "inverse crime" we solve the forward problem on a finer mesh, then we add some noise to the computed membrane displacement and we project it on the (coarser) mesh used for the solution of the inverse problem. In other words, we use the measures given by

$$\eta_m = \Pi\, \eta_{f,h} + ||\eta_{f,h}||_\infty \xi e$$

where Π is a projection from the fine to the coarse mesh, $\eta_{f,h}$ is the displacement computed on the fine mesh, $e \sim \mathcal{U}(-1,1)$ is a random vector, and ξ is the noise level, reciprocal of the SNR.

The Young modulus used to generate the measures is $E = [1.3, 1.8, 1.3] \times 10^6\,\text{dyn/cm}^2$, assuming a piecewise constant profile along the axial direction. In particular, E_1 is the value of the Young modulus for the first quarter of the torus, E_2 is the value for the second quarter, and E_3 is the value in the cylindrical part. For the generation of the POD basis, we use the sample $S = \{E \in \mathbb{R}^3 : E_i \in \{1,2\} \times 10^6\,\text{dyn/cm}^2\}$. In Table 6.6 we report the dimension N_u, N_η of the velocity and displacement POD basis for different choices of the POD threshold τ.

Table 6.6 Dimension of the fluid velocity and membrane displacement POD basis for different values of the POD threshold for the idealized aortic arch test case

τ	0.9	0.95	0.99
N_u	5	8	22
N_η	5	7	12

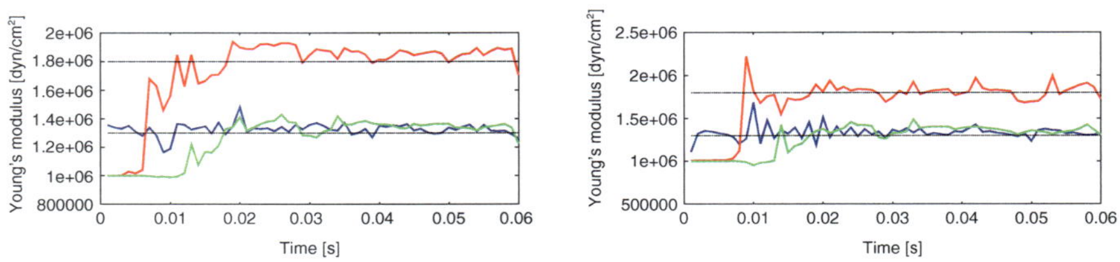

Fig. 6.14 History of the Young modulus estimates for SNR = 10. On the *left*, the estimates obtained by solving the inverse problem on the full finite element space. On the *right*, the estimates obtained by first projecting the problem on the reduced space. The *colors* refer to the different components of the vector E (E_1 (*blue*), E_2 (*red*), E_3 (*green*))

Table 6.7 Comparison between Full Space (FS) and Reduced Space (RS) performance for the idealized aortic arch test case

	FS	RS
E	$[1.33, 1.84, 1.31]\times10^6$	$[1.34, 1.80, 1.33]\times10^6$
Rel. error	1.91 %	2.01 %
Exec. time	3,176 s	277 s
NS solves	492 s	480 s

Table 6.8 Time average of the estimates and relative error for different values of the POD threshold for the idealized arch test case

	$\tau = 0.9$	$\tau = 0.95$	$\tau = 0.99$
E	$[1.37, 1.81, 1.32]\times10^6$	$[1.34, 1.80, 1.32]\times10^6$	$[1.30, 1.78, 1.29]\times10^6$
Rel. error	2.83 %	1.97 %	0.87 %

In this test we compare the reduced space approach with the full space approach (i.e., the minimization in the full Finite Element space). The history of the estimates at each time step for the case SNR = 10 and $\tau = 0.95$ is shown in Fig. 6.14, while in Table 6.7 we report their performance. The optimal estimate for the Young's modulus is computed by averaging all but the first 10 time steps estimates, which are clearly significantly affected by the initial guess.

We can see in addition that the reduced space approach estimates are as good as the full space approach. Moreover, the error on the estimates is remarkably smaller than the intensity of the noise in the measures, for both the approaches, dropping from 10 % to about 2 %, that shows also how DA filters the noise in the measures. Regarding the behavior of the estimates with respect to the POD threshold, in Table 6.8 we report the time averages (excluding the first ten time steps) and the corresponding relative error for three different POD thresholds.

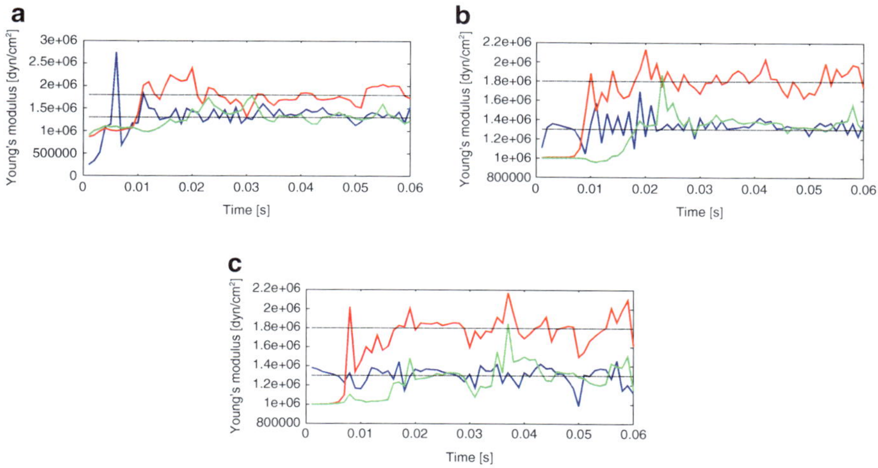

Fig. 6.15 History of the Young modulus estimates for SNR $= 5$ for different values of the POD threshold for the aortic arch test case. (**a**) $\tau = 0.9$ ($N_u = 5$, $N_\eta = 5$), (**b**) $\tau = 0.95$ ($N_u = 8$, $N_\eta = 7$), (**c**) $\tau = 0.99$ ($N_u = 22$, $N_\eta = 12$). The *colors* refer to the different components of the vector E (E_1 (*blue*), E_2 (*red*), E_3 (*green*))

Finally, in Fig. 6.15 we show the history of the Young modulus estimates for different choices of the POD threshold in the case of SNR $= 5$. It is interesting to notice that, despite the fact that the level of the noise is as large as 20 % of the intensity of the signal, the average estimates are still close to the correct values. In particular, even when using a low dimensional size for the reduced model, the optimization procedure clearly detects that the Young modulus in the second region is larger than in the other two regions.

A Kalman-Based Parameter Estimation Approach

Let us consider the FSI system after time–space discretization and linearization that we write as

$$\boldsymbol{U}^k = A^{k-1}\boldsymbol{U}^{k-1} + \boldsymbol{F}^{k-1},$$

where $\boldsymbol{U}^k \in \mathbb{R}^N$ is the vector of velocity and pressure degrees of freedom. In order to estimate the parameter $\boldsymbol{E} \in \mathbb{R}^p$ the augmented state approach is used. Define $\boldsymbol{X}^{(k)} := [\boldsymbol{U}^{(k)}, \boldsymbol{E}^{(k)}]$, then the system becomes

$$\boldsymbol{X}^{(k)} = A_X^{(k-1)}\boldsymbol{X}^{(k-1)} + \boldsymbol{F}_X^{(k-1)}, \qquad A_X^{(k)} = \begin{bmatrix} A^{(k)} & 0 \\ 0 & I \end{bmatrix}, \qquad \boldsymbol{F}_X^{(k)} = \begin{bmatrix} \boldsymbol{F}^{(k)} \\ 0 \end{bmatrix}. \tag{6.81}$$

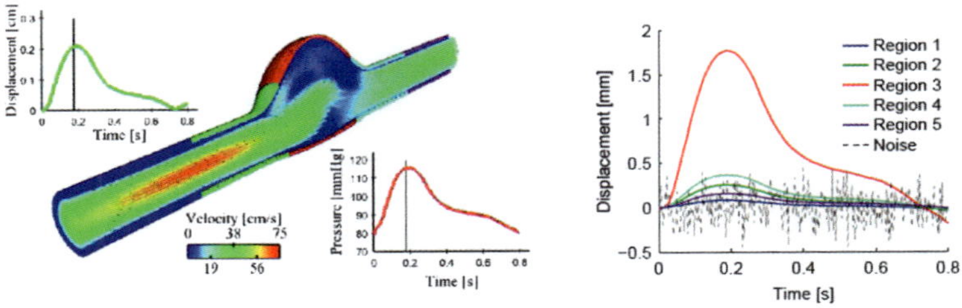

Region	SNR
1	0.43
2	1.25
3	8.8
4	1.8
5	0.75

Fig. 6.16 *Left*: Idealized abdominal aortic aneurysm geometry with subregions and fluid velocity field; displacement and pressure fields at the outlet as a function of time. *Right*: Noise compared to the typical wall displacements in the five regions and signal to noise ratios. Adapted from [6], with permission of © J. Wiley & Sons, 2012

The initial state is assumed to be $\boldsymbol{X}^{(0)} = [\boldsymbol{U}^{(0)}, \boldsymbol{E}_{\text{ref}} + \boldsymbol{\theta}^{(0)}]$, and $\boldsymbol{F}^{(k)} = \boldsymbol{0}$, i.e., the initial velocity and displacement are assumed to be known without uncertainty and the model is considered exact. The variables to be estimated are denoted by $\boldsymbol{\theta}$. The measures of the displacement are affected by a white noise $\boldsymbol{\nu}^{(k)}$, i.e.,

$$\boldsymbol{\eta}_{\text{meas}}^{(k)} = H_k \boldsymbol{X}^{(k)} + \boldsymbol{\nu}^{(k)}.$$

Since the problem is nonlinear, an UKF approach (see Sect. 6.2.5) is used where $(N + p + 1)$ sample points (for details see [6]) are needed to approximate the average and the covariance of the evolving state. As explained in Sect. 6.2.5, the predictor phase consists in evaluating $\boldsymbol{X}_i^{(k)}$ for each sample $\boldsymbol{X}_i^{(k-1)}$, which requires the solution of the FSI problem $(N + p + 1)$ times at each time step. This is computationally prohibitive, therefore a model reduction is performed. The idea is to exploit the fact that the initial covariance is given by

$$\Lambda^{(0)} = \begin{bmatrix} 0 & 0 \\ 0 & \text{Cov}\left(\boldsymbol{\theta}^{(0)}\right) \end{bmatrix},$$

and to use a factorized formulation of the UKF. In this way [6] it is possible to use only $p + 1$ sample points, which significantly reduce the computational cost of the method when $p \ll N$, i.e., when the number of parameters is much smaller than the dimension of the state.

Consider the idealized 3D geometry of an abdominal aortic aneurysm showed in Fig. 6.16, left. The structure is divided, a priori, into five regions featuring different values of the Young modulus E, corresponding to different colors in Fig. 6.16. The typical displacements and noise recorded in the five regions are shown in Fig. 6.16, right.

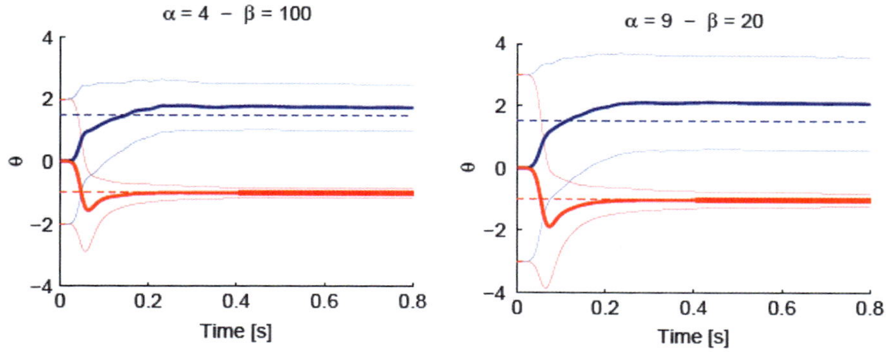

Fig. 6.17 Mean values (*thick solid lines*) and plus/minus standard deviations (*thin solid lines*) of the logarithm of estimated Young modulus, for $\alpha = 4$, $\beta = 100$ (*left*) and $\alpha = 9$, $\beta = 20$ (*right*). Adapted from [6], with permission of © J Wiley & Sons 2012

Numerical Results

Consider the idealized 3D geometry of an abdominal aortic aneurysm showed in Fig. 6.16, left. The length of the geometry, the minimum and maximum diameters are 23 cm, 1.7 cm, and 5 cm, respectively. The Poisson ratio, density, and viscosity of the structure are 0.46, 1.2 g/cm^3 and 10^{-3} s, respectively. The fluid density and viscosity are 1 g/cm^3 and 0.035 Po, respectively. A Windkessel boundary condition is used at the outflow (see Fig. 6.16 for details). We assume to have displacement measures at each grid point of the mesh and that $\boldsymbol{v}^{(k)} \sim \mathcal{N}(0, \sigma^2 I)$. In analogy with the variational approach, comparing the FE discretization of the cost functional (6.71) and the cost functional (6.72), we assume the covariance matrix R_k to be inversely proportional to the M_k^{Σ}, the mass matrix on Σ. In particular we take

$$R_k^{-1} = \beta\sigma^{-2}\frac{\tau_m}{T_{\text{ref}}}\frac{M_k^{\Sigma}}{|\Sigma|},$$

where τ_m is the time sampling of the measurements, T_{ref} is a reference time, and β is a positive scalar used to weight the importance of the measurements. Also we assume that $\theta^{(0)} \sim N(0, \alpha I)$. Figure 6.17, shows the reconstructed Young modulus in the different regions, as a function of α and β. The coefficient β represents the level of confidence attributed to the displacement measures, whereas α is the a priori covariance. As expected, the sensitivity with respect to β is higher when α is larger and the sensitivity with respect to α is higher in regions with smaller SNR. Together with the estimated parameters, the Kalman filter provides also their covariances, which is an important index to evaluate the confidence we should have in the results. The results are in fact more (less) reliable when the covariances are small (large). For more results see [6].

With respect to the variational method, the filtering approach has the advantages that only the solution of the forward problem is needed and that it provides an estimate of the covariance of the parameters. Also, it is computationally cheaper when the parameter space is much smaller than the state space. However, the nonlinearities are not solved accurately and this can lead to a suboptimal estimate of the parameters. Also, when the space of the parameters is large (e.g., $\boldsymbol{E}$ is a finite element field with as many DOFs as the number of grid points), the Kalman approach may become expensive.

6.5 Conclusions

Cardiovascular Mathematics is nowadays a mature discipline not only for understanding and improving basic knowledge of diseases but also for supporting the clinical practice, with an accurate quantitative estimate, prediction, identification of optimal therapies. In particular, the common denominator of this exciting perspective is the presence of *inverse problems*, where problems related to blood flow and FSI, traditionally per se challenging, need to be solved several times, assimilated to available measures, analyzed with probabilistic tools. This is true not only for DA but also for the identification of the optimal realization of a therapy or, more specifically, of a surgical intervention. For instance, in [49, 50, 56] the identification of the optimal placement of leads for optimizing pacemaking action in the heart is addressed; the computation of a personalized patient-specific peritoneal dialysis is addressed in [63, Chap. 7]. In Fig. 6.18 we report the aortic blood flow simulated with different options of a Left Ventricular Assisted Device (LVAD) implant; in particular, the emphasis is on the location of the cannula from the pump. The identification of the optimal location is still an open problem whose solution certainly depends on the patient-specific morphology.

This process bringing complex quantitative analyses from the computer to the bedside requires a strong integration with available data, shifting the goal of performing a patient-specific computation to the patient-specific "assimilation" [71]. This is a crucial step for improving reliability of numerical elaborations, reducing uncertainty and eventually the risks of failure.

Several methods can be pursued to this goal and extensive investigation is required to establish the most appropriate approach for the different problems. A genuinely numerical-statistical research is necessary for understanding how to reduce the computational costs to be able to tackle challenges presented by clinical problems that typically feature short timelines and large number of patients.

This chapter intended to offer a short introduction with a special emphasis on FSI problems to some possible methods and to their interplay. Far to be a conclusive and exhaustive presentation, we aimed at turning on interest for the emerging topic of Inverse Cardiovascular Mathematics, with the final—ambitious but possible—goal of introducing mathematically advanced methods in the clinical practice to improve doctors activity and—more importantly—patients healthcare.

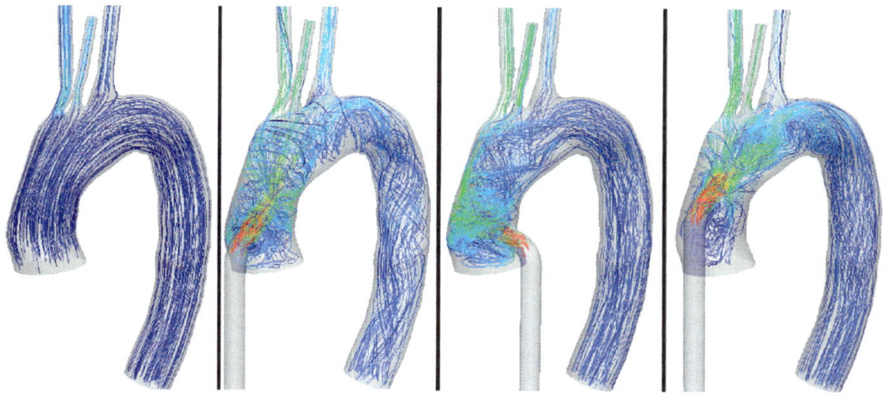

Fig. 6.18 Simulation of different locations of the cannula of an LVAD in a real aorta. *Leftmost*: Pre-op fluid dynamics. Images reproduced with permission of © D. Gupta, Emory University Hospital, image processing and simulations in collaboration with M. Piccinelli (Radiology, Emory University) and T. Passerini (Math & CS, Emory University)

Acknowledgements The authors wish to thank Tiziano Passerini (Siemens, Princeton, NJ, USA) and Marina Piccinelli (Emory University, Department of Radiology) for several contributions in the development of methods and codes used for the topics considered in the chapter.

References

1. H. Abou-Kandil, G. Freiling, V. Ionescu, G. Jank, *Matrix Riccati Equations: In Control and Systems Theory* (Springer, Berlin, 2003)
2. H.T. Banks, *A Functional Analysis Framework for Modeling, Estimation and Control in Science and Engineering* (Taylor & Francis, London, 2012)
3. H.T. Banks, K. Kunisch, *Estimation Techniques for Distributed Parameter Systems.* (Birkhauser, Boston, 1989)
4. P.E. Barbone, A.A. Oberai, Elastic modulus imaging: some exact solutions of the compressible elastography inverse problem. Phys. Med. Biol. **52**, 1577 (2007)
5. P.E. Barbone, C.E. Rivas, I. Harari, U. Albocher, A.A. Oberai, Y. Zhang, Adjoint-weighted variational formulation for the direct solution of inverse problems of general linear elasticity with full interior data. Int. J. Numer. Methods Eng. **81**(13), 1713–1736 (2010)
6. C. Bertoglio, P. Moireau, J.-F. Gerbeau, Sequential parameter estimation for fluid–structure problems: application to hemodynamics. Int. J. Numer. Methods Biomed. Eng. **28**(4), 434–455 (2012)
7. L. Biegler, G. Biros, O. Ghattas, M. Heinkenschloss, D. Keyes, B. Mallick, L. Tenorio, B. Waanders, K. Willcox, Y. Marzouk, *Large-Scale Inverse Problems and Quantification of Uncertainty*. Wiley Series in Computational Statistics (Wiley, Chichester, 2011)
8. J. Blum, F.-X. Le Dimet, I. Michael Navon, Data assimilation for geophysical fluids, in *Handbook of Numerical Analysis*, vol. 14, ed. by P.G. Ciarlet (Elsevier, Amsterdam, 2009), pp. 385–441
9. P.B. Bochev, Analysis of least-squares finite element methods for the navier-stokes equations. SIAM J. Numer. Anal. **34**, 1817–1844 (1997)

10. P.B. Bochev, M.D. Gunzburger, Accuracy of least-squares methods for the Navier-Stokes equations. Comput. Fluids **22**, 549–563 (1993)
11. P.B. Bochev, M.D. Gunzburger, *Least-Squares Finite Element Methods* (Springer, Berlin, 2009)
12. P.T. Boggs, J.W. Tolle, Sequential quadratic programming. Acta Numer. **4**, 1–51 (1995)
13. D. Calvetti, E. Somersalo, *An Introduction to Bayesian Scientific Computing: Ten Lectures on Subjective Computing*. Surveys and Tutorials in the Applied Mathematical Sciences (Springer Science+Business Media, New York, 2007)
14. I. Campbell, W. Robert Taylor, Flow and atherosclerosis, in *Hemodynamics and Mechanobiology of Endothelium* (World Scientific, Hackensack, 2010)
15. D. Chapelle, A. Gariah, J. Sainte-Marie, Galerkin approximation with proper orthogonal decomposition: new error estimates and illustrative examples. ESAIM: Math. Model. Numer. Anal. **46**, 731–757 (2012)
16. M. D'Elia, A. Veneziani, Uncertainty quantification for data assimilation in a steady incompressible navier-stokes problem. ESAIM: Math. Model. Numer. Anal. **47**, 1037–1057 (2013)
17. M. D'Elia, L. Mirabella, T. Passerini, M. Perego, M. Piccinelli, C. Vergara, A. Veneziani, Some applications of variational data assimilation in computational hemodynamics, in *Modelling of Physiological Flows*, ed. by D. Ambrosi, A. Quarteroni, G. Rozza. MS&A Series (Springer, Berlin, 2011), pp. 363–394
18. M. D'Elia, M. Perego, A. Veneziani, A variational data assimilation procedure for the incompressible navier stokes equations in hemodynamics. J. Sci. Comput. **52**(2), 340–359 (2012)
19. H. Delingette, M. Sermesant, R. Cabrera-Lozoya, C. Tobon-Gomez, P. Moireau, R.M. Figueras i Ventura, K. Lekadir, A. Hernandez, M. Garreau, E. Donal, C. Leclercq, S.G. Duckett, K. Rhode, C.A. Rinaldi, A.F. Frangi, R. Razavi, D. Chapelle, N. Ayache, S. Marchesseau, Personalization of a cardiac electromechanical model using reduced order unscented kalman filtering from regional volumes. Med. Image Anal. **17**, 816–829 (2013)
20. J. Donea, S. Giuliani, J.P. Halleux, An arbitrary lagrangian-eulerian finite element method for transient dynamic fluid-structure interactions. Comput. Methods Appl. Mech. Eng. **33**(1–3), 689–723 (1982)
21. R.P. Dwight, Bayesian inference for data assimilation using least-squares finite element methods, in *IOP Conference Series: Materials Science and Engineering*, vol. 10 (IOP Publishing, Bristol, 2010), p. 012224
22. B. Einarsson, *Accuracy and Reliability in Scientific Computing*, vol. 18 (Society for Industrial Mathematics, Philadelphia, 2005)
23. H.W. Engl, M. Hanke, A. Neubauer, *Regularization of Inverse Problems*. Mathematics and its Applications (Springer, Berlin, 1996)
24. L. Formaggia, A. Veneziani, C. Vergara, A new approach to numerical solution of defective boundary value problems in incompressible fluid dynamics. SIAM J. Numer. Anal. **46**(6), 2769–2794 (2008)
25. L. Formaggia, A. Quarteroni, A. Veneziani, *Cardiovascular Mathematics: Modeling and Simulation of the Circulatory System*, vol. 1 (Springer, Berlin, 2009)
26. L. Formaggia, A. Veneziani, C. Vergara, Flow rate boundary problems for an incompressible fluid in deformable domains: formulations and solution methods. Comput. Methods Appl. Mech. Eng. **9**(12), 677–688 (2010)
27. P.C. Franzone, L.F. Pavarino, A Parallel Solver for Reaction-Diffusion Systems in Computational Electrocardiology, Math. Model. Methods in Appl. Sci. **14**(6), 883–911 (2004) doi:10.1142/s0218202504003489
28. B. Fristedt, N. Jain, N.V. Krylov, *Filtering and Prediction: A Primer*, STML vol. 38, AMS, Providence, RI (2007)
29. K. Funamoto, T. Hayase, Reproduction of pressure field in ultrasonic-measurement-integrated simulation of blood flow. Int. J. Numer. Methods Biomed. Eng. **29**(7), 726–740 (2013)

30. G.P. Galdi, A.M. Robertson, R. Rannacher, S. Turek, *Hemodynamical Flows: Modeling, Analysis and Simulation*. Oberwolfach Seminar Series, vol. 37, Birkhauser Verlag AG, Basel (2008)
31. J.F. Gerbeau, D. Lombardi, Reduced-order modeling based on approximated lax pairs. Technical Report RR 8137, INRIA. arXiv:1211.4153v1 (November 2012)
32. E. Gilboa, P.S. La Rosa, A. Nehorai, Estimating electrical conductivity tensors of biological tissues using microelectrode arrays, in *Engineering in Medicine and Biology Society (EMBC), 2012 Annual International Conference of the IEEE* (2012), pp. 1040–1044
33. R. Glowinski, J.L. Lions, Exact and approximate controllability for distributed parameter systems. Acta Numer. **3**, 269–378 (1994)
34. R. Glowinski, J.L. Lions, Exact and approximate controllability for distributed parameter systems. Acta Numer. **4**, 159–328 (1995)
35. R. Glowinski, J.-L. Lions, J. He, *Exact and Approximate Controllability for Distributed Parameter Systems: A Numerical Approach (Encyclopedia of Mathematics and its Applications)*, 1st edn. (Cambridge University Press, New York, 2008)
36. G.H. Golub, C.F. Van Loan, *Matrix Computations*, vol. 3 (Johns Hopkins University Press, Baltimore, 1996)
37. L.S. Graham, D. Kilpatrick, Estimation of the bidomain conductivity parameters of cardiac tissue from extracellular potential distributions initiated by point stimulation. Ann. Biomed. Eng. **38**(12), 3630–3648 (2010)
38. M.D. Gunzburger, *Perspectives in Flow Control and Optimization*, vol. 5 (Society for Industrial Mathematics, Philadelphia, 2003)
39. P.C. Hansen, *Rank-Deficient and Discrete Ill-Posed Problems*. SIAM Monographs on Mathematical Modeling and Computation (Society for Industrial and Applied Mathematics, Philadelphia, 1998)
40. J.J. Heys, T.A. Manteuffel, S.F. McCormick, M. Milano, J. Westerdale, M. Belohlavek, Weighted least-squares finite elements based on particle imaging velocimetry data. J. Comput. Phys. **229**(1), 107–118 (2010)
41. K. Hinsch, 3-Dimensional particle velocimetry. Meas. Sci. Technol. **6**, 742–753 (1995)
42. T.J.R. Hughes, W.K. Liu, T.K. Zimmermann, Lagrangian-eulerian finite element formulation for incompressible viscous flows. Comput. Methods Appl. Mech. Eng. **29**(3), 329–349 (1981)
43. J. Humpherys, P. Redd, J. West, A fresh look at the kalman filter. SIAM Rev. **54**(4), 801–823 (2012)
44. S.J. Julier, J.K. Uhlmann, A new extension of the kalman filter to nonlinear systems, in *Proceedings of SPIE 3068, Signal Processing, Sensor Fusion, and Target Recognition VI, 182* (1997), pp. 182–193
45. S.J. Julier, J.K. Uhlmann, Unscented filtering and nonlinear estimation. Proc. IEEE **92**(3), 401–422 (2004)
46. T. Kailath, *Lectures Notes on Wiener and Kalman Filtering* (Springer, Berlin, 1981)
47. J. Kaipio, E. Somersalo, *Statistical and Computational Inverse Problems (Applied Mathematical Sciences)*, vol. 160, 1st edn. (Springer, Berlin, 2004)
48. R.E. Kalman, A new approach to linear filtering and prediction problems. Trans. ASME J. Basic Eng. **82**, 35–45 (1960)
49. K. Kunisch, M. Wagner, Optimal control of the bidomain system (iii): existence of minimizers and first-order optimality conditions. ESAIM: Math. Model. Numer. Anal. **47**, 1077–1106 (2013)
50. K. Kunisch, M. Wagner, Optimal control of the bidomain system (ii): uniqueness and regularity theorems for weak solutions. Annali di Matematica Pura ed Applicata **192**, 1–36 (2012)
51. P. Lancaster, L. Rodman, *Algebraic Riccati Equations* (Oxford Science Publications, New York, 1995)
52. T. Lassila, A. Manzoni, A. Quarteroni, G. Rozza, A reduced computational and geometrical framework for inverse problems in hemodynamics. Int. J. Numer. Methods Biomed. Eng. **29**(7), 741–776 (2013)

53. J. Modersitzki, *FAIR: Flexible Algorithms for Image Registration*. Fundamentals of Algorithms (Society for Industrial and Applied Mathematics, Philadelphia, 2009)
54. P. Moireau, D. Chapelle, Reduced-order unscented kalman filtering with application to parameter identification in large-dimensional systems. ESAIM: Control Optim. Calc. Var. **17**(02), 380–405 (2011)
55. A.M. Mood, F.A. Graybill, D.C. Boes, *Introduction to the Theory of Statistics* (McGraw-Hill, New York, 1974)
56. C. Nagaiah, K. Kunisch, G. Plank, Numerical solutions for optimal control of monodomain equations. PAMM **9**(1), 609–610 (2009)
57. F. Nobile, C. Vergara, An effective fluid-structure interaction formulation for vascular dynamics by generalized Robin conditions. SIAM J. Sci. Comput. **30**(2), 731–763 (2008)
58. J. Nocedal, S. Wright, *Numerical Optimization* (Springer, Berlin, 2000)
59. M. Perego, A. Veneziani, C. Vergara, A variational approach for estimating the compliance of the cardiovascular tissue: an inverse fluid-structure interaction problem. SIAM J. Sci. Comput. **33**(3), 1181–1211 (2011)
60. K.B. Petersen, M.S. Pedersen, The matrix cookbook. Technical report, http://matrixcookbook.com (2008)
61. M. Piccinelli, L. Mirabella, T. Passerini, E. Haber, A. Veneziani, 4d image-based cfd simulation of a compliant blood vessel. Technical report, Technical Report TR-2010-27, Department of Mathematics & CS, Emory University, www.mathcs.emory.edu (2010)
62. A. Quarteroni, R. Sacco, F. Saleri, *Numerical Mathematics*. Texts in Applied Mathematics Series (Springer GmbH, Berlin, 2000)
63. A. Quarteroni, L. Formaggia, A. Veneziani, *Complex Systems in Biomedicine* (Springer, Berlin, 2007)
64. G. Rozza, K. Veroy, On the stability of the reduced basis method for stokes equations in parametrized domains. Comput. Methods Appl. Mech. Eng. **196**(7), 1244–1260 (2007)
65. G. Rozza, D.B.P. Huynh, A.T. Patera, Reduced basis approximation and a posteriori error estimation for affinely parametrized elliptic coercive partial differential equations. Arch. Comput. Methods Eng. **15**(3), 229–275 (2008)
66. S. Salsa, *Partial Differential Equations in Action: From Modelling to Theory* (Springer, Berlin, 2008)
67. O. Scherzer, The use of morozov's discrepancy principle for tikhonov regularization for solving nonlinear ill-posed problems. Computing **51**(1), 45–60 (1993)
68. R. Todling, Estimation theory and foundations of atmospheric data assimilation. DAO Office Note 1:1999 (1999)
69. F. Tröltzsch, *Optimal Control of Partial Differential Equations: Theory, Methods, and Applications*, vol. 112 (American Mathematical Society, Providence, 2010)
70. K. Urban, A.T. Patera, A new error bound for reduced basis approximation of parabolic partial differential equations. C. R. Math. **350**(3–4), 203–207 (2012)
71. A. Veneziani, C. Vergara, Inverse problems in cardiovascular mathematics: toward patient-specific data assimilation and optimization. Int. J. Numer. Methods Biomed. Eng. **29**(7), 723/725 (2013). Editorial of the special issue "Inverse Problems in Cardiovascular Mathematics"
72. C.R. Vogel, *Computational Methods for Inverse Problems*. Frontiers in Applied Mathematics (Society for Industrial and Applied Mathematics, Philadelphia, 2002)
73. E.A. Wan, R. Van der Merwe, The unscented kalman filter for nonlinear estimation, in *Adaptive Systems for Signal Processing, Communications, and Control Symposium 2000. AS-SPCC. The IEEE 2000* (2000), pp 153–158
74. H. Yang, A. Veneziani, Variational estimation of cardiac conductivities by a data assimilation procedure. Technical Report TR-2013-007, Math&CS, Emory University (July 2013)

A. Veneziani (✉) • L. Bertagna
Department of Mathematics and Computer Science, Emory University, Atlanta, GA, USA
e-mail: avenez2@emory.edu; lbertag@emory.edu

M. D'Elia • M. Perego
Computer Science Research Institute, Sandia National Laboratories, Albuquerque, NM, USA
e-mail: mdelia@sandia.gov; mperego@sandia.gov

Chapter 7
Mathematical Models for Blood Coagulation

Tomáš Bodnár, Antonio Fasano, and Adélia Sequeira

Abstract This chapter presents an overview and introduction to blood coagulation models. The historical exposure of the development of classical coagulation modeling theories is followed by a basic overview of blood coagulation biochemistry. The recent developments of cell-based models are explained in detail to demonstrate the current shift from the classical cascade/waterfall models. This phenomenological overview is followed by a survey of available mathematical concepts used to describe the blood coagulation process at various spatial scales including some of the related biophysical phenomena. A comprehensive survey of basic literature is provided for each of these topics.

Keywords Blood • Coagulation • Coagulation cascade • Coagulation model • Chemical reaction • Kinetics rate • Platelet • Red blood cell • Thrombin • Thrombosis

MSC2010: 76Z05, 76V05, 74F10, 80A32, 92C35, 74F25, 80A30, 92C45, 92C40

7.1 Introduction

Blood coagulation is a very complex process that plays an indispensable role in wide variety of living organisms. Better understanding of this process is both important and complicated. The extreme complexity of blood coagulation has several origins.

Complexity of blood. Blood itself, even without considering its biochemistry and coagulation, is an extremely complicated fluid. It exhibits many unique properties and its behavior is not yet fully described and understood.

Sensitivity. Both, the external and internal stimuli can alter and heavily affect the blood coagulation process. This sensitivity is a major problem in repeatability of experiments and their interpretation.

T. Bodnár et al. (eds.), *Fluid-Structure Interaction and Biomedical Applications*, Advances in Mathematical Fluid Mechanics, DOI 10.1007/978-3-0348-0822-4_7,

Variability. Despite many similarities, there are important differences in coagulation processes between various biological species. Non-negligible variances also exist between individuals of the same species, depending on a great number of factors.

Multiscale nature. Blood coagulation process has a multiscale nature in both, space and time. This is not only very demanding when it comes to experimental studies, but the multitude of spatial and temporal scales is also hard to include within a single mathematical model.

Multidisciplinary nature. Blood coagulation is a typical example of a multidisciplinary problem requiring detailed knowledge of many areas of physics, chemistry, biology, medicine, and other branches of science. It's impossible to get a full picture of the coagulation process from a narrow point of view of a single scientific discipline.

These are few from many problems that make the studying and modeling of blood coagulation so challenging. Within this chapter we will try to introduce and address some of the most important concepts forming the basis of current theories and models of blood coagulation. The beginning of the chapter focuses on the historical background and evolution of the traditional, biochemical approach to blood coagulation. Special attention is paid to recent discoveries and consequent revisions of historical models. The second part of the chapter is more oriented towards the mathematical description of coagulation models. It shortly reviews various methods and models used to describe blood coagulation and related phenomena.

7.2 Historical Remarks

Despite the familiarity with the phenomenon of coagulation following bleeding, its scientific explanation had to go through the knowledge of blood composition and above all the development of biochemistry, which took place in relatively recent times. Actually, the process, which is not confined to external wounds, is so complicated that the theory of blood coagulation is still in full evolution. A great help came from the observation of how the process can fail. Roughly speaking, there are two ways blood clotting can go wrong: excessive coagulation and defective coagulation. The first case may produce thrombosis with the possible occlusion of blood vessels. Defective coagulation instead may end up with severe hemorrhages. Both kind of disorders can be lethal. Therefore, it is not surprising that the history of blood coagulation has much to do with the observation of clotting disorders. In this section we will shortly review the history preceding the twentieth century.

Thrombosis was one of the first reported clotting disorders, though just through its symptoms, namely limb swelling, most frequently in pregnant women.[1] For a review on pregnancy and post-partum related thrombosis, see [234]. Gender-specific studies on thrombosis are presented in [52, 252].

A very well-written historical review on thrombosis in general is [13], which contains many important references. Symptoms attributable to arterial thrombosis have been described in the old traditional Chinese medicine, whose development is attributed to the legendary *Yellow Emperor (Huang Di)*, who is supposed to have reigned between 2698 BC and 2599 BC.

Documents from Ancient Egypt, like the *Edwin Smith Papyrus* (written around 1700 BC, but based upon much earlier material[2]) and the *Ebers Papyrus* (existing in a copy of the sixteenth century BC, but probably having its roots back to 3000 BC) make no reference to blood coagulation, nor does the readable part of the *Kahun Gynecological Papyrus* (nineteenth century BC).

Thus our history jumps to *Hippocrates* (ca. 460 BC–370 BC). He was a central character in the medicine of the classic world[3] and he coined the term *leucophlegmatia* to denote limb swelling, though this term was later used to indicate symptoms of various origin. The Greek *Galen of Pergamon* (131–201), who acquired great fame in Rome where he became known as *Aelius (Claudius) Galenus*,[4] was the first to

---

[1]Pregnancy may increase the risk of thrombosis in various ways. The swollen uterus can compress pelvic vessels, reducing blood circulation in the legs. Hormonal changes can also produce hypercoagulability by increasing the concentration in blood of pro-coagulant factors and reducing the concentration of anticoagulant factors. A more serious condition sometimes related to pregnancy is the Antiphospholipid Antibody (or sticky blood) Syndrome, due to the autoimmune production of antibodies against a cell membrane substance called phospholipid causing platelets aggregation.

[2]The most famous physician of ancient Egypt was *Imhotep* (a semi-divine character, he lived during the twenty-seventh century BC and is supposed to be the legendary author of this papyrus). The practice of mummifying corpses must have taught much to Egyptians about the human body, but the papyrus (mainly dealing with wounds healing) can hardly be considered a scientific document and the suggested remedies could easily be fatal to the patients because they could produce infections.

[3]He was the author of the *humoral theory*, according to which four *humors* (blood, phlegm, black bile, and yellow bile) had to be in a proper balance in healthy individuals. The theory, somehow anticipated by *Alcmaeon of Croton* (fifth century BC), parallels the contemporary claim by Empedocles that four elements (air, water, fire, earth) are the basic constituents of the world and which may have had a much older origin. Hippocrates tremendous authority (and the immense reputation of *Galen*, who took his legacy to the Roman world and passed it on to the next era) prevented the development of medicine on a scientific basis for centuries, thanks to the blindness of his followers. The humoral theory found its way through Islamic medicine: the Persian *Avicenna* (*Ib Sīnā*, 980–1037) based his *Canon of Medicine* (1025) on Hippocrates' and Galen's theories. It was instead opposed by another Persian, *Razi* (*Muhammad ibn Zakariyā Rāzī*, 865–925) an eclectic scientist, very famous in his times, who explicitly questioned several of Galen claims on the basis of his own experimental observations.

[4]He adopted and propagated Hippocrate humoral theory, adding his own theory of *four temperaments* (choleric, melancholic, sanguine, phlegmatic), resulting from the combinations of the humors with four *qualities* of (cold, warm, moisty, dry). He sketched an erroneous scheme of the

use the term *thrombosis* in connection with limb swelling (from Greek *thrombos*, meaning clot).

Throughout middle ages and even for most of the eighteenth century, still under Galen's influence, humors were considered responsible for limb swelling.[5] Particularly interesting was the point of view of *Aristotle* (384–322 BC) about blood clotting, which he claimed to be caused by *heat loss*. This is of course a very naïve claim, but he also said that a *fibrous material* was necessary for clotting.[6]

Coming closer to our times, while a lot of humoral trash was still invading the medical literature, fundamental discoveries were made in the seventeenth century, laying the foundation for a scientific approach to biology.[7] These were mainly related to the development of microscopy, particularly in the Netherlands. *Anthony Leeuwenhoek* (1632–1723) can be considered the initiator of microbiology. With the help of his instruments he studied Red Blood Cells[8] (1674), previously identified by *Marcello Malpighi* (1628–1694).[9]

Clearly, the knowledge of blood composition is indispensable to formulate any theory on coagulation, but that was just a first step. The famous French surgeon *Jean-Louis Petit* (1674–1750) recognized that blood clotting was part of the process of hemostasis. A description of vein occlusion by blood clots was provided (1676) by the celebrated English surgeon *Richard Wiseman* (1601–1686). Much before the so-called Virchow triad became famous (see below) he recognized two different causes of thrombosis: stasis and hypercoagulability.

---

circulatory system. We had to wait until the famous treatise *Exercitatio anatomica de motu cordis et sanguinis in animalibus* (1628) by *William Harvey* (1578–1657) for a correct systematic description of blood circulation (limited to great vessels: microcirculation was a later discovery). It is worth mentioning at this point the important contributions later given by the English eclectic scientist *Stephen Hales* (1677–1761), who determined the blood volume, the heart output and who first measured arterial blood pressure. For completeness we recall the revolutionary work of *Andreas Vesalius* (Latinized from *Andries van Wesel*) (1514–1564), who opened a new era in physiology. It is interesting to note that Vesalius studied Rāzī's books and that he based his famous treatise *De humani corporis fabrica libri septem* (1543) on direct observation of dissected human bodies. He pointed out several of Galen's mistakes (particularly in the description of circulatory system), indifferent to the harsh criticism of Galen's followers.

[5]Explanations given to pregnancy or post-partum related limb swelling by various authors during the seventeenth and eighteenth centuries, largely based on humors look today simply ridiculous. See the paper [13].

[6]Such a fibrous component was isolated much later by *Marcello Malpighi* (1628–1694), the Italian physician famous above all for his studies on kidneys.

[7]A new course in the medical studies was set by the book *The Philosophical Principles of Medicine* (1725) by *Thomas Morgan*.

[8]Cells were first observed at the microscope by the physicist *Robert Hooke* (1635–1703) (the founder of the theory of elasticity) in a thin sample of cork (1665). He did not know what cells were, but he called them that way because of their particular and regular arrangement in the sample, resembling the one of monks cells.

[9]Malpighi first described RBC as fat corpuscles (1663). Actually RBCs had been observed earlier (1658) by the Dutch *Jan Swammerdam* (1637–1680). Malpighi was also the discoverer of capillaries (1661).

Attention towards blood composition in the interpretation of the coagulation process was brought by *William Hewson* (1739–1774), who isolated a "coagulation lymph" responsible for the process (identifiable with what is known today as *fibrinogen*).

In the first part of the nineteenth century it became clear that limb swelling had to be attributed to veins obstruction, though the struggle against the humoral theory was still going on![10] One of the most revered physician of the nineteenth century was *Rudolph Virchow* (1821–1902), today still remembered in the field of blood coagulation for his *Virchow Triad*, emphasizing three elements contributing to thrombosis. In the modern language we can list them as hypercoagulability, alterations of hemodynamics (stasis), and endothelial injuries.[11] He introduced the term *embolia*. Of course there were many other physicians who had similar views, but since this is not a historical paper it is more urgent to deal with the discovery of *platelets*.

However, before we come to that, let us briefly mention some historical remark about bleeding disorders. It is natural that such phenomena have been important to people practicing circumcision.[12] Babies whose blood could not coagulate properly could bleed to death. In the second century AD Rabbi *Judah haNasi* exempted babies from circumcision when two sons of the same mother had previously died after the operation (Babylonian Talmud). The Islamic physician *Albucasis* (*Abu al-Qasim Khalaf ibn al-Abbas Al-Zahrawi*, 936–1013, dates are uncertain) who lived in Andalusia, reported cases of what today is known to be hereditary hemophilia. Accounts of bleeding disorders of hereditary type can be found in many later sources and with different names. The term hemophilia was used for the first time by Friedrich Hopff in his 1828 treatise *Über die haemophilie oder die erbliche Anlage zu todlichen Blutungen* (Zurich). We know today that some classes of disorders known as hemophilia (namely hemophilia A and B) are due to a defective gene in the X-chromosome. That explains why it is extremely rare in women (possessing two X-chromosomes),[13] who can however carry the illness (only one chromosome being defective) without symptoms and transmit it through their genealogic tree.

---

[10] It is really amazing that a well-known scientist like the Swedish *Robin Fåhraeus* (1888–1968), still quoted today, for instance, in the field of blood rheology, believed to have found a confirmation of Galens theories on the basis of the observation that blood coagulates in four layers with different colours, corresponding to the famous humors [121].

[11] Virchow described the mechanism of thromboembolism [240], a phenomenon that was by no means clear at his time (inflammation was considered by many physicians the real cause of thrombosis: this was the subject of a famous dispute with the French pathologist Jean Cruveilhier). Curiously, he did not formulate the famous triad which for some reason found a firm place in the literature much after his death (apparently not before 1950!). See the interesting review [17].

[12] Circumcision is a very old practice, already found in the ancient Egypt and that was widely adopted also in the Islamic world. Its origin in ancient Egypt was probably as an initiation practice to religious offices. The *Book of the Dead* describes self-circumcision by the sun-god Ra: *Blood fell from the phallas of Ra after he had finished cutting himself*.

[13] In numbers, 1 over 10,000 men is hemophilic. The probability that a woman is hemophilic is the square of that number.

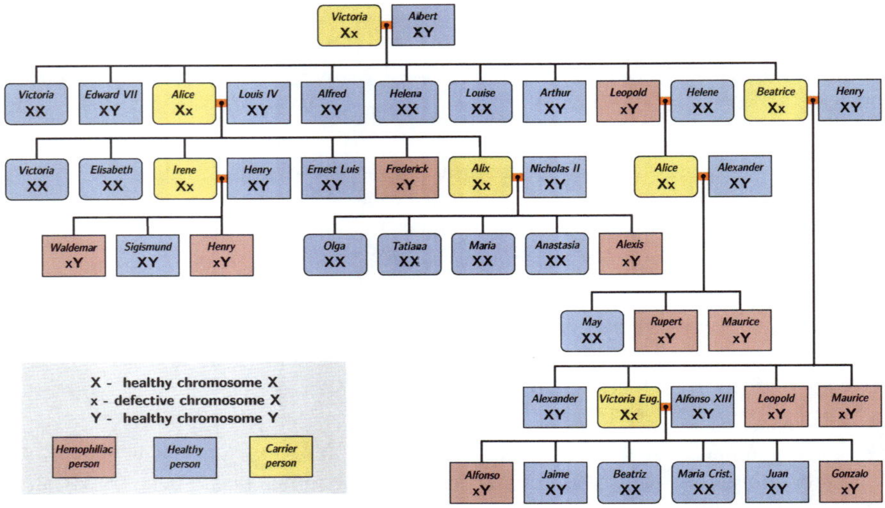

Fig. 7.1 Queen Victoria's family tree, showing hemophiliac descendants

We refer to the important review paper [117], containing a large section on the Royal hemophilia. It is well known that Queen Victoria was a carrier of the disease and that members of many royal families got from her the defective chromosome (Fig. 7.1), a fact that was going to have important historical consequences [139,227]. Hemophilia used to be a life-threatening illness and any minor injury could be fatal. Later on patients could benefit transfusions, but at the insurgence of HIV infection a great number of them died.

Relationship between cancer and thromboembolism was recognized already in the 1860 by *Armand Trousseau* (1801–1867), who had the chance to diagnose in that way his own fatal illness.

But let us finally come to platelets. Due to their smallness, their discovery was delayed until sufficiently powerful microscopes became available. In 1865 *Max Johann Sigismund Schultze* (1825–1874) made an accurate description of tiny cells that he recognized as normal constituents of blood. However, there were previous observations: in 1836 by *Hermann Nasse*, and in 1842 by the French *Alfred Donné* (1801–1878), frequently referred to as the discoverer of platelets, and many others (see the review paper [51]). It was however *Giulio Bizzozero* (1846–1901) who understood their role in blood coagulation (1881) as *Fibrin* producers.[14] He coined the Italian name *Piastrine* and the German *Blutplächtten*. This was the starting point

[14]In this connection the name of the eminent French hematologist *Georges Hayem* (1841–1933) has to be remembered as one of the founders of modern hematology. He performed the first count of platelets. In 1882 he illustrated the effects of *thrombocytopenia* (low platelets count).

of modern investigations on blood coagulation. The reader is referred to the papers [38, 88, 203].

Of course that was just the beginning, since platelets, in spite of their smallness, are a very complex universe, still offering new surprises.

7.3 Cells and Proteins Intervening in the Formation and Dissolution of Clots

Clotting and fibrinolysis are the result of *cascades* of chemical reactions, characterized by strong positive feedback, with the active participation of platelets and of endothelial cells, particularly those making the blood vessels (see [1] for a review on their specialization according to the vessel size, and on thrombosis in general). In this section we will list the many substances involved in clotting and fibrinolysis, grouping them in classes according to their role and nature. The exposition will be only moderately technical, since all we need to know is the role these elements have in the process, not, for instance, the details about their chemical structure, nor how the reactions actually take place.[15] The reader may be discouraged by the length of this section, but at least a quick glance at it is necessary before to proceed.

7.3.1 Blood Cells

1. **Platelets**

Platelets are cells with no nucleus produced in the bone marrow, having a diameter of 2–4 μm and a lifespan of 5–9 days.[16] Their average concentration in blood is 1.5–$4 \times 10^5/\text{mm}^3$. In the rest state their shape is discoid, but in response to various stimuli they may become *star shaped* (rolling over blood vessels wall to inspect its integrity), or they may assume a very irregular shape. The latter transformation characterizes *activated* platelets, with the emission of *philopodia* (or *pseudopods*). In this stage they can bind both among themselves and (in the fastest stage of the coagulation process) to the growing *fibrin network*.

Despite their smallness, platelets can perform an incredible number of actions, interacting with the environment by means of the many *receptors* on their membrane (Fig. 7.2). Concerning this subject we quote the review paper [207].

Figure 7.2 shows the specialization of each class of receptors, emphasizing the ability of platelets to interact with specific elements. Engaging receptors provide stimulus for platelet *activation*. The substances recognized by receptors will be

[15] Large molecules like proteins have specific sites which are engaged in specific reactions.

[16] All data concerning human blood are subjected to large variations, according to sex, body weight, and health conditions.

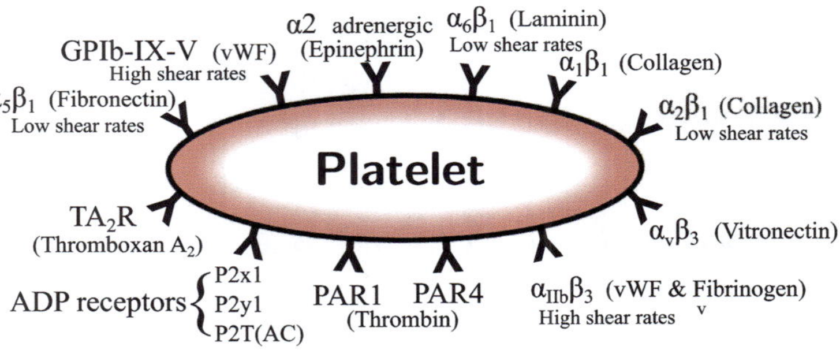

Fig. 7.2 Schematic representation of main platelets receptors

described in the sequel. Drugs neutralizing some of the receptors are administered to prevent blood clotting, particularly after heart surgery. The most known among a long list are *Plavix* (Clopidogrel), acting on the ADP and other receptors, and *Aspirin*, inhibiting the *Thromboxane A2* production. A natural platelet inhibitor, synthesized in the endothelium, is NO (*nitric oxide*). Platelets exhibit also an intense internal activity, since they possess corpuscles, called *α-granules* and *dense* (or *δ*) *granules*. The *α-granules* secrete, under activation, the binding factors vWF and *Fibrinogen* (see Coagulation Factors) in addition to *Platelet Factor 4* (another platelet activator) and other proteins having different functions. For instance, *Thrombospondin 1*, a multifunctional protein (e.g., antiangiogenic), and *Nexin II*, whose role will be discussed later. Platelets also release *Growth Factors* which help repairing the damaged tissue. The *δ-granules* produce platelet activators like the already mentioned *Thromboxane A2* and ADP, thus triggering positive feedback, but also *Serotonin* (5-hydroxytryptamine), a vasoconstrictor. Activated platelets are also able to synthesize some of the coagulation factors.

An important role in platelets (and other cells activity) is played by the *membrane*. Not only through the receptors, but also by exposing negative charges to the exterior, which become sites of attraction of active coagulation factors.

Finally, we recall that in most cases the presence of the ion *Ca++* is necessary for receptors to perform their action. Such an ion is also contained in the δ-granules. Drugs containing citrates have the effects of lowering Calcium concentration, inducing some anticoagulant effect. Therefore their use must be avoided while assuming potent anticoagulants.

Literature on platelets is extremely large and includes several books (e.g., [171], whose foreword [50] contains a huge bibliography). An accurate review paper on platelets receptors is [207], also with numerous references.

2. **White Blood Cells (WBC)**

Also called *Leukocytes*, they are distinguished in various classes (*Granulocytes*: neutrophils, eosinophils, basophils; *Lymphocytes*, including T-cells; *Monocytes*, further differentiated in *Macrophages*) possessing nuclei with different shapes.

Their concentration in blood ranges between 4,500 and 11,000/mm^3 (neutrophils making 50–70 % of the whole population) and they have generally short life (up to about a week, with the exception of Lymphocytes who can live some weeks). Their diameter can be up to ~20 μm. They are highly specialized cells performing many important functions, particularly in the immune system (T-cells). Their importance in the coagulation process consists in the ability of monocytes to produce *Tissue Factor* (defined below), which we will see to be a triggering factor for clotting, see, e.g., [53].

3. **Red Blood Cells (RBC)**

They make ~45 % of blood, with a concentration of 5–6 × 10^6/mm^3, lifespan ~120 days. They have a diameter of ~8 μm and no nucleus. Normally they have a discoid shape, but they are very flexible and they bent to flow in capillaries. In conditions of low shear rate they may aggregate forming *rouleaux*. This fact is responsible for the *shear-thinning* property of blood. They are considered just as passive elements in blood coagulation, being trapped in the fibrin network and providing a large part of the clot volume. However, the discovery that they can synthesize NO [131], which we know to down regulate platelets activity, may open new perspectives.

7.3.2 Platelet Regulators

Here we list stimulators or inhibitors of platelets activity. Some of them have been already mentioned as a comment to Fig.7.2.

1. **Activators**

- TXA_2 (*Thromboxane A2*). It is produced by activated platelets, thus triggering a positive feedback. It favors platelets aggregation. It degrades to *Thromboxane B2*, eliminated through urine.
- ADP (*Adenosine-Di-Phosphate*). A very important molecule,[17] contained also in RBCs. For this reason hemolysis (RBCs disruption, which may occur under excessive mechanical stress[18]) can liberate ADP in the bloodstream, activating platelets.
- *Serotonin* is a vasoconstrictor, thus favoring hemostasis.
- *Platelet Factor 4*[19] is an antagonist of heparin (see Sect. 7.3.5(2))

[17]AMP, ADP, ATP contain 1 (Mono-), 2 (Di-), 3 (Tri-) atoms of phosphorus and they are obtained in that sequence by addition of a P atom (a process called *phosphorylation*). ATP has a vital importance in cells metabolism.

[18]This condition can be produced by arterial stenosis, possibly as a consequence of clotting itself, or due to the mechanical action of implanted devices (rigid artificial heart valves).

[19]Platelet Factors 1–3 actually regulate interactions with the Coagulation Factors IIa (thrombin), V, X.

2. **Inhibitors**

- NO (*Nitric Oxyde*), produced by endothelial cells, is a platelet inhibitor. Its main function in the body is vasodilation via relaxation of the vessel smooth muscle.
- *Prostacyclin* (*or Prostaglandin I2, PGI_2*), also released by endothelial cells, prevents platelets adhesion and also counteracts the effects of TXA_2 binding.
- *Ecto-ADP-ase*, synthesized by the endothelium,[20] neutralizes ADP.

7.3.3 The Coagulation "Factors"

1. **Von Willebrand Factor (vWF)**

The Finnish physician ***Erik Adolf von Willebrand*** (1870–1949) studied a severe, often fatal, bleeding disorder (which later took his name) affecting several individuals in a remote Finnish village, recognizing that it was different from hemophilia (indeed also girls had the symptoms).

He published his studies in 1926, though he could not provide any explanation of the disease. We know today that the cause of that disease is deficiency or dysfunction of the so-called von Willebrand Factor, which is a large multimeric molecule stored in cytoplasmic granules (*Weibel–Palade bodies*) of many cells (and in the platelet δ-granules, as we said), but also circulating in blood (see the subsection Complexes). vWF molecules of smaller size are produced by cleavage of larger molecules by the enzyme ADAMTS13. The fragmentation of the very large molecules is essential for vWF to work correctly. The configuration of vWF molecules is sensitive to stress, which tends to unfold them thus exposing more binding sites to platelets. vWF stress-induced self-association is a property having a special influence on platelets adhesion to fibrin (see [211]). The papers [209, 210] are excellent sources of information about vWF.

2. **Tissue Factor (TF)**

Frequently called *Thrombokinase* or *Thromboplastin*.[21] The chemical cascade leading to the final formation of the fibrin network is initiated by the exposition of endothelial TF to blood, following an injury (for instance the rupture of a plaque). Thus TF has a fundamental role in the clotting process. We have mentioned that TF can be produced by monocytes, so that blood born TF is available. This fact is not a recent discovery (see [104]). Instead, what is really new is the evidence that *platelets can synthesize TF* (see [58]). To our knowledge this fact has not yet been

---

[20]Ecto-enzymes act at the exterior of cells. For more details about this enzyme see, [87].

[21]Though there is some disagreement on the exact meaning of this name, it is very frequently attributed to TF.

taken into account in any mathematical model, but it may deeply modify the whole picture of the process.

3. **The numbered Factor pairs**

The path which leads from TF exposure at the injury site to the formation of the fibrin network is basically a cascade of chemical reactions in which a *zymogen* is modified to an *enzyme* (a *serine protease*),[22] which is said to be the active form of its precursor. In the cascade the enzyme will perform a similar operation on another zymogen (what is called an *enzymatic reaction*). Thus many factors come in an activated and in a non-activated form. This is the basic structure of the biochemical model that will be illustrated in Sect. 7.4, though the actual cascade requires the intervention of platelets and of *complexes* (combinations of factors) and is also accompanied by antagonist reactions. Some of the factors once activated do not possess an enzymatic activity, but they bind in a very active complex with another activated factor, to which, so to speak, they are subsidiary. For this reason they are called *cofactors*. In 1954 the International Committee for the Nomenclature of Blood Clotting Factors was created with the aim of standardizing the names of the *coagulation factors pairs* that had been discovered at that time[23] (some of the former names are still in use). The reader can find a short report of the activity this Committee in [255, 256]. In 1957 the Committee met for the first time in Rome.

Roman numbers were attributed following the chronological order of discovery, following the trend already established by Paul Morawitz at the beginning of the twentieth century, when only the first four were known, and continued by the Norwegian Paul Owren, who called Factor V the one he discovered in 1944 [188]. The final list is the following (alternative names in italic)[24]:

- FI *Fibrinogen*[25] /FIa *Fibrin*, resulting from polymerization of Fibrinogen
- FII *Prothrombin*/FIIa *Thrombin* (the main enzyme in the cascade, deserving its name)[26]
- FIII is nothing but *Tissue Factor*
- FIV identifiable with Ca^{++}
- FV/FVa (formerly *proaccelerin* and *accelerin*, resp.), FVa is cofactor of FXa
- FVI, later recognized to be identical to FVa
- FVII/FVIIa (formerly *proconvertin–convertin*)
- FVIII/FVIIIa, FVIIIa is cofactor of FIXa

[22] Serine proteases are a large class of enzymes including the amino acid *serine*. The list of serine proteases is impressively long. See http://biochem.wustl.edu/~protease/ser_pro_help.html.

[23] Simultaneous and independent discoveries had produced a great confusion in nomenclature.

[24] Actually there were many more names: see [208] for a complete list.

[25] Fibrinogen is not just the precursor of Fibrin, but it has also other specific functions, illustrated in this chapter. It is also known to stimulate RBCs aggregation (forming the so-called *rouleaux*), a phenomenon of some importance in blood rheology.

[26] In order to prevent coagulation and keep blood flowing through the wound it produces, the *Hirudo Medicinalis* (leech) secretes *Hirudin*, a natural and very effective inhibitor of thrombin.

- FIX *Christmas Factor*[27] / FIXa
- FX *Stuart–Prower Factor*[28]/ FXa
- FXI (formerly *Plasma Thromboplastin Antecedent, PTA*)/FXIa
- FXII *Hageman Factor*[29]/FXIIa
- FXIII *Laki–Lorand Factor*[30]/FXIIIa

Factors VIII, IX, XI are called *antihemophilic factors* A, B, C, respectively. We will explain why in the section about bleeding disorders. A review of the intense activity that led to the discovery of Coagulation Factors can be found in the papers [15,208] and [228].

4. **Other Proteins involved in blood coagulation**

One more zymogen–serine protease pair is

- *Protein C* (PC), rarely called FXIV, with its *activated version APC*. Activation is performed by the *Thrombin–Thrombomodulin complex* (see below). Its action is contrasting coagulation, since it *inactivates* FVa and FVIIIa.

The just mentioned action of APC is mediated by

- *Protein S*[31] (PS) as a cofactor.

A protein structurally related to serine proteases is

- *Protein Z* (PZ) has a role in the degradation of FXa.

This is a very appropriate place to mention

- *Vitamin K*[32] (actually a family of vitamins produced in the liver, as PC is),

since many coagulation factors (FVII, FIX, FX) and the Proteins C, S, Z are *vitamin K-dependent*. It means that they cannot perform their action in the absence of vitamin K. Hence its fundamental importance. Drugs inhibiting vitamin K are widely used as anticoagulants. Since the most rapid action is on the anticoagulant proteins PC, PS, PZ, they first favor clotting, while the strong anticoagulant effect

[27]Stephen Christmas was the first patient diagnosed with FIX deficiency (hemophilia B) (1952) at the age of five. He died in 1993 by AIDS. Many of the transfusion-dependent patients were infected by the HIV virus before blood screening became obligatory. A case which became emblematic was the one of Ryan Wayne White, affected by hemophilia A, who became discriminated when he was diagnosed with AIDS. He died still a teenager in 1990.

[28]Named after the patients Rufus Stuart and Audrey Prower.

[29]Named after Ratnoff's patient John Hageman (1955).

[30]Laki and Lorand suggested its existence in 1948 [138].

[31]A very important function of Protein S in the organism is to facilitate phagocytosis of apoptotic cells by macrophages. Discovered in 1979 in Seattle, takes its name after that city.

[32]Denominated after the German name *Koagulationvitamin*. Discovered in the 1930 a Nobel prize was attributed in 1943 for studies on it, though its real action in the coagulation process became clear only in the 1970.

is seen with some delay. For this reason *heparin* (see next section) is simultaneously administered. Drugs neutralizing vitamin K are on the market with various trademarks. The most known are *Warfarin* (in the USA)[33] and *Coumadin*.

Activated factors have their own *inhibitors*, that will be described below (Sect. 7.3.5).

5. **Complexes**

The following *Factor Complexes* have an important role in the coagulation process:

- *Complex FVIII-vWF* is the main carrier of inactive FVIII in blood.[34] Under the action of FIIa it dissociates and FVIII is rapidly activated.
- *Complexes FVII-TF, FVIIa-TF* intervene in the initiation phase of the cascade.
- *Complex FVIIIa-FIXa* ($+Ca^{++}$) is called *Tenase*[35] because it activates FX.
- *Complex FVa-FXa* ($+Ca^{++}$) is called *Prothrombinase* because it promotes the transition from FII to FIIa. We will discuss its pivotal role in the cascade.
- *Complex Thrombin–Thrombomodulin*. Thrombomodulin is a protein expressed by endothelial cells. This complex induces the activation of Protein C and turns TAFI (see Sect. 7.3.4) into its active form, providing protection to Fibrin.

6. **More Coagulation Factors**

According to the *Cell-Based Model* (Sect. 7.4), all the Factors listed above enter the chemical cascade leading to Fibrin production, *except FXII*. This is the main discrepancy with the *3-pathway Cascade Model* (Sect. 7.6), used until recently. In the latter, activation of FXII is the triggering event of the *intrinsic pathway* of coagulation (i.e., a process originated within blood, independently of exposure to TF, which is of *extrinsic* nature). Since there is evidence of clotting of intrinsic origin, and also of the fact that FXII can become activated when blood comes into contact with artificial materials, the intrinsic pathway, though not endorsed in the Cell-based Model, still is worth being considered. It is also called *contact activation pathway*. Here we list the Factors which, in addition to FXII, take part in it.

- *Prekallikrein* (PK), also known as *Fletcher Factor*,[36] complexes with *High Molecular Weight Kininogen* (HMWK) by contact with collagen, in the presence of FXII.

---

[33]Patented in 1948 as a rat poison and used as anticoagulant for humans since 1954. It was isolated in 1941 by a group at the University of Wisconsin after a 6-year work investigating a widespread hemorrhagic disease that affected cattle in the USA, the so-called sweet clover disease (the research was funded by WARF, i.e., Wisconsin Alumni Research Foundation). See [248].

[34]Already in the 1950 it was known that deficiency of vWF was accompanied by a deficiency of FVIII (see [208]).

[35]The symbol X-ase is sometimes used.

[36]Also referred to as *Williams Factor* or *Flaujeac Factor*.

- *High Molecular Weight Kininogen*[37] (HMWK), also known as *Fitzgerald Factor*.
- *Kallilkrein*,[38] the active form of PK, following the formation of the complex PK-HMWK. In turn, Kallikrein is a fast activator of FXII.

7.3.4 Fibrinolysis Factors

Fibrinolysis is the process eventually destroying the clot. It goes through positive feedback too, with the intervention of the following factors.

- *Plasminogen* (a zymogen)
- *Plasmin* (a serine protease), the active form of its precursor, attacks fibrin, gradually destroying the clot. Plasmin is also active on vWF and other proteins.
- *Tissue Plasminogen Activator* (tPA), a serine protease catalyzing the transition from plasminogen to plasmin.
- *Urokinase*[39] (*urokinase-type Plasminogen Activator: uPA*) (a serine protease), another activator of plasminogen.[40]
- *Thrombin Activatable Fibrinolysis Inhibitor* (*Carboxypeptidase B2*), or *TAFI*, when activated (by the thrombin-thrombomodulin complex) is an enzyme which protects fibrin from the action of plasmin by slightly modifying its structure.

Besides tPA and uPA, plasminogen can be activated to plasmin by the Hageman factor FXII, a fact established long ago by Ratnoff himself and coworkers [96]. One more strong plasminogen activator is the enzyme streptokinase, used as a trombolytic.

7.3.5 Factors Inhibitors

Both coagulation and fibrinolysis factors have their inhibitors, as all proteases do. The delicate game played in the body by proteases and their inhibitors tells us that indeed health requires the equilibrium of a huge number of substances in permanent mutual conflict. In a sense this fact expands to an unthinkable scale the naïve idea of the equilibrium among the four Hippocratic humors!

---

[37]*Kininogens* are proteins which are precursors of *kinins* (see next footnote), such as *bradikinin* and *kallidin*, which are vasodilator.

[38]Here we refer to *Plasma Kallikrein*, distinct from the numerous group of *Tissue Kallikreins*, which are enzymes performing various actions. Discovered in 1934, it was named after the Greek words *kalli* (sweet, in this context) and *krein* (flesh) referring to pancreas tissue. Plasma Kallikrein (like some of its tissue analogs) liberates kinins from the kininogens. The so-called *kinin-kallikrein system* has a role in regulating blood pressure, owing to the vasodilation action.

[39]First isolated in the urine.

[40]Also Kallikrein and FXIIa can activate plasminogen.

Serine Protease Inhibitors make a large group of proteins neutralizing specific serine proteases. Of course their list is as long as the one of serine proteases. The acronym *serpin* is frequently used to denote any of these proteins. For our purposes we are interested in the following ones.

1. **Serpins Neutralizing Pro- or Anti-coagulation Factors**
   - *Antithrombin*. Most commonly referred to as *Antithrombin III* (ATIII),[41] it inhibits most of the activated coagulation factors (FIXa, FXa, FXIa, and FIIa) and the FVIIa-TF complex. Its action is enormously enhanced by *Heparin*.[42]
   - *Tissue Factor Pathway Inhibitor* (TFPI) can inhibit thrombin (FIIa) and also FXa by forming a complex with it. In turn the latter complex can inhibit the FVIIa-TF complex (the initiator of coagulation). It is released by endothelial cells and also by platelets.
   - *Alpha 1-antitrypsin*. This serpin interacts with many proteases and therefore is particularly important.[43] It is known to inhibit APC.
   - *Protein C inhibitor* limits the expression of Protein C.
   - *Protein Z-related protease inhibitor*. It neutralizes FXa in the presence of PZ. It also inhibits FXIa.
   - *Kallistatin*, an inhibitor of Kallikrein.
   - *Heparin cofactor II* rapidly inhibits thrombin in the presence of heparin.

   Another protease inhibitor, not frequently quoted in the literature in the context of blood coagulation, is

   - *Nexin II*, an inhibitor of FXIa, secreted by activated platelets (see [244]).
2. **Serpins Neutralizing Pro- or Anti-fibrinolytic Factors**
   - *Plasminogen activator inhibitor-1* (PAI1) and *Plasminogen activator inhibitor-2* (PAI2)[44] inactivate both tPA and urokinase.
   - *Neuroserpin* inhibits tPA and urokinase.
   - *Alpha 2-antiplasmin* is an inhibitor of plasmin.
   - *Alpha 2-Macroglobulin* is another multifunction serpin. It inhibits plasmin and Kallikrein.

---

[41] AT I–IV are also found in the literature, with specific targets.

[42] Discovered in 1918 [109], though isolated in 1916 in canine liver tissue [167]. There has been some controversy about heparin discovery (see [248] and [162]]). It is a large polymer, also naturally produced by endothelial cells (as *heparan sulfate*). A side effect can be a strong reduction of platelets count (*Heparin Induced Thrombocytopenia*, HIT), see [128]. HIT can be sometimes observed in patients undergoing hemodialysis, during which heparin is supplied to prevent clotting (after passing through the dialyzer and before being returned to the patient, *protamine sulfate* is added, which neutralizes heparin's action). Platelet Factor 4 contrasts the action of heparin on platelets.

[43] Its deficiency leads to degradation of tissues, particularly in the lungs, causing emphysema. Smoke is believed to inactivate this serpin, thus causing additional damage to lungs.

[44] PAI2 is detectable only in pregnant women, a fact that may justify the increased risk of thrombosis during pregnancy.

We hope that this long list has not been too tedious. We believe it can be very useful not only for understanding the rest of the paper, but also as a kind of glossary to move around in the field of blood coagulation. Having introduced all the characters, let us now see how they play the game.

7.4 The Cell-Based Model for Secondary Hemostasis

During the last decade it became apparent that the *Cascade* (or *Waterfall*) model for secondary hemostasis elaborated during the 1960s and long considered the correct explanation of the complex blood coagulation process was in need of a deep revision. In Sect. 7.6 we will illustrate both the model and the reasons for its final rejection. The model that took its place is known as the *Cell-Based Model* and has its roots in many papers (see [56, 67, 106, 160, 161, 206, 216, 219] and the literature quoted therein). The very synthetic exposition in [22] is interesting because it contains some details not frequently recalled elsewhere.

According to the cell-based model, the fibrin production process goes through four steps (*initiation, amplification, propagation, termination*). Fibrinolysis goes in parallel and becomes visible over a longer time scale. The necessity of having two separate time scales comes from the fact that the clot has to be formed soon, but it has to dissolve slowly. Indeed it has to stay in position long enough to allow wound healing (at least partial) and it must not break into pieces which would be dangerously released in the bloodstream.

7.4.1 Secondary Hemostasis

1. **Initiation** (Fig. 7.3)

Once the tissue factor has become exposed to blood at the injury site, the complex FVII-TF is readily formed, and the tiny amount of FVIIa which normally circulates in blood gives rise to the FVIIa-TF complex.[45]

The latter activates FVII-TF and (*at a low rate*) FIX, FX. Now we have a small amount of FIXa, FXa (in turn able to activate more of the complex FVII-TF). In particular, at this stage, a small quantity of FVa is produced as the result of the action of FXa on FV. The ability of FXa to activate FV has been proved long ago (in [176], see also [82]). Though the main activator of FV is thrombin and the activation rate by FXa is orders of magnitude less, the production of FVa even in very small quantity

[45]Some FVIIa can reach TF in nonvascular tissues even in the absence of a lesion [279], thus making FIXa and FXa accidentally available. However, coagulation does not start because it requires, for instance, the intervention of platelets, which are not available out of the bloodstream.

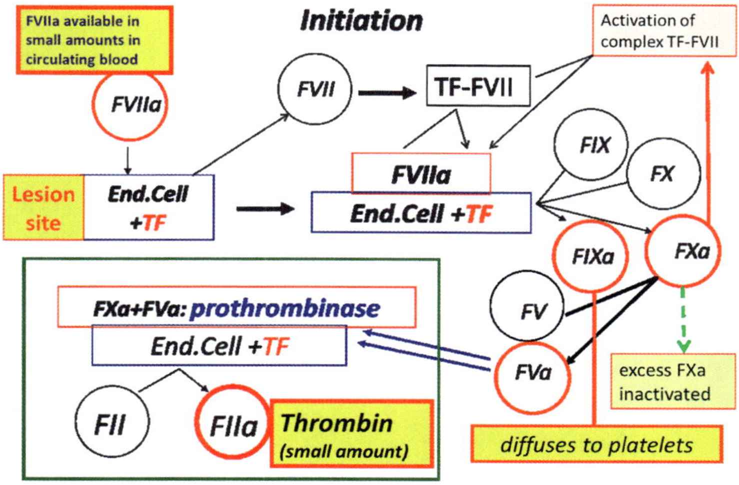

Fig. 7.3 Sketch of initiation phase

is absolutely crucial in order to give rise to complex FVa-FXa (prothrombinase), which transforms FII (prothrombin) to FIIa (thrombin). Thus the initiation stage has the fundamental task of making some thrombin available. We must not forget that in the meanwhile platelets keep accumulating at the lesion site.

It is now important to say that Fig. 7.3 tells only part of the story. Indeed two additional considerations are in order.

- FXa and thrombin leak from the clotting site and are carried by blood. However, they are not going to trigger clotting downstream, because FXa is neutralized by ATIII (more rapidly than thrombin, and the same happens to FIXa) and by endothelium produced TFPI, while thrombin reaching endothelium combines with thrombomodulin (TM), losing its procoagulant activity. Moreover, the thrombin–thrombomodulin complex is an activator of Protein C, and APC strongly inhibits FVa on the surface of endothelial cells, thus helping to keep the coagulation process confined close to the initiation site.
- In addition, the complex FXa-TFPI has an inhibiting action on the complex FVIIa-TF (see [185]). Therefore we can say that, due to the formation of FXa-TFPI, and of APC, during this stage a regulatory mechanism is present, somehow delaying the exit to the next stage.[46]

---

[46]The proteins responsible for this regulatory action have a fundamental role in eventually halting the clot growth. They are inevitably produced at this initial stage too, but it is known that FVa inhibition by APC is far less efficient than on the surface of endothelial cells [106, 219]. One can wonder whether, besides the clotting confining action, the simultaneous slowing down of the initial process may have a precise aim, for instance letting the platelet plug become thicker.

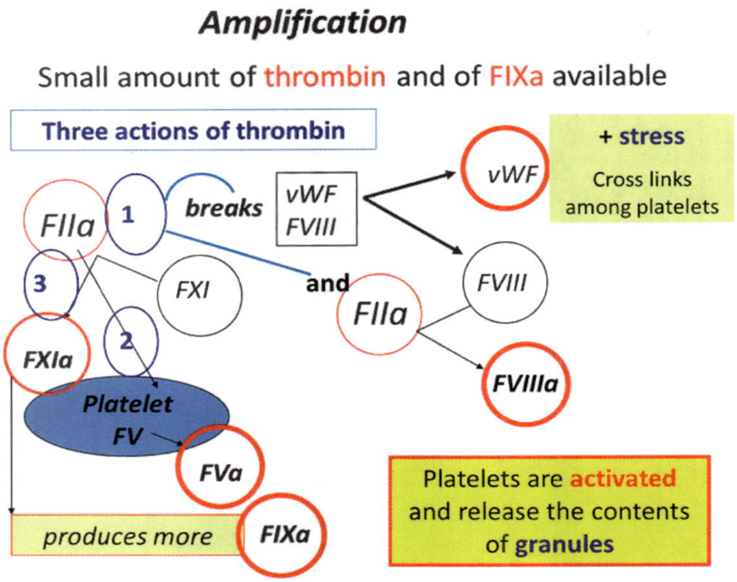

Fig. 7.4 Sketch of amplification phase

2. **Amplification** (Fig. 7.4)

Thrombin produced in the previous stage exerts many actions both at this and at later stages. We list them as (Thr. n):

- (Thr. 1). It dissociates the complex FVIII-vWF, at the same time activating FVIII,
- (Thr. 2). It activates FV, stimulating platelets to produce more of FV,
- (Thr. 3). It activates FXI.

These actions have important consequences. FXIa is a fast activator of FIX, so that the amplification stage makes both FIXa and FVIIIa available. Moreover vWF can promote further platelets aggregation (if stress conditions are favorable). Finally, platelets become fully activated, releasing the granules content.

3. **Propagation** (Fig. 7.6)

The coagulation machine is now ready to produce the *thrombin burst* (during which 95 % of thrombin is produced [160]). The surface of activated platelets provides the ideal site for the combination of FVIIIa, FIXa into the complex *tenase*, activating FX very rapidly. FXa combines with the available FVa (still on activated platelets surface) yielding the *prothrombinase* complex. From now on activation of prothrombin occurs at a large speed and the processes already described in the previous stage trigger an enormously effective positive feedback. The onset of the propagation phase is indeed recognizable by the sudden and marked increase of

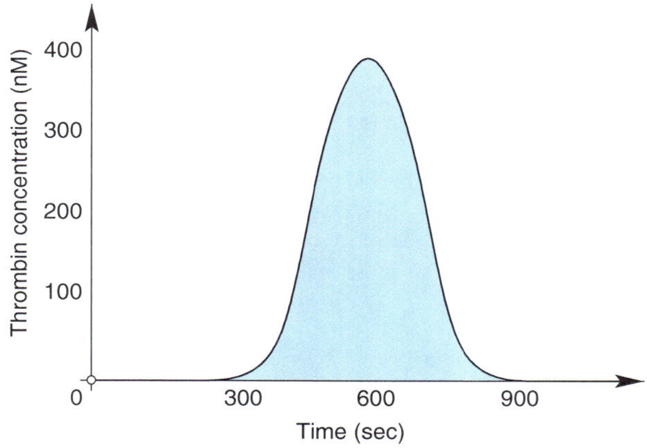

Fig. 7.5 Sketch of the thrombin concentration vs. time

thrombin production rate shown in the sketch in Fig 7.5 (for experimental evidence of this behavior, see, e.g., [40, 161, 186]).

Besides continuing actions (Thr. 1–3), thrombin performs two more important tasks:

- (Thr. 4) transition from FI (*fibrinogen*) to polymer FIa (*fibrin*),
- (Thr. 5) activation of FXIII.

Fibrinogen polymerization occurs at a high rate. From now on the fibrin network traps blood constituents and we may say that the clot progression is *fibrin dominated*. Fibrin generation is just the first step of the network production, which results from the aggregation of polymer chains into fibers [79]. The role of FXIIIa is to *consolidate* the fibrin network by forming cross links among fibers. Clots lacking *cross links* are unstable, as proved by the fact that a nontrivial bleeding disorder is associated with deficiency of FXIII.

A further remark is about the fact that, while FVa is effectively inactivated by APC (Activated Protein C) when not included in the prothrombinase complex, it appears to be protected from this action if combined with FXa. Protection of FVa from APC exerted by FXa was established in [181, 243].

4. **Termination** (Fig. 7.7)

As we have said several times, the biological processes of clot formation and dissolution are the result of an unbalance between contrasting elements which happen to be first in one direction and then in the opposite. If starting and accelerating the growth of a thrombus is important for arresting bleeding, terminating it is absolutely essential to prevent vessel occlusion. We have seen how thrombin helps in confining clotting. By means of the very same reaction it actually starts the termination mechanism:

- (Thr. 6) The thrombin-TM complex activates PC.

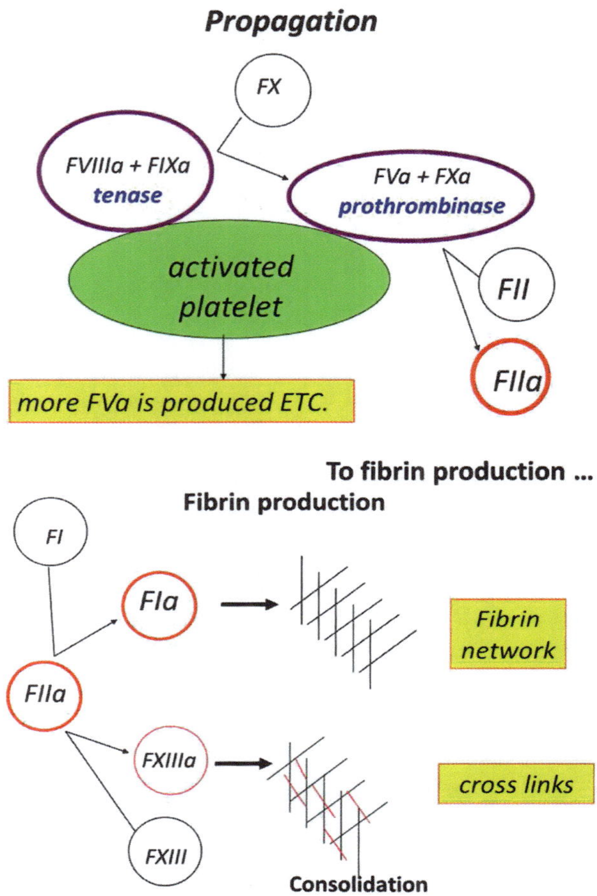

Fig. 7.6 Sketch of propagation phase

APC (with the cofactor PS) inhibits both cofactors FVa and FVIIIa, thus switching off the production of tenase and of prothrombinase. Surviving FIXa, FXa, FXIa, and thrombin are inactivated by ATIII (this action is greatly accelerated by heparin), and TFPI complexes with FXa, inhibiting FVIIa-TF at the same time.

The clot has grown to its maximum size and it is now time to take care of its gradual dissolution, which, however, needs to be delayed. Once more thrombin has a leading role:

- (Thr. 7) thrombin activates TAFI, which makes fibrin more resistant to the attack of plasmin.

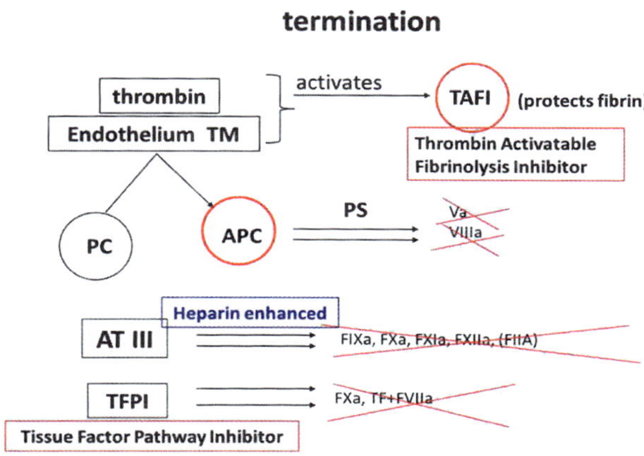

Fig. 7.7 Sketch of termination phase

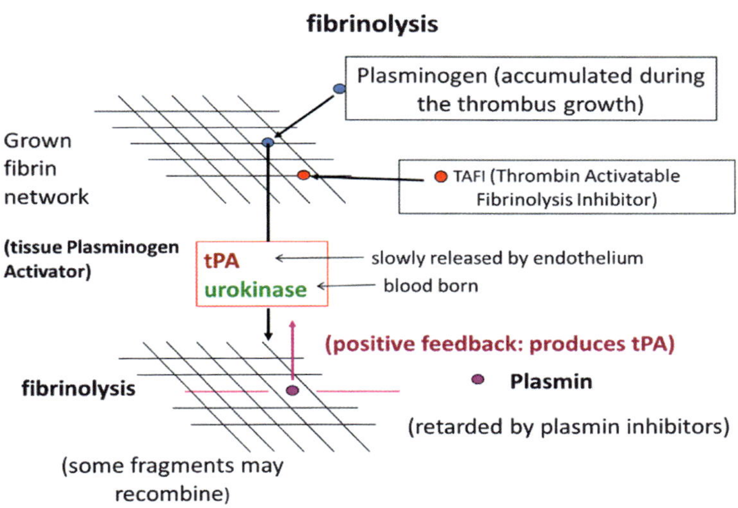

Fig. 7.8 Sketch of fibrinolysis pattern

7.4.2 *Fibrinolysis (Fig. 7.8)*

As we said, fibrinolysis goes in parallel to clotting, but it is regulated so that the fibrin network can develop sufficiently rapidly and that the clot dissolves slowly and gradually, without delivering dangerous fragments in the bloodstream. Since the clot is basically impervious to new plasma and all diffusion processes (particularly of heavy molecules) are considerably slow in it, the fibrinolytic elements have to be stored within the clot during its growth. We have listed them in Sect. 7.3.

Plasminogen, synthesized by the liver, is available in circulating blood, so it is naturally trapped in the clot. We know that it is activated to Plasmin by the serine proteases tPA and urokinase (and also by Kallikrein and FXIIa). We recall that plasmin and plasminogen activators have their own serpins, making sure that those proteases are eliminated once they have performed their task. We have seen that TAFI, accumulated during clot formation, reduces plasmin action, retarding the process. Secretion of additional tPA by the endothelium is stimulated by the clot itself. At the early stage of fibrinolysis fragments of the fibrin network may recombine.

A product of fibrin degradation is the so-called *D-dymer*, whose presence in blood reveals ongoing thrombosis.

7.5 Bleeding Disorders

There are two classes of *coagulopathies*:

(1) *hypocoagulability*, generally with *bleeding diathesis*, i.e., spontaneous bleeding,
(2) *hypercoagulability* or *thrombophilia*, producing thrombosis.

Such phenomena can occur with different levels of intensity and consequently with a different impact on health. This is largely an open field for mathematical modeling.

We can review bleeding disorders according to their origin. So a natural way of exposing this material is to follow the same articulation as Sect. 7.3.

7.5.1 Platelet-Related Bleeding Disorders

Since platelets are so complicated there are several types of dysfunctions that may produce bleeding disorders.

- **Thrombocytopenia** Literally it means scarcity of thrombocytes (i.e., platelets). It is diagnosed if the platelet count drops below 1/3 of the minimum (i.e., less than $50{,}000/\text{mm}^3$). There are very many possible causes of different gravity. For instance, the spleen may start storing too many platelets, sequestering them from the bloodstream, or the immune system may produce antibodies attacking platelets (*Immune Thrombocytopenia*), or it can be related to other pathological conditions. It can also be drug induced. A striking and puzzling example in the latter class is
  - *Heparin-Induced Thrombocytopenia* (HIT). Thrombotic episodes (also of a massive character) were observed in patients under anticoagulant treatments with heparin. This counterintuitive phenomenon has been the object of many controversial studies. Today it is believed to result from an immune reaction triggered by heparin, which induces platelets activation and clustering. We recommend the reading of [128].

Thrombocytopenia is frequently accompanied by the appearance of bruises on the skin associated with subcutaneous bleeding (*purpura*) [47] and some forms of the disorder are associated with this name:

- *Idiopathic* [48] *Thrombocytopenic Purpura* (ITP). It can sometimes be of immune (rather than unknown) origin (in that case the I in the acronym can stand for Immune). It can be a very serious illness. Treatments include the surgical removal of spleen. The interesting paper [39] develops a mathematical model helping clinicians to decide whether or not the spleen removal can be advantageous, depending on the measurements of specific parameters.
- *Thrombotic Thrombocytopenic Purpura* (TTP). This is in a sense the opposite, since it produces microscopic clots in the microcirculation. It is a rare disorder, actually only indirectly related to platelets, since it is caused by inhibition, or dysfunction or deficiency[49] of the enzyme ADAMTS13 (see the subsection on vWF, Sect. 7.3), which results in an excessive interaction between platelets and the uncleaved large vWF multimers.

- **Thrombocytosis** is the abnormally high platelet count (say, twice the maximum). It can be due to excessive production by the bone marrow, or to spleen dysfunction, or to medical treatments. It predisposes to thrombophilia.
- **Platelets receptors dysfunctions.** We have seen the crucial importance of the various platelets receptors both for primary and secondary hemostasis. They are all rare disorders.
  - *Glanzmann's thrombasthenia* [94] involves receptors of the Gp-family, reducing the ability of binding with fibrinogen, vWF, etc.
  - *Bernard Soulier syndrome* (*Giant Platelets Syndrome*) [25] also relates to inability of binding to vWF because of defective GpIb receptors. It has an influence on platelet morphology.
  - *Scott syndrome* is a defect in a delicate mechanism occurring on the membrane which prevents the formation of complexes tenase and prothrombinase, while the typical functions in primary hemostasis are not altered.
  - *Stormorken syndrome*, still related to wrong membrane reactions, but with the opposite consequence of a self-activation of platelets, inducing thrombophilia.
- **Granules dysfunctions.** Also rare disorders. We just mention
  - *Gray platelet syndrome* consists in α-granules deficiency with the consequent lack of the proteins there synthesized.

---

[47]The Latin word for *purple*. *Purpura* denotes spots in the range 3–10 mm, smaller spots are called *petechiae*, and those more extended are called *ecchymoses*.

[48]The Latin equivalent of the Greek derived word *idiopathic* is *sui generis*. In this context it means of no specific origin.

[49]Congenital deficiency accounts for a small fraction of TTP cases and is known as *Upshaw–Schülman syndrome*.

– *Quebec Platelet Disorder*: the plasminogen stored in platelets is abnormally converted to plasmin, destroying proteins in the α-granules.

Actually, the list of diseases due to defective platelets is much longer. Besides [49], a good review is [100].

7.5.2 Disorders Due to Coagulation Factors Deficiency

- **von Willebrand Disease (vWD).** We have already illustrated the circumstances of its discovery in the mid-1920s. Many children in the Åland islands were affected by life-threatening bleeding diathesis (consanguineous mating was an important factor, revealing the hereditary character of the disease). Both sexes were affected (girls could die by menorrhagia). Von Willebrand recognized it was not one of the disorders known at that time. Since vWF is known to be a carrier of FVIII in blood (as a complex), its deficiency can be accompanied by FVIII deficiency. A recent study [202] indicates that vWF deficiency is accompanied by reduced platelets activation. We refer to [20] and once more the basic reviews [209, 210].
- **Hemophilia.** In the historical section we have already said much about hemophilia. There are three types:

  – *Hemophilia A*, namely FVIII deficiency,
  – *Hemophilia B*, or FIX deficiency,
  – *Hemophilia C*, or FXI deficiency,[50]

  justifying the alternative names of anti-hemophilic factors A,B,C for FVIII, FIX, FXI, respectively. The first two types are ascribable to defective genes in an X chromosomal branch and are therefore extremely rare in women. Hemophilia C instead affects both sexes with equal probability and is a less severe form. If one inspects the role of FXI in the cell-based model, the latter statement may look very surprising, since FIX is activated by FXIa, suggesting that deficiency of FXI should prevent the activation of FIX, with effects completely similar to Hemophilia B. The explanation can be found in the ability of platelets to synthesize FXI, thus compensating the deficiency of the blood born factor (see [86, 195]).
- **FV Leiden mutation.** First identified in the homonymous Dutch city (1994), it is probably the most frequent cause of thrombophilia. This mutated version of FV can do the same job of the normal protein, but the corresponding protease FVa is not attacked by Activated Protein C. Other (rarer) mutations of FV (*FV Cambridge*, *FV Hong Kong*) have similar effects.

[50]More common among Ashkenazi Jews.

- **More disorders due to factors deficiency or malfunction.** They all cause thrombophilia and are usually rare or very rare. We omit listing them for the sake of brevity.

7.5.3 Disorders Due to Proteins C, S, Z or to Vitamin K Deficiency

- *Protein C deficiency.*
- *Protein S deficiency.*
- *Protein Z deficiency.*

Due to the anticoagulant actions of PC, PS, PZ deficiency in each of these proteins predisposes to thrombophilia. They are rare. We quote [57].

- *Vitamin K deficiency.* Induced by warfarin treatment or (rarely) occurring naturally. We have already discussed its consequences.

We omit mentioning disorders linked to other serpins deficiencies and to fibrinolysis factors deficiencies, since they are very rare.

7.5.4 Deep Vein (or Venous) Thrombosis (DVT)

Blood stasis or significantly reduced circulation in limbs is known to be responsible of clot formation in deep veins, particularly in the large veins in the legs (femoral, popliteal, saphenous, etc.).[51] It can also affect deep veins of the pelvis and more rarely in the arms (succlavian or axillary veins).[52] It causes swelling and it may cause pulmonary embolism. For that reason it has to be treated as soon as possible with strong anticoagulants.[53] A typical consequence is a permanent damage of vein valves.[54] Altered circulation can produce conditions favorable to clotting particularly in the valves pocket.

Clot formation in the absence of lesions and in an environment of reduced stress is difficult to explain on the basis of the cell-based model, since the process has a clear intrinsic origin. It is out of question that fibrin production requires

[51]Because immobilization is a frequent cause, DVT is also called the *economy class syndrome*, since many cases have been reported in passengers after long flights.

[52]In that case it is known as *Paget–Schrötter disease*.

[53]Not with fibrinolytic proteins (like tPA or UPA), because they could fragment rather than gradually dissolve the clot. Fibrinolytic therapies are instead used to attack arterial thrombosis in the heart or the brain.

[54]Major veins are provided with valves preventing flux inversion, thus helping circulation in the presence of reduced pressure gradients.

the prothrombinase complex, thus the activation of both FV and FX. Rather than appealing to the contact activation pathway, involving FXII (see next section), the modern view is to identify TF delivery as a triggering cause. An interesting discussion about DVT in different veins can be found in [1], where some hypotheses are presented for the TF source (see also [44]). It can be secreted by the endothelium (particularly the endothelial cells lining the valves) into the blood under abnormal condition, or it can be attributed to leucocytes (which can even happen to remain trapped in the valves). Hypoxia[55] is also an element to be considered as a possible stimulus for TF release. In addition we know that platelets (that can suffer hypoxia and can also be trapped in the valves) are able to produce TF. To these considerations we may add that RBCs severe hypoxia (worsen by RBCs aggregation) could lead to the release of ADP, activating platelets. DVT can also be induced by other pathological conditions. For instance, it was noted long ago [235] that DVT occurrence was more frequent in patients with tumors. Today this is attributed to an overexpression of TF [205].[56]

7.5.5 Heart Arrhythmia and Thromboembolism

In the family of disorders related with altered blood flow we must include the possible formation of clots caused by atrial fibrillation.[57] To prevent the risk of strokes, patients suffering from atrial fibrillation are frequently given an anticoagulant therapy.

7.5.6 Coagulation on Artificial Surfaces

This is a very important subject, because the implant of artificial bodies (heart valves, stents, joints, etc.) is likely to be followed by blood coagulation. Clotting can be caused by high shear stress (also possibly causing hemolysis). Some materials have affinity to fibrinogen, which can produce platelet aggregation on the body (similarly to vWF in primary hemostasis), and/or trigger immune reaction, by aggregating leukocytes.[58] Cell-born TF does the rest. Hence the particular attention

---

[55]Reduced oxygen concentration is more marked in valves, since, differently from veins and other blood vessel, they do not possess their own vessels (*vasa vasorum*).

[56]This paper is an extensive study on the role of TF and of thrombin in promoting angiogenesis and contains a large bibliography. Excessive TF production may be accompanied by upregulated expression of VEGF (the angiogenic factor) and downregulated expression of thrombospondin 1 (see Sect. 7.3).

[57]Clots are mostly originated in the left atrium and more precisely in an area called *left atrial appendage*.

[58]See, e.g., http://courses.washington.edu/overney/NME498_Material/NME498_Lectures/\Reading_on_Adsorption_Kinetics.pdf.

in designing anticoagulant coatings. Traditionally this kind of coagulation used to be attributed to self-activation of FXII on *artificial surfaces* (a fact established by the discoverer of FXII, Ratnoff [201]), initiating the so-called intrinsic pathway (see next section)). This can be a concomitant process, but, as we shall see, the intervention of TF is anyway required.

7.5.7 *Disseminated Intravascular Coagulation*

This disorder consists in a deep dysregulation of the entire coagulation-fibrinolysis system, which may be caused by other pathologies (sepsis, tumors, etc.). It occurs when abnormal, critical conditions stimulate TF secretion in blood, e.g., by leukocytes, resulting in the production of circulating thrombin. As a consequence microthrombi are diffusely generated, sequestering platelets from blood. Therefore, paradoxically, hypocoagulability sets in, with resultant bleeding in various parts of the body. Blood perfusion of vital organs is altered which may induce failure.

7.6 The 3-Pathway Cascade Model

Figure 7.9 shows a typical sketch for the 3-pathway (or *Cascade* or *Waterfall*) model, proposed independently in 1964 by Davie and Ratnoff [59] and by Macfarlane [158]. For more than four decades it has been theİ coagulation model, though it had eventually to be replaced because of its inability to explain hemophilia A (see [206]). Let us comment this important aspect. The extrinsic origin starts with the formation of the FVIIa-TF complex, exactly like in the cell-based model. The intrinsic origin requires the activation of the Hageman factor FXII.

The original scheme was conceived so to provide two independent ways of leading to the FVa-FXa complex.

The extrinsic pathway leads directly from the formation of complex FVIIa-TF to the activation of FX. Then, entering the common pathway, the subsequent production of prothrombinase takes place via the activation of FV and its complexing with FXa and Ca++. Thus the extrinsic pathway is sufficient to eventually produce thrombin and fibrin.

In the intrinsic pathway a small amount of FXII gets activated in contact with negatively charged surfaces, like collagen or the membrane of activated platelets (but also, e.g., of foreign bacteria). This phenomenon is mediated by HMWK and induces the Prekallikrein to Kallikrein conversion. Kallikrein is now a fast activator of FXII, so that a positive feedback sets in. FXIIa activates FXI, and FXIa activates FIX. At this point the former scheme by Davie and Ratnoff (1964) was as follows[59]

[59]The scheme is inspired to an original document held by Ratnoffs heirs (see [208]).

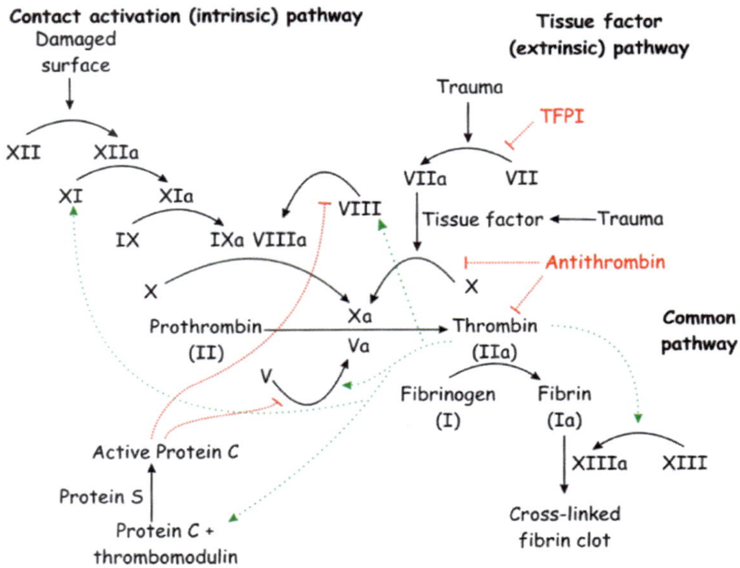

Fig. 7.9 Scheme of the 3-pathway cascade model. Extrinsic pathway is also called TF-pathway. An alternative name for the intrinsic pathway is contact activation pathway

$$
\begin{aligned}
&FXII \overset{?}{\rightarrow} FXIIa \Rightarrow \\
&\quad FXI \rightarrow FXIa \Rightarrow \\
&\qquad FIX \overset{Ca++}{\rightarrow} FIXa \Rightarrow \\
&\qquad\quad FVIII \overset{Ca++}{\rightarrow} FVIIIa \Rightarrow \\
&\qquad\qquad FX \overset{Ca++}{\rightarrow} FXa \Rightarrow \\
&\qquad\qquad\quad FV \rightarrow FVa \Rightarrow \\
&\qquad\qquad\qquad FII \rightarrow FIIa \Rightarrow \\
&\qquad\qquad\qquad\quad FI \rightarrow FIa
\end{aligned}
$$

Fig. 7.10 Sketch of the *waterfall* model of Davie and Ratnoff

(it includes the question mark concerning the activation of FXII) (Fig. 7.10). From the appearance of FIXa onward the sequence in the sketch above is rather different from the one we know (in which FVIII is activated by thrombin, FXa is a product of the tenase complex, and thrombin is generated by prothrombinase) and in this form it is autonomous from the extrinsic pathway.

The sketch of Fig. 7.9 has instead the correct chemical correlations, letting products from extrinsic pathway (thrombin, FXa) enter the intrinsic pathway, which becomes so to speak subsidiary. The purely intrinsicİ pathway is instead based on the assumption that FVIII is directly activated by FIXa (as in the original scheme), which makes it really independent of the parallel extrinsic pathway.

Two main experimental observations clash with the former 3-pathway cascade model:

- The TF-pathway *bypasses FVIII*, but is nevertheless sufficient to produce clotting,
- FXII deficiency, as we have seen in the previous section, produces no significant bleeding disorder.

So the troubles about the 3-pathway cascade model come from the position occupied by FVIII, whose deficiency we know to be associated with Hemophilia A. If the TF-pathway alone would be enough for clotting, FVIII deficiency would be inessential. Moreover, FXII deficiency would exclude FVIII from the process, causing the same effects as FVIII deficiency. None of these facts are true, hence the necessity of a model, like the cell-based model, which can produce the clot without FXII and recognizes the crucial role of FVIII.

Nevertheless, it would be wrong to dismiss the role of FXII altogether. Its ability of becoming activated on artificial surfaces can enhance clotting on implanted bodies. A paper recently devoted to FXII and its possible alternative functions is [212].

7.7 Mathematical Description of Coagulation Models

Mathematical models of blood coagulation are at the same time the main tools as well as outputs to study the coagulation processes. Solutions of these models are used for comparison with experimental observations to test our understanding of the underlying processes. The wide range of physical phenomena to be considered and the complexity of their interactions are responsible for a high number of different models being developed for specific use.

There is a large number of coagulation models that can be applied for various specific purposes [266]. They differ in many aspects like physical features to be considered, scales of the phenomena, or the biochemistry model to be coupled to biomechanics. Important differences are also in the mathematical formulation of the coagulation models and in the numerical methods to solve them.

In order to give a brief overview of some of the existing mathematical models of blood coagulation, we will classify the models in two ways. First, the models will be grouped together according to the (spatial) scales they are able to describe and resolve. Second, the models can be classified according to the physical features they are including. These two classification schemes are not independent, as some physical phenomena can only be described at appropriate scales. Nevertheless, these classification schemes will be useful to demonstrate some of the main trends in the past and current evolution of mathematical models of blood coagulation.

7.7.1 Scale-Based Classification of Coagulation Models

Blood coagulation is a highly complex process that involves various mechanisms acting at a wide range of spatial scales. It is hardly possible to include all of them in one single model. The main problem is not just in the number of processes to be considered, but also in the need of using different scale-specific mathematical formulations and numerical solvers. This led to the development of a whole range of models, each one of them treating specific coagulation problems from the point of view of a specific spatial scale. This approach is reflected in the hereinafter adopted models classification.

1. **Sub-microscale Models**

By the *sub-microscale* (or nanoscale) in the coagulation models we understand the *sub-cellular scale* with characteristic dimensions up to the order of 0.1 μm. At this scale the internal structure of various proteins and macromolecules plays a major role. The interaction between them, their activity, and sensitivity with respect to external chemical, mechanical, or electrical stimuli has to be taken into account. The information about the three-dimensional structure of proteins and the functional activity of molecules can be gained using a wide range of computational methods and tools commonly associated under the label of *Structural Bioinformatics*. This rather new scientific area is quite rapidly developing together with the increasing power of modern computers. This sub-microscale approach is somehow at the edge of the scope of the present survey and thus only a very brief note with few references is given here to point out where the blood coagulation models come from and where the most detailed models are able to go in terms of prediction accuracy.

For a detailed understanding of the molecular mechanism of protein functions it is essential to know their 3D structure. Although some experimental tools are available,[60] an important insight can be gained using computational simulations. The underlying mathematical models are mainly based on some elementary physical principles including Newtonian laws of motion, classical electrostatics, or thermodynamics.

There are now several mathematical methods used for modeling at the sub-microscale biochemistry level. Starting from the atomistic scale, the most detailed information can be derived from the molecular dynamics models.

Molecular Dynamics (MD). These models were developed in the late 1950s [2], used later, e.g., in inorganic chemistry [198] and quickly entering in biochemical applications [166]. They are based on solving the Newtonian equations of motion for point masses (particles, i.e., atoms, molecules, etc.), following the Newton's second law:

$$m_i \frac{d^2 \boldsymbol{r}_i}{dt^2} = \boldsymbol{f}_i \qquad i = 1, \ldots, N \tag{7.1}$$

[60] e.g., Nuclear Magnetic Resonance (NMR) spectroscopy, X-ray crystallography.

The set of N particles is described by their three-dimensional position vectors $\boldsymbol{r}_i$. The acceleration of i-th particle is then driven by the corresponding force $\boldsymbol{f}_i$. This is a net force summarizing the effects of inter- and intra- molecular forces acting on each particle. These forces are usually expressed in the form of potentials $\mathcal{V}_i$ depending on the positions of all particles in the system:

$$\boldsymbol{f}_i = -\nabla \mathcal{V}_i(\boldsymbol{r}_1, \boldsymbol{r}_2, \ldots, \boldsymbol{r}_N) \tag{7.2}$$

For each particle the potential $\mathcal{V}_i$ accounts for all important non-bonded (e.g., external, the Lennard–Jones or the Coulomb potential) and bonded (e.g., related to stretching, angular or torsion intra-molecular deformations) interactions. The numerical evaluations of these potentials and of the corresponding forces are the most time-consuming part of the whole numerical solution procedure. Thus the potentials should be kept as simple as possible, but capturing all the major physical characteristics of interactions. Using these methods it is possible to study the spatial structure, kinematic behavior (e.g., folding) as well as some other functions (or dysfunctions) of enzymes [126, 141]. This kind of information concerning blood coagulation factors is of key importance in the mathematical modeling and experimental investigation of the coagulation process [238, 239].

MD models are very detailed, capable of highly accurate predictions, but computationally extremely expensive. So far they can only be applied to spatially very small systems (typically $N = 10^3$–10^8 particles, i.e., to regions with dimensions of nano-meters) to study their short-term (typically 100 ps–100 ns) evolution.[61] To avoid these severe spatio-temporal limits a number of simplified techniques have been developed and used so far. The family of ***Coarse-Grained MD*** models is based on the idea of creating larger pseudo-particles (e.g., whole molecules, their parts, or ensembles) with defined mean properties. By solving the equations of motion for these coarse-grained pseudo-particles, a significant amount of computational effort can be saved at the price of losing some micro-scale details in the simulations [247].

Besides the coarse-grained models, many further levels of simplifications are possible on the way from the most detailed fully deterministic atomic Molecular Dynamics simulations up to stochastic simulations of bulk properties. Among these intermediate simplified and hybrid models we mention the following ones:

Steered Molecular Dynamics (SMD). One of the techniques used to overcome the simulation time limitations relies on applying an external force to the studied model system and analyzing its response [118, 229]. This method was used to study some elementary elasticity properties of the clot in [147].

Poisson–Boltzmann (PB) model. A completely different approach is based on the (continuous) theory of electrostatic interactions [61, 107]. The key point in this method is to solve a Poisson type equation for electrostatic potential and use the result to reconstruct the spatial structure of macromolecules.

[61] See [237] for historical evolution and future trends in MD simulations complexity with respect to computational power growth.

Langevin Dynamics (LD). The molecules of the solvent (in which the molecules or proteins of interest are dissolved) are approximated using additional force terms expressing drag and random collisions associated with the thermal motions of the solvent molecules [19, 81]. The modified molecular dynamics equation can now be written as

$$m_i \frac{d^2 \boldsymbol{r}_i}{dt^2} = \boldsymbol{f}_i - \gamma \frac{d \boldsymbol{r}_i}{dt} + R(t) \qquad i = 1, \ldots, N_p \tag{7.3}$$

where the first additional term reflects the friction force while the second, time-dependent term $R(t)$, approximates the stochastic collision force. The exclusion of the solvent molecules from the simulated particle set reduces the number of numerical degrees of freedom of the problem and saves a significant part of the computational time.

Brownian Dynamics (BD). This method can be seen as a viscous limit of the Langevin Dynamics. The viscosity is assumed to be large and the inertial effects are suppressed [85].

Normal Mode Analysis (NMA). This method is based on the harmonic analysis of the system (e.g., protein) oscillations about its local minimum energy state. Based on experimental observations it is assumed that only the slowest harmonic modes of protein oscillations have some functional consequences. In the original implementation of NMA, the force field is required as in MD simulations. However, instead of solving the Newtonian equations of motion, an harmonic analysis is performed to find the slow, most important oscillation modes. For further details on this method, see, e.g., [84, 145, 156, 218]. Some simplifications were introduced using *Coarse Grained NMA* [18], *Elastic Network Model* [102], or *Essential Dynamics* [5, 236].

2. **Microscale Models**

In this case the coagulation models consider *microscale* objects, i.e., *cellular scale* matter with dimensions of the order of 1–10 μm. The need for modeling blood flow and coagulation at microscale has at least two different motivations.

Micromechanics of blood clotting. The importance of the role of RBCs, platelets, and other microparticles in the blood coagulation process can be better understood and captured by models that are actually resolving all these objects. The motion, deformation, aggregation, and adhesion dynamics as well as other microscopic behavior of blood cells can only be accurately described in this scale. So the first motivation is to gain detailed and high-resolution simulation results.

Blood coagulation in microvessels. The blood flow and thus also blood coagulation in microvessels, which are vessels with diameter comparable with the size of RBCs, cannot be accurately described by any macroscopic (continuous or statistical) model. The continuum hypothesis is no more valid at this scale and thus the use of models explicitly taking into account the blood cells is inevitable [70]. The second motivation is thus the necessity of adjusting the scale of the resolved object to the spatial size of the domain of interest.

The microscale models are typically built on Lagrangian tracking of red blood cells, platelets, and other cellular matter within the blood flow. As such, they are often closely related to blood rheology and cellular mechanics. Many of these models are used as microscale components of more general hybrid multiscale models that include also components from macroscopic continuum models. Here we only focus on some of the methods often used at the microscopic scale. Although the models are in principle very similar to each other, at least three major groups can be distinguished among them.

Euler–Lagrangian Particle Tracking methods (ELPT). Within this subclass of microparticle tracking methods we classify those where Lagrangian methods are used for tracing particles, while an independent Eulerian description is used for the fluid flow field. The particles are typically only coupled to the fluid flow by a *one-way coupling* scheme, i.e., particle trajectories depend on the fluid velocity field, which however is not affected by the presence and motion of the particles.

The fluid field is described by the Navier–Stokes like equations for incompressible, but possibly non-Newtonian fluid:

$$\nabla \cdot \boldsymbol{u} = 0 \tag{7.4}$$

$$\rho\left(\frac{\partial \boldsymbol{u}}{\partial t} + \boldsymbol{u} \cdot \nabla \boldsymbol{u}\right) = -\nabla p + \operatorname{div} \boldsymbol{T} \tag{7.5}$$

The stress tensor $\boldsymbol{T}$ is given by an appropriate rheological constitutive relation. The fluid velocity $\boldsymbol{u}(\boldsymbol{x}, t)$ and pressure $p(\boldsymbol{x}, t)$ are first computed (independently of particles' motion) and further used to evaluate the forces acting on each particle that follows the trajectory being governed by the Newtonian second law of motion:

$$m_p \frac{d\boldsymbol{v}}{dt}(\boldsymbol{x}', t) = \boldsymbol{F}^D + \boldsymbol{F}^P + \boldsymbol{F}^G \tag{7.6}$$

This means that the particle at position $\boldsymbol{x}'$ in time t having the velocity $\boldsymbol{v}(\boldsymbol{x}', t)$ and mass m_p accelerates due to the action of the drag force $\boldsymbol{F}^D$, pressure gradient force $\boldsymbol{F}^P$, and gravity force $\boldsymbol{F}^G$. These forces can be expressed in terms of particle parameters and actual fluid velocity field. For spherical particles with diameter d_p and density ρ_p the drag force can be expressed as

$$\boldsymbol{F}^D = \frac{1}{8}\pi \rho_p d_p^2 C_D (\boldsymbol{u}(\boldsymbol{x}', t) - \boldsymbol{v}(\boldsymbol{x}', t))|\boldsymbol{u}(\boldsymbol{x}', t) - \boldsymbol{v}(\boldsymbol{x}', t)| \tag{7.7}$$

Here the fluid velocity at the position $\boldsymbol{x}'$ is denoted by $\boldsymbol{u}(\boldsymbol{x}', t)$ while $\boldsymbol{v}(\boldsymbol{x}', t)$ stands for the velocity of the particle, as mentioned above. The drag coefficient C_D depends on the local Reynolds number Re_p of the relative motion of the particle as:

$$C_D = \frac{24}{\mathrm{Re}_p}\left(1 + 0.15\mathrm{Re}_p^{0.687},\right) \quad \text{where} \quad \mathrm{Re}_p = \frac{\rho|\boldsymbol{u} - \boldsymbol{v}|d_p}{\mu} \tag{7.8}$$

Using the local pressure gradient $\nabla p(\boldsymbol{x}', t)$ the corresponding force can be computed as

$$\boldsymbol{F}^P(\boldsymbol{x}', t) = -\frac{1}{6}\pi d_p^3 \nabla p(\boldsymbol{x}', t) \tag{7.9}$$

In a similar way the gravity force (due to gravity acceleration $\boldsymbol{g}$) acting on a particle with density ρ_p in a fluid with density ρ can be computed according to

$$\boldsymbol{F}^G(\boldsymbol{x}', t) = -\frac{1}{6}\pi d_p^3 (\rho - \rho_p)\boldsymbol{g} \tag{7.10}$$

Based on these forces and corresponding particle velocity $\boldsymbol{v}(\boldsymbol{x}', t)$, their positions can be updated from

$$\frac{d\boldsymbol{x}'}{dt} = \boldsymbol{v}(\boldsymbol{x}', t) \quad . \tag{7.11}$$

A simple implementation of this method was used, e.g., in [21]. A particle transport and deposition study for different variants of the carotid artery bifurcation was presented in [115]. For blood coagulation and atherosclerosis trajectories of particles and their near-wall residence times this method gives important results. These quantities were simulated using the above described approach, e.g., in [154, 155] or [132]. Slightly different modeling assumptions, based on Stokes background flow (see [165]), have been, used e.g., in [270] or [178].

Immersed Boundary methods (IB). This is probably the most commonly used method in biological multiphase microfluid mechanics. Similarly to the ELPT methods the basic idea is to solve the fluid phase (Eulerian) equations in parallel with the solid phase (Lagrangian) equations and perform a *two-way* coupling of these two phases via an extra solid–liquid (fluid–structure) force(s). For blood flows, the generic IB algorithm starts from the classical Navier–Stokes equations describing the flow of a viscous incompressible Newtonian[62] fluid.

$$\nabla \cdot \boldsymbol{u} = 0 \tag{7.12}$$

$$\rho\left(\frac{\partial \boldsymbol{u}}{\partial t} + \boldsymbol{u} \cdot \nabla \boldsymbol{u}\right) = -\nabla p + \mu \Delta \boldsymbol{u} + \boldsymbol{f}. \tag{7.13}$$

[62]This can easily be applied to non-Newtonian fluids, if necessary. However, for blood flows we usually assume the fluid phase being the blood plasma which is a Newtonian fluid under physiological conditions.

As usual ρ denotes the (constant, in this case) fluid density, $\boldsymbol{u}(\boldsymbol{x},t)$ is the fluid velocity depending on spatial coordinate $\boldsymbol{x}$ and time t, p is the pressure and μ stands for the dynamical viscosity of the fluid. The solid-to-fluid coupling force $\boldsymbol{f}(\boldsymbol{x},t)$ is expressed using the structural force $\boldsymbol{F}(\boldsymbol{x}',t)$ assigned to the solid phase point $\boldsymbol{x}'$ and interpolated to the fluid domain using a Dirac function[63] $\delta(\boldsymbol{x}-\boldsymbol{x}')$ as

$$\boldsymbol{f}(\boldsymbol{x},t) = \int_{\Omega_S} \boldsymbol{F}(\boldsymbol{x}',t)\,\delta(\boldsymbol{x}-\boldsymbol{x}')d\boldsymbol{x}'. \tag{7.14}$$

The integration over Ω_S denotes the domain occupied by the solid phase. This domain can either be a volume or just a membrane represented by fictitious particles connected by elastic links. Important is the (numerical implementation of) Dirac function that allows the interpolation of the force from the solid phase (moving) grid to the fluid phase (fixed) grid. The key point and most tricky part in this method is thus the appropriate parametrization of the structure force[64] $\boldsymbol{F}(\boldsymbol{x}',t)$.

In the inverse direction, i.e., for the fluid-to-solid coupling the link is established by assuming that the solid phase velocity $\boldsymbol{v}(\boldsymbol{x}',t)$ can be interpolated from the background fluid velocity field $\boldsymbol{u}(\boldsymbol{x},t)$

$$\boldsymbol{v}(\boldsymbol{x}',t) = \int_{\Omega_F} \boldsymbol{u}(\boldsymbol{x},t)\,\delta(\boldsymbol{x}-\boldsymbol{x}')d\boldsymbol{x}. \tag{7.15}$$

The integral over the whole fluid domain Ω_F is restricted by the Dirac function in order to take into account only the fluid velocity in the close proximity of the material point $\boldsymbol{x}'$. The solid phase particles positions $\boldsymbol{x}'$ are then simply updated using the velocity $\boldsymbol{v}(\boldsymbol{x}',t)$

$$\frac{d\boldsymbol{x}'}{dt}(t) = \boldsymbol{v}(\boldsymbol{x}',t) \qquad . \tag{7.16}$$

There exist numerous variants and different implementations of this basic method. The *Immersed Finite Element Method (IFEM)* was used in [150, 276] to model a wide range of biological systems at various scales. A lattice Boltzmann method implementation of the IB method was used, e.g., in [54, 258, 277].

In the context of blood microflows and coagulation, the IB method was used, e.g., in [190, 191] to simulate leukocytes rolling and adhesion. The erythrocytes aggregation and deformation have been studied using IFEM method in [148, 149]. Platelets interactions were studied, e.g., in [192]. The lattice Boltzmann version of IB method was recently used to simulate flexible fiber suspensions in [257] or

---

[63] approximated numerically in the simulations.

[64] See, e.g., [192] for platelet simulations.

to investigate platelet activation and blood damage in [259]. Platelets were also investigated using an IB model in [215], where different versions of immersed boundary representations of platelets are tested or as a part of a complex multiscale model in [78]. Aggregation of RBCs under shear flow was studied using IB in [268] or [16].

Discrete-Particle Methods (DPM). This subclass of Lagrangian tracking-based methods is characterized by using both real and fictitious particles to describe the whole fluid mechanics problem including the fluid, the immersed structure, or domain boundaries. In these methods the mesh generation is unnecessary for typical Eulerian fluid description. The motion of continuum is reconstructed from the assembly behavior of discretized particles carrying the information about physical quantities such as position, velocity, pressure, and density. Many different implementations of DPM are used for specific applications [37, 66, 135, 269]. Only few examples will be mentioned here to demonstrate the basic modeling principles and possible biomedical applications with special focus on blood coagulation.

(a) *Dissipative Particle Dynamics (DPD).* In this method the *dissipative particles* represent mesoscopic portions (e.g., clusters of molecules) of a real fluid [108]. These fictitious coarse-grained particles interact with the surrounding particles through elementary pair-wise forces. The particles motion is governed by the Newton's second law. For a given set of N particles having mass m_i, positions $\boldsymbol{r}_i$ and velocity $\boldsymbol{v}_i$, this can be written as:

$$\frac{d\boldsymbol{r}_i}{dt} = \boldsymbol{v}_i \qquad i = 1, \ldots, N \tag{7.17}$$

$$m_i \frac{d\boldsymbol{v}_i}{dt} = \sum_{j=1, j \neq i}^{N} \left(\boldsymbol{F}_{ij}^C + \boldsymbol{F}_{ij}^D + \boldsymbol{F}_{ij}^R\right) + \boldsymbol{F}_i^E. \tag{7.18}$$

The three main interaction forces applied in DPD are the conservative force $\boldsymbol{F}_{ij}^C$, the dissipative force $\boldsymbol{F}_{ij}^D$, and the random force $\boldsymbol{F}_{ij}^R$. The external force $\boldsymbol{F}_i^E$ represents, e.g., the contributions of the pressure gradient or gravity forces.

Let's denote by $\boldsymbol{r}_{ij} = (\boldsymbol{r}_i - \boldsymbol{r}_j)$ the vector connecting the particle j with the particle i, $r_{ij} = (\boldsymbol{r}_{ij} \cdot \boldsymbol{r}_{ij})^{1/2}$ being the magnitude of this vector and $\hat{\boldsymbol{r}}_{ij} = \boldsymbol{r}_{ij}/r_{ij}$ the corresponding unit vector. Similarly $\boldsymbol{v}_{ij} = (\boldsymbol{v}_i - \boldsymbol{v}_j)$ denotes the relative velocity of particles i and j. The forces are assumed to act only within a spherical cut-off[65] region with the characteristic radius r_c.

---

[65] In a similar way as in *Smoothed Particle Hydrodynamics (SPH)* [146,175] where the interpolation kernel is usually truncated to have a compact support. The SPH method differs significantly from many other particle methods because the equations of motion for the fictitious particles in SPH are derived directly from the partial differential equations of fluid mechanics by integration using an interpolation kernel [66].

Conservative force. This is the *repulsive* potential force acting between particles i and j.

$$\boldsymbol{F}_{ij}^{C}(r_{ij}) = \begin{cases} a\left(1 - \frac{r_{ij}}{r_c}\right)\hat{\boldsymbol{r}}_{ij} & r_{ij} \le r_c \\ \mathbf{0} & r_{ij} > r_c \end{cases} \tag{7.19}$$

Here a denotes the repulsive force coefficient corresponding to the maximum of this force at distance $r_{ij} = r_c$.

Dissipative force. Represents the *friction* force approximated by

$$\boldsymbol{F}_{ij}^{D}(r_{ij}) = -\gamma\, w_D(r_{ij})\,(\hat{\boldsymbol{r}}_{ij} \cdot \boldsymbol{v}_{ij})\hat{\boldsymbol{r}}_{ij} \tag{7.20}$$

where γ is the friction coefficient and w_D stands for the dissipative force amplitude weight function

$$w_D(r_{ij}) = \begin{cases} \left(1 - \frac{r_{ij}}{r_c}\right)^{2k} & r_{ij} \le r_c \\ 0 & r_{ij} > r_c \end{cases} \tag{7.21}$$

The exponent $k = 1/2$ is used in the standard DPD method, however other values are possible (e.g., $k = 0.25$ see [222]) to modify the diffusivity of the method.

Random force. Based on the Brownian motion of particles, the random forces can be expressed as

$$\boldsymbol{F}_{ij}^{R}(r_{ij}) = \sigma\, w_R(r_{ij})\,\xi_{ij}\, dt^{-1/2}\hat{\boldsymbol{r}}_{ij}, \tag{7.22}$$

where ξ_{ij} is a random variable with normal distribution, zero mean, and unit variance satisfying the symmetry $\xi_{ij} = \xi_{ji}$. The random force coefficient is linked to friction coefficient γ and absolute temperature T via the Boltzmann constant k_B by the relation $\sigma = \sqrt{2\gamma k_B T}$. The random force weight function is in this case given by

$$w_R(r_{ij}) = \begin{cases} \left(1 - \frac{r_{ij}}{r_c}\right)^{k} & r_{ij} \le r_c \\ 0 & r_{ij} > r_c \end{cases} \tag{7.23}$$

The random and dissipative forces form must satisfy the fluctuation–dissipation theorem so that the DPD model preserves the equilibrium temperature. This leads to the condition $w_D = w_R^2$ which is satisfied in the above case. Parametrization for external forces $\boldsymbol{F}_i^E$ as well as for various additional structural (bonding, torsion, adhesion, etc.) forces can be found in specialized literature.

The DPD method has been successfully used for simulations of blood coagulation related phenomena. For details on the implementation of the DPD model of thrombosis, see, e.g., [75,233]. The influence of RBCs on platelets aggregation was studied using DPD in [193] or more recently in [222].

(b) *Fluid Particle Model (FPM).* This method can be seen as an extension of the DPD method, i.e., the interactions in the FPM are modeled by forces of a finite range. In comparison with the DPD method the FPM allows the fluid particles to rotate in space and also the interaction range for FPM is usually shorter due to a more realistic interaction forcing model [65]. Some information about the theoretical background of this method can be found, e.g., in [184]. The blood coagulation related applications of this method can be found, e.g., in [36] simulating fibrin aggregation and blood flow in capillaries. The RBCs aggregation in capillary vessels was studied using FPM e.g., in [65].

(c) *Moving Particle Semi-implicit (MPS) method.* This method has been introduced in [135]. Its formulation is based on the Lagrangian form of the Navier–Stokes equations for a viscous incompressible fluid

$$\frac{D\rho}{Dt} = 0 \tag{7.24}$$

$$\rho \frac{D\boldsymbol{u}}{Dt} = -\nabla p + \mu \Delta \boldsymbol{u} + \boldsymbol{f}\,. \tag{7.25}$$

External forces (including inter- and intra-cellular bonding) are summed up in $\boldsymbol{f}$. The spatial gradients and Laplacians of quantities in MPS formalism are approximated in a specific way. Let's consider a scalar quantity ϕ assigned to a particle with index i. The gradient of this quantity is approximated by[66]:

$$\nabla \phi_i = \frac{\dim}{n_0} \sum_{j=1, j \neq i}^{N} \frac{\phi_j - \phi_{\min}}{r_{ij}} \hat{\boldsymbol{r}}_{ij} w(r_{ij}) \tag{7.26}$$

where dim is the spatial dimension of the considered model, i.e., $\dim = 2,\ 3$ in practical simulations. The initial (reference) particle number density is denoted by n_0. The $\phi_{\min}$ stands for the local discrete minimum of the quantity ϕ among the particles surrounding the one with index i (within the radius of influence r_c). The kernel (weighting) function $w(r_{ij})$ is defined by the following formula

$$w(r_{ij}) = \begin{cases} \left(\frac{r_c}{r_{ij}} - 1\right) & r_{ij} \leq r_c \\ 0 & r_{ij} > r_c \end{cases} \tag{7.27}$$

with r_c being the cut-off distance (radius of influence) assuring the compact support for particle interaction forces. Similarly, the Laplacian can be expressed as

---

[66] We use the same notation of particles position vectors as in the description of the DPD method.

$$\Delta\phi_i = \frac{2\,\text{dim}}{\lambda_i n_0} \sum_{j=1, j\neq i}^{N} (\phi_j - \phi_i)\, w(r_{ij}) \tag{7.28}$$

The coefficient λ_i is expressing the increase in the statistical diffusion for the distribution of physical quantities. It is calculated from:

$$\lambda_i = \frac{\sum_{j=1, j\neq i}^{N} (\phi_j - \phi_i)^2\, w(r_{ij})}{\sum_{j=1, j\neq i}^{N} w(r_{ij})} \tag{7.29}$$

These approximations are applied to velocity and pressure gradients and to Laplacians in a time-advancing algorithm with pressure-correction semi-implicit coupling [182]. For a more detailed description of the computational algorithm, see, e.g., [123].

From the blood flow and coagulation simulation point of view, the crucial part of the model is the realistic representation of platelets and RBCs. To simulate these blood cells, a specific form of the inter-particle external force $\boldsymbol{f}$ can be adopted

$$\boldsymbol{f} = \underbrace{\boldsymbol{f}^{\text{el}} + \boldsymbol{f}^{\text{ad}} + \boldsymbol{f}^{\text{ag}}}_{\text{platelets}} + \underbrace{\boldsymbol{f}^{s} + \boldsymbol{f}^{b}}_{\text{RBCs}} \tag{7.30}$$

to represent the platelets *elasticity* force $\boldsymbol{f}^{\text{el}}$, *adhesion* bonding force $\boldsymbol{f}^{\text{ad}}$, *aggregation* force $\boldsymbol{f}^{\text{ag}}$ and similarly for the RBCs elastic membrane *stretching* and *bending* forces $\boldsymbol{f}^{s}$ and $\boldsymbol{f}^{b}$.

For example, using the simple mechanical spring model, the platelets elasticity force[67] is simulated using the linear spring model defined by

$$\boldsymbol{f}_{ij}^{\text{el}} = k_{\text{el}} \left(r_{ij} - r_{ij}^{\text{el}} \right) \hat{\boldsymbol{r}}_{ij} \tag{7.31}$$

where k_{el} is the adhesion elastic (spring) force coefficient and r_{ij}^{el} is the natural length of the spring connecting particles i and j.

A similar model is used to evaluate the platelet-to-wall adhesion force, but with an important limitation. The adhesion of a platelet to an injured wall is only activated when a platelet comes within the distance r_{ad} from the injured wall.

$$\boldsymbol{f}_{ij}^{\text{ad}} = \begin{cases} k_{\text{ad}} \left(r_{ij} - r_{ij}^{\text{ad}} \right) \hat{\boldsymbol{r}}_{ij} & r_{ij} \leq r_{\text{ad}} \\ \boldsymbol{0} & r_{ij} > r_{\text{ad}} \end{cases} \tag{7.32}$$

[67] The net force acting on the particle i is $\boldsymbol{f}_i = \sum_{i\neq j} \boldsymbol{f}_{ij}$.

where again k_{ad} is the adhesion elastic (spring) force coefficient and r_{ij}^{ad} is the natural length of the spring.

Further details on the platelets and RBCs models within the MPS method framework can be found, e.g., in [123] or [124], where the mechanical interaction between a thrombus and red blood cells was studied. A more detailed MPS-based model of platelets adhesion dynamics under shear flow taking into account various receptors bonding forces was recently published in [125]. A specific problem of malaria infected RBCs flow in capillaries was studied using an MPS model in [116] or [134].

Cellular Potts Model (CPM). The CPM is a cell-level lattice model based on energy-minimization[68] following the ideas of [95]. The effective energy E of the system sums up the true energies, like the cell-to-cell adhesion associated energy, with other energy-like contributions, e.g., the effects of blood flow on cell or virtual energies arising from dimensions constraints and chemotaxis (see, e.g., [262, 263]).

$$E = E_{\mathrm{Adhesion}} + E_{\mathrm{Flow}} + E_{\mathrm{Dim}} + E_{\mathrm{Chem}} \tag{7.33}$$

From an effective energy the resulting cell motion can be calculated using algorithms based on the Monte–Carlo Boltzmann acceptance rule. The CPM uses integer indices defined on a lattice to describe cells. The value of the index (kind of marker) at a specific lattice site (in the position (i, j, k) within the cartesian lattice) is equal to C if the site lies in the cell C. The sets of lattice sites having the same index represent cells. The cell is thus treated as a set of discrete sub-cells that can rearrange to form a cell motion or shape changes.

The CPM is capable to predict, e.g., microscopic cell motion, aggregation and interaction of cells, their adhesion, differentiation or division. The CPM was used in [45–47] as a component of a multiscale model in three-dimensional morphogenesis simulations. Thrombus development was studied using CPM in [262, 263].

3. **Mesoscale Models**

The *mesoscale* in this context refers to models at the *sub-continuum scale*. They are classified in this group mainly due to their common modeling principles based on statistical methods, rather than to their spatial resolution. The particulate matter in blood is treated in ensembles, described, e.g., using particle probability densities and their spatial integrals. This statistical approach requires spatial control volumes being much larger than typical particle sizes or inter-particular distances. The mesoscale methods typically use information about microscopic particle interactions to provide information about macroscopic quantities.

Lattice Boltzmann Method (LBM). This method is often used as a discretization tool for numerical simulations of Navier–Stokes like models in fluid mechanics.

---

[68] It uses similar principles as the Monte Carlo method in sub-microscale models.

LBM simulates the flow using the evolution of fictitious microscopic particles living on a set of discrete lattice nodes. Their dynamics depend only on interactions between particles in neighboring lattice points. The LBM tracks particle distribution functions at points of a lattice, instead of tracking discrete particles. The LBM can also be seen as a discrete approximate solution method for a specific form of the continuous Boltzmann equation known in statistical mechanics [54].

The fictitious particles in LBM only live in lattice nodes, thus only discrete particle velocity (shift) vectors are allowed. These discrete velocities are denoted by $\boldsymbol{e}_i$. The spatio-temporal distribution of the particles having velocity $\boldsymbol{e}_i$ is described by a scalar function $f_i(\boldsymbol{x}, t)$. The evolution in time of these particle distribution functions is governed by:

$$f_i(\boldsymbol{x} + \delta t\, \boldsymbol{e}_i, t + \delta t) = f_i(\boldsymbol{x}, t) - \frac{1}{\tau}\big[f_i(\boldsymbol{x}, t) - f_i^{eq}(\rho, \boldsymbol{u})\big] \tag{7.34}$$

Here δt is the time-step, τ is the relaxation time. The equilibrium distribution $f_i^{eq}(\rho, \boldsymbol{u})$ depends on the macroscopic fluid velocity $\boldsymbol{u}$, on the density ρ, and of course on the structure of the computational lattice, resp. on the discrete velocities $\boldsymbol{e}_i$, see, e.g., [54, 258]. The macroscopic density $\rho(\boldsymbol{x}, t)$ and the velocity $\boldsymbol{u}(\boldsymbol{x}, t)$ at point $\boldsymbol{x}$ and time t can be evaluated using the following simple relations:

$$\rho(\boldsymbol{x}, t) = \sum_i f(\boldsymbol{x}, \boldsymbol{e}_i, t) \tag{7.35}$$

$$\rho(\boldsymbol{x}, t)\boldsymbol{u}(\boldsymbol{x}, t) = \sum_i f(\boldsymbol{x}, \boldsymbol{e}_i, t)\boldsymbol{e}_i \tag{7.36}$$

A more detailed description of the LBM method can be found, e.g., in [197]. In most cases the LBM is only used for (blood) flow simulations. Some applications also include coagulation related phenomena. For example, the platelet motion induced by RBCs was simulated in [54]. A study of fully resolved blood flow (including blood cells representation with IB method) through aneurysmal vessels using LBM was shown in [180]. Red blood cell aggregation and dissociation in shear flows is simulated by the lattice Boltzmann method in [278]. An interesting extension of LBM for deformable particles and flexible fibers was used in [257, 258]. A simple LBM-based model of thrombosis in intracranial aneurysms was used in [187]. A clotting initiation mechanism based on *particle residence time* was adopted in [101] and [27]. A blood damage model using the LBM approach was developed in [172].

Statistical methods for simulations in (bio)chemistry are much less common. An important contribution in this field was provided in [89, 90] where a model based on the exact stochastic simulation of coupled chemical reactions was introduced for well-stirred chemical systems.[69] A very comprehensive review of these stochastic

[69]See [93] for the relation between this stochastic approach and continuous deterministic reaction rate equations.

techniques can be found in [91] with further details in [92]. This approach forms a starting point, e.g., for the class of ***Kinetic Monte Carlo (KMC)*** methods. Variants of this technique were used, e.g., in [151] to simulate the blood coagulation initiation or in [140] to study the aggregation kinetics of platelets.

A mechanical model of fibrin polymerization based on probabilistic approach was used in [174]. It focuses on the polymerization processes under shear flow and thus it is assumed that platelets have already been activated and have released thrombin into the blood plasma, initiating fibrinogen polymerization into fibrin fibers. The density of connected fibrin links $n(\boldsymbol{x}, t)$ is described by the following transport equation

$$\frac{\partial n}{\partial t} + \boldsymbol{v}_{ri} \cdot \nabla n = K C_T f(r) \cdot (1 - n) - \overline{K} C_P \cdot n \tag{7.37}$$

The left-hand side of Eq. (7.37) describes an advection process governed by a relative (slip-like) velocity $\boldsymbol{v}_{ri}$ at the interface between fixed and moving parts of the polymeric network (resp. i.e., clot and blood). The K and $\overline{K}$ are the chemical reaction rates for cross-link formation and dissociation, respectively, and C_T, C_P are the concentrations of thrombin and plasmin. The $f(r)$ is a probability function (based on random walk theory) that describes the probability of finding a fibrin fiber link at distance r. More details can be found in [174], mentioning [127] as a basic reference.

Another important class of methods fitting into the mesoscale group is based on the ***Population Balance Equation (PBE)***. It is a general mathematical concept used to describe various aggregation and disaggregation processes in physics and technology. A simple population balance model of coagulation can be described by an integro-differential equation[70] for particle size distribution [136]. The density $s(v, t)$ of particles of size[71] v at time t changes according to:

$$\begin{aligned} \frac{\partial s(v,t)}{\partial t} &= \frac{1}{2} \int_0^v K(v-w, w) s(v-w, t) s(w, t) dw \\ &- s(v,t) \int_0^\infty K(v, w) s(w, t) dw \end{aligned} \tag{7.38}$$

The first integral on the right-hand side follows the idea that a particle of size v can only be created by aggregation during collision of particles with sizes $(v - w)$ and w. The second integral represents the situation when a particle of size v aggregates with another particle and thus it increases its size and has to be withdrawn from the particle count of v-sized particles. The aggregation rate K of these particles is described by the so-called *aggregation kernel* $K(v - w, v)$. It is usually a symmetric function of particle sizes of colliding particles. The specific form of this kernel

[70] This equation is also known as Smoluchowski coagulation equation [221].

[71] The size can be, e.g., particle volume, mass, or dimension.

depends on the physical nature of the collision process. It has a different form, e.g., for shear flow than for Brownian motion or gravitational settling. More details can be found, e.g., in [62, 63, 220] or in the review [3].

Such coagulation model based on coagulation Eq. (7.38) was used, e.g., in [24] for the analysis of shear-induced platelet aggregation. The coagulation kernel used in this model has a specific shear-dependent form

$$K(v, w) = \frac{G}{\pi}\left[v^{1/3} + w^{1/3}\right]^3 \tag{7.39}$$

with G being a shear rate and v, w the particles (platelet aggregates) volumes. Another implementation of the PBM model was, e.g., used to study the aggregation of tumor cells in [157].

This simplest version of the model describes a spatially homogeneous process of pure coagulation, i.e., just the aggregation of particles without any breakup or disaggregation. These processes can be added to the model equation, e.g., considering a *fragmentation kernel*[72] $F(v, w)$ and extending the model (7.38) by two additional integrals [83]:

$$\begin{aligned}\frac{\partial s(v,t)}{\partial t} &= \frac{1}{2}\int_0^v K(v-w, w)s(v-w, t)s(w, t)dw \\ &\quad - s(v,t)\int_0^\infty K(v, w)s(w, t)dw \\ &\quad + \int_0^\infty F(v, w)s(v+w, t)dw - s(v,t)\int_0^w F(v-w, w)dw\end{aligned} \tag{7.40}$$

The first breakup integral expresses the increase of the number of particles of size v due to disaggregation of larger particles of size $(v + w)$. The second integral reduces the count of v-sized particles due to their breakup into smaller particles of size $v - w$. This general breakup model can be simplified to the form investigated earlier in [169, 170]. This simplified version was used in the platelet aggregation studies [111–113].

Further generalization of population balance models leads to their spatially nonhomogeneous versions taking into account also the advection of particles in flow.[73] This extension of the coagulation balance equation is described, e.g., in [64] and the references therein.

Besides the above described continuous model a discrete variant is often used as in [140] or [3].

[72] Also called breakage function.

[73] A basis for this approach comes from atmospheric science where it was used to describe cloud formation [28].

4. Macroscale Models

The class of *macroscale* models is based on the *continuum* hypothesis and macroscopic description of blood with typical scales being larger than 100 μm. Although scales are similar to the previously introduced mesoscale models, the modeling assumptions, formalisms, and methods have a completely different (more macroscopic-oriented) basis. It is assumed that at this scale the single particles are not resolved, particulate character of blood is not essential and thus continuum description can be used without introducing large errors. The blood flow and the coagulation process are described using macroscopic variables and parameters like velocity, concentration, viscosity, or diffusivity. The simplest models within this class are dealing with concentrations of coagulation factors and other chemicals described using coupled systems of ODEs. At a more complex level the diffusivity and advection effects should be taken into account. The coagulation model is then written in the form of a large coupled system of nonlinear advection–diffusion–reaction equations. If blood flow is simulated alongside with the coagulation process, it can be described using Navier–Stokes like models. A very detailed review of recent developments and use of these models and of the corresponding numerical methods can be found, e.g., in [71] or [48].

Most of the information concerning the biochemical basis of macroscopic models was already included in the historical introduction part of the chapter. The mathematical description is covered in great detail in Sect. 7.7.2 on feature-based classification of coagulation models. This gives us a chance and space to present here some results of numerical simulations based on macroscopic models to demonstrate the kind of results and outputs that can be expected and obtained from the contemporary methods of this type. As examples, two models from recent works of the authors will be used.

Coagulation Cascade and Viscoelastic Flow Model

A typical modern macroscopic blood coagulation model was presented in [8, 9] and solved in a simplified form in [32, 34]. It is a model describing a coupled problem of blood flow and biochemistry of thrombus development in a three-dimensional vessel with simulated arterial wall damage. Basic elements of this model are described at several places in this chapter.

Blood Flow Model. The flow is described by a nonlinear shear-thinning viscoelastic model following the thermodynamic framework established in [199] and extended for blood flow in [7]. The set of governing equations is based on the conservation of mass (reduced to divergence-free constraint) and conservation of linear momentum for an incompressible fluid.

$$\nabla \cdot \boldsymbol{u} = 0. \tag{7.41}$$

$$\rho\left(\frac{\partial \boldsymbol{u}}{\partial t} + \boldsymbol{u} \cdot \nabla \boldsymbol{u}\right) = \nabla \cdot \left(-p\mathbf{1} + \eta\, \boldsymbol{B}_{\kappa_p(t)} + \mu_1 \boldsymbol{D}\right) \tag{7.42}$$

where $\boldsymbol{D} = (\boldsymbol{L} + \boldsymbol{L}^T)/2$ denotes the symmetric part of the velocity gradient tensor $\boldsymbol{L}$ and $\mathbf{1}$ stands for the identity tensor. The upper-convected time derivative of the elastic stretch tensor $\boldsymbol{B}_{\kappa_{p(t)}}$ is given by:

$$\overset{\nabla}{\boldsymbol{B}}_{\kappa_{p(t)}} = -\frac{1}{\tau(\boldsymbol{B}_{\kappa_{p(t)}})}\left[\boldsymbol{B}_{\kappa_{p(t)}} - \lambda\mathbf{1}\right] \tag{7.43}$$

Here the $\tau = \tau(\boldsymbol{B}_{\kappa_{p(t)}})$ defined by Eq. (7.44) has the dimension of time and plays a role similar to the relaxation time in the classical Oldroyd-B (Maxwell) model [33].

$$\frac{1}{\tau(\boldsymbol{B}_{\kappa_{p(t)}})} = 2K\left(\mathrm{tr}(\boldsymbol{B}_{\kappa_{p(t)}}) - 3\lambda\right)^n \tag{7.44}$$

The coefficient λ depends on the trace of the inverse of the tensor $\boldsymbol{B}_{\kappa_{p(t)}}$ according to

$$\lambda = \frac{3}{\mathrm{tr}\left(\boldsymbol{B}^{-1}_{\kappa_{p(t)}}\right)} \quad . \tag{7.45}$$

The remaining model coefficients for blood are taken exactly from [7]:

$$\mu_1 = 0.01\,\mathrm{Pa\,s}; \quad \eta = 0.1611\,\mathrm{N/m^2}; \quad n = 0.5859;$$
$$K = \left(\frac{\eta}{\alpha}\right)^{1+2n} = 58.0725\,\mathrm{s^{-1}}$$

Using the definition of the upper-convected time derivative,[74] the left-hand side can be rewritten in a more conventional form

$$\frac{\partial \boldsymbol{B}_{\kappa_{p(t)}}}{\partial t} + (\boldsymbol{u}\cdot\nabla)\,\boldsymbol{B}_{\kappa_{p(t)}} = -\frac{1}{\tau}\left[\boldsymbol{B}_{\kappa_{p(t)}} - \lambda\mathbf{1}\right] + \left[\boldsymbol{L}\boldsymbol{B}_{\kappa_{p(t)}} + \boldsymbol{B}_{\kappa_{p(t)}}\boldsymbol{L}^T\right] \tag{7.46}$$

where the coefficients λ and τ are scalar functions of the tensor $\boldsymbol{B}_{\kappa_{p(t)}}$ and its invariants.

Biochemistry Model. The biochemistry model is based on a coupled set of advection–diffusion–reaction (ADR) equations. It has been originally developed in [8] and further extended in [11]. It describes the spatio-temporal evolution of concentrations $[C_i]$ of 23 chemical constituents (enzymes, zymogens, proteins, etc.) and takes the form

$$\frac{\partial [C_i]}{\partial t} + \boldsymbol{u}\cdot\nabla[C_i] = \nabla\cdot(D_i\nabla[C_i]) + R_i \tag{7.47}$$

[74]See p. 555 for details.

The nonlinear chemical reaction terms R_i are mainly based on second order or Michaelis–Menten kinetics. As an example we mention the reaction term R_{Ia} in the equation for fibrin[75]:

$$R_{Ia} = \frac{k_1[IIa][I]}{K_{1M} + [I]} - \frac{h_1[\text{PLA}][Ia]}{H_{1M} + [Ia]} \tag{7.48}$$

The concentrations of *thrombin* ($[IIa]$), *fibrinogen* ($[I]$), *fibrin* ($[Ia]$), and *plasminogen* ([PLA]) are used to evaluate the reaction term R_{Ia}. The chemical kinetics rates k_1, h_1 and constants K_{1M}, H_{1M} are known (taken from [11]). The values of the diffusion parameters D_i and the exact form of the reaction terms R_i are given in [32], where the model has been for the first time implemented and used in 3D simulations. The clot formation is initiated by flux boundary conditions imposed within a small region on the internal surface of the vessel. No-flux, homogeneous Neumann conditions are imposed elsewhere.

Coupling Strategy. The coupling between blood flow and biochemistry is based on the fibrin concentration. It is described in detail elsewhere in this chapter.[76] The fluid viscosity μ_1 is multiplied by a non-dimensional factor $\tilde{\mu}_1$ that locally depends on fibrin concentration $[Ia]$.

$$\tilde{\mu}_1([Ia]) = \min\left\{1 + \frac{\mu^* - 1}{C_{\text{clot}}}[Ia], \mu^*\right\} \tag{7.49}$$

In this case the values $\mu^* = 100$ and $C_{\text{clot}} = 1{,}000\,\text{nM}$ were used. In practice, when the fibrin concentration is sufficiently large such a high viscosity is generated to immobilize the clot.

Numerical Methods. The system of governing equations is rather complex and highly nonlinear. A semi-discretization approach is adopted to first discretize the PDEs in space and then integrate in time the resulting system of ODEs. The same discretization is employed for flow variables, viscoelastic stress tensor, and concentrations in the biochemistry model.

We adopted a *space discretization* based on a simple central finite-volume discretization on a structured grid with hexahedral cells. The multiblock grid topology with wall-fitted cells was used. The viscous fluxes were also discretized using finite volumes over diamond-shaped cells adjoint to primary control volumes faces. See, e.g., [31, 33, 35] for more details.

The *time integration* was performed using a Runge–Kutta (RK) multistage scheme. A specific advection–diffusion optimized RK method has been used to reduce the computational cost. The basic idea behind this subclass of RK methods is to split the space-discretization operator into its inviscid and viscous parts. The

---

[75]The subscript Ia refers to the chemical notation for fibrin.

[76]See p. 554 for details.

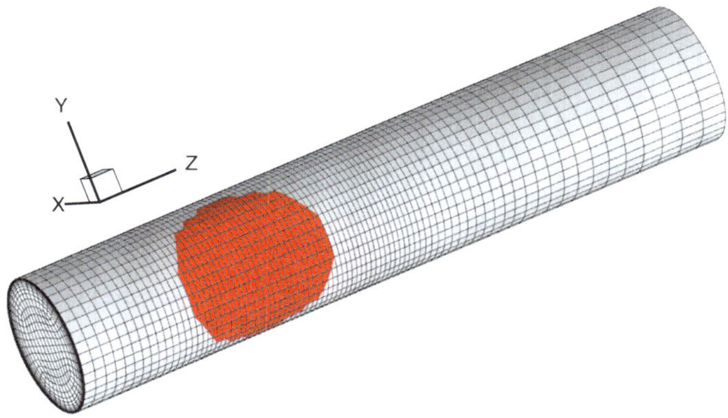

Fig. 7.11 Grid structure and clotting surface position

inviscid part is evaluated at every stage of the RK method, while the viscous fluxes are only evaluated in few stages. This corresponds to an operator splitting technique with different RK methods (coefficients) used for the advection step and another for the diffusion step. This allows to save several (very expensive) evaluations of diffusive fluxes per time-step while retaining the rather large stability region of the RK method. For details see [119, 120] or [33, 35].

Along with these two basic components of the numerical solver a specific *stabilization* technique was used to avoid nonphysical numerical oscillations arising from high solution gradients and due to the use of central discretization. The nonlinear TVD filter [68, 217] was used to smooth the concentration fields, as reported recently in [30].

Numerical Results. The numerical test case follows almost exactly the setups used in [32] and [34] where we refer for the complete parameter set for this simulation. The geometry represents a straight section of a blood vessel with diameter 6.2 mm and length 31 mm with grid shown in Fig. 7.11. The clotting surface is simulated in a region that is formed by the intersection of a sphere (with radius 3.1 mm) with the blood vessel wall.

The evolution in time of some of the coagulation factors can be observed in Fig. 7.12. The concentration is visualized in a single point located in the center of the clotting surface on the vessel wall. These graphs show the nature of the coagulation process, initially very fast, with a rather slow long-term evolution in the later phase.

The spatial extent of the clot can be shown using the contours of fibrin. Figure 7.13 shows the contours of fibrin concentration on the surface of the blood vessel. The velocity field is affected by the clotting which results in flow deceleration and reduction of velocity in the regions of high fibrin concentrations. A snapshot of the flow velocity reduction δ is shown in Fig. 7.14. It is evident that the core of the clot is rather stable in time, while the downstream concentration field varies in time as a consequence of the interaction with the flow of blood.

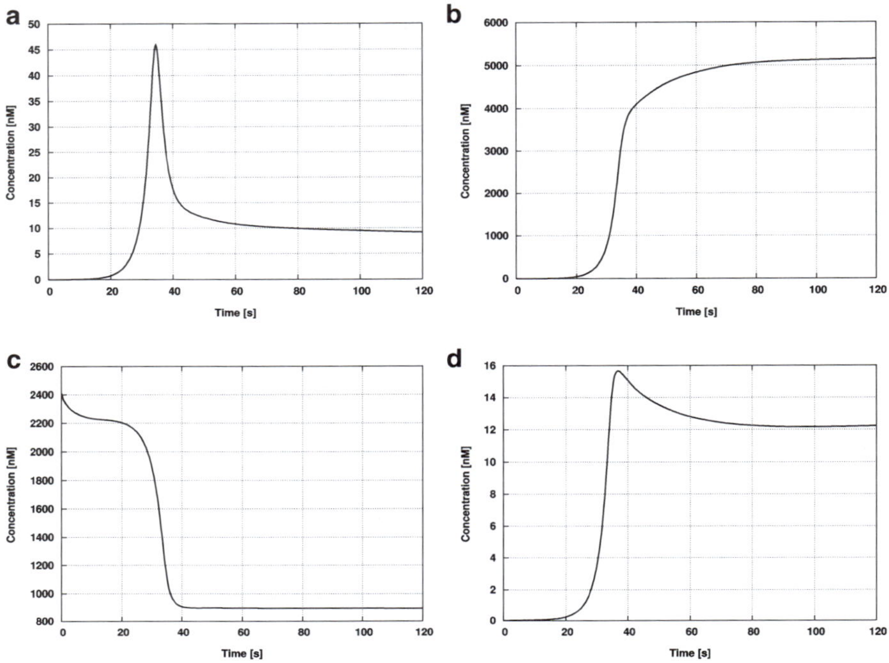

Fig. 7.12 Time evolution of selected concentrations in the center of the clotting surface during the initial 120 s. (**a**) Thrombin, (**b**) Fibrin, (**c**) Anti-Thrombin III, and (**d**) Tissue Plasminogen Activator

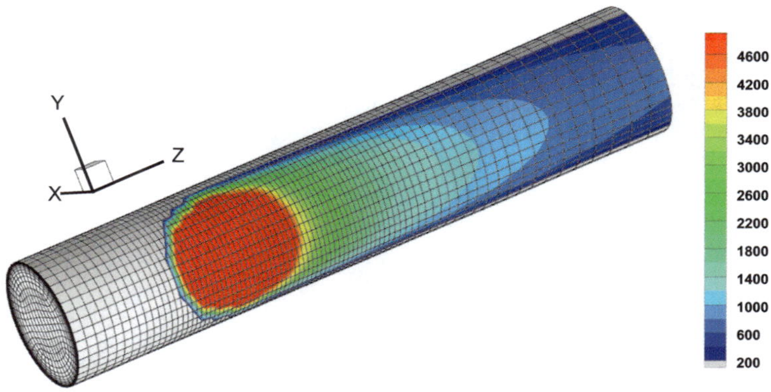

Fig. 7.13 Fibrin concentration contours at the vessel wall at the time 120 s

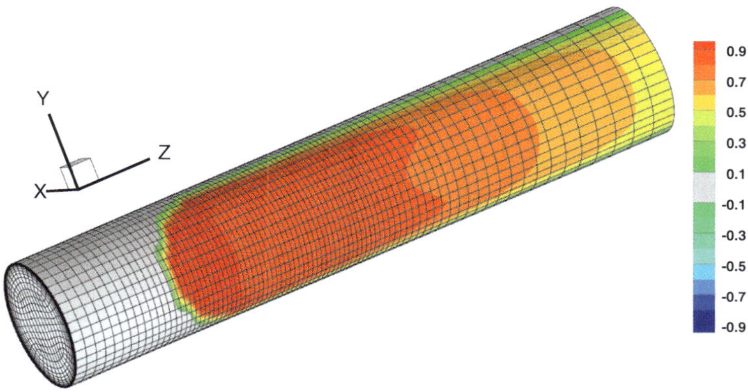

Fig. 7.14 Near-wall velocity magnitude reduction δ at the time 120 s

A longer time evolution of the clot and its effect on blood flow can be seen from Figs. 7.15–7.18. Besides the fibrin concentration on the deployed vessel surface,[77] a non-dimensional velocity reduction δ is defined

$$\delta(\boldsymbol{x}, t) = \frac{q(\boldsymbol{x}, 0) - q(\boldsymbol{x}, t)}{q(\boldsymbol{x}, 0)} \tag{7.50}$$

Here $q(\boldsymbol{x}, t)$ is the local velocity magnitude $q = \sqrt{u^2 + v^2 + w^2}$. The velocity reduction is evaluated in the first near-wall grid node, because evidently the no-slip condition is imposed on the vessel wall.

The time evolution of the clot is depicted in Fig. 7.17. The resulting blood flow velocity reduction is visualized in Fig. 7.18.

Simplified Cell Based Coagulation Model

Another macroscopic coagulation model was recently developed in [72]. Although it uses the same mathematical basis, i.e., a coupled set of ADR equations linked to a non-Newtonian blood flow model, as the previously mentioned complex coagulation model, its biochemical and biomechanical foundations are different. In comparison with the above described model the model mentioned here differs notably in several points:

- The biochemistry model follows the *cell-based model* formalism, rather than the traditional three-pathway scheme.
- The model is *simplified* in terms of number of ADR equations being solved (13 equations compared to 23 in the previous model).

[77]The dimensions are normalized using the vessel radius (half-diameter) $R = 3.1$ mm.

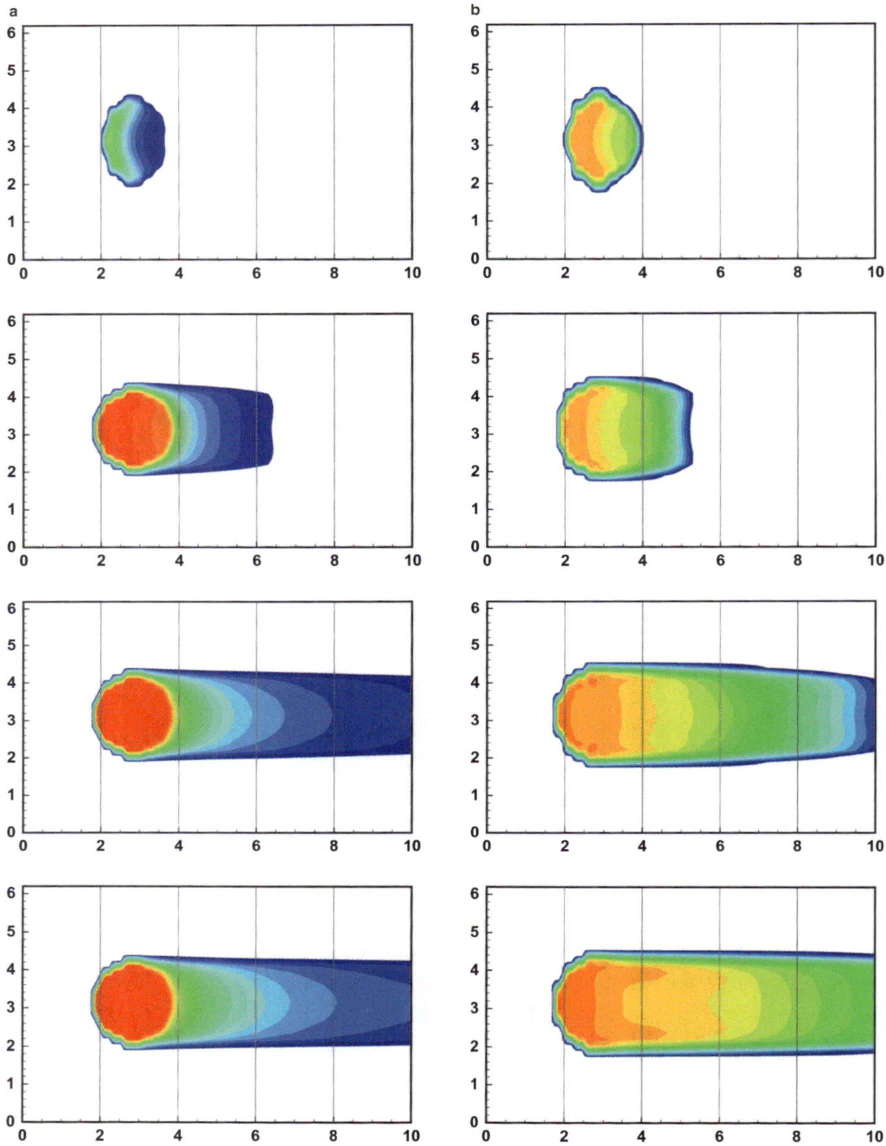

Fig. 7.15 Clot evolution during the initial 120 s. The snapshots at the times $t = 30, 60, 90, 120$ s are shown. (**a**) Fibrin concentration $[Ia]$, (**b**) blood velocity reduction δ

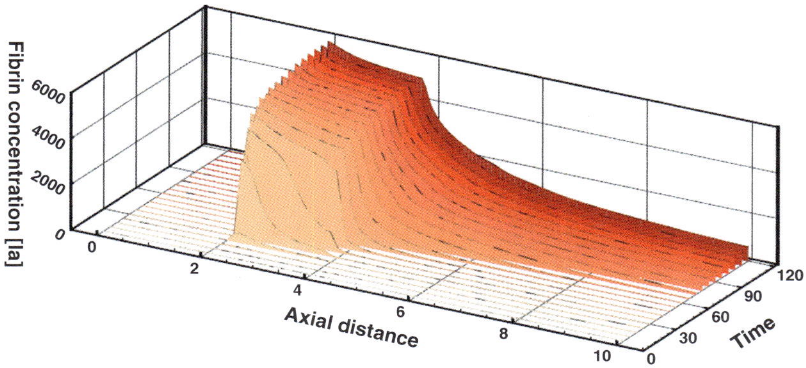

Fig. 7.16 Fibrin concentration evolution along an axial surface line during the initial 120 s

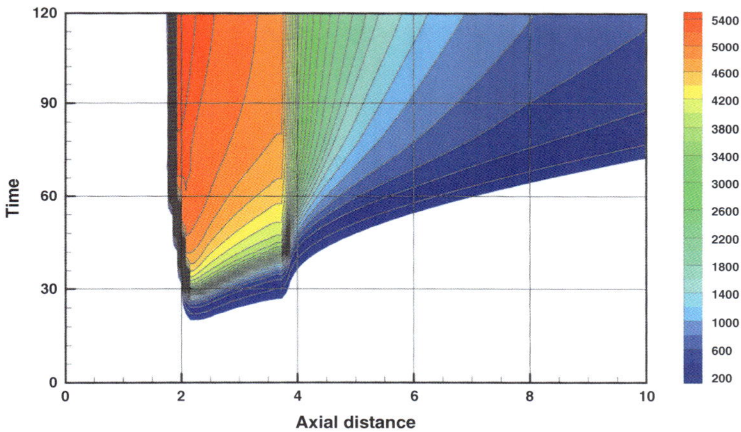

Fig. 7.17 Fibrin concentration $[Ia]$ along the axial surface line

- Coupled flow model is *generalized-Newtonian* (i.e., with variable, shear-dependent viscosity), compared to the more complex nonlinear viscoelastic model used in previous case.
- *Slip boundary conditions* are used here, showing that blood slip influence can be very important to the whole blood coagulation process, due to the consequent supply of activated platelets in the clot region. Physical motivations for the use of slip conditions in blood dynamics (usually disregarded in the recent literature) are illustrated in [72].

The type of results produced by this model is similar to those of the previous model, so only the basic features of this new model will be shown here.

Biochemistry Model. The biochemistry model can formally be written as a set of 13 ADR equations describing the *propagation* and *termination* phases of

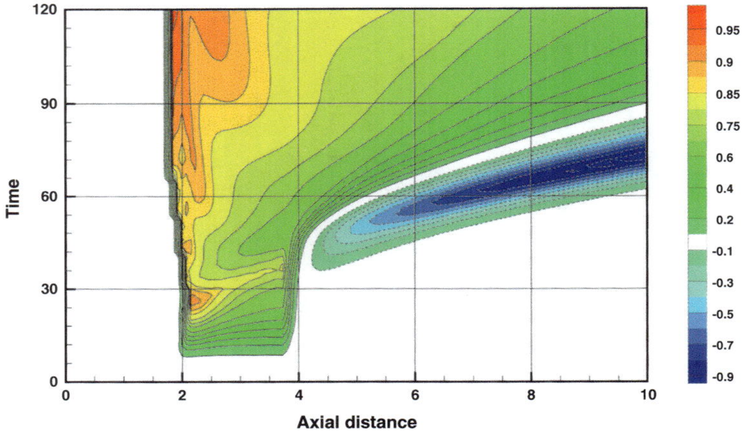

Fig. 7.18 Flow velocity reduction δ in the near-wall layer along the axial direction

coagulation as well as of the clot *lysis*.

$$\frac{\partial[C_i]}{\partial t} + \boldsymbol{u} \cdot \nabla[C_i] = \nabla \cdot (D_i \nabla[C_i]) + R_i \qquad i = 1, \ldots, 13 \tag{7.51}$$

The specific form of the corresponding reaction terms R_i is summarized below in Eqs. (7.52)–(7.64). These equations can be grouped according to which coagulation phase they describe. The *initiation* and *amplification* phases are modeled explicitly using an equivalent chemical reaction (7.52). This equation for the prothrombinase[78] concentration $[W]$ does not represent a chemical reaction in the classical sense, but rather a *virtual reaction* synthesizing the positive feedback loop prothrombinase-thrombin-prothrombinase in combination with the thrombin production described below in (7.53).

$$R_W = k_W C_P [IIa]\left(1 - \frac{[IIa]}{[IIa]_{\max}}\right) - h_{1W}[\text{APC}][W] - h_{2W}[ATIII][W] \tag{7.52}$$

The rate constants in (7.52) have to be carefully adjusted to reproduce results comparable with the experimentally observed behavior of the realistic biochemical cascade.

[78]We should clarify that the simplification introduced assigning a pivotal role to prothrombinase and summarizing in one equation the complex process leading to its production makes sense only in the framework of the normal physiological process. If we have to consider any type of pathology referring, e.g., to a defective or missing factor, the model has to incorporate the equations involving the dynamics related to that specific factor. In other words, the model is conceived in an elastic way, adapting the number of equations to the complexity that needs to be taken into account.

The consequent *propagation* phase is described by Eqs. (7.53)–(7.56):

$$R_{II} = -\frac{k_2[W][II]}{K_{2M} + [II]} \tag{7.53}$$

$$R_{IIa} = -R_{II} - h_2[IIa][ATIII] \tag{7.54}$$

$$R_I = -\frac{k_1[IIa][I]}{K_{1M} + [I]} \tag{7.55}$$

$$R_{Ia} = -R_I - \frac{h_{1a}[\text{PLA}][Ia]}{H_{1M} + [Ia]} \tag{7.56}$$

The *termination* of clotting is described by (7.57)–(7.60):

$$R_{ATIII} = -h_2[IIa][ATIII] - h_{2W}[W][ATIII] \tag{7.57}$$

$$R_{\text{PC}} = -\frac{k_{\text{PC}}[IIa][I]}{K_{\text{PCM}} + [I]} \tag{7.58}$$

$$R_{\text{APC}} = -R_{\text{PC}} - h_{\text{PC}}[\text{APC}][\alpha_1 AT] - h_{1W}[\text{APC}][W] \tag{7.59}$$

$$R_{\alpha_1 AT} = -h_{\text{PC}}[\text{APC}][\alpha_1 AT] \tag{7.60}$$

while the clot *lysis* is modeled by (7.61)–(7.64)

$$R_{tPA} = 0 \tag{7.61}$$

$$R_{\text{PLS}} = -\frac{k_{\text{PLA}}[tPA][\text{PLS}]}{K_{\text{PLAM}} + [\text{PLS}]} \tag{7.62}$$

$$R_{\text{PLA}} = -R_{\text{PLS}} - h_{\text{PLA}}[\text{PLA}][\alpha_2 AT] \tag{7.63}$$

$$R_{\alpha_2 AT} = -h_{\text{PLA}}[\text{PLA}][\alpha_2 AT] \tag{7.64}$$

For the full description of the biochemistry model, see the original paper [72] and the references therein.

Flow Model. The flow is simulated using the generalized Newtonian model similar to the one described in Sect. 7.7.2 using Eqs. (7.118), (7.119). With respect to the classical Newtonian fluid case the model is altered to account for shear rate $\dot{\gamma}$ and fibrin concentration $[Ia]$ dependence in both, viscosity ρ and density μ. The corresponding modified momentum equations can be written as:

$$\rho([Ia])\left(\frac{\partial \boldsymbol{u}}{\partial t} + \boldsymbol{u}\cdot\nabla\boldsymbol{u}\right) = -\nabla p + \nabla\cdot\left[\mu(\dot{\gamma},[Ia])\left(\nabla\boldsymbol{u} + \nabla\boldsymbol{u}^T\right)\right] \tag{7.65}$$

The shear-rate $\dot{\gamma}$ dependent (shear-thinning) fluid viscosity $\mu(\dot{\gamma})$ can be defined, e.g., using the *generalized Cross model* (7.119). The change in values of ρ and μ depending on the fibrin concentration $[Ia]$ is governed by a Heaviside jump function $H([Ia])$:

$$H([Ia]) = \begin{cases} 0 & \text{for} \quad [Ia] \le [Ia]^* \\ 1 & \text{for} \quad [Ia] > [Ia]^* \end{cases} \tag{7.66}$$

The step change[79] between blood and clot values appears when the fibrin concentration reaches a certain critical value $[Ia]^*$:

$$\rho([Ia]) = \rho_{\text{Blood}} + (\rho_{\text{Clot}} - \rho_{\text{Blood}})H([Ia]) \tag{7.67}$$

$$\mu([Ia]) = \mu_{\text{Blood}} + (\mu_{\text{Clot}} - \mu_{\text{Blood}})H([Ia]) \tag{7.68}$$

One of the original features of this model is the choice of *slip boundary conditions* on vessel walls (and clot boundaries). This choice is quite unusual in classical continuum (bio-)fluid mechanics, but it is quite natural, given the particulate microstructure of blood resembling rather particle laden flows, well known from industrial and environmental applications. This kind of boundary condition can be characterized by a vanishing wall-normal component of the velocity on the wall, while the tangent velocity is proportional to the corresponding stress component. This can be summarized as:

$$\boldsymbol{u} \cdot \hat{\boldsymbol{n}} = 0 \qquad \text{normal component} \tag{7.69}$$

$$\big(\boldsymbol{S}(\boldsymbol{u}, p)\hat{\boldsymbol{n}}\big) \cdot \hat{\boldsymbol{t}} = \beta \boldsymbol{u} \qquad \text{tangential component} \tag{7.70}$$

Here the $\boldsymbol{S}(\boldsymbol{u}, p)$ is the stress tensor including the spherical part coming from pressure p. The normal and tangent unit vectors are denoted by $\hat{\boldsymbol{n}}$ and $\hat{\boldsymbol{t}}$. The adjustable slip coefficient is denoted by β.

The slip condition is responsible for faster near-wall flow and thus the enhanced supply of platelets needed to build the clot as mentioned above. This fact, together with the explicit simulation of the clot initiation allows to impose no-flux[80] boundary conditions for concentrations of all simulated chemicals at the whole vessel wall. Additional arguments supporting this choice can be found again in [72].

5. **Multiscale Models**

The computational requirements associated with the fundamental micro- and sub-microscale models are so far very high. This is one of the main arguments in trying to couple some of the computationally cheaper, larger scale models with those very expensive finer scale models. Typical multiscale computational methods couple several elementary modules, such as fluid mechanics, coagulation cascade,

[79]In contrast to a linear, continuous change in the previous model.

[80]I.e. a homogeneous Neumann condition in the context of this model.

cell mechanics, and receptor–ligand binding, where each of these modules focuses on a specific range of the length scales.[81]

Multiscale models became standard in mathematical modeling of blood coagulation. Most of the currently developed models use some of the forms of multiscaling. There is an extensive body of literature published on this subject in the last couple of years. We don't aim to give a complete overview in this section, but rather point out some of the recent models and associated techniques with few basic references.

An overview of multiscale models with application to platelet adhesion and thrombogenesis was published recently, e.g., in [76, 246], and [267]. There are several examples of successful application of multiscale models in practical applications. In [262] a model of thrombus growth was built from *macroscopic* models for flow dynamics (Navier–Stokes equations) and chemical coagulation cascade (advection–diffusion–reaction equations) while cells are simulated using *mesoscopic* a stochastic model (Cellular Potts Model). This model was used to study the effects of pulsatile and non-Newtonian behavior of flow on growing thrombus [263]. A porous media model was added to simulate the macroscopic effects of thrombus on flow in [264, 265]. Another class of multiscale models was already discussed in the section devoted to *immersed boundary* and Lagrangian *particle tracking* microscale methods (see the references therein). For example, a model of thrombus formation composed from *macroscopic* continuum-based fluid flow module coupled with *microscopic* particle (blood cells) tracking sub-model was published, e.g., in [21] or [233]. A completely different implementation based on the same sub-model splitting was used in [168] where the Lattice Boltzmann Method (LBM) was applied to account for the plasma motion, and a modified molecular dynamics scheme was used for the cellular motion. An extensive review of other particle-based multiscale methods for optimization of cardiovascular implants can be found in [173, 269]. Some theoretical aspects of such atomistic-continuum models coupling were discussed in [241]. The *microscale* discrete stochastic Cellular Potts Model was coupled with a macroscale advection–diffusion–reaction model in [45]. An interesting three-scale algorithm starting from *sub-microscale* molecular dynamics model with *microscale* dissipative particle dynamics and macroscopic Navier–Stokes models was presented in [73].

Many other references can be given here concerning multiscale models. The above list is, however, sufficiently representative and convincing to see that the multiscale models are the most comprehensive mathematical modeling tools available for investigation of blood flow and coagulation. They represent not only a rational compromise between computational cost and accuracy, but they are also very efficient and versatile tools capable to focus on specific sub-problems in the blood coagulation process.

---

[81] We don't speak here about time-scales explicitly, however it is clear that each of the sub-models, depending on its resolved spatial scale, has an associated time scale that is able to treat (within a reasonable computational time and accuracy).

7.7.2 Feature-Based Classification of Coagulation Models

Different coagulation models include different features, i.e., they take into account different physical phenomena which results in more or less complete mathematical models. Besides models for the biochemistry, additional models for the flow, structure and other physical phenomena can be included. The reason for including (resp. omitting) certain physical phenomena in a full model is mainly due to economical reasons, i.e., based on the computational efficiency and accuracy of the considered model. To some extent it also reflects the historical evolution of coagulation models with visible trend to include more features and in the most recent state-of-the-art models. In this section we will mainly, but not exclusively, focus on the *macroscopic continuum models*, as many of the other biochemistry-only sub-micro and microscale models were already described.

1. **Biochemistry (Only) Models**

Within this class, two separate subgroups of models can be recognized, based on whether they allow just for temporal or also spatial variations of chemical fields.

Spatially Homogeneous Models

This class of models deals only with the chemical part of the coagulation process. The stoichiometric equations used in classical chemistry are translated into the mathematical language of ordinary differential equations describing the evolution in time of concentrations $C_i(t)$ of various components taking part in the coagulation process.

$$\frac{dC_i}{dt} = R_i(C_1, C_2, \dots, C_N) \quad i = 1, \dots, N \tag{7.71}$$

These equations are called the ***Reaction Rate Equations (RRE)***. The corresponding models are sometimes referred to as zero-dimensional[82] models, because all spatial variations of concentrations are excluded and thus no diffusion (or convection) processes are taken into account. The functions R_i are in general nonlinear and can depend on all considered concentrations. Typical forms are the *first-order*, *second-order*, or *Michaelis–Menten* kinetics terms:

$\alpha C i$	first order kinetics
$\alpha C i\ C_j$	second order kinetics
$\dfrac{\alpha C i\ C_j}{\beta + C_j}$	Michaelis–Menten kinetics

The rate constants α, β in all reaction terms have to be determined experimentally.

[82]In contrast with spatially two- or three-dimensional models.

One of the first modern coagulation models of this type was created in 1964 in [158] based on the waterfall enzymatic cascade. This model was later formulated mathematically in [144] and studied analytically as a biochemical amplifier system. The resulting cascade of linear ODEs is built on very simple assumptions. A finite sequence of N reactions, where at the ith stage the proenzyme (zymogen) with concentration y_i is converted to an active enzyme with concentration y_{ia} using the enzyme concentration $y_{(i-1)a}$ produced in the previous $(i-1)$th stage of the cascade. The rate at which the active enzyme is produced is denoted by k_i while the same enzyme is destroyed at a rate K_i.

$$\begin{aligned} \frac{dy_{1a}}{dt} &= k_1 y_1 I(t) \\ \frac{dy_{2a}}{dt} &= k_2 y_2 y_{1a} - K_2 y_{2a} \\ &\vdots \\ \frac{dy_{Na}}{dt} &= k_N y_N y_{(N-1)a} - K_N y_{Na} \end{aligned} \tag{7.72}$$

The initial impulse for this reaction cascade is provided by the time-dependent function $I(t)$. This model is so simple that it can be (under certain assumptions) solved analytically. This basic variant was later generalized to allow for time-dependent concentrations of proenzymes [179], or by adding negative feedback loops [163].

As a follow-up of these pioneering linear models a new wave of more realistic nonlinear coagulation models was started by Khanin and Semenov [129]. This simple nonlinear coagulation model represents the extrinsic coagulation pathway consisting of a cascade of four chemical reactions with a single nonlinear feedback loop.

This model can be written in the form of a system of ODEs for the concentrations[83] of coagulation factors $VIIa$, Xa, Va, IIa:

$$\frac{d[VIIa]}{dt} = k_1\alpha - K_1[VIIa] \tag{7.73}$$

$$\frac{d[Xa]}{dt} = k_2[VIIa] - K_2[Xa] \tag{7.74}$$

$$\frac{d[Va]}{dt} = k_3[IIa] - K_3[Va] \tag{7.75}$$

$$\frac{d[IIa]}{dt} = k_4[Xa]\frac{[Va]}{K_a + [Va]} - K_2[IIa] \tag{7.76}$$

[83] Square brackets are used to distinguish the concentrations of the corresponding chemicals from their names.

The coagulation cascade is initiated by a chemical substance[84] with concentration α, called stimulation intensity.[85] It is possible to show that there exists a threshold value α_{THR} for this stimulation intensity. For sub-threshold stimulation there only exists a single steady solution in which all concentrations vanish, with the exception of $[VIIa]$, while for higher stimulation an extra nonzero (i.e., positive) steady state solution exists. The threshold value α_{THR} is fully determined by the model coefficients:

$$\alpha_{\text{THR}} = K_a \frac{K_1 K_2 K_3 K_4}{k_1 k_2 k_3 k_4} \tag{7.77}$$

Many other models of this type have been developed in the past and are still used nowadays. After the success of the above simple nonlinear model, a more complicated one was published in [253] with six equations combining linear first order kinetics with second order and Michaelis–Menten kinetics reaction terms. This model was capable to simulate the whole coagulation pathway starting from extrinsic part up to fibrin production. The paper also provided valuable comparison with experimental data. The same part of coagulation cascade was simulated later in [122] using already 20 coupled ODEs. Even a more complicated model for Tissue Factor activated coagulation was presented in [142] including already 36 equations to simulate the serine protease inhibition. To study the contact activation (intrinsic initiation) of blood coagulation, a model with nine equations was proposed and theoretically studied in [196]. A special mathematical model with 35 equations was proposed in [133] for the analysis of Activated Partial Thromboplastin Time (APTT) commonly used as a laboratory test for diagnosis of blood coagulation disorders. To include also the role of platelets a simple model with 6 equations was developed in [260] assuming that the concentration of platelets is a function of thrombin concentration. One of the most commonly used models was published in [105] for the study of stoichiometric regulation of blood coagulation in extrinsic (and common) pathway cascade. It describes the evolution of 34 species and contains 42 rate constants. This model was used, e.g., in [43] to evaluate the significance of the circulating Factor IXa, in [40] for Factor Xa generation by computational modeling, or in [41] to model thrombin generation and risk of disease.

The above list of models is incomplete and still open. Many other models are used in their spatially nonhomogeneous version or as a part of more complex multi-scale or multi-phenomena models. Some of those models are mentioned in other parts of this overview.

---

[84]released from an injured vessel wall (Tissue Factor TF)

[85]From the biochemical point of view the first term on the right-hand side of (7.73) has a dubious interpretation. Actually VIIa is present in small quantities but its production has to come from VII and is mediated by VIIa itself (positive loop), in addition to the Xa coming out at the initiation stage. All this is replaced by a constant stimulus α. Thus it should be kept in mind that this model is a shortcut. Actually, it bypasses the action of the tenase complex, which in turn involves VIIIa and IXa.

In addition to the progress in the medical understanding of coagulation processes via mathematical models, an important effort has also been devoted to the study of the corresponding mathematical models. These theoretical aspects of the coagulation models are studied, e.g., in [23, 55, 99, 179, 196, 260, 261] or [213].

The kinetic models based on RRE have proven to be a valuable tool in studying the blood coagulation chemistry. One of the major drawbacks of these models lies in their high requirement when it comes to rate constants. These are numerous and have to be determined from data obtained in laboratory experiments. So the relatively low computational cost is balanced by the high amount of laboratory experiments to assembly and calibrate the model. A critical review of this class of models has been recently published in [103].[86]

Some of these disadvantages concerning the number of parameters to be supplied to such RRE-based models can be solved by using ***stochastic simulations*** of coupled chemical reactions. These are, however, computationally much more expensive and still not suitable for simulations of spatially nonhomogeneous large-scale problems. They can, however, help in deeper understanding of the chemistry and can replace some of the laboratory supplied data by outputs from microscale simulations.

Spatially Nonhomogeneous Models

The above described zero-dimensional spatially homogeneous models can be extended for variable in space concentration fields $C_i(\boldsymbol{x}, t)$. This allows to take into account the clot growth including diffusion phenomena[87] in two- or three-dimensional models. Instead of ODEs, a set of coupled reaction–diffusion PDEs can be built:

$$\frac{\partial C_i}{\partial t} = R_i(C_1, C_2, \dots, C_N) + \nabla \cdot \big(D_i \nabla C_i\big) \qquad i = 1, \dots, N \tag{7.78}$$

In the diffusion part of the model it is assumed that the diffusive flux is proportional to the concentration gradient ∇C_i by the diffusion coefficient D_i. This very simple generalization of the purely chemical system leads to many interesting results concerning the clotting process including the information about clot growth and local structure.

One of the earliest models of this type was published in [273, 274] where a model of the intrinsic coagulation pathway was built on tracking the spatio-temporal dynamics of activation of factors XI, IX, X, II, I, VIII, V, and protein C. In the spatially homogeneous version the model is described by the following set of

---

[86]This paper by H.C. Hemker et al. is entitled *Is there value in kinetic modeling of thrombin generation? No (unless...)*. This review presents a summary of models showing the number of reactions they take into account, using the number of rate constants, based on the amount of cited papers. The rather critical point of view from this paper is balanced by another paper in the same issue of the journal written by K.G. Mann, under an almost identical title *Is there value in kinetic modeling of thrombin generation? Yes* [159].

[87]in one or more spatial dimensions

8 ODEs for concentrations of factors IXa, Xa, IIa (thrombin), II (prothrombin), VIIIa, Va, APC (activated protein C), and Ia (fibrin).

$$\frac{\partial [IXa]}{\partial t} = k_9[XIa] - K_9[IXa] = R_{IXa} \tag{7.79}$$

$$\frac{\partial [Xa]}{\partial t} = k_{10}[IXa] - K_{10}[Xa] - \overline{k}_{10}[Z] = R_{Xa} \tag{7.80}$$

$$\frac{\partial [IIa]}{\partial t} = k_2[Xa]\frac{[II]}{k_{2m} + [II]} - K_2[IIa] + \overline{k}_2[W]\frac{[II]}{\overline{k}_{2m} + [II]} = R_{IIa} \tag{7.81}$$

$$\frac{\partial [II]}{\partial t} = -k_2[Xa]\frac{[II]}{k_{2m} + [II]} - \overline{k}_2[W]\frac{[II]}{\overline{k}_{2m} + [II]} = R_{II} \tag{7.82}$$

$$\frac{\partial [VIIIa]}{\partial t} = k_8[IIa] - K_8[VIIIa] - k_a[\mathrm{APC}]([VIIIa] + [Z]) = R_{VIIIa} \tag{7.83}$$

$$\frac{\partial [Va]}{\partial t} = k_5[IIa] - K_5[Va] - k_a[\mathrm{APC}]([Va] + [W]) = R_{Va} \tag{7.84}$$

$$\frac{\partial [\mathrm{APC}]}{\partial t} = k_{apc}[IIa] - K_{apc}[\mathrm{APC}] = R_{\mathrm{APC}} \tag{7.85}$$

$$\frac{\partial [Ia]}{\partial t} = k_1[IIa] = R_{Ia} \tag{7.86}$$

Concentrations of Z (tenase) and W (prothrombinase) were calculated from

$$[Z] = \frac{k_{8,9}[VIIIa][IXa]}{K_{8,9} + k_a[\mathrm{APC}]} \tag{7.87}$$

$$[W] = \frac{k_{5,10}[Va][Xa]}{K_{5,10} + k_a[\mathrm{APC}]} \tag{7.88}$$

In the original paper this model was implemented in a one-dimensional form, with a single spatial variable x. But it can easily be written in a multi-dimensional form using the coupled set of equations (7.78), where the reaction terms R_i correspond to the right-hand side of Eqs. ((7.79)–(7.86)). For example, the equation for the spatio-temporal evolution of factor IXa takes the form:

$$\frac{\partial [IXa]}{\partial t} = k_9[XIa] - K_9[IXa] + \nabla \cdot \left(D_{IXa}\nabla[IXa]\right) \tag{7.89}$$

The impulse for the initiation of the clotting cascade is provided by flux boundary conditions[88] written in the form of homogeneous/nonhomogeneous Neumann

[88]For homogeneous models the *Initial Value Problem* is solved subject to initial data. By adding spatial variability to concentrations, the *Initial Boundary Value Problem* has to be solved using both, initial and boundary conditions.

boundary conditions simulating healthy/injured vessel wall. According to [273], the activation stimulus is provided via an extra diffusion equation for factor XIa:

$$\frac{\partial[XIa]}{\partial t} = k_{11} + \nabla \cdot \big(D_{XIa}\nabla[XIa]\big) \tag{7.90}$$

supplemented by boundary conditions

$$\frac{\partial[IXa]}{\partial \hat{\boldsymbol{n}}} = \begin{cases} A & \text{on the clotting surface} \\ 0 & \text{elsewhere} \end{cases} \tag{7.91}$$

where $\hat{\boldsymbol{n}}$ is the boundary surface outer normal unit vector and the number (or function) A is the prescribed activation.

The diffusion–reaction (without advection) models are usually studied to clarify the spatio-temporal dynamics of clotting and pattern formation in quiescent blood. To do so, the coagulation models are often simplified to allow at least partial analytical solution or mathematical analysis [14, 275]. Some generalizations were also introduced, e.g., to take into account spatially nonuniform diffusion coefficients in a simple model of fibrin polymerization [153]. On the other hand, a much more complicated model of this type was published in [189]. This model was developed to describe the spatial propagation and localization of blood coagulation and their regulation in intrinsic and protein C pathways. The whole model, however, consists of almost 30 differential equations containing more than 100 parameters. This model was used, e.g., in [230] to study the spatial dynamics of contact-activated fibrin clot formation.

The diffusion coefficients D_i enter the model as "tuning parameters" and usually have to be determined from experiments, simplified analytical models or microscopic numerical simulations. The diffusion coefficients for proteins were estimated, e.g., in [272] using correlation based on Stokes–Einstein relation leading to a simple expression

$$D = 8.34 \times 10^{-8} \frac{T}{\mu M^{1/3}} \qquad (\text{cm}^2\,\text{s}^{-1}) \tag{7.92}$$

depending just on the absolute temperature T, the solvent dynamic viscosity μ, and the molecular weight M. This approximation was developed under the assumptions of relatively large (with respect to solvent) protein molecules with globular shape. Another estimate relaxing these shape restrictions was published in [42]. There are many other methods and models developed for prediction of diffusion coefficients. For a detailed study the above papers and references therein can be used as a starting point.

2. **Biochemistry and Flow models**

It is evident that the biochemical reactions of blood coagulation are in physiological situations strongly affected by the flow of blood. Most of the chemicals and particulate matter necessary to build up the clots (or thrombus) are supplied by the flowing blood. On the other hand, the flow itself is finally affected by the presence and growth of the clot. Flow-induced shear stress (to which von Willebrand factor is sensitive) is also known to affect the initiation of clotting as well as the structure and mechanical properties of the resulting clot. The flow variables or fields can be incorporated into the coagulation models in several ways.

(a) *Parametric.* A characteristic flow velocity or shear rate is used in some of the model parameters to represent the effects of flow on blood coagulation. This is typically only one way coupling, i.e., the coagulation has no direct effect on the flow field that is usually considered as fixed (i.e., a priori prescribed). This simplest approach was used, e.g., in the model published in [137]. This is a rather complex, spatially homogeneous model of surface-mediated control of blood coagulation described by reaction rate equations. It consists of 59 coupled ODEs and besides the chemistry it also includes the role of binding site densities and platelet deposition. This model is coupled to flow via variable (flow dependent) chemistry kinetics-like coefficients[89]:

$$k_c^{\text{flow}} = \frac{3}{4}\left(\frac{V^2 D}{R^2 L^2}\right)^{1/3} \tag{7.93}$$

This coefficient depends on flow velocity V, molecular diffusion coefficient D, vessel size (radius) R, and injury length L. It is derived from steady state advection–diffusion problem assuming a parabolic velocity profile and chemical boundary layer over a finite-length surface source. Using this coefficient a first order kinetics source/sink term is built

$$\pm\, k_c^{\text{flow}}(C - C_{\text{out}})$$

where C is the local concentration and C_{out} is the concentration far from the injury. This extra term is used to take into account the combined effect of advection and diffusion on the chemical transport to/from the coagulation site.

The flow parameters play also an important role in the diffusion processes during coagulation. Quite common is to consider shear-dependent diffusion coefficients. A correlation for the enhancement of diffusion depending on the characteristic shear rate and hematocrit in concentrated erythrocyte suspensions was described, e.g., in [245]. Another such shear rate-dependent diffusion

[89] See [80] for details of derivation and use.

coefficients formulae has the form:

$$D = \alpha \dot{\gamma}_{\max} \qquad \text{where} \qquad \alpha = 7.0 \cdot 10^{-9}\text{cm}^2 \tag{7.94}$$

It is based on [280] and uses the maximum shear rate $\dot{\gamma}_{\max}$ as a characteristic flow parameter. It was applied, e.g., in a mechanical model of acute platelet accumulation in thrombogenic stenoses [254].

(b) *One-way coupling*. The local velocity field enters the coagulation model mainly in its advection part. The velocity field, steady or unsteady, is prescribed, e.g., as an output from an external model or experimental measurement. The velocity field affects, but is not affected by the blood coagulation process and the growth and the presence of the clot. This approach is typical for some simple implementations of Lagrangian particle tracking methods.[90] As an example let's mention the *Stokesian dynamics* based on the additivity of velocities [178, 270]. The velocity $\boldsymbol{v}_i$ of a ith particle (blood cell) is expressed as a sum of fluid velocity $\boldsymbol{u}$ (background flow velocity) at particle position $\boldsymbol{x}_i$ and additional velocity contribution due to binding and other non-hydrodynamic forces.

$$\boldsymbol{v}_i = \boldsymbol{u}(\boldsymbol{x}_i) + \frac{1}{6\pi a \mu}\boldsymbol{F}_i + \cdots \tag{7.95}$$

Here a denotes the particle radius and μ stands for the dynamic viscosity of the solvent. The non-hydrodynamic force $\boldsymbol{F}_i$ can, e.g., be considered as a sum of *binding* force (acting between particles due to coagulation factors) and *collision* force. This model was used, e.g., in [178, 270] to analyze thrombus formation and destruction under flow.

Another example of one-way coupling is the model proposed in [69]. It is a simple continuous macroscopic coagulation model that is coupled to a (prescribed, fixed) flow field.

(c) *Two-way coupling*. In this case the flow-field is not a priori known and can be affected by the spatio-temporal evolution of the clot. This approach requires (in addition to the above described one-way coupling mechanisms) to have a corresponding flow model, e.g., Navier–Stokes equations, coupled together with the biochemistry model. The exact choice of this flow model depends mainly on the considered rheological behavior of blood (e.g., shear-thinning, viscoelastic, etc.). This kind of two-way coupling is evidently the most realistic way of blood coagulation modeling. It is, however, also the most complicated and expensive. At a microscopic level, e.g., the above described Immersed Boundary method belongs to this category. An example of a macroscopic continuum model with two-way coupling will be given later.

Most of the modern computational coagulation models include in some way the effects of blood flow. Some of these models were already introduced in the previous

[90] See the paragraph on Euler–Lagrangian Particle Tracking methods (ELPT) in the section on microscale coagulation methods.

sections and classified according to their characteristic scale. Here we will mention some additional references emphasize some interesting flow-related phenomena in blood coagulation models.

The simplest, chemistry-free way to simulate some of the flow effects on coagulation process is to use a *residence time* model. It is based on the assumptions that clotting occurs after a certain time since the blood activation. A passive scalar, governed by an advection–diffusion–reaction equation can be used as a tracer to estimate the residence time of activated fluid. The local concentration of the tracer can be used as a threshold parameter within the clotting (solidification) model. Such model within the lattice Boltzmann framework was used, e.g., in [26] to investigate the clotting in an idealized stenosed artery.

Very simple coagulation models were proposed in [152] and [69] to demonstrate the combined effects of advection and diffusion on the spatial extent and structure of the thrombus. For example, the model used in [152] consists only of three chemical species (thrombin, protein C, and fibrin) and is described by the following set of PDEs:

$$\frac{\partial \theta}{\partial t} + \boldsymbol{u} \cdot \nabla \theta = \frac{\alpha \theta^2}{\theta + \theta_0} - \gamma \theta \varphi - K_\theta \theta + \nabla \cdot (D_\theta \nabla \theta) \tag{7.96}$$

$$\frac{\partial \varphi}{\partial t} + \boldsymbol{u} \cdot \nabla \varphi = \beta \theta \left(1 - \frac{\varphi}{C}\right)\left(1 - \frac{\varphi^2}{\varphi_0^2}\right) - K_\varphi \varphi + \nabla \cdot \left(D_\varphi \nabla \varphi\right) \tag{7.97}$$

$$\frac{\partial \psi}{\partial t} = K_\varphi \theta \tag{7.98}$$

In this model θ, φ, ψ are concentrations of the *activator*, *inhibitor*, and *fibrin* (at point $\boldsymbol{x}$ and time t). The diffusion coefficients for activator and inhibitor are denoted by D_θ and D_φ. The remaining constants and chemical kinetics rates are determined by fitting to experimental data. The first two equations for θ and φ are of the advection–diffusion–reaction type. The last equation only describes the conversion of activator θ to the final product ψ (fibrin) by a first order chemical reaction. The velocity field $\boldsymbol{u}(\boldsymbol{x}, t)$ is prescribed and is not affected by coagulation. The model solutions lead in the 1D case to the formation of various patterns of fibrin due to the propagation of traveling and pulsing waves of the activator and inhibitor. The same effects were also observed in 2D simulations. Similar problems were studied in [152] using a simplified biochemistry model with only three ADR equations for thrombin (IIa), activated factor XI (XIa), and activated protein C (APC).

The above described simple model belongs to a large family of continuum-based macroscopic models that can be written in the form of ***Advection–Diffusion–Reaction (ADR)*** equations:

$$\frac{\partial C_i}{\partial t} + \boldsymbol{u} \cdot \nabla C_i = R_i(C_1, C_2, \ldots, C_N) + \nabla \cdot \left(D_i \nabla C_i\right) \qquad i = 1, \ldots, N \tag{7.99}$$

A model of that kind, consisting of seven ADR equations was used, e.g., in [224, 225] for the computational simulation of platelet deposition and activation. The pre-computed velocity field is used in the advection part of the biochemistry model and also in the shear-enhanced diffusion coefficients due to the motion of RBCs according to

$$D = 0.18\, d_{\mathrm{rbc}}^2 \dot{\gamma}/4 \tag{7.100}$$

where $d_{\mathrm{rbc}} = 7\mu\mathrm{m}$ is the RBC diameter and $\dot{\gamma}$ is the characteristic shear rate (e.g., at the vessel wall) of the flow.

A much more complex model of this type was proposed in [8,9,11] consisting of ADR equations for more than 20 chemicals and a simple two-way coupling strategy [6] applied to a nonlinear viscoelastic model for blood flow. It includes various mechanical and biochemical factors underlying clot formation and dissolution in flowing blood. This model was used in [32] to simulate clot evolution in a straight three-dimensional vessel segment. Some theoretical aspects of the stability of the underlying biochemical model were presented in [213] together with clotting simulations in 3D axisymmetric stenosed vessel. Another biochemistry model with 18 ADR equations was used in [29] for finite-element simulation of vortical flow structures affecting the biochemistry in formation of intra-luminal thrombus in abdominal aortic aneurysms. A non-Newtonian shear-thinning flow model of blood was used to calculate the background flow (i.e., one-way coupling was used). A two-way coupling mechanism was adopted in [216] where about 40 ADR equations (originally proposed in [189]) were coupled to the Navier–Stokes flow solver to perform a detailed 2D study of the mechanisms controlling the initiation of coagulation in the presence of flow. It has been found that when sufficiently strong blood flow is present, factor Xa is rapidly removed, and the rate of factor X production becomes insufficient to create a fibrin clot.

Among the flow-coupled coagulation models, the most complex are probably some of the multi-scale models. A detailed description of a model of this type can be found, e.g., in [77]. It consists of several coupled blocks (sub-models) where a set of ADR equations is used to describe macroscopic concentrations of non-activated and activated platelets as well as the corresponding activator concentration. An extra equation describes the transport of platelet–platelet elastic bonds which is used to couple the whole platelet thrombosis model to the flow solver.

3. **Biochemistry and Flow and Structure Models**

The blood clot itself can typically be seen as a kind of solid (or elastic, viscoelastic) material being bound to the vessel wall or to other clotting surface. From the mechanical, structural point of view, it might be important to consider the deformation or rupture of thrombi, the change of mechanical properties (e.g., elasticity) of vessel wall due to clotting, and the flow deviations resulting from the presence of clots. To account for these effects an appropriate structural model has to be used and coupled to the rest of the coagulation and flow models.

Pure fluid–structure (without biochemistry) interaction models are quite often used to evaluate possible effects of existing thrombus/embolus on surrounding blood flow and consequent blood vessel deformations. Some of the flow–structure interaction aspects are naturally embedded in small-scale coagulation models dealing with separate blood cells within the blood flow. Several examples can be found in micro-scale ([123] using the MPS method, [193] with DPD method) or multi-scale models ([262, 265] using coupled continuous and CPM models) discussed in previous sections of this text. These methods are capable to distinguish between blood, thrombus, and vessel wall points or particles. This allows to use different rheological and structural models for each of these distinct components of biological system.

Much less common are the macroscopic continuum-based flow–structure interaction multiphase coagulation models. A simple model of this type was developed and tested in [249, 250] and [251] solving a free boundary problem modeling thrombus growth using the level set method. The mathematical model is composed of the incompressible Navier–Stokes equations for blood flow description, coupled with a single scalar transport equation for platelet concentration. The flow–structure (thrombus) interaction is provided by tracking in time the spatial evolution of thrombus and adjusting the blood/thrombus boundary. This moving boundary problem is solved using the ***Level–Set Method (LSM)***. The flow model can be written down as:

$$\nabla \cdot \boldsymbol{u} = 0 \tag{7.101}$$

$$\rho \left(\frac{\partial \boldsymbol{u}}{\partial t} + \boldsymbol{u} \cdot \nabla \boldsymbol{u} \right) = -\nabla p + \mu \Delta \boldsymbol{u} \tag{7.102}$$

This system is solved[91] simultaneously with the platelet transport (advection–diffusion) equation

$$\frac{\partial c}{\partial t} + \boldsymbol{u} \cdot \nabla c = \nabla \cdot (D\, \nabla c) \tag{7.103}$$

Here $c(\boldsymbol{x}, t)$ is the platelet concentration and D is the corresponding diffusion coefficient. Considering wall-bounded flow, the platelet boundary condition can be of Neumann-type:

$$-\,D \frac{\partial c}{\partial \hat{\boldsymbol{n}}} = \begin{cases} k(s)c & \text{on clotting (injured) surface} \\ 0 & \text{on impermeable (healthy) wall} \end{cases} \tag{7.104}$$

Here the surface adhesion of platelets is described as a first order chemical reaction with rate $k(s)$ depending on the local surface shear rate $s = |\boldsymbol{S}\hat{\boldsymbol{n}} \cdot \hat{\boldsymbol{t}}|/\mu$ with $\boldsymbol{S}$ being the stress tensor and $\hat{\boldsymbol{t}}$, $\hat{\boldsymbol{n}}$ the clotting surface tangent, resp. normal unit vectors. When the volume growth of the thrombus is non-negligible, a level set function $\phi(\boldsymbol{x}, t)$ can be introduced to distinguish between fluid and structure (blood and thrombus) regions. The function ϕ is initialized as a signed-distance[92] from the

[91] Subject to appropriate initial and boundary conditions.

[92] Being positive, e.g., in the fluid (blood) and negative in the solid (thrombus).

clotting surface (thrombus–blood boundary) which is further evolved according to the transport equation:

$$\frac{\partial \phi}{\partial t} + \boldsymbol{v} \cdot \nabla \phi = 0 \tag{7.105}$$

The thrombus surface "advection" velocity can be approximated from the platelet concentration field as $\boldsymbol{v} = \alpha \nabla c$, i.e., proportional to the platelets density gradient, neglecting the platelets rolling and other minor transport effects. Taking into consideration the definition of the surface platelet flux (7.104) postulating that $D(\nabla c \cdot \hat{\boldsymbol{n}}) = k(s)c$, the level-set function transport equation can be rewritten as

$$\frac{\partial \phi}{\partial t} + \left(\frac{\alpha k(s)}{D} c\, \hat{\boldsymbol{n}} \right) \cdot \nabla \phi = 0 \tag{7.106}$$

which can further be modified using the definition of unit normal vector $\hat{\boldsymbol{n}} = \nabla \phi / \|\nabla \phi\|$ where $\|\nabla \phi\|^2 = \nabla \phi \cdot \nabla \phi$

$$\frac{\partial \phi}{\partial t} + \frac{\alpha k(s)}{D} c \, \|\nabla \phi\| = 0 \quad . \tag{7.107}$$

Using the function ϕ, the blood/thrombus regions can be determined according to the sign of ϕ with the interface being characterized by $\phi = 0$. Distinct material parameters can be prescribed in each one of these regions. Typically the velocity is set to vanish inside the (wall attached) thrombus and also the diffusivity can be set different for blood/thrombus regions. This model can easily be extended using more complex biochemistry as it was done, e.g., in [231] where a level-set method has been used coupled with a coagulation model taking into account concentrations of activated and resting platelets in blood as well as in the aggregate, transport of RBCs and ADP.

4. **Biochemistry and Other Features Models**

There are many important aspects affecting and being affected by blood coagulation. This is why in certain models some features not previously mentioned have also been implemented to deal with some specific effects. From the long list of more or less important features to be included in coagulation models we will focus on platelets activation due to mechanical stress and on the possible rheological effects of blood coagulation and blood cells aggregation.

Stress-Induced Activation of Platelets

As already observed, platelets play an essential role in the whole coagulation process. One of the important mechanisms of platelet activation is related to the exposition of platelets to excessive stress. It is not just the local instantaneous stress value, but it is the history of stress acting on the platelet that plays a crucial role. This aspect of platelet activation and formation of clotting surfaces is especially important in evaluating the artificial mechanical devices coming into contact with blood (e.g., blood pumps or mechanical heart valves).

Several models have been developed to quantify the blood damage and platelet activation due to flow-induced stress. The most widely used approach defines the *Blood Damage Index* $\mathcal{D}$ using an empirical power law relation

$$\mathcal{D} = C\, \tau^{\alpha} t_{\mathrm{exp}}^{\beta} \tag{7.108}$$

where τ is a scalar measure of stress and t_{exp} is the exposure time. The constants C, α, and β have to be determined experimentally depending on the type of solvent, blood cells (RBCs or platelets), and other typical flow parameters. For an overview of available model parameters see the summary in [114] and related references therein. This model was used to integrate the blood damage along particle trajectories in computational analysis of a blood pump in [223], or for numerical investigation of the effects of channel geometry on platelet activation and blood damage in [259]. A similar model, but a bit simpler in its form is the blood damage index proposed in [204]

$$\mathcal{D} = \frac{C}{\mu}\, \tau^2\, t_{\mathrm{exp}} \tag{7.109}$$

On the other hand, a more complicated differential model for this quantity was proposed in [200]

$$\frac{d\mathcal{D}}{dt} = \frac{C}{(1-\mathcal{D})^{\delta}}\, \tau^2 \tag{7.110}$$

Here C and δ are case-dependent constants to be determined experimentally. A model of this type[93] was used, e.g., in [4] to study the flow-induced platelet activation and damage accumulation in a mechanical heart valve. This approach was generalized in [271]:

$$\frac{d\mathcal{D}(t)}{dt} = \dot{\mathcal{D}}_0 + F(\mathcal{D}, \tau) + \hat{F}(\dot{\tau}) \tag{7.111}$$

where $\dot{\mathcal{D}}_0$ is the *constant* activation/damage rate, $F(\mathcal{D}, \tau)$ is the *stress*-dependent part, and $\hat{F}(\dot{\tau})$ is the *stress rate*-dependent contribution. A similar approach was recently adopted in [222] to develop a mathematical model of activation and sensitization of platelets subjected to dynamic stress histories.

These simple empirical laws are suitable for situations that can easily be characterized by a single stress level and exposure time. This is, however, seldom the case of realistic blood flow simulations. In order to be able to handle also complex flows a more general Lagrangian approach is often adopted. The one introduced for blood (erythrocyte) damage in [97] and [98] was extended for platelets activation in

[93]In Lagrangian particle tracking form.

[183]. It is based on *Platelet Activation State (PAS)* associated with the kth platelet at time t:

$$\mathcal{PAS}_k(t) = \int_{t_0}^{t} Ca\left[\int_{t_0}^{\psi} \tau(\xi)^{b/a} d\xi + \frac{\mathcal{PAS}_k^{1/a}(t_0)}{C}\right]^{a-1} \tau(\psi)^{b/a} d\psi \qquad (7.112)$$

Here $\mathcal{PAS}_k(t_0)$ is the value of activation of the platelet at the starting time of observation t_0, $\tau = \tau(t)$ is time-dependent (local, along platelet path) scalar value of stress.[94] The constants $a = 1.3198$, $b = 0.6256$, and $C = 10^{-5}$ are tuning parameters of the model obtained from experimental data fitting. This model was used, e.g., in [164] to evaluate shear-induced platelet activation and its relationship with blood flow in a model of stenosed carotid bifurcation, or in [177] to predict blood damage in prosthetic heart valves.

A different approach to quantify platelet damage and activation in complex flows was described in [6, 8]. It evaluates a *platelet activation function* $\mathcal{A}$ as an integral over the platelet stressing history

$$\mathcal{A}(t) = \mathcal{A}(t_0) + \frac{1}{A_0}\int_{t_0}^{t} \exp\left[k\left(\frac{\tau(\xi)}{\tau_c} - 1\right)\right] H(\tau(\xi) - \tau_c) d\xi \qquad (7.113)$$

where $\tau = \tau(t)$ is again a scalar measure of stress with τ_c being the critical value of stress above which the platelets get potentially activated. The integral expresses the accumulated stress acting on the platelets between the instants t_0 and t. The Heaviside function H is defined as

$$H(\tau - \tau_c) = \begin{cases} 1 & \text{for} \quad \tau \geq \tau_c \\ 0 & \text{for} \quad \tau < \tau_c \end{cases} \qquad (7.114)$$

to assure that only supra-critical stresses are taken into consideration, i.e., contributing to platelet activation. Lower, physiologically standard values have no effect on activation. The values of the activation function $\mathcal{A}(t)$ are compared against the platelet activation threshold A_{act} and damage threshold A_{dam} to decide, whether the platelet remains resting or gets activated. More details including the full activation algorithm can be found in [8], where a complex blood flow and biochemistry model is built for thrombus formation and lysis. In the abovementioned implementation, the model is capable of platelet activation prediction for each separate platelet/fluid-parcel at the price of Lagrangian tracing and damage integration for every such particle. This is useful when effects of delayed platelet activation[95] or damage have to be included, i.e., the values of $\mathcal{A}(t - t_{\text{act}})$ have to be considered when deciding whether the platelets get activated. For immediate activation, i.e., negligible time t_{act}

[94] Usually some kind of norm of the stress tensor. See, e.g., [214, 222].

[95] Platelet does not get activated immediately, but only after a certain time t_{act}.

the above Langrangian model for activation function $\mathcal{A}(t)$ can be rewritten to more computational friendly Eulerian form for function $\mathcal{A}(\boldsymbol{x}, t)$:

$$\frac{\partial \mathcal{A}}{\partial t} + \boldsymbol{u} \cdot \nabla \mathcal{A} = C_0 \exp\left[k\left(\frac{\tau(\boldsymbol{x}, t)}{\tau_c} - 1\right)\right] H(\tau(\boldsymbol{x}, t) - \tau_c) \tag{7.115}$$

This transport equation has a source term on the right-hand side, that only contributes to the activation function $\mathcal{A}(\boldsymbol{x}, t)$ when $\tau(\boldsymbol{x}, t) \geq \tau_c$. Otherwise the values of $\mathcal{A}$ are only advected by the flow, i.e., remembered without modification.

Another non-Lagrangian platelet activation model was implemented in the lattice Boltzmann framework in [172] based on the so-called Virtual Particle Integration. The principle reminds the above Eulerian activation model. The Lagrangian integration of a quantity is simulated using source terms proportional to the quantity that should otherwise be integrated along the particle pathway.

Blood Rheology Changes

One of the essential phenomena in blood rheology is the aggregation of RBCs forming *rouleaux*. This aggregation is made possible by the presence of macromolecules in blood plasma. Among them the fibrinogen is probably the most important. It plays an essential role in both, blood coagulation and aggregation. This is just one of the examples showing that blood rheology is closely related to coagulation and deserves special attention.

Many of the methods used to resolve microstructure of blood and the process of aggregation and adhesion of blood cells are also capable to predict the most notable rheological properties of blood, such as shear-thinning, viscoelasticity, or thixotropy. This is why blood rheology comes naturally into play when considering blood coagulation models. Recent microscopic blood flow simulations have clearly demonstrated that macroscopic rheological parameters of blood can be derived from relatively simple assumptions imposed on micro-structural inter- and intra-cellular bonds. For more details on this aspect of coagulation and aggregation models, see the corresponding section in this chapter, devoted to micro and multi-scale models of blood coagulation. Some more details and references can be found in recent reviews [242] or [74]. By nature, the microscopic cell-oriented models are suitable for small-scale simulations (e.g., [16, 232]) and cannot be efficiently used for macroscopic problems. This is where the classical, macroscopic rheological models have to be employed.

At a macroscopic continuum level the rheology of blood is often interpreted using bulk material properties, such as viscosity in both Newtonian and non-Newtonian models, yield-stress in thixotropic models or relaxation and retardation times in viscoelastic models. There is an extensive amount of specialized literature on blood rheology and its mathematical modeling. This section will just point out some of the models that can be used within the coagulation models to simulate thrombus and its interaction with blood flow.

Porous media models are often applied to describe the flow and transport phenomena within the clot. Under the assumption of highly viscous flow with negligible inertia a simplified flow model can be built starting from the Navier–Stokes equations. The resulting approximate model is known as the *Darcy's law*. In its generic form it leads to a velocity field inside the clot defined as

$$\boldsymbol{u} = -\frac{k}{\mu}\nabla p \tag{7.116}$$

The permeability k is related to the clot porosity ϵ and to the diameter of fibrin fibers D_f by

$$k = \frac{D_f^2}{64(1-\epsilon)^{3/2}\,(1.0 - 56(1-\epsilon)^3)} \tag{7.117}$$

This model was used in [264, 265] as a component of a multi-scale model of clot formation. A similar model with only slightly different coefficients was also used in [60]. The same kind of permeability formulae, but with *Brinkmann model* instead of the Darcy model was used, e.g., in [130] to study fibrin networks regulating protein transport during thrombus[96] development. The Brinkmann model adds an extra Laplacian-like term to Darcy's law (7.116) in low porosity (permeable external clot) regions where viscous dissipation of momentum is no more negligible.

Generalized-Newtonian models are the most common non-Newtonian models used to describe the macroscopic behavior of blood flow. They capture its shear-thinning behavior by introducing a variable (local shear rate dependent) apparent viscosity instead of the classical constant viscosity used for Newtonian fluids. This type of non-Newtonian models was used in [32] for both, blood and clot in a complex coagulation model adapted from [8,11]. The basic principle is very simple. The concentration of one of the coagulation products, e.g., fibrin in the described case, is used as a tracer (indicator) to distinguish between blood and clot regions. The fluid (blood/clot) viscosity is adjusted according to this tracer concentration. The resulting model equations can be written in the form *generalized Navier–Stokes equations*:

$$\rho\left(\frac{\partial \boldsymbol{u}}{\partial t} + \boldsymbol{u}\cdot\nabla\boldsymbol{u}\right) = -\nabla p + \nabla\cdot\left[\tilde{\mu}([Ia])\mu(\dot{\gamma})\left(\nabla\boldsymbol{u} + \nabla\boldsymbol{u}^T\right)\right] \tag{7.118}$$

The shear-rate[97] dependent (shear-thinning) fluid viscosity $\mu(\dot{\gamma})$ can be defined using *generalized Cross model*:

$$\mu(\dot{\gamma}) = \mu_\infty + (\mu_0 - \mu_\infty)\frac{1}{\left(1 + (\lambda\dot{\gamma})^b\right)^a} \tag{7.119}$$

---

[96] The thrombus model employs a composite structure with an impermeable core (activated platelets and fibrin) and a permeable shell (fibrin cap).

[97] The *shear-rate* is defined as $\dot{\gamma} = 2\sqrt{\boldsymbol{D}:\boldsymbol{D}}$ where $\boldsymbol{D} = (\nabla\boldsymbol{u} + \nabla\boldsymbol{u}^T)/2$ is the symmetric part of velocity gradient.

Model adjustable parameters are taken from [143]:

$$\mu_0 = 1.6 \cdot 10^{-1}\,\text{Pa s} \qquad \mu_\infty = 3.6 \cdot 10^{-3}\,\text{Pa s}$$
$$a = 1.23,\ b = 0.64 \qquad \lambda = 8.2\,\text{s}$$

The symbols μ_0 and μ_∞ denote the zero- and infinity-shear rate limits of viscosity. The apparent viscosity is multiplied by a non-dimensional factor $\tilde{\mu}$ that locally depends (linearly, up to a certain saturation value μ^*) on fibrin concentration $[Ia]$.

$$\tilde{\mu}([Ia]) = \min\left\{1 + \frac{\mu^* - 1}{C_{\text{clot}}}[Ia], \mu^*\right\} \tag{7.120}$$

where $\mu^* = 100$ and $C_{\text{clot}} = 1{,}000\,\text{nM}$ was used in [32].

In this model the clot is simulated as a highly viscous fluid (up to μ^* times more viscous than blood). The region occupied by the clot represents an obstacle to the flow of blood having a much higher viscosity. This viscosity amplification effect is even significantly magnified due to the shear-thinning non-Newtonian behavior of blood, leading to a further increase of fluid viscosity in regions of low shear. The biochemistry (i.e., fibrin)-induced changes in viscosity results into modifications of the local flow field. These flow deviations consequently affect the concentration field that lead to further changes in the viscosity. Using this algorithm the two-way biochemistry-flow coupling is enforced.

Viscoelastic models are used in some recent blood flow and coagulation models to account for more complex viscoelastic effects in the rheology of blood. One of the simplest models of this kind, being implemented as a component of a coagulation model, is the shear-thinning viscoelastic model proposed in [6]. It is a mathematical model describing the change in the constitutive character of blood due to platelet activation. The considered stress-induced activation of platelets is described by the function $\mathcal{A}$ defined previously in (7.113). According to the local value of this activation function the actual status of platelets (activated/resting) is decided and corresponding parameters of the rheological model are adjusted.

In its simplest form the model can be written as a generalized Oldroyd-B model with local shear-rate and platelet activation-dependent viscosity. As in the Newtonian case for the Navier–Stokes model the equations of fluid motion are based on the conservation of mass and linear momentum for an incompressible fluid:

$$\nabla \cdot \boldsymbol{u} = 0 \tag{7.121}$$

$$\rho\left(\frac{\partial \boldsymbol{u}}{\partial t} + \boldsymbol{u} \cdot \nabla \boldsymbol{u}\right) = -\nabla p + \nabla \cdot \boldsymbol{T} \tag{7.122}$$

The (viscoelastic) stress tensor is in this case defined by:

$$\boldsymbol{T} + \lambda_1 \overset{\triangledown}{\boldsymbol{T}} = 2\mu\left(\boldsymbol{D} + \lambda_2 \overset{\triangledown}{\boldsymbol{D}}\right) \tag{7.123}$$

The parameters λ_1 resp. λ_2 denote the relaxation, resp. retardation times. Upper-convected (time) derivative[98] of tensors is marked by the symbol ∇. The viscosity μ (being constant in the original Oldroyd model) is defined as a product of a shear-dependent apparent viscosity $\mu(\dot{\gamma})$ and an activation dependent factor $\tilde{\mu}(\mathcal{A})$, i.e., $\mu = \mu(\dot{\gamma})\tilde{\mu}(\mathcal{A})$. For the shear-thinning part of viscosity, the model of Yeleswarapu is suggested:

$$\mu(\dot{\gamma}) = \mu_\infty + (\mu_0 - \mu_\infty)\frac{1 + \ln(1 + \Lambda\dot{\gamma})}{1 + \Lambda\dot{\gamma}} \tag{7.124}$$

where $\dot{\gamma}$ is the shear-rate and Λ is an adjustable parameter. The platelet activation-dependent multiplicative factor $\tilde{\mu}(\mathcal{A})$ is defined as[99]

$$\tilde{\mu}(\mathcal{A}) = \begin{cases} 1 & \text{for resting platelets} \\ \mu^* \gg 1 & \text{for activated platelets} \end{cases} \tag{7.125}$$

A much more complex macroscopic coagulation biochemistry model, described by a set of more than 20 coupled ADR equations, was introduced in [8]. To describe the shear-thinning viscoelastic character of blood a nonlinear model was used based on earlier works considering a novel thermodynamically consistent framework for rate type fluid models [199]. Within this model the viscoelastic stress tensor $\boldsymbol{T}$ is split as:

$$\boldsymbol{T} = \eta\, \boldsymbol{B}_{\kappa_p(t)} + \mu_1 \boldsymbol{D} \tag{7.126}$$

where $\boldsymbol{D}$ denotes the symmetric part of the velocity gradient tensor and $\boldsymbol{B}_{\kappa_p(t)}$ stands for the elastic stretch tensor.[100] The upper-convected time derivative of the tensor $\boldsymbol{B}_{\kappa_p(t)}$ is then given by:

$$\overset{\nabla}{\boldsymbol{B}}_{\kappa_p(t)} = -2K\left(\mathrm{tr}(\boldsymbol{B}_{\kappa_p(t)}) - 3\lambda\right)^n \left[\boldsymbol{B}_{\kappa_p(t)} - \lambda\mathbf{1}\right] \tag{7.127}$$

where the coefficient λ depends on the trace of the inverse of the tensor $\boldsymbol{B}_{\kappa_p(t)}$ according to

$$\lambda = \frac{3}{\mathrm{tr}\left(\boldsymbol{B}^{-1}{}_{\kappa_{p(t)}}\right)} \quad . \tag{7.128}$$

[98]The upper-convected derivative $\overset{\nabla}{\boldsymbol{M}}$ of a tensor $\boldsymbol{M}$ is defined using the classical material time-derivative $\dot{\boldsymbol{M}}$ and the symmetric resp. skew-symmetric parts of the velocity gradient $\boldsymbol{D}$ resp. $\boldsymbol{W}$ as $\overset{\nabla}{\boldsymbol{M}} = \dot{\boldsymbol{M}} - \boldsymbol{W}\boldsymbol{M} + \boldsymbol{M}\boldsymbol{W} - (\boldsymbol{D}\boldsymbol{M} + \boldsymbol{M}\boldsymbol{D})$.

[99]See [6] for the complete algorithm of platelet activation depending on the values of $\mathcal{A}$.

[100]The subscript $\kappa_p(t)$ is used to emphasize that the stretch is expressed with respect to natural (time-dependent) configuration $\kappa_p(t)$. This notation follows exactly the original papers [7, 12, 199] and [34] where the model has been introduced and used.

The remaining model coefficients η, μ_1, K and n are chosen depending on the considered fluid. This model is able to capture both, the viscoelastic and shear-thinning effects. In the coupled blood flow & coagulation model two distinct parameter sets are used for the blood and the clot. To distinguish between the blood

		Blood	Clot
η	[Pa]	0.0227	4.9
μ_1	[Pa s]	0.01	0.1
K	[s^{-1}]	1.2056	2.2
n	[1]	0.7525	0.1

and the clot a combined chemo-mechanical indicator is used depending on the fibrin concentration $[Ia]$ and platelet activation function $\mathcal{A}$. For details see [8–11].

7.8 Conclusions and Remarks

One of the aims of this chapter was to present the complexity of the blood coagulation process from different perspectives. These different points of view are provided by a wide range of mathematical models and concepts being used to describe and study blood coagulation. This very complicated picture is framed by a historical time-line of the evolution of coagulation theories based on underlying experimental discoveries.

There are several lessons to be learned from this chapter:

- The historical evolution of coagulation theories exhibits many breaking points and bifurcations. An example is the recent shift from waterfall motivated models to cell-based theories leading to major changes in the whole concept of macroscopic biochemistry models of blood coagulation. Such a change of the point of view not only generates new mathematical models, but it also leads to new interpretation of many older experiments.
- It is evident that the traditional, purely macroscopic, and exclusively chemical description of the coagulation process is not able to provide a complete picture of blood coagulation. In the contemporary literature many efforts have been made to include more of the physics involved in the mathematical models of blood coagulation (fluid mechanics, structural mechanics, molecular dynamics, etc.).
- Recent rapid evolution of molecular chemistry and mechanics is mainly driven by the increased availability of massive computational resources. This leads to more information being obtained about the sub-micro and micro-scale nature of blood coagulation. The corresponding mathematical models can not only provide these very specialized microscopic level simulations, but more importantly they can add more details to meso- and macro-scale models. This multiscale coupling

is probably the most promising from the current trends in blood coagulation modeling.

At this very end of the chapter it is fair to note that the theory and mathematical modeling of blood coagulation is so complex, that despite our effort of producing a comprehensive overview, our presentation is necessarily incomplete. There can be different ways of organizing this large material, according to different viewpoints and also to give more or less emphasis to the various subjects. Since the topics treated here are in a stage of rapid evolution, as more and more is known about key elements intervening in the process (e.g., platelets, anticoagulant therapies, etc.), this chapter should be understood as a starting point for a more detailed and specialized investigation in this very specific, challenging and fascinating scientific area.

Acknowledgements The financial support for the present project was partly provided by the *Czech Science Foundation* under the *Grant No.201/09/0917* and by the *Portuguese Science Foundation* under The Research Center CEMAT-IST and under the *Project EXCL/MAT-NAN/0114/2012*.

References

1. W.C. Aird, Vascular bed-specific thrombosis. J. Thromb. Haemost. **5**(Suppl. 1), 283–291 (2007)
2. B.J. Alder, T.E. Wainwright, Phase transition for a hard sphere system. J. Chem. Phys. **27**(5), 1208–1209 (1957)
3. D.J. Aldous, Deterministic and stochastic models for coalescence (aggregation and coagulation): a review of the mean-field theory for probabilists. Bernoulli **5**(1), 3–48 (1999)
4. Y. Alemu, D. Bluestein, Flow-induced platelet activation and damage accumulation in a mechanical heart valve: numerical studies. Artif. Organs **31**(9), 677–688 (2007)
5. A. Amadei, A.B.M. Linssen, H.J.C. Berendsen, Essential dynamics of proteins. Proteins Struct. Funct. Genet. **17**(4), 412–425 (1993)
6. M. Anand, K.R. Rajagopal, A mathematical model to describe the change in the constitutive character of blood due to platelet activation. C. R. Mec. **330**(8), 557–562 (2002)
7. M. Anand, K.R. Rajagopal, A shear-thinning viscoelastic fluid model for describing the flow of blood. Int. J. Cardiovasc. Med. Sci. **4**(2), 59–68 (2004)
8. M. Anand, K. Rajagopal, K.R. Rajagopal, A model incorporating some of the mechanical and biochemical factors underlying clot formation and dissolution in flowing blood. J. Theor. Med. **5**(3–4), 183–218 (2003)
9. M. Anand, K. Rajagopal, K.R. Rajagopal, A model for the formation and lysis of blood clots. Pathophysiol. Haemost. Thromb. **34**(2–3), 109–120 (2005)
10. M. Anand, K. Rajagopal, K.R. Rajagopal, A viscoelastic fluid model for describing the mechanics of a coarse ligated plasma clot. Theor. Comput. Fluid Dyn. **20**(4), 239–250 (2006)
11. M. Anand, K. Rajagopal, K.R. Rajagopal, A model for the formation, growth, and lysis of clots in quiescent plasma. A comparison between the effects of antithrombin III deficiency and protein C deficiency. J. Theor. Biol. **253**(4), 725–738 (2008)
12. M. Anand, J. Kwack, A. Masud, A new generalized Oldroid-b model for blood flow in complex geometries. Int. J. Eng. Sci. **72**, 78–88 (2013)
13. S.T. Anning, The historical aspects of venous thrombosis. Med. Hist. **1**(1), 28–37 (1957)

14. F.I. Ataullakhanov, G.T. Guria, V.I. Sarbash, R.I. Volkova, Spatiotemporal dynamics of clotting and pattern formation in human blood. Biochim. Biophys. Acta Gen. Subj. **1425**(3), 453–468 (1998)
15. F. Bachmann, The discovery of factor X: a personal reminiscence. Thromb. Haemost. **98**(1), 16–19 (2007)
16. P. Bagchi, Mesoscale simulation of blood flow in small vessels. Biophys. J. **92**, 1858–1877 (2007)
17. C.N. Bagot, R. Arya, Virchow and his triad: a question of attribution. Br. J. Haematol. **143**(2), 180–190 (2008)
18. I. Bahar, A.J. Rader, Coarse-grained normal mode analysis in structural biology. Curr. Opin. Struct. Biol. **15**, 586–592 (2005)
19. F. Bai, Z. Wu, J. Jin, P. Hochendoner, J. Xing, Slow protein conformational change, Allostery and Network Dynamics, in *Protein-Protein Interactions - Computational and Experimental Tools* (InTech, Croatia, 2012), pp. 169–188
20. L. Baronciani, P.M. Manucci, The molecular basis of von Willebrand disease, in *Molecular Hematology*, chap. 19, 3rd edn., ed. by D. Provan, J.G. Gribben (Wiley-Blackwell, London, 2010), pp. 233–245
21. C. Basciano, C. Kleinstreuer, S. Hyun, E.A. Finol, A relation between near-wall particle-hemodynamics and onset of thrombus formation in abdominal aortic aneurysms. Ann. Biomed. Eng. **39**(7), 2010–2026 (2011)
22. R.C. Becker, Cell-based models of coagulation: a paradigm in evolution. J. Thromb. Thrombolysis **20**(1), 65–68 (2005)
23. E. Beltrami, J. Jesty, Mathematical analysis of activation thresholds in enzyme-catalyzed positive feedbacks: application to the feedbacks of blood coagulation. Proc. Natl. Acad. Sci. USA **92**(19), 8744–8748 (1995)
24. T.K. Belval, J.D. Hellums, Analysis of shear-induced platelet aggregation with population balance mathematics. Biophys. J. **50**(3), 479–487 (1986)
25. J. Bernard, J.P. Soulier, Sur une nouvelle variété de dystrophie thrombocytaire hémorragipare congénitale. Sem. Hôp. Paris **24**, 3217–3223 (1948)
26. J. Bernsdorf, S.E. Harrison, S.M. Smith, P.V. Lawford, D.R. Hose, Concurrent numerical simulation of flow and blood clotting using the lattice Boltzmann technique. Int. J. Bioinform. Res. Appl. **2**(4), 371–380 (2006)
27. J. Bernsdorf, S.E. Harrison, S.M. Smith, P.V. Lawford, D.R. Hose, Applying the lattice Boltzmann technique to biofluids: a novel approach to simulate blood coagulation. Comput. Math. Appl. **55**(7), 1408–1414 (2008)
28. E.X. Berry, A mathematical framework for cloud models. J. Atmos. Sci. **26**, 109–111 (1969)
29. J. Biasetti, P.G. Spazzini, J. Swedenborg, T. Christian Gasser, An integrated fluid-chemical model toward modeling the formation of intra-luminal thrombus in abdominal aortic aneurysms. Front. Physiol. **3**, 1–16 (2012)
30. T. Bodnár, On the use of non-linear TVD filters in finite-volume simulations, in *Algoritmy 2012 Proceedings of Contributed Papers and Posters*, Bratislava. Faculty of Civil Engineering, Slovak University of Technology, pp. 190–199 (2012)
31. T. Bodnár, J. Příhoda, Numerical simulation of turbulent free-surface flow in curved channel. J. Flow Turbulence Combust. **76**(4), 429–442 (2006)
32. T. Bodnár, A. Sequeira, Numerical simulation of the coagulation dynamics of blood. Comput. Math. Methods Med. **9**(2), 83–104 (2008)
33. T. Bodnár, A. Sequeira, Numerical study of the significance of the non-Newtonian nature of blood in steady flow through a stenosed vessel, in *Advances in Mathematical Fluid Mechanics*, ed. by R. Rannacher, A. Sequeira (Springer, Berlin, 2010), pp. 83–104
34. T. Bodnár, K.R. Rajagopal, A. Sequeira, Simulation of the three-dimensional flow of blood using a shear-thinning viscoelastic fluid model. Math. Model. Nat. Phenom. **6**(5), 1–24 (2011)
35. T. Bodnár, A. Sequeira, M. Prosi, On the shear-thinning and viscoelastic effects of blood flow under various flow rates. Appl. Math. Comput. **217**(11), 5055–5067 (2011)

36. K. Boryczko, W. Dzwinela, D.A. Yuen, Modeling fibrin aggregation in blood flow with discrete-particles. Comput. Methods Programs Biomed. **75**, 181–194 (2004)
37. K. Boryczko, D.A.Yuen, W. Dzwinel, Modeling mesoscopic fluids with discrete-particles – methods, algorithms, and results, in *Finely Dispersed Particles: Micro-, Nano-, and Atto-Engineering* (CRC Press, West Palm Beach, 2005), pp. 715–778
38. D.B. Brewer, Max Schultze (1865), G. Bizzozero (1882) and the discovery of the platelet. Br. J. Haematol. **133**(3), 251–258 (2006)
39. L. Brugnano, F. Di Patti, G. Longo, An "incremental" mathematical model for Immune Thrombocytopenic Purpura (ITP). Math. Comput. Model. **42**(11–12), 1299–1314 (2005)
40. K.E. Brummel-Ziedins, T. Orfeo, M. Gissel, K.G. Mann, F.R. Rosendaal, Factor Xa generation by computational modeling: an additional discriminator to thrombin generation evaluation. PLoS ONE **7**(1), e29178 (2012)
41. K. Brummel-Ziedins, Models for thrombin generation and risk of disease. J. Thromb. Haemost. **11**(Suppl.1), 212–223 (2013)
42. D. Brune, S. Kim, Predicting protein diffusion coefficients. Proc. Natl. Acad. Sci. USA **90**(9), 3835–3839 (1993)
43. S. Butenas, T. Orfeo, M.T. Gissel, K.E. Brummel, K.G. Mann, The significance of circulating factor IXa in blood. J. Biol. Chem. **279**(22), 22875–22882 (2004)
44. M.S. Chatterjee, W.S. Denney, H. Jing, S.L. Diamond, Systems biology of coagulation initiation: kinetics of thrombin generation in resting and activated human blood. PLoS Comput. Biol. **6**(9), 1–24 (2010)
45. R. Chaturvedi, C. Huang, B. Kazmierczak, T. Schneider, J.A. Izaguirre, T. Glimm, H.G.E. Hentschel, J.A. Glazier, S.A. Newman, M.S. Alber, On multiscale approaches to three-dimensional modelling of morphogenesis. J. R. Soc. Interface **2**(3), 237–253 (2005)
46. T.M. Cickovski, C. Huang, R. Chaturvedi, T. Glimm, H.G.E. Hentschel, M.S. Alber, J.A. Glazier, S.A. Newman, J.A. Izaguirre, A framework for three-dimensional simulation of morphogenesis. IEEE/ACM Trans. Comput. Biol. Bioinform. **2**(4), 273–287 (2005)
47. T. Cickovski, K. Aras, M.S. Alber, J.A. Izaguirre, M. Swat, J.A. Glazier, R.M.H. Merks, T. Glimm, H.G.E. Hentschel, S.A. Newman, From genes to organisms via the cell a problem-solving environment for multicellular development. Comput. Sci. Eng. **9**(4), 50–60 (2007)
48. S. Cito, M.D. Mazzeo, L. Badimon, A review of macroscopic thrombus modeling methods. Thromb. Res. **131**(2), 116–124 (2013)
49. K.J. Clementson, Platelets disorders, in *Molecular Hematology*, chap. 20, 3rd edn., ed. by D. Provan, J.G. Gribben (Wiley-Blackwell, London, 2010), pp. 246–258
50. B.S. Coller, A brief history of ideas about platelets in health and disease, in *Platelets* (Academic, New York, 2007), pp. xxiii–xlii
51. B. Cooper, Osler's role in defining the third corpuscle, or "blood plates". Proc. (Bayl. Univ. Med. Cent.) **18**(4), 376–378 (2005)
52. J.M. Coutinho, J.M. Ferro, P. Canh ao, F. Barinagarrementeria, C. Cantú, M.-G. Bousser, J. Stam, Cerebral venous and sinus thrombosis in women. Stroke **40**(7), 2356–2361 (2009)
53. K.J. Croce, M. Sakuma, D.I. Simon, Platelet-leukocyte-endothelial cross talk, in *Platelets in Hematologic and Cardiovascular Disorders: A Clinical Handbook* (Cambridge University Press, Cambridge, 2007), pp. 106–123
54. L.M. Crowl, A.L. Fogelson, Computational model of whole blood exhibiting lateral platelet motion induced by red blood cells. Int. J. Numer. Method Biomed. Eng. **26**, 471–487 (2010)
55. S. Cunha Orfao, G. Jank, K. Mottaghy, S. Walcher, E. Zerz, Qualitative properties and stabilizability of a model for blood thrombin formation. J. Math. Anal. Appl. **346**, 218–226 (2008)
56. B. Dahlbäck, Blood coagulation and its regulation by anticoagulant pathways: genetic pathogenesis of bleeding and thrombotic diseases. J. Inter. Med. **257**(3), 209–223 (2005)
57. B. Dählbeck, A. Hillarp, Molecular coagulation and thrombophilia, in *Molecular Hematology*, chap. 17, 3rd edn., ed. by D. Provan, J.G. Gribben (Wiley-Blackwell, London, 2010), pp. 208–218

58. G. Davì, C. Patrono, Platelets activation and atherothrombosis. N. Engl. J. Med. **367**, 2482–2494 (2007)
59. E.W. Davie, O.D. Ratnoff, Waterfall sequence for intrinsic blood clotting. Science **145**(3638), 1310–1312 (1964)
60. S.L. Diamond, S. Anand, Inner clot diffusion and permeation during fibrinolysis. Biophys. J. **65**(6), 2622–2643 (1993)
61. F. Dong, B. Olsen, N.A. Baker, Computational methods for biomolecular electrostatics, in *Biophysical Tools for Biologists, Volume One: In Vitro Techniques, Volume 84 of Methods in Cell Biology* (Elsevier, Amsterdam, 2008), pp. 843–870
62. R.L. Drake, A general mathematical survey of the coagulation equation, in *Topics in Current Aerosol Research (Part 2), Volume 3 of International Reviews in Aerosol Physics and Chemistry*, ed. by G.M. Hidy, J.R. Brock (Pergamon, Oxford, 1972), pp. 201–376
63. R.L. Drake, The scalar transport equation of coalescence theory: moments and kernels. J. Atmos. Sci. **29**, 537–547 (1972)
64. P.B. Dubovskii, Mathematical theory of coagulation. Technical report, Seoul National University, Research Institute of Mathematics, Global Analysis Research Center (1994)
65. W. Dzwinel, K. Boryczko, D.A. Yuen, A discrete-particle model of blood dynamics in capillary vessels. J. Colloid Interface Sci. **258**, 163–173 (2003)
66. W. Dzwinel, D.A. Yuen, K. Boryczko, Bridging diverse physical scales with the discrete-particle paradigm in modeling colloidal dynamics with mesoscopic features. Chem. Eng. Sci. **61**, 2169–2185 (2006)
67. K.-E. Eilertsen, B. Østerud, The role of blood cells and their microparticles in blood coagulation. Biochem. Soc. Trans. **33**(2), 418–422 (2005)
68. B. Engquist, P. Lötstedt, B. Sjögreen, Nonlinear filters for efficient shock computation. Math. Comput. **52**(186), 509–537 (1989)
69. E.A. Ermakova, M.A. Panteleev, E.E. Shnol, Blood coagulation and propagation of autowaves in flow. Pathophysiol. Haemost. Thromb. **34**(2–3), 135–142 (2006)
70. A. Farina, A. Fasano, J. Mizerski, A new model for blood flow in fenestrated capillaries with application to ultrafiltration in kidney glomeruli. A.M.S.A. (2014, to appear)
71. A. Fasano, R.F. Santos, A. Sequeira, Blood coagulation: a puzzle for biologists, a maze for mathematicians, in *Modeling of Physiological Flows, Volume 5 of MS&A – Modeling, Simulation and Applications* (Springer, Milan, 2012), pp. 41–75
72. A. Fasano, J. Pavlova, A. Sequeira, A synthetic model for blood coagulation including blood slip at the vessel wall. Clin. Hemorheol. Microcirc. **54**(1), 1–14 (2013)
73. D.A. Fedosov, G.E. Karniadakis, Triple-decker: interfacing atomistic–mesoscopic–continuum flow regimes. J. Comput. Phys. **228**(4), 1157–1171 (2009)
74. D.A. Fedosov, H. Noguchi, G. Gompper, Multiscale modeling of blood flow: from single cells to blood rheology. Biomech. Model. Mechanobiol. 1–20 (2013). doi:10.1007/s10237-013-0497-9
75. N. Filipovic, M. Kojic, A. Tsuda, Modelling thrombosis using dissipative particle dynamics method. Philos. Trans. R. Soc. A **366**, 3265–3279 (2008)
76. M.H. Flamm, S.L. Diamond, Multiscale systems biology and physics of thrombosis under flow. Ann. Biomed. Eng. **40**(11), 2355–2364 (2012)
77. A.L. Fogelson, R.D. Guy, Platelet-wall interactions in continuum models of platelet thrombosis: formulation and numerical solution. Math. Med. Biol. **21**(4), 293–334 (2004)
78. A.L. Fogelson, R.D. Guy, Immersed-boundary-type models of intravascular platelet aggregation. Comput. Methods Appl. Mech. Eng. **197**, 2087–2104 (2008)
79. A.L. Fogelson, J.P. Keener, Toward an understanding of fibrin branching structure. Phys. Rev. E **81**(5), 051922-1–051922-9 (2010)
80. A.L. Fogelson, N. Tania, Coagulation under flow: the influence of flow-mediated transport on the initiation and inhibition of coagulation. Pathophysiol. Haemost. Thromb. **34**(2–3), 91–108 (2006)
81. C. Forrey, M. Muthukumar, Langevin dynamics simulations of genome packing in bacteriophage. Biophys. J. **91**, 25–41 (2006)

82. W.B. Foster, M.E. Nesheim, K.G. Mann, The factor Xa-catalyzed activation of factor V. J. Biol. Chem. **258**(22), 13970–13977 (1983)
83. S.K. Friedlander, On the particle size spectrum of a condensing vapor. Phys. Fluids **3**(5), 693–696 (1960)
84. S. Fuchigami, S. Omori, M. Ikeguchi, A. Kidera, Normal mode analysis of protein dynamics in a non-Eckart frame. J. Chem. Phys. **132**, 104109-1–104109-7 (2010)
85. R.R. Gabdoulline, R.C. Wade, Brownian dynamics simulation of protein-protein diffusional encounter. Methods **14**(3), 329–341 (1998)
86. D. Gailani, A. Zivelin, D. Sinha, P.N. Walsh, Do platelets synthesize factor XI? J. Thromb. Haemost. **2**(10), 1709–1712 (2004)
87. R.B. Gayle, C.R. Maliszewski, S.D. Gimpel, M.A. Schoenborn, R.G. Caspary, C. Richards, K. Brasel, V. Price, J.H. Drosopoulos, N. Islam, T.N. Alyonycheva, M.J. Broekman, A.J. Marcus, Inhibition of platelet function by recombinant soluble ecto-ADPase/CD39. J. Clin. Investig. **101**(9), 1851–1859 (1998)
88. V. Gazzaniga, L. Ottini, The discovery of platelets and their function. Vesalius **VII**(1), 22–26 (2001)
89. D.T. Gillespie, A general method for numerically simulating the stochastic time evolution of coupled chemical reactions. J. Comput. Phys. **22**(4), 403–434 (1976)
90. D.T. Gillespie, Exact stochastic simulation of coupled chemical reactions. J. Phys. Chem. **81**(25), 2340–2361 (1977)
91. D.T. Gillespie, Stochastic simulation of chemical kinetics. Annu. Rev. Phys. Chem. **58**, 35–55 (2007)
92. D.T. Gillespie, Simulation methods in systems biology, in *Formal Methods for Computational Systems Biology, Volume 5016 of Lecture Notes in Computer Science*, ed. by M. Bernardo, P. Degano, G. Zavattaro (Springer, Berlin, 2008), pp. 125–167
93. D.T. Gillespie, Deterministic limit of stochastic chemical kinetics. J. Phys. Chem. B **113**(6), 1640–1644 (2009)
94. W.E. Glanzmann, Hereditäre hämorrhägische Thrombasthenie. ein Beitrag zur Pathologie der Blutplättchen. Jahrb. Kinderheilkund. **88**(1–42), 113–141 (1918)
95. J.A. Glazier, F. Graner, Simulation of the differential adhesion driven rearrangement of biological cells. Phys. Rev. E **47**(3), 2128–2154 (1993)
96. G.H. Goldsmith, H. Saito, O.D. Ratnoff, The activation of plasminogen by Hageman factor (factor XII) and Hageman factor fragments. J. Clin. Investig. **62**(1), 54–60 (1978)
97. M. Grigioni, C. Daniele, U. Morbiducci, G. D'Avenio, G. Di Benedetto, V. Barbaro, The power-law mathematical model for blood damage prediction: analytical developments and physical inconsistencies. Artif. Organs **28**(5), 467–475 (2004)
98. M. Grigioni, U. Morbiducci, G. D'Avenio, G. Di Benedetto, C. Del Gaudio, A novel formulation for blood trauma prediction by a modified power-law mathematical model. Biomech. Model. Mechanobiol. **4**(4), 249–260 (2005)
99. K.G. Guria, A.R. Gagarina, G.T. Guria, Instabilities in fibrinolytic regulatory system. theoretical analysis of blow-up phenomena. J. Theor. Biol. **304**, 27–38 (2012)
100. R.I. Handin, Inherited platelet disorders. Am. Soc. Hematology **2005**(1), 396–402 (2005)
101. S.E. Harrison, S.M. Smith, J. Bernsdorf, D.R. Hose, P.V. Lawford, Application and validation of the lattice Boltzmann method for modelling flow-related clotting. J. Biomech. **40**(13), 3023–3028 (2007)
102. S. Hayward, B.L. de Groot, Normal modes and essential dynamics, in *Molecular Modeling of Proteins, Volume 443 of Methods in Molecular Biology* (Humana Press, Clifton, 2008), pp. 89–106
103. H.C. Hemker, S. Kerdelo, R.M.W. Kremers, Is there value in kinetic modeling of thrombin generation? No (unless…). J. Thromb. Haemost. **10**(8), 1470–1477 (2012)
104. O. Hetland, A.B. Brovold, R. Holme, G. Gaudernack, H. Prydz, Thromboplastin (tissue factor) in plasma membranes of human monocytes. Biochem. J. **228**(3), 735–743 (1985)
105. M.F. Hockin, K.C. Jones, S.J. Everse, K.G. Mann, A model for the stoichiometric regulation of blood coagulation. J. Biol. Chem. **277**(21), 18322–18333 (2002)

106. M. Hoffman, Remodeling the blood coagulation cascade. J. Thromb. Thrombolysis **16**(1–2), 17–20 (2003)
107. B. Honig, A. Nicholls, Classical electrostatics in biology and chemistry. Science **268**(5214), 1144–1149 (1995)
108. P.J. Hoogerbrugge, J.M.V.A. Koelman, Simulating microscopic hydrodynamic phenomena with dissipative particle dynamics. Europhys. Lett. **19**(3), 155–160 (1992)
109. W.H. Howell, E. Holt, Two new factors in blood coagulation: heparin and pro-antithrombin. Am. J. Physiol. **47**, 228–241 (1918)
110. L. Hsieh, D. Nugent, Factor XIII deficiency. Haemophilia **14**, 1190–1200 (2008)
111. P.Y. Huang, J.D. Hellums, Aggregation and disaggregation kinetics of human blood platelets: part I. Development and validation of a population balance method. Biophys. J. **65**(1), 334–343 (1993)
112. P.Y. Huang, J.D. Hellums, Aggregation and disaggregation kinetics of human blood platelets: part II. Shear-induced platelet aggregation. Biophys. J. **65**(1), 344–353 (1993)
113. P.Y. Huang, J.D. Hellums, Aggregation and disaggregation kinetics of human blood platelets: part III. The disaggregation under shear stress of platelet aggregates. Biophys. J. **65**(1), 354–361 (1993)
114. S.J. Hund, J.F. Antaki, M. Massoudi, On the representation of turbulent stresses for computing blood damage. Int. J. Eng. Sci. **48**(11), 1325–1331 (2010)
115. S. Hyun, C. Kleinstreuer, J.P. Archie Jr., Computational particle-hemodynamics analysis and geometric reconstruction after carotid endarterectomy. Comput. Biol. Med. **31**, 365–384 (2001)
116. Y. Imai, H. Kondo, T. Ishikawa, C.T. Lim, T. Yamaguchi, Modeling of hemodynamics arising from malaria infection. J. Biomech. **43**, 1386–1393 (2010)
117. G.I. Ingram, The history of hemophilia. J. Clin. Pathol. **29**(6), 469–479 (1976)
118. B. Isralewitz, J. Baudry, J. Gullingsrud, D. Kosztin, K. Schulten, Steered molecular dynamics investigations of protein function. J. Mol. Graph. Model. **19**, 13–25 (2001)
119. A. Jameson, Time dependent calculations using multigrid, with applications to unsteady flows past airfoils and wings, in *AIAA 10th Computational Fluid Dynamics Conference*, Honolulu, Paper 91-1596, June 1991
120. A. Jameson, W. Schmidt, E. Turkel, Numerical solutions of the Euler equations by finite volume methods using Runge-Kutta time-stepping Schemes, in *AIAA 14th Fluid and Plasma Dynamic Conference*, Palo Alto, Paper 81-1259, June 1981
121. I. Johansson, N Lynöe, *Medicine & Philosophy: A Twenty-First Century Introduction* (Ontos Verlag, Frankfurt, 2008)
122. K.C. Jones, K.G. Mann, A model for the tissue factor pathway to thrombin. II. A mathematical simulation. J. Biol. Chem. **269**(37), 23367–23373 (1994)
123. H. Kamada, K.I. Tsubota, M. Nakamura, S. Wada, T. Ishikawa, T. Yamaguchi, A three-dimensional particle simulation of the formation and collapse of a primary thrombus. Int. J. Numer. Methods Biomed. Eng. **26**, 488–500 (2010)
124. H. Kamada, Y. Imai, M. Nakamura, T. Ishikawa, T. Yamaguchi, Computational analysis on the mechanical interaction between a thrombus and red blood cells: possible causes of membrane damage of red blood cells at microvessels. Med. Eng. Phys. **34**, 1411–1420 (2012)
125. H. Kamada, Y. Imai, M. Nakamura, T. Ishikawa, T. Yamaguchi, Computational study on thrombus formation regulated by platelet glycoprotein and blood flow shear. Microvasc. Res. **89**, 95–106 (2013)
126. M. Karplus, J. Kuriyan, Molecular dynamics and protein function. Proc. Natl. Acad. Sci. USA **102**(19), 6679–6685 (2005)
127. J. Keener, J. Sneyd, Muscle, in *Mathematical Physiology, Volume 8 of Interdisciplinary Applied Mathematics*, chap. 18 (Springer, New York, 1998), pp. 542–578
128. J.G. Kelton, T.E. Warkentin, Heparin-induced thrombocytopenia: a historical perspective. Blood **112**(7), 2607–2616 (2008)
129. M.A. Khanin, V.V. Semenov, A mathematical model of the kinetics of blood coagulation. J. Theor. Biol. **136**(2), 127–134 (1989)

130. O.V. Kim, Z. Xu, E.D. Rosen, M.S. Alber, Fibrin networks regulate protein transport during thrombus development. PLoS Comput. Biol. **9**(6), e1003095 (2013)
131. P. Kleinbongard, R. Schulz, T. Rassaf, T. Lauer, A. Dejam, T. Jax, I. Kumara, P. Gharini, S. Kabanova, B. Ozüyaman, H.G. Schnürch, A. Gödecke, A.A. Weber, M. Robenek, H. Robenek, W. Bloch, P. Rösen, M. Kelm, Red blood cells express a functional endothelial nitric oxide synthase. Blood **107**(7), 2943–2951 (2006)
132. C. Kleinstreuer, J.R. Buchanan, M. Lei, G. A. Truskey, Computational analysis of particle hemodynamics and prediction of the onset of arterial diseases, in *Biomechanical Systems, Techniques and Applications, Volume II. Cardiovascular Techniques* (CRC Press, West Palm Beach, 2001)
133. A.E. Kogan, D.V. Kardakov, M.A. Khanin, Analysis of the activated partial thromboplastin time test using mathematical modeling. Thromb. Res. **101**(4), 299–310 (2001)
134. H. Kondo, Y. Imai, T. Ishikawa, K.-I. Tsubota, T. Yamaguchi, Hemodynamic analysis of microcirculation in malaria infection. Ann. Biomed. Eng. **37**(4), 702–709 (2009)
135. S. Koshizuka, Y. Oka, Moving-particle semi-implicit method for fragmentation of incompressible fluid. Nucl. Sci. Eng. **123**(3), 421–434 (1996)
136. M. Kostoglou, Extended cell average technique for the solution of coagulation equation. J. Colloid Interface Sci. **306**(1), 72–81 (2007)
137. A.L. Kuharsky, A.L. Fogelson, Surface-mediated control of blood coagulation: the role of binding site densities and platelet deposition. Biophys. J. **80**(3), 1050–1074 (2001)
138. K. Laki, L. Lóránd, On the solubility of fibrin clots. Science **108**(2802), 280 (1948)
139. N. Lannoy, C. Hermans, The 'royal disease'- haemophilia A or B? A haematological mystery is finally solved. Haemophilia **16**(6), 843–847 (2010)
140. I.J. Laurenzi, S.L. Diamond, Monte Carlo simulation of the heterotypic aggregation kinetics of platelets and neutrophils. Biophys. J. **77**(3), 1733–1746 (1999)
141. E.H. Lee, J. Hsin, M. Sotomayor, G. Comellas, K. Schulten, Discovery through the computational microscope. Structure **17**, 1295–1306 (2009)
142. R.J. Leipold, T.A. Bozarth, A.L. Racanelli, I.B. Dicker, Mathematical model of serine protease inhibition in the tissue factor pathway to thrombin. J. Biol. Chem. **270**(43), 25383–25387 (1995)
143. A. Leuprecht, K. Perktold, Computer simulation of non-Newtonian effects on blood flow in large arteries. Comput. Methods Biomech. Biomed. Eng. **4**(2), 149–163 (2001)
144. S.N. Levine, Enzyme amplifier kinetics. Science **152**(3722), 651–653 (1966)
145. M. Levitt, C. Sander, P.S. Stern, Protein normal-mode dynamics: trypsin inhibitor, crambin, ribonuclease and lysozyme. J. Mol. Biol. **181**, 423–447 (1985)
146. L.D. Libersky, A.G. Petschek, T.C. Carney, J.R. Hipp, High strain lagrangian hydrodynamics: a three-dimensional SPH code for dynamic material response. J. Comput. Phys. **109**(1), 67–75 (1993)
147. B.B.C. Lim, E.H. Lee, M. Sotomayor, K. Schulten, Molecular basis of fibrin clot elasticity. Structure **16**, 449–459 (2008)
148. Y. Liu, W.K. Liu, Rheology of red blood cell aggregation by computer simulation. J. Comput. Phys. **220**, 135–154 (2006)
149. Y. Liu, L. Zhang, X. Wang, W.K. Liu, Coupling of Navier-Stokes equations with protein molecular dynamics and its application to hemodynamics. Int. J. Numer. Methods Fluids **46**(12), 1237–1252 (2004)
150. W.K. Liu, Y. Liu, D. Farrell, L. Zhang, X.S. Wang, Y. Fukui, N. Patankar, Y. Zhang, C. Bajaj, J. Lee, J. Hong, X. Chen, H. Hsu, Immersed finite element method and its applications to biological systems. Comput. Methods Appl. Mech. Eng. **195**, 1722–1749 (2006)
151. K. Lo, W.S. Denney, S.L. Diamond, Stochastic modeling of blood coagulation initiation. Pathophysiol. Haemost. Thromb. **34**(2–3), 80–90 (2006)
152. A.I. Lobanov, T.K. Starozhilova, The effect of convective flows on blood coagulation processes. Pathophysiol. Haemost. Thromb. **34**(2–3), 121–134 (2006)
153. A.I. Lobanov, A.V. Nikolaev, T.K. Starozhilova, Mathematical model of fibrin polymerization. Math. Model. Nat. Phenom. **6**(7), 55–69 (2011)

154. P.W. Longest, C. Kleinstreuer, Comparison of blood particle deposition models for non-parallel flow domains. J. Biomech. **36**, 421–430 (2003)
155. P.W. Longest, C. Kleinstreuer, J.R. Buchanan, Efficient computation of micro-particle dynamics including wall effects. Comput. Fluids **33**, 577–601 (2004)
156. J. Ma, Usefulness and limitations of normal mode analysis in modeling dynamics of biomolecular complexes. Structure **13**, 373–380 (2005)
157. Y. Ma, J. Wang, S. Liang, C. Dong, Q. Du, Application of population dynamics to study heterotypic cell aggregations in the near-wall region of a shear flow. Cell. Mol. Bioeng. **3**(1), 3–19 (2010)
158. R.G. MacFarlane, An enzyme cascade in the blood clotting mechanism, and its function as a biochemical amplifier. Nature **202**(4931), 498–499 (1964)
159. K.G. Mann, Is there value in kinetic modeling of thrombin generation? Yes. J. Thromb. Haemost. **10**(8), 1463–1469 (2012)
160. K.G. Mann, S. Butenas, K. Brummel, The dynamics of thrombin formation. Arterioscler. Thromb. Vasc. Biol. **23**(1), 17–25 (2003)
161. K.G. Mann, K. Brummel-Ziedins, T. Orfeo, S. Butenas, Models of blood coagulation. Blood Cells Mol. Dis. **36**(2), 108–117 (2006)
162. J.A. Marcum, The origin of the dispute over the discovery of heparin. J. Hist. Med. Allied Sci. **55**(1), 37–66 (2000)
163. F. Martorana, A. Moro, On the kinetics of enzyme amplifier systems with negative feedback. Math. Biosci. **21**(1–2), 77–84 (1974)
164. D. Massai, G. Soloperto, D. Gallo, X.Y. Xu, U. Morbiducci, Shear-induced platelet activation and its relationship with blood flow topology in a numerical model of stenosed carotid bifurcation. Eur. J. Mech. B Fluids **35**, 92–101 (2012)
165. M.R. Maxey, B.K. Patel, Localized force representations for particles sedimenting in Stokes flow. Int. J. Multiphase Flow **27**(9), 1603–1626 (2001)
166. J.A. McCammon, B.R. Gelin, M. Karplus, Dynamics of folded proteins. Nature **267**(5612), 585–590 (1977)
167. J. McLean, The discovery of heparin. Circulation **19**(1), 75–78 (1959)
168. S. Melchionna, A model for red blood cells in simulations of large-scale blood flows. Macromol. Theory Simul. **20**(7), 548–561 (2011)
169. Z.A. Melzak, A scalar transport equation. Trans. Am. Math. Soc. **85**(2), 547–560 (1957)
170. Z.A. Melzak, A scalar transport equation ii. Mich. Math. J. **4**(3), 193–206 (1957)
171. A.D. Michelson, *Platelets*, 2nd edn. (Academic, New York, 2007)
172. G. Moiseyev, P.Z. Bar-Yoseph, No need for particle tracing: from accumulating fluid properties to novel blood coagulation model in the lattice Boltzmann method. J. Biomech. **43**(5), 864–870 (2010)
173. G. Moiseyev, P.Z. Bar-Yoseph, Computational modeling of thrombosis as a tool in the design and optimization of vascular implants. J. Biomech. **46**, 248–252 (2013)
174. G. Moiseyev, S. Givli, P.Z. Bar-Yoseph, Fibrin polymerization in blood coagulation-a statistical model. J. Biomech. **46**(1), 26–30 (2013)
175. J.J. Monaghan, Smoothed particle hydrodynamics. Annu. Rev. Astron. Astrophys. **30**(1), 543–574 (1992)
176. D.D. Monkovic, P.B. Tracy, Activation of human factor V by factor Xa and thrombin. Biochemistry **29**(5), 1118–1128 (1990)
177. U. Morbiducci, R. Ponzini, M. Nobili, D. Massai, F.M. Montevecchi, D. Bluestein, A. Redaelli, Blood damage safety of prosthetic heart valves. Shear-induced platelet activation and local flow dynamics: a fluid–structure interaction approach. J. Biomech. **42**, 1952–1960 (2009)
178. D. Mori, K. Yano, K.-i. Tsubota, T. Ishikawa, S. Wada, T. Yamaguchi, Computational study on effect of red blood cells on primary thrombus formation. Thromb. Res. **123**, 114–121 (2008)
179. A. Moro, A.T. Bharucha-Reid. On the kinetics of enzyme amplifier systems. Math. Biosci. **5**(3–4), 391–402 (1969)

180. L. Mountrakis, E. Lorenz, A.G. Hoekstra, Where do the platelets go? A simulation study of fully resolved blood flow through aneurysmal vessels. Interface Focus **3**(2), 20120089 (2013)
181. M.E. Nesheim, W.M. Canfield, W. Kisiel, K.G. Mann, Studies of the capacity of factor Xa to protect factor Va from inactivation by activated protein C. J. Biol. Chem. **257**(3), 1443–1447 (1982)
182. K.C. Ng, Y.L. Ng, M.Z. Yusoff, Development of a Lagrangian meshless flow solver based on the moving particle semi-implicit (MPS) method, in *4th International Conference on Energy and Environment 2013 (ICEE2013), Volume 16 of IOP Conference Series: Earth and Environmental Science*, p. 012151 (2013)
183. M. Nobili, J. Sheriff, U. Morbiducci, A. Redaelli, D. Bluestein, Platelet activation due to hemodynamic shear stresses: damage accumulation model and comparison to in vitro measurements. ASAIO J. **54**(1), 64–72 (2008)
184. P. Espa nol, Fluid particle model. Phys. Rev. E **57**(3), 2930–2948 (1998)
185. Th. Orfeo, S. Butenas, K.E. Brummel-Ziedins, K.G. Mann, The tissue factor requirement in blood coagulation. J. Biol. Chem. **280**, 42887–42896 (2005)
186. T. Orfeo, M. Gissel, S. Butenas, A. Undas, K.E. Brummel-Ziedins, K.G. Mann, Anticoagulants and the propagation phase of thrombin generation. PLoS ONE **6**(11), e27852 (2011)
187. R. Ouared, B. Chopard, B. Stahl, D.A. Rüfenacht, H. Yilmaz, G. Courbebaisse, Thrombosis modeling in intracranial aneurysms: a lattice Boltzmann numerical algorithm. Comput. Phys. Commun. **179**(1–3), 128–131 (2008)
188. P. Owren, Parahaemophilia. Haemorrhagic diathesis due to absence of a previously unknown clotting factor. Lancet **249**(6449), 446–448 (1947)
189. M.A. Panteleev, M.V. Ovanesov, D.A. Kireev, A.M. Shibeko, E.I. Sinauridze, N.M. Ananyeva, A.A. Butylin, E.L. Saenko, F.I. Ataullakhanov, Spatial propagation and localization of blood coagulation are regulated by intrinsic and protein c pathways, respectively. Biophys. J. **90**(5), 1489–1500 (2006)
190. V. Pappu, P. Bagchi, 3D computational modeling and simulation of leukocyte rolling adhesion and deformation. Comput. Biol. Med. **38**, 738–753 (2008)
191. V. Pappu, S.K. Doddi, P. Bagchi, A computational study of leukocyte adhesion and its effect on flow pattern in microvessels. J. Theor. Biol. **254**, 483–498 (2008)
192. I.V. Pivkin, P.D. Richardson, G. Karniadakis, Blood flow velocity effects and role of activation delay time on growth and form of platelet thrombi. Proc. Natl. Acad. Sci. USA **103**(46), 17164–17169 (2006)
193. I.V. Pivkin, P.D. Richardson, G.E. Karniadakis, Effect of red blood cells on platelet aggregation. IEEE Eng. Med. Biol. Mag. **28**(2), 32–37 (2009)
194. A. Pizard, C. Richer, N. Bouby, N. Picard, P. Meneton, M. Azizi, F. Alhenc-Gelas, Genetic deficiency in tissue kallikrein activity in mouse and man: effect on arteries, heart and kidney. Biol. Chem. **389**(6), 701–706 (2008)
195. A. Podmore, M. Smith, G. Savidge, A. Alhaq, Real-time quantitative PCR analysis of factor XI mRNA variants in human platelets. J. Thromb. Haemost. **2**(10), 1713–1719 (2004)
196. A.V. Pokhilko, F.I. Ataullakhanov, Contact activation of blood coagulation: trigger properties and hysteresis. J. Theor. Biol. **191**(2), 213–219 (1998)
197. D. Raabe, Overview of the lattice Boltzmann method for nano- and microscale fluid dynamics in materials science and engineering. Model. Simul. Mater. Sci. Eng. **12**(6), R13–R46 (2004)
198. A. Rahman, Correlations in the motion of atoms in liquid argon. Phys. Rev. **136**(2A), A405–A411 (1964)
199. K.R. Rajagopal, A.R. Srinivasa, A thermodynamic frame work for rate type fluid models. J. Non-Newtonian Fluid Mech. **88**(3), 207–227 (2000)
200. J.M. Ramstack, L. Zuckerman, L.F. Mockros, Shear-induced activation of platelets. J. Biomech. **12**(2), 113–125 (1979)
201. O.D. Ratnoff, J.M. Rosenblum, Role of Hageman factor in the initiation of clotting by glass. Evidence that glass frees Hageman factor from inhibition. Am. J. Med. **25**(2), 160–168 (1958)

202. R.L. Reddick, T.R. Griggs, M.A. Lamb, K.M. Brinkhous, Platelet adhesion to damaged coronary arteries: comparison in normal and von Willebrand disease swine. Proc. Natl. Acad. Sci. USA **79**(16 I), 5076–5079 (1982)
203. D. Ribatti, E. Crivellato, Giulio Bizzozero and the discovery of platelets. Leuk. Res. **31**(10), 1339–1341 (2007)
204. E. Richardson, Applications of a theoretical model for haemolysis in shear flow. Biorheology **12**(1), 27–37 (1975)
205. F.R. Rickles, S. Patierno, P.M. Fernandez, Tissue factor, thrombin, and cancer. CHEST **124**, 58S–68S (2003)
206. J.P. Riddel Jr., B.E. Aouizerat, C. Miaskowski, D.P. Lillicrap, Theories of blood coagulation. J. Pediatr. Oncol. Nurs. **24**(3), 123–131 (2007)
207. J. Rivera, M.L. Lozano, L. Navarro-Núñez, V. Vicente, Platelet receptors and signaling in the dynamics of thrombus formation. Haematologica **94**(5), 700–711 (2009)
208. H.R. Roberts, Oscar Ratnoff: his contributions to the golden era of coagulation research. Br. J. Haematol. **122**(2), 180–192 (2003)
209. Z.M. Ruggeri, Perspectives series: cell adhesion in vascular biology – von Willebrand factor. J. Clin. Investig. **99**(4), 559–564 (1997)
210. J.E. Sadler, Biochemistry and genetics of von Willebrand factor. Annu. Rev. Biochem. **67**, 395–424 (1998)
211. B. Savage, J.J. Sixma, Z.M. Ruggeri, Functional self-association of von Willebrand factor during platelet adhesion under flow. Proc. Natl. Acad. Sci. USA **99**(1), 425–430 (2002)
212. A.H. Schmaier, G. LaRusch, Factor XII: new life for an old protein. Thromb. Haemost. **104**(5), 915–918 (2010)
213. A. Sequeira, R.F. Santos, T. Bodnár. Blood coagulation dynamics: mathematical modeling and stability results. Math. Biosci. Eng. **8**(2), 425–443 (2011)
214. S.C. Shadden, S. Hendabadi, Potential fluid mechanic pathways of platelet activation. Biomech. Model. Mechanobiol. **12**, 467–474 (2012)
215. V. Shankar, G.B. Wright, A.L. Fogelson, R.M. Kirbya, A study of different modeling choices for simulating platelets within the immersed boundary method. Appl. Numer. Math. **63**, 58–77 (2013)
216. A.M. Shibeko, E.S. Lobanova, M.A. Panteleev, F.I. Ataullakhanov, Blood flow controls coagulation onset via the positive feedback of factor VII activation by factor Xa. BMC Syst. Biol. **4** (2010)
217. W. Shyy, M.-H. Chen, R. Mittal, H.S. Udaykumar, On the suppression of numerical oscillations using a non-linear filter. J. Comput. Phys. **102**, 49–62 (1992)
218. L. Skjaerven, S.M. Hollup, N. Reuter, Normal mode analysis for proteins. J. Mol. Struct. THEOCHEM **898**, 42–48 (2009)
219. S.A. Smith, The cell-based model of coagulation: State-of-the-Art Review. J. Vet. Emerg. Crit. Care. **19**(1), 3–10 (2009)
220. D.J. Smith, M.J. Hounslow, W.R. Paterson, Aggregation and gelation - I. Analytical solutions for CST and batch operation. Chem. Eng. Sci. **49**(7), 1025–1035 (1994)
221. M. V. Smoluchowski, Drei Vorträge über Diffusion, Brownsche Bewegung und Koagulation von Kolloidteilchen. Z. Phys. **17**, 557–585 (1916)
222. J.S. Soares, J. Sheriff, D. Bluestein, A novel mathematical model of activation and sensitization of platelets subjected to dynamic stress histories. Biomech. Model. Mechanobiol. **12**, 1127–1141 (2013)
223. X. Song, A.L. Throckmorton, H.G. Wood, J.F. Antaki, D.B. Olsen, Computational fluid dynamics prediction of blood damage in a centrifugal pump. Artif. Organs **27**(10), 938–941 (2003)
224. E.N. Sorensen, G.W. Burgreen, W.R. Wagner, J.F. Antaki, Computational simulation of platelet deposition and activation: I. Model development and properties. Ann. Biomed. Eng. **27**(4), 436–448 (1999)

225. E.N. Sorensen, G.W. Burgreen, W.R. Wagner, J.F. Antaki, Computational simulation of platelet deposition and activation: II, results for Poiseuille flow over collagen. Ann. Biomed. Eng. **27**(4), 449–458 (1999)
226. E. Stavrou, A.H. Schmaier, Factor XII: what does it contribute to our understanding of the physiology and pathophysiology of hemostasis & thrombosis. Thromb. Res. **125**(3), 210–215 (2010)
227. R.F. Stevens, The history of haemophilia in the royal families of europe. Br. J. Haematol. **105**(1), 25–32 (1999)
228. H. Stormorken, *Paul A. Owren and the Golden Era of Haemostasis* (Gazettebok, Sandvika, 2001)
229. O. Taboureau, O.H. Olsen, Computational study of coagulation factor VIIa's affinity for phospholipid membranes. Eur. Biophys. J. **36**, 133–144 (2007)
230. A.A. Tokarev, Yu.V. Krasotkina, M.V. Ovanesov, M.A. Panteleev, M.A. Azhigirova, V.A. Volpert, F.I. Ataullakhanov, A.A. Butilin, Spatial dynamics of contact-activated fibrin clot formation in vitro and in silico in haemophilia b: effects of severity and ahemphil b treatment. Math. Model. Nat. Phenom. **1**(2), 124–137 (2006)
231. A. Tokarev, I. Sirakov, G. Panasenko, V. Volpert, E. Shnol, A. Butylin, F. Ataullakhanov, Continuous mathematical model of platelet thrombus formation in blood flow. Russ. J. Numer. Anal. Math. Model. **27**(2), 191–212 (2012)
232. A. Tosenberger, V. Salnikov, N. Bessonov, E. Babushkina, V. Volpert, Particle dynamics methods of blood flow simulations. Math. Model. Nat. Phenom. **6**(5), 320–332 (2011)
233. A. Tosenberger, F. Ataullakhanov, N. Bessonov, M. Panteleev, A. Tokarev, V. Volpert, Modelling of thrombus growth in flow with a DPD-PDE method. J. Theor. Biol. **337**, 30–41 (2013)
234. S.D. Treadwell, B. Thanvi T.G. Robinson, Stroke in pregnancy and the puerperium. Postgrad. Med. J. **84**(991), 238–245 (2008)
235. A. Trousseau, Phlegmatia alba dolens, in *Clinique médicale de l'Hôtel-Dieu de Paris*, vol. 3 (J.B. Balliére et Fils, Paris, 1865), pp. 654–712
236. D.M.F. Van Aalten, B.L. De Groot, J.B.C. Findlay, H.J.C. Berendsen, A. Amadei, A comparison of techniques for calculating protein essential dynamics. J. Comput. Chem. **18**(2), 169–181 (1997)
237. W.F. Van Gunsteren, D. Bakowies, R. Baron, I. Chandrasekhar, M. Christen, X. Daura, P. Gee, D.P. Geerke, A. Glättli, P.H. Hünenberger, M.A. Kastenholz, C. Oostenbrink, M. Schenk, D. Trzesniak, N.F.A. Van Der Vegt, H.B. Yu, Biomolecular modeling: goals, problems, perspectives. Angew. Chem. Int. Ed. **45**(25), 4064–4092 (2006)
238. B.O. Villoutreix, Structural bioinformatics: methods, concepts and applications to blood coagulation proteins. Curr. Protein Pept. Sci. **3**, 341–364 (2002)
239. B.O. Villoutrei, O. Sperandio, In silico studies of blood coagulation proteins: from mosaic proteases to nonenzymatic cofactor inhibitors. Curr. Opin. Struct. Biol. **20**, 168–179 (2010)
240. R.L.K. Virchow, Thrombose und Embolie. Gefässentzündung und septische Infektion, in *Gesammelte Abhandlungen zur wissenschaftlichen Medicin* (Von Meidinger & Sohn., Frankfurt am Main, 1856), pp. 219–732
241. G.J. Wagner, W.K. Liu, Coupling of atomistic and continuum simulations using a bridging scale decomposition. J. Comput. Phys. **190**, 249–274 (2003)
242. C. Wagner, P. Steffen, S. Svetina, Aggregation of red blood cells: from rouleaux to clot formation. C. R. Phys. **14**(6), 459–469 (2013)
243. F.J. Walker, P.W. Sexton, C.T. Esmon, The inhibition of blood coagulation by activated protein C through the selective inactivation of activated factor V. Biochim. Biophys. Acta **571**(2), 333–342 (1979)
244. A.H. Schmaier, L.D. Dahl, A.J. Rozemuller, R.A. Roos, S.L. Wagner, R. Chung, W.E. Van Nostrand, Protease nexin-2/amyloid beta protein precursor. A tight-binding inhibitor of coagulation factor IXa. J. Clin. Invest. **92**(5), 2540–2545 (1993)
245. N.-H.L. Wang, K.H. Keller, Augmented transport of extracellular solutes in concentrated erythrocyte suspensions in Couette flow. J. Colloid Interface Sci. **103**(1), 210–225 (1985)

246. W. Wang, M.R. King, Multiscale modeling of platelet adhesion and thrombus growth. Ann. Biomed. Eng. **40**(11), 2345–2354 (2012)
247. H. Wang, C. Junghans, K. Kremer, Comparative atomistic and coarse-grained study of water: what do we lose by coarse-graining? Eur. Phys. J. E **28**, 221–229 (2009)
248. D. Wardrop, D. Keeling, The story of the discovery of heparin and warfarin. Br. J. Haematol. **141**(6), 757–763 (2008)
249. F.F. Weller, Modeling, analysis, and simulation of thrombosis and hemostasis. Ph.D. thesis, Ruprecht–Karls–Universität Heidelberg, 2008
250. F.F. Weller, Platelet deposition in non-parallel flow: influence of shear stress and changes in surface reactivity. J. Math. Biol. **57**(3), 333–359 (2008)
251. F.F. Weller, A free boundary problem modeling thrombus growth: model development and numerical simulation using the level set method. J. Math. Biol. **61**(6), 805–818 (2010)
252. N.K. Wenger, Clinical characteristics of coronary heart disease in women: emphasis on gender differences. Cardiovasc. Res. **53**(3), 558–567 (2002)
253. G.M. Willems, T. Lindhout, W.T. Hermens, H.C. Hemker, Simulation model for thrombin generation in plasma. Haemostasis **21**(4), 197–207 (1991)
254. D.M. Wootton, C.P. Markou, S.R. Hanson, D.N. Ku, A mechanistic model of acute platelet accumulation in thrombogenic stenoses. Ann. Biomed. Eng. **29**(4), 321–329 (2001)
255. I.S. Wright, Nomenclature of blood clotting factors. Can. Med. Assoc. J. **80**(8), 659–661 (1959)
256. I.S. Wright, Nomenclature of blood clotting factors. Can. Med. Assoc. J. **86**, 373–374 (1962)
257. J. Wu, C.K. Aidun, A method for direct simulation of flexible fiber suspensions using lattice Boltzmann equation with external boundary force. Int. J. Multiphase Flow **36**, 202–209 (2010)
258. J. Wu, C.K. Aidun, Simulating 3D deformable particle suspensions using lattice Boltzmann method with discrete external boundary force. Int. J. Numer. Methods Fluids **62**(7), 765–783 (2010)
259. J. Wu, B.M. Yun, A.M. Fallon, S.R. Hanson, C.K. Aidun, A.P Yoganathan, Numerical investigation of the effects of channel geometry on Platelet activation and blood damage. Ann. Biomed. Eng. **39**(2), 897–910 (2011)
260. C.Q. Xu, Y.J. Zeng, H. Gregersen, Dynamic model of the role of platelets in the blood coagulation system. Med. Eng. Phys. **24**(9), 587–593 (2002)
261. C. Xu, X.H. Xu, Y. Zeng, Y.W. Chen, Simulation of a mathematical model of the role of the TFPI in the extrinsic pathway of coagulation. Comput. Biol. Med. **35**(5), 435–445 (2005)
262. Z. Xu, N. Chen, M.M. Kamocka, E.D. Rosen, M. Alber, A multiscale model of thrombus development. J. R. Soc. Interface **5**, 705–722 (2008)
263. Z. Xu, N. Chen, S.C. Shadden, J.E. Marsden, M.M. Kamocka, E.D. Rosen, M. Alber, Study of blood flow impact on growth of thrombi using a multiscale model. Soft Matter **5**(4), 769–779 (2009)
264. Z. Xu, J. Lioi, M. Alber, J. Mu, X. Liu, D.Z. Chen, M.M. Kamocka, E.D. Rosen, Combined experimental and simulation study of blood clot formation, in *TIC-STH'09: 2009 IEEE Toronto International Conference - Science and Technology for Humanity*, pp. 357–362 (2009)
265. Z. Xu, J. Lioi, J. Mu, M.M. Kamocka, X. Liu, D.Z. Chen, E.D. Rosen, M. Alber, A multiscale model of venous thrombus formation with surface-mediated control of blood coagulation cascade. Biophys. J. **98**, 1723–1732 (2010)
266. Z. Xu, M. Kamocka, M. Alber, E.D. Rosen, Computational approaches to studying thrombus development. Arterioscler. Thromb. Vasc. Biol. **31**, 500–505 (2011)
267. Z. Xu, O. Kim, M. Kamocka, E.D. Rosen, M. Alber, Multiscale models of thrombogenesis. Wiley Interdiscip. Rev. Syst. Biol. Med. **4**(3), 237–246 (2012)
268. D. Xu, E. Kaliviotis, A. Munjiza, E. Avital, C. Ji, J. Williams, Large scale simulation of red blood cell aggregation in shear flows. J. Biomech. **46**, 1810–1817 (2013)

269. T. Yamaguchi, T. Ishikawa, Y. Imai, N. Matsuki, M. Xenos, Y. Deng, D. Bluestein, Particle-based methods for multiscale modeling of blood flow in the circulation and in devices: challenges and future directions. Ann. Biomed. Eng. **38**(3), 1225–1235 (2010)
270. K. Yano, D. Mori, K. Tsubota, T. Ishikawa, S. Wada, T. Yamaguchi, Analysis of destruction process of the primary thrombus under the influence of the blood flow. J. Biomech. Sci. Eng. **2**(1), 34–44 (2007)
271. K.K. Yeleswarapu, J.F. Antaki, M.V. Kameneva, K.R. Rajagopal, A mathematical model for shear-induced hemolysis. Artif. Organs **19**(7), 576–582 (1995)
272. M. E. Young, P. A. Carroad, R. L. Bell, Estimation of diffusion coefficients of proteins. Biotechnol. Bioeng. **22**(5), 947–955 (1980)
273. V.I. Zarnitsina, A.V. Pokhilko, F.I. Ataullakhanov, A mathematical model for the spatio-temporal dynamics of intrinsic pathway of blood coagulation. I. The model description. Thromb. Res. **84**(4), 225–236 (1996)
274. V.I. Zarnitsina, A.V. Pokhilko, F.I. Ataullakhanov, A mathematical model for the spatio-temporal dynamics of intrinsic pathway of blood coagulation. ii. Results. Thromb. Res. **84**(5), 333–344 (1996)
275. V.I. Zarnitsina, F.I. Ataullakhanov, A.I. Lobanov, O.L. Morozova, Dynamics of spatially nonuniform patterning in the model of blood coagulation. Chaos **11**(1), 57–70 (2001)
276. L. Zhang, A. Gerstenberger, X. Wang, W.K. Liu, Immersed finite element method. Comput. Methods Appl. Mech. Eng. **193**, 2051–2067 (2004)
277. J. Zhang, P.C. Johnson, A.S. Popel, An immersed boundary lattice Boltzmann approach to simulate deformable liquid capsules and its application to microscopic blood flows. Phys. Biol. **4**, 285–295 (2007)
278. J. Zhang, P.C. Johnson, A.S. Popel, Red blood cell aggregation and dissociation in shear flows simulated by lattice Boltzmann method. J. Biomech. **41**, 47–55 (2008)
279. R.F.A. Zwaal, P. Comfurius, E.M. Bevers, Surface exposure of phosphatidylserine in pathological cells. Cell. Mol. Life Sci. **62**(9), 971–988 (2005)
280. A.L. Zydney, C.K. Colton, Augmented solute transport in the shear flow of a concentrated suspension. PCH. Physicochem. Hydrodyn. **10**(1), 77–96 (1988)

T. Bodnár (✉)
Faculty of Mechanical Engineering, Department of Technical Mathematics, Czech Technical University in Prague, Prague, Czech Republic
e-mail: Tomas.Bodnar@fs.cvut.cz

A. Fasano
Dipartimento di Matematica “Ulisse Dini”, Università degli Studi di Firenze, Florence, Italy
e-mail: fasano@math.unifi.it

A. Sequeira
Departamento de Matemática, Instituto Superior Técnico, Lisbon, Portugal
e-mail: adelia.sequeira@math.ist.utl.pt

Printed by Printforce, the Netherlands